Biology
A Functional Approach

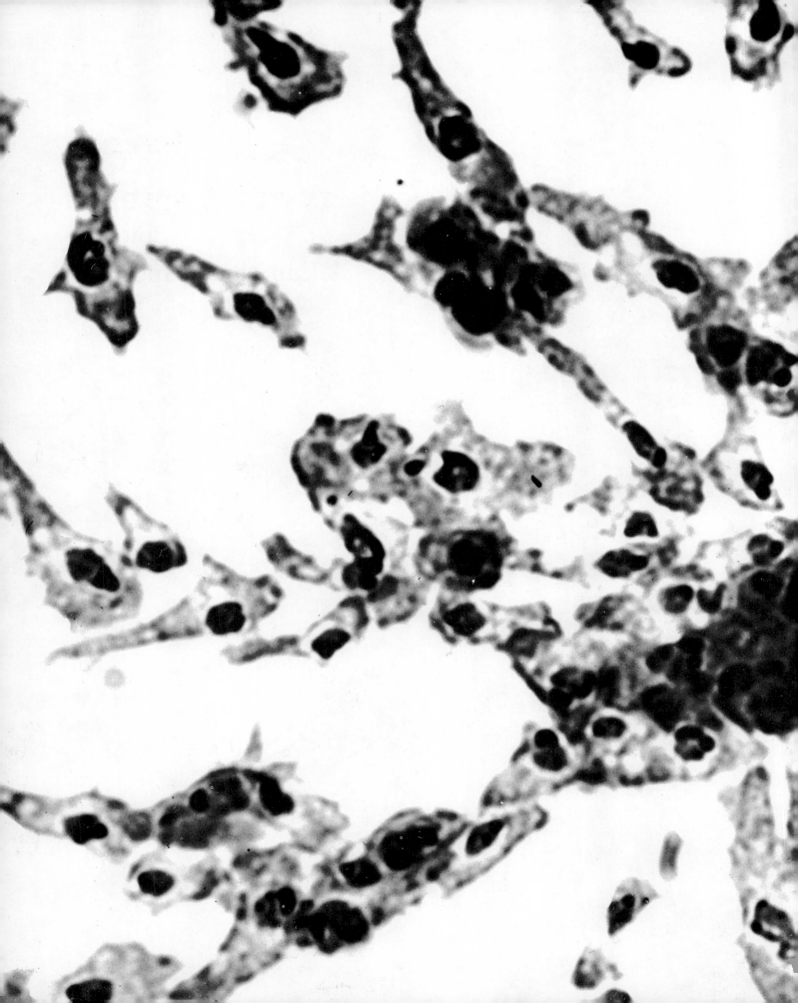

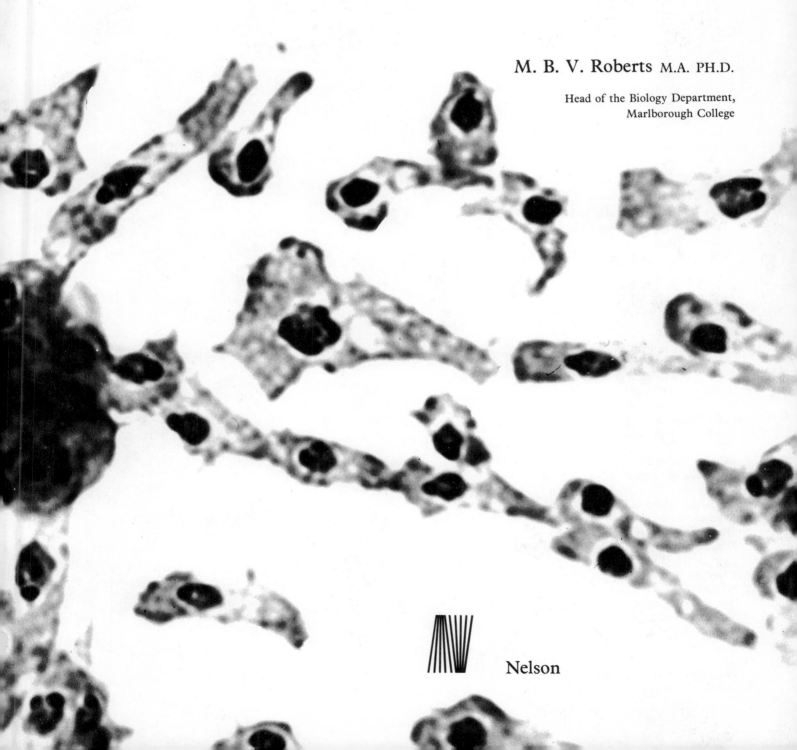

Biology
A Functional Approach

M. B. V. Roberts M.A. PH.D.

Head of the Biology Department,
Marlborough College

Nelson

To my Mother and Father

Thomas Nelson and Sons Ltd.
36 Park Street London W1Y 4DE
PO Box 18123 Nairobi Kenya
Thomas Nelson (Australia) Ltd.
171-175 Bank Street South Melbourne Victoria 3205
Thomas Nelson (Nigeria) Ltd.
PO Box 336 Apapa Lagos

© M. B. V. Roberts 1971
First published 1971
Sixth impression 1974

School Edition ISBN 0 17 448001 6
Trade Edition ISBN 0 17 143025 5

Art Editor Robin Fior
Designer Ken Campbell
Diagrams Colin Rattray and associates
Diagrams © Thomas Nelson & Sons Ltd. 1971

The cover shows the molecular model of
DNA (by courtesy of Professor M. H. F. Wilkins).
The title page shows aggregation of amoe-
boid cells of the slime mold Dictyostelium
(photo by courtesy of Professor J. T. Bonner,
Princeton University).

Filmset in Lumitype Plantin by St. Paul's Press Ltd., Malta
Printed in Great Britain by Cox and Wyman Ltd., London, Fakenham and Reading

Preface

Until recently sixth form biology courses centred on the structure and physiology of a series of animal and plant types. For the most part modern biology was given a raw deal. There was virtually no cell biology, no functional genetics, no experimental embryology, very little biochemistry and no molecular biology.

In the last five years things have changed. Influenced by the Nuffield programme and the American BSCS curricula, and encouraged by some of the examining boards, schools have been swinging away from the descriptive approach and placing greater emphasis on investigation. Six of the examining boards have already rewritten their Advanced level syllabuses to take into account modern developments, and there has been a general shift from the memorization of facts to the understanding of principles.

Most teachers and students have welcomed these changes but they have brought problems. How far should one take the modern discoveries? What should one do about the traditional topics?

This book attempts to strike the right balance between the two. I have tried to reassess traditional topics such as anatomy and organ physiology in the light of recent advances and to combine both into a modern functional framework. Modern work in biology, particularly cell structure and function, is making it possible to draw together seemingly disconnected threads into a series of unifying principles, and it is these that I try to emphasize. I have kept the burden of descriptive facts to the minimum consistent with an understanding of these principles.

Within limits the chapters can be taken in any order. The grouping of the various topics and the overall sequence gives a thematic development that I personally favour, but I have intentionally made each chapter as self-contained as possible so as to give teachers and students maximum freedom and flexibility in using the book.

I have not included laboratory schedules or examination questions. To have done so would have made the book unwieldy, and I feel that such material is best dealt with separately. But biology is an experimental science and I have taken pains to give evidence for stated facts wherever possible, particularly where it would be difficult or impossible for the student to verify them for himself. This has obviously taken up space, but I have allowed room for it by omitting any systematic treatment of the animal and plant kingdoms. I decided at the outset that to include a superficial survey of the animal and plant kingdoms would be of little value and indeed would run contrary to the philosophy behind the book. Nevertheless I firmly believe that the 'whole organism' should occupy a central position in any basic biology text. I have therefore drawn examples from a wide range of organisms, describing each in sufficient detail to illustrate the basic principle under discussion.

This book is intended to be a pre-university text for prospective biologists, medical students and agriculturalists. At the same time I hope it may provide a useful introduction for the increasing number of students who choose to read biological science at university having done no formal biology at school. I have assumed an elementary knowledge of physics and chemistry, but I have tried to write the book in such a way that it is intelligible to a student who has not done biology before. I have not stuck rigidly to any particular syllabus but the content and approach should make the book suitable for those following the Advanced level syllabuses of the University of London, the Joint Matriculation Board, the Oxford and Cambridge Schools Examination Board, the Associated Examining Board, the University of Cambridge Local Examinations Syndicate and the Oxford Delegacy of Local Examinations.

Acknowledgements

It is a pleasure to acknowledge the debt I owe to the many friends and colleagues who have helped me in producing this book. Various parts of the manuscript have been read by Dr H. W. Lissmann, F.R.S., Professor R. A. Hinde, F.R.S., Professor J. A. Ramsay, F.R.S., Dr Richard Bainbridge, Dr P. K. Tubbs, Dr A. V. Grimstone, Professor R. R. A. Coombs, F.R.S., Dr B. L. Gupta, Dr C. B. Goodhart, Dr Sydney Smith, Dr T. ap Rees, Dr P. J. Grubb and Dr Rufus Clarke, all of the University of Cambridge; Dr B. E. Juniper and Dr J. B. Land of the University of Oxford; Professor Bernard John of the University of Southampton; and my former colleagues in the University of California, Dr Garrett Hardin and Dr J. L. Walters. To all of them I owe my thanks for giving me the benefit of their expertise. Dr R. Gliddon of Clifton College kindly undertook the formidable task of reading the whole of the manuscript and I am most grateful to him for his frank and helpful comments. I also owe a debt of gratitude to my colleagues at Marlborough College, Mr Malcolm Hardstaff, Mr J. H. Halliday, and Mr John Emmerson for reading parts of the manuscript and assisting me in sundry other ways; working closely with them over the years has done much to shape my own ideas. I am also grateful to my colleagues Dr T. E. Rogers and Dr F. R. McKim for assistance on matters pertaining to chemistry and physics, and Mr S. W. Hockey for advice on units. In thanking all these people I must emphasize that I am fully responsible for any shortcomings that remain.

A substantial proportion of the text was written during tenure of a Fellow Commonership at Corpus Christi College, Cambridge, and I owe grateful thanks to the Master and Fellows for their hospitality, and to Professor T. Weis-Fogh for laying the facilities of the University Zoology Department at my disposal.

The names of the many people who have so kindly supplied photographs are given in the captions, but I would specially like to mention Mr J. F. Crane of the Cambridge University Anatomy School who prepared many of the photomicrographs, and my colleague Mr Beverley Heath whose skill as a photographer speaks for itself.

Many of my pupils have had a hand in the production of the book, particularly in the preparation of the illustrations and the checking of the typescript. I owe particular thanks to my former pupil Mr Peter Saugman for his sterling editorial work on an earlier draft, and to my technicians Mr C. R. Hughes and Mr Michael Ward for assistance with apparatus. The latter also prepared some of the more difficult drawings and assisted with the reading of proofs, as did my pupil Mr Guy Northridge. I also owe grateful thanks to the good ladies who typed the manuscript, particularly Mrs Berenice Loney, Miss Hermione Budge, and Mrs Jeanette Radford. The last not only typed much of the book but managed to maintain a semblance of order in my office during the final stages of its preparation.

Finally I am indebted to my publishers, Thomas Nelson and Sons Ltd., particularly Mr W. T. Cunningham and Mr D. R. Worlock, for their patience and encouragement. No author could have wished for a more sympathetic treatment of his manuscript.

M. B. V. Roberts
Marlborough, February 1971

Contents

Part I

Organization in Cells and Organisms

Part II
The Maintenance of Life

Part III
Adjustment and Control

Part IV
Response and Coordination

Part V
Reproduction, Development and Heredity

Part VI

Ecology and Evolution

Appendixes

1 Introducing Biology

Biology, the study of life and living organisms, is a branch of natural science. It is an enormous subject, involving many other disciplines such as chemistry, physics, mathematics, geology and psychology. Medicine and agriculture are really applied biology. It is only necessary to glance through the current issues of the British scientific journal *Nature*, or its American equivalent *Science*, to appreciate the extensiveness and ramifications of biology, both pure and applied. Over one million original papers are published in the biological sciences each year, ranging from descriptions of new species to analyses of complex chemical reactions in organisms. Moreover the volume of information grows at a rate that makes it impossible for any one person to keep up with it all. Of course this great burst of scientific activity is by no means peculiar to biology: it is happening in other sciences as well. In fact the explosion in scientific knowledge prompted a former President of the Royal Society to remark that nowadays the only item on the agenda which all Fellows of the Society can be guaranteed to understand is the statement that tea will be served in the lounge at 4 p.m. With such a wealth of information pouring out of research laboratories it is sometimes difficult for a beginner to see the wood for the trees.

The purpose of this book is to present the broad sweep of the subject in a reasonably integrated manner. Throughout the book fundamental concepts are stressed: concepts such as how life is maintained, how organisms adjust to changes in the environment, and so on. But one cannot really appreciate such concepts without first being familiar with the facts on which they are based. However, not all facts are necessary, or even desirable. It is a matter of separating the important from the less important ones. In this book we shall look at a large number of specific biological facts within a framework of concepts and basic principles. The specific facts must come first, just as the parts of a jigsaw come before the completed picture.

THE DIFFERENT BRANCHES OF BIOLOGY

Biology means the science of life (Greek: *bios* – life, *logos* – knowledge). Traditionally the subject has been divided into **zoology**, the study of animals, and **botany**, the study of plants. A third subdivision, **microbiology**, embraces that vast assembly of microscopical organisms many of which do not fit neatly into either the animal or plant kingdoms. Within this no-man's land come such subjects as **bacteriology** and **virology**, the study of bacteria and viruses respectively.

Fifty years ago biologists were mainly preoccupied with describing the structure and general form of animals and plants – **anatomy** and **morphology**. But in more recent times there has been a shift of interest towards the way organisms function, resulting in the development of animal and plant **physiology**. During the last thirty years or so such functional studies have become more and more chemical, resulting in the growth of **biochemistry**, now a subject in its own right.

Biochemical studies are showing us that in many respects the traditional division of biology into botany and zoology is an unnatural one. At the chemical level animal and plant cells are remarkably similar. The **cell** is the fundamental structural unit of which organisms are made. Research on cells, the study of **cytology**, has shown that in their structure as well as their functioning, animal and plant cells share much in common. This similarity is also seen in the way an organism's characteristics are transmitted to its offspring, the study of **heredity** or **genetics**. Genetics comes from the word **gene**, the term used to describe the particles that are transmitted from parents to offspring. Recent studies on the structure and properties of molecules found in cells have helped us to understand how genes exert their action. This new field, in which spectacular advances have been made in recent years, is known as **molecular biology** and, like biochemistry, is now an established subject in its own right.

So, looking back over the last fifty years, there has been a gradual shift of interest from anatomical description of the whole organism to functional studies at the cellular and molecular levels. This does not of course mean that all biologists are engaged in cellular or molecular research, or even that all modern biologists work in these fields. The older, more traditional studies still continue and have a very important part to play in the building of an integrated conceptual biology.

From the earliest times men have been interested in observing the ways and habits of animals and plants: **natural history** as one would call it. Today the naturalist, still enthusiastic in the pursuit of his hobby, is adopting a more analytical and experimental approach, studying **animal behaviour** and organisms in relation to their environment.

Behaviour itself is becoming a more experimental science, involving increasingly the techniques of **neurophysiology**. Environmental studies constitute **ecology** and are important if only because they affect the future wellbeing of man. Man's hopes of solving the world food problem and of making better use of his environment and natural resources will depend in large measure on advances in this field. Man is beginning to realize the importance of not destroying his natural environment. Ecological studies are demonstrating how man can derive material benefits from the environment without destroying it: **conservation** as opposed to exploitation.

WHAT IS LIFE?

The honest answer is that we cannot say categorically what life is. The best we can do is to list those attributes of a living organism that distinguish it from non-living matter. These are as follows:

(1) **Movement.** It is characteristic of organisms that they, or some part of them, are capable of moving themselves. Even plants, which at first sight appear to be an exception, display movements within their cells. For example, if you examine the hairs on the stamens of *Tradescantia* under a

microscope you will notice granules circulating round the periphery of the cells, a phenomenon which cannot be explained by purely physical forces.

(2) Responsiveness. All organisms, including plants, react to stimulation. Such responses range from the growth of a plant towards light to the rapid withdrawal of one's hand from a hot object. Even the simplest organisms respond to stimuli, as you will know if you have ever looked at unicells like *Amoeba* or *Paramecium* under the microscope.

(3) Growth. A crystal grows by new material being added at the surface. Organisms grow from within by a process which involves the intake of new materials from outside and their subsequent incorporation into the internal structure of the organism. This is called **assimilation** and it necessitates some kind of feeding process.

(4) Reproduction. All organisms are able to reproduce themselves. At its simplest, reproduction involves the replication of certain giant molecules (macromolecules). Advances in molecular biology enable us to say that these molecules are **nucleic acids**, generally **deoxyribonucleic acid (DNA)**, sometimes **ribonucleic acid (RNA)**. The presence of these complex molecules in all organisms from viruses to man qualifies them as an essential characteristic of life.

(5) Release of energy. To stay alive an organism must be able to release energy in a controlled and usable form. This it does by breaking down a compound called **adenosine triphosphate (ATP)**. The energy for building up ATP beforehand is obtained from the breakdown of sugar (**respiration**). One of the interesting facts to emerge from biochemical research is that ATP appears to be of universal occurrence in all living cells.

One might sum all this up by saying that a living organism is a self-reproducing system capable of growing and of maintaining its integrity by the expenditure of energy. Life is the sum total of all these things.

INVESTIGATIONS IN BIOLOGY: THE SCIENTIFIC METHOD

Most biological investigations start with an **observation**. What is observed may be a structure, a physiological process, a behaviour pattern, or the occurrence of an organism in a particular habitat. Whatever it is, a scientist's natural reaction is to ask questions about it. What is this structure for? How does this physiological process take place? Why does the animal behave in this way? Why does this particular organism occur here and not there? To arrive at an answer to such questions, the scientist gathers as many relevant facts as possible and then formulates a **hypothesis**.

An examination candidate once wrote that a hypothesis is 'better than a hunch but not as good as a theory'. This was his way of saying, quite correctly, that a hypothesis is a tentative theory – an intelligent guess. Hypotheses are the lifeblood of science. In formulating them the scientist shows his originality and intuition. It is here that, to use Einstein's analogy, he plays the detective. The hypothesis may not come to him until after many months or even years of careful observation, gathering of data and intensive thought. But when it does come, it enables him to go forward to the next stage in the argument.

If the hypothesis is to lead anywhere it must be possible to make **predictions** from it. If it is true that the DNA molecule is shaped like a double helix, then it follows that X-ray diffraction analysis of it should give a particular pattern. In other words a prediction is a logical deduction from the hypothesis, and does not necessitate any kind of experimentation.

Generally a number of different predictions can be made from a single hypothesis. The next step is to **test** each prediction to find out whether or not the hypothesis is correct. In testing the predictions we are therefore testing the hypothesis. This is where the scientist shows his prowess as an experimenter, for the testing of a hypothesis generally involves performing one or more crucial **experiments**. To revert to DNA, if we have predicted that the X-ray diffraction pattern should show certain characteristic features, then we must obviously carry out an X-ray analysis to show whether or not this is true. So the testing of a prediction, unlike its formulation, involves experimentation, i.e. it is an empirical process. If all the predictions turn out to be correct, then the chances are that our hypothesis is sound. If so much as a single prediction turns out to be wrong, the hypothesis must be false.

Before proceeding further let us consider an example of this scientific process in action. We are working in a medical laboratory on diphtheria. In a preliminary experiment we find that if we inject diphtheria bacilli into a mouse, the mouse develops certain symptoms and dies. However, we observe that the bacteria, instead of spreading through the body, remain localized at the site of the injection. From this observation we tentatively conclude that the disease was caused, not by the bacteria themselves, but by a toxic chemical produced by them. From this hypothesis several predictions may be made, one of them being that if we could isolate this chemical substance and inject it on its own into a mouse, the mouse should develop diphtheria. So we set about the appropriate experiment: we prepare bacterial cultures and extract fluid from them. On injection these prove, as predicted, to be powerfully toxic, inducing death with all the symptoms of diphtheria. Our original hypothesis would therefore seem to be correct. (Would it be fair to say that it was proved?) This is an actual case history: it was carried out in Germany in the 1880s and led to the development of immunization against diphtheria.

Of course it is not always possible to confirm a hypothesis as unequivocally as in the case recorded above. There are many instances in biology where the validity, or otherwise, of a hypothesis rests on the gradual accumulation of indirect evidence. As more and more evidence comes to hand, the hypothesis gains increasing acceptance and eventually is promoted to the rank of a **theory**. A theory is a set of scientific assumptions consistent with one another and supported by evidence, but not fully proved. A good example is the theory of evolution: though supported by an overwhelming body of evidence, it would be a bold man indeed who would assert that it was proved beyond all reasonable doubt.

The procedure described above is often labelled **the scientific method**, as if it were endowed with some special mystique. But, when one reflects, it is really what any intelligent person would do when confronted with a practical problem. Essentially the same procedure is adopted by a doctor in making a diagnosis, and a police detective when investigating a crime. Indeed, it is what most people do when they try to discover what is wrong with their car, record player, or washing machine. In other words 'the scientific method' is really nothing more than common sense.

The important thing for any scientist, or aspiring scientist, to do is to observe accurately and constantly to ask questions. Only by doing this can he put himself in a position to formulate hypotheses.

Although the formulation of hypotheses is at the very heart of the scientific process, it is beset with dangers. In particular one must beware of clinging to disproved hypotheses. The hypothesis is the scientist's 'brain child', the product of his creative thought. This may sometimes make him reluctant to abandon a hypothesis, even after contrary evidence has been produced. How tempting when you have done 99 experiments all of which support your hypothesis, not to bother about the hundredth. And yet this one experiment could disprove the hypothesis. Cases of omission are not uncommon: this is the theme of *The Search* by C. P. Snow, a book which might interest many aspiring scientists. Less frequent, but nevertheless on record, are cases where contrary results have been suppressed. This is not necessarily intentional: it is sometimes difficult to be objective when conducting and interpreting an experiment in the light of what one believes the answer should be. It is nonsense to think of the scientist as an impersonal robot who never lets his emotions creep into his work. Scientific laboratories would be very dull places if all scientists were like that!

Nevertheless, over-subjectivity should be avoided as far as possible. Clinging to outmoded hypotheses is an occupational hazard in those branches of biology where it is difficult, perhaps impossible, to test predictions experimentally. Such is the case with palaeontology, the study of fossils, and certain branches of animal behaviour. In such subjects the validity of hypotheses rests mainly on the gradual accumulation of indirect evidence.

CONDUCTING BIOLOGICAL EXPERIMENTS

A hypothesis stands or falls on the experiments that are carried out to test it. It is therefore essential to do the right experiment in the correct way. The predictions made from the hypothesis should tell the investigator exactly what experiments need to be done. This might seem obvious, but it is surprising how often a student, having formulated a very sensible hypothesis, goes on to devise experiments which do not test it at all. Often a single crucial experiment is enough to settle the matter. The diphtheria story illustrates this well. Here is another example.

On the basis that all cells have one, we conclude that the nucleus is essential for life. From this we predict that a cell without a nucleus will die. The experimental test is obvious: assuming that it is technically feasible, we must remove the nucleus from a cell and see what happens.

Enucleation of cells is possible, at any rate with a large cell like *Amoeba*. The nucleus is carefully sucked out with a very fine pipette. It is found, as predicted, that the cell dies. But can we be certain that it is absence of the nucleus which causes death, and not the damage inflicted on the cell during the operation? To find the answer to this we must set up a **control**: a second *Amoeba* has its nucleus removed in exactly the same way, and then immediately put back again. The same amount of damage is done to the cell but it is not permanently deprived of its nucleus. Apart from this the two *Amoebae* must be kept in exactly the same conditions. The control *Amoebae* do in fact survive, suggesting (proving?) that it is the absence of the nucleus which causes death. Setting up appropriate controls is an essential part of any investigation. It provides a standard with which the experimental situation can be compared.

One further point before we leave this topic. It is not sufficient to do the experiment on only one pair of *Amoebae*, even though they may give the expected results. The experiment must be repeated many times, and consistent results obtained, before the prediction can be regarded as confirmed. The quickest way of doing this in the case described above is to set up a group of enucleated *Amoebae*, perhaps fifty in all (the **experimental group**). At the same time we prepare a **control group** of non-enucleated specimens. If we are lucky it may turn out that all the individuals in the experimental group die, whereas all those in the control group survive. But in biological experiments it often turns out that only a proportion of the tests give the expected results. In such cases it may be necessary to repeat the experiment over and over again, and subject the results to statistical analysis, to establish whether the results are significant or merely the result of chance.

ORGANISMS

Although nowadays great emphasis is placed on the cell, there is no getting away from the fact that biology is concerned with organisms. In looking at organisms one is struck, first and foremost, by the great variety of living things. The number of different types of organism is prodigious: for example, there are almost two million different kinds of animals. Yet, despite this great diversity, basic similarities exist to a greater or lesser extent between them. For purposes of identification it is obviously important that each should have a precise name or some equivalent designation. It is also necessary that they should be placed in groups on the basis of similarities and differences. This is an aid to rapid identification and creates order out of what would be chaos. Think of the uselessness of an unclassified library of two million books.

The branch of biology concerned with naming organisms and placing them in groups is known as **taxonomy** or **systematics**. Taxonomic studies have led to the development of a system of classification which divides all living things into two, or three, large groups called **kingdoms**, each of which is divided into a series of major sub-groups called **phyla** (sing. **phylum**). Each phylum is further divided into a series of successively smaller groups known as **classes, orders, families, genera** (sing. **genus**) and finally **species**. An organism's proper name is designated by the genus to which it belongs (**generic name**), followed by the name of the species (**specific name**). There is generally only one kind of organism in a species (though sometimes there may be several **varieties** of the same species), so the specific name is usually a precise indication of the identity of an organism. As a matter of custom the generic name always begins with a capital letter, the specific name with a small letter. When printed, generic and specific names are always put in italics (when handwritten or typed they are underlined). Thus the proper name of the common earthworm is *Lumbricus terrestris*, the meadow buttercup *Ranunculus bulbosus*, and man *Homo sapiens*. You may well feel that all this is rather pedantic and that the common names would do equally well. However, common names are sometimes ambiguous since they may embrace more than one species which in some instances may be totally unrelated. There are also many instances of several different names being used for the same species. Can you think of some examples of such ambiguities? Despite these obvious dangers common names are easier to remember than the latinized proper names and, unless confusion is likely to result, they will be frequently used in this book. This is permissible when dealing with phenomena that apply to organisms generally as opposed to particular species.

The naming of organisms with a generic and specific name was first introduced by the Swedish naturalist Carl Linnaeus in 1735. Known as the **binomial system** it has been used ever since, and is the basis of modern taxonomy.

This book does not include a systematic treatment of the **animal** and **plant kingdoms**. However, many different organisms are brought in to illustrate the various topics discussed. As far as possible I have tried to confine examples to organisms that are likely to be familiar to the reader. This is not always possible, and to aid the reader to quickly see where a particular organism fits into the scheme of things an outline classification of the animal and plant kingdoms is included in the Appendix.

Certain organisms cannot be described as either animal or plant. These include **bacteria** and **viruses,** both micro-organisms. There is some reason to suppose that bacteria may be closely related to the plant kingdom, indeed they are often included with the **Fungi,** but the viruses are more problematical. Viruses are the simplest things to display the fundamental properties of life. Although they have no means of propelling themselves, they can reproduce and have characteristics which are transmitted from one generation to the next. However, their chemical simplicity makes it impossible for them to do this unaided: they can only reproduce inside the living cell of a 'host'. In so doing they destroy the host cell. For this reason viruses are always associated with disease: smallpox, measles, poliomyelitis, etc., and numerous plant diseases as well. Viruses are extremely small and can only be seen with an electron microscope (see p. 16). For this reason they will pass through a filter whose pores are small enough to retain bacteria. In fact this is how they were discovered: it was found that if the juice extracted from a tobacco plant suffering from mosaic disease was passed through a very fine filter, the filtrate when applied to the leaves of a healthy plant induced the disease – despite the fact that the filter was too fine to let through all known bacteria. This was the first indication that disease-causing particles of submicroscopical dimensions existed, and it led subsequently to the isolation and characterizing of viruses.

SOME BASIC CONCEPTS

If there is one over-riding concept in biology it is **survival,** not merely of the individual but of the species. To this end organisms are **adapted** to the environment, i.e. their structure and physiology fits them to live successfully in their particular situation. Fig 1.1 shows a rather spectacular example, but in fact every organism is adapted to its environment to a greater or lesser extent: if this were not so it would perish.

What specifically is an organism adapted for? To begin with, it must be adapted in such a way that it can reproduce its kind, defend itself from attack by other organisms, and compete successfully for food and other essential commodities. It must also be able to respond quickly and efficiently to changes in the environment. Above all, it must ensure that its cells always enjoy those conditions which are optimum for efficient functioning. In short, the organism must be able to maintain all those processes which we listed earlier as fundamental to life. Much of this book will be devoted to exploring how this is achieved.

From the principle of adaptation a further concept emerges, namely that the structure of organisms and their component parts is closely related to the

A

Fig 1.1 (*opposite*) Adaptation and survival. A sundew plant *Drosera rotundifolia*, feeding on a fly. The insect-catching device is a modified leaf with tentacles projecting from the upper surface and edges. The tentacles secrete a sticky substance which adheres to the insect's body, holding it fast. The tentacles bend over and press the insect against the flat part of the leaf whose cells secrete a protein-digesting enzyme. Its ability to devour small animals enables sundew to thrive in nitrogen-deficient soils. (*B. J. W. Heath, Marlborough College*)

Fig 1.2 The leaf of the giant water lily of the Amazon, *Victoria amazonica*, illustrates the close relationship between structure and function. The vertical margin round the edge of the large floating leaf prevents flooding over the surface, and thick fibrous ribs on the underside prevent it from collapsing. Reminiscent of fan vaulting, this strong construction is said to have been the inspiration for the design of the Crystal Palace in London. (A: *J. H. Gerard, courtesy of National Audubon Society;* B: *Walter Singer, New York*)

B

functions they are called on to perform. Guts, kidneys, gills, lungs, stems, leaves all demonstrate this close correlation between structure and function (see Fig. 1.2 for example). This concept is by no means restricted to organization at the level of whole organisms and organs; it also applies to individual cells as we shall see in Chapter 2.

One final point. How has the great variety of life that we see on this planet today come into being? Current belief is that organisms have arisen by a process of slow, gradual change from simpler forms that existed in the geological past. In Chapter 34 we shall consider the evidence for this theory of **evolution**. For the moment we should merely be aware of the fact that in the animal and plant kingdoms we see a progression from the simple to the complex.

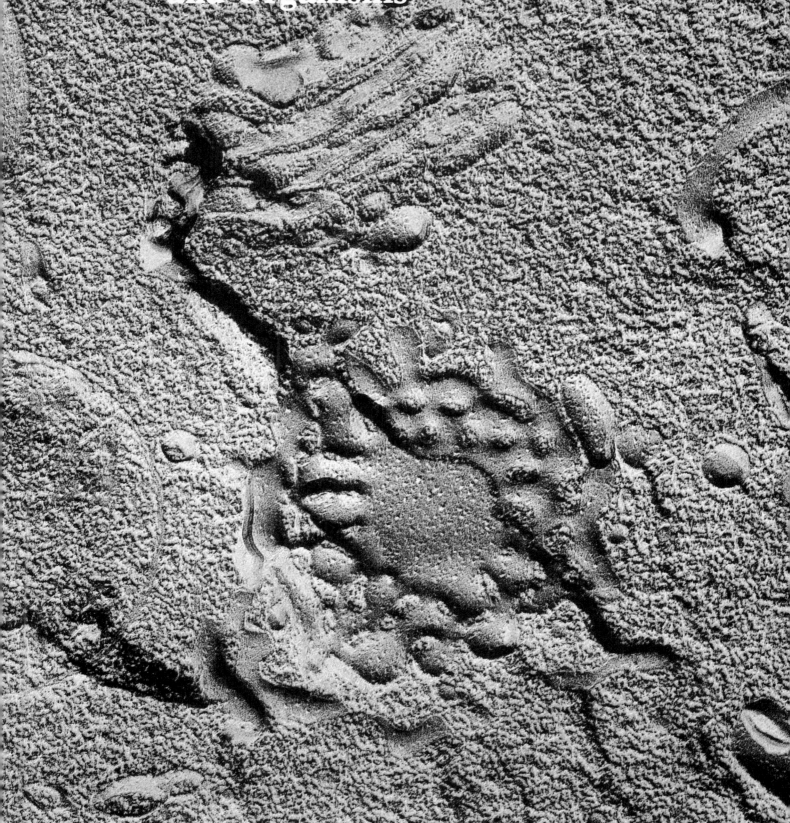

Part I   Organization in Cells
and Organisms

An organism can be studied at a number of different levels: at the level of the whole organism and populations, at the level of organs and tissues, or at the level of individual cells, and their constituent chemical substances and molecules. In Part I we shall be mainly concerned with organization at the cellular and chemical levels.

The structure of cells and the functions of their visible components are discussed in Chapter 2, together with their diversity in organisms as a whole. Cells do not generally exist in isolation but are massed together to form tissues and organs. The construction of tissues and organs is discussed in Chapter 3. To carry out its varied functions a cell must be in contact with its surroundings. How materials get in and out of cells is discussed in Chapter 4. The two remaining chapters deal with organization at the chemical level. In Chapter 5 the principal chemical substances found in organisms are surveyed from a functional standpoint; Chapter 6 discusses the basic principles underlying the chemical reactions in which they play a part.

Electron micrograph of a replica of the Golgi body in a root tip cell of the onion. *(Daniel Branton, University of California.)*

2 Structure and Function in Cells

The bodies of all but the simplest organisms are made up of **cells**, tiny micro-scopical units which collectively carry out the processes that make the organism a living entity. Even the simplest organisms are composed of cellular materials, i.e. molecules found inside the cells of more complex organisms. The cell, then, would seem to be a basic structural and functional unit of an organism, and so it is not surprising that biologists have devoted a great deal of attention to its structure and the processes which go on inside it.

THE CELL THEORY

Cells were first described in 1665 by Robert Hooke, a scientist of great talent and versatility who was an accomplished technician as well as a biologist. He designed one of the earliest optical microscopes with which he examined, amongst other things, thin sections of cork. He discovered that cork is com-posed of numerous box-like structures which we now know to be cells. Though Hooke coined the word cell for these structures he did not realize their significance. As more and more material was examined under the microscope it gradually became apparent that the great majority of organisms are composed of cells, a fact which is embodied in the **cell theory**. First proposed by M. J. Schleiden and Theodore Schwann in 1839, the cell theory states that cells are of universal occurrence and are the basic units of an organism. Like many generalizations, the cell theory has been greatly overworked to the point that it is taken by some biologists to mean that the cell is the most important unit, the whole organism being little more than a collection of independent though co-operating cells. As a reaction to this extreme view a rival idea has grown up, the **organismal theory**, which proposes that the whole organism is the basic entity and the cells merely incidental sub-units. An analogy may be useful: the cell theory (at its extreme) looks upon an organism as a kind of republic of semi-independent cells; the organismal theory on the other hand, regards the organism as a utopia of inter-dependent cells whose functions are dictated by the needs of the whole organism. We will not consider the evidence for these two viewpoints at the moment but as you discover more about cells and organisms, try to decide which you consider approximates more closely to the truth.

AN ANIMAL CELL AS SEEN WITH THE LIGHT MICROSCOPE

You will probably examine your first cells with the **optical** or **light micro-scope** and it is therefore best to start by considering the structure of a cell

Fig 2.1 A modern light microscope. Light rays from the object on the stage pass through two glass lenses, the objective and ocular lenses, which, depending on their strength, give magnifications of up to 800 times. Higher magnifications can be achieved without loss in resolution by using special objective lenses in conjunction with an immersion fluid such as cedar-wood oil (oil immersion). The resolving power of the instrument shown here is approximately 0.2 μm. Even in the best conditions it is not possible to achieve magnifications of more than 1,500 times without loss of detail. (*B. J. W. Heath, Marlborough College*)

Fig 2.2 The structure of typical animal and plant cells as seen under the light microscope. The Golgi body and mitochondria shown in the right hand animal cell can only be detected by using special staining techniques.

as seen with this instrument (Fig. 2.1). Needless to say the light microscope has had a profound influence on the development of biology, but it has severe limitations. In theory it might seem possible to magnify an object indefinitely by means of glass lenses in series, but in practice this cannot be done without losing detail. The limited resolution of the light microscope, imposed by the wavelength of light, means that little can be gained by magnifying an object more than 1,500 times at the most. This puts a limit on the amount of structural detail that can be detected within individual cells.

The structure of a typical animal cell is illustrated in Fig 2.2A. The whole cell has a diameter of about one hundredth of a millimetre (10 μm).[1] It is bounded by a thin **cell membrane** or **plasma membrane**, which encloses the protoplasm. The latter consists of the **nucleus** in the centre surrounded by **cytoplasm**. On first examination the cytoplasm appears to be a uniformly homogeneous substance, but closer inspection of cells stained with special dyes shows it to contain numerous granules and inclusions. Food materials are stored in the cytoplasm and it is here that complex chemical reactions take place, building up materials and supplying energy for the cell's activities.

The nucleus is bounded by a **nuclear membrane** and contains a dense body called the **nucleolus** together with small **chromatin granules**, the precursors of the **chromosomes** which make their appearance when the cell undergoes division. The chromosomes carry hereditary material determining the organism's characteristics and transmitting these to subsequent generations.

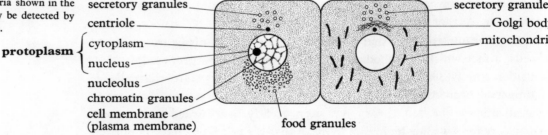

A) Animal Cells

secretory granules
centriole
protoplasm { cytoplasm
nucleus
nucleolus
chromatin granules
cell membrane (plasma membrane)
10 μm
food granules
secretory granules
Golgi body
mitochondria

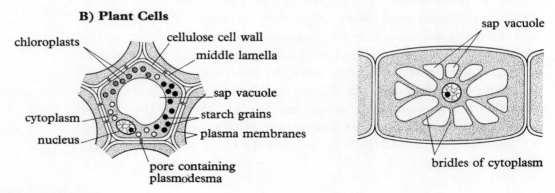

B) Plant Cells

chloroplasts
cellulose cell wall
middle lamella
sap vacuole
starch grains
cytoplasm
nucleus
plasma membranes
pore containing plasmodesma
sap vacuole
bridles of cytoplasm

[1] If you are not familiar with the units used in microscopy you should consult Appendix III, p. 607. The symbol μm means micrometre, i.e. a thousandth of a millimetre.

Later we shall have much to say about the functions of the nucleus. For the moment you should appreciate that it is vital for the continued life of the cell. This has been shown by removing it. Certain cells are large enough to permit this sort of operation to be performed successfully. For example, using special instruments it is possible to extract the nucleus from *Amoeba*, a single-celled animal which may be as large as a pinhead. When this is done the enucleated cell eventually dies. If, however, the nucleus is replaced sufficiently soon the cell revives and will continue to live indefinitely. It is therefore clear that although important chemical reactions take place in the cytoplasm, the nucleus is essential for directing activities and maintaining the life of the cell.

The **centriole**, found just outside the nuclear membrane plays an important part in the formation of **cilia** and **flagella**, slender motile hairs that project from the surface of certain cells. The behaviour of the centriole is also related to the way cells divide during multiplication.

A PLANT CELL AS SEEN WITH THE LIGHT MICROSCOPE

As you can see in Fig 2.2B, most of the structures found in an animal cell also occur in plant cells. However, a typical plant cell has certain additional features. The centre of the cell is taken up by a large **vacuole** filled with a solution containing sugars and salts, the **cell sap**. The cell is bounded by a comparatively thick wall made of the polysaccharide carbohydrate **cellulose**. This is tough but slightly elastic. We shall see later that the sap vacuole and cellulose wall play a major part in maintaining the shape and form of the cell.

The presence of a sap vacuole results in the cytoplasm being pushed to the sides of the cell. In most plant cells the nucleus is located somewhere in this peripheral cytoplasm, but not uncommonly it is suspended in the middle of the vacuole by slender strands of cytoplasm.

Another consequence of the vacuole is that there are two plasma membranes, one lining the outer surface of the cytoplasm (in contact with the cellulose wall), and the other lining the inner surface bordering the vacuole. The protoplasmic material sandwiched between these two membranes is referred to as the **protoplast**.

The cell wall is laid down during development of the cell, and starts as a thin layer of **calcium pectate** beneath which cellulose, secreted by the outer layer of cytoplasm, is laid down (**primary wall**). Further cellulose constitutes the **secondary wall**. The point of demarcation between one cell and the next, known as the **middle lamella**, represents the fused pectate walls of the two adjoining cells. Although each cell appears to be enclosed in a box of cellulose it is by no means isolated from its neighbours. The cellulose wall is interrupted at intervals by narrow pores carrying fine strands of cytoplasm which join the cells to one another. These are called **plasmodesma strands** or **plasmodesmata** and are thought to facilitate movement of materials between cells. They also play an important part in the deposition of cellulose during thickening of the secondary cell wall.

Granules and inclusions found in the cytoplasm include hollow ovoid or spherical bodies called **plastids**. These are of two main types. The **leucoplasts** are colourless and contain starch: **starch grains** are found very widely in plant cells and they represent the major form of storage carbohydrate, equivalent to glycogen in animal cells. The other main type of plastid, the **chloroplast**, contains the green pigment **chlorophyll** about which we shall have much to say later. This plays a crucial rôle in photosynthesis, the process by which plants manufacture food materials.

This brief account of cell structure, based on information gained from the optical microscope, demonstrates some of the important differences between animal and plant cells. These are related to the different tasks which each type of cell has to perform. But in this chapter we shall be concerned with the structures which they have in common: cytoplasm, nucleus and plasma membrane.

It is clear that the optical microscope does not tell us very much about the detailed structure of these three parts of the cell. It is true that with appropriate staining techniques chromosomes can be detected in the nucleus and granules in the cytoplasm, but the plasma membrane appears to be no more than a thin line and no detailed organization can be seen in the cytoplasm.

THE ELECTRON MICROSCOPE

In recent years our entire concept of the cell has been revolutionized by the development of the **electron microscope** (Fig. 2.3). This instrument uses an electron beam instead of light and electro-magnets instead of glass lenses. Freed from the limitations imposed by the wavelength of light, the electron microscope has a resolving power five hundred times as great as the optical microscope. A good optical microscope can only magnify an object without loss of detail about 1,500 times. The electron microscope can achieve magnifications of over 500,000 times. It is important to appreciate what this means in practice: with the electron microscope an object the size of a pinhead can be enlarged without loss of detail to the point at which it has a diameter of well over a kilometre; a cell with a diameter of 10 micrometres finishes up with a diameter of five metres. It is impossible to exaggerate the impact which this instrument has had on biology. Materials which were formerly described as structureless have been shown to have an elaborate internal organization, and so-called homogeneous fluids are now known to contain a variety of complex structures. The electron microscope has opened up a new world of structure whose existence was barely realized twenty years ago.

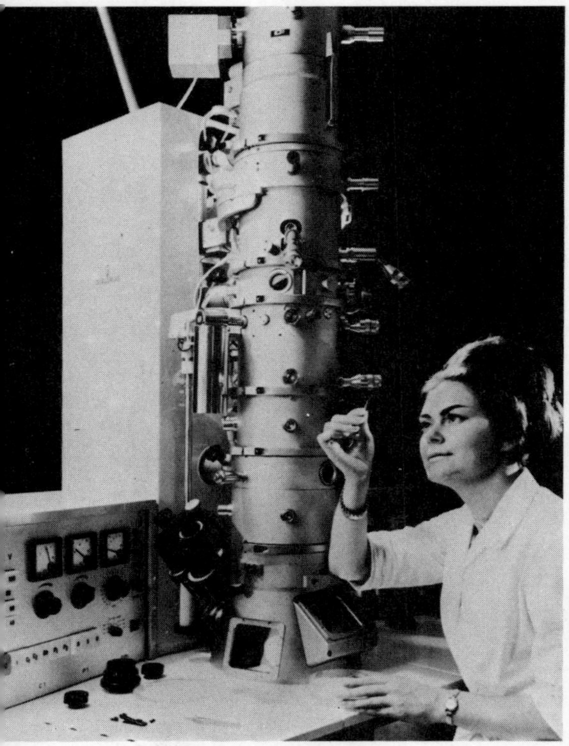

But there are problems. One snag is that the material for examination has to be mounted in a vacuum, and is therefore dead, before it can be viewed. This, coupled with the preliminary treatment to which it has to be subjected, may cause distortion of the delicate structures inside cells. The electron-microscopist is always on the look-out for such artifacts and uses every means to prevent them occurring. To some extent these problems can be overcome by using other types of microscopy in addition to the electron microscope. The **phase-contrast microscope**, for example, enables transparent objects to be seen, and is ideal for studying unstained living cells. Special illumination techniques can also be employed for increasing the contrast between the object and its background; for example, **dark-ground illumination**, in which the object is illuminated from above against a dark background, enables tiny structures inside cells to be seen clearly. Another technique is to examine the object in polarized light: the **polarizing microscope** is useful for differentiating between different types of material embedded in another substance.

The electron microscope, combined with these other kinds of microscopy, has provided us with an incredibly detailed picture of cell structure.

Fig 2.3 A high-resolution electron microscope currently used in biological research. This instrument has a resolution of approximately 1 nm and a magnification of up to 280,000 times. The operator is holding in her forceps the tiny copper grid on which the specimen is mounted before being examined in the microscope. (*Siemens (U K) Ltd.*)

At the same time biochemists have been busy investigating the functions carried out by many of the structures revealed by the cytologists. Using high-speed centrifuges it is possible to separate, isolate and purify the constituents of cells and investigate their chemical properties. These techniques have yielded much information on the processes that occur inside cells.

Fine Structure of the Cell

The organization in an individual cell as revealed by the electron microscope is known as its **fine structure,** or **ultra-structure.** The specific entities making up this organization are called **organelles.** In a sense organelles stand in relation to the cell as organs do to the whole organism. In fact sub-cellular organization is in many respects analogous to the kind of organization which exists in the body of a complex multicellular organism, specific functions being carried out in certain parts of the cell, and all these functions being integrated in an orderly manner.

Fig. 2.4 is an electron micrograph of the central part of an animal cell.

Fig 2.4 Electron micrograph of part of a pancreas cell. Notice pores in the nuclear membrane, endoplasmic reticulum encrusted with ribosomes, and prominent mitochondrion with cristae. The dense spherical structure in the nucleus is the nucleolus. (*K. R. Porter, Harvard University*).

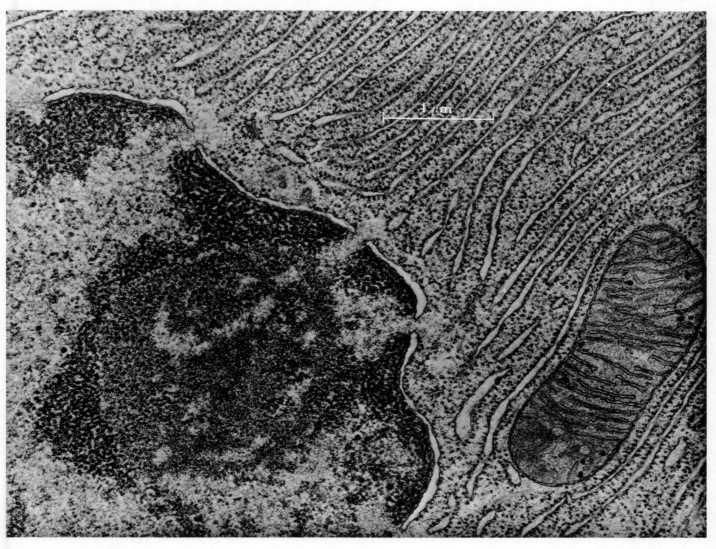

Fig. 2.5 is a diagram of a generalized animal cell based on detailed examination of numerous electron micrographs. Refer to these illustrations as you read the following description of the various components of the cell.

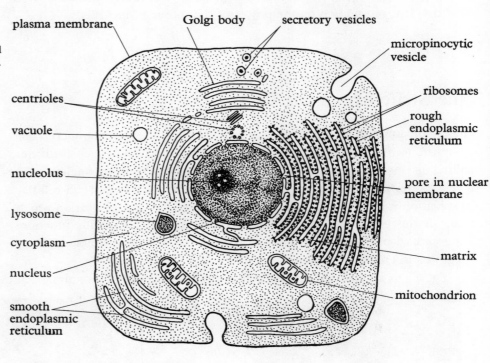

Fig 2.5 Fine structure of a generalized animal cell based on studies with the electron microscope. (*modified after Brachet*)

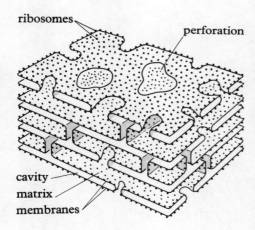

Fig 2.6 The endoplasmic reticulum consists of a series of parallel membranes, encrusted with ribosomes, enclosing a system of interconnected flattened cavities.

¹nm is the symbol for nanometre, a thousandth of a micrometre. See Appendix III.

ENDOPLASMIC RETICULUM

You will see immediately that, far from being homogeneous, the cytoplasm is a highly organized material consisting of a **matrix** containing a system of parallel flattened cavities lined with a thin membrane about 4 nm thick[1]. This system is known as the **endoplasmic reticulum**, or ER for short. The cavities are interconnected as shown in Fig 2.6 and the lining membranes are continuous with the nuclear membrane. Attached to the matrix side of the membranes are numerous granules which, because they are rich in ribonucleic acid, are called **ribosomes.** These are the sites where proteins are synthesized in the cell. The bulk of the ER in most cells is encrusted with ribosomes and accordingly is known as **rough endoplasmic reticulum**. Its general function is to isolate and transport the proteins which have been synthesized by the ribosomes. Many of these proteins are not required by the cell in which they are made but are for 'export', i.e. they are **secreted** by the cell. Such proteins include digestive enzymes and hormones. The ER is thus a kind of intracellular transport system facilitating movement of materials from one part of the cell to another. In this connection it is interesting that the nuclear membrane is pierced by tiny pores which are continuous with the ER, thus providing a route by which materials might move from nucleus to cytoplasm and vice versa. These pores are seen in section in Fig 2.4 and in surface view in Fig 2.7.

In certain parts of the cell the ER is not encrusted with ribosomes and accordingly is known as **smooth endoplasmic reticulum**. This is not continuous with the rough ER and its cavities are tubular rather than flattened sacs. It is thought to be concerned with the synthesis and transport of lipids and steroids.

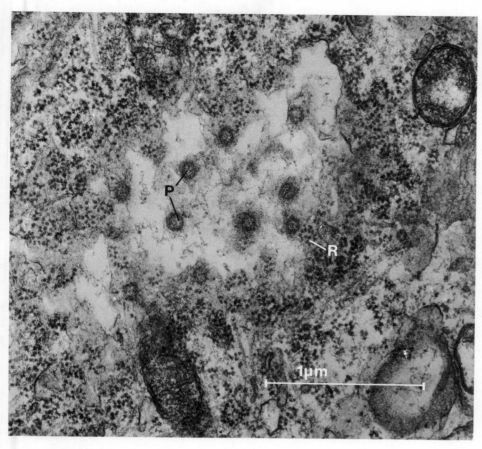

Fig 2.7 Section of a cell showing pores (**P**) in the nuclear membrane in surface view. This unusual electron micrograph also shows the way ribosomes are arranged in small groups to form polyribosomes or polysomes (**R**) (see p. 495) (*A. R. Lieberman, University College, London.*)

GOLGI BODY

A structure where smooth ER is particularly prominent is the so-called **Golgi body**. With special staining techniques this can be detected under the optical microscope as a particularly dense region of the cytoplasm, and as such it has exercised the minds of cytologists ever since it was first discovered by the Italian physician Camillo Golgi at the turn of the century. In the electron microscope it is seen to consist of stacks of flattened cavities lined with smooth ER close to which are numerous vesicles containing secretory granules (Fig 2.8).

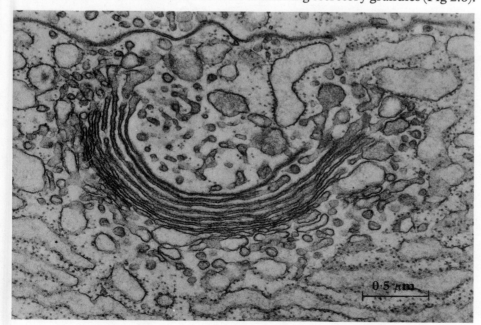

Fig 2.8 Golgi body of a specialized unicell. Notice the vesicles pinched off from the ends of the flattened cavities. (*B. L. Gupta, University of Cambridge*)

Much controversy has centred on the function of the Golgi body. The fact that secretory granules are closely associated with it suggests that it may be concerned with the production of substances by the cell. Recently this hypothesis has received support from experiments in which the distribution of radioactively labelled substances taken up by the cell is followed by autoradiography (see p. 82). Nearly all cell secretions are glycoproteins, i.e. proteins conjugated with a carbohydrate. Newly-synthesized proteins are found in the channels of the rough ER. From here they move to the Golgi body where the carbohydrate is added to them. Enveloped in part of the Golgi membrane, they then leave the cell. Thus the Golgi apparatus is an assembly point through which raw materials for secretion are funnelled before being shed from the cell.

How does this movement of materials take place? We are a long way from knowing the full answer to this question but certain facts are clear. In the first place the rough ER is not continuous with the Golgi body. The latter is formed by the fusion of vesicles which are pinched off from the cavities of the rough ER. Vesicles containing the secretory molecules then get pinched off from the channels of the Golgi body. These vesicles then move to the surface of the cell and discharge their contents to the exterior. Thus, to summarize, the function of the Golgi body is to add the carbohydrate component to the protein and package the finished product before it leaves the cell (Fig. 2.9).

Fig 2.9 Schematic diagram illustrating the formation and function of the Golgi body. Protein and carbohydrate derived from the channels of the rough endoplasmic reticulum are combined in the Golgi body to form a glycoprotein secretion which is discharged from the cell as shown.

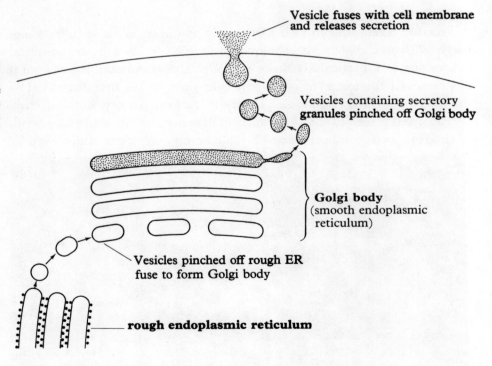

Vesicle fuses with cell membrane and releases secretion

Vesicles containing secretory **granules pinched off Golgi body**

Golgi body (smooth endoplasmic reticulum)

Vesicles pinched off rough ER fuse to form Golgi body

rough endoplasmic reticulum

MITOCHONDRIA

Embedded in the matrix of the cytoplasm are found variable numbers of **mitochondria**. Under the phase-contrast microscope, or under the light microscope with dark-ground illumination, they appear as minute rods, but in the electron microscope their internal structure becomes apparent. It has been estimated that a typical cell contains about a thousand mitochondria, though some have many more than this. Their shape and size vary, but generally they are sausage-shaped with a diameter of approximately 1.0 μm and a length of about 2.5 μm. The wall of the mitochondrion consists of two thin membranes separated by an extremely narrow fluid-filled space. The inner membrane is highly folded giving rise to an irregular series of partitions, or **cristae**, which project into the interior (Fig. 2.10). The latter contains an organic matrix consisting of numerous chemical compounds.

Fig **2.10** Structure of mitochondrion. A) Electron micrograph of a large mitochondrion in longitudinal section from the oöcyte of a bird. Notice the cristae projecting into the hollow interior. (*Ruth Bellaires, University College, London*). B) Cutaway view showing the inside of a mitochondrion. Most of the chemical reactions involved in the release of energy (respiration) take place in the matrix and on the cristae. Tiny stalked granules attached to the surface of the cristae, of diameter approximately 0.8 nm, are thought to be the site of respiratory enzymes.

A

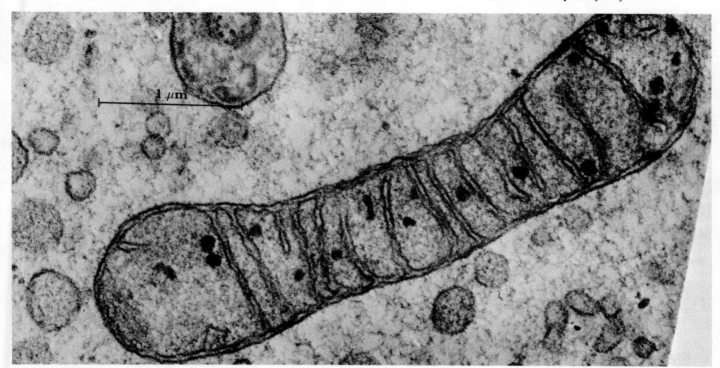

The mitochondrion is one of the cell's most important organelles, for it is here that the chemical reactions of respiration take place, by which energy is generated for the needs of the cell. These reactions are accelerated by **enzymes**, organic catalysts, which are thought to be located on the inner walls and cristae of the mitochondrion. The cristae have the effect of increasing the surface area for attachment of enzymes so that more can be packed into a comparatively small space. It is interesting that cells whose function requires them to expend particularly large amounts of energy contain unusually large numbers of mitochondria. These are often packed close together in the part of the cell where the energy is required. This is seen dramatically in spermatozoa where the mitochondria are tightly packed at the base of the motile tail. They are also found in great abundance alongside the contractile fibrils in muscle, particularly the flight muscle of insects, and at the surface of cells in which active transport occurs. In some cases the cristae may be very close together thus increasing the surface area within each mitochondrion.

B

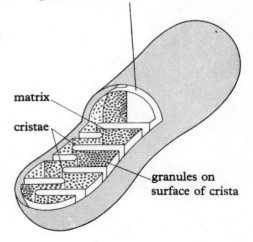

space between the two membranes lining mitochondrion

matrix

cristae

granules on surface of crista

LYSOSOMES

Also prominent in the cytoplasm of most cells are dark spherical bodies called **lysosomes**. The word 'lysis' means to break up, and there is evidence

Fig 2.11 Diagrams to show how lysosomes destroy unwanted organelles in a cell. In this case two mitochondria are broken down.

The two mitochondria are about to be destroyed by the lysosomes on the left.

A membrane forms round the mitochondria enclosing them in a sac, and the lysosomes move towards the sac.

The lysosomes discharge their contents into the sac.

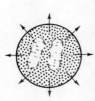

The lysosome enzymes break down the two mitochondria and the products of this digestive process are absorbed into the surrounding cytoplasm.

Fig 2.12 Plasma membranes (**M**) of two adjacent epithelial cells. Notice that each plasma membrane has the triple structure characteristic of the unit membrane. The space (**S**) between the two plasma membranes contains intercellular material. (*B. L. Gupta, University of Cambridge.*)

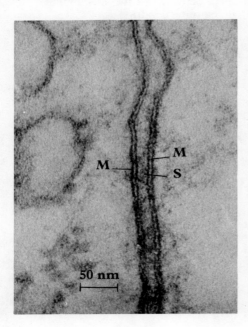

50 nm

that the lysosomes contain potent enzymes responsible for splitting complex chemical compounds into simpler sub-units. Digestion is carried out in a membrane-lined vacuole into which several lysosomes may discharge their contents. This is discussed in more detail on p. 56.

Another important function of lysosomes is that they destroy worn-out organelles within the cell (Fig 2.11). The unwanted structures, they may be mitochondria or part of the endoplasmic reticulum, are enveloped in a membrane which forms a bag round them. Into this bag several lysosomes discharge their contents. The organelles are broken down by the lysosome enzymes, and the products of digestion absorbed into the surrounding cytoplasm to be used in the construction of new organelles.

Sometimes lysosomes may destroy the entire cell. In this case the lysosome membrane ruptures or dissolves thereby liberating the enzymes which proceed to digest the contents of the cell, killing it in the process. This may seem rather disastrous but in certain instances it may be desirable, as for example when old, worn-out cells have to be replaced by new ones. Lysosomes thus act as 'suicide bags', destroying cells after they have done their job.

PLASMA MEMBRANE

The thin line seen under the light microscope turns out to be more complex when viewed in the electron microscope. From Fig 2.12 you will see that the plasma membrane is made up of three layers: two dark layers, separated by a lighter region. The total thickness of the membrane is approximately 7.0 nm.

Examination of the plasma membrane with the polarizing microscope has thrown light on the molecular structure of these layers. This, coupled with chemical analysis, has shown that the two dark layers are made of protein and the light region in between consists of lipid. The plasma membrane is therefore a thin sandwich of lipid contained between two layers of protein. Beyond the plasma membrane is a glycoprotein coat of variable thickness. Cells abutting against each other in a multicellular tissue are separated by this intercellular material. Through it various materials pass as they flow in and out of cells.

The membranes of the endoplasmic reticulum, the nuclear membrane, the membrane surrounding the sap vacuole and plastids of plant cells, and the membranes surrounding the mitochondria and other organelles all share the characteristic three-layered structure typical of the plasma membrane. This kind of structure, wherever it occurs, is known as the **unit membrane**. On account of their basic similarity of appearance, it has been suggested that the internal membranes of the cell may have been formed by evagination of the nuclear membrane or possibly by infolding of the plasma membrane. However, the various membranes in the cell are of different thicknesses and have different biochemical properties and it is possible that they have had an independent origin.

Various structures are associated with the surface of the cell. Of these the most universal in occurrence are **micropinocytic vesicles**, flask-like invaginations of the plasma membrane. The neck of the flask eventually closes up so that the vesicle is sealed off from the outside and becomes entirely enclosed within the cell. Now known as a **vacuole**, it moves about freely in the cytoplasm. An animal cell may contain numerous small vacuoles which are constantly being formed by **pinocytosis** at the cell surface. As we shall see later this may provide an important means by which large molecules are taken into the cell. It is thought that the large sap vacuole characteristic of mature plant cells is formed by the fusion of numerous small vacuoles derived from pinocytosis.

MICROVILLI

It has already been mentioned that one function of the plasma membrane is to permit the entry of materials into the cell. Plainly, the greater the surface area of the plasma membrane the greater will be the exchange of materials across it. To this end the membrane of many cells is folded to form minute, hair-like projections called **microvilli**. Each microvillus is a very thin, finger-like process about 1.0 μm long and 0.08 μm wide. It is lined with plasma membrane and filled with cytoplasm which is continuous with the main body of the cell. Microvilli are only visible in the electron microscope though larger ones, if densely packed, may show up under the optical microscope as a fuzzy line at the cell surface—the so-called **brush border** or **striated border**. Two places where microvilli abound are the lining of the convoluted tubules in the kidney and the epithelial lining of the small intestine. In both cases their function would appear to be to increase the surface area of the epithelium thereby aiding absorption of materials from the enclosed cavity. It has been estimated that a single epithelial cell may have 3,000 microvilli and that in one square millimetre of intestinal lining there may be as many as 200,000,000, giving an estimated twenty-fold increase to the surface area. However, microvilli are not invariably associated with absorption and the possibility that they perform other functions cannot be ruled out.

CILIA AND FLAGELLA

The surface of certain cells is drawn out to form fine hair-like processes called **cilia** or **flagella** according to their length. Although they are found only in certain cells, they are of sufficiently wide occurrence to merit discussion in this chapter. Cilia and flagella are fundamentally similar but the former are usually shorter and more numerous than flagella. Both have the propensity to lash back and forth, and their rôle depends on this fact. Many unicellular organisms move by means of cilia or flagella: the surface of *Paramecium*, for example, is covered with cilia which beat in a coordinated fashion, driving the animal through the water in which it lives; in contrast *Euglena* has a single flagellum, much longer than a cilium, which propels the organism by means of rapid undulations passing from the base to the tip. Sometimes cilia occur on the underside of comparatively large animals like flatworms and marine snails where their rapid beating, aided by muscular contractions, enables the animal to glide on smooth surfaces. Flagella are nearly always associated with locomotion, but cilia, which are of much wider occurrence, perform other functions as well. For example, they are often found lining ducts and tubules and other specialized surfaces along which materials are wafted by means of their rapid and rhythmical beatings.

Until the advent of the electron microscope very little was known about cilia and flagella. The reason for this was their small size. Although fairly long (an average flagellum has a length of about 100 μm, a cilium 5 to 10 μm), they are extremely fine, being generally less than 0.3 μm in diameter. They are therefore barely visible under the optical microscope. However the electron microscope has enabled cytologists to analyse their detailed structure (Figs 2.13 and 2.14).

Cross-sections of a cilium show that it contains eleven hollow fibres or tubules which run longitudinally along its length. These are arranged in a precise way: there are two in the centre surrounded by a ring of nine. This arrangement is described as the **9 + 2 pattern**. The bundle of fibres is enclosed within a membrane which is continuous with the plasma membrane. At the base of the cilium is an elaborate attachment apparatus consisting of a **basal body** from which **rootlet fibres** may be seen to penetrate into the deeper layers of the cytoplasm. The basal body is composed of nine peripheral fibres continuous with those in the cilium itself. However, the two central fibres are absent.

Fig 2.13 Fine structure of cilia and flagella. A) Electron micrograph of cilia of a trochophore larva (see p. 559). This striking three-dimensional picture was obtained using a scanning electron microscope: the specimen is covered with a thin film of conducting metal and scanned by an electron beam in a manner essentially similar to that used in a television set. The reflected electrons produce an image on the cathode ray screen. (*V. I. Barber, Patricia Holborow and M. S. Laverack, Universities of Bristol and St. Andrews*). B) Transverse section through the flagella of the unicell *Trichonympha*. Notice the 9 + 2 arrangement of tubular fibres within each flagellum. C) Longitudinal section through the centre of a single flagellum of *Trichonympha* (B and C: *A. V. Grimstone, University of Cambridge*)

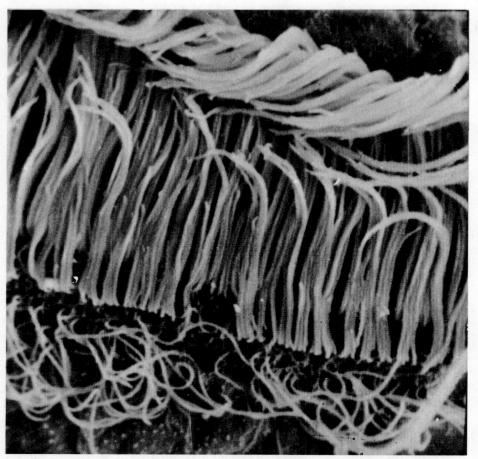

A

Little is known about how cilia and flagella move. It might be surmised from their internal structure that the fibres are contractile and that shortening of them causes bending. Alternatively they may slide relative to each other. As yet there is no proof of either of these hypotheses. Indirect evidence that the two central fibres play a vital rôle in the process comes from the observation that in non-motile cilia and their derivatives they are absent. What does seem certain is that ciliary movement will only occur if the basal body is intact. This has been shown by several different experiments. For example the cilia lining the gills of the fresh-water mussel stop beating if they are stripped from their cell attachments. The flagellum of *Euglena* will continue to undulate when separated from the organism provided that the basal bodies are still attached to it. Isolated sperm tails are reported to behave similarly.

It appears, therefore, that while shortening or sliding of the fibres may bring about bending of the cilium or flagellum, the process is initiated from the base and is transmitted towards the tip possibly along the ciliary membrane. That ATP provides the energy for ciliary movement has been shown by the fact that isolated cilia separated from their basal bodies bend in the presence of ATP. Furthermore chemical analysis of isolated fragments of cilia has shown that

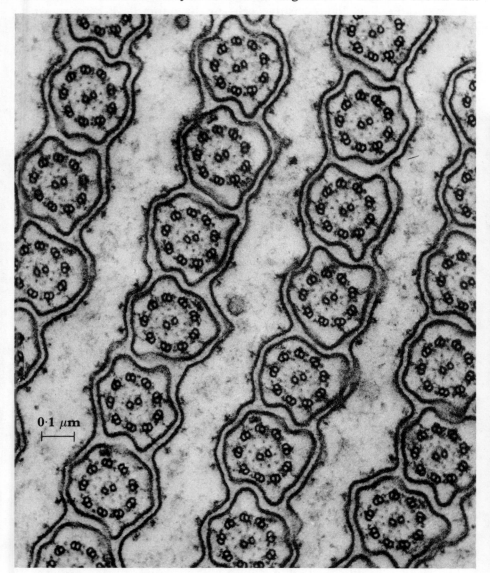

B

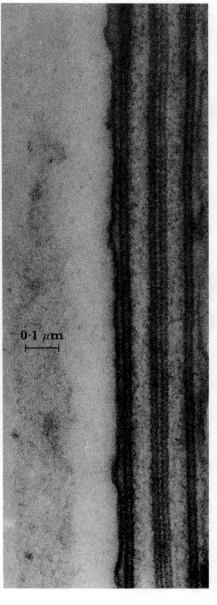

C

an enzyme which splits ATP is found in great abundance in short arm-like processes which project from the peripheral fibres. These arms thus represent the site of energy-production, but how the energy is utilized in bending the cilium is quite unknown.

OTHER STRUCTURES WITH THE SAME ORGANIZATION AS CILIA

As you can see in Fig 2.5 an animal cell has two rod-like **centrioles** situated at right angles to each other. Under the optical microscope these appear as a single unit. Their behaviour in cell division is discussed in Chapter 23 and for the moment we will concern ourselves only with their structure.

A discovery of great interest is that the centrioles have an internal structure similar to that of the basal body of cilia and flagella, each rod

Fig 2.14 The detailed structure of a cilium based on electron micrographs. Note that the nine peripheral fibres penetrate into the basal body where they may be attached to collagen-like root-let fibres (not always present). The little arm-like processes projecting from the peripheral fibres shown in the cross-section in B are thought to be the site of ATP breakdown where energy is released for shortening of the fibres.

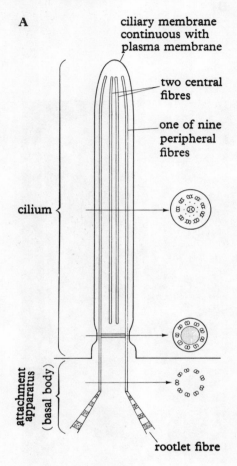

A

ciliary membrane continuous with plasma membrane

two central fibres

one of nine peripheral fibres

cilium

attachment apparatus (basal body)

rootlet fibre

B

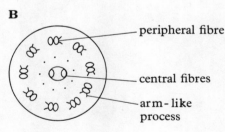

peripheral fibre

central fibres

arm-like process

containing nine peripheral fibres. This is no coincidence, for they are now known to be responsible for the formation of cilia and flagella. It is possible that the basal body is derived in evolution from a centriole which migrated towards the edge of the cell and became associated with the cell membrane.

Another surprising place where ciliary structure appears is in the eye of vertebrates. The light-sensitive cells in the retina have been shown to contain nine peripheral fibres. What is more they also possess basal bodies and rootlet fibres just like ordinary cilia. These cells are certainly not concerned with movement but their tell-tale internal structure suggests that they have evolved from ciliated cells which lost their motility and became adapted for the reception of light stimuli. What all this suggests is that the 9 + 2 pattern of fibres probably arose at an early stage in evolution and, despite much modification, is now a basic feature of many cells.

MICROTUBULES

The tubular nature of the ciliary fibres is very clear in Fig 2.13B. Since their discovery in cilia and flagella similar tube-like structures have been found in many other cells, both animal and plant. The similarity may, of course, be superficial but they do look remarkably alike and have been given the general name of **microtubules**. They are widely distributed in the cytoplasm where they may occur singly or in bundles. The spindle fibres involved in cell division (see Chapter 23), are also microtubules. At the present time it is not possible to say anything positive about their general function but they seem to be associated with cellular movements of one sort and another.

FINE STRUCTURE OF PLANT CELLS

Although we have been looking at the fine structure of animal cells, all the organelles described so far are also found in plant cells. The only qualification is that cilia, flagella and centrioles are confined to the more primitive plant groups such as the algae and are absent in higher plants.

However, plant cells possess certain additional features of their own, the most important of which are the **chloroplasts** and **cell wall**. The fine structure of the chloroplast is discussed in Chapter 10 and will therefore not be described here. With regard to the cell wall, the electron microscope has confirmed earlier light microscope studies that this is not a uniform structure. In the early stages of its development a plant cell has a thin **primary wall** composed of calcium pectate and some cellulose. Later, further cellulose is laid down in a series of layers on the inner surface of the primary wall, giving rise to the thick **secondary wall** typical of a mature cell. Now in places the secondary wall is absent altogether, giving rise to a **pit**. Where pits occur two adjacent cells may be separated only by the primary wall (Fig 2.15). Presumably the thinness of the wall in these regions facilitates movement of molecules between adjoining cells. This function also applies to the **plasmodesmata**. The electron microscope has shown that the plasmodesma strands contain endoplasmic reticulum which is therefore continuous from cell to cell (Fig 2.16).

Molecular Structure of the Plasma Membrane

The plasma membrane separates the cell from its surroundings and undoubtedly it is important in controlling what enters and leaves the cell. For this reason much attention has been focused on its structure.

Being so thin and difficult to see, one has to resort to an indirect approach in working out the structure of the plasma membrane. For example, it is possible to make certain predictions from its physico-chemical properties. One of its most interesting properties is that it is penetrated particularly rapidly by substances which dissolve in oil, suggesting that it contains lipid. This is supported by other lines of evidence including the observation that its permeability properties are greatly influenced by treatment with lipid solvents.

Assuming that the plasma membrane does contain lipid, the question arises as to how the lipid molecules are arranged. This is a matter of some importance as it could determine the functional properties of the membrane. A possible answer to this question is provided by the properties of lipids.

In certain lipids (the phospholipids for example) the long hydrocarbon chains which project from the glycerol part of the molecule are insoluble in water whereas the glycerol end of the molecule is water-soluble, that is it contains polar groups. Now when such a lipid is allowed to spread over the surface of pure water, the water-soluble ends of the lipid molecules are drawn

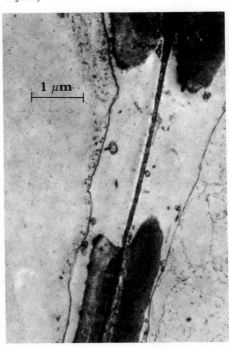

Fig 2.15 Electron micrograph of a pit between two cortical cells in the root of the pea plant *Pisum*. Notice the complete absence of the secondary cell wall (cellulose), only the primary wall being present. (*B. E. Juniper, University of Oxford.*)

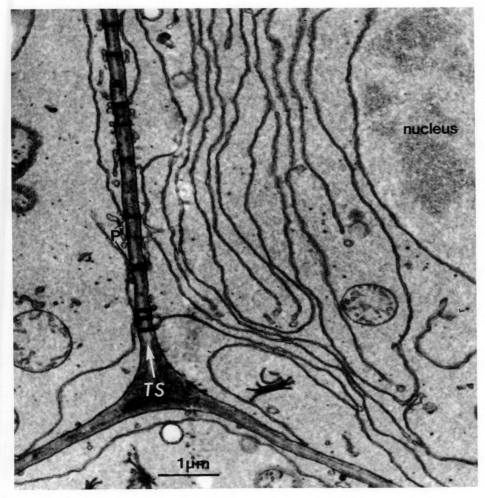

Fig 2.16 Plasmodesmata (**P**) linking adjacent cells in a root cap cell of maize. Notice that the plasmodesmata are connected to the endoplasmic reticulum and are confined to the transverse wall (**TS**) of the cell (*B. E. Juniper, University of Oxford*)

into the water and the insoluble hydrocarbon chains, if the lipid molecules are sufficiently tightly-packed, point directly away from the surface of the water. Thus we get a single layer of lipid molecules with their hydrocarbon chains orientated at right angles to the surface: a so-called **monomolecular film** or **monolayer**. The lipid component of the cell membrane cannot in fact be a monolayer for this is only formed where there is a water surface in contact with air, and the cell membrane generally has water in contact with both sides. However, we know that when the non-polar sides of two monomolecular films are brought into contact, the non-polar ends of the lipid molecules are attracted to each other to form what is called a **bimolecular leaflet** (Fig 2.17).

Fig 2.17 Diagrams illustrating the formation of a bimolecular leaflet. The polar ends of the lipid molecules are shown as solid dots, the hydrocarbon chains as bold lines (*after J. A. Ramsay, Experimental Basis of Modern Biology, Cambridge University Press*).

A

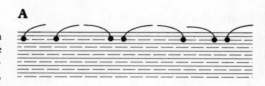

A) A polar lipid spreads out over the surface of water. The polar ends of the molecules, being soluble in water, enter the water while the insoluble hydrocarbon chains stay outside. If there is a large surface available and comparatively few lipid molecules, the latter lie parallel to the surface as shown.

B

B) But if a large number of lipid molecules are packed together on a restricted surface, the hydrocarbon chains project from the water at right angles to the surface. This gives a monolayer of lipid at the air-water interface.

C

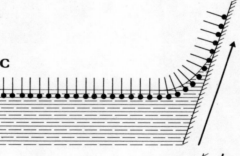

C) A clean glass surface is drawn slowly out through the lipid monolayer. The lipid molecules adhere to the glass as shown.

D

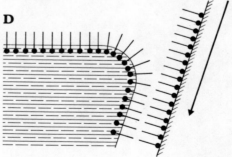

D) If the glass surface, after drying, is then pushed back through the lipid monolayer, the two layers of lipid molecules arrange themselves as shown, forming a bimolecular leaflet. This has a number of properties in common with the plasma membrane.

The bimolecular leaflet could very well be the basis of the lipid part of the plasma membrane, but this on its own would be insufficiently strong as a cell membrane and there would have to be something to hold it together and give it mechanical strength. This function, it has been argued, could be performed by a protein framework on each side of the lipid layer, a suggestion which is supported by the elasticity of the membrane as well as its surface tension properties.

In the late 1930s J. F. Danielli and H. Davson put forward a theory of membrane structure based on the kind of arguments outlined above. The Danielli-Davson hypothesis proposes that the plasma membrane is made up of three layers: a bimolecular layer of lipid sandwiched between two layers of protein, the lipid molecules being set at right angles to the surface. From the speeds at which various molecules penetrate the membrane they predicted the lipid layer to be about 6.0 nm in thickness, and each of the protein layers about 1.0 nm, giving a total thickness to the membrane of about 8.0 nm.

In recent years the Danielli-Davson hypothesis has received support from more direct methods of analysis. Electron micrographs of the plasma membrane show it to consist of two dark (electron-dense) layers separated by a lighter area in between (see Fig 2.12). The total thickness of the membrane is about 7.5 nm, a figure which agrees very closely with that proposed by Danielli and Davson some twenty years earlier. While the electron microscope has confirmed the dimensions and triple-layered structure of the cell membrane, X-ray diffraction studies have gone a long way towards confirming the orientation of the constituent lipid and protein molecules. A diagram of the cell membrane based on such studies is shown in Fig 2.18.

Fig 2.18 Diagram of the cell membrane based on the Davson-Danielli hypothesis and electron microscopy. The polar (electrically charged) ends of the protein and lipid molecules are shown as open and filled in circles respectively.

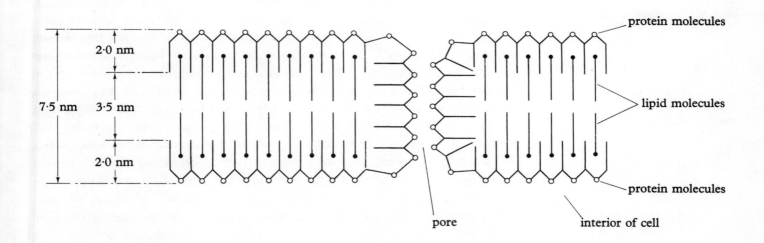

PORES IN THE CELL MEMBRANE

In Fig 2.18 a pore is shown in the plasma membrane. It is envisaged that such pores perforate the membrane at regular intervals. That such pores exist was predicted many years ago on the grounds that certain molecules, insoluble in lipid and therefore unable to get in between the lipid molecules, are still capable of penetrating the membrane. The pores are thought to be lined with hydrophilic groups (the polar groups of the protein molecules) and are therefore penetrated readily by water-soluble substances. By measuring the resistance of the membrane to the passage of such molecules, together with other methods of determining the diameter of small pores, it has been established that the pores must be less than 1.0 nm wide. As such they are too small to be seen even with the electron microscope, and so the truth of their existence must rest, for the time being at any rate, on experimental evidence.

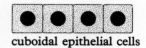

cuboidal epithelial cells

white blood cell

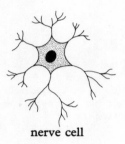

nerve cell

smooth muscle fibre

motile tail

spermatozoon

pigment

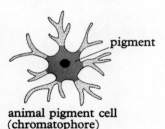

animal pigment cell
(chromatophore)

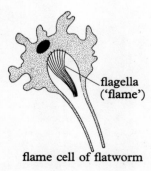

flagella
('flame')

flame cell of flatworm

The Diversity of Cells

So far we have considered the basic features of cells in general. But this should not be taken to imply that all cells are identical. Structures like chromosomes, mitochondria, endoplasmic reticulum and ribosomes are common to virtually all cells, but the shape, form and contents of individual cells show much variation. The structural characteristics of a particular cell are closely correlated with its functions. This will be amply illustrated in later chapters, but it is as well to appreciate from the start the striking diversity of cells in both the animal and plant kingdoms. This diversity is seen between different species, and within a single species. Thus the cells of *Hydra* differ considerably from those of man despite the fact that they share the same basic features and to some extent perform the same functions. Within the body of *Hydra*, a comparatively simple animal, there are seven different types of cell, each more or less specialized to perform a particular task. In man this differentiation is carried much further. In the human body there are numerous different types of cell, each highly specialized and adapted for a particular function. Thus **epithelial cells** (of which there are several different kinds) possess a shape and form that makes them most suitable for lining the surface of the body and the organs and cavities within it. **Glandular cells** are responsible for producing some kind of secretion, for example mucus. Such cells have a particularly prominent Golgi body, an indirect piece of evidence implicating the Golgi apparatus in secretion. **Fibroblasts** secrete the protein fibres that make up connective tissue binding tissues and organs together. **Chondroblasts** and **osteoblasts** produce cartilage and bone. Blood contains several kinds of specialized cells. **Erythrocytes** (**red blood cells**), biconcave, disc-shaped, and loaded with the red pigment haemoglobin, are responsible for conveying oxygen about the body; they are one of the few cells to lack a nucleus. **Leucocytes** (**white blood cells**), generally amoeboid, play an important part in defending the body against disease. One of the most specialized of all animal cells is the **nerve cell** or **neurone** whose slender arm-like processes transmit electrical impulses through the nervous system. **Sensory cells** are also capable of electrical activity which is generated by specific kinds of stimulation: touch, sound, light, heat or chemical substance as the case may be. Also capable of electrical activity are **muscle fibres**, elongated cells in which electrical activity is accompanied by contraction.

Muscle fibres are not the only cells capable of movement. Certain white blood cells can undergo amoeboid movement by streaming of the cytoplasm, but the most spectacular case of mobility is seen in the **spermatozoon,** a single cell with an elongated tail containing the 9 + 2 arrangement of fibres characteristic of flagella. Capable of powerful undulations, the tail propels the spermatozoon forward in its watery medium. Muscle cells and spermatozoa, with their high energy requirements, contain large numbers of densely packed mitochondria which, as we have seen, are the site where energy-producing chemical reactions take place.

The cells described so far all occur in the mammal. Other specialized cells are found in animals below the level of mammals. For example the skin of animals such as amphibians and reptiles contain **chromatophores,** stellate cells containing a pigment which can be concentrated or expanded, thereby changing the colour of the body. **Flame cells,** so called because of the 'flickering' of flagella in the cavity within them, are found in flatworms where they play an important part in osmoregulation. One of the most remarkable animal cells is the **nematoblast** or **stinging cell** of sea anemones, jelly fish and *Hydra*. Its eversible thread, containing a toxic fluid, is capable of piercing and poisoning prey.

Turning to plants, **photosynthetic cells** of various shapes and forms, packed with chloroplasts, perform the task of building up complex molecules. **Parenchyma cells,** packed tightly together, fill up spaces between tissues. **Epidermal cells,** whose outer walls possess a waxy cuticle, protect the plant from excessive water-loss. **Guard cells** control the opening and closing of air pores. **Root hairs** absorb water and mineral salts from the soil. **Collen-chyma cells** and **sieve tubes** are both elongated cells, the former concerned with strengthening the plant, particularly the young stem, the latter with conducting soluble food materials from one part of the plant to another. One of the most striking specializations is seen in lignified **vessels, tracheids** and **fibres.** These play an important part in strengthening the stems of higher plants and, in the case of vessels and tracheids, conducting water and mineral salts from the roots to the leaves. These cells are enormously elongated, their cellulose walls being impregnated with the complex polysaccharide lignin. What results is a non-living tube through which water can be conducted. In this case specialization involves the total loss of protoplasm from the cell. A more drastic modification can hardly be imagined.

Some of the cells mentioned in this section are illustrated in Fig 2.19. This figure also includes a few representative single-celled organisms such as *Amoeba* and *Euglena*. Though generally regarded as occupying a lowly position in the evolutionary hierarchy, some of these **unicells** are surprisingly elaborate. This matter will be taken up again in Chapter 3.

So it is clear that cells are structurally specialized to perform particular tasks. In extreme cases specialization may entail loss of protoplasm, or some important constituent of it; but in the majority of cases it involves modification of the shape and form of the cell, its basic features (nucleus, mitochondria, endoplasmic reticulum etc.) remaining unchanged. The flame cell of a flatworm and a nerve cell in the brain look very different, and may perform quite different functions, but in their fundamental structure and chemistry they are remarkably alike. This is not surprising when we bear in mind that both are living entities and, whatever else they do, they must perform those functions which are necessary for the maintenance of life.

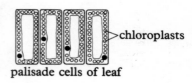
palisade cells of leaf

parenchyma cells of plant

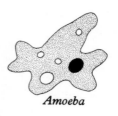
Amoeba

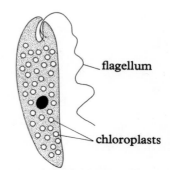
Euglena, a green flagellate

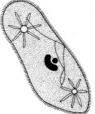
Paramecium, a ciliated unicell

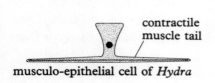
musculo-epithelial cell of *Hydra*

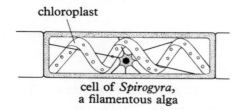
cell of *Spirogyra,* a filamentous alga

Fig 2.19 A variety of cells. Despite differences in shape and form, all these cells are fundamentally similar in possessing nucleus and cytoplasm together with many of the organelles described earlier. Nucleus is shown solid.

3 Tissues, Organs and Organization

In the last chapter we talked about cells as if they exist in isolation, lacking any kind of functional contact with one another. However, in multicellular organisms cells of a particular type are generally massed together to form a **tissue** whose function is determined by the kind of cell of which it is composed. Furthermore, in more complex organisms, particularly animals, different tissues are united to form **organs**. In this chapter we will see how tissues and organs are built up.

CLASSIFICATION OF TISSUES

There are various ways of classifying tissues. For convenience we will divide animal tissues into **epithelial, connective, skeletal, blood, nerve, muscular** and **reproductive**. Plant tissues will be divided into **meristematic, epidermal, parenchyma, photosynthetic, mechanical** and **vascular**. The purpose of this chapter is not to systematically describe all these tissues in detail, but to look at the principles underlying their construction.

EPITHELIUM: A SIMPLE ANIMAL TISSUE

One of the simplest animal tissues is **epithelium**, which deserves detailed attention because it demonstrates how individual cells can be built up into tissues of varying complexity. Epithelium is lining tissue: in its simplest form it consists of a single layer of cells covering the surface of the body and the organs within it. It also lines various spaces and tubes, in which situation it is usually referred to as **endothelium**. Typically the individual cells, firmly attached to each other, rest on a **basement membrane** and have a free surface. Clearly the main function of epithelial tissue is protective.

There are six main types of epithelial tissue which depend on the detailed structure and function of the cells comprising them (Fig 3.1). Least specialized is **cuboidal epithelium** whose cells are cubical in shape. One of the best places to see this kind of epithelium is in the thyroid gland where it forms the lining of the vesicles. Many other glands and their ducts are lined with cuboidal epithelium, as are certain of the tubules in the kidney. In the latter case numerous **microvilli** project from the free surface of the cells. Their function is to increase the surface area for the reabsorption of materials from the fluid in the tubules.

In **pavement (squamous) epithelium** the cells are flattened, giving a crazy paving appearance in surface view. The resulting sheet of cells is thin and delicate, often less than 2.0 μm thick. It is found in places where the protective covering needs to be readily permeable to molecules

A) Cuboidal

free surface

nucleus

cytoplasm

basement membrane

free surface

b

a

a and b equal.

a single cell

B)

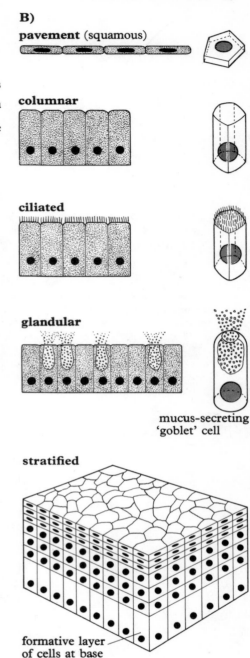

pavement (squamous)

columnar

ciliated

glandular

mucus-secreting
'goblet' cell

stratified

formative layer
of cells at base

in solution, for example the lining of capillaries and alveoli in the lungs (Fig 3.2).

Columnar epithelium, on the other hand, consists of cells elongated at right angles to the basement membrane; in other words they appear to be tall and column-like when viewed in section. Columnar epithelium is found lining the intestine where it plays an important part in supporting other types of cells and absorbing soluble food material. Like the cells in the kidney, the columnar epithelial cells lining the small intestine possess microvilli on their free surfaces for increasing the surface area over which absorption can take place.

A specialized form of lining tissue is **ciliated epithelium**. Usually columnar in shape, the free surface of each cell bears numerous cilia capable of beating rapidly and rhythmically. In the mammal ciliated epithelium lines tubes and cavities in which materials have to be moved. For example cilia lining the respiratory tract are responsible for expelling small particles of dust and other foreign materials which have got caught up in the mucus adhering to the cells. In certain animals, flatworms for example, cilia lining the underside of the body play an important part in locomotion.

Epithelial cells are frequently interspersed with secretory cells. The resulting tissue, known as **glandular epithelium**, secretes materials into the cavity or space which it happens to be lining. A good example is seen in the lining of the mammalian intestine. Amongst the columnar epithelial cells are large numbers of **goblet cells** which secrete mucus into the lumen of the intestine (Fig. 3.3). The lubricating action of the mucus facilitates the movement of solid matter along the intestine.

In the intestine the goblet cells are supported by columnar epithelial cells. However, certain glandular epithelia contain so many densely packed secretory cells that the tissue consists of little else. Such epithelia may be folded in various ways to increase the surface area from which secretion takes place. Folding of glandular epithelia results in the formation of **glands** whose sole function is secretion. Fig 3.4 illustrates how different kinds of glands are formed by epithelial folding.

In the types of epithelium considered so far the tissue consists of a single layer of cells. **Stratified epithelium**, however, is made up of a series of layers and is, therefore, much thicker than ordinary epithelia. Tough and impervious, it comprises the epidermis of skin where its function is protective. The multi-layered nature of stratified epithelium derives from the fact that the cells at its base, the formative layer, maintain the capacity to

Fig 3.1 Structure of epithelial tissue. A) Cuboidal epithelium. This is one of the simplest forms of tissue. The height of each cell is approximately equal to its width so that when viewed in section the cells appear square. Viewed from the free surface they are polygonal in shape. Each cell is thus a small prism.   B) Five other types of epithelium.

divide repeatedly along a plane parallel to the basement membrane. As multiplication continues the daughter cells get pushed gradually outwards as new cells are formed beneath them. As the cells move outwards they become flattened and eventually flake off, to be replaced by new ones from beneath. In skin the cells towards the surface are transformed into a tough, non-living layer of keratin, a feature that greatly enhances its efficiency as a protective covering.

A modified form of stratified epithelium, called **transitional epithelium**, is found lining cavities and tubes that are subject to stretching, for example the bladder. Like normal stratified epithelium, it has several layers

Fig 3.2 Electron micrograph showing flattened endothelial cells (**E**) lining a small blood vessel (**B**). (*A. R. Akester, University of Cambridge*)

Fig 3.3 (*opposite*) Mucus-secreting goblet cells in the lining of the mammalian small intestine. The section has been specially stained to show up the secretory region in each goblet cell (dark blobs). (*J. F. Crane*)

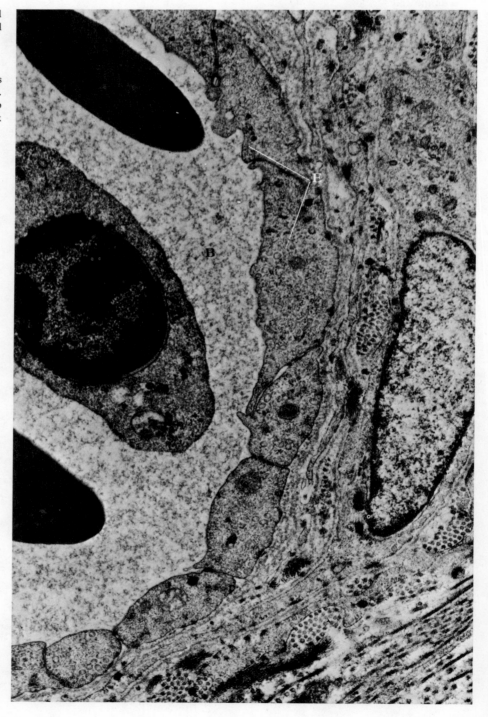

surface epithelium

non-secretory
neck of gland

secretory
part of gland

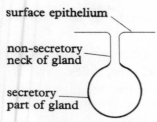

(a) **Simple saccular gland** e.g.
mucus glands in the skin of
the frog and other amphibia.

(b) **Simple tubular gland** e.g. crypts of
Lieberkühn in the wall of the
mammalian small intestine.

(c) **Coiled tubular gland** e.g.
sweat glands in the skin
of man.

(d) **Simple branched tubular gland**
e.g. Brunner's glands in the wall of
the mammalian small intestine and
gastric glands in the wall of the
stomach.

(e) **Simple branched saccular gland**
e.g. oil- secreting sebaceous glands
in mammalian skin.

(f) **Compound tubular glands** e.g.
salivary glands.

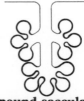

(g) **Compound saccular gland** e.g.
the part of the pancreas which
secretes digestive enzymes, and
the mammary glands.

Fig 3.4 Diagrams summarizing the different
types of glands found amongst the vertebrates.
In all cases a patch of glandular epithelium
invaginates to form a sac or tube which may show
various degrees of branching. Branching has the
effect of increasing the surface area from
which secretion can take place.

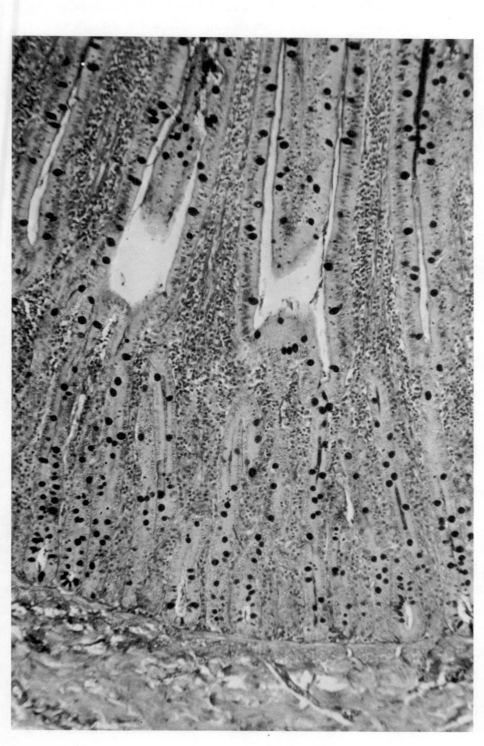

of cells, but the cells are all approximately the same size, do not flake off, and can change their shape according to circumstances.

To summarise: epithelial tissue forms the lining of structures, cavities and tubes. There are several different types whose functions, collectively, are protection, absorption, movement (by means of cilia), and secretion.

CONNECTIVE TISSUE

The tissues and organs in an animal's body must be supported and held in position. This function is performed by **connective tissue** which binds organs and tissues together. It follows from its function that connective tissue must be strong. It consists of a **matrix** or **ground substance** in which a variety of structures may be embedded.

The most fundamental type of connective tissue is **areolar tissue** found all over the body: beneath the skin, connecting organs together, and filling the spaces between adjacent tissues. Areolar tissue consists of a gelatinous glycoprotein **matrix** or **ground substance** containing four types of cell and two types of protein fibre (Fig 3.5). The cells include large flat **fibroblasts**

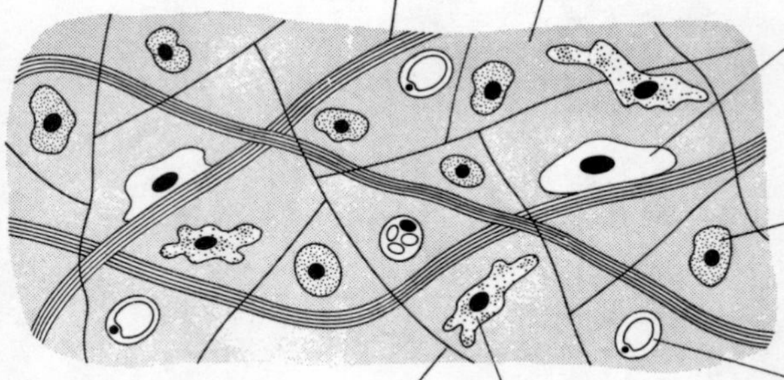

Fig 3.5 Diagram of areolar tissue as it appears in microscopical section. This type of connective tissue is found in the interstices between adjacent tissues and has the general function of binding tissues and organs together.

Collagen fibres, flexible but very strong and non-stretchable.

Ground substance in which the various fibres and cells are embedded.

Fibroblast: long flat cell which produces the collagen and elastic fibres.

Mast cell secretes the ground substance and an anticoagulant.

Fat cell which stores fat

Elastic fibres form a loose network of stretchable fibres.

Macrophage: a large amoeboid cell which ingests a wide variety of foreign particles and is important in defending the body against disease.

which synthesize the fibres, amoeboid **mast cells** which secrete the matrix, a variable number of **fat-filled cells** and phagocytic **macrophages**. It is the fibres which give areolar tissue its strength and toughness. These are of two types: unbranched **collagen fibres** (**white fibres**) running parallel to each other in bundles; and branched **elastic fibres** (**yellow fibres**) which form a dense network in the matrix.

Three other types of connective tissue are recognized. All are derivatives of areolar tissue, differing from each other in the structures present in the matrix. **Collagen tissue** (**white fibrous tissue**) consists of a glycoprotein matrix containing nothing but densely-packed collagen fibres. This tissue is comparatively inelastic and has great tensile strength; **tendons,** by which muscles are attached to bones, are composed of collagen fibres. **Elastic tissue** (**yellow elastic tissue**) is composed of a glycoprotein matrix containing only elastic fibres. Combining strength with elasticity, this tissue is found in ligaments which bind bones together. In **adipose tissue** (**fatty tissue**) the matrix contains nothing but closely-packed fat-filled cells and is important in storage. In the dermis of the skin adipose tissue insulates the body from heat loss.

Collagen and elastic fibres are non-living derivatives of cells, both fibrous proteins. Their great importance justifies the interest which has been shown in them in recent years. X-ray analysis of collagen has shown it to consist of three polypeptide chains coiled round each other to form a triple helix. The chains are interlinked by hydrogen bonds and the whole structure is tough and inextensible, like a plaited rope. Elastin, on the other hand, consists of interconnected globules within which the polypeptide chains are arranged randomly, giving a more pliant structure.

Connective tissue can be summarized by describing it as a mixture of fibres in different proportions. Its efficiency in binding structures together is achieved by the molecular configuration of the protein molecules. The particular type and abundance of fibre present depends on the stresses and strains to which the tissue is normally subjected.

SKELETAL TISSUE

Closely related to connective tissue is **skeletal tissue**, responsible for supporting the body and providing it with a rigid framework. Like connective tissue it consists of cells embedded in an organic matrix but in this case the matrix is comparatively hard. Two kinds of skeletal tissue occur in the vertebrates: **cartilage** (**gristle**) and **bone**. The skeleton of the more primitive fishes, such as dogfish, sharks and rays, is composed entirely of cartilage. The mammal, with its predominantly bony skeleton, has cartilage at the joints and in the discs between successive vertebrae.

Cartilage consists of an organic matrix, **chondrin**, in which are embedded groups of spherical cells called **chondroblasts**. The latter secrete the matrix.

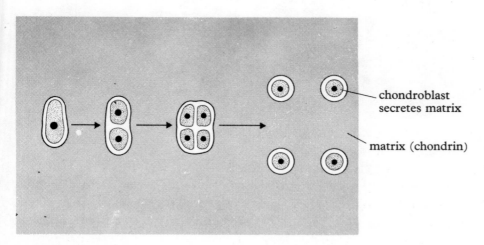

chondroblast secretes matrix

matrix (chondrin)

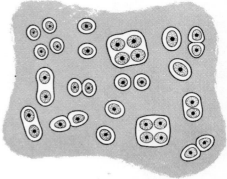

Hyaline cartilage as it appears in a microscopical preparation.

The simplest form, **hyaline cartilage**, consists of nothing but chondrin and chondroblasts (Fig 3.6). Other more complex forms are strengthened by the presence in the matrix of connective tissue fibres. For example, the cartilage of the intervertebral discs contains collagen fibres, whilst that in the nose and pinna of the ear contains elastic fibres.

Bone is much harder than cartilage. It consists of an organic matrix impregnated with calcium salts, mainly calcium phosphate. These salts confer

Fig 3.6 Cartilage consists of an organic matrix secreted by cells called chondroblasts. The chondroblasts divide repeatedly into small groups of cells which gradually get separated from one another as each secretes matrix material. Cartilage is much softer than bone but may be strengthened by the presence of elastic or collagen fibres. Simple cartilage lacking such inclusions is known as hyaline cartilage and is found, for example, in the wall of the trachea.

upon bone its property of extreme hardness. They are secreted by stellate cells called **osteoblasts** which arrange themselves in concentric rings round nerves and blood vessels, thereby giving bone tissue its characteristic pattern of holes and lamellae (Fig 3.7).

A

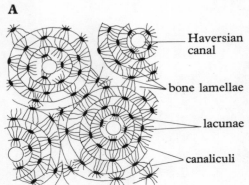

Haversian canal

bone lamellae

lacunae

canaliculi

(i) Transverse section

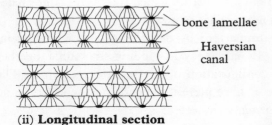

bone lamellae

Haversian canal

(ii) Longitudinal section

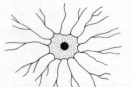

single bone-secreting osteoblast.

Fig 3.7 A) Diagram showing structure of compact bone in transverse and longitudinal section. In development osteoblasts arrange themselves in concentric rings around a series of Haversian canals, each of which contains an artery and vein. The osteoblasts secrete an organic matrix which later becomes greatly hardened by impregnation with mineral salts: calcium phosphate plus some calcium carbonate. Because of the concentric arrangement of the osteoblasts, the matrix is laid down in a series of layers or lamellae encircling the Haversian canal. In a microscopical preparation of compact bone the lacunae represent the spaces occupied by the central region of the osteoblasts, the canaliculi are the fine channels that contained their slender arm-like processes. B) Low power photomicrograph of compact bone in transverse section. (*Gene Cox, Micro-Colour International*)

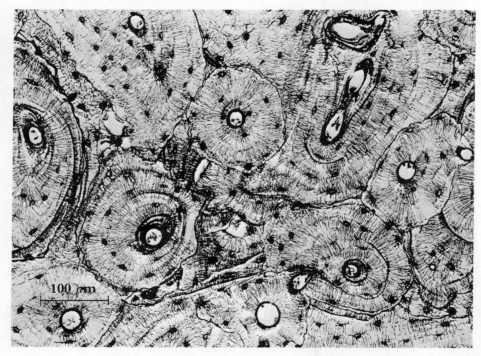

100 μm

B

OTHER ANIMAL TISSUES

We have discussed epithelial, connective and skeletal tissues in detail because they are of general occurence in animals and we shall frequently refer to them. Only brief mention will be made of other animal tissues at this point because they are described in detail in later chapters.

Blood is a circulating tissue consisting of three types of cell suspended in a fluid (**plasma**). The most numerous cells are the **red blood cells** (**erythrocytes**), as many as five million per cubic millimetre, whose function is to transport oxygen. Less numerous are the **white blood cells** (**leucocytes**) of which there are several different types. Collectively they combat disease by destroying pathogenic micro-organisms, bacteria and viruses, which have got into the body. Finally, blood contains minute cell fragments called **platelets** which play an important part in the process by which blood clots when exposed to air. So blood is a complex tissue whose main function is transport and defence.

By now you will appreciate that cells are not isolated units, but are structurally and functionally integrated. Nowhere can this be better seen than in **nervous tissue** whose function is to transmit electrical messages from one part of the body to another. To this end nerve cells are elaborately interconnected, the processes of one cell linking up with adjacent cells. Nervous tissue is an intricate network of interconnected cells whose function is to transmit, and sometimes to store, information.

There are three types of **muscular tissue: visceral, cardiac** and **skeletal.** Visceral (also known as **smooth** or **unstriated**) muscle consists of sheets of densely-packed, elongated fibres running parallel to each other and bound together by connective tissue. Each muscle fibre is a single cell containing a nucleus and numerous fine contractile fibrils. Controlled by the involuntary part of the nervous system, visceral muscle contracts, and fatigues, comparatively slowly. It is found in places where slow, involuntary movements take place, such as the walls of the gut, bladder, blood vessels, and the ducts of glands. It is also found in the dermis of the skin.

Skeletal muscle, as the name implies, is associated with the skeleton, cardiac muscle with the heart. Both appear characteristically **striated** when viewed under the microscope, and both contain contractile fibrils. Skeletal muscle is controlled by the voluntary part of the nervous system and provides the basis of all voluntary movement in the body such as locomotion and movement of the limbs. Since they are described in detail in later chapters nothing further will be said about them now.

Finally, **reproductive tissue**. Associated with ovaries and testes, this is concerned with production of gametes: eggs and sperm respectively. Reproductive tissue is composed of developing gametes in the process of division and differentiation, together with cells that provide support and nourishment.

PLANT TISSUES

Structurally the simplest plant tissue is **meristematic tissue**. This occurs at the apical growing points of a plant, for example the tip of the stem and root. Meristematic cells are small, thin walled, and lack sap vacuoles and chloroplasts. But they contain undifferentiated plastids whose membranes are the site of intense synthetic activity. Their important feature is their ability to divide and subsequently differentiate into specialized cells. Of course cell division and differentiation also occur in animals, but there is no separate meristematic tissue as such because growth does not continue throughout life.

A plant's equivalent to epithelium is **epidermal tissue** located at the surface of the plant. Somewhat flattened and often irregular in shape, epidermal cells form a protective layer covering the more delicate tissues beneath. They lack chloroplasts and the outer cellulose walls are frequently thick and impregnated with waxy materials, forming a **cuticle** that is impermeable to water and prevents excessive evaporation.

The principal feature distinguishing a plant from an animal is its method of feeding: photosynthesis. This function is performed by cells containing chloroplasts. These cells are massed together to form **photosynthetic tissue** which is particularly abundant in leaves.

A plant must have support. Several tissues contribute towards this (Fig 3.8). **Parenchyma** is packing tissue. Consisting of rounded cells packed tightly together, its function is to fill the spaces between other tissues and, assuming that the cells are turgid, to maintain the shape and firmness of the plant, particularly the leaves. Parenchyma cells frequently contain starch and therefore also function as **storage tissue**. Further support is provided by a variety of **mechanical tissues**. These are mainly composed of tightly packed **collenchyma** cells and **sclerenchyma** fibres. The former have unevenly thickened cellulose walls, whilst the latter have uniformly thick lignified walls (Fig 3.8 B and C). These woody fibres are mainly found in stems and the midrib of leaves

Fig 3.8 The principal mechanical elements found in the stem of a flowering plant. Such tissues are also found in the midrib and veins of leaves where their function, as in the stem, is to provide strength and support.

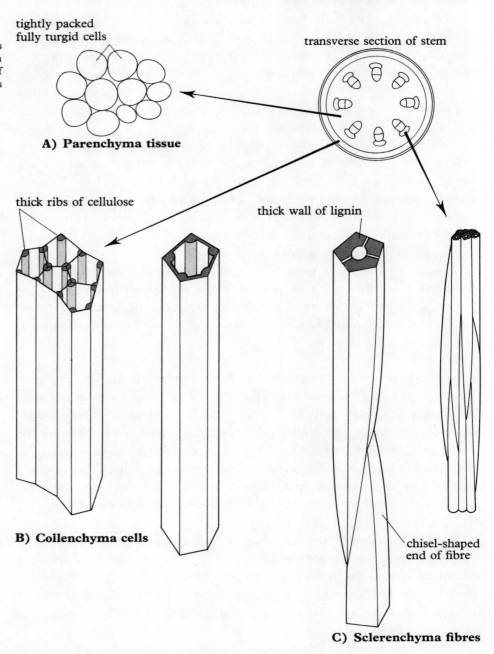

tightly packed
fully turgid cells

transverse section of stem

A) Parenchyma tissue

thick ribs of cellulose

thick wall of lignin

B) Collenchyma cells

chisel-shaped
end of fibre

C) Sclerenchyma fibres

where they contribute towards their toughness and rigidity. They are less abundant in roots, except in specialized roots that are involved in providing support (Fig 3.9). Sclerenchyma fibres are functionally equivalent to bone.

Can you detect any similarities between them that might help to explain their ability to support heavy loads?

Fig 3.9 The famous Banyan tree in the grounds of the Abdin Palace in Cairo. In the course of its growth large aerial roots have been let down from its larger branches. Well endowed with mechanical tissue, these prop roots form, in effect, secondary trunks, providing extra support and enabling the tree to spread over a considerable area. (*Paul Popper Ltd*)

Vascular tissue is concerned with transport, and is functionally equivalent to the circulatory system in animals. There are two types: **xylem** and **phloem**. Xylem tissue consists of elongated, lignified tubes called **vessels** and **tracheids**, which conduct water and mineral salts from the roots to the leaves; phloem tissue consists mainly of unlignified living cells called **sieve tubes** whose function is to convey food materials from the leaves to other parts of the plant.

THE DIFFERENCES BETWEEN ANIMAL AND PLANT TISSUES

Animals and plants are thought to be the products of two distinct lines of evolution which diverged from a common ancestor many millions of years ago. Evidence for this is considered in Chapter 36, but let us for a moment consider the implication of this as far as cells and tissues are concerned. If animals and plants share a common ancestry we would expect to find fundamental similarities in the structure and functioning of their cells; and indeed we do find this, particularly at the biochemical and molecular levels. But the divergence would lead us to suppose that the detailed structure of their cells, particularly the more specialized ones, would be markedly different. Here again our supposition is borne out by the facts: the specialized cells and tissues of animals and plants are very dissimilar.

If we get down to the core of the problem these dissimilarities hinge on one basic difference, namely nutrition. Plants, unlike animals, can synthesize their own organic food. This of course relates to the fact that plants possess photosynthetic tissue whilst animals do not. But it explains other facts as well. For example, since animals cannot manufacture their own organic food materials, they have to obtain it in ready-made form, and this often means searching for it. Sensitivity and movement are thus essential attributes of most animals, and this explains why they possess muscle and nerve tissue, as well as specialized epithelia for the digestion and absorption of food. Photosynthesis renders such tissues unnecessary in plants.

All this can be summarized by saying that their method of nutrition imposes upon animals the necessity to move and respond rapidly, and this demands a greater variety of specialized tissues than are necessary in plants. In their general organization animals are thus more complex than plants. It is a curious paradox that the root cause of this is that they lack the ability to perform a chemical process which plants are capable of performing, namely photosynthesis.

ORGANS AND ORGAN SYSTEMS

A further indication of the complexity of animals is seen in the fact that all but the simplest ones possess **organs**. An organ is a structurally distinct part of the body which usually performs a particular function. Typically it is made up of several types of tissue which have a highly organized structural relationship with each other. The mammalian kidney, for example, contains blood, smooth muscle, connective tissue and several different types of epithelium, all organized into a complex system of interrelated structures whose combined function is the elimination of unwanted chemical substances from the body.

Just as cells and tissues cannot operate in isolation, organs too are interrelated to form **organ systems**. An organ system is made up of several organs which together perform a specific function. The digestive system, for example, consists of the gut plus various accessory organs such as the pancreas and liver. Sometimes an organ belongs to more than one system: the pancreas, for example, secretes hormones in addition to digestive enzymes, and therefore belongs to the endocrine as well as the digestive system.

In the development of organs and organ systems the mammalian body is unsurpassed in the animal kingdom, but most animals have developed organs to some extent. Plants have them too, though they are generally less elaborate and fewer in number than in animals. One of the most prominent plant organs is the leaf whose basic function is photosynthesis. The flower is a reproductive organ, and structures such as bulbs and corms are organs of perennation and vegetative reproduction.

LEVELS OF ORGANIZATION

In this account of cells, tissues and organs we have considered mainly higher animals and plants. How do other organisms fit into this picture?

Three grades of organization are recognized. The great majority of animals, like the mammal, are on the **organ level**: that is, their body functions are carried out for the most part by organs and organ systems. Naturally simpler forms, such as the earthworm, have comparatively simple organs, but nevertheless their organization is also based on the possession of interrelated organs.

At the other end of the scale are single-celled organisms representing the **unicellular level** of organization. In these **unicells** or **protists** the functions, which in higher forms are carried out by organs and organ systems, must be performed by organelles within the single cell. In some protists, particularly ciliates like *Paramecium*, intracellular organization may be surprisingly elaborate (Fig 3.10). *Paramecium* is a comparatively large fresh water ciliate

A

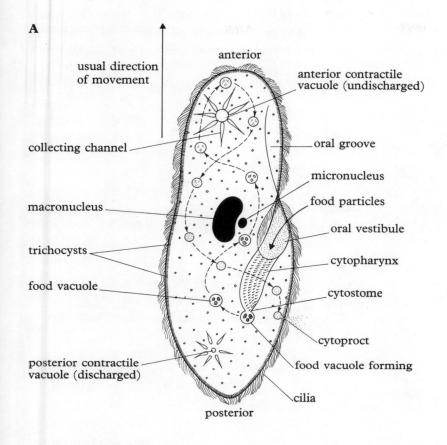

B

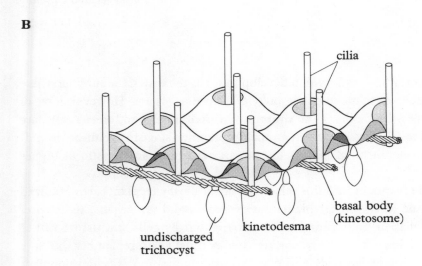

Fig 3.10   As a unicellular organism, *Paramecium* has to carry out within one cell all the functions that in a multicellular animal are performed by differentiated tissues and organs. A) diagram showing the structures visible under the light microscope. The macronucleus exercises control over metabolic functions, including growth, whilst the micronucleus is necessary for sexual reproduction. The cilia are responsible for locomotion, the contractile vacuoles for osmoregulation. The trichocysts, tiny explosive sacs containing a needle-shaped thread, are used in some species for defence. Food particles, swept into the oral vestibule by ciliary action, are taken up into a food vacuole formed at the base of the cytopharynx. Food vacuoles circulate through the cytoplasm before discharging undigested remains to the exterior through the cytoproct. B) diagram of the surface region based on electron micrographs. The cilia beat in a fully coordinated manner in which each performs its backstroke slightly before the one immediately in front (metachronal rhythm). The basal bodies (kinetosomes) are interconnected by a system of threads (kinetodesmata) situated immediately beneath the elaborately sculptured pellicle. Most of the structures seen in *Paramecium* have readily identifiable counterparts in a multicellular animal like man: thus the oral vestibule and cytopharynx are equivalent to the mouth and pharynx, the cytoproct to the anus, and the contractile vacuoles to the kidneys. (*diagrams modified after Vickerman and Cox*, The Protozoa, *John Murray*)

whose internal structure can be observed under the microscope. In addition to the usual organelles, it has a specialized apparatus for the intake of food; two contractile vacuoles with collecting channels for expelling excess water; and an elaborate system of threads interconnecting the basal bodies of its cilia. So although they are unicellular, protists are by no means simple.

Between these two extremes are primitive multicellular animals at the **tissue level of organization**. Apart from reproductive organs they have few organs as such, physiological processes being carried out mainly by isolated cells and tissues. Such animals include coelenterates, like *Hydra* (Fig 3.11). The wall of *Hydra's* simple sac-like body is composed of seven different types of cell arranged in two sheet-like layers, the **ectoderm** and **endoderm**. Apart

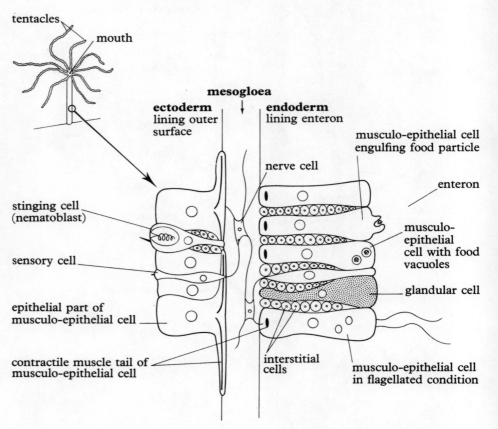

tentacles

mouth

mesogloea

ectoderm
lining outer
surface

endoderm
lining enteron

musculo-epithelial cell
engulfing food particle

nerve cell

enteron

stinging cell
(nematoblast)

musculo-
epithelial
cell with food
vacuoles

sensory cell

glandular cell

epithelial part of
musculo-epithelial cell

contractile muscle tail of
musculo-epithelial cell

interstitial
cells

musculo-epithelial cell
in flagellated condition

Fig 3.11 Structure of the body wall of *Hydra*. *Hydra* is on the tissue level of organization: its cells are inter-dependent and to some extent integrated by a nervous system. For example, in its feeding behaviour, food (usually small water fleas) is immobilized by the stinging cells and pulled to the mouth by the tentacles. This is achieved by contraction of the muscle tails which are controlled by the nerve net in the mesogloea. Digestion of the food starts in the enteron by enzymes secreted from the glandular cells, and is completed inside the musculo-epithelial cells. These cells take up particles phagocytically, and then form flagella which help to circulate the semi-fluid contents of the enteron.

from the ovary and testis which develop towards the end of the summer, there are no structures that can be remotely described as organs. However, most of the cell types, particularly the **musculo-epithelial cells** and **nerve cells**, are integrated to form tissues. Such animals represent a stage in evolution preceding the development of organs and organ systems characteristic of higher forms.

How do plants fit into this scheme? For the most part plants are organized on the tissue level. Higher plants possess leaves and sex organs, to be sure, but most of their body functions are carried out by cells and tissues rather than organs. In seeking the reason for this we must again remember the basic difference between animals and plants, namely nutrition. The development of organ systems in animals is associated with increased complexity resulting from their feeding habits. In a plant, feeding by photosynthesis, elaborate organ systems are not necessary; it simply does not need them.

COLONIAL ORGANIZATION

Although *Hydra* is constructed on a tissue basis, its constituent cells, with certain notable exceptions, are not functionally isolated units but are inte-

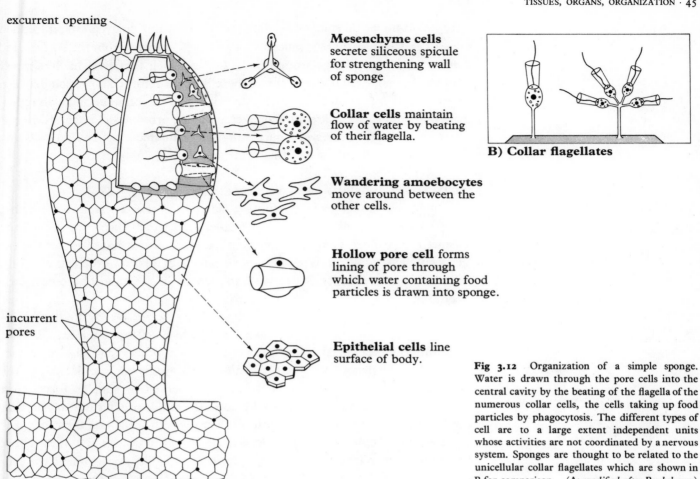

excurrent opening

Mesenchyme cells secrete siliceous spicule for strengthening wall of sponge

Collar cells maintain flow of water by beating of their flagella.

B) Collar flagellates

Wandering amoebocytes move around between the other cells.

incurrent pores

Hollow pore cell forms lining of pore through which water containing food particles is drawn into sponge.

Epithelial cells line surface of body.

A) Simple sponge

Fig 3.12 Organization of a simple sponge. Water is drawn through the pore cells into the central cavity by the beating of the flagella of the numerous collar cells, the cells taking up food particles by phagocytosis. The different types of cell are to a large extent independent units whose activities are not coordinated by a nervous system. Sponges are thought to be related to the unicellular collar flagellates which are shown in B for comparison. (A: *modified after Buchsbaum*)

grated at least to a degree. This is most clearly shown by the musculo-epithelial cells which are controlled by the nerve net, which in turn is connected to sensory cells in the ectoderm. This means that *Hydra*, though simple in organization, behaves as a fully coordinated individual.

Some organisms, however, are composed of cells that are functionally much more isolated from one another. Such is the case with sponges. These organisms are best regarded as **colonies of single cells**, rather than multi-cellular individuals. A simple sponge, such as the one illustrated in Fig 3.12, consists of five types of cell: flattened **epithelial cells** and hollow **pore cells** at the surface, numerous flagellated **collar cells** lining the internal cavity, spicule-secreting **mesenchyme cells**, and **amoeboid cells** (**amoebocytes**) which move about in the spaces between the other cells. This is not the place to go into the detailed biology of sponges. Suffice it to say that these types of cell are functionally independent of one another; they can exist on their own or in small isolated groups, and there is no trace of a nervous system to coordinate their activities. Their colonial status is confirmed by the fact that the collar cells are almost identical in their structure and behaviour with a group of protistans called **collar flagellates** which can exist either as single individuals or as small groups. It seems likely that sponges have arisen from collar flagellates in evolution.

Another case of colonial organization is seen in the *Volvocales*, a group of incredibly beautiful microscopical green algae. These organisms consist of varying numbers of simple flagellated cells, containing chloroplasts, embedded in spheres of transparent mucilage. The flagella project from the surface and drive the colony along through the water in which they live. In the simpler forms the cells are all identical and there is no structural connection between them. In *Volvox*, however, a small proportion of the cells specialize in reproduction, and the cells are interconnected by fine protoplasmic strands. In this case the colony behaves in a surprisingly coordinated manner. The cells are, as it were, beginning to lose their independence.

THE ADVANTAGES OF THE MULTICELLULAR STATE

A single cell cannot grow indefinitely. When it reaches a certain size it either stops growing, or divides into two smaller cells which then grow. Indefinite growth appears to be limited by the nucleus. It seems that any one nucleus can only exert control over a certain volume of cytoplasm. In terms of evolution this means that for an organism to increase in size it must become **multicellular**.

Becoming multicellular, then, allows increase in size. With this comes the possibility of specialization: instead of every cell carrying out all tasks, certain cells become specialized for one function, others for a different function. This division of labour permits greater efficiency and enables the organism to exploit environments that are denied to simpler forms. But although it is an undeniable advantage to the whole organism, it means that individual cells are unable to exist on their own: the cells lose their independence and become increasingly dependent on one another's activities. Think how interdependent the cells are, even in *Hydra*. For nourishment the cells depend on the explosive discharge of the nematoblasts, but these would be of little use without the coordinated action of the muscle tails after the prey has been paralysed. And none of this prey-catching apparatus would be of any use if the endodermal cells were unable to digest and absorb the food.

Although the multicellular state allows greater specialization, the increased size that accompanies it creates difficulties that have to be solved if the organism is to survive. One of these difficulties concerns the acquisition of respiratory gases and food materials by the constituent cells. How this difficulty has been overcome is discussed in the next chapter.

4 Movement in and out of Cells

We must now turn our attention to the processes by which materials get in and out of cells. Such exchanges take place by five main processes: **diffusion**, **osmosis**, **phagocytosis**, **pinocytosis** and **active transport**. Each depends on certain properties of the cell membrane. We will consider them in turn and discuss their consequences on the functioning of cells.

Diffusion

If a crystal of potassium permanganate is dropped into a beaker of water it dissolves, and gradually the purple colour of the permanganate spreads until eventually it is uniformly distributed. What causes the potassium permanganate molecules to behave in this way? The answer depends on the fact that the molecules are in a state of random motion. Although they can move in any direction, the fact that initially there are far more of them in the immediate vicinity of the crystal increases the probability of their moving away from the crystal. In other words there is a *net* movement of molecules away from the crystal. This process is **diffusion**, and it is defined as the movement of molecules from a region where they are at a comparatively high concentration to a region where they are at a lower concentration. Diffusion will always proceed whenever such a **concentration gradient** exists, and it will continue until eventually the molecules are uniformly distributed throughout the system, at which time **equilibrium** is said to be reached.

Diffusion is very important in the movement of molecules and ions in and out of cells. Think, for example, of a cell in your own body. Because it is continually being used up in respiration, the concentration of oxygen inside the cell will be lower than it is in the blood and tissue fluids. The concentration gradient results in oxygen molecules diffusing into the cell from outside. With carbon dioxide the reverse is true: its concentration is highest inside the cell, where it is continually being formed. This results in carbon dioxide molecules diffusing out of the cell. Anything that increases the concentration gradient will favour diffusion. This is one of the functions of a **circulatory system**. By rapidly carrying away the diffused substance the circulation maintains a steep concentration gradient, thus encouraging further diffusion. Similarly, the conversion of the diffused substance into a new product, for example sugar into another carbohydrate, also has the effect of maintaining a concentration gradient, favouring continued diffusion.

If diffusion is to take place freely, any membranes or partitions in the system must be fully permeable to the molecules or ions in question. Such is the case with the cell membrane in relation to oxygen and carbon dioxide: it is fully permeable to both these gases.

DIFFUSION AND THE STRUCTURE OF ORGANISMS

The fact that organisms rely on diffusion in the fulfilment of their physiological needs has had a profound effect on their structure. Consider, for example, the way gaseous exchange relates to the size of an organism. The organism's oxygen requirements (its needs) are proportional to its volume, i.e. the bulk of respiring tissue which it contains. Its exchanges, however, are proportional to the surface area over which diffusion of oxygen can take place. In any organism the effective surface area must be sufficient to fulfil the needs of the respiring tissues. Now it is a simple mathematical rule that as an object increases in volume the ratio of its total surface area to its volume decreases; in other words, the larger the object the smaller its **surface: volume ratio**. The significance of this is as follows. In a small organism like a protist, or even an earthworm, the surface:volume ratio is sufficiently great for diffusion over the general body surface to satisfy its respiratory needs. But in larger forms the surface:volume ratio is too small for this to be the case. Such organisms have got round this problem by developing a special **respiratory surface** for gaseous exchange. The latter is generally highly folded so as to increase the surface area over which diffusion can take place, for example lungs and gills. The general strategy of increasing the surface area for diffusion is also shown by individual cells in the development of microvilli.

The fact that lungs and gills generally have a good blood supply illustrates another important principle. If there is no circulatory system the organism is dependent on passive diffusion for conveying oxygen from the respiratory surface to the innermost tissues. In this case its size (and shape) are limited by the law of diffusion which states that the rate of diffusion is inversely proportional to the distance over which it has to take place. In practical terms this means that the thicker the tissue the slower is the rate at which oxygen reaches the cells furthest from the surface. In small organisms the diffusion distance is short and there is therefore no problem. In larger organisms the problem has been overcome either by developing a circulatory system with an oxygen-carrying pigment, or by reducing the diffusion distance so that none of the cells are far removed from the surface. The latter has been achieved in various ways:

(1) The body may be flattened, thus reducing the distance between the two surfaces (e.g. flatworms, and the leaves of plants).

(2) The body may be constructed in such a way that the tissues are thin (e.g. *Hydra* and sea anemones).

(3) There may be a system whereby the external medium is brought into the body so that it comes into intimate association with all the tissues (e.g. the tracheal system of insects, see p. 122).

In addition to respiration, diffusion affects two other processes, namely the distribution of food materials and elimination of toxic waste (excretion). In the majority of animals the circulatory system takes on the job of transporting food materials and toxic waste. But what about those animals that lack a circulatory system but are too large for diffusion alone to suffice? This is not the place to go into details of the various adaptations which have evolved but one example will be given to illustrate the principle. In planarian worms the gut, instead of being a straight tube, as it is in most animals, gives off numerous blindly-ending branches which ramify throughout the body penetrating deep into the tissues. Digested food is forced into these branches by muscular action and is thus conveyed directly to the cells. Planarians have no circulation and this extraordinary alternative means that they do not have to depend on diffusion alone for distributing their digested food.

Osmosis

Although the plasma membrane of a cell is fully permeable to respiratory gases, it is by no means permeable to all substances. The porous nature of the membrane means that only those molecules that are small enough will diffuse through it unimpeded. Larger molecules either penetrate slowly or not at all. The plasma membrane is thus **semipermeable**, permitting the passage of some substances but not of others. To appreciate the significance of this we must examine the properties of semipermeable membranes.

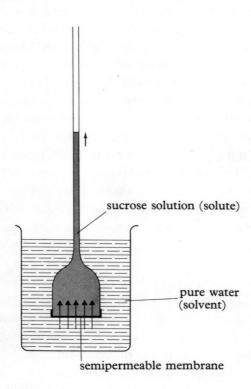

sucrose solution (solute)

pure water (solvent)

semipermeable membrane

Fig 4.1 A simple osmometer. The arrows indicate the passage of water molecules (solvent) into the sucrose solution (solute). The semipermeability of the membrane prevents sucrose molecules moving into the solvent.

Consider the situation shown in Fig 4.1. The mouth of a thistle funnel is covered with a semipermeable membrane such as cellophane or dead pig's bladder. The funnel is then filled with a strong sugar solution and immersed in a beaker of pure water. What happens? Quite quickly the level of the solution in the tube starts to rise. Analysis of this solution shows that it gradually becomes more dilute, indicating that water is passing into it from the surrounding beaker.

In seeking for an explanation of this it is necessary to appreciate that the membrane is permeable to the water molecules but impermeable (or relatively so) to the much larger sugar molecules. With this in mind consider the situation on each side of the membrane. In the beaker there are nothing but water molecules; in the funnel there are water molecules plus sugar molecules. The presence of sugar molecules means that the concentration of water molecules in the funnel is less than in the beaker. The result is a net movement of water molecules from the beaker into the funnel. This movement of molecules across a semipermeable membrane is known as **osmosis.**

Osmosis is really a special case of diffusion: it involves the passage of water molecules from a region of high concentration (the beaker) to a region of lower concentration (the funnel). It is the presence of sugar molecules in the funnel which makes the concentration of water molecules lower on that side of the semipermeable membrane. As water molecules continue to move from beaker to funnel a hydrostatic pressure develops in the latter. This presses down on the semipermeable membrane and eventually stops further net movement of water molecules across it. The head of pressure which can be developed by a solution when permitted to take up water by osmosis is known as the **osmotic pressure**.

A solution consists of the molecules of one substance (**the solute**) dissolved in another (**the solvent**). Osmosis will occur whenever solvent is separated from solute (or solution) by a semipermeable membrane. The membrane does not have to be completely impermeable to the solute molecules. Many semipermeable membranes allow the passage of solute and solvent, though not to the same extents. All that is required for osmosis to occur is that the solvent molecules move more rapidly than the solute molecules.

Plainly the osmotic pressure a solution is capable of developing will depend on the concentration of its solute molecules. To revert to Fig 4.1, a much greater osmotic pressure will be developed if the sugar solution is saturated (100 per cent) than if it is, say, only 10 per cent. It also follows that osmosis will be reduced if solute molecules are present in the beaker as well as in the funnel. If the concentration of solute is the same on both sides of the membrane osmosis ceases altogether; and if the solute concentration is greater in the beaker than the funnel the solvent will move in the opposite direction – from funnel to beaker.

CELLS AS OSMOMETERS

Cells owe many of their properties to the fact that the plasma membrane is semipermeable. If a cell is surrounded by pure water, or a solution whose solute concentration is lower than that of the cell's contents, water molecules pass into the cell by osmosis, and the cell swells up. In this case the osmotic pressure of the external solution is lower than that of the cell, and accordingly the solution is said to be **hypotonic** to the cell. On the other hand, if the cell is surrounded by a solution whose solute concentration, and hence osmotic pressure, exceeds that of the cell, water passes out of the cell, which therefore shrinks. In this case the external solution is said to be **hypertonic** to the cell. If the solute concentration of the cell and its surrounding medium are the same, osmosis does not occur in either direction and the external solution is said to be **isotonic** with the cell.

All this can be readily demonstrated with human red blood cells. A solution of 0.9 per cent sodium chloride is isotonic with human cells, and if red blood cells are placed in such a solution they will neither shrink nor swell. However, if they are placed in a stronger (i.e. hypertonic) salt solution (say 1.2 per cent) they shrink perceptibly, the cell membrane crinkling as shown in Fig 4.2. This is known as **laking** or **crenation**. If placed in a weaker (i.e. hypotonic) solution they swell and may even burst, a phenomenon known as **haemolysis** (meaning literally blood-splitting).

It follows that if a cell is to maintain its proper shape it must either exist permanently in an isotonic environment or, failing that, it must have special mechanisms enabling it to survive in a hypertonic or hypotonic medium. These special mechanisms are all part of the business of **osmoregulation** which is immensely important in the lives of organisms. They are discussed in detail in Chapter 14, but one example will be mentioned at this stage to clarify the principle involved. The unicellular animal *Amoeba* has both marine and fresh water species. The marine species is isotonic with sea water with the result that there is no net loss or gain of water by the cell. The fresh water species, on the other hand, is markedly hypertonic to its surrounding medium. The result is that water continually passes into the cell across its semipermeable plasma membrane. Undoubtedly the animal would swell up and burst, just like our red blood cell in distilled water, were it not for the **contractile vacuole**, a spherical sac which collects water as fast as it enters, and periodically discharges it to the outside. By this means the cell is prevented from swelling, and its solute concentration, and hence osmotic pressure, is kept constant.

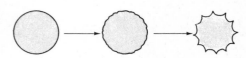

Fig 4.2 Three stages in the crenation of a red blood cell. The cell is placed in a hypertonic solution: water is drawn out by osmosis leading to shrinkage of the cell and crinkling of the membrane.

OSMOSIS AND PLANT CELLS

Plant cells generally have a solute concentration which is markedly higher than that of their surroundings. The solutes are located mainly in the sap vacuole which is surrounded by two semipermeable membranes, one on each side of the protoplast. For our present purposes we can regard the entire protoplast as a single semipermeable membrane. When the cell is surrounded by a hypotonic solution, water is drawn into the sap vacuole by osmosis (Fig. 4.3A).

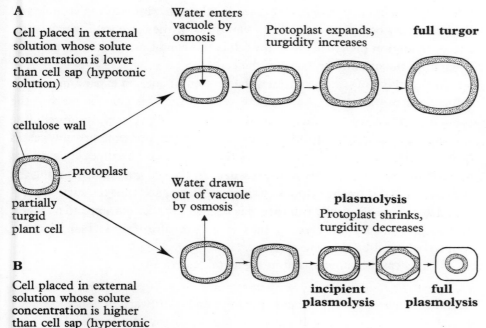

A

Cell placed in external solution whose solute concentration is lower than cell sap (hypotonic solution)

Water enters vacuole by osmosis

Protoplast expands, turgidity increases

full turgor

cellulose wall

protoplast

partially turgid plant cell

B

Cell placed in external solution whose solute concentration is higher than cell sap (hypertonic solution)

Water drawn out of vacuole by osmosis

plasmolysis
Protoplast shrinks, turgidity decreases

incipient plasmolysis

full plasmolysis

Fig 4.3 Summary of the events that ensue if a partially turgid plant cell is placed in A) a hypotonic solution, B) a hypertonic solution. The protoplast acts as a semipermeable membrane; the cellulose wall being fully permeable. See explanation on p. 52.

The result is that the cell swells. However, it does not burst because of the cellulose wall which, once fully stretched, resists any further expansion of the cell.

As a plant cell takes up water by osmosis the protoplast presses against the cellulose wall and an internal pressure is developed. This is called **turgor pressure** and is equal to the back pressure exerted by the cellulose wall against the cell's contents. Turgor pressure reaches its maximum when the cellulose wall can be stretched no more. At this point **full turgor** is said to be achieved, or, to put it another way, the cell becomes **fully turgid**. Turgidity plays a very important part in supporting plants and maintaining their shape and form. The stems of herbaceous plants are held erect by being filled with fully turgid cells packed tightly together. The same thing is responsible for holding leaves in a flat, horizontal position. If the cells lose water, as they often do on hot dry days, turgidity is reduced and the plant visibly droops. This phenomenon is called **wilting**.

Let us now consider what happens to a plant cell immersed in a solution which is strongly hypertonic to the cell sap. This can be observed by mounting a piece of plant epidermal tissue in a strong sugar solution and watching it under a microscope (Fig. 4.3B). First the volume of each cell decreases as water is drawn out of its sap vacuole. Within a matter of minutes the protoplast shrinks to such an extent that it pulls away from the cellulose wall, leaving a perceptible gap between the wall and the outer membrane of the protoplast. Shrinkage of the protoplast continues until eventually the sap vacuole is no longer visible and the protoplast rounds off into a ball. This process is called **plasmolysis**. Only in special circumstances does it occur under natural conditions but, when induced by experiment, it can help us to understand the rôle of the semipermeable protoplast in determining the physical conditions inside the cell.

SUCTION PRESSURE

Let us imagine that we have just plasmolysed a plant cell by immersing it in a hypertonic solution. The cell is now taken out of the solution and placed in pure water. What happens? Water immediately rushes into the sap vacuole and the protoplast begins to expand. We can speak of the force with which water enters as **suction pressure**. With the cell in this condition the suction pressure is equal to the osmotic pressure of the cell sap.[1]

As the influx of water continues the protoplast goes on expanding until it comes into contact with the cellulose wall. When this point is reached the suction pressure starts being opposed by the inward pressure of the cellulose wall acting against the expanding protoplast. The suction pressure now ceases to be equal to the osmotic pressure of the cell sap, and becomes equal to the osmotic pressure minus this **wall pressure**. As the cell goes on expanding the resistance offered by the cellulose wall gets greater, and thus the suction pressure decreases. Eventually full turgor is reached: the cell can expand no more; suction pressure drops to zero and the osmotic pressure of the cell sap is exactly counterbalanced by the wall pressure.

All this can be summarized in a general statement:

SP	=	**OP**	−	**WP**
Suction pressure of cell		Osmotic pressure of cell sap		Inward pressure exerted by cellulose wall

[1] In some respects the term 'suction pressure' is misleading since it implies that the cell actually sucks water in. For this reason some biologists prefer to use the term **water diffusion potential** instead of suction pressure. The use of the term 'osmotic pressure' applied to the contents of a cell is also questionable: in the absence of an opposing force the cell sap can hardly be said to exert an actual pressure. To overcome this the term **osmotic potential** is sometimes adopted. In this book the terms suction pressure and osmotic pressure are used, but you should be aware of their shortcomings.

The suction pressure is therefore the net tendency of a cell to draw in water from outside. In a cell which is plasmolysed to the extent that the protoplast loses contact with the cellulose wall, the wall pressure is zero and SP = OP. At full turgor the wall pressure is equal and opposite to the osmotic pressure, the suction pressure being zero. These relationships are summarized in Fig 4.4.

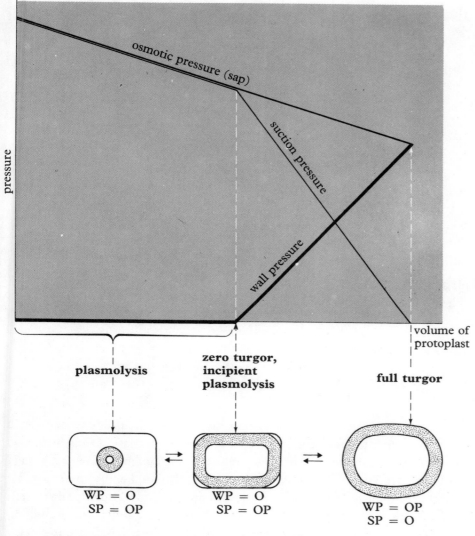

Fig 4.4 Graph illustrating the relationship between the suction pressure, osmotic pressure and wall pressure at different stages of plasmolysis of a plant cell.

It is sometimes necessary to measure the suction pressure and osmotic pressure of a plant tissue. To measure the suction pressure, pieces of tissue of known weight or volume are placed in a series of solutions of different solute concentration. The solution which produces no change in weight or volume has a solute concentration, and therefore osmotic pressure, equal to the suction pressure of the tissue. Some class results are shown in Fig 4.5.

Measuring the osmotic pressue is slightly more difficult because one has to eliminate effects caused by wall pressure. One method is to find that strength of external solution which causes the cells to plasmolyse so that the protoplasts just lose contact with the cellulose walls. This condition is described as **incipient plasmolysis** (Fig 4.4). With the cellulose walls no longer pressing in on the cells' contents, we can assume that the osmotic pressure of the external solution is equal to the osmotic pressure of the cell sap. In actual practice the cells tend to plasmolyse at different rates, and for practical purposes incipient

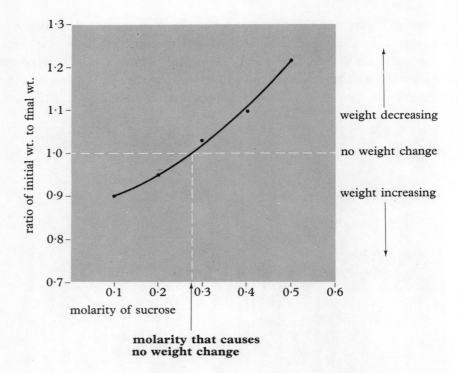

Fig 4.5 Results of an experiment to determine the suction pressure of the cell sap of potato tuber cells. Samples of tissue were placed in a series of sucrose solutions of different molarities and the change in weight measured. The molarity of sucrose causing no weight change is 0.27 which corresponds to a suction pressure of 0.74 N/mm².

plasmolysis is taken as the point when half the cells are visibly plasmolysed. What criticisms could you level against this method?

Active Transport

Diffusion is a purely physical process in which molecules or ions move from a region of higher to lower concentration. However, there are certain biological situations where the reverse happens: molecules or ions move from a region of low concentration to a region of higher concentration, i.e. they move against a concentration gradient. A spectacular example of this is provided by certain seaweeds which take up iodine so vigorously that it is more than two million times as concentrated inside the cells than in the surrounding water. This movement against a concentration gradient is called **active transport**.

Active transport cannot be accounted for by purely physical forces. It will only take place in a living system that is actively producing energy by respiration. It has been found that variables such as temperature and oxygen concentration, which influence the rate of respiration, also influence the rate of active transport. Further evidence that active transport is linked with energy production is supplied by biochemical studies. Energy in cells comes from the breakdown of adenosine triphosphate (ATP). Anything that inhibits the formation of ATP, or prevents it from being used, stops active transport from taking place. Cyanide, for example, which prevents ATP from being synthesized, inhibits active transport. Another interesting correlation is that certain cells that are known to indulge in active transport on a large scale have an unusually large number of mitochondria.

It is not yet certain how active transport takes place. One hypothesis (Fig

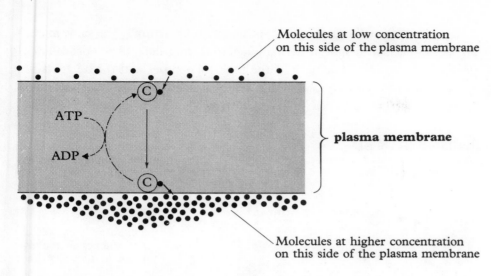

Molecules at low concentration
on this side of the plasma membrane

plasma membrane

Molecules at higher concentration
on this side of the plasma membrane

Fig 4.6 Schematic diagram depicting how molecules might be actively transported across the plasma membrane against a concentration gradient. A carrier (**C**) picks up the molecule on one side of the membrane and deposits it on the other side. It then returns and repeats the process. ATP is known to be required for active transport.

4.6) proposes that there are **carriers** in the plasma membrane which attach themselves to the molecules or ions at the outer surface of the membrane and convey them to the inner surface where they are deposited in the cytoplasm. The carrier then returns to the outer surface and repeats the process. None of these proposed carriers has been positively identified, but it does seem certain that to return to its original position and pick up another molecule, energy from the breakdown of ATP is required.

The plasma membrane plays a vital rôle in controlling not only what goes into the cell but also what comes out of it. This is seen in the fact that many cells accumulate potassium but expel sodium ions. Whether separate carriers are involved is not known but clearly these two ions are being moved in opposite directions within one and the same cell membrane. Moreover this situation can be altered by circumstances. For example, when a nerve cell transmits an impulse the extrusion of sodium ceases, and sodium ions pour into the cell by diffusion.

It is also important to realize that the membrane of a given cell may permit the passage of some molecules or ions but not of others. Furthermore the molecules or ions absorbed may be taken up at different rates. Later we shall have occasion to study the uptake of minerals by plant cells (p. 192) and we shall see that various ions (sodium, potassium, calcium, magnesium, chloride) are all absorbed against a concentration gradient but at markedly different rates.

All this leads us to suppose that the plasma membrane is not merely a passive barrier, but a dynamic interface between the cell's contents and its immediate surroundings.

Phagocytosis

So far we have only discussed how individual molecules traverse the plasma membrane. How do larger particles get into cells? The majority of cells are actually incapable to taking in large particles, i.e. particles visible under the light microscope. However, certain specialized cells can do so by a process called **phagocytosis** which literally means 'cell-eating'. This was first observed in white blood cells which take up bacteria, and later in *Amoeba*, which feeds on a variety of small organisms. What happens is essentially the same in both cases: the plasma membrane invaginates to form a flask-like depression enclosing the particles. The neck of the flask then closes and the invagination becomes sealed off as a **phagocytic vesicle** or **food vacuole** which migrates towards the centre of the cell (Fig 4.7). The particles are now digested by enzymes

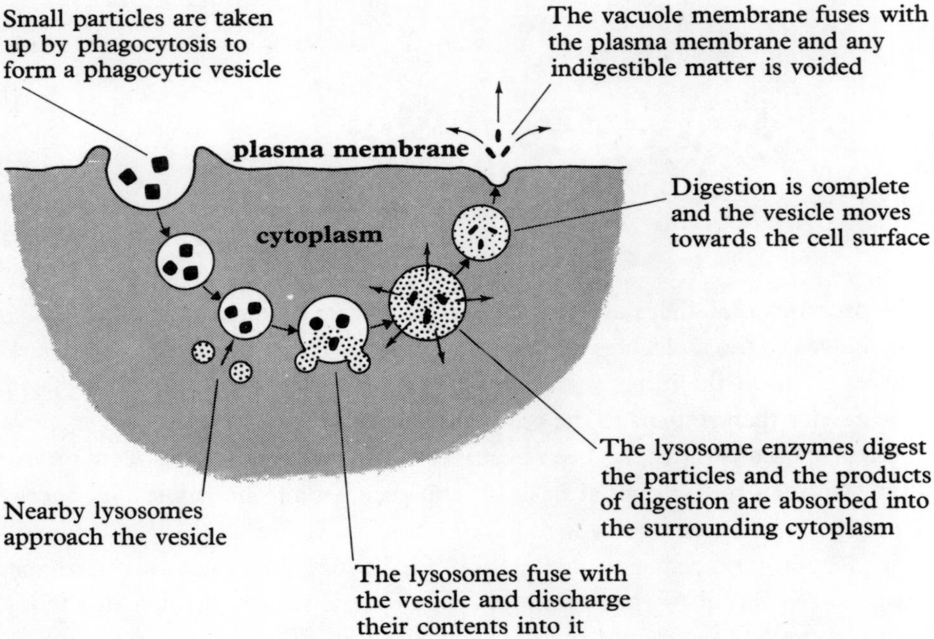

Fig 4·7 Diagram showing the sequence of events in the phagocytic digestion of food particles inside a cell (intracellular digestion.)

Small particles are taken up by phagocytosis to form a phagocytic vesicle

The vacuole membrane fuses with the plasma membrane and any indigestible matter is voided

plasma membrane

cytoplasm

Digestion is complete and the vesicle moves towards the cell surface

The lysosome enzymes digest the particles and the products of digestion are absorbed into the surrounding cytoplasm

Nearby lysosomes approach the vesicle

The lysosomes fuse with the vesicle and discharge their contents into it

secreted into the vesicle from lysosomes which fuse with it. The soluble products of digestion (sugars, amino acids and the like) are then absorbed into the surrounding cytoplasm. In *Amoeba* any indigestible material is voided by the vesicle moving to the edge of the cell and fusing with the plasma membrane, essentially the reverse of the process by which it was formed. Phagocytosis is a selective process, the cell discriminating between different kinds of particle. *Amoeba*, for example, will ingest particles of nutritional value but generally fails to take up particles that are of no food value. Similarly phagocytic white blood cells will only attack certain bacteria.

Pinocytosis

Pinocytosis means 'cell-drinking' and is essentially similar to phagocytosis on a much smaller scale. The light microscope has shown that in, for example, *Amoeba*, tiny **pinocytic channels** are continually being formed at the cell surface by invagination of the plasma membrane. From the inner end of each channel small vacuoles are pinched off, and these move towards the centre of the cell. Smaller vesicles may then be pinched off each vacuole and these migrate to different parts of the cell. Pinocytic channels provide a means by which liquids can be brought into the body of the cell, and their breaking up into numerous vacuoles aids distribution and increases the surface area across which absorption can take place.

The membrane lining the vacuoles, derived from the plasma membrane, remains intact. Any materials absorbed into the cytoplasm from a vacuole must traverse the lining membrane first before it can be regarded as fully inside the cell. Pinocytosis is therefore not a substitute for transport across the plasma membrane but a supplementary process facilitating it.

Pinocytosis, as described above, can be seen under the light microscope. However, much smaller invaginations of the plasma membrane are detectable in electron micrographs. A mere 0.1 μm in diameter, these **micropinocytic vesicles** appear to become sealed off from the exterior of the cell, forming minute vacuoles. The fact that this process can be induced by the attachment of certain materials to the cell surface has led to the suggestion that it may provide a means by which macromolecules are selectively taken up into cells. Judging from electron micrographs, micropinocytic vesicles appear to be of universal occurrence in cells. The vacuoles so formed may fuse together in groups forming larger vacuoles into which enzymes are then discharged from surrounding lysosomes. Alternatively the vacuole membrane may break down, thus releasing the enclosed macromolecules which may then be incorporated into the cytoplasm.

One cannot help being struck by the apparent ease with which the cell membrane invaginates to form phagocytic or pinocytic vesicles. The plasma membrane, together with various internal membranes including those lining phagocytic and pinocytic vesicles, are all unit membranes with the characteristic three-layered structure (see p. 22). It is a striking property of the unit membrane that more membrane can be added to it, or taken away from it, with apparent ease. Thus vesicles pinch off from the plasma membrane, lysosome membranes fuse with vacuole membranes, vacuole membranes fuse with the plasma membrane, and so on.

5 The Chemicals of Life

Cells, tissues and organs are composed of chemicals, many of which are identical with those found in non-living matter. Others are unique to living organisms. The study of chemical compounds found in living systems, and the reactions in which they take part is known as **biochemistry**. Studies on the structure and behaviour of individual molecules constitute **molecular biology**, a subject in which spectacular advances have been made in recent years. If the 'secret of life' is to be found anywhere it is in these molecules.

Chemical compounds are conventionally divided into two groups: **organic** and **inorganic**. Under the organic heading are included all complex compounds of carbon. All other compounds are classified as inorganic. Both are found in living things. Until the German chemist Friedrich Wöhler synthesized urea in 1828, it was thought that organic compounds could only be formed in living materials. In fact organic matter was thought to be unique to living systems, synonymous with life itself. Since Wöhler's experiment hundreds of organic compounds have been synthesized and it is clear that they are by no means unique to living organisms. None the less, organic compounds make up a goodly proportion of the average cell, comprising about 20 per cent of the total; second only to water. The principal organic compounds found in organisms are **carbohydrates, fats, proteins** and **nucleic acids**. Of the inorganic constituents **acids, bases, salts** and **water** are amongst the more important.

In this chapter we shall review the structure and functions of these cell constituents. We shall see that at the chemical level the relationship between structure and function is no less apparent than it is when dealing with visible organelles. In fact the chemical structure of a particular compound is often directly related to its function in the cell.

Water

Water is by far the most abundant component of organisms. An individual cell in man contains approximately 80 per cent water, and the whole body is made up of over 60 per cent. As J. B. S. Haldane used to say, even the Archbishop of Canterbury is 65 per cent water. Life originated in water, and today numerous organisms make their home in it. Water provides the medium in which all biochemical reactions take place and has played a major rôle in the evolution of biological systems.

The importance of water as a medium for life springs from four of its properties: its **solvent properties**, **heat capacity**, **surface tension** and **freezing**. Let us look at each of these in turn.

WATER AS A SOLVENT

Water's properties as a solvent depend on the fact that it is a **polar molecule**. This means that the distribution of electric charge is permanently such that centres of positive and negative charge are separated by a short distance. This is because of the configuration of the water molecule; instead of being arranged in a straight line the hydrogen and oxygen atoms are situated asymmetrically, thus:

The oxygen part of the molecule has a net negative charge and the hydrogen parts a net positive charge. Thus the molecule as a whole shows **polarity**.

Now consider what happens if a crystal of sodium chloride is placed in water. In the crystal the sodium and chloride ions are held together by their being strongly attracted to one another. But in the presence of water this attraction is greatly reduced, with the result that the ions part company and go into solution. Water effectively weakens the attraction between ions of opposite charge because, having positive and negative charges itself, it attracts both. Moreover, once the ions have separated they are prevented from rejoining by water molecules clustering round each of them.

Water is therefore a good solvent, ionic solids and polar molecules readily dissolving in it. This is of great biological importance because all the chemical reactions that take place in cells do so in aqueous solution.

The polarity of water molecules also explains their association with each other. A positive hydrogen atom of one molecule may be attracted to the negative oxygen atom of another, leading to the formation of small clusters of water molecules. The hydrogen-oxygen attraction holding the molecules together is known as **hydrogen bonding**. We shall see later that, although it is not strong, the hydrogen bond plays an important part in holding organic molecules together.

WATER'S THERMAL PROPERTIES

Heat capacity may be defined as the amount of heat required to raise the temperature of 1 gramme by 1°C. Now water has a very high heat capacity compared with other liquids. In other words a large increase in heat results in a comparatively small rise in temperature of the water. This means that water is good at maintaining its temperature irrespective of fluctuations in the temperature of the surrounding environment. The biological importance of this is that the range of temperatures in which biochemical processes can proceed is narrow, and most organisms cannot tolerate wide variations in temperature. The high thermal capacity of water is important in keeping the temperature constant, thus making it an ideal environment for animal and plant life.

One further point related to water's thermal properties: water has a remarkably high boiling point for a substance of such low molecular weight. Were it not for this fortuitous accident it is likely that liquid water would never have existed on this planet, and most would probably have been lost to outer space. And with no water life as we understand it could not have evolved.

SURFACE TENSION

Surface tension is the force that causes the surface of a liquid to contract so that it occupies the least possible area. It is caused by inward-acting cohesive forces between the molecules at the surface, this being caused by the polarity of water already referred to. Now water has the highest surface tension of any known liquid and this is of great biological importance. Molecules dissolved in water lower its surface tension and tend to collect at the interface between its liquid phase and other phases. This may have been important in the development of the plasma membrane, and certainly plays an important part in the movement of molecules across it. The high surface tension of water, together with the strong cohesion forces which exist between water molecules, also plays an important part in the movement of water up the capillary-like vessels and tracheids in the stems of plants.

WATER'S FREEZING PROPERTIES

Most liquids decrease in volume and increase in density as the temperature drops. When such a liquid freezes the molecules become densely packed together and the resulting ice sinks. Now with water the reverse is true: as water is cooled below a certain temperature its volume increases and its density decreases. This means that ice tends to float rather than sink. Moreover, when the temperature drops the coldest water is at the surface and, being less dense than the slightly warmer water lower down, tends to remain at the surface. So ice forms at the surface first, and the bottom later. Organisms which live towards the bottom of fresh-water lakes are therefore protected from freezing, a factor which has undoubtedly contributed towards their survival. It has been calculated that if it were not for ice floating on water the oceans would all be frozen solid except perhaps for a thin layer at the top.

Acids and Bases

You will remember from chemistry that an **acid** is a compound which, when it dissociates in water, liberates a hydrogen ion. For example:

$$HCl \longrightarrow H^+ + Cl^-$$

Hydrochloric acid   Hydrogen ion   Chloride ion

The strength of an acid is determined by the extent to which it dissociates in aqueous solutions. Hydrochloric acid is a strong acid because it dissociates completely into hydrogen and chloride ions. This is why in the above equation the arrow points in only one direction viz. $\longrightarrow$ On the other hand acetic acid is comparatively weak, only a proportion of its molecules dissociating into hydrogen and acetate ions in aqueous solution.

The acidity of a solution is expressed as its **pH**. This is the negative logarithm to the base 10 of the hydrogen ion concentration in moles per dm³ of solution: in other words it is a measure of the hydrogen ion concentration of the solution. A pH of 7.0 represents neutrality. A solution with a pH of less than 7.0 is acidic, and the lower the figure the higher the acidity, i.e. the greater is the hydrogen ion concentration. A solution whose pH is greater than 7.0 is basic or alkaline, and the higher the figure the more basic is the solution.

A **base** is a compound which can combine with hydrogen ions liberated by dissociation of an acid. For example, sodium bicarbonate readily dissociates in solution into sodium and bicarbonate ions:

$$NaHCO_3 \longrightarrow Na^+ + HCO_3^-$$

Sodium   Sodium   Bicarbonate
bicarbonate   ion   ion

The bicarbonate can then accept a free hydrogen ion, if one is available, to form carbonic acid:

$$HCO_3^- + H^+ \rightleftharpoons H_2CO_3$$

Bicarbonate   Hydrogen   Carbonic
ion   ion   acid

Clearly the sodium bicarbonate can remove hydrogen ions from an aqueous solution and thereby lower the latter's acidity. In so doing it is acting as a **buffer**. This is a compound which behaves in such a way as to resist changes in pH on dilution or addition of moderate amounts of acid or alkali[1].

In the case of increased acidity sodium bicarbonate acts as a buffer by combining with free hydrogen ions. If the alkalinity is increased it can react with the free hydroxyl ions to form carbonate ions and water:

$$HCO_3^- + OH^- \longrightarrow CO_3^{2-} + H_2O$$

Sodium bicarbonate can therefore resist either a decrease or an increase in pH by mopping up the hydrogen or hydroxyl ions as appropriate.

In the human body bicarbonates play a relatively minor role as buffers. Much more important are **phosphate salts** like potassium and sodium phosphate both of which play an important part in suppressing the hydrogen ion concentration in the blood. The phosphate combines with free hydrogen ions to form the dihydrogen phosphate.

$$HPO_4^{2-} + H^+ \rightleftharpoons H_2PO_4^-$$

Certain organic compounds, for example proteins and haemoglobin, can also accept hydrogen ions and are therefore important as buffers. The biological importance of all these compounds is this. Cells and tissues can only function properly at or around neutrality; they cannot tolerate fluctuations in pH of more than a unit or two. For this reason it is essential that the pH of the body fluids should be kept as constant as possible. Any tendency for the acidity to increase is counteracted by the buffers which mop up the excess hydrogen ions, thereby helping to maintain constant conditions in the cells.

Salts

In the body some of the most common solutes found dissolved in water are

[1] Although sodium bicarbonate acts as a buffer on its own, in most cases two or more compounds interact to form a 'buffer system'.

mineral salts, compounds of a metal with a non-metal or non-metallic radical. Sodium chloride is an example. The previous section will have made it clear that when a salt dissolves in water it dissociates into its constituent ions. Thus:

$$\text{NaCl} \longrightarrow \text{Na}^+ + \text{Cl}^-$$

Sodium chloride   Sodium ion   Chloride ion

Salts are therefore ionic compounds, and since they are decomposed by electricity and their free ions enable the passage of electric currents, they are classed as electrolytes. Although sodium and chloride are two of the most common ions in cells, many others are found too. Important **cations** are: sodium (Na^+), potassium (K^+), calcium (Ca^{2+}), magnesium (Mg^{2+}), copper (Cu^{2+}) and iron (Fe^{2+} or Fe^{3+}). Important **anions**: chloride (Cl^-), bicarbonate (HCO_3^-), phosphate ($H_2PO_4^-$), sulphate (SO_4^{2-}) and iodide (I^-). In addition to these **major mineral elements** additional ones are often present in minute amounts. These **trace elements** include molybdenum (generally as (MoO_4^-) and cobalt (Co^{2+}). Some are required by animals, some by plants; a few by both. Though needed in the smallest quantities, absence of them can cause disease and death. Some examples of the effects of mineral deficiencies in plants are shown in Fig 5.1.

A

B

Fig 5.1 The importance of mineral elements in plants. A) The effect of deficiency of three major mineral elements on the growth of barley. The plant on the left is growing in a complete solution; the others are from left to right: nitrogen-free, iron-free and sulphur-free. B) The effect of molybdenum deprivation on the growth of tomato plants. Molybdenum is a trace element required in extremely small amounts. Deficiency results in poor growth of both shoots and roots. (*G. Bond and E. J. Hewitt, University of Glasgow*)

All these mineral salts have specific rôles to fulfil in organisms. Thus, sodium and potassium ions play an important part in the transmission of nerve impulses, calcium and phosphate enter into the composition of bone, and magnesium is a constituent of the prosthetic group in chlorophyll. Many of them are discussed in detail in later chapters. However, one function applies to them all: mineral salts, together with other solutes, determine the osmotic pressure of the body fluids (see Chapter 14). Like pH, the osmotic pressure must not be allowed to fluctuate and an organism's physiology is to a large extent directed towards preventing this.

Organic Compounds

Organic compounds are immensely important in living things: they constitute the structures of the body and regulate the chemical processes going on inside it. It is their structural complexity which enables them to do this, and it is important to understand the chemical basis of this complexity.

THE CONSTRUCTION OF ORGANIC MOLECULES

Organic compounds owe their complexity to the **carbon atom**. Carbon has a valency of four which means that in its compounds every carbon atom forms four covalent bonds. Carbon's ability to use these bonds to build up long chains of atoms is responsible for the unique variety and complexity of organic molecules. It explains why more carbon compounds are known than all the compounds of all other elements put together.

To make this clearer, think of the way an organic molecule is built up. Starting with a single carbon atom one can visualise adding on further carbon atoms (or other atoms for that matter) in any of four different directions. In this way elaborate three-dimensional molecules can be constructed, the carbon atoms forming a skeleton to which other atoms (hydrogen, oxygen, nitrogen and so on) are attached. Sometimes the carbon skeleton consists of a single chain, a kind of backbone: such is the case with fatty acids for example. Sometimes it is branched, as in amino acids; or it may be a ring as in simple sugars. In Fig. 5.2 you will notice that the four bonds projecting from it are arranged 3-dimensionally, giving the carbon atom a tetrahedral shape. For this reason organic chains are not straight but 'crinkled'. This is difficult to show in a structural formula and will therefore be ignored in the descriptions to follow. The important point is that the variety, complexity and sheer size of organic molecules are caused by the carbon atom. It is doubtful if without it life could ever have evolved as we know it, and its absence may have been a major factor preventing the development of life on other planets.

The other principal elements found in organic molecules are **oxygen** and **hydrogen**. Oxygen has a valency of two which makes it suitable for linking carbon atoms together and forming side-chains. Hydrogen, with its valency of one, can only occupy terminal positions in the molecule. **Nitrogen** is a constituent of some organic compounds, proteins and nucleic acids for instance. It has a valency of three and enters into the skeleton of certain important ring compounds such as porphyrins, steroids and the organic bases found in nucleic acids.

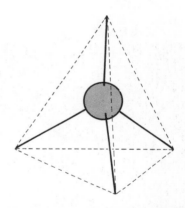

Fig 5.2 The carbon atom has four potential bonds capable of linking up with other atoms. These bonds are oriented in such a way that if their ends were joined up (dotted lines) the resulting shape would be a tetrahedron.

Carbohydrates

Carbohydrates include **sugars** and **starch**. On combustion they yield energy, and their principal function in organisms is the production, and storage, of this vital commodity. They also enter into the composition of important structures, particularly the cellulose wall of plant cells.

The elements present in a carbohydrate are carbon, hydrogen and oxygen. A simple sugar like glucose has the general formula $(CH_2O)_n$, the n signifying that the total number of carbon, hydrogen and oxygen atoms varies from one carbohydrate to another. If n = 6 the resulting carbohydrate has the formula $C_6H_{12}O_6$ and is known as a **hexose sugar**. If n = 5 the carbohydrate is $C_5H_{10}O_5$, a **pentose sugar**. And if n = 3 we get a **triose sugar** with the formula $C_3H_6O_3$.

Fig 5.3 The basic unit of all carbohydrates is the simple sugar molecule $C_6H_{12}O_6$. Below is shown the structural formula of α glucose (A) in full, (B) slightly simplified. The molecule consists of five carbon atoms arranged in a ring. An oxygen atom links carbon atoms 1 and 5. The ring lies at right angles to the plane of the paper, the thick bonds lying in front of the thinner ones behind. The various side groups stick obliquely up and down.

A

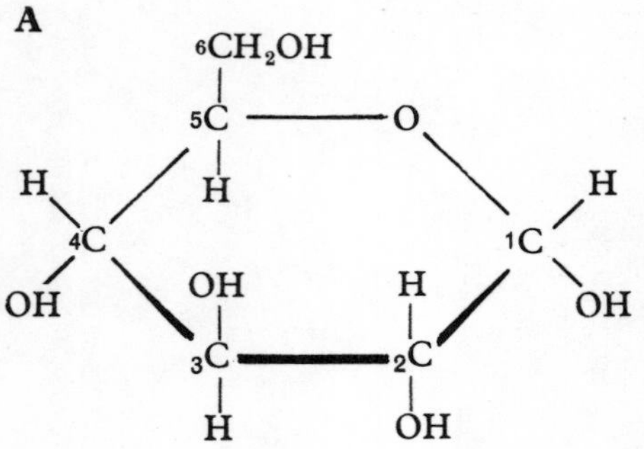

B

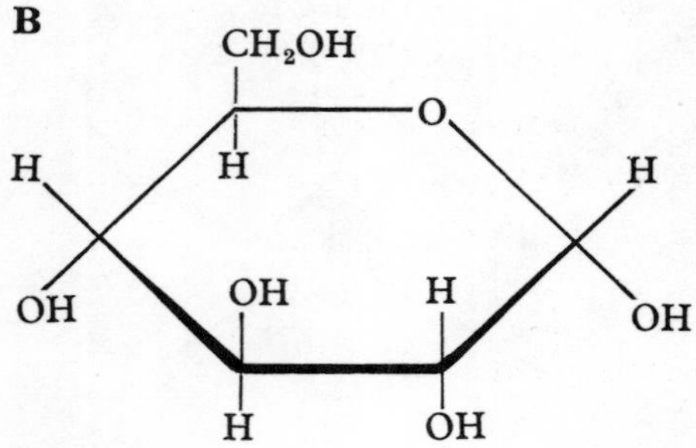

THE HEXOSE SUGARS

The best known sugars are the hexoses, of which glucose is the most common in the mammalian body. To say that glucose has the formula $C_6H_{12}O_6$ does not convey much information about its structure. It tells us that there are six carbon, twelve hydrogen and six oxygen atoms in each molecule, but it gives no information as to how these are related to each other in space. This information is embodied in the structural formula shown in Fig 5.3.

Let us consider this formula for a moment since, as we shall see later, it is important in understanding the functions of carbohydrates. First, notice the shape of the molecule: it is a ring whose sides are formed by five carbon atoms and one oxygen atom. Second, notice the side branches. Some of these terminate as hydrogen atoms, others as OH (hydroxyl) groups, and one of them as a CH_2OH group. Each group occupies a particular position with respect to the carbon atoms of the ring. This is made clear in the structural formula by numbering the carbons 1 to 6 in a clockwise direction, starting with the one on the extreme right.

It is this specific relationship between the carbon atoms and the side groups which determines the nature of the sugar and its properties. The type of glucose whose structural formula is shown in Fig 5.3 is called α **glucose**. If however the H and OH groups are interchanged at position 1, another sugar with slightly different properties results: β **glucose**. (Fig 5.4). Another sugar is formed by interchanging the CH_2OH group at position 6 with the H at position 1, and interchanging the H and OH groups at positions 2 and 3: this is **fructose**. There would be little point in multiplying examples; the important thing is that the type of sugar is determined by the positioning of the side groups. Some implications of this will be discussed later.

BUILDING UP COMPLEX CARBOHYDRATES

Hexose sugars are known as **monosaccharides**, literally 'single sugars'. All are sweet, soluble in water, and crystallizable. Monosaccharides are the building blocks of more complex carbohydrates. Thus two of them may combine to form a **disaccharide** or 'double sugar' (Fig 5.5). The process is one of **condensation** involving the loss of water. A single molecule of water is removed from a pair of monosaccharide molecules, a hydrogen atom coming from each of the two hydroxyl groups at positions 1 and 4 respectively, and the oxygen atom from one of the two hydroxyl groups. As a result a covalent bond (known in this case as a **glycosidic bond**) is established joining the two monosaccharide molecules at positions 1 and 4. This gives us a 1–4 linked disaccharide with the general formula $C_{12}H_{22}O_{11}$. The reaction is reversible, and under suitable conditions the disaccharide can be hydrolysed into its monosaccharide sub-units.

The particular disaccharide formed depends on the monosaccharides that enter into its composition. **Sucrose**, for example, is formed by the union of glucose and fructose; it is abundant in plant cells particularly in the stem of the sugar cane and the root of sugar beet which provide the sources of commercial sugar (Fig 5.6). **Lactose**, the sugar found in milk, is formed by the union of glucose and galactose. And **maltose**, which occurs in certain germinating seeds (barley, for example), is an intermediate in the formation of starch, and is formed by the union of two α glucose molecules. Like monosaccharides all disaccharides are sweet, soluble in water and crystallizable.

Fig 5.4 All hexose sugars have the same basic ring structure; they differ from each other in the arrangement of the various side groups as can be seen directly below by comparing α glucose, β glucose and fructose.

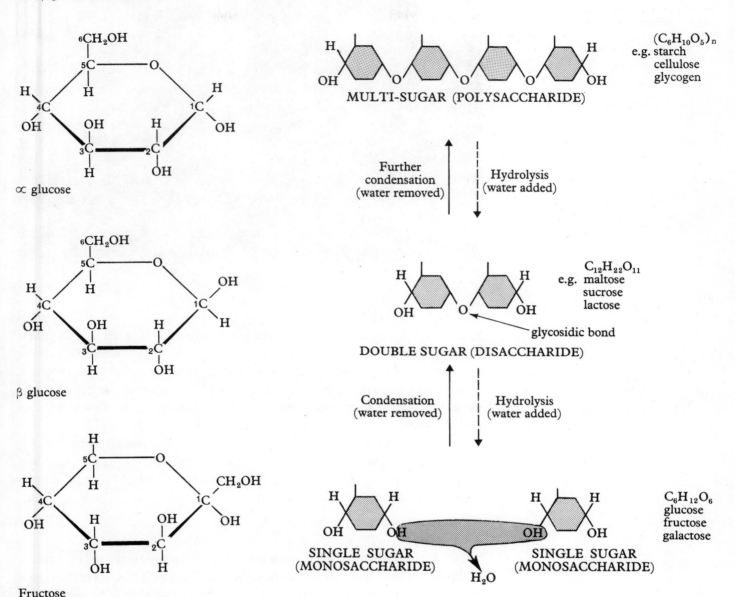

$\propto$ glucose

β glucose

Fructose

MULTI-SUGAR (POLYSACCHARIDE)

$(C_6H_{10}O_5)_n$
e.g. starch
cellulose
glycogen

Further condensation (water removed) | Hydrolysis (water added)

DOUBLE SUGAR (DISACCHARIDE)

glycosidic bond

$C_{12}H_{22}O_{11}$
e.g. maltose
sucrose
lactose

Condensation (water removed) | Hydrolysis (water added)

SINGLE SUGAR (MONOSACCHARIDE)

SINGLE SUGAR (MONOSACCHARIDE)

H_2O

$C_6H_{12}O_6$
glucose
fructose
galactose

Under appropriate conditions disaccharides link up through glycosidic bonds to form a **polysaccharide** or 'multi-sugar'. In its final form a polysaccharide consists of a long chain which may be folded or branched and in which the total number of monosaccharide units is variable. The general formula is therefore $(C_6H_{10}O_5)_n$. Polysaccharides are insoluble in water, unsweet, and cannot be crystallized.

It will be apparent at once that the compact structure of a polysaccharide makes it ideal as a storage carbohydrate. By building up free monosaccharide molecules into an insoluble polysaccharide, the sugar can be stored in a compact form in which it cannot diffuse out of the cell and exerts no osmotic action within the cell. When occasion demands it can be hydrolyzed into free monosaccharide molecules which may then be used for the production of energy.

Fig 5.5 Removal of a molecule of water from two monosaccharide molecules results in the formation of a disaccharide. Further condensation leads to the formation of a polysaccharide.

Fig 5.6 Sugar cane is the main source of sucrose for commercial sugar. Here a harvesting machine in Puerto Rico is cutting cane at the rate of 78.4 tonnes per hour. This particular cane, known as PR-1016, is a high-yielding variety developed by an American experimental station. (*US Department of Agriculture*)

STARCH AND GLYCOGEN

Starch and glycogen are both storage polysaccharides. **Starch** is the storage carbohydrate of plants: it consists of long chains of α glucose molecules with branches in places. The chains are folded and packed together in spherical plastids to form **starch grains**. Starch grains are particularly abundant in those parts of a plant which are specially concerned with storage such as potato tubers.

Glycogen, sometimes called animal starch, is the storage carbohydrate of animals. It consists of long, profusely branched chains of α and β glucose molecules linked by 1–4 or 1–6 glycosidic bonds. Glycogen is more soluble than starch and exists in the cytoplasm as tiny granules.

CELLULOSE

We have seen how polysaccharides can be built up for storage. This process of linking sugar molecules can be taken further with the formation of a material of great structural importance, **cellulose**. Cellulose is found in plant cell walls, and we shall see later that the efficient functioning of plants is very much bound up with the properties of their cellulose walls. Cellulose is a polysaccharide consisting of long chains of β glucose molecules linked by glycosidic bonds. The way the sugar molecules are orientated means that OH groups stick outwards from the chain in all directions. These can form hydrogen bonds with neighbouring chains, thereby establishing a kind of three-dimensional lattice. This is in marked contrast to the way starch molecules are built up. Like cellulose, starch consists of a long chain of glucose molecules but the chain is coiled into a helix and the orientation of the glucose sub-units is such that most of the OH groups potentially capable of forming hydrogen bonds project inwards. So there are no cross-linkages in starch, and this is one reason why it lacks the structural properties possessed by cellulose. The structure of starch and cellulose are compared in Fig 5.7.

A) STARCH

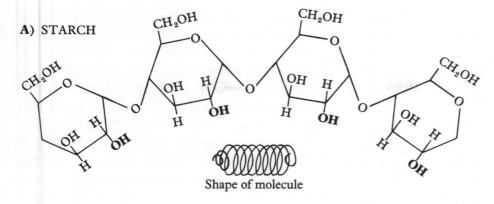

Shape of molecule

B) CELLULOSE

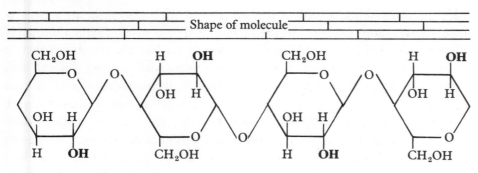

Shape of molecule

Fig 5.7 Starch and cellulose compared. A) The starch molecule consists of a long chain of 1-4 linked α glucose units of which four are shown. The chain is coiled into a helix forming, in effect, a cylinder in which most of the OH groups capable of forming cross linkages project into the interior. These are shown in bold. There are six glucose units for every complete turn of the spiral. B) The cellulose molecule consists of a straight chain of β glucose units joined together in such a way that the OH groups project from both sides of the chain. These are capable of forming hydrogen bonds with neighbouring OH groups resulting in the formation of bundles of cross-linked parallel chains.

A single cellulose chain may contain as many as 10,000 sugar units with a total length of 5 μm. The strength of the glycosidic bonds, together with the cross-links between adjacent chains, makes it tough like rubber. In the cell wall, groups of about 2,000 cellulose chains are massed together to form ribbon-like **microfibrils** each between 10 and 30 nm in diameter. The electron microscope has shown that these are laid down in layers, the microfibrils of each layer running roughly parallel with each other but at an angle to those in other layers. The microfibrils of successive layers are frequently interwoven as can be seen in Fig 5.8.

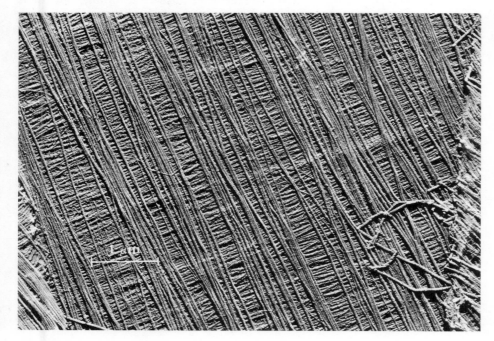

Fig 5.8 Electron micrograph of cellulose microfibrils (shadowed) from the cell wall of the alga *Chaetomorpha*. (*R. D. Preston, University of Leeds*)

In the cell wall the cellulose microfibrils are embedded in a gel-like organic matrix consisting of polymerised sugars and various other substances. The result is a material of great strength which in its structural organization has been likened to reinforced concrete: the matrix is equivalent to the concrete, the microfibrils to the metal framework within the concrete. Moreover there is evidence of chemical bonding between the cellulose microfibrils and surrounding matrix, thus further enhancing its mechanical properties.

Despite its strength cellulose is fully permeable to water and solutes. This is because the matrix is riddled with minute water-filled channels through which free diffusion of salts, sugars and the like can take place. This is important in the functioning of plant cells.

LIGNIFICATION

Certain plant cells, notably those concerned with providing strength and conducting water, become **lignified** (converted into wood) during their development. In this process **lignin**, a polymer of various alcohol derivatives, is deposited in the spaces between the sugar molecules, making the cell wall much more rigid, and rendering it impermeable. It thus serves as a waterproof cement. Once lignification is complete the protoplasm can no longer absorb materials from outside the cell, which therefore dies. Hence lignified tissue is always dead. Its function of providing mechanical strength is entirely due to its ligno-cellulose composition. Its ability to transport water and mineral salts is due to the fact that lignification involves loss of the protoplasm resulting in the formation of a hollow tube.

THE COMMERCIAL IMPORTANCE OF CELLULOSE

Cellulose provides the raw material for the manufacture of many commercially important products such as cellophane, celluloid, rayon and various plastics. Cellulose derivatives such as cellulose nitrate are used in the manufacture of lacquers, films and explosives. Cotton is almost pure cellulose and requires comparatively little treatment before being used in the manufacture of fabrics and other materials. But the main source of cellulose for commercial use is wood pulp obtained mainly from spruce. The first stage in the preparation of the cellulose involves the removal of the lignin and other non-cellulose materials. This can be done by treating it at high temperature and pressure with sulphur dioxide and calcium bisulphite. The purified pulp, containing up to 97 per cent pure cellulose, is then dissolved and the cellulose regenerated as synthetic fibres or sheets. The regenerated cellulose has enormous tensile strength as well as protective properties. This makes cellophane, for example, ideal for wrapping purposes. Rayon is even stronger and is used in the manufacture of industrial belts and tyre cords.

INTERCONVERSION OF CARBOHYDRATES

We have seen that carbohydrates can be classified into monosaccharides, disaccharides and polysaccharides which are interconvertible. The relationship between these different carbohydrates is summarized in Table 5.1.

In the laboratory it is possible to convert one carbohydrate into another by chemical means. For example a polysaccharide like starch can be split into free monosaccharides by heating it with strong acid. But in organisms these conversions take place much more quickly (and gently!) under the influence of

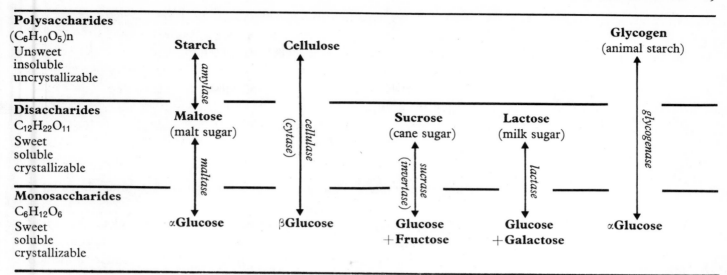

Polysaccharides
$(C_6H_{10}O_5)n$
Unsweet
insoluble
uncrystallizable

Starch　Cellulose　Glycogen (animal starch)

amylase　*cellulase (cytase)*　*glycogenase*

Disaccharides
$C_{12}H_{22}O_{11}$
Sweet
soluble
crystallizable

Maltose (malt sugar)　Sucrose (cane sugar)　Lactose (milk sugar)

maltase　*sucrase (invertase)*　*lactase*

Monosaccharides
$C_6H_{12}O_6$
Sweet
soluble
crystallizable

αGlucose　βGlucose　Glucose + Fructose　Glucose + Galactose　αGlucose

organic catalysts called **enzymes** (see p. 87). Each reaction shown in Table 5.1 is reversible and catalysed by a specific enzyme. The direction in which the reaction goes depends on the equilibrium conditions.

OTHER SUGAR COMPOUNDS

Simple sugars such as hexoses and pentoses can readily link up with other molecules to form more elaborate compounds. One of the most important associations is that between sugar and phosphoric acid. Phosphorylation of hexose sugar is a necessary first step in the oxidative breakdown of sugar in respiration. Phosphorylation is also involved in the formation of **nucleotides**: in this case a pentose sugar links up with an organic base at position 1 and one or more phosphoric acid molecules at position 5 (condensation again). Nucleotides are the building blocks of **nucleic acids** such as deoxyribonucleic acid (DNA), but some of them have important functions in their own right. In Chapter 7 we shall come across several that play a vital rôle in respiration.

Table 5.1 Chart illustrating how one carbohydrate can be converted into another by condensation or hydrolysis. All the reactions are reversible; the direction in which the reaction proceeds depends on the equilibrium conditions. The names in italics refer to the enzymes, organic catalysts responsible for speeding up the reactions.

Lipids

Lipids include natural **fats** and **oils**, distinguished from each other simply by the fact that at normal temperatures fats are in the solid state and oils in the liquid state. Like carbohydrates, fats contain carbon, hydrogen and oxygen but a fat molecule contains a much smaller proportion of oxygen than a molecule of carbohydrate.

Natural fats and oils are compounds of **glycerol** and **fatty acids**, and it is impossible to understand their composition without first considering the structure of these two constituents.

The formula of glycerol is $C_3H_8O_3$, and the arrangement of the carbon, hydrogen and oxygen atoms is shown in the following structural formula:

$$H$$
$$|$$
$$H - C - OH$$
$$|$$
$$H - C - OH$$
$$|$$
$$H - C - OH$$
$$|$$
$$H$$

There is only one kind of glycerol: its molecular configuration shows no variation and it is exactly the same in all lipids.

Fatty acids, however, show considerable variation and the nature of a particular lipid depends on the fatty acids which it happens to contain. The general formula of a saturated fatty acid is $C_nH_{2n}O_2$. A more informative way of writing the formula is $R - (CH_2)_n COOH$ where R stands for a variety of different hydrocarbon groups (e.g. CH_3, C_2H_5, C_3H_7, etc.), and n can be any number from 1 to 16. To take an example: in **stearic acid**, a common constituent of adipose tissue, the R group is CH_3 and there are 16 CH_2 groups. The formula of stearic acid is therefore $CH_3(CH_2)_{16}COOH$. The full structure of this fatty acid is shown in Fig 5.9 and you will see that it consists of a single chain of carbon atoms to which hydrogen atoms are attached, a **hydrocarbon chain**. The molecule terminates in the form of an acid carboxyl (COOH) group. As there are no polar groups in the chain, fatty acids are insoluble in water.

Fig 5.9 The general formula for a fatty acid is $R(CH_2)_n COOH$. The fatty acid depicted on the right is stearic acid in which $R = CH_3$ and there are 16 CH_2 groups making up the hydrocarbon chain.

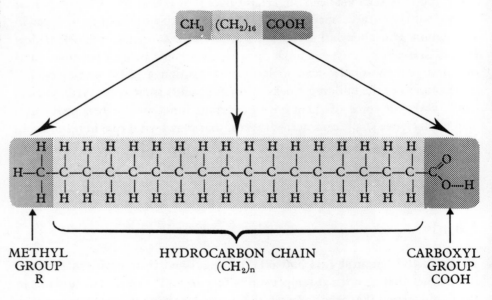

METHYL GROUP R HYDROCARBON CHAIN $(CH_2)_n$ CARBOXYL GROUP COOH

A glance at the structure of stearic acid will tell you that it contains the maximum possible number of hydrogen atoms: all the bonds are used. Such fatty acids are said to be **saturated**. Some, however, do not contain the full number of hydrogen atoms and therefore have a double bond connecting two of the carbon atoms in the chain. Such fatty acids are **unsaturated**. For example, **oleic acid**, found in both animal and plant fats, has a double bond between two carbons right in the middle of the chain:

$$CH_3 (CH_2)_7 - \overset{\overset{\displaystyle H}{\displaystyle |}}{C} = \overset{\overset{\displaystyle H}{\displaystyle |}}{C} - (CH_2)_7 \; COOH.$$

THE BUILDING UP OF A LIPID

Fatty acids and glycerol are the sub-units of a lipid. In the synthesis of a fat or oil three fatty acid molecules combine with one molecule of glycerol to form a **triglyceride**. As with the construction of complex carbohydrates the process involves condensation in which water is lost. As you can see in Fig 5.10 each of the OH (hydroxyl) groups in the glycerol molecule reacts with the COOH (carboxyl) group of a fatty acid. In this reaction water is removed and an oxygen bond (known in this case as an **ester bond**) is established between the glycerol and fatty acid. As the glycerol possesses three hydroxyl groups, three fatty acids attach themselves to the glycerol and three molecules of water are removed.

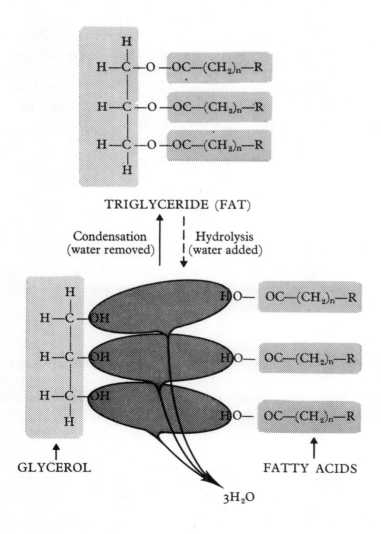

TRIGLYCERIDE (FAT)

Condensation (water removed) Hydrolysis (water added)

GLYCEROL   FATTY ACIDS

$3H_2O$

Fig. 5.10 The removal of water from a molecule of glycerol and three fatty acid molecules results in the formation of a triglyceride. Most naturally occurring fats are triglycerides similar in general plan to the one illustrated here.

The particular fat or oil resulting from this process depends on the fatty acids involved. The fatty acids may be all the same or different, saturated or unsaturated, complex or comparatively simple. It is this that determines the characteristics of individual fats and oils.

THE FUNCTIONS OF LIPIDS

Although carbohydrates provide the most important direct source of energy in organisms, fats too contain energy. In fact weight for weight they yield more energy on combustion than carbohydrates. However, the main function of fat is storage. Like starch and glycogen, fat is compact and insoluble and provides a convenient form in which energy-yielding molecules (the fatty acids) can be stored for use when occasion arises. Excess carbohydrate can be readily converted into fat, as many of us know to our cost. The fat deposits beneath the skin and elsewhere represent potential sources of energy which can be drawn upon when required (Fig 5.11).

Fats also enter into the composition of certain structures in cells, for example the middle layer of the unit membrane. In this case the fat is combined with phosphoric acid to form a **phospholipid**. The orientation of the phospholipid molecules is important in determining the functional properties of the plasma membrane (see p. 27). Another situation in which fat is structurally important is the mammalian skin. The dermis contains large deposits of **sub-cutaneous fat** which is predominantly for storage but also serves to insulate the body from excessive heat loss. It is extremely extensive in animals such as the polar bear which live in particularly cold climates.

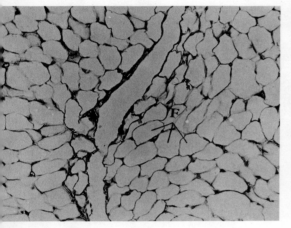

Fig 5.11 Section of adipose (fatty) tissue showing closely-packed fat cells (**F**). The large clear region within each cell represents the fat. In some of the cells the nucleus can be seen pushed against the cell membrane. (*Gene Cox, Micro-Colour International*)

Proteins

Proteins play a vital rôle in the formation of structures in organisms. They differ from carbohydrates and fats in that they always contain nitrogen as well as carbon, hydrogen and oxygen. In addition sulphur is often present and sometimes phosphorus and other elements.

THE BUILDING UP OF PROTEINS

Proteins are built up from **amino acids**. These are the sub-units of proteins in much the same way that monosaccharide sugars are the sub-units of poly-saccharides. About 20 naturally-occurring amino acids have been identified with certainty and they all have the same basic structure:

$$H_2N - \overset{\displaystyle R}{\underset{\displaystyle H}{C}} - C\overset{\displaystyle O}{\underset{\displaystyle OH}{}}$$

All amino acids have an NH_2 (amino) group and a COOH (carboxyl) group, as shown above.

They differ in the nature of the R group. The simplest amino acid is **glycine** in which R is a hydrogen atom. In **alanine** R is CH_3, in **valine** it is

C_3H_7, and so on. The names of the 20 amino acids are listed in Appendix V together with their R groups. There is no virtue in remembering the names and formulae of all the amino acids, but it is important to appreciate the principle on which their structure is based.

Amino acids unite to form proteins in much the same way that monosaccharide sugars combine to form polysaccharides, and fatty acids and glycerol to form fats. The first step in this process involves the combination of two amino acids (Fig 5.12). A reaction occurs between the amino group of

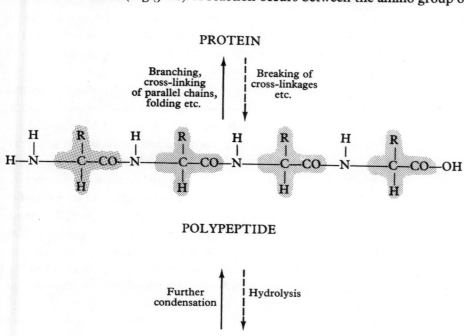

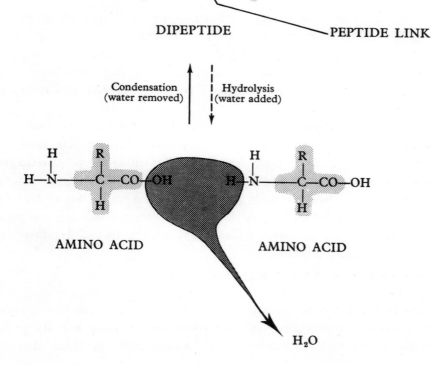

Fig 5.12 The removal of a molecule of water from two amino acid molecules results in the formation of a dipeptide. Further condensation leads to the formation of a polypeptide. Proteins are formed by the branching, folding and cross-linking of polypeptide chains.

one amino acid and the carboxyl group of another: a molecule of water is removed (condensation) and the two amino acids become joined by a **peptide** bond to form a **dipeptide**. Continued condensation leads to the addition of further amino acids resulting in the formation of a long chain called a **polypeptide**. A particular polypeptide may be composed of as many as a hundred amino acids. Further elaboration of the polypeptide chain results in the formation of the protein.

A typical protein consists of several adjacent polypeptide chains which may be folded, branched and cross-linked at intervals. The cross-linkages may be of several sorts: hydrogen bonds, salt links or sulphur bridges. The sulphur bridges are the strongest and contribute to the great toughness of certain proteins.

Looked upon as chains of amino acids, the structure of proteins is simple enough. Although there are only about 20 naturally occurring amino acids, the number of possible ways in which they can be combined is almost infinite. The individuality of a particular protein is determined by the sequence of amino acids comprising its polypeptide chains, together with the pattern of branching, folding and cross-linkages. Some proteins contain only a proportion of the known amino acids; others contain all of them. The total number of amino acids in the protein molecule varies from a few to several thousand. The latter are very large molecules and have correspondingly high molecular weights: serum **globulin**, for example, has a molecular weight of 140,000. So large are some protein molecules that they can even be detected in the electron microscope and this, combined with other techniques, has enabled their size and shape to be determined with accuracy.

On account of the large size of their molecules, proteins do not go into true solution but form **colloidal suspensions**. Collectively the colloidal particles have a very large surface area and, since they have a strong capacity to absorb water and other substances, they are important in holding molecules in position within the cell and maintaining molecular organization in protoplasm.

Another feature of proteins related to their molecular structure is that they have both acidic and basic properties. This is due to the presence of amino (basic) and carboxyl (acidic) groups at the free ends of the polypeptide chains. This makes it possible for the protein to combine with basic or acidic substances, a fact of importance when it comes to building up materials in the body.

In fact many structures found in organisms consist of a protein combined with another compound to form a **conjugated protein**. The non-protein component to which the protein is attached is called the **prosthetic group**. Egg yolk, mucus, haemoglobin and other pigments, are all examples of conjugated proteins. In egg yolk the prosthetic group is phosphoric acid; in mucus it is a carbohydrate; in haemoglobin it is an iron-containing pigment called haem. Nucleic acids, discussed in Chapter 30, are usually found combined with a protein as nucleoproteins.

THE FUNCTIONS OF PROTEINS

The shape of a protein molecule is directly related to its function. In general proteins fall into two groups: **globular** and **fibrous**. In the globular proteins, such as globulin itself, the polypeptide chains are tightly folded to form a spherical shape. Such proteins are relatively soluble and readily go into colloidal suspension. They occur in blood plasma and in all cells where they play an important part in maintaining the composition of protoplasm.

An extremely important function of globular proteins is the formation of **enzymes,** organic catalysts whose function is to speed up chemical reactions in organisms. Enzymes are always proteins and modern research has indicated that the functioning of an enzyme is directly related to its molecular configuration, and depends on the fact, already mentioned, that a protein will readily combine with other substances. The polypeptide chains are folded in such a way that the surface of the enzyme molecule has depressions into which other molecules can fit prior to reacting together. This is discussed in more detail in Chapter 6.

Fibrous proteins are insoluble and consist of long parallel polypeptide chains cross-linked at many points along their length. In this form they enter into the composition of various structures in the body. These structural proteins include **keratin** found in hairs, nails, hooves, horns and feathers, and the connective tissue proteins, **elastin** and **collagen** (Fig 5.13).

Fig 5.13 A) Shadowed collagen fibres from the neck tendon of a bird. B) Single collagen fibre from the tail of a rat negatively stained with uranyl acetate. The banding pattern is characteristic of collagen. (*Sir John Randall, King's College, London.*)

A

B

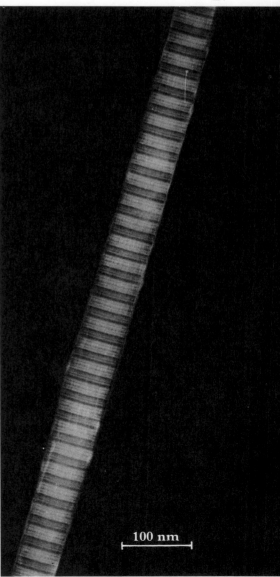

100 nm

A further property of proteins that helps to explain their functional characteristics is that they are irreversibly destroyed by heat. If the temperature rises above a critical level (about 40°C) the protein becomes denatured. It seems that although the peptide linkages remain intact, the molecular configuration and shape of the molecule become irreversibly altered. This is particularly important in the case of enzymes whose efficiency depends on the shape of their molecules.

THE ANALYSIS OF PROTEINS

In this account of protein structure we have visualized the gradual building up of a protein from its constituent amino acids. This is a perfectly justifiable way of approaching protein structure but a biochemist has to tackle the problem the other way round. He is confronted with a protein, perhaps a very complex one containing hundreds of amino acids, and his job is to identify the amino acids and determine their sequence in the polypeptide chains that make up the protein. In other words he has got to take the protein to pieces. How is this done?

It is not particularly difficult to identify the different amino acids. The protein can be hydrolyzed by appropriate enzymes into its constituent amino acids which are then separated and identified by **paper chromatography**. This technique depends on the fact that the different amino acids have different solubility properties. A small amount of solvent is put into the bottom of a glass column. A strip of absorptive paper, with a concentrated spot of the mixed amino acids near the bottom, is hung in the glass column so that its end dips into the solvent. The solvent moves slowly up the strip carrying the amino acids with it. Owing to their different solubilities, the latter travel at different speeds and thus separate from each other. Subsequent development of the **chromatogram** with a suitable reagent enables the different amino acids to be detected and identified. Many proteins have been analyzed in this way. But although it tells us what amino acids are present, and their relative proportions, this process gives us no information on how they are arranged within the protein molecule.

Determination of the amino acid sequence involves progressive hydrolysis of the protein: knocking off the amino acids one at a time, as it were, and then identifying them. The protein is systematically treated with different hydrolyzing enzymes which break it into a series of polypeptide chains of progressively shorter lengths. Each chain has a terminal amino acid at each end: one of these has a free amino (NH_2) group whilst the other has a free carboxyl (COOH) group. After each hydrolysis the polypeptide chains are reacted with various substances which attach themselves to either the amino or carboxyl groups, thereby 'labelling' the terminal amino acids for subsequent identification. Further hydrolysis with specific enzymes breaks off these labelled amino acids which can then be identified and their position noted. This process is continued until the protein has been split into all its constituent amino acids.

Needless to say this procedure is easier said than done, and it was not until 1954 that the first protein was fully analyzed. In that year Frederick Sanger of Cambridge University published the structure of **insulin**, the hormone produced by the pancreas. The insulin molecule contains a total of 51 amino acids, and 17 of the known amino acids are represented in the molecule. Sanger succeeded in demonstrating their precise sequence within the molecule.

Fig 5.14 Frederick Sanger, discoverer of the sequence of amino acids in the hormone insulin, an achievement for which he was awarded a Nobel Prize in 1958. (*Frederick Sanger, University of Cambridge, photo by K. Harvey*)

WORKING OUT THE SHAPE OF A PROTEIN MOLECULE

But there is still a further question, namely the three-dimensional shape of the molecule. Here again spectacular progress has been made in recent years. In 1913 Sir Lawrence Bragg worked out the structure of solid sodium chloride by bombarding crystals of this salt with X-rays and analyzing the way the X-rays were scattered. Since then this technique of **X-ray crystallography** has been used in the analysis of many different molecules, including proteins.

The technical problems of X-ray analysis are formidable but the principle is simple enough. A beam of X-rays is fired at the protein crystal. The X-rays are reflected by the atoms in the protein molecules and the way in which they are scattered, the X-ray diffraction pattern, is recorded on a photographic plate located behind the crystal. Between each firing the crystal is rotated slightly so that the X-rays hit all sides of it, thus enabling a complete analysis to be made. Difficulties arise not so much in obtaining the X-ray pictures as in interpreting them. We have already seen that proteins are large molecules composed of thousands of atoms. Their X-ray diffraction patterns are correspondingly complex, and this makes analysis of the data a difficult and laborious business. However, with the aid of high-speed computers, it is now possible to derive from X-ray diffraction patterns the precise positioning of the atoms and the shape of the whole molecule. It should be pointed out, however, that hydrogen atoms cannot be located by this technique. Their positions are inferred from the arrangement of the other atoms.

So far as protein structure is concerned, two major advances have been made as a result of X-ray analysis: the establishment of the helical configuration of many proteins, and the shape of globular proteins like haemoglobin. Let us look at these in turn.

THE α HELIX

It is now established that fibrous proteins such as keratin and collagen are built on a helical plan. This was first shown conclusively by Linus Pauling of the California Institute of Technology who demonstrated that the keratin molecule consists of a greatly elongated polypeptide chain twisted into a helix, the so-called α **helix**, rather like an extensible telephone cord. Successive turns of the helix are linked together by weak hydrogen bonds situated between the amino groups of one turn and the carboxyl groups of the next. This is what makes hair and wool so stretchable. The hydrogen bonds can be readily broken and the helical strand stretched out so that it becomes a straight untwisted cord. Collagen is also helical, in this case a triple helix. The molecule consists of three polypeptide chains coiled round each other rather like three-stranded electric light flex. Hydrogen bonds connect the adjacent polypeptide strands: this makes for a much more stable form of protein whose shape is less readily altered. Double helices are also known: for example, certain nucleic acids, such as deoxyribonucleic acid (DNA).

It is now known that the α helix is found in globular as well as fibrous proteins, and the question naturally arises: What is its function? Undoubtedly it contributes to the general toughness of a fibrous protein like collagen: the three polypeptide chains, twisted round each other and held together by cross-connections, form a strong cord rather like a plaited rope. But a more general function of the helical arrangement, and one which applies to globular as well as fibrous proteins, is that it helps to maintain the shape of the molecule. The

α helix, with its links between successive twists, is a much more stable structure than a straight untwisted polypeptide chain would be. There is little doubt
that this is of great importance in the biological functioning of proteins, particularly enzymes whose efficiency depends on the molecules maintaining a
particular shape.

HAEMOGLOBIN AND MYOGLOBIN

The second important advance made possible by X-ray crystallography involves
the globular proteins **haemoglobin** and **myoglobin**. Our knowledge of these
two molecules is largely due to the work of two Cambridge scientists, Max
Perutz and John Kendrew. Intense interest has centred on haemoglobin and
myoglobin because they are both blood pigments, the former in the general
circulation, the latter in muscle. Both have a remarkable tenacity for oxygen
with which they are able to combine reversibly. Only by understanding their
molecular structure are we likely to understand how they perform this unique
function.

Myoglobin is the simpler of the two (Fig 5.15). It contains over 2500 atoms
which show up as groups of spots on X-ray photographs. On analysis it turns
out that the molecule consists of a single polypeptide chain made up of 153
amino acids. The polypeptide chain is coiled to form an α helix, and this in turn
is folded on itself into a roughly spherical shape. Various kinds of chemical
bond, together with electrostatic attraction, keep the folds of the chain together
and help to maintain the shape of the molecule. Myoglobin is an example of a
conjugated protein: attached to the polypeptide chain is a flat group of atoms,
the prosthetic group, consisting of a central iron atom surrounded by rings

Fig 5.15 Three dimensional analysis of a
protein is here illustrated by the myoglobin
molecule. A) X-ray diffraction pattern of a
crystal of myoglobin. The pattern gives a total
of more than 50,000 reflections, represented
by dark spots in the X-ray photographs. Only a
very small part of the overall pattern is therefore
shown in this one photograph. B) three-
dimensional contour map showing positions of the
various atoms in part of the myoglobin molecule,
derived from analysis of numerous X-ray
photographs like the one shown in A. The black
disc (left of centre) is the haem group seen
sideways-on; this is surrounded by a helical
polypeptide chain, an α helix, a section of which
can be seen in side view at the bottom of the
photograph, and end-on towards the top right-
hand corner. C) model of the myoglobin molecule based on contour maps similar to the one
shown in B. The black disc at the top is the
haem group, and the spherical object attached to
it represents an oxygen molecule. (*J. C.
Kendrew, MRC Laboratory of Molecular Biology,
Cambridge*)

A

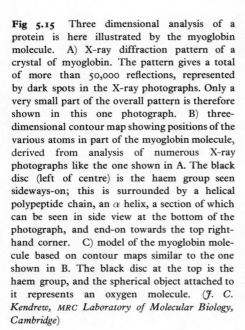

C

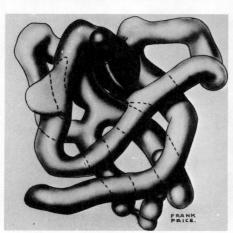

B

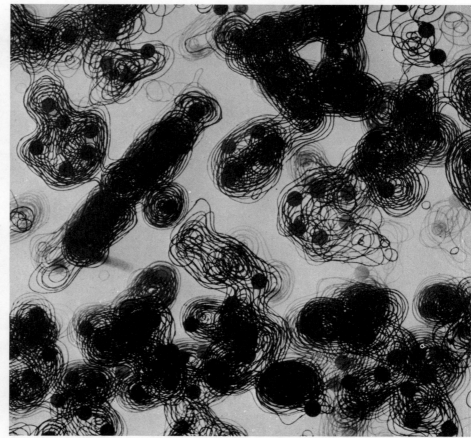

of carbon and nitrogen atoms. The prosthetic group is **haem** and it is to the iron atom in the middle of it that the oxygen molecule becomes attached.

Haemoglobin is a larger and more complex molecule than myoglobin. It contains 574 amino acids in all, over three times as many as in myoglobin. The molecule is composed of four polypeptide chains arranged round four haem groups. Each polypeptide chain is an α helix and this is folded and held together in much the same way as the single polypeptide chain is in myoglobin. The presence of four haem groups means that a single molecule of haemoglobin can carry four molecules of oxygen, thus giving it a greater capacity for this vital respiratory gas.

PORPHYRINS

Haem belongs to a class of organic compounds known as the **porphyrins**. They are of great biological importance, occurring in a variety of pigments, viz. blood pigments, chlorophyll and the cytochromes. These last are coenzymes required in respiration (see p. 104). The general structure of a porphyrin is illustrated in Fig 5.16. Porphyrins are one of many different types of molecule with which protein may be conjugated.

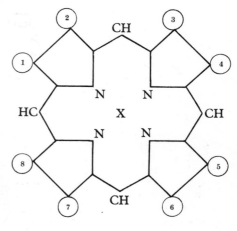

Fig 5.16 The general structure of a porphyrin. Four pyrrole rings are linked by CH bridges to form the complex molecule illustrated above. A variety of side groups can exist at positions 1 to 8, so many different isomers can be formed. Porphyrins can form a complex with a heavy metal at X. In chlorophyll the metal is magnesium, in haemoglobin iron, and so on.

THE GENERAL FUNCTIONS OF PROTEINS

As you will have inferred the main function of proteins is to form the structures of the body: cell membranes, the wall of cell organelles, connective tissue, muscle fibres, skeletal materials and so on. Protein forms the framework of cells and organisms and has been aptly described as the 'stuff of life'.

Like carbohydrate and fat, protein can be broken down with the release of energy. However, this only takes place in conditions of dire starvation. This is hardly surprising because the body is unlikely to break down its own tissues except when there is no other source of energy and even then it is always the less essential tissues such as the skeletal muscles and gut wall which are used first, the really vital structures such as the heart and brain remaining unaffected.

Proteins also function as regulators in that they control the chemical reactions and metabolic processes which occur in organisms. Many hormones are proteins, and these regulate physiological processes such as growth and reproduction. But it is as enzymes that proteins perform their most important regulatory function. This is discussed in Chapter 6.

Vitamins

Many other organic compounds, besides those already described, are found in living organisms. These include the **vitamins**, a mixed assortment of chemical compounds grouped together not because of any chemical affinity between them but because they share the same functional attributes. Vitamins are organic constituents of food required in very small amounts for a variety of metabolic purposes. Plants synthesize them, but animals require them in their diet, though there are exceptions to this. The exceptions include vitamins synthesized by bacteria in the gut; and those manufactured by the animal itself. Vitamin D_3, for example, can be synthesized in the human body and is then activated in the skin by ultra-violet rays.

The importance of vitamins can be appreciated from the ill-effects which ensue if an organism is deprived of one of them. Indeed this was how they were discovered. Around the turn of the century a number of observations made it quite clear that a diet consisting of only carbohydrates, fats, proteins and salts was not enough. The word vitamine (meaning 'vital amine') was coined to cover the additional food factors required. We now know that, with one exception, vitamins are not amines, but they are essential for healthy life and the word has been retained though the 'e' has been dropped. Two of the important deficiency diseases caused by lack of particular vitamins are illustrated in Fig 5.17.

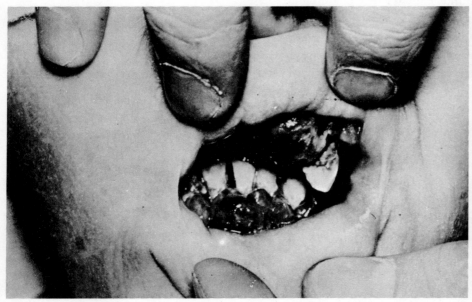

A

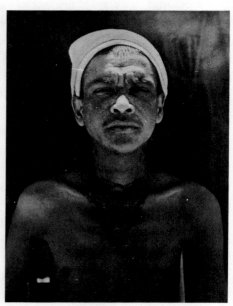

B

Fig 5.17 Two examples of 'deficiency diseases' caused by lack of vitamins in the diet. A) Bleeding of the gums is one of the symptoms of scurvy, caused by deficiency of vitamin C, (ascorbic acid), an ingredient of citrus fruits. Although its medical aspects have been much studied, little is known of the biochemical processes in which this vitamin takes part. B) Lack of the vitamin nicotinic acid results in the deficiency disease pellagra, one of whose symptoms is pigmentation in the exposed parts of the chest ('Casal's necklace'). Other symptoms include abdominal pains, diarrhoea, and skin lesions. Acute deficiency results in mental disorder and delirium. Nicotinic acid, a water-soluble vitamin, is found in a variety of foods including meat, fish, wheat and maize. It is an essential component of the hydrogen-carrying coenzyme nicotinamide adenine dinucleotide (NAD) required in cell respiration. (*Wellcome Museum of Medical Science*).

What is the rôle of vitamins in metabolism? In some cases we simply do not know. In other cases vitamins have been implicated in various metabolic pathways. To take an example, **vitamin B$_2$, (riboflavine)** combines with certain phosphate compounds to form **flavine adenine dinucleotide** (FAD), an important coenzyme concerned with the transport of hydrogen atoms in cell respiration (see p. 104). Another important vitamin is **nicotinic acid**, whose amide is also involved in hydrogen transport. Nicotinic acid enters into the composition of two coenzymes: **nicotinamide adenine dinucleotide** (NAD), and **nicotinamide adenine dinucleotide phosphate** (NADP). In both of these the nicotinic acid amide forms the effective group which carries hydrogen atoms. Both occur in cell respiration, but NADP also occurs in photosynthesis where it is involved in taking up hydrogen atoms derived from water for the subsequent reduction of carbon dioxide.

Little would be gained by multiplying examples. The important principle is that many vitamins form coenzymes and other essential metabolites without which the chemical reactions that occur in cells would be unable to take place. As such they are equally important to animals and plants, not to mention micro-organisms such as bacteria.

6 Chemical Reactions in Cells

In the last chapter we reviewed the structure and functions of the different chemical components found in cells. We now turn to the reactions in which they take part.

The chemical reactions that take place in cells constitute **metabolism** and the participating molecules are termed **metabolites**. Some of these metabolites are synthesized within the organism, while others have to be taken in as food. Metabolism is a basic characteristic of all living systems; indeed it is the metabolic reactions, particularly those that produce energy, which keep cells alive. Even seemingly inert structures like bone, cartilage and connective tissue contain living cells engaged in metabolic activities. It is only the truly dead parts of an organism, such as the hair and nails of mammals, the shells of molluscs and the lignified fibres of plants, which do not metabolize; and indeed it is because they do not metabolize that they are dead.

It is characteristic of metabolic reactions that they occur in small steps. Instead of a metabolite being converted into products in one large explosive reaction, it is gradually transformed through a series of gentle reactions which together comprise a **metabolic pathway**. Each reaction, though small in itself, brings the raw materials closer to the end product.

There are two main reasons why it is important that metabolism should proceed in small steps. First, large reactions, particularly of a violent kind, would endanger the cell and probably kill it. Second, long metabolic pathways involving numerous small steps enable the cell to derive maximum benefit from the reactions. It is like dismantling a building brick by brick so that the bricks can be used again, rather than blowing it up. Needless to say, the orderliness of these metabolic pathways requires a fantastic degree of structural and functional organization in the cell. Much of this organization is achieved by **enzymes** about which we shall have more to say later.

ANALYSING METABOLIC PATHWAYS

One of the main problems of biochemistry is to analyse the metabolic pathways that occur in cells and establish the individual steps by which they proceed. This is a difficult task partly because of the small size of the average cell, and also because so many reactions take place inside it. It has been estimated that as many as 1,000 different reactions occur in an individual cell, an object only 10 μm in diameter. This makes a biochemist's life something of a nightmare.

Biochemists resort to a number of strategies to get round these difficulties. One is to grind up organs and tissues, chunks of liver for example, and extract the juices from them. The cell-free extract is centrifuged so as to separate the different structural components of the juice, and the different fractions are then analysed. The various chemical constituents of each fraction can be subsequently purified and their properties investigated. This kind of direct chemical analysis is a necessary prerequisite to any sort of experimental approach.

Each step in a metabolic pathway is catalysed by a specific enzyme. If a particular enzyme is inactivated by means of a specific poison the molecules on which the enzyme normally acts will accumulate. These particular molecules, whatever they happen to be, can then be identified by the kind of analysis outlined above. This approach has proved invaluable in identifying the intermediates formed, for example, in the oxidative breakdown of sugar, discussed in Chapter 7.

The approach which has been most generally fruitful in tracing metabolic pathways is that of **isotope labelling**. This technique depends on the fact that the atoms of a particular element are not all identical but exist in several different forms which differ from each other in the structure of their nucleus. These **isotopes,** as they are called, share the same atomic number and chemical properties but are distinguished from each other by their physical properties and atomic weights. Moreover, some of them are radioactive, emitting characteristic radiations such as alpha particles and gamma rays. Radioactive carbon ^{14}C is an example of such a radioactive isotope, the 14 indicating that it has an atomic weight of 14 as compared with the normal carbon atom ^{12}C with its atomic weight of 12. Such isotopes can be detected, and the amount of radiation measured, by **Geiger counters** and other monitoring devices sensitive to the radiations emitted by them.

Other isotopes are stable and do not emit radiations. The only way of distinguishing this kind of isotope from the normal one is by the fact that it has a greater mass. Such isotopes are usually detected and measured very accurately by means of a **mass spectrometer**, an instrument which separates them according to their atomic weights.

The development of the cyclotron and nuclear reactor has made available to biologists an artificial source of isotopes which can be used in biochemical research. The organism is supplied with a specially prepared compound in which one of the elements is replaced by its radioactive isotope. For example, plants can be put into an atmosphere containing $^{14}CO_2$, i.e. carbon dioxide in which the normal carbon (^{12}C) is replaced by the radioactive isotope (^{14}C). The radioactive carbon is described as a **tracer**, and we say that the carbon dioxide has been **labelled** or **tagged**.

Subsequent analysis can be carried out to determine which parts of the plant have come to contain the radioactive element. To do this the tissue is placed on a photographic plate which is exposed to the radiations wherever they happen to occur. Subsequent development of the plate reveals the exact whereabouts of the radioactive material. Such photographs are called **autoradiographs** and an example is shown in Fig 6.1.

If it is desired to identify the chemical compounds into which the radioactive isotope has become incorporated, the tissue is ground up and the chemical constituents separated by chromatography (Fig 6.2). If paper chromatography is used the chromatogram is exposed against a photographic plate, and the radioactive compounds identified in the resulting autoradiograph.

Fig 6.1 An example of the use of radioactive isotopes in biological research. A eucalyptus with dead areas in some of its leaves was supplied with water containing radioactive phosphorus ^{32}P. A) photograph of some of the leaves showing the dead areas. B) autoradiograph showing that ^{32}P does not enter the dead areas during uptake. (*Australian Atomic Energy Commission*)

A

B

Fig 6.2 For detailed chemical analysis, compounds labelled with radioactive isotopes may be separated from non-labelled compounds by means of chromatography. This photograph shows chromatographic separation in an ion-exchange column of a peptide labelled with radioactive carbon ^{14}C (*UK Atomic Energy Commission*)

This kind of technique has been used to trace the fate of various elements in metabolism. The organism is supplied with a labelled compound, the choice of which depends on the particular element being investigated, and analysis is carried out at various times after the beginning of the experiment, to determine where the radioactive material has got to, and which compounds contain it. In this way metabolic pathways can be reconstructed.

In later chapters we shall come across examples of metabolic pathways which have been worked out by painstaking research using radioactive tracers. Our present knowledge of photosynthesis and cell respiration is largely the result of such techniques.

TYPES OF CHEMICAL REACTIONS IN CELLS

Two types of chemical reaction occur in cells: **synthetic** and **breakdown**. Synthetic reactions are those in which molecules are linked together to form a more complex compound:

$$A + B \rightarrow AB$$

A and B are the **reactants** or **substrate** molecules; AB represents the **product**. We saw examples of this kind of reaction in Chapter 5: the linking up of sugar molecules to form a disaccharide, or of fatty acid and glycerol to form a lipid.

Breakdown reactions are those in which a complex compound is split into simpler molecules:

$$AB \rightarrow A + B$$

In this case AB is the substrate, and A and B are the products. The hydrolysis of a fat molecule into fatty acid and glycerol is an example.

Both kinds of reaction occur in metabolism and both happen in cells all the time. Synthetic reactions comprise **anabolism**, and breakdown reactions **catabolism**. The important difference between them is that anabolic reactions require energy whereas catabolic reactions produce it. Energy-absorbing reactions are termed **endergonic**; energy-liberating reactions **exergonic**. The relationship between these two types of reaction is summarized in Fig 6.3.

Fig 6.3 The fate of organic food materials in a generalized animal cell. Can you construct a comparable scheme for a generalized plant cell?

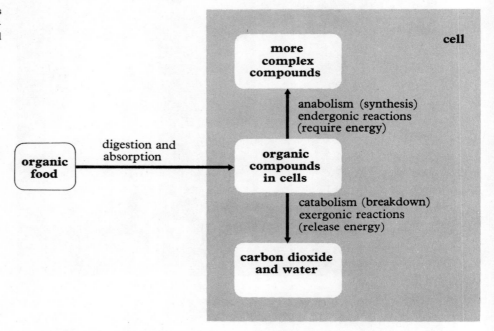

Anabolic reactions are concerned with building up structures, storage compounds and complex metabolites in the cell. Starch, glycogen, fats and proteins are all products of anabolic pathways. Plants and certain bacteria build up these complex organic molecules from simple inorganic sources. They have much greater synthetic powers, and their anabolic pathways are therefore much more extensive, than those of animals. But even animals have to build them up from their simpler organic sub-units, i.e. polysaccharides have to be built up from monosaccharides, fats from fatty acid and glycerol, and proteins from amino acids.

The function of catabolic reactions is to liberate energy. This is required for three main purposes:

(1) To drive the cells' anabolic reactions, i.e. synthesis of proteins, storage compounds, etc.;

(2) For work: for example contraction of muscles, transmission of nerve impulses, secretion of glands.

(3) For maintenance: for example maintenance of a constant internal environment, and of the tissues and organs in a state of health and functional efficiency.

Energy is therefore of vital importance to an organism, and it is no wonder that much of metabolism is concerned with its production.

Energy

Let us now take a closer look at energy from the biological point of view. One of the most important metabolic pathways yielding energy is the breakdown of sugar;

$$C_6H_{12}O_6 + 6O_2 \rightarrow 6CO_2 + 6H_2O + Energy$$

What this greatly over-simplified equation tells us is that the breakdown of sugar is an **oxidation** process: the sugar reacts with oxygen in some way and is eventually broken down into carbon dioxide and water with the liberation of energy. The sugar molecule can be looked upon as containing **potential energy** which is freed only when the molecule is split. The potential energy is locked in the chemical bonds which join the atoms in the sugar molecule together, and only when these bonds are broken is the energy released. The energy released is capable of doing useful work. As an analogy think of a boulder sitting at the top of a hill (Fig 6.4). While at rest the boulder is full of potential energy. If, however, it is pushed so that it rolls down the hill, its potential energy is converted into the kinetic energy of motion. The energy, formerly locked up in the boulder, is released.

The boulder analogy can usefully be taken a step further. When the boulder reaches the bottom of the hill and comes to rest, it contains less potential energy than it did at the top of the hill. In order to restore its potential energy it must be pushed back to its former position, and this requires the expenditure of energy. Similarly there is far less potential energy in the carbon dioxide and water resulting from the breakdown of sugar than there is in the much larger and more complex sugar molecule. The energy content can only be restored if the carbon dioxide and water molecules are built up again into sugar, a chemical feat which requires energy and can only be performed by green plants and certain micro-organisms.

This illustrates a further point, namely that chemical materials can be cycled. The carbon dioxide produced by the respiration of animals and plants can be resynthesized by plants into sugars; some of the latter will in turn be consumed by animals that eat the plants. It is therefore clear that animals are absolutely dependent upon plants, as they alone can resynthesize sugars from carbon dioxide and water.

Matter can in fact be cycled with what may be regarded as complete efficiency: none is lost and none need be gained from outside sources. Energy, however, cannot be cycled with such efficiency. Only a proportion of the energy released by the breakdown of sugar is capable of being used to resynthesize sugar. Much of it is lost as heat, in which form it cannot be reconverted into the energy of organic molecules. The bulk of the energy required for the building up of sugars comes from an outside source, the sun. The trapping of

Fig 6.4 The concept of energy. Just as it requires energy to push the boulder over the hump before it can roll down the hill, so also the sugar molecule must be supplied with 'activation energy' before it can be broken down into carbon dioxide and water. (*modified after G. G. Simpson and W. S. Bede*, Life; An Introduction to Biology, *Harcourt, Brace and Jovanovich Inc. 1965*)

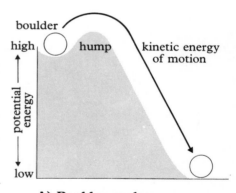

A) Boulder analogy

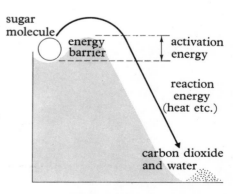

B) Oxidation of sugar

this solar energy by green plants requires the participation of **chlorophyll**. Chlorophyll is therefore of enormous importance in the living world. The process by which plants use the energy of sunlight in order to build up sugars from carbon dioxide and water is **photosynthesis.**

In the case of the boulder rolling down the hill, the energy released is mechanical energy. In cells useful energy is contained in chemical bonds and is released when these bonds are broken. Anabolic processes like photosynthesis are concerned with building up bonds which can subsequently be broken for energy release.

ENERGY CONVERSIONS

You will remember from the laws of thermodynamics that different forms of energy (chemical, light, heat and so on) are interconvertible. This is a very obvious fact in our daily lives: for example, every time we light a match chemical energy is converted into heat and light energy. This is also important biologically because it means that, given the right conditions, an organism (or certain specialized cells within it) can transform one type of energy into another. The most obvious example is the photosynthetic cell of a green plant which can convert light energy from the sun into chemical energy. The latter can in turn be converted into a variety of other forms of energy: in muscle tissue it is converted into the mechanical energy of contraction, in luminescent organisms like the glow-worm it is converted into light energy, in electric fishes it is converted into electrical energy, and so on.

The **First Law of Thermodynamics** states that when one form of energy is converted into another there is no net loss or gain. In other words the total amount of energy resulting from a conversion is equal to the amount we start off with. However, this does not mean that all the energy is in the same form. As our boulder rolls down the hill only a proportion of the energy is converted into the mechanical energy of motion; the rest is used to overcome various frictional forces and is dissipated as heat. This is summed up in the **Second Law of Thermodynamics** which states that when one form of energy is converted into another, a proportion of it is turned into heat. This, too, is an obvious fact of everyday life. Electricity is used to drive the motor of an electric train, but inevitably some of the energy is lost as heat. This unavoidable fact is of great importance in biology. It means that when sugar is broken down not all the energy liberated can be used for establishing new chemical bonds and so forth. A proportion of it is dissipated as heat. Of course it does not follow that this heat is of no value to the organism, indeed in warm-blooded animals it contributes to the maintenance of a constant body temperature, but from the thermodynamic standpoint the energy has been changed into a form in which it has no direct metabolic use.

ACTIVATION ENERGY

The boulder analogy illustrates another important concept. Imagine that the boulder resides in a slight depression at the top of the hill (Fig 6.4). It is obviously necessary to push it over the hump before it can start rolling down the hill, and this requires the expenditure of energy. The hump represents an **energy barrier**, and the energy required to push the boulder over this barrier can be called the **activation energy**. Much the same applies to our sugar molecule: a small amount of activation energy must be supplied before the process of oxidative breakdown can get under way.

Clearly any factor, physical or chemical, that helps the sugar molecule over its energy barrier will facilitate the process of oxidative metabolism.

THE SPEED OF BIOLOGICAL REACTIONS.

Consider the type of chemical reaction in which two molecules A and B react to form a compound AB, the sort of reaction that unites two monosaccharide molecules to form a disaccharide. What kind of factors influence the speed of such a reaction? To answer this we must remember that the molecules of A and B are in a state of continual random motion. Only when they collide and come into contact can they react. Clearly any factor that increases the frequency of collision will increase the speed of the reaction. In general three factors achieve this effect:

(1) the concentration of the substrate molecules;
(2) the temperature;
(3) the presence of a catalyst.

Let us consider each in turn.

The first two need little explanation. It is obvious that the more concentrated the molecules, i.e. the more densely packed they are, the more likely they are to collide and react. Raising the temperature speeds up the random motion of the molecules, thereby increasing the probability of their colliding.

Any factor which, directly or indirectly, raises the concentration of the substrate molecules will obviously speed up the reaction. Increasing the pressure has this effect, as does removal of molecules of water in which the substrate molecules are dissolved. But the most effective way of concentrating the substrate is to supply a catalyst. The substrate molecules are adsorbed onto the surface of the catalyst where, having been brought into close proximity, they react. The product leaves the surface of the catalyst, which is quite unchanged by the process and may be used again.

Many inorganic catalysts are known, for example, iron, platinum, sulphuric acid and so on. Catalysis is also important in metabolism, but in this case the catalysts are always organic and operate in a slightly different way from the surface catalysis described above. The catalysts found in living systems are called **enzymes,** to which we now turn.

Enzymes

Enzymes were discovered by the German chemist Eduard Buchner towards the end of the nineteenth century. His discovery is an example of one of those fortuitous accidents by which advances are sometimes made in science. Buchner had been endeavouring to extract from yeast a fluid of medicinal use. One of his difficulties was to prevent the extract going bad. To this end he tried adding sugar to it and found to his astonishment that the sugar was converted into alcohol: in other words it fermented. Now there was nothing new about his discovery that yeast promotes fermentation; in fact this had been demonstrated some twenty years before by Louis Pasteur, but Pasteur believed that fermentation was brought about by the living yeast cells. Buchner showed that the living cells were not responsible for fermentation, but that the juice extracted from them was. The word **enzyme** was coined for the active ingredient in the juice that promotes fermentation. Enzyme literally means 'in yeast', but it is now used as the collective name for the many hundreds of compounds that have since been extracted from cells and shown to have a catalytic action on specific chemical reactions.

Enzymes are organic catalysts which speed up chemical reactions in organisms. Before discussing their properties one must appreciate why their presence is so important. In the first place the reactions in cells would, in the absence of enzymes, be so slow as virtually not to proceed at all. But they do more than merely speed up metabolic processes. They also control them. It was mentioned earlier that as many as 1,000 different reactions take place in an individual cell, a mere 10 μm in diameter. The functional organization which this demands is achieved by each individual reaction being catalysed by a specific enzyme. It is this which ensures that metabolism proceeds by gentle steps in an orderly fashion.

Fig 6.5 Demonstration of enzyme action. A) a piece of liver (approximately 35 g) is dropped into a beaker of hydrogen peroxide. B) the result after about 3 minutes. Liver contains catalase, one of the fastest-acting enzymes known. (*B. J. W. Heath, Marlborough College*)

A

B

NAMING AND CLASSIFYING ENZYMES

Enzymes can be divided into two main groups: **intracellular** and **extracellular**. The former occur inside cells where they control metabolism. The latter are produced by cells but achieve their effects outside the cell: they include the digestive enzymes that break down food in the gut.

Normally an enzyme is named by attaching the suffix -ase to the name of the substrate on which it acts. Thus maltase acts on maltose, lipase on lipid, urease on urea, and so on. Pepsin and trypsin, both found in the mammalian gut, attack proteins and are hence known as proteases.

Enzymes are grouped according to the type of reaction they catalyse. In many cases the reaction involves the transfer of atoms, or groups of atoms, from one molecule to another: it may be a hydrogen atom, phosphate group, amino group, and so on. Enzymes catalysing such reactions are known as **transferases** and they play an important part in energy-production and other metabolic processes in cells.

Enzymes that specifically catalyse the removal of hydrogen atoms from a substrate are called **dehydrogenases,** and those that catalyse the addition of oxygen to hydrogen are **oxidases.** Both of these are involved in the final steps in respiration. Also important in respiration are the **phosphokinases** which catalyse the addition of phosphate groups to a compound. Phosphorylation is a necessary prerequisite to the splitting of sugar for energy production. Very often, if a compound is to serve a useful purpose, certain of its component atoms must be rearranged so that its structural formula conforms to a particular pattern. Enzymes catalysing such transfer of atoms from one part of a molecule to another are called **isomerases.**

Other important types of enzymes are **dehydrases** which remove water, **decarboxylases** which remove carboxyl (COOH) groups, and **deaminases** which transfer amino (NH$_2$) groups. In what situations would you expect to find such enzymes?

THE PROPERTIES OF ENZYMES

Enzymes are always proteins, and their characteristics therefore reflect the properties of proteins. Their main properties are as follows:

(1) They generally work very rapidly. The speed of action of an enzyme is expressed as its **turnover number**. This is the number of substrate molecules which one molecule of the enzyme turns into products per minute. The turnover numbers of different enzymes vary from 100 to several million; for the majority it is around several thousand. The fastest known enzyme is **catalase**. Found in tissues where it speeds up the decomposition of hydrogen peroxide into water and oxygen, catalase has a turnover number of 6 million. Its action can be demonstrated by dropping a small piece of liver into a beaker of hydrogen peroxide: the fizzing that ensues as oxygen is given off is a dramatic demonstration of an enzyme in action (Fig 6.5). In their speed of action enzymes are much more efficient than inorganic catalysts. Finely divided platinum or iron filings will also speed up the decomposition of hydrogen peroxide but nothing like as effectively as a piece of liver. The reason is that the enzyme achieves a greater lowering of the activation energy than can be brought about by inorganic catalysts.

(2) Enzymes are not destroyed by the reactions they catalyse and so can be used again. The explanation of this will become clear when we discuss how enzymes work. This is not to say that a given molecule of enzyme can be used indefinitely for enzymes are very unstable and are readily inactivated by heat, acids etc. In this respect they differ from inorganic catalysts which are completely stable and can be used over and over again indefinitely.

(3) An enzyme can work in either direction. Metabolic reactions are reversible and the direction in which they proceed depends on the relative amounts of substrate and products present. The reaction will proceed from left to right until an equilibrium between substrate and products is reached.

$$A + B \rightleftharpoons C$$

If for some reason a large amount of C happens to be present the reverse reaction occurs, C being split into A and B until again equilibrium is re-established. The enzyme responsible for accelerating this reaction will catalyse it in either direction depending on conditions. The enzyme has no effect on the equilibrium point; it merely speeds up the reaction until equilibrium is reached.

(4) Enzymes are inactivated by excessive heat. This property of enzymes relates to the fact that they are proteins. Fig 6.6 shows the effect of temperature on the rate of an enzyme-controlled reaction. Up to about 40°C the rate increases smoothly, a ten degree rise in temperature being accompanied by an approximate doubling in the rate of the reaction. Above this temperature the rate begins to fall off, and at about 60°C the reaction ceases altogether. This is because proteins (and therefore enzymes) at high temperatures are **denatured**. For this reason few cells can tolerate temperatures higher than approximately 45°C. Organisms living in situations where the temperature exceeds 45°C either have heat-resistant enzymes or are capable of regulating their body temperature. A striking example of the former is provided by certain blue-green algae that live in hot springs at temperatures of 100°C.

(5) Enzymes are sensitive to pH. Every enzyme has its own range of pH in which it functions most efficiently. Most intracellular enzymes function best at or around neutral. Excessive acidity or alkalinity renders them inactive. On the other hand certain digestive enzymes prefer a distinctly acidic or alka-

Fig 6.6 The effect of temperature on the rate of an enzyme-controlled reaction. The concentration of enzyme and substrate were kept constant at all the temperatures investigated.

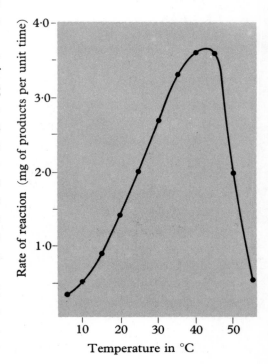

Rate of reaction (mg of products per unit time)

Temperature in °C

line environment. Thus the protein-splitting enzyme pepsin only functions in an acid medium at a pH of about 2.0 and accordingly is found in the stomach where conditions are markedly acidic. Trypsin, on the other hand, only functions in an alkaline medium at about pH 8.5, and is found in the duodenum where conditions are alkaline.

(6) Enzymes are specific in the reactions they catalyse, much more so than inorganic catalysts. Normally a given enzyme will catalyse only one reaction, or type of reaction. However, the degree of specificity varies from one enzyme to another. Most intracellular enzymes only work on one particular substrate, but certain digestive enzymes work on a comparatively wide range of related substrates. Thus catalase will only split hydrogen peroxide and is ineffective on any other natural substrate, but pancreatic lipase is less specific and will digest a variety of different fats.

HOW ENZYMES WORK

The only way we can adequately explain the properties of enzymes is to assume that when an enzyme-controlled reaction takes place the enzyme and substrate molecules combine with each other. This hypothesis is supported by various lines of evidence. Curve A in Fig 6.7 shows the effect on the rate of an enzyme-controlled reaction of increasing the concentration of the substrate, the enzyme concentration being kept constant. As you can see, the rate of the reaction rises at first and then levels off. These results are consistent with the idea that substrate molecules collide with the usually much larger enzyme molecules and then combine with them. Obviously the more substrate molecules there are, the greater the chances of substrate and enzyme molecules colliding in the right way. The plateau in the graph can be explained by assuming that when the substrate concentration reaches a certain level the system becomes saturated: all the enzyme molecules are working flat out. When this point has been reached the only way to increase the rate of the reaction is to raise the concentration of the enzyme, and curve B in Fig 6.7 shows the result of doing just this.

Experiments of this sort, together with other lines of evidence, suggest that in an enzyme-controlled reaction the substrate molecules combine with the enzyme to form an **enzyme-substrate complex**. With their various bonds held in relation to each other by the enzyme, the substrate molecules react together to form an enzyme-product complex. This splits into the enzyme and product. The enzyme, unchanged by the reaction, can then be used again.

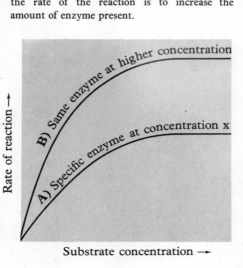

Fig 6.7 The effect of substrate concentration on the rate of an enzyme-controlled reaction. Assuming an optimum temperature, when a plateau is reached the only way of increasing the rate of the reaction is to increase the amount of enzyme present.

Rate of reaction →

B) Same enzyme at higher concentration

A) Specific enzyme at concentration x

Substrate concentration →

Enzyme + Substrate $\rightleftharpoons$ Enzyme- $\rightleftharpoons$ Enzyme- $\rightleftharpoons$ Enzyme + Product
molecules   substrate   product   molecule.
      complex   complex

The high degree of specificity shown by enzymes suggests that the combination of substrate and enzyme is very exact. It is thought that each enzyme molecule has a precise place on its surface, the **active site**, to which the substrate molecules become attached. In Chapter 5 we saw that every protein has a particular shape, and this applies to enzymes no less than to other proteins. We can picture the active site of an enzyme molecule as having a distinctive configuration into which only certain specific substrate molecules will fit (Fig 6.8). The shape of the active site, and the positions of the different chemical groups within it, ensure that only those substrate molecules with a complementary structure will combine with the enzyme. Thus we have an explanation of the specificity of enzymes, enzyme and substrate fitting together like two bits of a jigsaw puzzle. This explanation of enzyme action is known as the **lock-and-key hypothesis**: the substrate can be represented by a padlock and the enzyme by the key. Only one key will fit the lock. The key will open or close the lock just as an enzyme will split or unite substrate molecules.

The lock-and-key hypothesis also explains why enzymes are inactivated by excessive heat. Heating denatures the protein, bringing about changes in shape that prevent the substrate molecules fitting into the active site.

ENZYME INHIBITORS

Certain substances inhibit enzymes, thereby slowing down or stopping enzyme-controlled reactions. These **inhibitors** have excited the interest of biologists partly because of their medical and agricultural applications, but also because they go some way towards confirming the lock-and-key hypothesis.

One of the steps in the series of reactions which release energy in cells involves the oxidation of succinic acid. The enzyme catalysing this particular reaction is succinic dehydrogenase. Now malonic acid has a molecular configuration similar to that of succinic acid and if it is added to the system the rate of the reaction is reduced. What seems to happen is that the malonic acid molecule is so similar to the succinic that it fits into the active site of the enzyme. It therefore competes with the normal substrate for the active site. While it is attached to the enzyme molecule it prevents the normal substrate from doing so. This is known as **competitive inhibition**. The degree of inhibition depends on the relative concentrations of substrate and inhibitor. If the inhibitor is sufficiently concentrated a high proportion of the enzyme molecules will combine with it and the reaction will be slowed.

A quite different type of inhibition is that in which the inhibitor attaches itself permanently to the active site of the enzyme thereby excluding any possibility of the normal substrate taking up its rightful place. In this case the extent of the inhibition depends entirely on the concentration of inhibitor and cannot be varied by changing the amount of substrate present (note why). As substrate and inhibitor are not competing with each other for the active site, this kind of inhibition is known as **non-competitive** (Fig 6.9).

Arsenic, and other metals such as mercury and silver, owe their toxic effects to the fact that they are enzyme inhibitors, in this case of the non-competitive type. Other well known poisons such as the sulphonamide drugs are competitive enzyme inhibitors. The sulphonamide drugs are used against pathogenic bacteria and exert their effect by releasing compounds that combine with the bacteria's enzymes. Other enzyme inhibitors include the phosphate insecticides and the nerve gases developed during the Second World War. Both act in the same way: they attack the enzyme cholinesterase which plays a vital rôle in the transmission of nerve impulses (see p. 272). Cyanide is another potent inhibitor: it combines with the cytochrome enzymes responsible for the transfer of hydrogen atoms during cellular respiration and thus blocks the production of energy in cells.

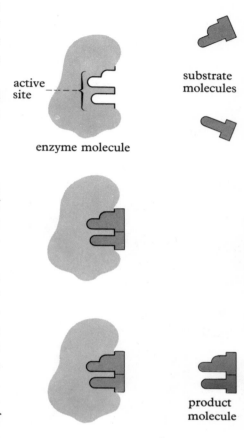

Fig 6.8 Hypothesis explaining enzyme action. The substrate molecules fit into the active site where they react together to form the product. The product then leaves the enzyme molecule.

Fig 6.9 In non-competitive inhibition of an enzyme, an inhibitor molecule, similar in structure to one of the normal substrate molecules, fits into the active site and stays there, thereby excluding the normal substrate molecules from their rightful place.

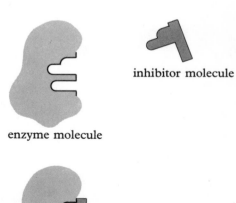

The terms competitive and non-competitive, though useful in understanding the general nature of enzyme inhibition, can be somewhat misleading, and nowadays biochemists prefer to use the terms **reversible** and **irreversible** for these two kinds of inhibition. Nor is inhibition confined to substances which combine with the active site. Some inhibitors combine with other parts of the enzyme molecule, altering its shape in such a way that the normal substrate no longer fits the active site. In general all that is required for a substance to qualify as an enzyme-inhibitor is that it should, somehow or other, prevent the substrate molecules combining with the enzyme. In other words the binding of inhibitor and substrate with the enzyme must be mutually exclusive.

The fact that poisons are often enzyme inhibitors should not give the impression that enzyme-inhibition is invariably unnatural or disastrous. Far from it. To a great extent moment-to-moment variations in the rate of cellular metabolism are caused by the control of enzyme action by inhibitor substances occurring naturally in the body.

It has been shown that in some metabolic pathways the end-product itself acts as an inhibitor. This happens when the end-product is in excess, as for instance when there is a surplus of a particular compound (or the raw materials required for its synthesis) in the diet. In these circumstances the end-product combines with one of the enzymes responsible for its own production so that the formation of further end-product is temporarily slowed down or stopped. This is an example of **negative feedback** and is important in the normal control of biochemical reactions. The general principles underlying negative feedback mechanisms are discussed in Chapter 13.

COENZYMES

Many enzymes only work in the presence of an appropriate **coenzyme**, the latter serving as a carrier for transferring chemical groups from one enzyme to another. A coenzyme of particular importance is **nicotinamide adenine dinucleotide (NAD)** whose function is discussed on p. 103. NAD works in conjunction with dehydrogenase enzymes. The latter catalyse the transfer of hydrogen atoms from a substrate to NAD. This coenzyme, and others like it, play a very important part in the final steps in respiration.

PROSTHETIC GROUPS

Sometimes the function of coenzymes is carried out not by a separate chemical substance but by a non-protein **prosthetic group** which is tightly bound to the enzyme, of which it is an integral part. For example, the enzyme **cytochrome oxidase**, which plays an important part in respiration, has a prosthetic group containing iron. A highly active part of the enzyme molecule, the function of the prosthetic group is to take up chemical groups from the protein part of the enzyme. It can therefore be looked upon as a kind of built-in coenzyme.

Part II   The Maintenance
          of Life

For an organism to maintain life energy is required. How, and in what form, this energy is made available is the subject of Chapter 7. An essential raw material for energy release is oxygen. Chapter 8 discusses how oxygen is obtained, particularly by animals in which specialized respiratory surfaces have been developed. The other essential raw material for energy release is food. In animals and certain other organisms this must be obtained in organic form, a process called heterotrophism which is described in Chapter 9. Green plants, however, can synthesize their own organic materials from inorganic sources, a process called autotrophism. Photosynthesis is the main method of autotrophic nutrition, and is discussed in Chapter 10.

An organism's problems do not end after food and oxygen have been taken into the body. They must then be transported, by active or passive means, to all its cells. How this is achieved in animals is the subject of Chapter 11. In plants uptake and transport of materials are inseparable as physiological topics and are therefore dealt with together in Chapter 12, the final chapter in this part of the book.

Longitudinal section of sieve tubes in the phloem of *Cucurbita. (Flatters and Garnett Ltd., Manchester.)*

7 The Release of Energy

We have seen that for cells to perform their vital activities energy is required. In this chapter we shall examine the process of energy production in more detail and its importance in organisms as a whole.

Energy is released by the metabolic breakdown of organic compounds, mainly carbohydrates. The reactions constitute **cell, tissue** or **internal respiration**. Oxygen is normally required, and carbon dioxide and water produced.

Respiration occurs in all organisms, and in every living cell, for the breakdown of organic compounds is the only way a cell can obtain usable energy. This applies no less to plants than to animals. The fact that a green plant can absorb energy from sunlight does not exempt it from the necessity to respire. A plant uses the energy of sunlight to build up complex organic compounds, the light energy being incorporated into their chemical bonds. The latter must then be broken down to release the energy for use by the cells.

THE GENERAL NATURE OF RESPIRATION

Cell respiration generally involves the oxidation of sugar. This equation summarizes the process:

$$\underset{\text{Sugar}}{C_6H_{12}O_6} + \underset{\text{Oxygen}}{6O_2} \longrightarrow \underset{\text{Water}}{6H_2O} + \underset{\substack{\text{Carbon} \\ \text{dioxide}}}{6CO_2} + \text{Energy}$$

This simplified equation is misleading for it gives the impression that respiration occurs in one chemical step, a notion which is far from the truth as we shall see later. But it is useful as an overall summary of the process, and its general validity can be demonstrated by simple experiments.

That oxygen is used can be demonstrated in man by breathing in and out of a sealed chamber. After a short time the air in the chamber can be shown to extinguish a candle flame, suggesting that oxygen has been used up. A similar experiment can be done on small animals and plants by enclosing them for a time in a sealed jar, and then testing for oxygen. If a plant is used the experiment must be done in the dark in order to prevent oxygen being produced by photosynthesis.

That expired air contains carbon dioxide can be demonstrated in man by bubbling air through lime water. For small animals and plants a more sensitive test is required: the organism is placed in a sealed chamber containing an indicator solution sensitive to small traces of carbon dioxide. A suitable indicator for this purpose is a standard mixture of a bicarbonate solution containing a substance such as bromothymol blue that changes colour according to the pH. The use of such a mixture for detecting carbon dioxide rests on the fact that the carbon dioxide in a sample of air is directly related to the pH of a bicarbonate solution in equilibrium with it. As the concentration of carbon dioxide increases the pH of the bicarbonate solution falls and the indicator changes colour.

THE COMPOSITION OF INSPIRED AND EXPIRED AIR

Quantitative information on what happens to air when it is taken into the lungs (inspired) can be obtained by analysing the gases present in atmospheric and expired air. The practical details of **gas analysis** vary with the type of apparatus used, but the principle is the same in all cases. A sample of the air is taken into a special gas burette and its volume noted. Potassium hydroxide is then introduced, absorbing any carbon dioxide present and causing the air to decrease in volume. The new volume is noted, and then alkaline pyrogallate is added. This absorbs any oxygen present, causing the air to shrink once more. Again the new volume is noted. Any residual air is nitrogen. The whole operation is conducted at constant temperature and pressure. From the shrinkage of the air when the hydroxide and pyrogallate are added, the percentages of carbon dioxide and oxygen in the sample can be calculated.

A comparison of the oxygen, carbon dioxide and nitrogen in atmospheric and expired air in a resting human subject at sea level is shown below.

	Atmospheric air (%)	*Expired air (%)*
Oxygen	20.95	16.4
Nitrogen	79.01	79.5
Carbon dioxide	0.04	4.1

Predictably, atmospheric air is found to contain more oxygen and less carbon dioxide than expired air. However, the relative proportions of oxygen and carbon dioxide in expired air are not always the same but vary with conditions. For example, in heavy muscular exercise more energy is needed and the production of carbon dioxide increases. In man the effect of changing conditions on the composition of expired air can be investigated by analysing the oxygen and carbon dioxide in a known volume of expired air collected over a given period of time. The air is collected in a **Douglas bag**, a large expandable sac impermeable to respiratory gases. The bag is connected to the subject's mouth by a flexible breathing tube fitted with valves so that air is breathed in from the atmosphere and out into the bag. Leading from the breathing tube is a short side tube from which samples of the expired air can be taken for analysis. Knowing the total volume of air produced, and the time taken to produce it, oxygen consumption and carbon dioxide production can be calculated. This is a useful method for investigating the composition of expired air in different conditions, for the Douglas bag is light and can be easily attached to a person's back, enabling expired air to be collected during various kinds of activity (Fig 7.1). Professor Douglas and his colleague J. B. S. Haldane used to run round the University Parks in Oxford collecting expired air for subsequent analysis in the laboratory in order to find the effects of exercise on respiratory exchange.

Fig 7.1 An athlete collecting his expired air in a Douglas bag for subsequent analysis in the laboratory. (*B. J. W. Heath, Marlborough College*)

STUDYING THE PATTERN OF RESPIRATION

In man the pattern of respiration can be studied by means of a **recording spirometer**. Spirometers come in various shapes and forms but they all operate on the same principle, namely that of a gasometer. In the model shown in Fig 7.2 the subject breathes in and out of an air-tight chamber consisting of

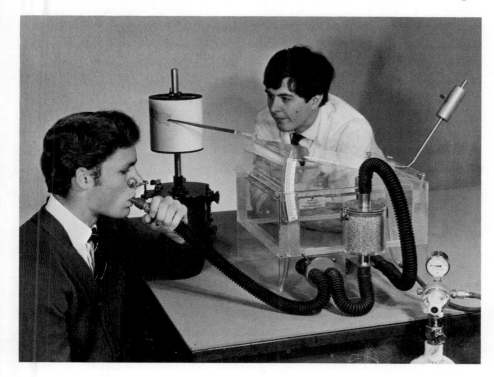

Fig 7.2 A subject's respirations being recorded by means of a spirometer. Notice the canister of soda lime which absorbs the carbon dioxide expired by the subject. Further explanation in text. (*B. J. W. Heath, Marlborough College*)

a light perspex 'lid' floating in water. As the subject breathes the 'lid' goes up and down in time with his respirations, its movements being recorded by a pen writing on a revolving drum. The chamber is first filled with oxygen; the subject is then connected to a flexible breathing tube which leads into the air chamber via a canister of soda lime. The purpose of the latter is to absorb all carbon dioxide exhaled by the subject. The spirometer 'lid' moves up and down with each respiration, and slowly sinks as the oxygen is used up. The recording paper is calibrated for time and volume so that the depth and frequency of the subject's respirations can be measured. From this the rate of oxygen consumption may be calculated.

Oxygen consumption may be related quantitatively to energy expenditure by getting the subject to operate a **bicycle joulometer**. An ordinary bicycle is fixed in a stationary stand, the back wheel being replaced by a flywheel working against a frictional resistance. In this way the mechanical work done can be measured while the subject continues to breathe in and out of the spirometer. From experiments of this sort it can be shown that oxygen consumption is directly proportional to the mechanical work done by the subject.

It is obviously important to any organism that it should convert the chemical energy of its food into the mechanical energy of muscular contraction with maximum efficiency, i.e. with minimum loss of energy as heat. Knowing the energy input and output, the organism's **efficiency** can be calculated. For man the efficiency of the body as a whole turns out to be about 23 per cent, a figure which compares favourably with, for example, motor cars whose efficiencies range from about 10–25 per cent.

MEASURING THE OXYGEN CONSUMPTION OF A SMALL ORGANISM

The oxygen consumption of a small organism can be measured by techniques of the sort shown in Fig. 7.3. The organism is placed in a chamber through which air is drawn by a filter pump. Before reaching the chamber, carbon dioxide and water are removed by passing the air-stream through soda lime and concentrated sulphuric acid respectively. After the air has been through the chamber it is again passed through these reagents. By weighing these before and after the experiment, any water and carbon dioxide produced by the organism can be estimated. The oxygen taken in can be determined indirectly by weighing the animal before and after the experiment. Knowing its change in weight, and the amount of water and carbon dioxide it has lost, the organism's oxygen consumption can be calculated.

Fig 7.3 The Haldane method of determining the oxygen consumption and carbon dioxide production by a small animal. This is explained in the text. What are the defects of this method? Can you suggest ways in which it might be improved?

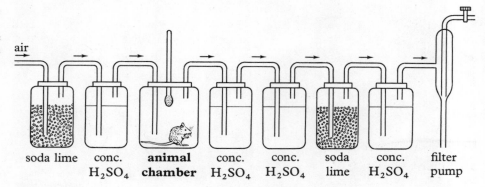

air

soda lime | conc. H_2SO_4 | **animal chamber** | conc. H_2SO_4 | conc. H_2SO_4 | soda lime | conc. H_2SO_4 | filter pump

This apparatus was first used by J. B. S. Haldane to measure the respiration rate of small animals. Many variations can be introduced to suit particular requirements. For example the carbon dioxide produced can be passed through a suitable absorber and then estimated by titration, a useful technique when only small amounts are produced. On the intake side, it is sometimes desirable to measure oxygen uptake directly, and this can be done by connecting the respiration chamber to a manometer. Carbon dioxide is absorbed by soda lime placed in the respiration chamber, so that any movement of fluid in the manometer is due to uptake of oxygen.

Modern methods of measuring oxygen consumption involve the use of sophisticated machines which measure the total volume of air expired whilst at the same time drawing off small samples for analysis. The samples are fed into a device which makes a continuous record of the percentage of oxygen and carbon dioxide present.

RESPIRATORY QUOTIENT

What is the point of these experiments? First and foremost they enable us to see how oxygen consumption and carbon dioxide production, i.e. the **metabolic rate**, vary in different conditions of diet, muscular activity, temperature and so on. They also enable us to work out the ratio of carbon dioxide produced to oxygen consumed. This gives us the **respiratory quotient (RQ)** which can provide useful information on the nature of the respiratory process.

The respiratory quotient is the amount of carbon dioxide produced, divided by the amount of oxygen used, in a given period of time:

$$RQ = \frac{\text{Carbon dioxide produced}}{\text{Oxygen used}}$$

Its main importance is that it can tell us what kind of food is being oxidized, i.e. the substrate being used in respiration. Theoretical RQs for the complete oxidation of carbohydrate, fat and protein can be worked out from the appropriate chemical equations. Carbohydrate gives an RQ of 1.0, fat 0.7 and protein 0.9. In theory we might expect an organism to give one of these three RQs, or a close approximation to it, depending on the type of food being respired. However, many factors influence the values obtained by experiment. For example, a respiratory substrate is rarely oxidized fully, and often a mixture of substrates is used. Most animals have an RQ in resting conditions of between 0.8 and 0.9. Man's is generally around 0.85. As protein is normally not used except in starvation, an RQ of less than 1.0 can be taken to mean that fat as well as carbohydrate is being respired.

As well as indicating the type of food being used, the RQ tells us what sort of metabolism is going on. For example, high RQs (exceeding 1.0) are often obtained from organisms, or tissues, which are short of oxygen. Under these circumstances they resort to **anaerobic respiration** (respiration without oxygen) with the result that the amount of carbon dioxide produced exceeds the oxygen used. We shall return to this later. High RQs also result from the conversion of carbohydrate to fat, because carbon dioxide is being liberated in the process. This is most pronounced in organisms that are laying down extensive food reserves, for example in mammals preparing to hibernate, and in fattening livestock.

A low RQ, on the other hand, may mean that some (or all) of the carbon dioxide released in respiration is being put to some sort of use by the organism. In plants it may be used for building carbohydrates (photosynthesis), in animals for the construction of calcareous shells, and so on.

MEASURING THE ENERGY OUTPUT OF THE BODY
Estimating the rate of oxygen consumption, or carbon dioxide production, gives us a method of measuring the metabolic rate. An alternative method is to determine the energy released by the organism in a given time. There are two main ways of doing this. The first depends on knowing how much heat is liberated when a molecule of a particular substrate (carbohydrate, fat or protein) is oxidized. If we know the rate of oxygen uptake, and the type of food being oxidized, the metabolic rate can be calculated. There is no difficulty in finding the rate of oxygen consumption: methods for doing this have already been described. The difficulty is knowing what sort of food is being oxidized, for this is not necessarily indicated by the diet. (Why?) However, a knowledge of the respiratory quotient can give us some indication of whether carbohydrate, fat or protein is being used. This is an indirect, and not very accurate, method of determining energy output but is adequate for most purposes.

A more direct method is to measure the heat production by the body. This can be done by putting the organism in a **calorimeter** in which it can live for several days. For human subjects the calorimeter chamber is the size of a small room, and has thermally insulated walls permitting no heat loss. The subject's heat production is determined by measuring temperature changes in a current of water circulated through the chamber.

THE METABOLIC RATE IN DIFFERENT CIRCUMSTANCES
What these experiments tell us is that the metabolic rate depends on the

amount of physical work done. The amount of energy released by the body may be increased more than ten times during heavy muscular activity. But even when a person is at rest and all voluntary movement ceases a certain amount of energy is still produced. This is used for maintaining the circulation, breathing, body temperature and so on: in short for maintaining the life of the cells. This is the minimum amount of energy on which the body can survive, and it represents the **basal metabolic rate** (BMR). The BMR varies with the age, sex and health of the individual but for a healthy young adult man it is about 167 kilojoules (kJ) per square metre of body surface, or about 4.2 kJ per kilogramme of body mass, per hour.[1] It is usual to express the BMR per unit body surface, but for general purposes it is sufficient to give the total heat output per day. For an average-sized man this comes to about 7,500 kJ, and for a woman about 5,850 kJ, per day.

The BMR does not remain constant throughout life, but changes as growth and development take place. In a new-born baby the BMR is low: about 100 kJ/m^2/h. It then rises rapidly, reaching the maximum of about 220 kJ/m^2/h by the end of the first year. This corresponds to the child's period of most rapid growth. The BMR then gradually declines as the rate of growth decreases, and continues to do so even after growth has ceased altogether.

FOOD AND ENERGY

The energy produced by the body comes from the oxidation of food, and it is therefore important to know the energy value of different foods. This can be found by measuring the amount of heat produced when a measured quantity of food is burned. A **bomb calorimeter** can be used for this. The food is placed in a strong steel chamber which is filled with oxygen and tightly sealed. The calorimeter is surrounded by a jacket containing a known volume of water whose temperature is recorded. The food is ignited by means of a small electric heating coil, and the amount of heat produced estimated by measuring the rise in temperature of the surrounding water.

From bomb calorimetry it has been estimated that 1 gramme of carbohydrate yields 17.2 kJ of heat; in other words the **energy value** of carbohydrate is 17.2 kJ. Fat has an energy value of 38.5 kJ, and protein 22.2 kJ. In the case of fat and protein rather less energy is released when these are oxidized in the body than when they are burned in a bomb calorimeter, the reason being that inside the cells they are not completely broken down.

Carbohydrates and fat are the body's primary sources of energy. Protein is required for growth and repair and is not normally used for energy production. However, if carbohydrate and fat are in short supply proteins from the diet or the tissues can be converted into carbohydrates and used for supplying energy.

Alcohol, as well as being socially and psychologically desirable, can be a valuable source of energy, yielding as much as 29.3 kJ per gramme. However, it cannot be stored and can therefore only be used as an immediate source of energy. In some countries an energy-poor diet may be made good with beer or wine.

If the total energy value of a person's food is estimated over a period of time it is found to be equal to the heat he produces plus the energy content of any materials lost from his body, such as urine and faeces. Thus his energy budget can be drawn up, the energy input balancing energy output. This is

[1] In the past it has been customary to express metabolic rates in kilocalories. 1 kilocalorie is approximately equal to 4.2 kilojoules.

not really surprising in view of the First Law of Thermodynamics, which states that energy is neither gained nor destroyed when it is converted from one form to another. Calorimetry confirms that this is as true of organisms as it is of non-living systems.

MAN'S ENERGY REQUIREMENTS

From what has been said it follows that the metabolic rate is limited by the energy value of the food taken into the body. The body cannot expend more energy than it takes in and, unless it is to draw heavily on its reserves, the necessary energy must be provided by the diet. What, then, are man's basic energy requirements?

We have already seen that to sustain life at its minimum level a basal metabolic rate of about 7,500 kJ/day is required. However, this figure applies to a man lying at rest. It does not even include the energy required for him to feed himself, and therefore as an indication of a person's normal energy requirements it is hardly applicable to everyday life. Attempts have been made to work out the energy requirements of an 'average man', one who spends a normal proportion of his time sleeping, walking, and performing the sundry activities that most people face in their daily lives. Ignoring individual variations, this works out at about 13,400 kJ/day (9.3 kJ/min) for a young man, and about 9,600 kJ/day (6.7 kJ/min) for a young woman. Corrections have to be made for age, weight and occupation as Table 7.1 makes clear. A woman's energy requirements increase during pregnancy, and a coal miner's are much greater than those of a high court judge.

The figures quoted above are based on those recommended by the United Nations Food and Agriculture Organization (FAO). They agree very closely with those put forward by other organizations such as the Medical Research Council (Great Britain), the British Medical Association, and the National Research Council of the USA. In view of these generally agreed figures it is disturbing to realize that over two-thirds of the world's population receive less energy than the recommended minimum, and in many parts of the world the daily energy intake is only just sufficient to maintain the basal metabolic rate.

On the other hand the daily energy intake of many people living in the affluent countries of the world is far more than they need.

Activity	Energy Expenditure (kJ/min)
Sleeping	4.2
Sitting (woman)	5.0
Sitting (man)	5.8
Light housework (woman)	14.6
Machine fitting (man)	17.6
Cross-sawing	37.6
Very heavy work	44.0–52.5
Exceedingly heavy work	over 52.5

Table 7.1 Energy expenditure in relation to different kinds of activity. (*after Christensen and others*).

THE IMMEDIATE SOURCE OF ENERGY: ATP

We have seen that the oxidation of food materials in the body yields energy. However, this does not directly supply the immediate needs of the body such as muscular contraction and so on. The immediate source of energy for these activities is a phosphate compound called **adenosine triphosphate** (ATP). This is a nucleotide consisting of a complex organic molecule, adenosine, to which a chain of three phosphate groups is attached: Adenosine $- P - P \sim P$. The outstanding functional importance of ATP is that when the terminal phosphate group is broken off, as it can be in the presence of the appropriate enzyme, a relatively large amount of free energy is released – about 34 kJ per mole altogether. Some of this energy is lost as heat, but a proportion of it can be used directly for biological activities. On account of the large quantity of useful energy released when it is broken, the terminal phosphate bond is described as **energy-rich**. Such bonds are usually designated by a wavy line,

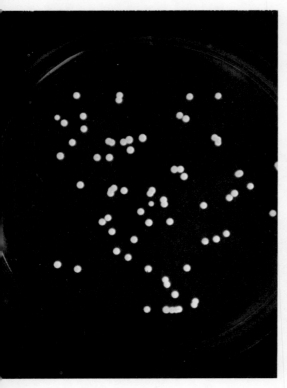

Fig 7.4 Photograph of colonies of luminescent bacteria on an agar plate, taken by their own light. (*William D. McElroy, Johns Hopkins University*).

thus $\sim$. They are by no means confined to ATP, but from the biological point of view ATP represents the most important source of useful energy. Driven by an enzyme, ATPase, the splitting of ATP is a hydrolysis reaction, yielding **adenosine diphosphate** (**ADP**) and phosphoric acid[1]:

$$\text{ATP} + \text{H}_2\text{O} \xrightarrow{\text{ATPase}} \text{ADP} + \text{H}_3\text{PO}_4 + 34\text{ kJ}$$

ATP was first isolated by K. Lohmann of the University of Heidelberg in the early 1930s. He extracted it from muscle tissue. Its function in muscles was subsequently demonstrated in America by the famous physiologist Albert Szent-Gyorgyi who showed that isolated muscle fibres contract when ATP is placed on them. Since then ATP has been shown to be the fuel driving many other biological processes: muscle contraction, nerve transmission, synthesis of many materials, luminescence, and so on (Fig 7.4). Luminescence is particularly interesting for it is used as a quantitative test for ATP. Ground-up firefly tails luminesce when solutions containing ATP are added to them, and the degree of luminescence, which can be accurately estimated, is used as a measure of the amount of ATP present. Preparations of firefly tails are now produced commercially for this purpose!

ATP is found in all cells and there is every reason to believe that it is the universal supplier of energy in all biological systems, plant and animal alike. It is therefore essential that there should always be a ready supply of it for use when required. This is where respiration comes in. The purpose of breaking down sugars in tissue respiration is to produce a continual stream of energy which can be used for synthesizing ATP. The final step in the synthesis involves attaching a phosphate group (derived from phosphoric acid) to ADP, the reverse of what happens when ATP is split. This is an endergonic reaction requiring as much energy as is produced when ATP is split. The energy for this reaction comes from the oxidation of sugars. The breakdown of sugars is thus coupled with the synthesis of ATP:

$$\text{Food} + \text{oxygen} \qquad \text{ADP} + \text{P}$$

$$\text{CO}_2 + \text{H}_2\text{O} \qquad \text{ATP}$$

LEVELS OF ENERGY

ATP represents the immediate source of energy in a cell; it can be split (and resynthesized) extremely quickly whenever and wherever it is needed. **Sugar** contains energy in a less readily available form; its breakdown is necessary for the synthesis of ATP. **Starch, glycogen** and **fats** contain energy in an even less readily available stored form which can be drawn upon when necessary. **Proteins** represent the remotest source of energy which is only used when there is no carbohydrate or fat available, as in extreme starvation.

A financial analogy may be useful. ATP is like the ready cash we carry in our pockets, available for immediate use at any time. Sugar is like a current account in the bank; the regular cashing of cheques ensures a continual supply of ready cash. Storage materials such as fat are like a deposit account

[1] Recently the traditional view that ATP is a high-energy compound whose hydrolysis yields energy for driving biological reactions has been challenged by Barbara Banks of University College, London (see *School Science Review*, Vol. 52, 1970). A reply to Banks' arguments is given by Linus Pauling and others in *Chemistry in Britain*, Vol. 6, 1970. The controversy is worth following.

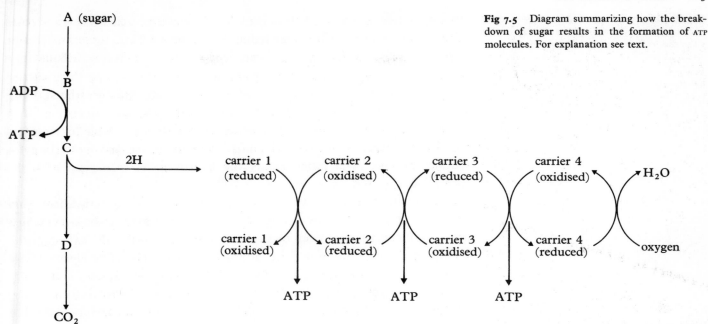

Fig 7.5 Diagram summarizing how the breakdown of sugar results in the formation of ATP molecules. For explanation see text.

which can be drawn upon when occasion arises. To take the analogy to its logical conclusion, protein is equivalent to one's property. Using protein for producing energy is like selling one's belongings, a course of action which most people resort to only in the unfortunate event of bankruptcy.

ENERGY FROM SUGAR

How exactly does the oxidation of sugar provide energy for the synthesis of ATP? This question can be answered without going into details of the metabolic pathways involved. The steps in the gradual breakdown of glucose can be represented by letters of the alphabet (Fig 7.5). Energy comes from two sources. First, some of the individual steps in the overall process are exergonic, releasing sufficient energy for a molecule of ATP to be synthesized from ADP and inorganic phosphate. In Fig 7.5 the conversion of B to C is depicted as being a reaction of this sort. However, comparatively few steps are capable of doing this, and the total number of ATP molecules built up in this way is relatively small.

THE HYDROGEN CARRIER SYSTEM

The second, and much the most important, source of energy depends on the fact that certain of the steps involve the removal of **hydrogen atoms**. This process is called dehydrogenation and is of the utmost importance in ATP synthesis. Under the influence of a dehydrogenase enzyme, two hydrogen atoms are removed from the intermediate compound (C in Fig 7.5) and taken up by a **hydrogen carrier** or acceptor, which is thereby reduced.

The hydrogen atoms then pass to a second carrier which is in turn reduced, the first carrier becoming reoxidized. In this transfer sufficient energy is released for the synthesis of a molecule of ATP. This process of oxidation-reduction is repeated with two further carriers, the hydrogen atoms finally combining with oxygen to form water. The chain of reactions results in the synthesis of a total of three molecules of ATP.

The first two carriers are both dinucleotides, derivatives of nucleic acid consisting of two nucleotides linked together. The first is **nicotinamide adenine**

Fig 7.6 Diagram illustrating the mode of action of the respiratory carriers. Note that at the FAD stage the hydrogen atoms split into protons and electrons which rejoin at the cytochrome oxidase stage. The cytochromes therefore receive and hand on only electrons. There may be other carriers in addition to the ones shown below.

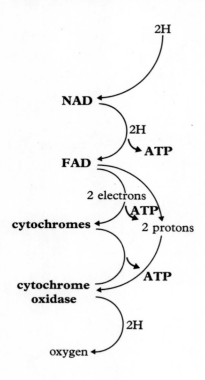

Fig 7.7 Warburg flask used in measuring oxygen-uptake by small samples of tissue. Inhibitor substances can be let into the flask from the side arm, and their effect on respiration investigated.

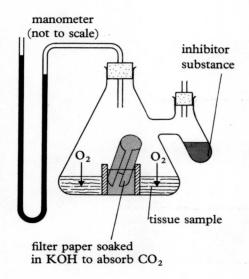

dinucleotide (NAD) derived from the vitamin nicotinic acid. When it accepts two hydrogen atoms it becomes reduced NAD, or $NADH_2$. The second carrier is **flavine adenine dinucleotide** (FAD) derived from vitamin B_2 (riboflavine). This is similarly reduced by accepting hydrogen atoms. The third carrier is a **cytochrome**, a protein pigment with an iron-containing prosthetic group similar to haemoglobin. The fourth is **cytochrome oxidase**, an enzyme. The first three carriers are **coenzymes** which are jointly responsible for transferring the hydrogen atoms from the initial dehydrogenase enzyme to the cytochrome oxidase. The latter hands the hydrogen on to oxygen with the formation of water.

Although hydrogen atoms are removed from the intermediate, and hydrogen atoms eventually join up with oxygen, it is mainly **electrons** which are actually transferred from one carrier to the next. After the FAD stage the hydrogen atoms split into their protons and electrons. The electrons are transferred to cytochrome and thence to cytochrome oxidase, and finally rejoin the protons at the last stage when oxygen is reduced to form water. The carriers are therefore more properly called **electron carriers**. They can be regarded as being at different energy levels. When the electrons are transferred from one to the next, energy is extracted from them for the synthesis of ATP. As can be seen from Fig 7.6 three molecules of ATP are formed every time a pair of electrons is transferred through the entire carrier system.

The process of electron transport just described involves oxidation in the sense that hydrogen is removed from a compound. It also involves the establishment of high-energy phosphate bonds in ATP. For these reasons the whole process is described as **oxidative phosphorylation**. It takes place at various stages in the gradual breakdown of glucose.

THE PATHWAY BY WHICH SUGAR IS BROKEN DOWN

Reconstructing the sequence of reactions which occurs as the sugar is broken down has been one of the triumphs of biochemistry over the last thirty years. Before continuing with the story let us look briefly at the kind of techniques which have helped to solve the problem.

Sugar is metabolized inside individual cells and to find out what happens it is necessary to study the respiration, not of whole organisms, but of isolated pieces of tissue. This can be done using a **Warburg manometer**. Designed by the German chemist Otto Warburg in the 1940s, this apparatus measures the respiratory rate of small pieces of tissue. The material is placed in a flask to which is attached a manometer (Fig 7.7). A compartment in the centre of the flask contains a small piece of filter paper soaked in caustic potash. This absorbs any carbon dioxide produced so that movement of the manometer fluid represents oxygen uptake. The whole apparatus is kept at a constant temperature by shaking it continuously in a thermostatically controlled water bath. The flask is equipped with one or two side arms from which various reagents and inhibitors can be added to the tissue in the course of an experiment. In this way different factors influencing the rate of respiration can be investigated.

How can this apparatus be used to reconstruct the steps in the metabolic breakdown of sugar? From a knowledge of the different organic compounds found in cells, it is possible to postulate various metabolic pathways that might conceivably occur. These can be tested by the use of inhibitors. The principle behind this is as follows. Suppose we have postulated that the following steps occur:

$$A \xrightarrow{\text{1}} B \xrightarrow{\text{2}} C \xrightarrow{\text{3}} D$$

and suppose we have shown that the enzyme which catalyses the conversion of B to C occurs in the cell. If our proposed sequence is correct inhibiting this enzyme should have certain predictable results, the most obvious of which will be an accumulation of B, a decline in the amounts of C and D, and a lowering in the rate of oxygen usage. Hypotheses of this sort can be tested with the Warburg apparatus. The inhibitor is let into the flask from one of the side arms, and its effect on the rate of oxygen consumption determined. The contents of different flasks can be analysed before and after inhibition to find out if a particular intermediate has accumulated.

Many of the results obtained by this kind of technique have been confirmed by the use of **radioactive isotopes**. Compounds labelled with the radioactive isotope of carbon ^{14}C are fed into metabolizing tissues, and the radioactive products subsequently identified. Combinations of these various techniques have given us a detailed picture of what happens in the metabolic breakdown of sugar. The main steps in the process are summarized in Fig 7.9.

GLYCOLYSIS

In the first stage of respiration sugar is broken down, step by step, to **pyruvic**

Fig 7.8 Sir Hans Krebs, discoverer of the citric acid cycle, with a Warburg apparatus. Notice the two manometers which are used to measure the uptake of oxygen by the tissue samples. The Warburg flasks are inside the cylindrical chamber. (*Sir Hans Krebs, University of Oxford*)

acid, a compound containing three carbon atoms. This sequence of reactions is known as **glycolysis**, meaning literally 'sugar splitting'.

Glycolysis begins with the **phosphorylation** of the sugar. In this process phosphate groups are added to the sugar. The reactions are endergonic and the necessary energy is derived from the splitting of ATP, which also supplies its terminal phosphate group for attachment to the sugar molecule.

Altogether two such reactions occur so that a total of two phosphate groups are added to each molecule of sugar, two molecules of ATP being used. This initial phosphorylation of the sugar is extremely important because it raises the energy level of the sugar molecules so that more useful energy can be extracted from them later.

In the next stage the phosphorylated sugar is split into two molecules of a **3-carbon** (**triose**) **sugar**. These are in equilibrium, and in normal circumstances each will be converted to pyruvic acid. In the first step two hydrogen atoms are removed from the triose sugar and taken up by NAD. As this happens in the cytoplasm outside the mitochondria where none of the other respiratory carriers are present, no useful energy can be derived from it. Instead of passing through the carrier system the hydrogen atoms are used for various reduction reactions in the cytoplasm. Although no energy comes from this initial dehydrogenation, the subsequent conversion of the triose sugar to pyruvic acid does yield some useful energy: enough for the synthesis of two molecules of ATP. This energy does not come from the transfer of hydrogen atoms but from two of the steps in glycolysis which are directly coupled with ATP synthesis.

What happens to the pyruvic acid depends on whether oxygen is available or not. If oxygen is present the pyruvic acid enters a mitochondrion where it is converted into a 2-carbon derivative of acetic acid called **acetyl coenzyme A** (acetyl-CoA). In this reaction carbon dioxide is given off, and the pyruvic acid loses two hydrogen atoms which are passed through the carrier system with the formation of three molecules of ATP. Acetyl-CoA is an extremely important intermediate in the breakdown of sugar; it links glycolysis with the next series of reactions and is also formed in the breakdown of fats and proteins. It is therefore a focal point in oxidative metabolism.

THE CITRIC ACID CYCLE

Acetyl-CoA has two carbon atoms and this now reacts with a 4-carbon compound, **oxaloacetic acid**, to form **citric acid** with six carbon atoms. There now follows a series of reactions in which the citric acid is gradually converted back to oxaloacetic acid step by step.

This cyclical series of reactions, worked out by Sir Hans Krebs and his colleagues, is known as **Krebs' citric acid cycle**. Two of the steps involve

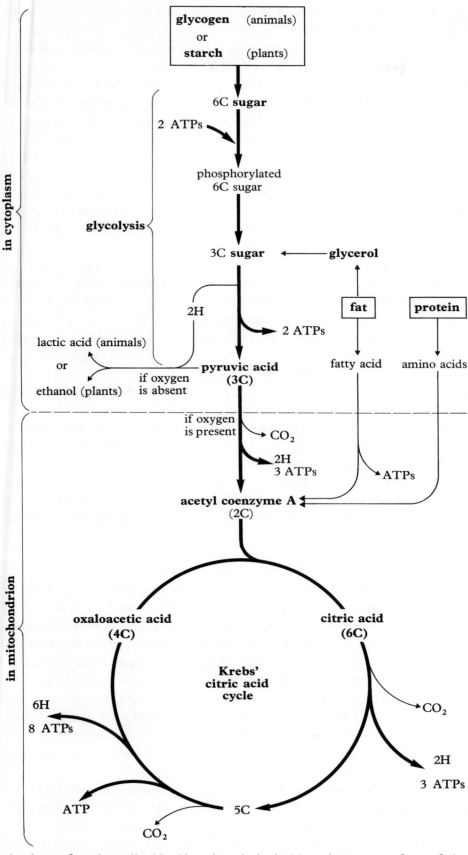

glycogen (animals)
or
starch (plants)

6C **sugar**

2 ATPs

phosphorylated
6C sugar

glycolysis

3C **sugar** ← **glycerol**

2H

2 ATPs

fat   **protein**

lactic acid (animals)
or
ethanol (plants)

if oxygen
is absent

**pyruvic acid
(3C)**

fatty acid   amino acids

if oxygen
is present

CO_2

2H
3 ATPs

ATPs

**acetyl coenzyme A
(2C)**

**oxaloacetic acid
(4C)**   **citric acid
(6C)**

**Krebs'
citric acid
cycle**

CO_2

6H
8 ATPs

2H
3 ATPs

ATP

5C

CO_2

in cytoplasm

in mitochondrion

Fig 7.9 Production of energy in cells. Most of the available energy in cells comes from the step-by-step breakdown of carbohydrate. This energy is then used for the synthesis of ATP molecules. In carbohydrate metabolism most of the energy for ATP synthesis comes from the process of dehydrogenation in which hydrogen atoms are removed and handed through the hydrogen carrier system. In the pathway shown here only the main steps in the breakdown of carbohydrate are included. In reality numerous intermediate steps are involved in the conversion of sugar to pyruvic acid and of citric acid to oxaloacetic acid. The scheme shows the stages at which hydrogen atoms are removed, together with the number of ATP molecules synthesized. Although carbohydrates provide the most important source of energy in cells, fats can also be broken down to provide energy for ATP synthesis as shown here. Proteins too can be broken down but this only happens in conditions of extreme starvation when all carbohydrate and fat has been used up.

the loss of carbon dioxide (decarboxylation). More important, four of the steps involve the removal of hydrogen atoms which are passed through carrier systems with the formation of ATP. Three of the carrier systems are like the one illustrated in Fig. 7.6, yielding three ATP molecules for every pair of hydrogen atoms transferred. The remaining one yields two molecules of ATP. In addition to the energy released by the carrier systems one of the steps in the cycle produces an ATP molecule direct.

In terms of ATP synthesis Krebs' cycle is the most important source of energy in metabolism. Careful auditing of the energy yield has shown that every time a molecule of acetyl-CoA goes through the citric acid cycle a total of 12 ATP molecules are produced, whereas the conversion of one molecule of triose sugar to pyruvic acid yields only 2 molecules of ATP. If we consider the complete oxidation of one molecule of 6-carbon sugar, glycolysis produces a net total of 2 ATP's, as compared with 30 produced by the citric acid cycle (including the formation of acetyl-CoA from pyruvic acid). Altogether a total of 32 ATP molecules are produced by the complete oxidation of one molecule of 6-carbon sugar.

Where do all these reactions take place? By centrifuging cells it is possible to separate the various organelles from each other, after which each fraction can be tested for its chemical properties. By techniques of this sort it has been shown that the formation of pyruvic acid (glycolysis) takes place in the general cytoplasm whereas the citric acid cycle, and transfer of hydrogen and electrons, take place in the mitochondria. In other words the mitochondrion is the site of ATP formation.

But whereabouts in a mitochrondrion do the reactions take place? The available evidence suggests that the reactions of the citric acid cycle take place in the matrix whilst the ATP-producing electron transfer reactions occur on the inner membrane and cristae. High-magnification electron micrographs show the inner membrane and cristae to be covered with densely-packed stalked globules. These, together with fragments of the membrane to which they are attached, have been separated from the rest of the mitochondrion by sonic oscillation followed by high-speed centrifugation and shown to be the site of the electron transfer system.

RESPIRATION WITHOUT OXYGEN

As you can see, the bulk of useful energy yielded by metabolism comes from the transfer of hydrogen atoms or electrons. For this to work oxygen must be available to accept the hydrogen atoms from the final carrier. However, certain organisms derive energy from breaking down sugar in the complete absence of oxygen, a process known as **anaerobic respiration** or **anaerobiosis**. Yeasts can respire anaerobically, as can various other fungi and bacteria. Certain worms living in oxygen-deficient mud can also do this, as can vertebrate skeletal muscle during heavy activity when oxygen cannot be delivered to the contracting muscles fast enough to meet their needs.

An organism that can respire anaerobically is known as an **anaerobe**. In general two types are recognized:

(1) Complete anaerobes live permanently in oxygen-deficient conditions and are completely independent of oxygen for respiration. Indeed in some cases they may be poisoned by oxygen, even in small concentrations.

(2) Partial anaerobes thrive in the presence of oxygen, but if oxygen happens to be absent or in short supply they can resort to anaerobic respiration. The majority of anaerobes fall into this category.

The main difference between aerobic and anaerobic respiration is that in the latter sugar is only partially broken down. Instead of being oxidized to carbon dioxide and water, it is converted into either **ethanol** (**ethyl alcohol**) or **lactic acid**: lactic acid is the end product of anaerobiosis in animals, ethanol in plants. The latter process is **alcoholic fermentation**.

Anaerobiosis in plants (**alcoholic fermentation**):

$$C_6H_{12}O_6 \longrightarrow 2CH_3CH_2OH + 2CO_2 \uparrow + 210 \, kJ$$
$$\text{ethanol}$$

Anaerobiosis in animals:

$$C_6H_{12}O_6 \longrightarrow 2CH_3CH(OH)COOH + 150 \, kJ$$
$$\text{lactic acid}$$

Aerobic respiration for comparison:

$$C_6H_{12}O_6 + 6O_2 \longrightarrow 6H_2O + 6CO_2 \uparrow + 2880 \, kJ$$

As the breakdown of the sugar is incomplete in anaerobiosis, less energy is released than when sugar is broken down in aerobic conditions. The meagre energy yield in anaerobic respiration contrasts sharply with the 2880 kJ produced in aerobic conditions.

In anaerobic respiration a considerable amount of energy remains locked up in the alcohol or lactic acid molecules. In animals this energy can be liberated by the subsequent conversion of the lactic acid back into pyruvic acid which is then oxidized in the usual way. This requires oxygen and if none is available the lactic acid must be excreted. In terms of energy production anaerobic respiration is therefore an inefficient process if carried out continuously. But as a short-term measure it is extremely useful. In muscle, for example, lactic acid that accumulates during prolonged activity can be subsequently broken down or reconverted to carbohydrate when oxygen becomes available.

Plants, however, cannot make use of the ethanol. It cannot be reconverted to carbohydrate, nor can it be further broken down in the presence of oxygen. As it is toxic it must not be allowed to accumulate, and this is the main reason why very few plants can be complete anaerobes. Many plants (or parts of plants) can indulge in anaerobiosis for a short period of time, germinating seeds for example, and roots living in waterlogged soil, but before the concentration of ethanol reaches a certain level they must revert to aerobic respiration.

This is even true of yeast which is often mistakenly believed to be a complete anaerobe. The fact that yeast can respire anaerobically to produce ethanol has long been used by man in brewing. But in fact yeast grows much better in aerobic conditions, and if too little oxygen is present the alcohol concentration rises so much that the yeast cells are killed. The secret in brewing is therefore not to let conditions become too anaerobic. At the same time it is obviously desirable to develop new strains of yeast that are tolerant to high concentrations of alcohol. This is one of the major occupations of microbiologists working for brewery companies.

In anaerobic conditions there is no oxygen available to accept hydrogen atoms as they come off the carrier system. This inevitably means that the carrier systems cannot work in the absence of oxygen. With no carrier systems there can be no Krebs' cycle, and accordingly it has been found that in anaerobic conditions none of the reactions which make up Krebs' cycle take place.

Glycolysis occurs in the usual way, and indeed is much speeded up, but the pyruvic acid, instead of being converted into acetyl-CoA and hence fed into Krebs' cycle, is converted into lactic acid or ethanol. How this conversion takes place provides yet another example of the neatness of biological systems.

You will remember that the first step in the conversion of the phosphory-lated triose sugar to pyruvic acid involves dehydrogenation: two hydrogen atoms are removed and taken up by NAD. Now in the absence of oxygen the hydrogen atoms are taken up by the initial hydrogen acceptor NAD and then handed to pyruvic acid which is thereby converted into lactic acid (Fig 7.10). The importance of this reaction is that it prevents the hydrogen atoms from accumulating as they would certainly otherwise do, for anaerobic respiration takes place at a very much higher rate than aerobic.

In the case of plants a slight qualification is necessary: the pyruvic acid is not converted directly into ethanol. It is first converted into acetaldehyde which is then reduced by the hydrogen atoms to form ethanol. In the conversion of pyruvic acid into acetaldehyde carbon dioxide is liberated. Alcoholic fermentation is therefore accompanied by the evolution of carbon dioxide, a fact which is made use of in baking: yeast is added to the flour and the escaping carbon dioxide causes the dough to rise.

Fig 7.10 Diagram showing how in anaerobic conditions, pyruvic acid is converted into either lactic acid or ethanol. The two hydrogen atoms that reduce the pyruvic acid or acetaldehyde come from the initial dehydrogenation of 3-carbon sugar at the commencement of glycolysis.

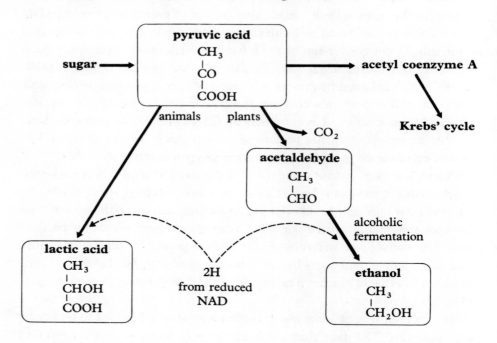

Thus in anaerobic respiration Krebs' cycle is omitted but glycolysis continues to take place at a greatly accelerated rate. The omission of Krebs' cycle means that far fewer ATP molecules are produced than in aerobic conditions. In fact the anaerobic breakdown of a single molecule of sugar yields only two molecules of ATP compared with the total of 32 molecules produced in aerobic conditions. The ability of an organism to utilize oxygen is therefore vitally important as far as energy production is concerned.

ENERGY FROM FAT

The scheme shown in Fig 7.9, as well as summarizing carbohydrate metabolism, also shows how fat can be utilized in releasing energy for the synthesis of ATP. When demands are great, or when carbohydrate is in short supply, fat may be split into **fatty acid** and **glycerol**. The latter is then converted into 3-carbon sugar, and thence either built up into glycogen or starch for temporary storage, or broken down to pyruvic acid with the production of energy. Meanwhile the fatty acid goes through a series of reactions in which carbon atoms are lost, two at a time. Each 2-carbon unit forms a molecule of acetyl-CoA which enters Krebs' cycle.

The process in which each 2-carbon unit is split off the fatty acid takes place in the mitochondria. It involves a series of steps some of which entail the removal of hydrogen atoms which are passed through the carrier system with the formation of ATP. When eventually a molecule of acetyl-CoA has been formed, the fatty acid, now containing two fewer carbon atoms than it did before, returns to the beginning and goes through the same series of reactions again. This process is repeated until the fatty acid is completely broken down. The whole sequence is productive of energy for ATP synthesis. Still more energy is released every time an acetyl-CoA molecule is fed into Krebs' cycle.

Exactly how many ATP molecules are produced by the complete oxidation of a fatty acid depends on the number of carbon atoms it contains. A fatty acid with a long chain of carbon atoms will obviously give more molecules of acetyl-CoA than one with a relatively short chain. The complete oxidation of a molecule of stearic acid yields a total of about 180 molecules of ATP, much more than is given by the oxidation of a single sugar molecule.

Clearly acetyl-CoA represents a kind of crossroads in metabolism. It is produced by the oxidation of both carbohydrate and fat, and it represents a common pathway through which the products of carbohydrate and fat metabolism are fed into Krebs' cycle. Fig 7.9 makes it clear how carbohydrate and fat metabolism are interlinked. With the exception of the conversion of pyruvic acid to acetyl-CoA all the reactions are reversible. This is important because it means that carbohydrate can be converted into fat for storage. The circumstances in which this takes place depend on numerous factors including diet, health and energy requirements of the body. In mammals the liver plays a central rôle in controlling these metabolic interactions, a topic which is discussed more fully in Chapter 13.

ENERGY FROM PROTEIN

Only in abnormal circumstances is protein used as a substantial source of energy, so only brief mention will be made of it here. To produce energy the protein must first be broken up into its constituent **amino acids**. Each amino acid is then **deaminated**; that is its amino (NH_2) group is split off to form ammonia which is excreted (see p. 206). The residual carbon compound then enters carbohydrate metabolism with the subsequent release of energy. In the simplest cases the residual carbon compound is itself identical with one of the intermediates in carbohydrate metabolism: pyruvic acid, or one of the acids of the Krebs cycle, for example. But frequently the process is more complex, the amino acid residue being converted into one of the carbohydrate intermediates by a series of steps which may themselves produce energy for ATP synthesis.

In this chapter we have seen that oxygen and organic food material are essential requirements for efficient energy release. How these raw materials are acquired by organisms is discussed in the following chapters.

8 Gaseous Exchange in Animals

In small animals, the surface:volume ratio is large enough for diffusion across the general body surface to satisfy their respiratory needs. But in larger animals, particularly active ones, the surface:volume ratio is too small for this to be so, and a special **respiratory surface** has to be developed. Animals that use the general body surface for gaseous exchange include protozoons, coelenterates, flatworms and many annelids including the earthworm. For the most part these animals are small enough not to require a special respiratory surface. Many are comparatively inactive, and their sluggishness considerably reduces their need for oxygen. In some the surface:volume ratio is increased by the whole body being flattened (e.g. flatworms), or by the development of flat surfaces inside the body (e.g. sea anemones). Both these adaptations can be looked upon as evolutionary strategies designed to solve the problem of gaseous exchange.

Fig 8.1 Diagrams showing the different kinds of repiratory surfaces found amongst animals. Respiratory surface shown in bold. Gut is labelled **G**.

(a)

Gaseous exchange across the **entire surface** of the body. Found in a wide range of small animals from protozoa to earthworms.

(b)

Gaseous exchange across the surface of a **flattened body**. Flattening increases the S:V ratio and also decreases the distance over which diffusion has to occur within the body, e.g. flatworms.

(c)

External gills. These increase the surface area but note that they are unprotected and therefore easily damaged. Gaseous exchange usually takes place across the rest of the body surface as well as the gills, e.g. lugworm.

(d)

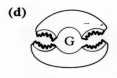

Highly vascularized **internal gills.** A ventilation mechanism draws water over the gill surfaces, e.g. fishes.

(e)

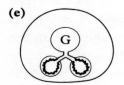

Highly vascularized **lungs.** The lungs are sacs connected to the pharynx. Air is drawn into them by a ventilation mechanism. Found in all air-breathing vertebrates.

(f)

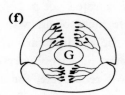

Gaseous exchange at the terminal ends of fine **tracheal tubes** which ramify through the body and penetrate into all the tissues. Found in insects and other arthropods.

SPECIALIZED RESPIRATORY SURFACES

Most other animals have developed special respiratory surfaces, a selection of which are shown diagramatically in Fig 8.1. In all cases they consist of numerous flat surfaces, sacs, or tubes, with a large surface area for gaseous exchange. The simplest are **external gills**, epidermal outgrowths from the body surface found in the lugworm and young tadpoles. In contrast, **internal gills** are enclosed in cavities within the body where they are protected from damage: in fishes they consist of folded epithelial sheets on either side of a series of pouches connecting the pharynx with the exterior. Air-breathing vertebrates have developed **lungs** which develop as sac-like outgrowths of the pharynx. A quite different arrangement is found in insects: air pores at the surface open into a system of branching **tracheal tubes** which ramify through the body, coming into close association with all the tissues. Although these various respiratory devices may seem rather different, they all have one essential feature in common: the exposure of a large surface area to whatever medium the animal happens to live in. We might note in passing that the leaves of a plant achieve the same effect.

To illustrate the basic principles involved in the structure and function of a respiratory surface, we will consider gaseous exchange in two quite different animals: man and fishes. Both have highly specialized internal surfaces connected to the outside world by a **respiratory tract**. In both cases a **ventilation mechanism** brings the external medium into contact with the respiratory epithelium. Let us start with man.

Fig 8.2 Respiratory apparatus of man. Expansion of the thorax draws air down the respiratory tract into the lungs. B) and C) show how the thorax expands. Expansion in an anteroposterior direction takes place by descent of the diaphragm (arrow 1); in a lateral direction by upward and outward movement of the ribs (arrow 2); and in a dorso-ventral direction by upward and forward movement of the sternum (arrow 3). Position of structures after inspiration shown in tone.

A) Respiratory tract and associated structures

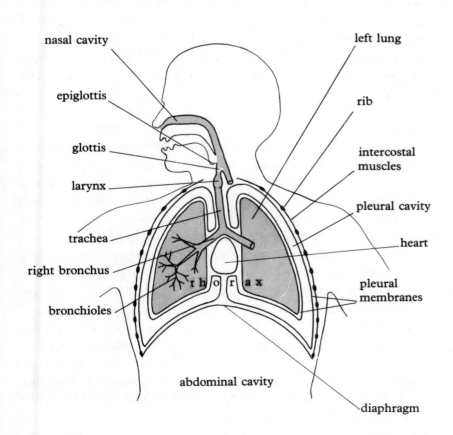

B) Ventral view of thoracic cage

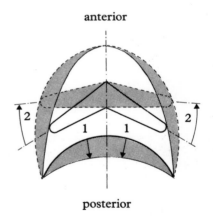

C) Side view of thoracic cage

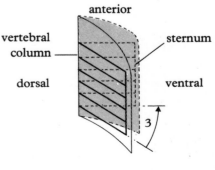

Gaseous Exchange in Man

Fig 8.3 Pressure and volume changes during the respiratory cycle in the mammal. Volume changes are measured by means of a recording spirometer, pressures by running a fine tube from the appropriate cavity to a suitable manometer. For practical purposes the pleural pressure can be estimated by measuring the pressure in the oesophagus.

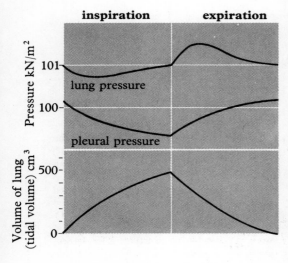

The structure of the human respiratory system is shown in Fig 8.2A. The **lungs** are located in the **thorax**, whose walls are formed by the **ribs** and **intercostal muscles,** and floor by the muscular **diaphragm**. The lungs are surrounded by a **pleural cavity** lined by **pleural membranes**. The pleural cavity contains a thin layer of fluid.

Air is drawn into the lungs via the **trachea** and **bronchi**. The expansion of the thoracic cavity is brought about by the upward and outward movement of the ribs, accompanied by flattening of the diaphragm (Fig 8.2B and C). The rib movements are achieved by contraction of the intercostal muscles, and the flattening of the diaphragm by contraction of its radial and circular muscles. All this constitutes **inspiration**. The process then goes into reverse, air being expelled from the lungs in the act of **expiration**. Expiration is a mainly passive process resulting from elastic recoil of the tissues that have been stretched during inspiration. However, in forced breathing, or when the respiratory tract is blocked, expiration is aided by contraction of the **abdominal muscles** which raises the pressure in the abdominal cavity, forcing the diaphragm upwards.

The pressure and volume changes that occur during the respiratory cycle are shown in Fig 8.3. At rest the pressure in the lungs is atmospheric, but because the lungs are elastic and tend to pull away from the walls of the thorax, the pressure in the pleural cavity is slightly less than atmospheric. During inspiration, when the walls and floor of the thorax are moving outwards and downwards respectively, the pleural pressure falls. This has the immediate effect of lowering the lung pressure to below atmospheric, so that air rushes into the lungs. This increases their volume and returns the lung pressure to atmospheric. On expiration the pressure of the thoracic wall and the diaphragm against the pleural cavity raises the pleural pressure. This is transmitted to the lungs whose pressure therefore increases and volume decreases as air is expelled.

STRUCTURE OF THE LUNGS

The lungs are sponge-like in texture, and consist of a tree-like system of tubes which ramify from the two **bronchi**. The tubes terminate as sac-like **atria** from which arise numerous **alveoli** (Fig 8.4). Although a certain amount of gaseous exchange can take place across the walls of the smaller tubes, it is the alveoli which play the leading role in this respect. The efficiency of the mammalian lung as a respiratory surface depends on the fact that a vast number of alveoli come into very close association with an extensive **capillary system**. In man the two lungs contain approximately 700 million alveoli, giving a total surface area of over 70 square metres; i.e. if the lungs were opened out into a continuous sheet they would just about cover the surface of a tennis court! The capillary network in the lungs has a total area of about 40 square metres. In the lungs, therefore, an enormous surface area for gaseous exchange is packed into a comparatively small space. This general principle also applies to other terrestrial vertebrates like amphibians and reptiles, except that in these more primitive vertebrates the total surface area relative to the size of the lungs is nothing like so great.

The relationship between the alveoli and the capillaries is an extremely intimate one. The walls of the capillaries and alveoli both consist of a single layer of flattened epithelial cells firmly applied to each other. The resulting barrier between the alveolar cavity and the blood is a mere 0.3 μm thick. As such, it offers minimum resistance to the diffusion of gases from one side to the other.

THE RESPIRATORY CYCLE

A person breathing normally at rest takes in, and expels, approximately 0.5 dm^3 (half a litre) of air during each respiratory cycle (Fig 8.5). This is known as his **tidal volume**, and it can be measured by means of a recording spirometer (see p. 97).

The rate of respiration can be expressed in terms of **ventilation rate**, the volume of air breathed per minute. Clearly: VENTILATION RATE = TIDAL VOLUME × FREQUENCY OF INSPIRATIONS. The ventilation rate changes according to circumstances: in muscular exercise, for example, both the frequency and depth increase, resulting in a greater ventilation rate. We shall return to this in Chapter 16. The important point is that the lungs have a much greater potential than is ever realized in resting conditions, and this permits the respiratory apparatus to adapt to changing needs. If you take a deep breath, you can take into your lungs about 3 dm^3 of air over and above the tidal volume. This is known as the **inspiratory reserve volume**, and is brought into use when required. If at the end of a normal expiration you expel as much air as you possibly can, the extra air expired amounts to about 1.0 dm^3, and is called the **expiratory reserve volume**. The total amount of air that can be expired after a maximum inspiration (i.e. the tidal volume plus inspiratory and expiratory reserve volumes) is known as the **vital capacity**. The vital

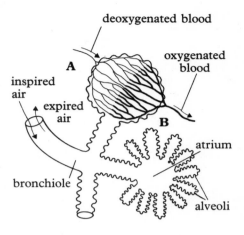

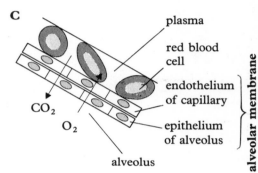

Fig 8.4 The bronchioles in the mammalian lung terminate as cavities across whose much folded and highly vascularized walls gaseous exchange takes place (A). The principal site of gaseous exchange are the numerous alveoli shown in B. The alveoli are separated from the bloodstream by a very thin alveolar membrane consisting of only two layers of pavement epithelial cells (C).

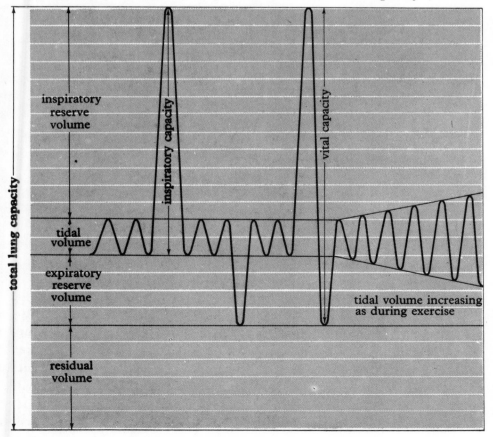

Fig 8.5 Diagram showing the different lung volumes in man. The tracings are based on spirometer recordings obtained from a large number of individuals. The horizontal divisions correspond to 250 cm^3. There is much variation between different individuals in the tidal volume and vital capacity. These tracings emphasize how little air is drawn into the lungs in resting conditions. The reserve volumes are brought into action only when necessary.

capacity of an average man lies between 4 and 5 dm³ but in a fit athlete it may exceed 6 dm³. Even after a maximum expiration, about 1.5 dm³ of air remains in the lungs. This is known as the **residual volume**.

How much of the air taken in is actually used in gaseous exchange? Of the half litre or so inspired in quiet breathing only about 350 cm³ gets into the parts of the lungs where gaseous exchange is possible. The rest remains in the trachea and bronchial tubes, collectively known as **dead space**, where no gaseous exchange takes places. The total capacity of the lungs is about 6 dm³, and it is therefore clear that in resting conditions only a small fraction of the total volume of air present in the respiratory apparatus is used in gaseous exchange.

At each inspiration during normal quiet breathing, about 350 cm³ of inspired air mixes with some 2.5 dm³ of air already present in the alveoli. With so little new air mixing with so much air already present, it is probable that the composition of the air in the depth of the lungs remains relatively constant in resting conditions. Through this air, situated between the inspired air and the blood, gases diffuse to and from the surface. Comparison of the composition of inspired and expired air gives some indication of the exchanges that take place in the lungs. Figures for this have already been given (p. 96), and at this point we should merely note that with continual gaseous exchange going on between the alveolar air and blood, it is the ventilation of the lungs which keeps the composition of the alveolar air more or less constant.

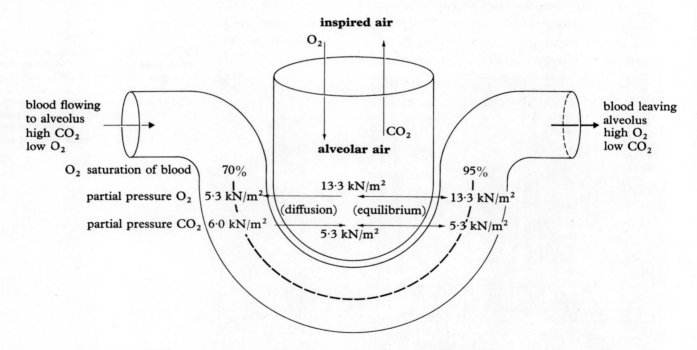

Fig 8.6 Diagram summarizing the exchange of oxygen and carbon dioxide that takes place as blood flows past an alveolus in the mammalian lung. By the time the blood leaves the alveolus it has the same partial pressure of oxygen and carbon dioxide as the alveolar air and it is almost fully saturated with oxygen.

EXCHANGES ACROSS THE ALVEOLAR SURFACE

Analysis of the blood flowing to and from the alveoli gives us an insight into what happens at the respiratory surface itself (Fig 8.6). Blood reaching the alveoli has a lower partial pressure of oxygen, and a higher partial pressure of carbon dioxide, than the alveolar air. There is thus a concentration gradient favouring the diffusion of these two gases in opposite directions. As blood flows past an alveolus, oxygen diffuses into it and carbon dioxide out, so that by the time the blood leaves the alveolus, it has the same partial pressure of oxygen and carbon dioxide as the alveolar air. During this equalization of partial pressures, the percentage saturation of the blood rises from about 70 per cent to over 95 per cent. The composition of alveolar air, however, remains relatively unchanged because of exchanges between it and the inspired air.

In fact the process is not quite as efficient as it seems because (1) some alveoli are inevitably under-ventilated, and (2) a proportion of the blood which goes to the lungs does not go through any alveolar capillaries and therefore never gets oxygenated. The result is that the blood leaving the lungs is not as fully oxygenated as it might be. Despite this limitation the respiratory mechanism is sufficiently good to allow mammals to lead highly active lives.

Gaseous Exchange in Fishes

In fishes gaseous exchange takes place between the water and highly vascularized **gills**. We will start by considering the dogfish in which the gills, and the blood vessels serving them, can be readily exposed by dissection. In the dogfish — and the same applies to other cartilaginous fishes such as sharks and rays — five **gills pouches** connect the **pharynx** with the exterior on each side of the body. The gills are situated between successive pouches (Fig 8.7). Water is drawn in through the **mouth** and **spiracles** by expansion of the pharyngeal cavity. This is achieved mainly by lowering the floor of the pharynx. This is brought about by the contraction of a series of **hypobranchial muscles** which run from a cartilaginous plate on the ventral side of the pharynx to the pectoral girdle. The suction exerted when the pharynx expands pulls the **branchial valves** tightly over the **gill slits** (Fig. 8.8A). This ensures

Fig 8.7 Gills (**G**), lying on either side of a series of gill pouches in the skate, a relative of the dogfish. Five afferent branchial arteries (**A**) convey blood to the gills from the heart (**H**). (*B. J. W. Heath, Marlborough College*)

that water enters only via the mouth and spiracles. The latter are lined with sensory epithelium which tests the incoming water. If undesirable, the water is expelled and the fish moves elsewhere. When the pharyngeal cavity is filled with water, the mouth and spiracles close and the floor of the pharynx is raised, forcing water through the gill pouches and over the respiratory epithelium.

STRUCTURE OF THE DOGFISH'S GILLS

Like the mammalian lung, gills are constructed so that a large surface area of highly vascularized epithelium is exposed to the respiratory medium (Fig 8.8B). Each gill is composed of about 70 leaf-like **lamellae** which project horizontally from a skeletal **branchial arch** orientated vertically in the wall of the pharynx. The lamellae are supported down the centre and held in position by a vertical **septum** which projects beyond the distal end of the gill as the **branchial valve**. On the upper and lower surfaces of each lamella are numerous vertical **gill plates** which greatly increase the total surface area of the gill. The gill can be regarded as a sheet of epithelium which is folded once to form the lamellae and then again to form the plates. At the base of each gill, close to the gill arch, is an **afferent branchial artery** which brings deoxygenated blood to the gill from the ventral aorta beneath the floor of the pharynx. The base of the gill also contains a pair of **efferent branchial arteries**, each derived from a loop vessel encircling the pouches on either side of the gill. An **epibranchial artery** conveys blood from each efferent loop to the dorsal aorta above the roof of the pharynx. The afferent and efferent arteries are interconnected within each lamella by an elaborate system of tiny

Fig 8.8 Respiratory apparatus of the dogfish. The gills, supported at the base by a series of skeletal branchial arches, are separated from one another by gill pouches through which the water flows on its way over the respiratory surfaces. Blood arriving at the base of the gill in the afferent branchial artery circulates through capillaries in the gill lamellae and plates before leaving via the efferent arteries. Gaseous exchange takes place in the lamellae and plates. (Deoxygenated blood indicated by solid line; oxygenated blood by broken line.)

A) Horizontal section through pharynx and gill region

water in

mouth cavity

pharynx

spiracle

branchial arch

gill

branchial valve

gill pouch

water out

gill slit

B) Part of one gill in detail

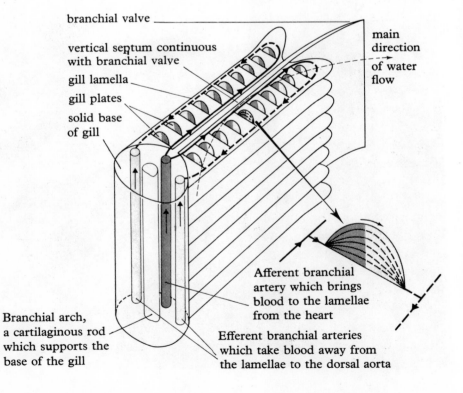

branchial valve

vertical septum continuous with branchial valve

gill lamella

gill plates

solid base of gill

main direction of water flow

Branchial arch, a cartilaginous rod which supports the base of the gill

Afferent branchial artery which brings blood to the lamellae from the heart

Efferent branchial arteries which take blood away from the lamellae to the dorsal aorta

vessels and capillaries. Gaseous exchange takes place as blood flows through the capillaries in the gill plates. Here the barrier between the blood and the water is only several cells in thickness, and offers little resistance to diffusion.

COUNTERFLOW AND PARALLEL FLOW

In any gill system, for maximum gaseous exchange to take place, it is best for the blood and water to flow in opposite directions (**counterflow system**). This ensures that as blood flows across the respiratory surface, it meets water which has had less and less oxygen extracted from it (Fig 8.9A). By the time the blood is about to leave the respiratory surface, it will have almost the same oxygen tension as the inhalent water. In other words the same steep concentration gradient is maintained throughout the respiratory surface, thereby maximizing gaseous exchange.

If the blood and water flow in the same direction, and at the same speed (**parallel flow**), the concentration gradient would be great at first, but would steadily decrease as the blood and water flowed together across the respiratory surface (Fig 8.9B). On leaving the respiratory surface, the blood would be in equilibrium with the water at a point well below its maximum possible saturation with oxygen. Parallel flow is therefore not as efficient as a counterflow system. It can, however, be improved if the flow of water is very rapid compared with that of the blood. This will ensure a higher saturation of the blood by the time it leaves the respiratory surface.

In the dogfish it is probably true that to some extent water flows between the gill plates in a direction opposite to that of the blood, but a really efficient counterflow system is prevented (1) by the fact that the main flow of water through the gill pouches is parallel to the lamellae, and (2) by the vertical septum which deflects the water so that it tends to pass over rather than between the gill plates. In view of this apparent inefficiency, it is interesting to find that in bony fishes the structural arrangement of the gills is such as to make the counterflow system more certain. This is made clear in the next section.

GASEOUS EXCHANGE IN TELEOST FISHES

In bony fishes, cod, whiting and the like, the entire gill region is flanked by a muscular flap of skin, the **operculum** (Fig 8.10A). This encloses an **opercular cavity** into which the gills project. The development of the operculum renders branchial valves unnecessary, with the result that the valves and septum have been completely lost. No longer bound together by tissue, the two sides of the gill part company and point obliquely outwards from the branchial arch, as shown in Fig 8.10B. The circulation of blood through each 'half-gill' is similar to the dogfish, but the change in orientation of the respiratory surface means that as water passes from the pharynx into the opercular chamber, it inevitably flows between the gill plates in a direction opposite to the blood flow. Moreover the free ends of adjacent gills touch each other, and this means that no water can avoid passing between the lamellae as it flows from the pharynx to the opercular chamber. The system is certainly more efficient than that of cartilaginous fishes, about 80 per cent of the oxygen being extracted from the water as it flows over the gills, compared with 50 per cent in the dogfish.

A) Counterflow system

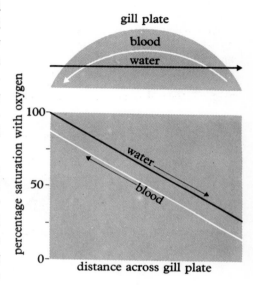

B) Parallel flow system

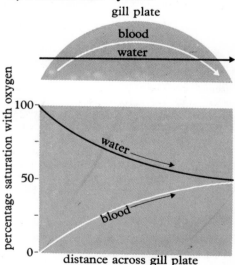

Fig 8.9 Diagrams illustrating counterflow and parallel flow across the gill plates of a fish. The graphs assume that the blood and water move at the same speed and have equal oxygen capacities. (*based on Hughes*)

A) Horizontal section through pharynx and gill region

B) Two gills in detail

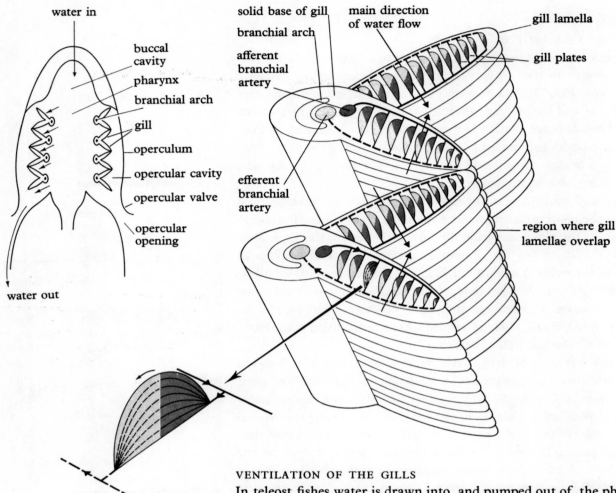

water in

buccal cavity

pharynx

branchial arch

gill

operculum

opercular cavity

opercular valve

opercular opening

water out

solid base of gill

branchial arch

afferent branchial artery

efferent branchial artery

main direction of water flow

gill lamella

gill plates

region where gill lamellae overlap

Fig 8.10 Respiratory apparatus of a teleost fish. A muscular operculum covers the entire gill region on either side of the body and a single pair of opercular valves performs the function which in the dogfish is carred out by the five pairs of branchial valves. The loss of the branchial valves and vertical septum in each gill allows the lamellae to project outwards from the base at approximately right angles to each other, an arrangement that allows for a more efficient counterflow system. Deoxygenated blood indicated by solid line; oxygenated blood by broken line. (*based on Hughes*).

VENTILATION OF THE GILLS

In teleost fishes water is drawn into, and pumped out of, the pharynx mainly by movements of the operculum (Fig 8.11). To achieve maximum efficiency it is obviously desirable that a continuous stream of water should be maintained over the gills. It used to be thought that when the pharynx was expanding the flow of water over the gills was temporarily suspended. However, thanks to the research of G. M. Hughes and G. Shelton, we now know that a current of water is maintained over the gills at all phases of the respiratory cycle. By means of an ingenious apparatus, Hughes and Shelton have recorded the pressure changes that take place in the buccal and opercular cavities of the trout (Fig 8.12). They have shown that a continuous flow of water is maintained by the combined action of the buccal cavity as a force pump and the opercular cavity as a suction pump. Throughout the respiratory cycle, except for one short period, the pressure in the buccal cavity is higher than that in the opercular cavity. The result is that there is an almost continuous flow of water from the former to the latter.

It is likely that much the same applies to cartilaginous fishes. There is no opercular chamber in the dogfish, but the so-called **parabranchial cavities**, enclosed by the flap-like branchial valves, function in the same way, sucking water through the gill pouches at the appropriate moment in the respiratory cycle (Fig 8.13).

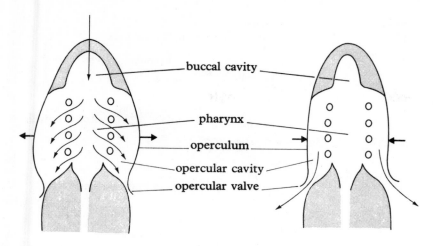

buccal cavity

pharynx

operculum

opercular cavity

opercular valve

Fig 8.11 Diagrams summarizing the ventilation of the gills in a teleost fish. Direction of water movements is shown by the fine arrows.

A) Inspiration

Water is sucked in through the mouth by expansion of the buccal cavity and thence into the opercular cavity by outward movement of the operculum accompanied by contraction of the buccal cavity.

B) Expiration

Water is expelled through the opercular openings by inward movement of the operculum together with the continued contraction of the buccal cavity.

Fig 8.12 Recording of pressures in the buccal cavity (**BC**) and opercular cavity (**OC**) of a trout. Notice that the pressure differences are such that a continuous flow of water from **BC** to **OC** is maintained throughout most of the respiratory cycle. At what stage of the cycle is this not so? (*based on data by Hughes and Shelton*)

1) BC expanding, acquires negative pressure; mouth valve opens and water enters from outside.

2) OC expanding, acquires negative pressure; opercular valve closes.

3) Negative pressure in OC falls below that of BC which has begun to contract. Result: water sucked into OC from BC

4) BC contracting, acquires positive pressure; mouth valve closes and water forced from BC to OC.

5) OC contracting, acquires positive pressure; opercular valve opens and water expelled.

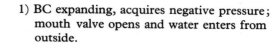

hypodermic tubing

electronic pressure recorder

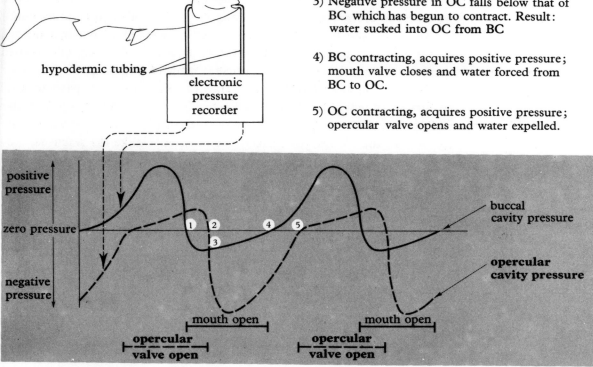

positive pressure

zero pressure

negative pressure

buccal cavity pressure

opercular cavity pressure

mouth open

mouth open

opercular valve open

opercular valve open

Fig 8.13 Ventilation of the gills in the dogfish. A) Water drawn into pharynx by lowering floor of mouth cavity and pharynx, brought about by contraction of hypobranchial muscles accompanied by relaxation of transverse muscles. Then water drawn into parabranchial cavities by outward movement of branchial valves. B) Water expelled from gill pouches by raising floor of mouth cavity and pharynx, brought about by relaxation of hypobranchial muscles and contraction of transverse muscles.

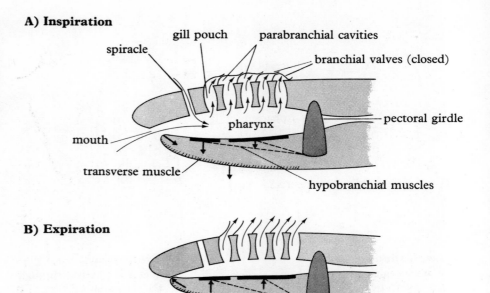

Gaseous Exchange in Insects

This would not be the place for a detailed study of comparative respiration, but brief mention must be made of the insect **tracheal system**, if only to demonstrate that terrestrial animals have ways of solving their respiratory problems other than by lungs.

In insects the integument on either side of the thorax and abdomen is perforated by a series of segmentally arranged pores or **spiracles** which open into a system of **tracheal tubes** or **tracheae** (singular **trachea**). The spiracles are guarded by valves or hairs to prevent excessive evaporation through them. The tracheae are arranged in a definite way, some of them running longitudinally, some transversely. The larger tracheae are about 1 mm in diameter, and are kept permanently open by spiral or annular thickenings of hardened chitin, the same material that the cuticle is made of (Fig 8.14).

The function of the spiracles and tracheae is to permit the passage of air to a further system of tubes, the **tracheoles**. These are very fine intracellular tubes, a mere 1 μm in diameter. They are extremely numerous and penetrate deep into all the tissues, particularly the muscles. Unlike the tracheae, they are not lined with a cuticle, gaseous exchange occurring freely across their walls.

The insect respiratory system is in marked contrast to most other animals. Instead of being picked up by blood at the respiratory surface, oxygen is conveyed direct to the tissues via this system of ramifying air passages.

How does the air get to the tissues? In most insects diffusion has to suffice, which may be one reason why insects are generally rather small. However, in some species diffusion is aided by rhythmical movements of the thorax or abdomen. Such ventilation is seen, for example, in the locust where it has been demonstrated that, by the differential opening and closing of the spiracles, air is drawn into the body through the thoracic spiracles, and leaves via the abdominal spiracles.

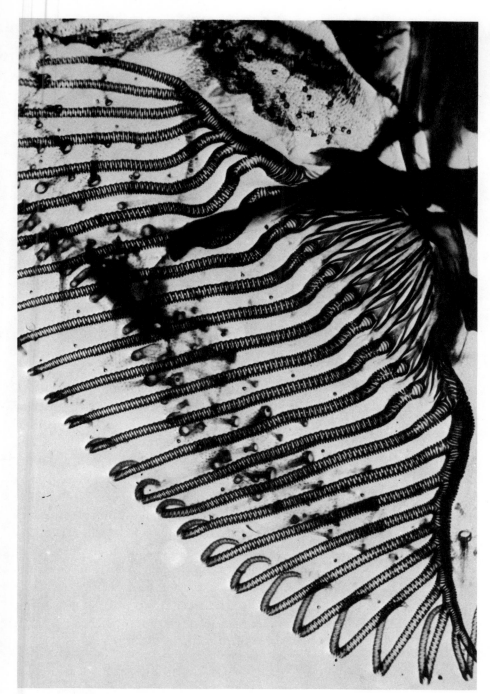

Fig 8.14 Pseudotracheae in the proboscis of the blow-fly *Calliphora*. These structures are connected to the pharynx and are used not for respiration but for feeding. However, the chitinous supports, characteristic of tracheae in general, are particularly clear. (*J. F. Crane*)

Moreover, insects can control the rate at which oxygen is delivered to the tissues. The mechanism depends on the fact that the tracheoles contain varying amounts of watery fluid. In severe muscular activity (flight, for example), lactic acid accumulates in the tissues. This raises the osmotic pressure of the tissue fluids, with the result that water is drawn out of the tracheoles. This has the effect of opening up the air passages, enabling more rapid diffusion of oxygen to the tissues.

The tracheoles are extraordinary structures whose pattern of growth can be modified according to the needs of the tissues. If a body segment is deprived of oxygen by cutting its main trachea, the tracheoles of neighbouring segments respond by growing towards the deprived segment. In this way uniform distribution of oxygen to the tissues is ensured.

9 Heterotrophic Nutrition

Organisms that are unable to synthesize their own food substances from simple inorganic raw materials must obtain their food in organic form. Such organisms include nearly all animals, all fungi, most bacteria and a few flowering plants. They are known as **heterotrophs**, and their mode of nutrition is **heterotrophic**. They obtain organic substances from various sources, which can ultimately be traced back to green plants that have manufactured them from inorganic materials.

Three types of heterotrophic nutrition are recognized. (1) **Holozoic nutrition**. This involves feeding on solid organic material obtained from the bodies of other organisms. Although this is an almost exclusively animal method of nutrition, specialized plants can do it too. (2) **Saprophytic nutrition**. This involves feeding on soluble organic compounds obtained from the remains of dead animals and plants. It is carried out by many fungi and bacteria. The activities of these saprophytes are extremely important in bringing about the decay of dead bodies and releasing from them elements that can subsequently be used by green plants to build up new organic substances. (3) **Parasitic nutrition**. This is indulged in by a variety of animals, flowering plants, fungi and bacteria. It involves feeding on the organic compounds present in the body of another living organism, the host. These organic compounds are often in solution, as in the case of a gut parasite such as the tapeworm, but sometimes the parasite feeds on solid tissue like a holozoic organism. Moreover, a parasite may kill its host and then continue to feed saprophytically on its dead remains. This shows that, as with so many other man-made categories, there are no hard and fast dividing lines between these three methods of nutrition.

General Considerations

The problem facing any heterotrophic organism is how to acquire and take in the organic substances it needs. This must occur before the organic materials can be assimilated into the tissues. For a heterotroph feeding on organic substances in solution (such as a gut parasite) there is no problem: all it has to do is to absorb the substances straight across its integument, or the lining of its gut. But an animal that feeds on solid organic material has to have means of obtaining food and rendering it into a suitable form for absorption into the body. For obtaining food a variety of structures are employed: teeth, tentacles, claws, pincers and so on. With the aid of these devices food is taken into a **gut** or **alimentary canal** or **gastro-intestinal tract** in the act of **ingestion**. It is then subjected to a combination of physical and chemical **digestion** which converts the solid food into soluble compounds capable of being **absorbed**. Finally the fluid products of digestion are absorbed across the walls of the gut, generally into some kind of **circulatory system** which distributes them to the various tissues where they are incorporated into the cells and either built up into complex materials (**assimilation**) or broken down for energy release (**cell respiration**).

PHYSICAL AND CHEMICAL DIGESTION

The physical part of digestion is achieved by the cutting and/or crushing action of teeth, or their equivalent, followed by rhythmical contractions of the gut which pound up and mix the semi-solid food. To this end the gut wall, particularly the mammalian stomach, is well endowed with muscles (Fig 9.1). These are also responsible for pushing the food along the gut in the process known as **peristalsis**. Physical action is accompanied by the secretion of **digestive enzymes**. Some of these are secreted by glands situated outside the gut, salivary glands and pancreas for example. Others come from glands located in the gut wall itself. Copious quantities of **mucus,** secreted along with the digestive enzymes, lubricate the gut and facilitate the passage of food along it. Variable quantities of acid or alkali are also secreted to provide the correct pH for optimum functioning of the enzymes.

The details of the digestive enzymes vary in different animals but in general they fall into three groups:

Fig 9.1 The gut exerts physical and chemical action on its contents. A) General structure of the wall of the mammalian gut showing the muscle layers and the different kinds of glands. (*after Ham*, Histology, *J. B. Lippincott Co*) B) The various movements of the gut brought about by contraction of the circular muscle.

A

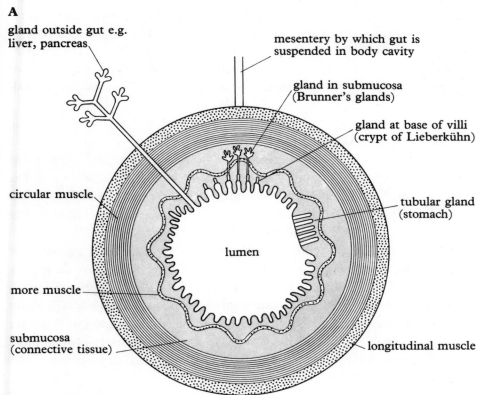

gland outside gut e.g. liver, pancreas

mesentery by which gut is suspended in body cavity

gland in submucosa (Brunner's glands)

gland at base of villi (crypt of Lieberkühn)

circular muscle

tubular gland (stomach)

lumen

more muscle

submucosa (connective tissue)

longitudinal muscle

B   **Peristalsis in the oesophagus**

Contractions of the small intestine

i) Localized constrictions

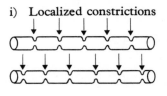

ii) Peristaltic wave

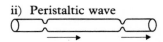

iii) Pendular constrictions

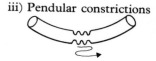

(1) **carbohydrases** which break down carbohydrates;
(2) **lipases** which break down fats; and
(3) **proteases** which break down proteins.

The physical and chemical processes of digestion go hand in hand. The chopping up of a solid lump of food before it is subjected to the action of digestive enzymes increases the surface area over which chemical digestion can subsequently take place.

It is unusual for any one enzyme to break a large complex organic molecule right down into its constituent sub-units, but between them this is what they do. Thus the various carbohydrases together break polysaccharides like starch into monosaccharides like glucose; lipases break down fats into fatty acids and glycerol; and proteases break down proteins into amino acids. In all these cases the action is one of hydrolysis, in which the larger molecule is split by the addition of water.

HOW DIGESTIVE ENZYMES WORK

The mode of operation of digestive enzymes can be illustrated by proteins. A protein is generally attacked first by enzymes that break the peptide links in the interior of the molecule. Such enzymes, called **endopeptidases**, have the effect of splitting proteins and large polypeptides into smaller polypeptides. In the human gut pepsin and trypsin are examples. The smaller polypeptides are then attacked by enzymes which break off their terminal amino acids. These enzymes are called **exopeptidases**. Some exopeptidases, known as **aminopeptidases**, will only attack that end of a polypeptide chain which has a free amino (NH_2) group. Others, known as **carboxypeptidases**, only attack the end of a polypeptide chain with a free carboxyl (COOH) group. Either way, the result is the liberation of free amino acids.

The same principles apply to the digestion of carbohydrates. Certain enzymes break the glycosidic bonds in the interior of polysaccharide chains, forming shorter polysaccharides; other enzymes then attack the ends of these polysaccharide chains, liberating free monosaccharides. Saliva and pancreatic juice both contain enzymes which break down starch first to dextrin, consisting of shorter polysaccharide chains, and then to the disaccharide maltose.

ABSORPTION AND ASSIMILATION

The end products of digestion (glucose, fatty acids, glycerol and amino acids) are all molecules small enough to be absorbed through the wall of the gut. The inner lining of the absorptive part of the intestine is generally greatly folded or bears numerous finger-like **villi** to increase its surface area (Fig 9.1). The absorptive surface may be made even greater by the individual epithelial cells bearing numerous **microvilli**. Aided by this large surface area, the products of digestion are absorbed through the wall of the gut for transport to the tissues. To this end the gut wall may be permeated by numerous capillaries. Through the bloodstream the soluble food substances are taken to the tissue cells where they are either assimilated or respired. In the meantime any indigestible material incapable of being broken down by the digestive enzymes continues into the posterior reaches of the gut, to be **egested** through the anus.

THE SITE OF DIGESTION

In the system just described, food is digested completely before being taken into the cells lining the gut. This is spoken of as **extracellular digestion** because chemical breakdown is completed outside the cells. In contrast to this is **intracellular digestion** where solid food particles are taken up into the cells by phagocytosis and then digested in food vacuoles within the cells. The most obvious example of this is *Amoeba*, a single-celled animal which has a very thin and flexible membrane enabling it to change its shape as a result of cytoplasmic streaming within the cell. When the cytoplasm streams towards one particular point a projection called a **pseudopodium** is formed. When it comes into contact with a small particle of nutritional value, a diatom or green flagellate for example, it responds by forming a cup-shaped pseudopodium which engulfs the food particle. Eventually the 'lips' of the cup seal, and the food becomes enclosed in a food vacuole. Digestive enzymes are now secreted into the vacuole and the products of digestion are absorbed into the surrounding cytoplasm. The entire process of digestion is therefore carried out inside the cell.

In some animals digestion is divided into extracellular and intracellular phases. Extracellular enzymes secreted into the gut break the food up into small particles. These are then taken up by phagocytosis into the cells lining the gut (or in some cases a special **digestive gland**) where digestion is completed by intracellular enzymes. In man and all other vertebrates digestion is exclusively extracellular. Intracellular digestion is more primitive, being associated with animals like *Hydra* and snails. In both these animals extracellular digestion is followed by intracellular. In the even more primitive sponges, which feed on small particles suspended in water flowing through them, digestion is entirely intracellular.

In evolution the tendency has been for intracellular to be replaced by extracellular digestion. It would appear that in the early stages of animal evolution the function of extracellular digestion was to break the food into particles small enough to be taken into the cells by phagocytosis. This would be particularly important to an animal like *Hydra* which has no physical means of breaking up its food. In the subsequent evolution of digestive systems, extracellular digestion has gradually replaced intracellular digestion.

Feeding and Digestion in Man

Let us now illustrate these general principles by considering what happens to a meal as it passes along the human gut. The anatomy of the alimentary canal is shown in Fig 9.2.

IN THE BUCCAL CAVITY

In the **buccal cavity** the food is subjected to the chewing action of the **teeth** (**mastication**). The teeth are ideally suited to this function, being exceedingly hard and differentiated (Fig 9.3). The **incisors** are for cutting the food and the **premolars** and **molars** for crushing it. The **canines** are poorly developed in man, but are well developed in other mammals such as the great cats where they are used for tearing flesh. While being chewed, the food is mixed with **saliva** secreted by the **salivary glands**. The flow of saliva is initiated by the sight, smell, taste or thought of food. It is a watery mixture of mucus and an

Fig 9.2 Diagram of the human alimentary canal and associated organs. (*based on Mackean*)

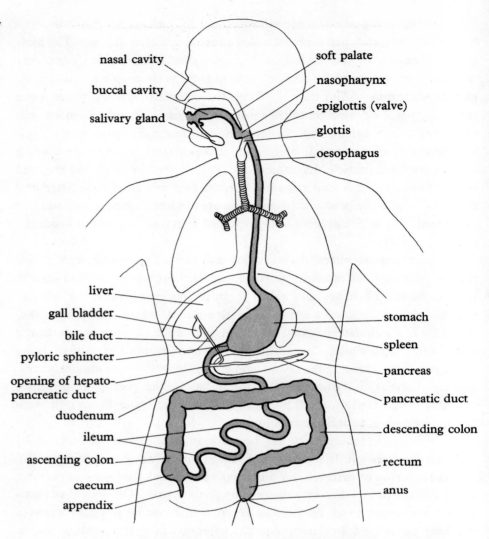

enzyme, **salivary amylase** or **ptyalin**, which hydrolyses the polysaccharide starch to the disaccharide maltose. Saliva is generally neutral or very faintly alkaline, this being the optimum pH for the action of the enzyme. The functions of saliva are twofold: it lubricates the food in preparation for its passage down the oesophagus, and it starts the digestion of starch. Towards the back of the buccal cavity the food is shaped by the action of the tongue into a bolus which is forced through the **pharynx** into the **oesophagus** in the act of **swallowing**. Triggered by tactile stimulation of the soft palate and wall of the pharynx, swallowing is a reflex in which contraction of the tongue forces the bolus against the soft palate thereby closing the nasal cavity. The opening into the larynx, the **glottis**, is closed by the valve-like **epiglottis**, so the bolus enters the oesophagus. While all this is happening respiration is momentarily inhibited. The centre responsible for controlling this swallowing reflex is located in the medulla of the brain.

DOWN TO THE STOMACH

The bolus is propelled down the oesophagus by **peristalsis**. Once in the stomach the food is acted on by **gastric juice** secreted by **gastric glands** situated in the thick stomach wall (Fig 9.4). Gastric juice contains three enzymes: **pepsin** which breaks down proteins into short polypeptide chains; **rennin** which coagulates caseinogen, the soluble protein of milk, forming an insoluble

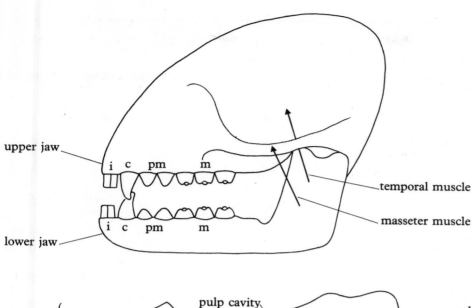

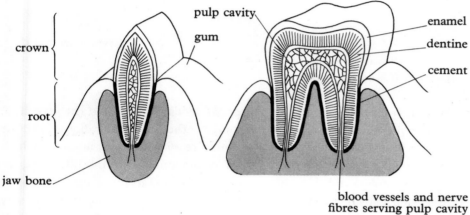

blood vessels and nerve
fibres serving pulp cavity

Fig 9.3 Skull and teeth of a generalized primate. The teeth are differentiated into cutting incisors (i) and canines (c), and crushing premolars (pm) and molars (m). In man the canines are shorter and less dagger-like than shown here. In man there are two incisors on either side of each jaw, one canine, two premolars and three molars (dental formula: $i\frac{2}{2}$, $c\frac{1}{1}$, $pm\frac{2}{2}$, $m\frac{3}{3}$.) The distinction between premolars and molars is less clear-cut than in other mammals. The teeth are very hard, which makes them well suited to their task of cutting and crushing food. The crown is covered by a layer of extremely hard enamel consisting of mineral salt crystals bound together by keratin. Beneath the enamel is a layer of dentine, similar to bone but with a higher mineral content thus making it harder. The dentine is perforated by fine canaliculae which contain the cytoplasmic processes of the tooth-forming cells (odontoblasts). The pulp cavity in the centre contains a plexus of blood capillaries and sensory nerve endings. The root is embedded in the jaw bone to which it is firmly attached by a layer of bone-like cement.

curd, casein, which is then attacked by pepsin; and **gastric lipase** which hydrolyses fat into fatty acid and glycerol. All these enzymes are secreted by the so-called **chief** (or **peptic**) **cells** in the wall of the gastric glands.

Pepsin functions best in an acid environment. To this end special **oxyntic cells** in the deeper regions of the gastric glands secrete **hydrochloric acid,** giving the gastric juice a pH of less than 2.0. Gastric juice also contains much mucus produced by cells situated towards the neck of the glands.

What brings about the production of gastric juice? Pavlov in his classical experiments on digestion in dogs, showed that gastric juice, like saliva, will flow as a result of the sight, smell, taste or expectation of food. But this reflex production is relatively slight compared with the copious secretion that occurs when the food arrives in the stomach. It is now known that mechanical and chemical stimulation of the stomach lining by the food itself causes secretion of a hormone called **gastrin** by the stomach wall. This then stimulates the gastric glands to produce gastric juice.

While the enzymes are working, rythmical contractions of the stomach pound the food into a semi-fluid state called **chyme.**

INTO THE DUODENUM

Peristaltic contractions of the stomach keep the chyme moving towards the **duodenum,** the first loop of the small intestine. The passage of food into the duodenum is controlled by a ring of muscle, the **pyloric sphincter,** situated

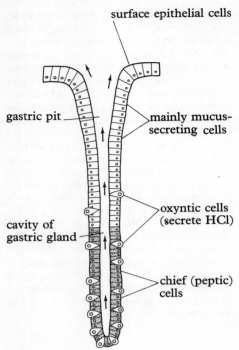

A) **Gastric gland in detail**

surface epithelial cells

gastric pit

mainly mucus-secreting cells

cavity of gastric gland

oxyntic cells (secrete HCl)

chief (peptic) cells

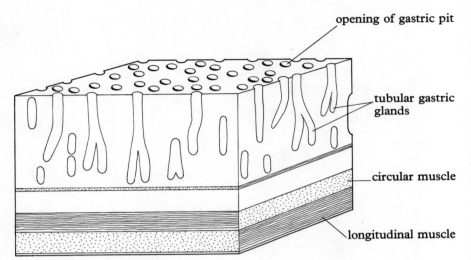

B) **Part of stomach wall**

opening of gastric pit

tubular gastric glands

circular muscle

longitudinal muscle

Fig 9.4 The thick wall of the stomach contains tubular glands lined with goblet, chief, and oxyntic cells which secrete mucus, enzymes and hydrochloric acid respectively. These cells are tightly packed in the epithelial lining of the glands thereby increasing the secretory surface. (B: *based on McCulloch*)

immediately between the pyloric end of the stomach and the duodenum. By alternately contracting and relaxing the pyloric sphincter can hold food back or let it through.

The duodenum is the main seat of digestion in the gut. The agents of diges-tion come from three sources: the **liver, pancreas** and **wall of the intestine.** The liver produces **bile** which, after storage in the gall bladder, flows down the bile duct into the duodenum. Bile is a mixture of substances, not all of which are concerned in digestion. The digestive components are the **bile salts,** so-dium taurocholate and glycocholate. These emulsify fats by lowering their sur-face tension, causing them to break up into numerous tiny droplets. In this way the total surface area of the fat is increased, thereby facilitating the digestive action of the enzyme lipase. It must be stressed that the bile salts are not enzymes. They are not proteins and have no chemical effect on the fat, only the physical effect of emulsifying them. Bile is also rich in sodium bicarbonate, which neutralises the acid from the stomach. The pH of the small intestine is therefore distinctly alkaline, which favours the action of the various enzymes.

Some of these enzymes are constituents of **pancreatic juice** which flows into the duodenum from the pancreas via the pancreatic duct. The three main pancreatic enzymes are: **pancreatic amylase** which breaks down starch to maltose; **trypsin** which breaks down proteins into short polypeptides; and **pancreatic lipase** which breaks down fat into fatty acids and glycerol. Trypsin is secreted as an inactive precursor **trypsinogen.** This is converted into trypsin by the action of the enzyme enterokinase, secreted from the wall of the small intestine. The products of amylase and trypsin digestion cannot be absorbed[1], but the fatty acids and glycerol can be. In fact pancreatic juice is the main source of lipase, and completes the digestion of fat. Pancreatic juice also con-tains **chymotrypsin** (secreted as an inactive precursor **chymotrypsinogen**) which is similar to trypsin but specifically attacks casein; and a **carboxypep-tidase** which releases free amino acids from polypeptide chains. However, protein digestion is completed mainly by enzymes produced in the wall of the intestine (Fig 9.5).

Secretory cells in the wall of the small intestine (duodenum and ileum) produce mucus and a variety of enzymes (**intestinal juice**) whose collective

[1] Trypsin does in fact liberate a small quantity of free amino acids, but the yield is negligible compared with that achieved by enzymes secreted from the intestinal wall.

function is to complete the digestion of the various compounds already started by the other secretions. These enzymes include **maltase** which hydrolyses maltose to glucose, thus completing the digestion of starch; a variety of **pep-**

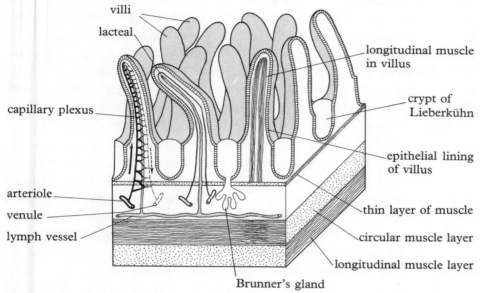

villi
lacteal
longitudinal muscle in villus
capillary plexus
crypt of Lieberkühn
epithelial lining of villus
arteriole
venule
lymph vessel
thin layer of muscle
circular muscle layer
longitudinal muscle layer
Brunner's gland

Fig 9.5 Stereogram of part of the wall of the mammalian small intestine showing glands and villi. Note Brunner's glands opening into crypts of Lieberkühn. The latter are present in the wall of both the duodenum and ileum, the former are restricted to the duodenum. Within each villus a dense capillary plexus surrounds a central lymph vessel. Movements of the small intestine are brought about by contraction of the circular and longitudinal muscle layers, and writhing of the villi by contraction of the longitudinal muscle fibres in each villus.

Fig 9.6 Microscopical structure of the wall of the small intestine. A) Section showing villi and crypts of Lieberkühn. B) Photomicrograph showing rich vascularization. The blood vessels have been injected with a dye that allows them to show up in the section. How many of the structures illustrated in Fig 9.5 can you see in these photomicrographs? (*Gene Cox, Micro-Colour International*)

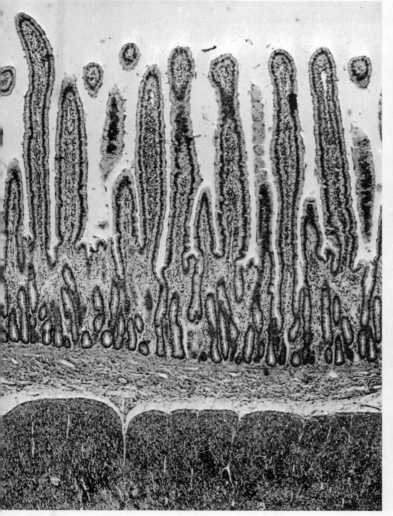

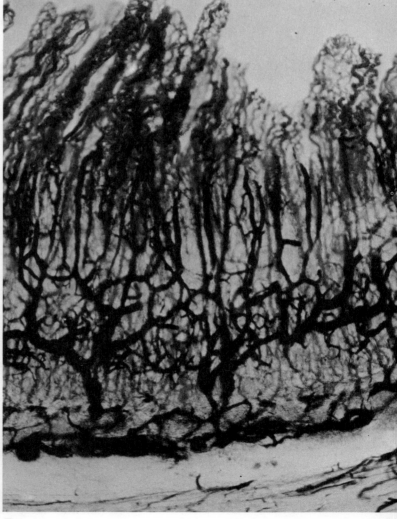

A B

tidases which break down polypeptides to free amino acids, thereby completing the digestion of proteins; **sucrase** which hydrolyses sucrose (cane sugar) to glucose and fructose; **lactase** which hydrolyses lactose (milk sugar) to glucose and galactose; and **enterokinase** which activates trypsin.

Precisely which cells produce these various secretions is uncertain. Alkaline fluid and mucus come mainly from **Brunner's glands** in the wall of the duodenum. The enzymes are produced principally by cells lining the **crypts of Lieberkühn** (Fig. 9.5). It is probable that the final breakdown of carbohydrates and proteins takes place on the membrane lining the microvilli of these cells. Some of the enzymes are known to be located here.

How is the flow of secretions into the small intestine controlled? In the case of bile and pancreatic juice, control is by hormones. The presence of acidified food in the small intestine stimulates the wall of the duodenum to secrete certain hormones into the bloodstream. These hormones are carried to the liver and pancreas where they stimulate the flow of bile and pancreatic juice respectively. The hormone that stimulates the pancreas is called **secretin**, the one stimulating the liver is **cholecystokinin**. The latter acts on the gall bladder.

In the case of the intestinal glands, the presence of food in contact with the lining of the intestine directly triggers the production of their secretions.

All the time the food is in the small intestine it is kept in continual motion by rings of contraction that mix it thoroughly with the various juices secreted into it. This facilitates rapid digestion. The contractions are stationary, or may be propagated along the intestine as peristaltic waves which push the food towards the large intestine. As a result of all these activities the food in the small intestine is converted into a watery emulsion called **chyle**. It is from this that the products of digestion are absorbed.

A summary of the enzymes present in the mammalian gut is given in Table 9.1.

ABSORPTION

Absorption takes place in the small intestine whose surface area is greatly increased by numerous finger-like **villi** that project into the lumen (Figs 9.5 and 9.6). The surface area is further increased by the fact that the cells lining the villi bear numerous **microvilli**. These tiny submicroscopic extensions of the individual cells are not to be confused with the villi, which are much larger multicellular structures. The latter contain smooth muscle enabling them to contract and expand, thus bringing them into contact with newly-digested food. Monosaccharide sugars (mainly glucose) and amino acids are absorbed, by a combination of diffusion and active transport, across the epithelial lining of the villi into the rich plexus of capillaries beneath. These capillaries drain into the **hepatic portal vein** whence the dissolved food is conveyed to the liver.

Fat is dealt with rather differently. Fatty acids and glycerol are absorbed into the columnar epithelial cells lining the villi and there resynthesized into neutral fat which is shed into the lymph vessels of the villi as a white emulsion of minute globules. This gives the lymph vessels a milky appearance, for which reason they are known as **lacteals**. As the lymphatic system ultimately opens into the veins, the fat eventually finds its way into the blood vascular system which distributes it round the body. Although this is the main way fat is absorbed, it may not be the only way. There is evidence that some is absorbed into the bloodstream rather than the lacteals, and a small proportion may be absorbed direct from the small intestine as tiny droplets of neutral fat rather than as fatty acid and glycerol.

Inorganic salts, vitamins and water are also absorbed in the small intestine. Water is extensively absorbed in the **colon** whose wall is much folded for this purpose. Thus by the time it reaches the rectum, indigestible food is in a semi-solid condition ready to be voided through the anus as faeces.

Table 9.1 Summary of the digestive enzymes secreted into the mammalian gut, together with their source, site of action, and functions. The products in italics are soluble and capable of absorption.

Secretion	Source	Site of action	Flow induced by	Enzymes etc.	Substrate	Products
Saliva (slightly alkaline)	Salivary glands	Mouth cavity	Expectation and reflex stimulation	Amylase (ptyalin)	Starch	Maltose
Gastric Juice (distinctly acid)	Stomach wall (gastric glands)	Stomach	Reflex stimulation and hormone (gastrin)	HCl (not an enzyme)		
				Pepsin	Protein	Polypeptides
				Rennin	Caseinogen (soluble)	Casein (insoluble)
				Lipase	Fats	*Fatty acids +glycerol*
Bile (alkaline)	Liver	Duodenum	Reflex action and hormone (cholecystokinin)	Bile salts (not enzymes)	Fats	Fat droplets (*chylomicrons*)
Pancreatic Juice (alkaline)	Pancreas	Duodenum	Hormone (secretin)	Amylase	Starch	Maltose
				Trypsin	Protein	Polypeptides + *amino acids*
				Chymotrypsin	Casein	Polypeptides
				Carboxypeptidase	Polypeptides	*Amino acids*
				Lipase	Fats	*Fatty acids + glycerol*
Intestinal Juice (alkaline)	Wall of small intestine	Small intestine	Mechanical stimulation of intestinal lining	Enterokinase	Trypsinogen (inactive)	Trypsin (active)
				Amylase	Starch	Maltose
				Maltase	Maltose	*Glucose*
				Sucrase (invertase)	Sucrose	*Glucose + fructose*
				Lactase	Lactose (milk sugar)	*Glucose + galactose*
				Peptidases	Polypeptides	*Amino acids*

Nutrition in other Heterotrophs

For descriptive purposes a convenient way of classifying heterotrophic

Fig 9.7 The feeding apparatus of three representative herbivores. In the case of the molar tooth, the wearing down of the cement-covered crown with continual use creates a series of enamel ridges giving an admirable serrated surface for grinding plant food. (*based on Ramsay*)

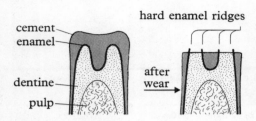

A Vertical section through molar tooth of herbivorous mammal e.g. horse

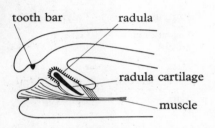

B Vertical section through head of snail to show radula.

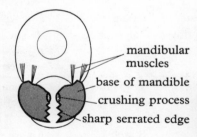

C Posterior view of head of locust to show mandibles.

organisms is in terms of the type of food they eat. On this basis we recognize **herbivores** which feed on plants, **carnivores** which feed on animals, **liquid-feeders** which consume a variety of animal and plant juices, and **microphagous feeders** which live on small particles suspended in water.

HERBIVORES

The problem facing all herbivores is that the nutritionally valuable components of a plant are enclosed within the cellulose walls of the cells. To obtain the protoplasm, the herbivore must grind up the tissues first. Only three groups of animals have evolved the necessary apparatus for doing this: mammals, molluscs, and insects (Fig 9.7).

Herbivorous mammals such as the horse or elephant use their **premolar** and **molar teeth**. As these teeth are worn down in the course of life, the different materials of the crowns are exposed. These wear away at different rates resulting in a flat serrated surface ideally suited to grinding up tough plant material (Fig 9.8).

Herbivorous molluscs such as the snail possess a rasping organ, the **radula**, located in a sac opening into the lower side of the buccal cavity. The radula is like a serrated conveyor belt which, by rubbing backwards and forwards against the hardened roof of the mouth, can tear plant food (Fig 9.9). Herbivorous insects like the locust have a pair of **mandibles** with a jagged edge for cutting through leaves and blades of grass. These, together with a set of manipulating devices, make up the **mouth parts** (Fig 9.10).

If maximum value is to be derived from plant food, the cellulose itself should be digested and absorbed. This requires the enzyme **cellulase**, a potent carbohydrase which splits cellulose into its constituent monosaccharides. A variety of micro-organisms, mainly bacteria and protozoons living in the intestine of other animals, are able to secrete cellulase but, apart from this, cellulase production is extremely rare in the animal kingdom.

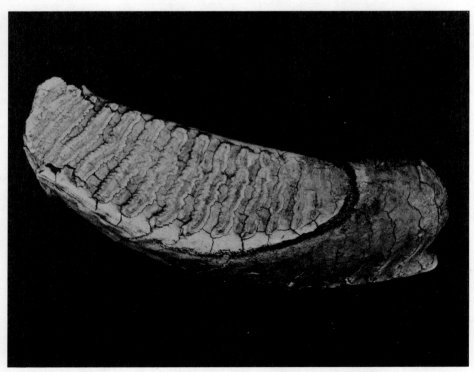

Fig 9.8 The molar surface of an elephant. Notice the ridges of enamel alternating with cement and dentine. Which is the cement and which the dentine? (*B. J. W. Heath, Marlborough College*)

One of the most striking examples of it is seen in the flagellate *Trichonympha* which lives in the intestine of wood-eating termites (Fig 9.11). The termites cannot themselves produce cellulase and for a long time it was a mystery how these insects could survive on a diet consisting of nothing but wood. We now know that the termites chew the wood into tiny fragments which are then taken up phagocytically by the flagellates in the gut. Into the food vacuoles so formed cellulase is secreted, wood fragments being broken down and the products absorbed into the surrounding cytoplasm. Not all the digestive products are used by the flagellates: some escape into the gut and are absorbed by the insect. The association between the insect and the flagellate is therefore beneficial to both organisms and is an example of **symbiosis** which is discussed in detail in Chapter 33.

A similar association is found in the gut of herbivorous mammals. Ruminant mammals like the cow harbour millions of symbiotic micro-organisms, mainly bacteria, in a complex four-chambered **rumen** situated between the oesophagus and the true stomach. Unchewed food is passed to the rumen, and later regurgitated bit by bit into the mouth for chewing. In the rumen fluid there may be as many as a billion cellulose-digesting bacteria per cm³, and these bring about the breakdown of cellulose by producing cellulase. Some of the products of digestion are absorbed by the micro-organisms, the rest by the host. The bacteria are anaerobic, and the acids produced by them are neutralized by an alkaline saliva secreted by the host, often in vast quantities. A cow, for example, may produce over 50 litres of saliva in a day! Although the association is generally regarded as symbiotic, the benefits reaped by the micro-organisms are relatively short-lived for eventually they pass with the rumen fluid into the true stomach where they are digested by the host's enzymes.

In non-ruminant herbivores such as the rabbit, cellulose digestion is effected by micro-organisms in the **caecum** and **appendix**, a blindly-ending

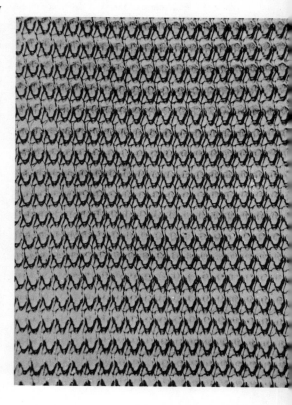

Fig 9.9 The radula of the garden snail *Helix aspersa*. Numerous pointed teeth project from a sheet of tissue. The radula is used for rasping and tearing plant food. (*J. F. Crane*)

Fig 9.10 Mouthparts of an orthopteran insect such as cockroach, locust or grasshopper. The mandibles are used for cutting and crushing plant food, the maxillae and labium for manipulating it. The salivary duct opens into the mouth cavity towards the tip of the hypopharynx, and the labrum serves as a protective 'upper lip'. (*based on Snodgrass*)

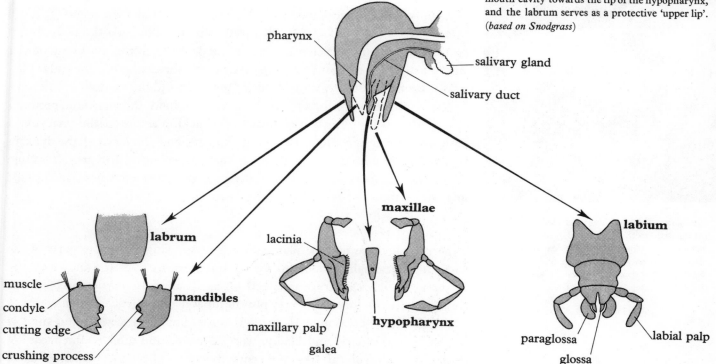

pharynx

salivary gland

salivary duct

labrum

maxillae

lacinia

labium

muscle

condyle

cutting edge

mandibles

crushing process

maxillary palp

galea

hypopharynx

paraglossa

glossa

labial palp

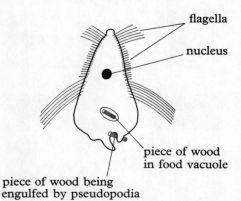

Fig 9.11 *Trichonympha*, an animal flagellate, lives symbiotically in the gut of wood-eating termites where it ingests pieces of wood that have been devoured by its host. Its ability to secrete the enzyme cellulase enables it to digest the wood, a feat which few other animals can perform.

flagella

nucleus

piece of wood in food vacuole

piece of wood being engulfed by pseudopodia

Fig 9.12 Skull of dog illustrating the way mammalian teeth may be adapted for a carnivorous diet. Notice the large carnassial teeth towards the back of the upper and lower jaws. What are the precise functions of the other teeth? (*B. J. W. Heath, Marlborough College*)

diverticulum of the gut situated at the point where the small intestine joins the large intestine. In non-herbivorous mammals like man, the caecum and appendix are functionless. Because plant material is so difficult to digest herbivorous mammals generally have much longer alimentary canals than carnivorous ones.

CARNIVOROUS ANIMALS

The problem here is not so much digesting the food as obtaining it. So we find that carnivores are adapted for catching prey. These adaptations take various forms: high speed locomotion and dagger-like **canine teeth** in the great cats, sucker-bearing **tentacles** in octopuses and squids, tentacles armed with **stinging cells** in *Hydra*, jelly fishes and other coelenterates.

Once captured, the prey is dealt with in one of three ways. It may be swallowed whole, chewed up and then swallowed, or digested externally outside the body and then ingested. Ingesting the food whole, without breaking it up first, is the method used by pythons and boaconstrictors, and also by *Hydra* and sea anemones. The trouble is that it puts a tremendous strain on the digestive system, which consequently takes a long time to break the food down into an absorbable state. In the python, for example, it may take several weeks for a complete rabbit to be digested and absorbed.

The majority of carnivores chew their prey first. This is achieved by the **teeth** in mammals and by **mouthparts** in carnivorous arthropods like crabs and crayfish. In the latter, the food, generally soft, is torn by the shredding action of the **maxillipeds**. Chewing is carried out by the **mandibles**. In mammals the sharp **incisor teeth** are used for biting pieces of flesh and tearing it off bone. In many mammals the last pair of premolars in the upper jaw and the first pair of molars in the lower jaw are enlarged and have sharp ridges for shearing flesh. As any dog-owner knows these **carnassial teeth** are very effective at scraping flesh off bones, and cutting through hard pieces of food (Fig 9.12).

Animals that digest food outside their bodies and then ingest it, do so in a variety of ways. Digestion is rarely completed outside the body, but it breaks the food up sufficiently for it to be drawn into the gut by suction or ciliary action. In certain free-living flatworms, for example, a muscular pharynx is everted from the ventral side of the body and buries itself in soft food. Digestive enzymes are then poured out onto the food, the semi-fluid products of digestion being drawn into the gut by sucking action of the pharynx. A basically similar technique is used by spiders and the larva of the dytiscus beetle. These animals insert a pair of sharp fangs into their prey. Through ducts opening near the tips of the fangs, digestive enzymes are injected into the victim. The resulting fluid is then sucked into the gut.

CARNIVOROUS PLANTS

Reports that there may be man-eating trees growing in remote parts of the world have gradually faded into legend, but there are over 400 known species of plants which can trap and feed on small animals, particularly insects. **Carnivorous (or insectivorous) plants** inhabit areas deficient in nitrates. A few such plants are found in Britain: for example butterwort and sundew of heaths and moors, and bladderwort of ponds and ditches. But most are tropical or sub-tropical, like the Venus fly-trap and pitcher plants. All have green leaves and obtain their carbohydrate by photosynthesis, but nitrogen they obtain from the bodies of their victims. The insect is attracted by colour, scent or sugary bait, then trapped, killed and digested by a fluid rich in proteolytic enzymes. The amino acids are absorbed into the plant. The method of trapping the insect varies from one species to another: sticky leaves in butterwort, an elaborate sac in bladderwort, and adhesive hairs in sundew (see p. 8). The Venus fly-trap (Fig 9.13A) has infolding leaves with spikes along the free edges and a hinge-like midrib. When an insect alights on a leaf, the two halves spring together, the spikes interlocking so that the unfortunate animal cannot escape. Pitcher plants have pitcher-shaped leaves which contain digestive fluid. Insects are attracted to the lips of the pitcher by nectar secreted near the brim. In trying to reach the nectar, the insect slips and falls into the fluid, there to be killed and digested (Fig 9.13B).

LIQUID FEEDERS

Animals that feed on liquid food fall into two groups, **wallowers** and **suckers**. Wallowers include gut parasites such as the tapeworm which inhabit the small intestine. They feed on the digested food of the host, absorbing it straight

Fig 9.13 The animal-trapping mechanism in two carnivorous plants. A) The leaf of the Venus flytrap, *Dionaea muscipula*. When an insect alights on it, the hinged leaf snaps shut as shown in A2. The insect's body is then broken down by digestive enzymes secreted by glandular cells at the surface of the leaf. B1) Side view of the pitcher of a pitcher plant. B2) Looking into the pitcher. Notice the digestive fluid. The downward-projecting hairs prevent the insect from clambering out. (*B. J. W. Heath, Marlborough College*)

A1

A2

B1

B2

through the integument, and therefore need no gut or digestive enzymes of their own. Other gut parasites such as *Ascaris lumbricoides*, a large round-worm inhabiting the human small intestine, have a muscular pharynx through which they suck the host's digested food into their own gut.

The sucking forms are mainly insects which feed on blood or plant juices. For this purpose their mouthparts consist of a sharp **proboscis** which can pierce the integument, and through which fluid is drawn by the suctorial action of the pharynx. In the mosquito the proboscis consists of a sucking tube surrounded by a group of sharp stylets whose cutting action helps the proboscis to pierce the skin. Having gained entrance, saliva flows down it into the blood which is then sucked up into the pharynx. The saliva contains an enzyme which prevents the blood from clotting and keeps it in a fluid state. Sucking mouthparts of a different kind are found in insects such as aphids and butterflies which feed on plant juices. Butterflies have a long flexible proboscis ideally adapted for probing into flowers. Some of these mouthparts are illustrated in Fig 9.14.

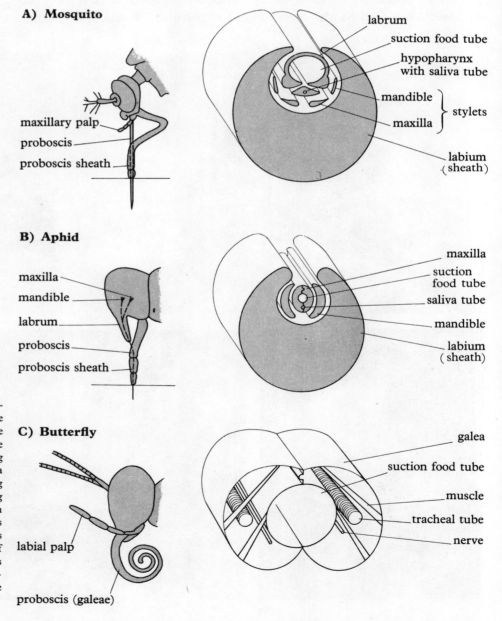

A) Mosquito

maxillary palp
proboscis
proboscis sheath

labrum
suction food tube
hypopharynx
with saliva tube
mandible
maxilla } stylets
labium
(sheath)

B) Aphid

maxilla
mandible
labrum
proboscis
proboscis sheath

maxilla
suction
food tube
saliva tube
mandible
labium
(sheath)

C) Butterfly

labial palp
proboscis (galeae)

galea
suction food tube
muscle
tracheal tube
nerve

Fig 9.14 Three examples of sucking mouth-parts in insects. The sharp proboscis of the mosquito and aphid is adapted for piercing the integument. The slender mandibles and maxillae in the mosquito form pointed 'stylets' for making a hole in the victim's skin. In both cases saliva flows down the salivary tube prior to food being sucked up the food tube. In the butterfly a long flexible proboscis is used for probing flowers in search of nectar: to this end the proboscis is equipped with muscles and nerves throughout its length. All three mouthparts are modifications of the basic pattern seen in orthopteran insects (Fig 9.10). Note that in each case different components have entered into the composition of the proboscis. (*based on various sources*)

MICROPHAGOUS FEEDERS

Microphagous feeders are always aquatic and feed on tiny particles suspended in water. Since their food is already 'broken up', digestion is a relatively simple matter. The problem facing a microphagous feeder is collecting, sorting and concentrating the particles, and conveying them to the mouth. Generally speaking, water is drawn towards the body either by the movements of appendages as in various crustaceans, or by the action of cilia. The water then goes through some kind of sieve which filters off the particles. For this reason microphagous feeders are also known as **filter feeders.**

The basic principles underlying filter-feeding can be illustrated by the fresh-water mussel *Anodonta* in which the gills are used for sorting and straining food particles (Fig 9.15). Four sheet-like gills, two on each side

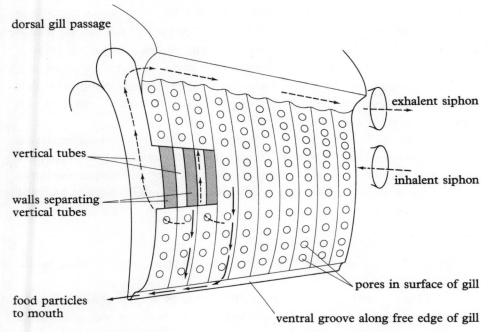

dorsal gill passage

vertical tubes

walls separating
vertical tubes

food particles
to mouth

exhalent siphon

inhalent siphon

pores in surface of gill

ventral groove along free edge of gill

Fig 9.15 Stereoscopic diagram of part of one gill of a fresh water mussel. There are four such gills, two on each side of the foot. The sheet-like gills hang into the mantle cavity and are used for respiration and also for filter feeding. The flow of water is shown by broken arrows, the movement of food particles by solid arrows.

of the muscular foot, hang into the mantle cavity. Each gill is a hollow structure made up of a series of **vertical tubes** which connect with a **dorsal gill passage**. The sides of the gill are perforated by numerous pores which open into the vertical tubes, and the free edge of the gill bears a **ventral groove**. The surface of the gills and the lining of the vertical tubes are covered with cilia. At the posterior end of the body are two apertures, the **inhalent** and **exhalent siphons**, which open into the mantle cavity.

Cilia on the gills draw water into the inhalent siphon and through the pores into the vertical tubes. As this happens particles suspended in the water get trapped in mucus on the surface of the gill. The water then passes up the vertical tubes, back along the dorsal gill passage, and out through the exhalent siphon. In the meantime the particles caught on the surface of the gill are drawn downwards to the free edge. Only small particles of nutritional value are treated thus. Larger particles of silt or sand drop off the gill and are wafted away by cilia lining the mantle cavity. On reaching the free edge of the gill, cilia lining the ventral groove carry the food particles to a pair of palps on either side of the mouth where further sorting occurs before the particles are taken into the mouth.

10 Autotrophic Nutrition

A dictum which we are all brought up to learn in elementary biology courses is that animals are utterly dependent on plants. This is in fact perfectly true for the simple reason that the food materials which animals eat are manufactured by plants. This reflects a basic difference between animals and plants: animals feed, and can only feed, on complex organic matter; plants, however, can feed on simple inorganic materials, building these up into complex organic molecules. All other differences between animals and plants stem from this fundamental difference of nutrition.

The type of feeding employed by plants is called **autotrophic nutrition** (or just **autotrophism**). It always involves the synthesis of organic compounds from inorganic raw materials and is sometimes called **holophytic nutrition** as it is typical of most plants. (Holophytism is a word of Greek origin meaning 'plant feeding'.)

DIFFERENT TYPES OF AUTOTROPHIC NUTRITION
Autotrophic organisms employ two methods of synthesizing organic from inorganic materials:

(1) Photosynthesis. This is the method used by all green plants. It is the synthesis of organic compounds, primarily sugars, from carbon dioxide and water using sunlight as the source of energy and chlorophyll (or some other closely related pigment) for trapping the light energy.

(2) Chemosynthesis. Used by certain bacteria, this is the synthesis of organic compounds from carbon dioxide and water but the energy instead of coming from light is supplied by special methods of respiration involving the oxidation of various inorganic materials such as hydrogen sulphide, ammonia and iron.

In this chapter we shall be mainly concerned with photosynthesis and will deal only briefly with chemosynthesis.

THE IMPORTANCE OF PHOTOSYNTHESIS
As a means of manufacturing sugar the contribution of photosynthesis is astounding. An acre (0.4 hectares) of corn can convert as much as 4000 kg of carbon from carbon dioxide into the carbon of sugar in a year, giving a total yield of 10,000 kg of sugar per year. Although it is difficult to arrive at a total world figure for photosynthesis, one biologist has calculated that 35×10^{15} kg of carbon are fixed by plants per year – probably a conservative estimate. It is therefore an exceedingly important process to man and other animals.

The way carbon compounds circulate in nature is shown in the **carbon**

cycle summarized in Fig 10.1. The organic compounds manufactured by plants are consumed by animals. The carbon dioxide subsequently released from the organic compounds in respiration is then built up again into organic compounds by the plants. From this you will appreciate the complete dependence of animals on plants.

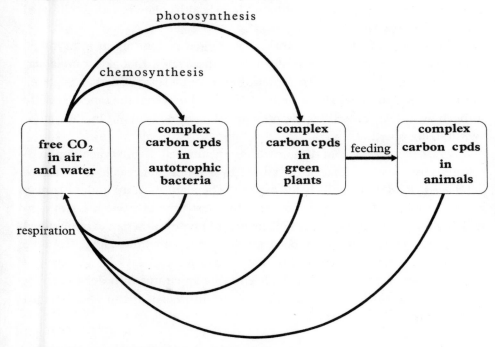

Fig 10.1 Simplified version of the carbon cycle to show the balance between photosynthesis (and chemosynthesis) and respiration. The cycle illustrates the dependence of animals on the photosynthetic ability of green plants. What other natural or artificial processes besides respiration contribute to the release of carbon dioxide into the atmosphere?

THE RAW MATERIALS OF PHOTOSYNTHESIS

The raw materials of photosynthesis are **carbon dioxide** and **water**. Experiments can be done to show that carbon dioxide is the only form in which a plant can take in carbon. A plant deprived of carbon dioxide will die even if it is grown in a culture solution containing soluble carbon compounds like calcium carbonate. Equally, water is the only source of hydrogen for reducing the carbon dioxide, the first step in the building up of carbohydrate. Mineral salts (sulphates, phosphates and nitrates) are also required, though these are needed not for the formation of carbohydrates but for their further conversion into proteins. It follows that if a plant is deprived of any of these essential raw materials it will die.

THE PRODUCTS OF PHOTOSYNTHESIS

The main product of photosynthesis is **monosaccharide sugar**. This can be built up into proteins for growth, broken down into carbon dioxide and water for energy production, or converted into starch for storage. Its fate depends on the requirements of the plant at the particular time.

Detecting the formation of starch is a convenient way of establishing whether or not a plant has been carrying out photosynthesis. The plant is first de-starched by placing it in the dark for not less than 12 hours. This prevents photosynthesis from taking place so the plant has to draw upon its starch reserves. After de-starching, one of the leaves should be tested for starch to check that none is present. The plant is now returned to the light, and after a period of time one of the leaves is re-tested for starch. Its presence indicates that photosynthesis has been going on. The test for starch is as follows: the leaf is boiled gently in water until flaccid and immersed in warm 90 per cent

Fig 10.2 Technique for demonstrating that oxygen is given off by a photosynthesising plant. The gas which collects at the top of the inverted test tube is tested for oxygen with a glowing splint. The rate of photosynthesis can be estimated from the number of bubbles given off per unit time. If it is necessary to find the amount of oxygen evolved the gas can be collected and analysed by a modification of the method described on p. 96.

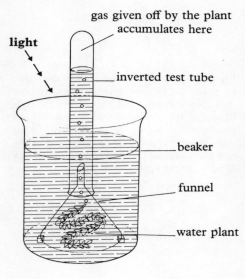

Fig 10.3 Experiment to show that carbon dioxide is required for photosynthesis. The whole plant is de-starched by putting it in darkness for 48 hours, and then set up as shown below. The flasks must be made completely airtight by smearing the split corks with vaseline. After 6 hours the two enclosed leaves are detached and tested for starch.

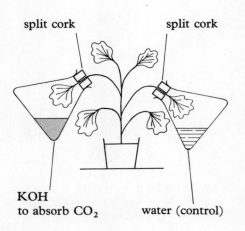

alcohol until decolorized. When placed in dilute iodine any parts of the leaf containing starch turn dark blue. Quantitative estimations of the amount of carbohydrate formed per unit time can be made to determine the rate of photosynthesis. This is generally expressed as the weight of dry matter formed by a square metre of leaf surface in the course of an hour. Naturally the rate varies tremendously but a typical figure for a healthy plant in well illuminated conditions might be around $1.0g/h/m^2$.

The other product of photosynthesis is **oxygen**. As long ago as 1772 Joseph Priestley, the father of modern chemistry, discovered that if a candle was allowed to burn in a sealed jar it went out. But if a sprig of mint was placed in the jar the candle could be re-kindled. We now know that the candle's ability to burn again was caused by the presence of oxygen given out by the photosynthesizing plant.

A rather better way of demonstrating the formation of oxygen in photosynthesis is to collect the gas given off by a water plant such as Canadian pondweed (*Elodea*) and test it for oxygen. The plant is set up under a funnel as shown in Fig 10.2. Gas is collected in a test tube and tested for oxygen with a glowing splint. Some idea of the rate of photosynthesis can be obtained by counting the number of bubbles given off per unit time, either from the whole plant or from the cut end of one of the stems.

There are various ways of modifying the apparatus in order to extract samples of the gas for chemical analysis: by this means it can be confirmed that the gas is rich in oxygen.

The Conditions required for Photosynthesis

Plainly for photosynthesis to take place a plant must be supplied with **carbon dioxide** and **water**. It also requires **light**, a **suitable temperature** and the presence of **chlorophyll**. The necessity for these five factors can be demonstrated by simple experiments either on whole plants or single leaves. In all cases the plant must first be de-starched by putting it in the dark for a period of time, and the necessary controls set up.

CARBON DIOXIDE

The necessity for carbon dioxide can be tested by the arrangement shown in Fig 10.3. One of the green leaves of a well-watered potted plant such as geranium is deprived of carbon dioxide by enclosing it in a flask containing a small volume of potassium hydroxide (caustic potash), which absorbs the carbon dioxide from the air in the flask. A second leaf, enclosed in a separate flask containing water, serves as a control. The plant is placed in a well lit place for several hours after which the two leaves are removed and tested for starch. The control leaf is usually found to have formed a significant quantity of starch, the other leaf little or none.

WATER

The necessity for water is difficult to demonstrate by a simple experiment. Depriving a plant of water will certainly kill it but this might be owing to any number of reasons, not necessarily connected with photosynthesis. The only way of showing unequivocally that water is required for photosynthesis is to trace what happens to it after it has been taken into the plant. This can be done by supplying a plant with the heavy isotope of water $H_2{}^{18}O$. This experiment, which is discussed later in connection with the chemistry of photosynthesis, has shown that water provides hydrogen for reducing the carbon dioxide.

LIGHT

The fact that a plant can be de-starched by putting it in the dark indicates that light is necessary for photosynthesis. The importance of light can be further demonstrated by covering part of a previously de-starched leaf with a strip of opaque paper and then exposing it to light for several hours. On testing the leaf with iodine it is found that the dark colour, signifying the presence of starch, is confined to the illuminated parts of the leaf. The covered part remains colourless, indicating the failure of this part of the leaf to form starch. The completed test gives a **starch print** (Fig 10.4), an elegant demonstration of the importance of light in photosynthesis.

Not all the wavelengths of light are absorbed by chlorophyll. This can be shown by projecting a beam of light through a solution of extracted chlorophyll and then through a prism which separates it into its different wavelengths. After passing through the prism the light is projected onto a screen and any colours absent from the normal spectrum are those that have been absorbed by the chlorophyll. This gives us an **absorption spectrum** for chlorophyll (Fig 10.5). It turns out that the red and blue ends of the spectrum disappear showing that light of these colours is absorbed to varying extents by the chlorophyll. However, the middle part of the spectrum, green light, is hardly absorbed at all: most of it is reflected, which is why chlorophyll looks green.

But are the wavelengths absorbed actually used in photosynthesis? That

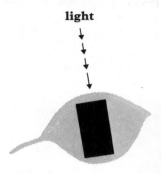

Fig 10.4 A simple experiment to show the necessity for light in photosynthesis. No starch is formed in the covered part of the leaf.

light

A) Black paper applied to previously de-starched leaf which is then illuminated for 6 hours.

B) Appearance of leaf after testing for starch with iodine.

Fig 10.5 (*below*) Absorption spectrum of chlorophyll. The black areas indicate those wavelengths of light which disappear after being passed through a general chlorophyll extract. Notice that most of the absorption is at the blue and red ends of the spectrum; there is relatively little in between.

blue green yellow orange red

450 500 550 600 650 700 750nm

Fig 10.6 (*left*) Comparison of absorption and action spectra of chlorophyll shows a close correspondence between the two, indicating that most of the wavelengths of light absorbed by chlorophyll are used in photosynthesis. The graphs are based on data obtained from the sea lettuce *Ulva taeniata*. In this particular plant efficiency in blue light is unusually high. The marked noncorrespondence of the two curves at **X** is because of absorption of this wavelength by carotenes which are not used in photosynthesis. (*from Haxo and Blinks, 1950*)

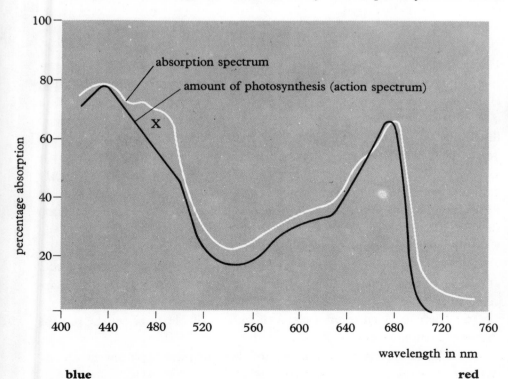

absorption spectrum

amount of photosynthesis (action spectrum)

X

percentage absorption

100
80
60
40
20

400 440 480 520 560 600 640 680 720 760

wavelength in nm

blue   red

Fig 10.7 A variegated geranium leaf before and after testing for starch with iodine. Notice that the dark colour indicating starch corresponds to the green areas where chlorophyll is located.

green part of leaf

yellow part of leaf

they are can be shown by exposing leaves to different coloured lights and then estimating the amount of carbohydrate formed in each case. This gives us an **action spectrum** for chlorophyll. Red and blue turn out to be the most effective wavelengths in photosynthesis; green is only used to a slight extent. There is thus a close correlation between the absorption and action spectra of chlorophyll, as can be seen in Fig 10.6.

TEMPERATURE

Photosynthesis proceeds by a series of chemical reactions controlled by enzymes. It can be shown by comparing a plant's rate of photosynthesis at different temperatures that the optimum temperature for photosynthesis is around 30°C. At lower temperatures the rate is slowed. If the temperature exceeds about 40°C the process stops altogether since the enzymes, which you will remember are proteins, become denatured.

CHLOROPHYLL

That chlorophyll is required for photosynthesis can be shown by studying the distribution of starch in a **variegated leaf**. A variegated leaf is one which lacks chlorophyll, either completely or in patches, giving it a yellow appearance wherever the green pigment is absent. This is common in such plants as privet, laurel, geranium and ivy. If a variegated plant is exposed to light and one of its leaves is then tested for starch, the dark colour develops only in those parts of the leaf that were green. The distribution of the dark colour corresponds exactly to the distribution of chlorophyll (Fig 10.7).

In the case of variegated ivy this experiment can be taken a step further. The leaves of this plant have four different shades of colour: dark green towards the centre, then two zones of progressively paler green and, at the edge, a yellow rim. If, after a period of illumination, one of these leaves is tested for starch the intensity of the blue colour follows exactly the same pattern as the chlorophyll.

INTERACTION OF FACTORS CONTROLLING PHOTOSYNTHESIS

Consider the following experiment. A plant is subjected to a series of increasing light intensities. The rate of photosynthesis is determined at each intensity, and the results plotted on a graph (Fig 10.8). The temperature and carbon dioxide concentration are kept constant, the temperature at 20°C, the carbon dioxide at 0.03 per cent, its normal value in the atmosphere. The experimental details need not concern us; let us concentrate on what happens and why.

As you can see from Fig 10.8, the rate of photosynthesis rises abruptly as the light intensity is increased, and then levels off as the process reaches its maximum rate. What causes the rate of photosynthesis suddenly to stop increasing?

There are three possible answers.

(1) The photosynthetic process is going at the fastest possible pace, and no amount of additional light will make it go any faster whatever the circumstances.

(2) There is insufficient carbon dioxide available to allow the process to speed up any further.

(3) The temperature is too low for the chemical reactions to go any faster.

How can we decide between these three possibilities? The simplest way is to raise either the temperature or carbon dioxide concentration and repeat the experiment. The results of doing this are shown in Fig 10.9. Curve (a) is the same one as we obtained before. If the experiment is now repeated at the same carbon dioxide concentration but a higher temperature (30°C instead of 20°C) curve (b) is produced which is virtually identical to curve (a). This shows that it cannot be temperature which is preventing the process going any faster. However, if the temperature is kept the same and the carbon dioxide concentration increased to 0.13 per cent, curve (c) is obtained: the rate of photosynthesis rises to a maximum which is more than double that achieved at the lower carbon dioxide concentration. This shows that our second hypothesis is the correct one, carbon dioxide limiting the rate of photosynthesis in the first experiment. What is limiting the process where curve (c) flattens out? Curve (d) shows that in this case the rate of photosynthesis is increased by raising the temperature, thus indicating that temperature is the limiting factor where curve (c) flattens out.

The facts demonstrated by this experiment can be put into a general statement, called the **law of limiting factors**. When a chemical process depends on more than one essential condition being favourable, its rate is limited by that factor which is nearest its minimum value. In Fig 10.9 light is the limiting factor where the curves are rising. When the curves flatten out we know that some other factor is limiting the process.

PHOTOSYNTHESIS AND THE ENVIRONMENT

Limiting factors are of great importance to plants in their natural surroundings. The concentration of carbon dioxide in the atmosphere does not vary much, but light and temperature fluctuate considerably both in the course of a day and from one season to another. On a warm summer day, light and temperature are generally well above their minimum value for plants living in the open, and carbon dioxide is the factor limiting photosynthesis. But in the cool of the early morning or evening, light or temperature may become limiting factors, as they do in winter. Habitat is also important: for plants living in shady places such as the floor of a forest or wood, light will be the limiting factor most, if not all, of the time.

Probably the most important single factor controlling photosynthesis is light: even a small change in illumination can make a considerable difference to the rate of photosynthesis. Plants often compete for light, endeavouring to put themselves in a situation which provides enough illumination for their activities. The most obvious adaptation for obtaining plenty of light is to be tall. This is most clearly seen in trees, but the sturdy, erect stems of herbaceous plants like nettles and willowherbs are also concerned with lifting the leaves into a well-lit position for photosynthesis. Their great height makes the taller trees like beech and oak the dominant plants of a woodland community, though for sheer size the jackpot must go to the Californian redwoods which can reach heights of over 100 metres.

Plants which live in the shadow of taller trees and shrubs resort to various strategies for obtaining sufficient light. Climbing plants like *Clematis*, honeysuckle and ivy have twining stems and/or tendrils enabling them to 'scramble' towards light. Certain woodland plants grow to maturity and flower in early spring before the leaves come out on the trees, good examples being dog's

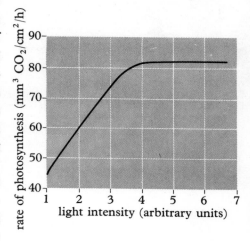

Fig 10.8 The result of an experiment showing the effect of different light intensities on the rate of photosynthesis of cucumber plants. The temperature was kept at 20°C and the carbon dioxide concentration at 0.03 per cent. The rate of photosynthesis was determined by measuring the volume of carbon dioxide given off per square centimetre of leaf per hour.

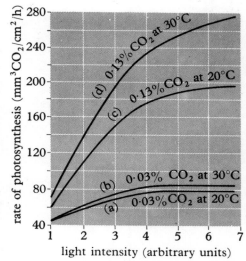

Fig 10.9 The results of an experiment investigating the effect of different light intensities on the rate of photosynthesis of cucumber plants at two temperatures and two carbon dioxide concentrations. (*data by Gaastra*)

Fig 10.10 Leaf mosaic in a rhododendron. Notice that the leaves arrange themselves in such a way that relatively few are in the shade. (*B. J. W. Heath, Marlborough College*)

mercury and bluebell. Some woodland plants, such as primroses and violets, can photosynthesize in conditions of very low illumination and can therefore survive in relatively dark places.

Even large plants have their difficulties, one being that the leaves may cast shadows on each other. This is overcome in some plants by a condition called **leaf mosaic**: the leaves are placed in such a way that they all fit neatly together into a sort of mosaic pattern, leaving few gaps between one leaf and the next. This is seen, for example, in beech trees and is the reason why beeches cast so much shade. Next time you are lying in a beech wood look up at the canopy and notice how dense it is. Leaf mosaic makes beech woods very dark and the ground flora is sparse as a result.

In order to survive a green plant must receive sufficiently intense light for sufficiently long to replenish its supplies of carbohydrates, which have been lost by respiration during dark periods. When photosynthesis and respiration proceed at the same rate so that there is no net loss or gain of carbohydrate, the plant is said to be at its **compensation point**. The time taken for a plant to reach its compensation point, having been in darkness, is called the **compensation period**. The length of this period varies with different plants and conditions. With their greater ability to utilize dim light, shade plants generally reach their compensation point earlier in the day than those requiring bright sunlight.

The Site of Photosynthesis

The distribution of starch in a variegated leaf demonstrates that photosynthesis can only take place in the green parts of a plant, and this suggests that the process is closely associated with **chlorophyll**. As chlorophyll is contained within chloroplasts it is logical to suppose that photosynthesis takes place in or close to the **chloroplasts**.

That this is so was first demonstrated by the German botanist T. W. Engelmann. Engelmann realized that it was no use trying to detect the site of photosynthesis in an ordinary cell filled with densely-packed chloroplasts. It was necessary to find a large cell containing a localized chloroplast. It so happens that the filamentous alga *Spirogyra* fills the bill very nicely. The filaments of *Spirogyra* are composed of comparatively large cylindrical cells placed end to end. Each contains a ribbon-like chloroplast which describes a spiral round the edge. Engelmann chose *Spirogyra* for his experiments and used the evolution of oxygen as an indication that photosynthesis was proceeding. His method of detecting the oxygen illustrates the ingenuity of this remarkable biologist. He had previously discovered that certain bacteria (*Pseudomonas*) move vigorously in the presence of oxygen, clustering together where the oxygen concentration is highest. A filament of *Spirogyra* was mounted on a microscope slide in a drop of water containing numerous bacteria. The slide was first put in darkness which prevented photosynthesis, stopped the evolution of oxygen and immobilized the bacteria. The slide was then exposed to light and viewed under a microscope. Motile bacteria were seen to cluster round the edge of the cells immediately adjacent to the chloroplast, indicating the evolution of oxygen at that point (Fig 10.11).

Engelmann's experiment illustrates an important general point, namely

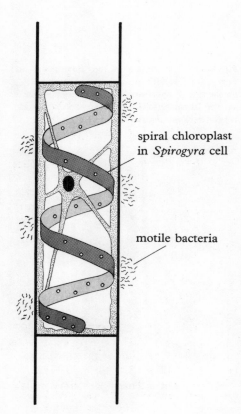

spiral chloroplast in *Spirogyra* cell

motile bacteria

Fig 10.11 In his classical experiment, Engelmann used bacteria of the genus *Pseudomonas* to demonstrate that oxygen is given off during photosynthesis by the filamentous alga *Spirogyra*. After oxygen deprivation motile bacteria are seen only in the immediate vicinity of the chloroplast.

the use that can be made of an organism to demonstrate a biochemical process. As a technique it is of limited application but occasions sometimes arise in the course of biological research where it may be the only way of solving the problem.

CHLOROPHYLL

What we have been calling chlorophyll is in fact a mixture of closely-related pigments of which chlorophyll is but one. This can be demonstrated by extracting the pigments from leaves with acetone and separating them by paper chromatography (Fig 10.12). Five pigments can be identified: **chlorophyll a** (blue-green), **chlorophyll b** (yellow-green), **xanthophyll** (yellow) and **carotene** (orange). The fifth pigment, **phaeophytin** (grey), is a breakdown product of chlorophyll. By making separate solutions of each pigment and determining the absorption spectrum of each it can be shown that chlorophyll a and b absorb light from both the red and blue/violet parts of the spectrum, whereas xanthophyll and carotene absorb light only from the blue/violet part (Fig 10.13). Chlorophyll a is the most abundant pigment and is of universal occurrence in all photosynthesizing plants. Its function is to absorb light energy and convert it into chemical energy. The other pigments do this too and then probably hand on the energy to chlorophyll a.

The chemical structure of chlorophyll is shown in Fig 10.14. It belongs to a group of organic compounds known as **porphyrins** (see p. 79) which also include haemoglobin and other respiratory pigments. One of the characteristic features of porphyrins is that they form complexes with heavy metals. In the case of chlorophyll the metal is magnesium, located at the centre of the molecule. What exactly makes chlorophyll so good at absorbing light is unclear but it certainly seems to be related to the presence of magnesium. The blood pigment haem, which is almost identical to chlorophyll but contains iron instead of magnesium, cannot absorb light to anything like the same extent.

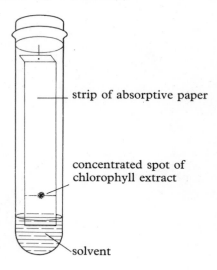

Fig 10.12 The different pigments present in chlorophyll can be separated by paper chromatography. A strip of absorptive paper carrying a concentrated spot of chlorophyll extract is dipped into a suitable solvent, for example a mixture of acetone and petroleum ether (A). The solvent rises up the paper sweeping the chlorophyll pigments with it (B).

— strip of absorptive paper

— concentrated spot of chlorophyll extract

— solvent

A

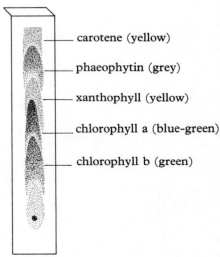

— carotene (yellow)

— phaeophytin (grey)

— xanthophyll (yellow)

— chlorophyll a (blue-green)

— chlorophyll b (green)

B

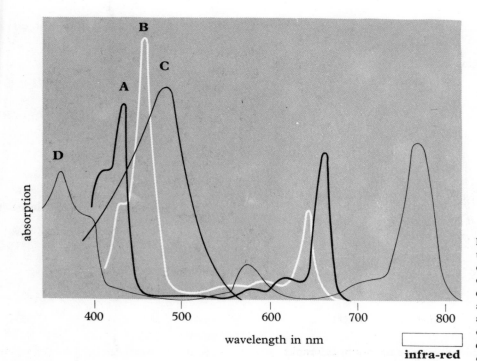

absorption

400 500 600 700 800

wavelength in nm

infra-red

Fig 10.13 Absorption spectra of various photosynthetic pigments: (A) chlorophyll a; (B) chlorophyll b; (C) xanthophyll; (D) bacteriochlorophyll of purple sulphur bacteria. Note differing abilities of the various pigments to absorb different wavelengths. Chlorophyll a absorbs at longer wavelengths than either chlorophyll b or xanthophyll, and bacteriochlorophyll can absorb infra-red. (*mainly after Clayton, 1965*)

Fig 10.14 Chlorophyll is a porphyrin with magnesium as its metallic radical. In chlorophyll a X is − CH₃; in chlorophyll b it is $-C {\overset{O}{\underset{H}{\lessgtr}}}$.

STRUCTURE OF THE CHLOROPLAST

With very few exceptions chlorophyll is contained within chloroplasts. A chloroplast of a higher plant is biconvex in shape, about 5 μm across in the widest part. Studies with the electron microscope show it to have a remarkably elaborate internal structure which can be readily related to its functions.

As you can see from Figs 10.15 and 10.16, the chloroplast is bounded by a double unit membrane within which are numerous sheet-like **lamellae** running from one end to the other. Each lamella consists of a pair of unit membranes close to each other with a narrow space between. The lamellae can be distinguished into two regions, the **granal** and **inter-granal regions**. Where

Fig 10.15 Structure of chloroplast. A) Section perpendicular to granal membranes. B) Stereogram of part of a granum. C) Section parallel to granal membranes. Notice the large surface area achieved by the lamellae. Further explanation in text.

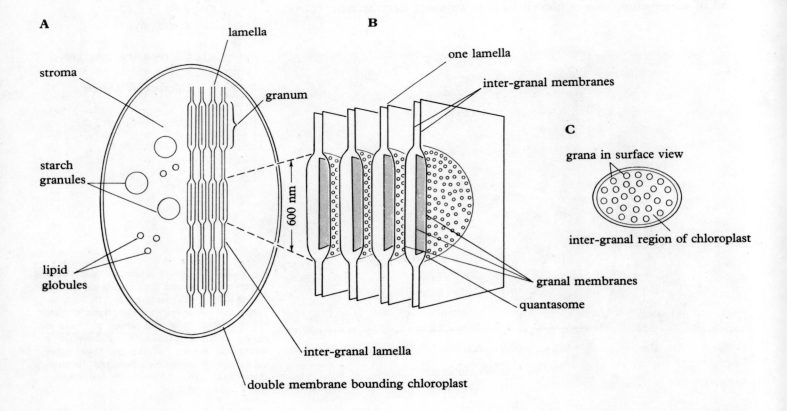

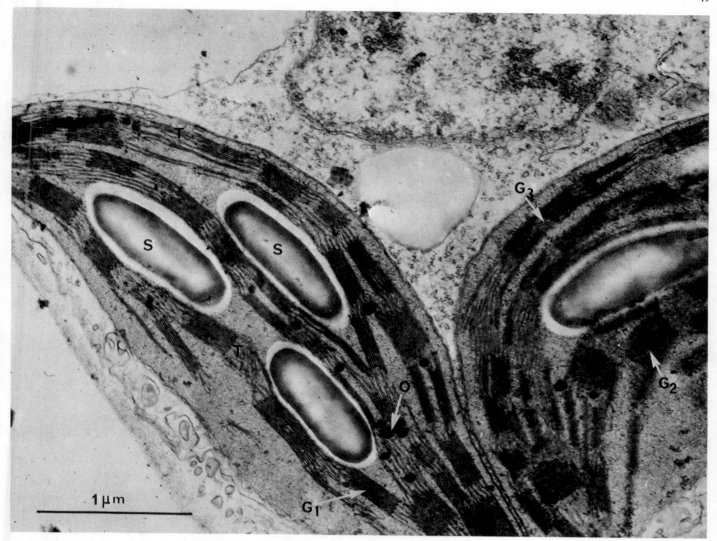

1 μm

the former occur the two membranes of each lamella are slightly further apart, enclosing another thin membrane. In these regions adjacent lamellae are neatly aligned, each group constituting a **granum**. As its constituent lamellae are disc-shaped in surface view, a granum is rather like a stack of coins. Its diameter is about 600 nm.

The function of the lamellae is to hold the chlorophyll molecules in a suitable position for trapping the maximum amount of light. In its internal organization the chloroplast appears to achieve this admirably. A typical chloroplast contains approximately 60 grana each consisting of about 50 lamellae. The chlorophyll molecules are, as it were, laid out on shelves stacked on top of each other with the greatest economy of space. This provides a large surface area without taking up too much room.

The lamellae are embedded in a watery matrix, the **stroma**. This contains, amongst other things, the enzymes responsible for the reduction of carbon dioxide and numerous starch granules, end-products of photosynthesis. From this it seems that while the absorption of light takes place in the lamellae, the subsequent building up of carbohydrates takes place in the stroma.

It has been found that isolated lamellae liberate oxygen when brightly illuminated, suggesting that photosynthesis (the part of it involving light at

Fig 10.16 Electron micrograph of parts of two chloroplasts from a leaf which had previously been in strong sunlight. Notice the conspicuous starch grains (**S**) between the bundles of lamellae (**T**). Granal lamellae (**G**) are seen in various planes: G_1 are cut at right angles to the plane of the section; G_2 are parallel with the plane of the section; and G_3 are oblique to the plane of the section. Lipid globules (**O**) can also be seen between the lamellae. (*B. E. Juniper, University of Oxford*)

Fig 10.17 Diagram illustrating possible arrangement of chlorophyll within an individual granum, based on studies with the polarizing and electron microscopes and chemical analysis of isolated grana. The chlorophyll molecules are thought to be located on one side of a series of unit membranes. (*after Hodge*)

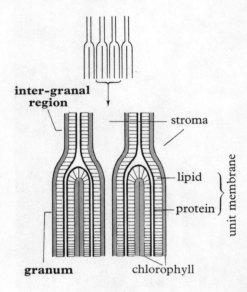

inter-granal region

stroma

lipid

protein

unit membrane

granum

chlorophyll

Fig 10.18 Electron micrograph showing quantasomes on the surface of lamellae of a spinach chloroplast. (*Daniel Branton, University of California*)

any rate) takes place in or on them. This and other evidence suggests that the chlorophyll molecules are located on the surface of the membranes (Fig 10.17). This idea is supported by high-resolution electron micrographs of ruptured grana which show that in these regions there are numerous granules attached to the surface of the membranes (Fig 10.18). These are called **quantasomes** and are believed to be rich in chlorophyll: it has been estimated that each quantasome contains 230 chlorophyll molecules. They may therefore represent the basic structural units of photosynthesis.

Our next problem is to understand where the chloroplasts are in relation to the whole plant and how light, carbon dioxide and water get to them. In some primitive plants like *Spirogyra* mentioned earlier, all the cells contain chloroplasts and carry out photosynthesis. But in higher plants structures specialized for photosynthesis have been developed. These are the leaves whose structure we must now consider.

STRUCTURE OF THE LEAF
Leaves are generally thin and flat and collectively present a large surface area

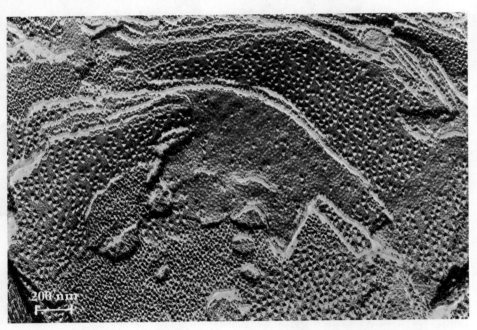

200 nm

to the light. Their thinness minimizes the distance over which diffusion of carbon dioxide has to take place. Of course being thin and flat makes them liable to sag, but their shape is maintained by the turgidity of the living cells and by the **midrib** and **veins** which are well endowed with strengthening tissue. Their large surface area also increases the chances of excessive water-loss but this is prevented by the **cuticle** (see page 230).

Most of the salient features of a flowering plant leaf (dicotyledonous type) are shown in Figs 10.19 and 10.20. You will see that the leaf is covered by a layer of cuticularized **epidermal cells**, the cuticle being generally thicker on the upper surface than the lower. The inside of the leaf is filled with cells which contain chloroplasts. These cells are of two types. Those immediately beneath the upper epidermis, called the **palisade cells**, are elongated with their long axes perpendicular to the surface. They are separated from each other by narrow air spaces and are densely packed with chloroplasts. The

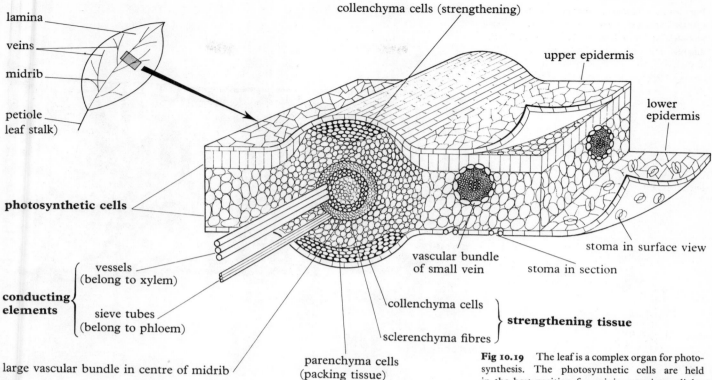

lamina
veins
midrib
petiole
(leaf stalk)

collenchyma cells (strengthening)

upper epidermis

lower epidermis

photosynthetic cells

stoma in surface view

stoma in section

vascular bundle
of small vein

**conducting
elements** {
vessels
(belong to xylem)

sieve tubes
(belong to phloem)

collenchyma cells

sclerenchyma fibres
} **strengthening tissue**

large vascular bundle in centre of midrib

parenchyma cells
(packing tissue)

Fig 10.19 The leaf is a complex organ for photosynthesis. The photosynthetic cells are held in the best position for gaining maximum light. Strengthening tissue maintains the shape of the leaf. Stomata allow the entry of carbon dioxide whilst the cuticularized epidermis prevents excessive water loss. Conducting elements in the midrib and veins bring water and mineral salts to the leaf, and remove the products of photosynthesis from it. *(based mainly on Brown)*

latter tend to arrange themselves in the part of the cell which receives maximum illumination, usually the upper part. The palisade cells collectively form the **palisade layer** which may be one or several cells in thickness.

Filling the leaf between the palisade layer and the lower epidermis are the **spongy mesophyll cells**. Irregular in shape and arrangement, they also contain chloroplasts but fewer than the palisade cells, which is why the underside of a leaf usually looks paler than the upper side. Their cellulose walls

Fig 10.20 *(below)* Photosynthetic cells as seen in a transverse section of a leaf. Note the intercellular air spaces allowing free diffusion of carbon dioxide, and the close proximity of the photosynthetic cells to the vascular tissues.

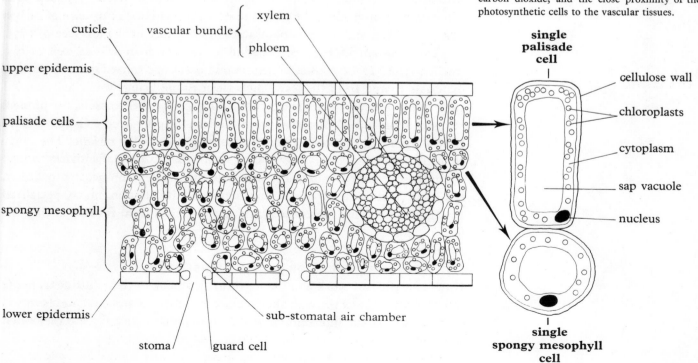

cuticle

vascular bundle {
xylem
phloem
}

upper epidermis

palisade cells

spongy mesophyll

lower epidermis

sub-stomatal air chamber

stoma

guard cell

**single
palisade
cell**

cellulose wall

chloroplasts

cytoplasm

sap vacuole

nucleus

**single
spongy mesophyll
cell**

Fig 10.21 Photomicrograph of the photosynthetic part of a leaf as seen in a transverse section. Can you relate the structures seen to the diagram in Fig 10.20? *(J. F. Crane)*

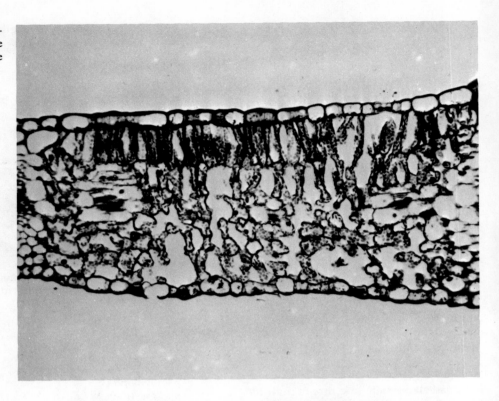

are thin and, assuming that the plant is well supplied with water, they are permanently saturated with moisture. Between the spongy mesophyll cells are large **air spaces** which communicate with each other and with those between the palisade cells. This system of air spaces allows gases to diffuse freely between the cells.

The lower epidermis is pierced by numerous pores, or **stomata** (sing. **stoma**), as many as 40,000 per cm^2 in some leaves. The upper epidermis may have some too but they are usually fewer than in the lower epidermis. Each with the intercellular air spaces described above. Bordered by **guard cells** which can open or close the pore, the stomata regulate the passage of carbon dioxide, oxygen and water vapour across the surface of the leaf (see p. 177).

The diagram in Fig 10.19 includes the central midrib as well as two smaller veins. These consist of conducting tissue specialized for transporting materials to and from the leaf. The **xylem tissue** brings water and mineral salts from the roots in elongated conducting tubes called **vessels**, the **phloem** removes soluble food materials from the leaf in specialized cells called **sieve tubes**. Xylem and phloem together constitute a **vascular bundle**. The xylem elements, being lignified (woody), also provide the flexible leaf with mechanical strength. The midrib is basically similar to the smaller veins except that there is a great abundance of conducting tissue and, in addition, much specialized strengthening tissue (**sclerenchyma** and **collenchyma**) for supporting the leaf.

THE LEAF AS AN ORGAN OF PHOTOSYNTHESIS

How does the leaf work as a photosynthetic organ? Carbon dioxide from the atmosphere diffuses through the stomata into the sub-stomatal air chambers and thence via the intercellular spaces to the chloroplasts in the mesophyll and palisade cells.

Water, drawn up to the leaves from the soil via the conducting tissues of the roots and stem, passes out of the xylem elements in the veins to the surrounding cells. Maintenance of this flow is discussed in Chapter 12. Its importance in the present context is that it supplies water for photosynthesis. With the water come mineral salts (nitrates, sulphates and phosphates) required for the synthesis of proteins. Oxygen and excess water vapour diffuse out of the leaf via the intercellular air spaces and stomata. Sugar and other products of photosynthesis are moved to other parts of the plant in the sieve tubes.

It is clear from what has been said that the stomata have a vital part to play in photosynthesis. When they are open the rate of photosynthesis may be ten or twenty times as fast as the maximum rate of respiration. Under these circumstances the plant will use the carbon dioxide from its respiration for photosynthesis but the bulk of its carbon dioxide must be brought in from the atmosphere. If the stomata are closed photosynthesis can still continue, using the carbon dioxide from respiration. In fact an equilibrium can be reached between photosynthesis and respiration, photosynthesis using carbon dioxide from respiration, and respiration using oxygen from photosynthesis. However, the rate of photosynthesis under these circumstances will be much slower than when an external source of carbon dioxide is available. The stomata cannot remain closed indefinitely, for open stomata are necessary in order to maintain the transpiration stream, which is the only way the leaf can obtain water (see Chapter 12).

The Chemistry of Photosynthesis

In the process of photosynthesis energy from sunlight is somehow trapped by chlorophyll and used in the manufacture of carbohydrate from carbon dioxide and water. The process can be summarized by the following equation:

$$CO_2 + H_2O \xrightarrow[\text{Chlorophyll}]{\text{Light energy}} CH_2O + O_2 \uparrow$$

Carbon dioxide   Water   Carbo-hydrate   Oxygen

Though useful as an overall summary of the process, this equation is a gross over-simplification, and indeed misleading because it gives the impression that the oxygen evolved comes from the carbon dioxide, which we now know is not true. It also suggests that photosynthesis takes place in one step whereas it occurs in two distinct stages each consisting of many steps.

PHOTOSYNTHESIS AS A TWO-STAGE PROCESS

If photosynthesis consisted only of photochemical reactions one would not expect the process to be influenced by temperature since photochemical reactions are temperature-insensitive. But in fact the rate of photosynthesis is strongly influenced by temperature, a $10°C$ rise in temperature approximately doubling the rate. This is typical of ordinary chemical reactions, and it would therefore seem that photosynthesis proceeds in two stages, one requiring light (**light stage**) and the other capable of occurring in the dark (**dark stage**).

Further evidence for this comes from the fact that the amount of carbo-hydrate formed is greater if a plant is provided with alternate light and dark periods than if it is given the same total amount of light continuously. It is envisaged that the light stage results in a product which is then fed into the dark stage thus:

$$A \longrightarrow B \text{ (Light stage)}$$
$$B \longrightarrow C \text{ (Dark stage)}$$

We may suppose that B builds up during a period of illumination and cannot be used up by the dark reactions as quickly as it is formed. Intermittent light ensures the complete conversion of B into C, and prevents B from accumulating in the plant.

We now know that in the light stage chlorophyll traps the radiant energy of sunlight and converts it into chemical energy. This is used for splitting water molecules into hydrogen and oxygen. The oxygen is evolved as oxygen gas and the hydrogen enters the dark stage in which carbon dioxide is reduced to form carbohydrate. These events are summarized in Fig 10.22. Let us now take each stage in turn and examine it in detail. Fig 10.23 should be consulted as you read the following account.

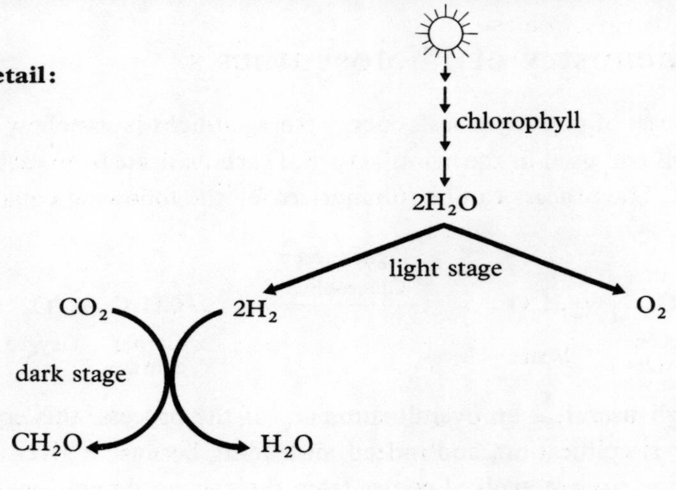

Fig 10.22 Summary of photosynthesis. The process consists of two stages, light and dark. The light stage involves the photochemical splitting of water. In the dark stage the hydrogen atoms formed in the light stage are used to reduce carbon dioxide with the formation of carbohydrate. This general scheme has been confirmed by the use of radioactive tracers. Note that the oxygen gas given off in photosynthesis comes from the water.

overall reaction:

$$CO_2 \quad + \quad 2H_2O \quad \xrightarrow[\text{chlorophyll}]{\text{light energy}} \quad CH_2O \quad + \quad H_2O \quad + \quad O_2 \uparrow$$

in more detail:

THE LIGHT STAGE

That the light stage involves the photochemical splitting of water has been ascertained by the use of isotope labelling (see p. 82). If the unicellular alga *Chlorella* is placed in water whose oxygen atom is replaced by the heavy isotope of oxygen ^{18}O, it is found by subsequent testing with a mass spectrometer that ^{18}O is present in the oxygen given off by the plant. If, however, the plant is given normal water but the carbon dioxide is labelled with ^{18}O, the oxygen liberated contains no ^{18}O, thus confirming that the oxygen formed in photosynthesis comes only from the water.

But the light stage involves more than just the splitting of water. It has been shown that isolated chloroplasts, as well as splitting water into hydrogen and oxygen, are capable of producing ATP.

The function of the light stage is therefore twofold:

(1) By the photochemical splitting of water it provides hydrogen atoms for the reduction of carbon dioxide; and

(2) by producing ATP it provides a source of chemical energy for the subsequent synthesis of carbohydrates.

How is the ATP produced? This is where the chlorophyll comes in. It will help us to understand this if we first recall how ATP is formed in cell respiration (see p. 103). In that process electrons are removed from organic compounds intermediate in the breakdown of sugar and transferred via a series of electron carriers to oxygen with the formation of water. The organic intermediates are electron donors and oxygen is the ultimate electron acceptor. As the electrons are handed from one carrier to the next, energy is released which is used for the synthesis of ATP. You will remember that this process is called oxidative phosphorylation on account of the fact that oxidation is involved and high-energy phosphate bonds are established.

Now ATP can be synthesized in the light stage of photosynthesis in essentially the same way. The main difference is that in photosynthesis chlorophyll functions as the donor and ultimate acceptor of electrons. What happens is something like this. When light strikes the chlorophyll molecule the energy level of an electron is raised. In this 'excited' state the electron is emitted and taken up by an electron carrier. The electron is then passed through a series of further carriers, the last one handing it back to the chlorophyll molecule. In losing an electron the chlorophyll becomes positively charged and unstable, but neutrality and stability are restored when the electron is returned. As in oxidative phosphorylation the electron carriers are at different energy levels, and as the electron is transferred from one carrier to the next energy is removed from it for the synthesis of ATP. Typically two molecules of ATP are formed every time a single electron goes through the cycle.

It goes without saying that for all this to happen ADP and inorganic phosphate must be supplied and the appropriate carriers must be present. If the carriers are not present no chemical energy can be obtained from the electron. Under these circumstances electrons emitted from the chlorophyll are returned to it unchanged, energy being released as heat and fluorescence. Pure chlorophyll from which the carriers have been removed will in fact fluoresce for just this reason.

The cyclical process outlined here is known as **photosynthetic phosphorylation** (or **photophosphorylation**) to distinguish it from oxidative phosphorylation which takes place during cell respiration. Both are concerned with establishing high-energy phosphate bonds but the source of energy is different in the two cases. In oxidative phosphorylation energy is derived from organic compounds. In photosynthetic phosphorylation it comes from sunlight via chlorophyll.

Let us turn now to the other function of the light stage of photosynthesis, the splitting of water. There is considerable argument over the details of how this is achieved but the main idea is reasonably clear. The process depends on the fact that water has a tendency to dissociate into hydrogen (H^+) and hydroxyl (OH^-) ions. This means that in a cell there will always be a certain number of these ions present. We have already seen that when light strikes chlorophyll an electron is emitted. Now not all the electrons are returned via the electron carrier system to chlorophyll. Sometimes an electron combines

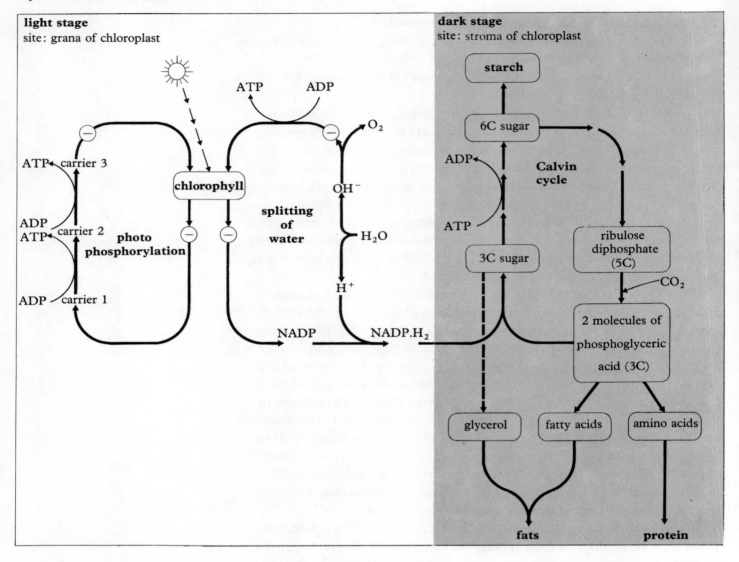

Fig 10.23 Summary of the chemistry of photosynthesis. See text for explanation.

with a hydrogen ion to form a hydrogen atom. The hydrogen atom is taken up by a hydrogen acceptor, **nicotinamide adenine dinucleotide phosphate** (**NADP**), which is thus reduced. The reduced NADP now enters the dark stage of photosynthesis, handing on the hydrogen which is then used in the reduction of carbon dioxide.

We will return to the dark reactions in a moment, but in the meantime the chlorophyll molecule has been deprived of an electron and is in an unstable state. How is its stability restored? This is where the hydroxyl ions come in. The hydroxyl ion donates an electron to chlorophyll and the OH resulting from this forms water and oxygen:

$$OH^- + \text{Chlorophyll}^+ \longrightarrow \text{Chlorophyll} + OH$$
$$4\,(OH) \longrightarrow 2\,H_2O + O_2\uparrow$$

By acting as an electron donor the hydroxyl ion restores the stability of the chlorophyll molecule. The oxygen is, of course, the oxygen gas which is given off in photosynthesis.

This is certainly not the whole story. Recent research has shown that ATP can be formed in the above process (photophosphorylation again), and the rôle of the different chlorophyll pigments in the transfer of electrons has been clarified. It appears that electrons are moved from one pigment to another, with the formation of ATP, and then to NADP. Chlorophyll a is the final pigment in the series, receiving electrons from other pigments and handing them on via electron carriers to NADP. However, the details are far from fully understood, largely because it is difficult to detect changes in the degree of oxidation and reduction in the presence of vast quantities of chlorophyll. You should refer to recent articles in scientific periodicals for a discussion of the more controversial issues.

THE DARK STAGE

Essentially the dark reactions involve the reduction of carbon dioxide to form carbohydrate. This is an endergonic process requiring energy. The energy is supplied by the splitting of the ATP formed in the light stage. The hydrogen for reducing the carbon dioxide is provided by the reduced NADP, also formed in the light stage.

The reduction of carbon dioxide, and subsequent synthesis of carbohydrates, takes place in a series of small steps, each controlled by a specific enzyme. The individual steps have been analyzed by Melvin Calvin and his associates at the University of California. They have done this by illuminating the unicellular alga *Chlorella* in the presence of carbon dioxide labelled with the radioactive isotope of carbon, ^{14}C (Fig 10.24). The plant is allowed to photosynthesize for a certain period of time after being given the labelled carbon dioxide. It is then quickly killed with boiling alcohol which inactivates all the enzymes and stops the reactions instantaneously. The radioactive compounds which have been formed are then extracted from the plant and separated by paper chromatography. Autoradiographs are made and the radioactive substances identified. The plants are killed at intervals after the initial fixation of the carbon dioxide, from a few seconds to several minutes. By identifying the intermediates formed after different periods of time the pathway through which carbon compounds are built up can be established.

From these investigations it has emerged that the chain of reactions is cyclical, known with justification as the **Calvin Cycle**. In the first step the carbon dioxide combines with a 5-carbon organic compound called **ribulose diphosphate**. This serves as a carbon dioxide acceptor and is responsible for fixing the carbon dioxide, i.e. incorporating it into the photosynthetic machinery of the plant. The combination of carbon dioxide with ribulose diphosphate gives an unstable 6-carbon compound which splits immediately into two molecules of a 3-carbon compound, **phosphoglyceric acid** (PGA). The next step is crucial: the PGA is reduced to form a 3-carbon sugar, **triose phosphate**. The hydrogen for this reduction comes from reduced NADP and the energy (for it is an endergonic reaction) from ATP. The triose phosphate is now built up through a series of intermediate steps to a 6-carbon sugar which can be converted into starch for storage. The necessary energy again comes from ATP.

Not all the 6-carbon sugar is converted into starch. Some of it enters a series of reactions which results in the regeneration of ribulose diphosphate. This is very important because only by ensuring a supply of ribulose diphosphate can the continued fixation of carbon dioxide take place.

That starch is not the only end-product of photosynthesis is shown by the fact that in Calvin's experiments radioactive carbon was eventually iden-

A

B

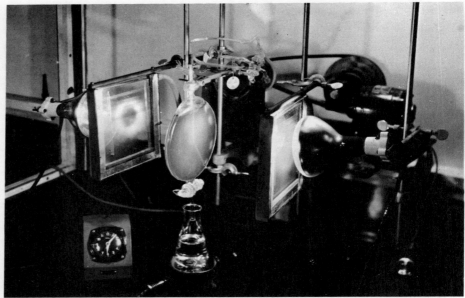

Fig 10.24 A) Melvin Calvin. B) Calvin's 'lollipop' apparatus. The 'lollipop' refers to the thin transparent vessel containing a suspension of algae. Carbon dioxide labelled with radioactive ^{14}C is bubbled through the suspension, and subsequent analysis of the radioactive compounds formed enables the path taken by carbon in photosynthesis to be traced. (*Lawrence Radiation Laboratory, University of California*)

tified in lipids and amino acids as well. These are the products of metabolic pathways leading from PGA. PGA is thus a kind of crossroads in photosynthesis, somewhat comparable to acetyl coenzyme A in respiration. From it carbohydrates, fats and proteins can ultimately be formed, depending on the requirements of the plant at the time.

THE EXACT SITES OF THE LIGHT AND DARK STAGES
We have seen that photosynthesis occurs in the chloroplasts but can we now be more specific about where the different stages take place?

By means of the high-speed centrifuge it has proved possible to separate the components of the chloroplast right down to the quantasomes. It has been shown that the quantasome is the smallest unit which can perform the light reactions. Soluble stroma material can be separated from the rest of the chloroplast structures. This is incapable of photochemical reactions but contains the enzymes necessary for reducing carbon dioxide and will do so if supplied with ATP and reduced NADP. So it appears that the light reactions are associated with the quantasomes and the dark reactions with the stroma. Not surprisingly when small fragments of lamella are added to soluble stroma material, the resulting mixture will both split water and reduce carbon dioxide.

Autotrophic Bacteria

Although in this account of photosynthesis we have concentrated on higher plants, most of what has been said about the basic chemical events applies to all plants. This is not however true of bacteria. Autotrophic bacteria are divided into two groups, **photosynthetic** and **chemosynthetic**. Both can build up organic compounds from simple inorganic raw materials. They differ in the way they obtain the necessary energy. Let us look at each in turn.

PHOTOSYNTHETIC BACTERIA
Like green plants these bacteria are able to build up carbon dioxide and water

into organic compounds using energy derived from sunlight. The energy is trapped by a pigment called **bacteriochlorophyll** which is similar to, though somewhat simpler than, chlorophyll. They differ from green plants in their source of hydrogen for reducing the carbon dioxide. Instead of obtaining it from water they get it from hydrogen sulphide, for which reason they are known as **sulphur bacteria**. They generally live in organic mud at the bottom of lakes and ponds where there is no shortage of hydrogen sulphide. The residual sulphur resulting from the splitting of hydrogen sulphide is deposited in the bacterial cells.

$$6CO_2 + 12H_2S \xrightarrow[\text{Bacteriochlorophyll}]{\text{Light}} C_6H_{12}O_6 + 12S + 6H_2O$$

Bacteriochlorophyll comes in two closely related forms, green and purple, giving the so-called green and purple sulphur bacteria respectively. The absorption spectrum of the purple form is included in Fig 10.13, from which it will be seen that it can absorb infra-red light.

CHEMOSYNTHETIC BACTERIA

The chemosynthetic bacteria can also synthesize organic from inorganic materials but instead of using sunlight they obtain the necessary energy from special chemical processes involving the oxidation of compounds other than sugar. Thus **iron bacteria**, living in streams that run over iron-containing rocks, oxidize ferrous salts to ferric. The **colourless sulphur bacteria** (not to be confused with the green and purple bacteria discussed in the last section) live in decaying organic matter and oxidize hydrogen sulphide to water and sulphur. There are even **hydrogen bacteria** which can oxidize hydrogen with the formation of water.

A particularly important group of chemosynthetic organisms are the **nitrifying bacteria** found in the soil. Through their metabolic activities they enrich the soil in available nitrogen, i.e. nitrogen in a form obtainable by plants. Some of these bacteria, specifically *Nitrosomonas* and *Nitrococcus*, obtain energy by oxidizing ammonia (formed by the breakdown of animal and plant proteins during decay) to nitrite. The conversion involves several steps. As soon as it is liberated into the soil the ammonia combines with carbon dioxide to form ammonium carbonate. This is then converted to nitrous acid under the influence of the bacteria referred to above:

$$(NH_4)_2CO_3 + 3O_2 \longrightarrow 2HNO_2 + CO_2 + 3H_2O + \text{Energy}$$

The nitrous acid immediately combines with, for example, calcium or magnesium salts to form the appropriate nitrite.

Another nitrifying bacterium, *Nitrobacter*, oxidizes nitrites to nitrates:

$$2Ca(NO_2)_2 + 2O_2 \longrightarrow 2Ca(NO_3)_2 + \text{Energy}$$

In all these cases the energy released is used for the synthesis of organic compounds.

These nitrifying bacteria do not exist in isolation but form part of a natural system in which nitrogen compounds are converted from one form to another step by step. Thus the ammonia released from the dead bodies of animals and plants by the activities of saprophytic bacteria and fungi is con-

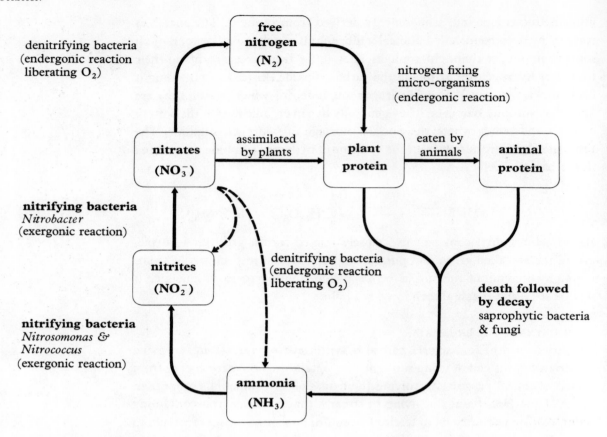

Fig 10.25 Scheme summarizing the cycling of nitrogen compounds in nature. Explanation in text.

verted by *Nitrosomonas* and *Nitrococcus* into nitrites. *Nitrobacter* then converts the nitrites into nitrates which are absorbed by plants. The full sequence of conversions constitutes the **nitrogen cycle** summarized in Fig 10.25. It shows how nitrogen compounds circulate in nature and demonstrates the inter-dependence of animals, plants and bacteria.

Looking at Fig 10.25 you will notice that certain bacteria convert nitrates, the only form of nitrogen which is directly available to plants, into nitrites, ammonia or even nitrogen. As these organisms deprive the soil of available nitrogen compounds they are known as **denitrifying bacteria**. The reason why they do this is of great interest. These bacteria tend to live in conditions of oxygen shortage, and in order to correct this deficiency they reduce nitrates to nitrites, ammonia, or nitrogen, thereby liberating oxygen. This reduction is an endergonic process, the energy coming from the anaerobic breakdown of carbohydrate. The oxygen liberated is then used for the aerobic breakdown of sugar, and the energy released is used for the synthesis of organic com-pounds – a roundabout process perhaps but a perfectly good way of obtaining energy.

FIXATION OF NITROGEN

An account of autotrophic nutrition would be incomplete without a reference to nitrogen-fixation, a process carried out by certain bacteria and primitive plants. These remarkable organisms are able to absorb atmospheric nitrogen and build it up into protein, a synthetic feat that commands our admiration. How they fit into the nitrogen cycle is shown in Fig 10.25.

The best known **nitrogen-fixing organisms** are bacteria that live in the roots of leguminous plants such as peas, beans and clover. It has long been known that such plants are able to thrive in soils deficient in nitrates and we now know that they owe this ability to the nitrogen-fixing bacteria in their roots. The bacteria enter the young plant through its root hairs, and they cause the cortical cells of the root to proliferate forming a swelling or **root nodule** (Fig 10.26). In the cells of the nodule the bacteria multiply rapidly, fixing atmospheric nitrogen and building it up into proteins.

It has been shown that the association between the bacteria and their host is mutually beneficial, both partners gaining from the relationship. Some of the products of the bacteria's nitrogen fixation pass into the host plant and are utilized by it: this has been confirmed by supplying an infected plant with the heavy isotope of nitrogen ^{15}N. Eventually the ^{15}N gets into the whole of the plant, not just the part containing the bacteria. The beneficial effect of the bacteria on the host is admirably illustrated in Fig 10.27.

What do the bacteria gain from the host apart from protection? From the host's photosynthesis they obtain carbohydrates which provide a source of carbon for the synthesis of proteins as well as energy for driving the endergonic reactions involved. An association of this sort, in which both partners benefit, is called **symbiosis** and is discussed in more detail in Chapter 33.

This is not to say that all organisms capable of fixing atmospheric nitrogen are found living inside the tissues of another organism. Many free-living examples are also known. But it is commonly associated with symbiosis, and far more plants harbour nitrogen-fixing micro-organisms than was originally thought to be the case.

Fig 10.26 Nodules in the root system of a 3-month old tree. The seeds were inoculated with an effective strain of nitrogen-fixing bacteria. (*O. N. Allen, University of Wisconsin*)

NOT INOCULATED   INOCULATED

Fig 10.27 The beneficial effects of nitrogen-fixing bacteria can be seen by comparing these two clover plants. The seed from which the right hand plant was grown was inoculated with an effective strain of nitrogen-fixing bacteria, whereas the one on the left received no such treatment. Both plants were planted at the same time and maintained under the same conditions. (*O. N. Allen, University of Wisconsin*)

11 Transport in Animals

In previous chapters we have seen how specialized surfaces are responsible for the absorption of oxygen and soluble food substances in animals. Once absorbed these materials must be distributed to the various tissues, some of them far removed from the absorptive surface. In small animals, such as protozoons, *Hydra* and flatworms, movement of respiratory gases and food materials takes place by diffusion. However, in larger animals the tissues are too bulky for diffusion alone to supply the body's needs, for which reason a transport system is necessary.

Transport systems range from ciliated water-filled canals in jellyfishes to the sophisticated **blood system** of the mammal. Between these two extremes all grades of complexity are found. Blood systems are present in all vertebrates and certain invertebrates, and in all cases they consist of a system of **blood vessels** containing **blood** which circulates round the body. This is why the blood system is also described as a **circulatory** or **blood vascular system**. (Vascular is derived from the Latin word *vasculum* meaning vessel.) The vessels need not necessarily be tubular as in ourselves: in some cases the blood flows through large cavities or **sinuses**. Generally the blood is circulated by muscular contractions of the vessels or a **heart**, the latter being essentially a greatly expanded and specialized blood vessel with a thick muscular wall.

In this chapter we shall be mainly concerned with transport in the mammal, considering first the composition and functions of blood, and then the structure of the circulatory system and the mechanism by which blood is conveyed through it.

Blood

Blood is a highly specialized tissue consisting of several types of cell suspended in a fluid medium called **plasma** (Fig 11.1). The cellular constituents consist of **red blood cells (erythrocytes)**, which carry respiratory gases; **white blood cells (leucocytes)**, which fight disease; and **platelets**, cell fragments which play an important part in the clotting of blood. So blood has a varied structure and performs a wide range of functions. So far as transport is concerned, its two important constituents are the red blood cells and plasma.

PLASMA

Plasma is mainly water containing a variety of dissolved substances which

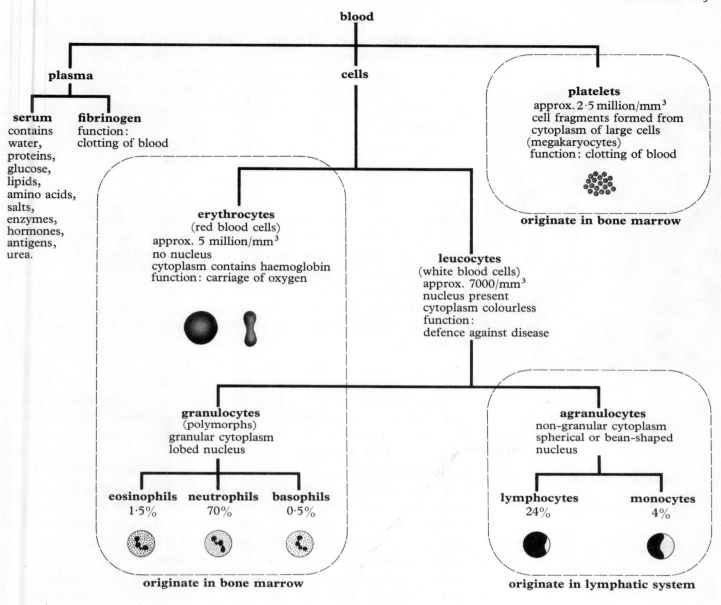

Fig 11.1 Chart summarizing the constituents of mammalian blood.

are transported from one part of the body to another. Thus food materials (glucose, lipids and amino acids) are conveyed from the small intestine to the liver; urea from the liver to the kidneys; hormones from various ductless glands to their target organs, and so on. Cells are constantly shedding things into the blood which flows past them, and removing things from it. Plasma provides the medium through which this continual exchange takes place.

RED BLOOD CELLS

The prime function of red blood cells is to carry oxygen from the respiratory organ to the tissues and their structure is modified accordingly. If you look at some of your own blood on a slide you will see that the red cells are small and numerous. There are approximately five million per cubic millimetre, each being about 8 μm in diameter and 3 μm thick in the widest part. There is no nucleus with the result that the cell is sunk in on each side giving it the shape of a **biconcave disc** (Fig 11.2). Surrounded by a thin elastic

Fig 11.2 Scanning electron micrograph of red blood cells showing their biconcave surfaces. The round white structure near the centre is a platelet. (*J. A. Clarke, St Bartholomew's Hospital, London*)

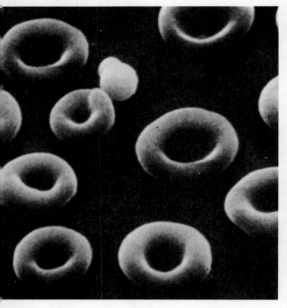

Fig 11.3 The chemical nature of haemoglobin A) Simple representation of human adult haemoglobin. The **H**'s stand for four haem groups, each of which contains an iron atom. α and β stand for two types of polypeptide chain each consisting of about 140 amino acids. Each chain is associated with one of the four haem groups. B) The structure of haem, the prosthetic group of haemoglobin. Haem, a porphyrin, is the part of the haemoglobin molecule that combines reversibly with oxygen to form oxyhaemoglobin. Note that the iron at the centre of the haem is in the divalent, i.e. ferrous, state. It remains in this state while oxygen is being transported from the lungs to the tissues.

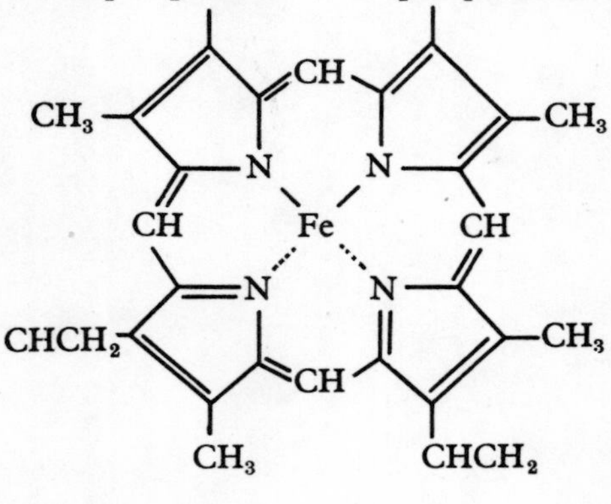

A

B

membrane, the whole of the interior of the cell is filled with the red pigment **haemoglobin**, a complex protein containing four iron haem groups (Fig 11.3). The red blood cell has a limited life span of about 120 days. Because of this the bone marrow manufactures new ones at the rate of about $1\frac{1}{2}$ million per second to replace those destroyed. Although the lack of a nucleus means a limited life span, it does permit more haemoglobin to be packed into the cell. It has been calculated that for its size the biconcave disc provides the highest possible surface : volume ratio for the absorption of oxygen.

CARRIAGE OF OXYGEN

Oxygen diffuses into the red blood cell across its thin membrane and combines with the haemoglobin to form **oxyhaemoglobin**. Each of the four iron-containing haem groups in the haemoglobin molecule can combine with a molecule of oxygen. The attachment of the oxygen does not involve chemical oxidation of the iron which remains in the ferrous state throughout the process. The union is a loose one, the oxygen molecules being quickly attached to the haemoglobin in the lungs and equally readily detached in the tissues. Thus:

$$Hb + 4O_2 \underset{tissues}{\overset{lungs}{\rightleftarrows}} HbO_8$$
$$\text{Haemoglobin} \qquad\qquad \text{Oxyhaemoglobin}$$

The ability of the blood to transport oxygen at a speed commensurate with the needs of the body is largely caused by the high affinity of haemoglobin for oxygen. This can be demonstrated experimentally by subjecting samples of blood to different oxygen tensions, and estimating the percentage saturation of the blood with oxygen in each case. In practice the blood samples are placed in a series of cylindrical glass containers into which air mixtures of known oxygen tension are introduced. Each sample of blood is given time to come into equilibrium with the air mixture, and then its percentage saturation with oxygen is determined. If the percentage saturation is then plotted against the oxygen tension, an **oxygen dissociation curve** is obtained. You will notice from Fig 11.4 that the curve is S-shaped. This is a clear indication of the efficiency of haemoglobin at taking up oxygen. To understand this, consider the situation when the oxygen tension is, say, $8\cdot0$ kN/m². At this tension the percentage saturation of the blood is over 90 per cent. If, however, the relationship between oxygen tension and percentage saturation was a linear one, as shown by the hypothetical broken line in Fig 11.4, the percentage saturation at the same oxygen tension would be only 60 per cent. In other words the S-shaped curve indicates that the blood can become fully saturated at relatively low oxygen tensions. This is what we mean when we say that haemoglobin has a high affinity for oxygen.

The oxygen dissociation curve, as well as facilitating the loading of haemoglobin with oxygen, also facilitates unloading. The steep part of the curve corresponds to the range of oxygen tensions found in the tissues. Over this part of the curve, a small drop in oxygen tension will bring about a comparatively large fall in the percentage saturation of the blood. In other words, if the oxygen tension falls as a result of the tissues utilizing oxygen at a faster rate, the haemoglobin responds by giving up more of its oxygen. The oxygen dissociation curve is therefore ideally suited to taking up oxygen in the lungs and releasing it in the tissues.

In the experiment just described all factors apart from oxygen tension were kept constant. Consider now what happens if the experiment is repeated at three different carbon dioxide tensions. The results are shown in Fig 11.5, curves A to C. You will notice that increasing the carbon dioxide tension has the effect of shifting the oxygen-dissociation curve towards the right. This is called the **Bohr effect** after the man who first discovered it. Under these circumstances the haemoglobin must be exposed to a higher oxygen tension in order to become fully saturated. But equally it will tend to release its oxygen at higher oxygen tensions. In other words at high carbon dioxide tensions the haemoglobin is less efficient at taking up oxygen, but more efficient at releasing it. The release of oxygen is therefore favoured in the tissues where the carbon dioxide tension tends naturally to be high as a result of its continual release from the respiring cells. On the other hand in the lungs the carbon dioxide tension is lower owing to its continual escape into the atmosphere, and this favours oxygen uptake.

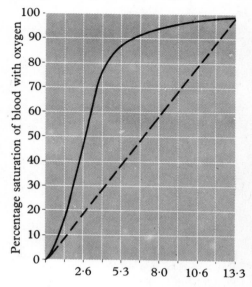

Fig 11.4 Oxygen dissociation curve for human haemoglobin. The sigmoid shape is indicative of haemoglobin's high affinity for oxygen.

MYOGLOBIN AND OTHER BLOOD PIGMENTS

From the discussion so far the fact emerges that the further an oxygen dissociation curve is to the left, the more tenaciously the pigment absorbs and holds onto its oxygen. There are certain types of blood pigment which readily take up oxygen even when the oxygen tension is very low. Such is the case with **myoglobin**, which has an oxygen dissociation curve situated well to the left of haemoglobin (Fig 11.5, curve D). Closely related to haemoglobin chemically, myoglobin is found in muscles where it remains fully saturated at tensions well below that required for haemoglobin to give up its oxygen. Myoglobin stores oxygen, only releasing it when the oxygen tension falls very low, as in severe muscular exertion. Myoglobin is responsible for the colour of 'red muscles', and is particularly abundant in active animals which are liable to suffer from oxygen shortage.

Functionally similar to myoglobin is the haemoglobin of animals like the lugworm which burrow into oxygen-deficient mud. The oxygen dissociation curve of lugworm blood is situated well to the left of human blood, and reflects its unusually high affinity for oxygen at low tensions. It is interesting, though inexplicable, that the haemoglobins of different species vary widely in their affinities for oxygen, and may even vary during the life history of a single individual.

The haemoglobin of the human foetus (**foetal haemoglobin**) has an oxygen dissociation curve situated to the left of adult haemoglobin. The reason for this is that foetal blood has got to pick up oxygen from the mother's blood across the placenta, and this can only take place if the foetal haemoglobin has a higher affinity for oxygen than the mother's haemoglobin.

Several other blood pigments are found in the animal kingdom, differing mainly in the nature of their prosthetic groups. Thus **chlorocruorin** and **haemoerythrin** both contain iron, as do haemoglobin and myoglobin. **Haemocyanin**, however, contains copper. These blood pigments are confined to lower animals (worms, molluscs and the like) and in some cases they are dissolved in the plasma rather than enclosed within cells. The affinity of these pigments for oxygen is comparable with haemoglobin, though their total oxygen capacity is generally lower.

An unfortunate property of haemoglobin is that it combines even more

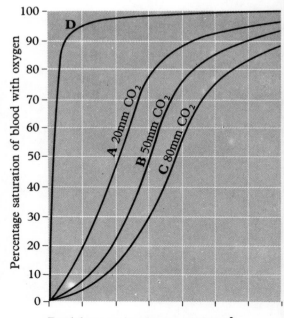

Fig 11.5 Curves **A**, **B** and **C** show the influence of three different carbon dioxide concentrations on the oxygen dissociation curve for haemoglobin. Notice that the effect of a high concentration of carbon dioxide is to shift the curve to the right, i.e. it lowers the affinity of haemoglobin for oxygen. Curve **D** is the oxygen dissociation curve for myoglobin, a pigment with an unusually high affinity for oxygen. Horizontal scale as in Fig 11.4 above.

readily with carbon monoxide than with oxygen. The result of this union is **carboxyhaemoglobin**. The carbon monoxide combines with the haemoglobin at the sites normally occupied by the oxygen molecules, thus preventing the latter from taking up their rightful position. This makes carbon monoxide a powerful respiratory poison.

The explanation of the Bohr effect, the shifting of the oxygen-dissociation curves by carbon dioxide, is to be found in the mechanism by which carbon dioxide is transported by the blood. To this topic we now turn.

CARRIAGE OF CARBON DIOXIDE

Carbon dioxide diffuses from the tissues into the red blood cells where it combines with water to form **carbonic acid**, H_2CO_3. This is normally a very slow reaction, but in the red blood cell it is greatly accelerated by the presence of the enzyme **carbonic anhydrase**. Because of this enzyme, most of the carbon dioxide is carried in the red blood cells rather than the plasma.

Fig 11.6 Summary of the main chemical events that take place in a red blood cell on reaching the tissues. The uptake of carbon dioxide results in the formation of hydrogen ions whose presence causes the oxyhaemoglobin to dissociate.

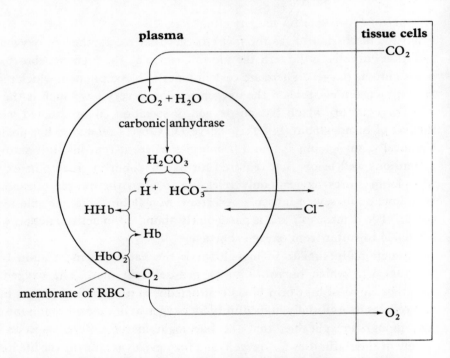

The carbonic acid then dissociates into bicarbonate and hydrogen ions. If the latter were allowed to accumulate they would raise the acidity of the cell and kill it. However, they are efficiently **buffered** by the haemoglobin itself. Their presence encourages the oxyhaemoglobin to dissociate into haemoglobin and oxygen. The latter diffuses out of the cell to the tissues, and the haemoglobin takes up the hydrogen ions forming a very weak acid, **haemoglobinic acid**, HHb. It is therefore clear that the Bohr effect is due not to carbon dioxide as such, but to the hydrogen ions resulting from its presence. These chemical events are summarized in Fig 11.6.

The carriage of carbon dioxide as just explained results in an accumulation of bicarbonate ions, upsetting the equilibrium conditions in the cell. As the membrane of the red blood cell is relatively impermeable to positive ions, equilibrium is restored by an outward movement of bicarbonate ions. Electroneutrality is maintained by an inward movement of chloride ions from the plasma, the so-called **chloride shift**. In this way the ionic and electrical conditions in the cell are kept on an even keel during the transport process.

Although most of the carbon dioxide is carried in this way, some of it combines with amino groups in the haemoglobin molecule, forming a **carbamino compound**, $HbCO_2$. A very small amount of carbon dioxide, probably not more than 5 per cent, never gets into the red blood cells at all but dissolves in the plasma to form carbonic acid. This dissociates, as in the red blood cell, into hydrogen and bicarbonate ions. In the absence of haemoglobin, the plasma proteins buffer the hydrogen ions, taking them up to form weak proteinic acids.

When the red blood cells reach the lungs they find themselves in a situation where the oxygen tension is high and the carbon dioxide tension low. With this sudden change in conditions all the reactions described above go into reverse. As a result oxygen is taken up by the red blood cells and carbon dioxide released.

The Mammalian Circulation

The general lay-out of the mammalian circulatory system is shown in Fig 11.7. Basically the muscular **heart** pumps the blood into a system of **arteries**, which split up within the tissues into **capillaries**. Here exchange of materials between blood and cells takes place. From the capillaries blood is collected up into a series of **veins** by which it is returned to the heart. The heart is divided into four chambers: **right** and **left atria** (**auricles**), and **right** and **left ventricles**. Blood returning to the heart enters the right atrium, whence it passes into the right ventricle, and then via the **pulmonary artery** to the lungs. This is deoxygenated blood, oxygen having been removed from it and carbon dioxide added to it during its passage through the tissues. As the blood flows through the capillaries in the lungs it sheds its carbon dioxide and takes up oxygen. The oxygenated blood now returns via the **pulmonary veins** to the heart, entering the left atrium. From this chamber it passes into the left ventricle, and thence to the **dorsal aorta**, the main artery of the body. From this numerous arteries, some median and some paired, convey blood to the capillary systems in the organs and tissues, where gaseous exchange takes place. Corresponding veins convey the deoxygenated blood to the **venae cavae** (**great veins**) by which it is returned to the right atrium. The walls of the arteries and veins are elastic, and the heart and veins are equipped with **valves** which prevent blood flowing in the wrong direction.

At one time it was thought that blood was pumped from the heart and subsequently drawn back into it in the same vessels, a sort of ebb-and-flow system. It is interesting to note that this kind of thing does happen in certain primitive animals, but not vertebrates. That the blood circulates was first discovered by the seventeenth century physician, William Harvey. By meticulous dissection and ingenious experiments, Harvey showed beyond all reasonable doubt that blood flows away from the heart in certain vessels (arteries), and returns to it in different vessels (veins). One of Harvey's experiments is shown in Fig 11.8 – a simple one to be sure, but a masterpiece of deductive reasoning[1].

[1] Although Harvey discovered the circulation, he was unable to demonstrate the existence of vessels connecting the arteries and veins. This was left to the Italian physiologist, Marcello Malpighi, who, towards the end of the seventeenth century, saw and described capillaries and demonstrated that they form the link between the arteries and veins.

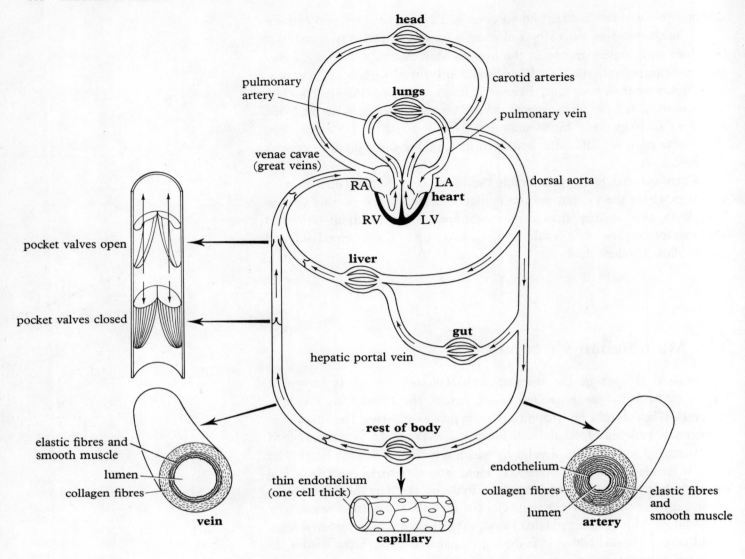

Fig 11.7 General plan of the mammalian circulation and the detailed structure of some of its component structures. The thick tough walls of the arteries, well endowed with connective tissue, are admirably suited to withstanding the pressures resulting from the pumping of the heart. The thinner-walled veins, equipped with pocket valves, minimize resistance to the flow of blood back to the heart after it has been through the narrow capillaries. The capillary walls are only one cell thick, thereby facilitating rapid diffusion of respiratory gases and soluble food materials between blood and tissues. **RA**, right atrium; **LA**, left atrium; **RV**, right ventricle; **LV**, left ventricle.

From a functional point of view the two most important parts of the circulatory system are the heart and capillaries. As the organ responsible for pumping the blood, the heart is of the utmost importance in maintaining the tissues in a state of health and efficiency. The capillaries represent the place where exchange of materials takes place, and as such provide the *raison d'etre* for the circulatory system. We will deal with these two parts of the circulation in turn.

THE HEART

The heart undergoes contraction (**systole**) and relaxation (**diastole**) rhythmically throughout the animal's life. In man its performance is prodigious: in the course of a normal life span it beats over 2.5×10^9 times, pumping more than 150 million dm^3 of blood from each ventricle. Although we shall mainly discuss the mammalian heart, it should be borne in mind that this basic function applies to any heart, be it of a whale, goldfish or lugworm.

Fig 11.9 shows the mammalian heart, and the passage of blood through it. Blood returning via the great veins enters the right atrium. The resulting pressure in this chamber forces open the flaps of the **atrio-ventricular valve** (also known as the **tricuspid valve** because it consists of three flaps) with the

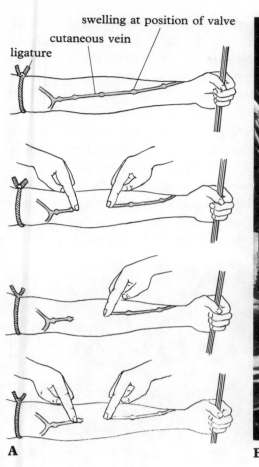

swelling at position of valve

cutaneous vein

ligature

A

B

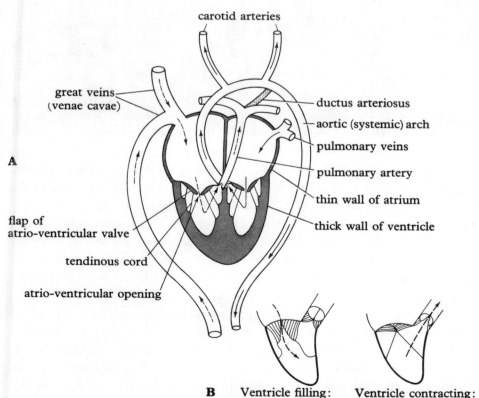

carotid arteries

great veins
(venae cavae)

ductus arteriosus

aortic (systemic) arch

pulmonary veins

pulmonary artery

thin wall of atrium

thick wall of ventricle

A

flap of
atrio-ventricular valve

tendinous cord

atrio-ventricular opening

B Ventricle filling:
atrio-ventricular
valve open;
pulmonary valve
closed.

Ventricle contracting:
atrio-ventricular
valve closed;
pulmonary valve
open.

Fig 11.8 (*above*) A) The technique used by William Harvey to demonstrate that blood flows towards the heart in the cutaneous vein of the arm. On tying a ligature round the upper arm, the valves in the veins show up as small swellings. If blood is pushed up to a point above one of these swellings, it fails to flow back even if pushed, indicating that its normal direction of flow must be towards the base of the arm. B) Harvey demonstrating his technique to a group of physicians in London. (*Parke, Davis and Co.*)

Fig 11.9 (*left*) Structure and action of the mammalian heart. A) Ventral view of heart and blood vessels connected to it. The flow of blood is indicated by arrows, broken for deoxygenated blood and solid for oxygenated blood. The four chambers of the heart are situated in relation to each other exactly as in Fig 11.7 on the opposite page. The atrio-ventricular valve on the right side of the heart consists of three flaps or cusps of which only two are shown (tricuspid valve); the atrio-ventricular valve on the left side consists of two flaps (bicuspid or mitral valve). The pulmonary valve guarding the entrance to the pulmonary artery, and the aortic valve at the entrance to the aortic arch, both consist of pocket-like flaps similar to the valves found in veins (see Fig 11.7). B) shows the action of the atrio-ventricular and pulmonary valves during filling and emptying of the right ventricle. The valves on the other side of the heart behave in the same way, the aortic valve opening when the left ventricle contracts.

result that blood flows through the atrio-ventricular opening into the right ventricle. When the atrium and ventricle are full of blood the atrium suddenly contracts, propelling the remaining blood into the ventricle. The contraction spreads from the right atrium over the rest of the heart. Atrial systole is comparatively weak but the ventricles, being particularly well endowed with muscle, contract much more powerfully. As a result blood is forced from the right ventricle into the pulmonary artery. It is prevented from flowing back into the atrium by the flaps of the atrio-ventricular valves, which close tightly over the atrio-ventricular aperture. They are prevented from turning inside out by tough strands of connective tissue, the **tendinous cords** (heart strings) which run from their undersides to the walls of the ventricle (Fig 11.10). Once in the pulmonary artery, blood is prevented from flowing back into the ventricle by **pocket valves** guarding the opening of the artery.

Fig 11.10 Atrio-ventricular valves and tendinous cords in sheep's heart. The tendinous cords prevent the valves being pushed inside out when the ventricle contracts. (*B. J. W. Heath, Marlborough College*)

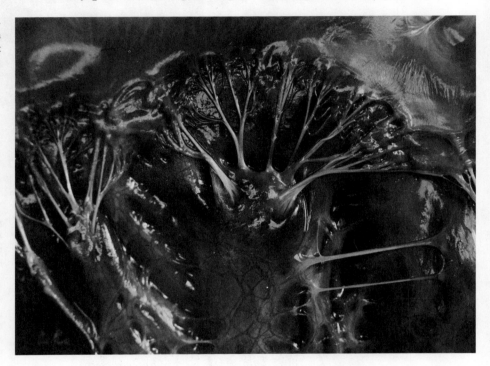

From the lungs oxygenated blood returns to the left atrium via the pulmonary veins. It is then conveyed to the left ventricle and so into the **aortic** (**systemic**) **arch**. This movement of blood takes place in the same way as on the right side of the heart. Although systole starts at the right atrium, it spreads quickly to the left so that the whole heart appears to contract synchronously. Thus deoxygenated blood is pumped from the right ventricle into the pulmonary artery at the same time as oxygenated blood is pumped from the left ventricle into the aortic arch.

Systole is followed by diastole during which the heart refills with blood again. The entire sequence of events is known as the **cardiac cycle**, and is accompanied by electrical activity in the wall of the heart and by 'sounds' corresponding to the closing of the various valves. The mammalian cardiac cycle is summarized in Fig 11.11, and it would be as well to ponder over this figure before reading further.

One of the most remarkable features of the heart is its ability to contract rhythmically without fatigue. It owes this property to its muscle. Known

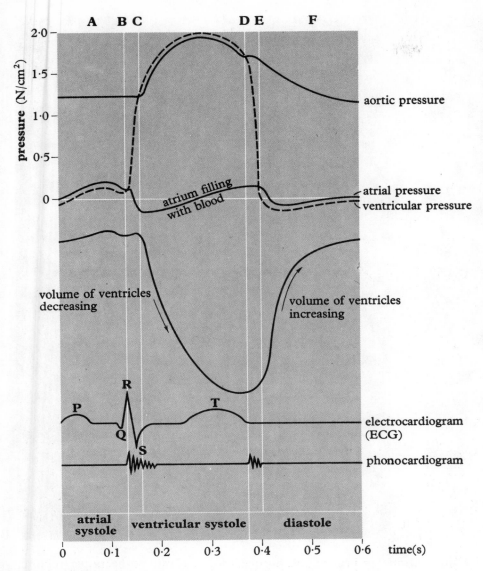

Fig 11.11 Graphs illustrating the pressure and volume changes that occur during the mammalian cardiac cycle (dog). Pressure changes were measured in the left atrium and ventricle, and the aorta. Volume changes were measured for both ventricles. The electrical activity in the heart wall (electrocardiogram) and heart sounds (phonocardiogram) as recorded in a human subject are also shown. The actions at different points on the graphs are as follows: A) atrium contracting: blood flows into ventricle.   B) ventricle starts to contract: ventricular pressure exceeds atrial pressure so atrio-ventricular valve closes.   C) ventricular pressure exceeds aortic pressure forcing open aortic valve; blood therefore flows from ventricle into aorta, and ventricular volume falls.   D) ventricular pressure falls below aortic pressure resulting in closure of aortic valve; ventricular volume starts to rise.   E) ventricular pressure falls below atrial pressure so blood flows from atrium to ventricle; ventricular volume rises rapidly.   F) atrium filling with blood from pulmonary vein; atrial pressure exceeds ventricular so blood flows from atrium to ventricle. Electrocardiogram (ECG): P wave corresponds to wave of excitation spreading over atrium; QRS and T waves correspond to wave of excitation spreading over ventricles. Phonocardiogram: the first and second heart sounds are due to sudden closure of atrio-ventricular and aortic valves respectively. (*after Winton and Bayliss*, Human Physiology, *Churchill*)

as **cardiac muscle**, it consists of a network of muscle fibres interconnected by bridges (Fig 11.12). The fibres are divided up into uninucleate cells containing fine longitudinal contractile fibrils. The muscle fibres show the same kind of cross-banding as skeletal muscle, and the mechanism of contraction is believed to be substantially the same (see Chapter 20). The bridges ensure a rapid and uniform spread of excitation throughout the wall of the heart, which in turn ensures a synchronous contraction.

THE BEATING OF THE HEART

What initiates the beating of the heart? Most muscles contract as a result of impulses reaching them from nerves. This is not, however, true of the heart, which will continue beating rhythmically even after its nerve supply has been severed. Indeed the heart will go on beating after it has been cut right out of the body. Cardiac muscle is, therefore, **myogenic**, its rhythmical contractions arising from within the muscle tissue itself.

What, then, initiates this rhythm? The mammalian heart has a specialized plexus of fine cardiac muscle fibres embedded in the wall of the right atrium close to where the great veins enter it. This is called the **sino-atrial node** (SAN), and experiments have shown that it serves as a **pacemaker**. If excised,

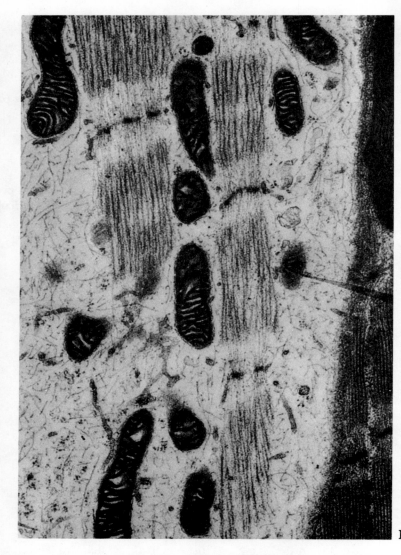

A

5μm

B

Fig 11.12 Microscopical structure of the heart wall. A) Low-magnification electron micrograph of cardiac muscle in longitudinal section. Notice the cross-connections linking adjacent fibres, and the numerous mitochondria for providing energy. The banding pattern in the fibres has the same basis as in skeletal muscle. B) Conducting tissue in the wall of the ventricle. This consists of modified cardiac muscle fibres which, instead of contracting, conduct electrical impulses like nerves. Compare this electron micrograph with the normal appearance of cardiac muscle seen in A. (*A. R. Akester, University of Cambridge*)

it will continue to beat at the normal rate of about 70 beats per minute. Other pieces of excised atrium will also beat on their own, but at a slower rate of about 60 beats per minute. Pieces of excised ventricle contract very much more slowly: about 25 beats per minute.

These experiments indicate that different parts of the heart are capable of beating at their own intrinsic rate. However, in the intact heart the beating of the ventricles is dependent on the atria, and the atria on the SAN. In other words the SAN, the region of the heart with the fastest intrinsic rhythm, sets the rate at which the rest of the heart beats.

Confirmation that the SAN is the pacemaker has come from recording electrical activity from various parts of the heart wall. It has been found that contraction of the heart is preceded by a wave of electrical excitation, similar to the nerve impulse discussed in Chapter 18. This starts at the SAN and then spreads over the two atria (Fig 11.13). When the wave reaches the junction between the atria and ventricles, it excites another specialized group of cardiac muscle fibres called the **atrio-ventricular node** (AVN). Continuous with the AVN is a bundle of modified cardiac muscle fibres, called **Purkinje tissue**, which runs down the interventricular septum and fans out over the walls of the ventricles where it breaks up into a sheet-like reticulum just beneath the endothelial lining (Fig 11.12B). When the AVN receives excitation from the atria, it sends impulses down the Purkinje tissue, and these then spread through the walls of the ventricles. Thus the pacemaker sends out rhythmical waves of electrical excitation which are transmitted first over the atria and then, via the AVN and Purkinje tissue, to the ventricles. This spread of excitation is accompanied by muscular contraction. The most remarkable aspect of the whole performance is that the rhythmical initiation of the excitatory waves is an intrinsic property of the pacemaker and is quite independent of nervous control.

INNERVATION OF THE HEART

The fact that the pacemaker initiates the rhythmical beating of the heart does not mean that it has no nerve supply. On the contrary it receives two nerves, a branch of the **sympathetic nervous system** and a branch of the **vagus nerve**. These do not initiate the beating of the heart, but can modify the activity of the pacemaker, thereby speeding up or slowing down the rate at which the heart beats. This can be demonstrated in an animal like the frog or turtle by attaching the heart to a lever that writes on a slowly revolving drum. The sympathetic and vagus nerves are hooked onto fine electrodes through which weak electrical shocks can be delivered. In this way impulses can be generated in one or other of the two nerves. The results of such an experiment are shown in Fig 11.14. If the sympathetic nerve is stimulated the heart speeds up; if the vagus is stimulated it slows down. The vagus and sympathetic nerves are thus antagonistic in their effects. This double innervation makes the animal's transport system much more adaptable than would otherwise be the case. It means that the speed at which respiratory gases are transported round the body can be modified to suit the needs of the animal as occasion demands. This matter will be taken up again in Chapter 16.

THE CAPILLARIES

As a transport system, the job of the circulation is to take up materials in one part of the body and deliver them to another. There must therefore be an intimate relationship between the circulatory system and the tissues. This

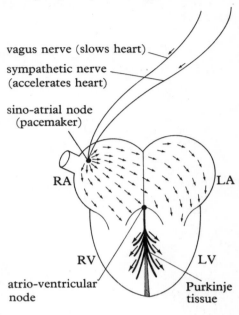

Fig 11.13 Ventral view of heart showing the spread of electrical excitation that accompanies contraction. The rhythmical beating of the heart is initiated by the pacemaker, the nerves merely serving to speed up or slow down its rate.

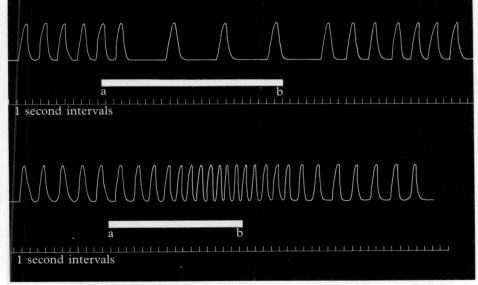

Fig 11.14 The effect on the turtle heart of stimulating (A) the vagus nerve, and (B) the sympathetic nerve with high frequency shocks. In each case the period of stimulation was from a to b. The recordings were made by attaching the heart to a lever which wrote on a revolving smoked drum. Note that impulses in the sympathetic nerve accelerate the heart, impulses in the vagus slow it.

Fig 11.15 Diagram of a capillary plexus. Note the sphincter muscle at the end of the arteriole which controls the blood flow through the capillaries. Some capillary systems, particularly those in the skin, have bypass vessels that constrict or dilate thus further controlling blood flow through the capillaries.

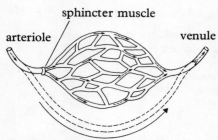

sphincter muscle

arteriole   venule

bypass vessel by which blood flows direct from arteriole to venule.

is achieved by the **capillaries**. Fig 11.15 shows a small part of a capillary network. In contrast to arteries and veins, the capillaries are narrow (an average of 10 μm in diameter) and thin-walled. The wall consists of a single layer of **pavement endothelium** which presents very little resistance to the diffusion of dissolved substances into or out of the capillary. The cells are bathed in tissue fluid derived from the blood plasma which provides a medium through which diffusion can take place. The close proximity between the capillaries and the tissue cells, and the thinness of the barrier between them, facilitates this exchange of materials.

The capillaries provide the means by which transport of materials can be regulated. Rings of muscle surround the capillaries at the points where they arise from the arterioles. Under the influence of nerves, hormones or local conditions, these **sphincter muscles** contract or relax, thereby decreasing or increasing the flow of blood through them. In some parts of the body large muscular vessels form a direct connection between arteries and veins, thereby bypassing the capillaries. By constricting or dilating, these **arteriovenous shunt vessels** can regulate the amount of blood which flows through a particular set of capillaries at any given time.

The capillaries are thus like a vast irrigation system different parts of which can be opened or closed according to local needs and conditions. This, coupled with the fact that the heart can vary its rate of beating, makes the mammalian circulation a highly adaptable transport system.

Single and Double Circulations

In Fig 11.16 the heart of the mammal is compared with the fish and frog. The fish has the simplest system: deoxygenated blood from the body is pumped to the gills whence it flows to various parts of the body and then returns to the heart. The heart has only one atrium and ventricle. As blood flows only once through the heart for every complete circuit of the body, this is spoken of as a **single circulation**.

The snag about this arrangement is that blood has to pass through two capillary systems, the capillaries of the gills and then those of the body, before returning to the heart. Capillaries offer considerable resistance to the flow of blood, and this means that in fishes there is a marked drop in blood pressure before the blood completes a circuit. For this reason the blood-flow tends to be sluggish on the venous side. In fishes this has been overcome to some extent by replacing the veins with large **sinuses** that offer minimum resistance to blood-flow. Nevertheless the problem of getting blood back to the heart is an acute one and probably imposes severe limitations on the activities of many fishes.

In mammals the problem has been overcome by the development of a **double circulation**. Blood is pumped from the heart to the lungs, whence it returns to the heart and is then re-pumped to the body. To prevent mixing of deoxygenated and oxygenated blood, the heart is divided into right and left sides, the right side dealing with deoxygenated and the left side with oxygenated blood. This we have already seen and there is no need to elaborate on it further.

The frog shows an interesting intermediate condition. It has a double circulation in that blood is returned to the heart after passing through the lungs, and there are two atria as in the mammal. However, the ventricle is completely undivided, as is the **conus arteriosus** into which the blood is pumped before entering the vessels leading to the lungs and body. On purely structural grounds it would seem that deoxygenated and oxygenated blood would be inextricably mixed in the single ventricle. Over the years, there has been much speculation on the extent to which this is so. This is certainly not the place to go into the details of so specialized a point. Suffice it to say that, despite the apparent anatomical shortcomings, separation of the two bloodstreams is surprisingly complete, mainly deoxygenated blood being sent to the lungs and oxygenated blood to the body. Exactly how this is achieved is not known but the various foldings in the wall of the ventricle, aided possibly by the **spiral valve** in the conus arteriosus, play an important part. Research on other vertebrates with incompletely divided hearts indicates the same thing: a high degree of separation of deoxygenated from oxygenated blood. The general principle which this illustrates is that if a double circulation is to be developed, it must be coupled with some kind of mechanism for keeping the deoxygenated and oxygenated bloodstreams apart.

Of course developing a double circulation is not the only way of overcoming the pressure problem seen in fishes. An alternative solution would be to have two separate hearts, one for pumping blood to the body, the other for pumping blood to the respiratory organs. This is precisely what has happened in octopuses and squids. Blood is pumped from the **main heart** to various parts of the body. It then flows through a system of sinuses to a pair of **branchial hearts** which pump it through the gills. The blood then returns to the main heart for distribution to the body. Octopuses and squids are on a quite different evolutionary line from vertebrates. But like vertebrates they are active creatures, and comparing their circulations shows us how the same physiological problem can be solved in two quite different ways.

Open and Closed Circulations

In the circulations considered so far the blood is confined to vessels, or in some cases sinuses, that are quite distinct from the general body cavity. However, in certain invertebrates, notably arthropods, the blood is contained in the body cavity. The reason lies in the way arthropods develop. In most animals the embryological cavity (coelom) that eventually becomes the main body cavity expands at the expense of the cavity (blastocoel) which eventually forms the blood vessels (see p. 418). In arthropods the reverse is the case. The coelomic cavities remain small and the blastocoel becomes the main body cavity. Since in this case the body cavity contains blood it is known as a **haemocoel**. The whole system is known as an **open circulation**, in contrast with the **closed circulations** of other animals.

In insects the haemocoel is divided by a transverse **pericardial membrane** into a **pericardial cavity** dorsally and a **perivisceral cavity** ventrally. The only blood vessel as such is a tubular **heart** which is suspended in the pericardial cavity by slender ligaments. The heart, which extends through the

Fig 11.16 Single and double circulations. For explanation see text. Note that the amphibian has a double circulation but the ventricle is not divided into right and left chambers, so oxygenated and deoxygenated blood are mixed before being pumped to the lungs and body. The octopus illustrates an alternative way of solving the pressure problem that faces an animal with a single circulation. It has developed a second pair of hearts which pump blood back to the main heart via the gills, after it has been to the body. **A**, atrium; **V**, ventricle; **R**, right; **L**, left.

Fish
single circulation

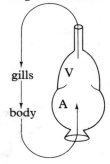

Amphibian
double circulation with partially divided heart

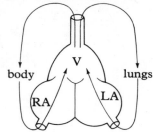

Mammal
double circulation with completely divided heart

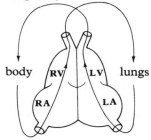

Octopus
separate hearts for body and gills

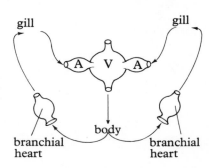

thorax and abdomen, is expanded in each segment to form a small chamber which is pierced by a pair of tiny holes or **ostia**. In the pericardial membrane is a series of **alary muscles** corresponding in position to the heart chambers. These structures are illustrated in Fig 11.17.

Blood is propelled forwards through the heart by waves of contraction (systole) which commence at the rear and work their way towards the anterior end. The blood then leaves the heart and enters the haemocoel through which it flows as shown in Fig 11.17A. How does the blood re-enter the heart? During

Fig 11.17 Diagrams illustrating the open circulation of an insect. Blood flows forward in the dorsal tubular heart whence it enters the haemocoel, to be returned to the heart via the ostia. The rôle of the heart ligaments and alary muscles in promoting the circulation of the blood is discussed in the text. (B: *after Ramsay,* A Physiological Approach to the Lower Animals, *Cambridge University Press*)

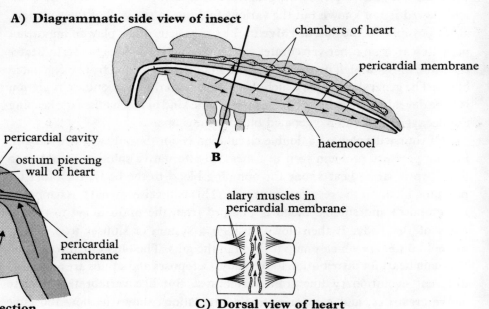

A) **Diagrammatic side view of insect**

chambers of heart

pericardial membrane

haemocoel

B

heart ligament

pericardial cavity

ostium piercing wall of heart

perforation in pericardial membrane

pericardial membrane

perivisceral cavity

B) **Transverse section**

alary muscles in pericardial membrane

C) **Dorsal view of heart**

systole the heart ligaments are stretched with the result that during diastole they pull the walls of the heart outwards. This results in blood being sucked into the heart through the ostia. The latter are equipped with valves that allow blood to enter, but not leave, the heart through them. Expansion of the heart is aided by contraction of the alary muscles which increases the tension on the ligaments. Contraction of the alary muscles also has the effect of pulling the pericardial membrane downwards, thereby raising the blood pressure in the perivisceral cavity and decreasing it in the pericardial cavity. This encourages the flow of blood from the former to the latter.

As we saw in Chapter 8 gaseous exchange in insects takes place through the tracheal system. The circulatory system is therefore not directly concerned with transport of respiratory gases. Accordingly it lacks a respiratory pigment, though it does contain phagocytes and plays an important part in the distribution of food substances and elimination of nitrogenous waste matter (see p. 226). The insect, with its tracheal system and open circulation, has solved the problem of transport by means that contrast sharply with those evolved by the vertebrates.

12 Uptake and Transport in Plants

To carry out photosynthesis a plant requires an adequate supply of carbon dioxide and water. It also requires mineral salts and oxygen. In primitive plants like the algae these materials, together with the products of photosynthesis, can move from cell to cell by diffusion and active transport, and there is no need for specialized transport tissues. In higher plants gases move by diffusion, but transport of water and mineral salts and the soluble products of photosynthesis takes place in specialized **vascular tissues**. We see the rudiments of such tissues in comparatively simple plants like mosses, but they reach their peak in ferns and flowering plants. But before we get on to these let us look briefly at the way carbon dioxide is taken into a plant.

Uptake of Carbon Dioxide

Aquatic plants obtain carbon dioxide from the surrounding water in which it is dissolved as carbonic acid. Land plants obtain it from the atmosphere. It diffuses into the plant through the **stomata** which were mentioned in Chapter 10 and will now be discussed in detail.

THE STOMATA

Stomata are pores perforating the epidermis of the leaves and stem. They are usually most numerous in the lower epidermis of the leaf where there may be as many as 400 per mm^2 (apple). There are generally fewer in the upper epidermis and fewer still in the stem. Their functions are:

(1) to allow exchange of carbon dioxide and oxygen between the inside of the leaf and the surrounding atmosphere;

(2) to permit the escape of water vapour from the leaf.

Stomata are important in several physiological processes, not just photosynthesis, but they also represent a hazard in that they may permit excessive evaporation from the leaf. In a sense a plant faces a conflict. If it opens its stomata it runs the risk of losing excessive water, particularly if it lives in a dry habitat. On the other hand if it closes them it may run short of carbon dioxide or oxygen. Plants resolve this problem by not opening them for longer than is necessary. This does not mean that a plant never loses more water than it can replace from the soil. The observation that plants frequently wilt in hot weather bears witness to the fact that they often do lose excessive water. However, wilting is not disastrous so long as the plant is given an opportunity

A) A stoma in surface view and section

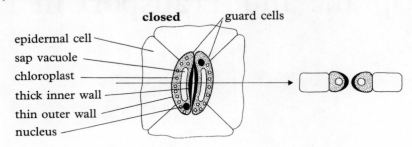

closed

guard cells

epidermal cell

sap vacuole

chloroplast

thick inner wall

thin outer wall

nucleus

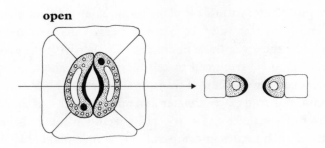

open

Fig 12.1 Structure and action of stomata.
A) Diagrams of a stoma in surface view and
section showing the method of opening. B)
Photomicrograph of epidermis of bindweed
Convolvulus showing several stomata in surface
view. C) Photomicrograph of a transverse
section of asparagus stem showing a stoma in
section. D) Electron micrograph of guard cells
of French bean *Phaseolus*. Notice the prominent
nuclei (**N**) and sap vacuoles (**S**). (B and C:
K. Esau and V. I. Cheadle, University of California.
D: *B. E. Juniper, University of Oxford*)

B D

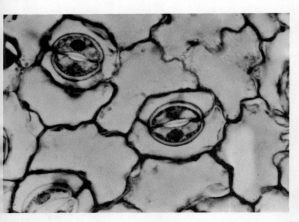

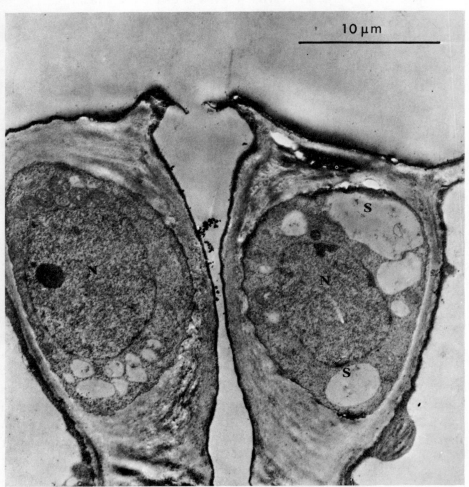

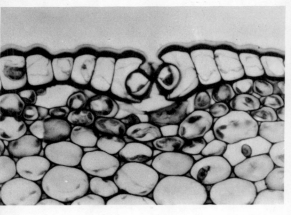

C

to recover later. The conflicting needs of the plant are resolved by the controlled opening and closing of the stomata.

In order to appreciate how the stomata are controlled we must first understand the mechanism by which they open and close, and this necessitates looking at their structure. You will notice in Fig 12.1 that the stomatal pore is bordered by a pair of **guard cells**. These are sausage-shaped and, unlike other epidermal cells, contain chloroplasts. There is a sap vacuole and, a point of great importance, the inner cellulose wall (i.e. the wall lining the pore itself) is thicker and less elastic than the thinner outer wall.

Stomatal opening and closure depends on changes in turgor of the guard cells. If water is drawn into the guard cells by osmosis the cells expand and their turgidity is increased. But they do not expand uniformly in all directions. The thick, inelastic inner wall makes them bend as shown in Fig 12.1A. The result is that the inner walls of the two guard cells draw apart from each other and the pore opens. The same effect can be achieved by blowing up a sausage-shaped balloon to which a strip of sellotape has been stuck down one side. As it is blown up it will bend over towards the sellotaped side. It is thought that in normal circumstances when a stoma opens the turgidity of the guard cells is increased by their taking up water from the surrounding epidermal cells. Isolated stomata will open when immersed in water, but if placed in a hypertonic solution they close.

When do the stomata open and close? This can be investigated by means of a **porometer**, an instrument for measuring the resistance to the flow of air through a leaf (Fig 12.2). If you attach a porometer to a leaf and take measurements of its resistance to airflow at intervals you will find that there is generally less resistance during daylight hours than at night. This is because the stomata open during the day and close at night.

THE MECHANISM OF STOMATAL OPENING AND CLOSURE

At first glance the mechanism causing this diurnal opening and closing might seem obvious. Unlike other epidermal cells, the guard cells have chloroplasts and at daybreak they start photosynthesizing; this leads to an accumulation of sugar in the guard cells whose osmotic pressure therefore increases. This in turn causes water to be drawn into them from the surrounding epidermal cells resulting in the opening of the pore. However, this theory is unsatisfactory. It is true that in the light sugar (sucrose mainly) accumulates in the guard cells, but the stomatal response is too rapid for it to be explained merely by a resumption of photosynthesis.

So we must look for an alternative explanation. One possible hypothesis depends on the fact that the enzymatic conversion of starch to sugar proceeds more readily when comparatively little acid is present (i.e., at a high pH). The conversion of sugar to starch on the other hand is favoured by a comparatively high concentration of acid (low pH). During the night carbon dioxide accumulates in the intercellular spaces of the leaf, and this raises the concentration of carbonic acid. The resulting drop in pH favours the conversion of sugar to starch in the guard cells, thereby decreasing their osmotic pressure and causing the stoma to close. In the morning the resumption of photosynthesis lowers the concentration of carbon dioxide. As a result the level of carbonic acid falls, the pH rises, starch is converted to sugar, the osmotic pressure of the guard cells increases, and the stoma opens.

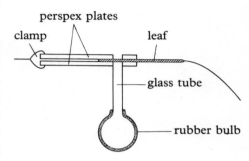

Fig 12.2 One of many types of porometer for measuring resistance to airflow through a leaf. The rubber bulb is squeezed then released and the time taken for it to reinflate is noted. All seals must be airtight: airtightness can be checked by using a thin glass plate in place of the leaf. This apparatus can be used to estimate the degree of openness or closure of stomata over a period of time.

perspex plates

clamp

leaf

glass tube

rubber bulb

Neat though it is, this theory leaves a number of facts unexplained. For example, starch is absent from the guard cells of certain plants; some guard cells lack chloroplasts but still open and close; and the stomatal movements of some plants may not necessarily be related to the time of day; in fact in some plants they open at night and close by day! We have to admit that there may be much more to the control of stomatal movements than is suggested above. One possibility is that closure may result from water being actively pumped out of the guard cells in response to specific stimuli. This idea is being explored at the present time.

When the stomata are open carbon dioxide from the atmosphere diffuses into the sub-stomatal air chambers and thence to the intercellular spaces between the mesophyll cells. When it comes into contact with the wet surface of a cell it goes into solution and diffuses into the cytoplasm. The fixation of carbon dioxide in the dark reactions of photosynthesis creates a concentration gradient so that carbon dioxide continues to diffuse into the leaf.

Uptake and Transport of Water

It is impossible to discuss the uptake and transport of water without first considering the way it is lost from the leaves. In fact the latter process is an integral part of the mechanism by which it is drawn up the stem from the roots.

TRANSPIRATION

The stomata, as well as permitting the entry of carbon dioxide, allow the evaporation of water from the plant, a phenomenon known as **transpiration**. Transpiration is by no means restricted to the leaves for there are generally a number of stomata in the stem epidermis as well. However, the leaves with their large surface area and abundant stomata represent the main source of water loss. To a small extent evaporation also takes place through the cuticle of the epidermal cells (**cuticular transpiration**) but this rarely exceeds 5 per cent of the total water loss.

The rate at which water is transpired from a plant may be considerable, particularly if the atmosphere is warm and dry. In one hour of a hot summer day a leaf may lose more water than it contains at any one moment. An oak tree may transpire as much as 680 dm^3 of water in a day. It has been estimated that a sunflower plant may transpire over 200 dm^3 of water during its life of six months. These figures convey some idea of the scale of transpiration. They also emphasize how essential it is for a plant to have an adequate system of water uptake, for the water which is lost must be replaced.

Transpiration can be demonstrated by means of absorptive paper saturated with **cobalt thiocyanate**. In the anhydrous state cobalt thiocyanate is blue, but when hydrated it turns pink. A piece of anhydrous cobalt thiocyanate paper is placed on each side of a leaf and sandwiched between two glass slides clamped together. After a period of time, which varies according to the type of leaf and the temperature and humidity, the paper turns pink, indicating that water has escaped from the leaf. Generally the pink colour develops more rapidly on the lower surface of the leaf than on the upper surface. (Why?).

MEASURING THE RATE OF TRANSPIRATION

The cobalt thiocyanate experiment is useful as a rough demonstration of transpiration but it gives no quantitative information on the rate. This can be estimated by measuring either the rate at which the plant loses weight, or the rate at which it takes up water. In the first case we have to assume that weight loss is due only to evaporation of water. In the second case we assume that water uptake is equal to water loss.[1]

The first experiment can be done by placing a potted plant on an automatic balance and measuring the weight changes over a given period of time. The whole of the pot must be enveloped in a polythene bag to prevent water evaporating from the soil. The latter must be well watered beforehand so that the plant has a plentiful supply of water for the duration of the experiment. Changes in the rate of transpiration are determined by recording the weight at intervals and plotting the results on a graph. The total leaf area is then estimated by tracing the outline of the leaves on squared paper and counting the number of squares within each leaf. In this way we obtain all the data necessary to express the rate of transpiration as weight loss per unit time per leaf area. As might be expected the results depend on numerous factors such as the type of plant used and the environmental conditions: temperature, humidity and so on. Typically water loss ranges between 15 and 250 g/h/m^2 by day, and between 1 and 20 g/h/m^2 by night.

To measure water uptake a **potometer** can be used. This has the advantage of being simple and portable but necessitates the cutting of the stem. One type of potometer is illustrated in Fig 12.3. The cut end of the stem is placed in a sealed vessel of water which is continuous with a capillary tube. There must be no air-locks in the system, the water in the stem, vessel, and capillary tube forming a continuous column. The rate of water uptake is measured by introducing an air bubble at the end of the capillary tube and timing how long it takes to move between successive divisions on a graduated scale. When one set of readings has been obtained the air bubble is returned to the beginning of the capillary tube by introducing water from the reservoir. The apparatus can be calibrated by estimating the volume of water represented by a given length of the capillary tube, and the leaf area is calculated with squared paper. This makes it possible for the uptake of water to be expressed as volume per unit time per leaf area. However, potometers are generally only used to measure changes in the rate of transpiration with various external conditions. For such purposes it is sufficient to express the results as the distance moved by the air bubble per unit time.

It is sometimes desirable to measure the evaporating power of the atmosphere to see how this correlates with water loss from a plant. For this an **atmometer** may be used (Fig 12.3). This is similar to a potometer but instead of a plant a porous pot is used as the evaporating surface. By running a potometer and an atmometer side by side one can compare changes in the rate of evaporation from a plant with that of a purely physical system.

THE EVAPORATING SURFACE

What emerges from these experiments is that the rate of transpiration is affected by a variety of internal and external factors. To understand how these operate we must first decide exactly where the evaporating surface is. Fig 12.4 shows that each stoma opens into a small **sub-stomatal air chamber**

[1] Are these assumptions justified?

Fig 12.3 A potometer (A) and a porous pot atmometer (B). In both cases the rate of water loss from the exposed surfaces is indicated by the speed at which the air bubble moves along the graduated capillary tube. When the bubble reaches the end of the capillary tube it can be sent back to the beginning by letting in water from the reservoir. The photograph (*right*) shows a potometer being used to measure the rate of water uptake by a leafy shoot of sycamore. (*B. J. W. Heath, Marlborough College*)

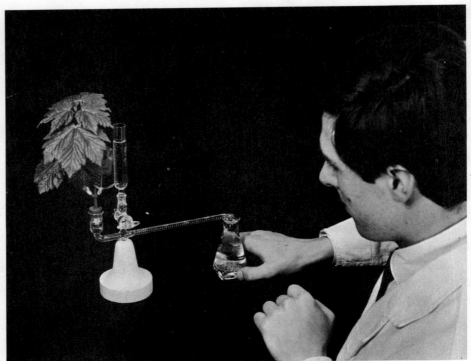

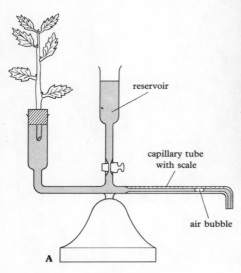

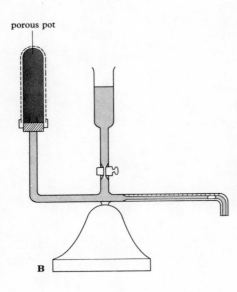

which is lined with **spongy mesophyll cells**. The effective evaporating surface of a leaf consists of the saturated walls of these spongy mesophyll cells. As evaporation proceeds, water vapour accumulates in the sub-stomatal air chambers and escapes through the stomata into the surrounding atmosphere. Any condition which directly or indirectly alters the rate of evaporation will change the rate at which water is drawn into the plant and transported through it.

FACTORS AFFECTING THE RATE OF TRANSPIRATION

The most important internal condition affecting transpiration is the state of the stomata: their number, distribution, structural features, and how open they happen to be. Any factor that influences the opening and closing of the stomata will obviously affect transpiration. In some plants the stomata are modified to prevent excessive water loss. This is discussed in Chapter 14.

Fig 12.4 (*right*) One of several routes by which water passes through a leaf from the vascular tissues to a sub-stomatal air chamber. Continual evaporation through the open stoma creates a gradient encouraging the further movement of water towards the surface of the leaf. The forces involved, and possible alternative pathways, are discussed in the text.

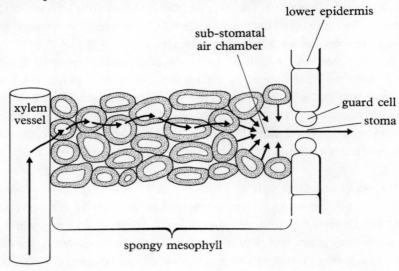

External conditions affecting transpiration include:

(1) Temperature. A high temperature provides latent heat of vaporization and therefore encourages evaporation of water from the mesophyll cells.

(2) Relative humidity, the degree to which the atmosphere is saturated with water vapour. This is important because it determines the **saturation deficit,** i.e. the humidity difference between the inside and outside of the leaf. As evaporation proceeds from the mesophyll cells the atmosphere in the sub-stomatal air chambers becomes more and more saturated. The lower the relative humidity of the surrounding atmosphere the greater will be the saturation deficit, and the faster will water vapour escape through the stomata.

Plants living in dry habitats such as the desert where the relative humidity is low have reduced their evaporating surface by cutting down the number of stomata, and protecting themselves from excessive water loss in various other ways. On the other hand plants living in humid situations such as a tropical jungle have numerous stomata. In such plants the humidity of the atmosphere may be so high that more water accumulates in the leaves than can be removed by evaporation. Under these circumstances water may ooze out of the stomata and drip from the leaves. This is known as **guttation** and even in temperate regions it may be observed on warm humid days.

(3) Air movements. In the last section it was pointed out that the atmosphere in the sub-stomatal air chambers becomes progressively more saturated as evaporation proceeds. This condition is not necessarily confined to the sub-stomatal chambers but may build up on the underside of the leaf as water vapour continues to diffuse out of the stomata. Obviously the atmosphere will be most highly saturated immediately outside each stoma and become progressively less saturated as water vapour diffuses away. The resulting layers of decreasing saturation are called **diffusion shells,** and if they are allowed to build up around the leaf the rate of evaporation from the mesophyll cells inevitably decreases. Air movements blow away these diffusion shells thereby increasing the rate of evaporation from the leaf.

(4) Atmospheric pressure. The lower the atmospheric pressure the greater the rate of evaporation. For this reason alpine plants are liable to have a high rate of transpiration, and many of them therefore have adaptations for preventing excessive water loss.

(5) Light. All the external conditions mentioned so far will influence the rate of evaporation from a porous pot atmometer in the same way as from a plant. But this is not the case with light. If light intensity is increased the rate of evaporation from an atmometer remains unchanged but the evaporation rate from a plant increases. The reason is not that light affects evaporation as such, but that it causes the stomata to open, thereby increasing water loss from the plant.

(6) Water supply. Transpiration depends on the walls of the mesophyll cells being thoroughly wet, and this in turn requires that the cells are kept fully turgid. For this to be so the plant must have an adequate water supply from the soil.

WATER MOVEMENT IN THE LEAF

When water evaporates from the walls of the spongy mesophyll cells into the sub-stomatal spaces, water molecules diffuse out of the sap vacuoles to replace those which have been lost to the atmosphere (Fig 12.4). This results in the solutes becoming more concentrated so that the osmotic pressure of the outermost cells exceeds that of the cells further in. Thus water is drawn into the surface cells from their next door neighbours. The latter will now acquire a higher osmotic pressure and will draw in water from their neighbours, and so on. Measurements have shown that there is in fact an **osmotic gradient** in the leaf, the cells at the edge having the highest osmotic pressure, and those at the centre, in the immediate vicinity of the xylem tissues, having the lowest osmotic pressure. This osmotic gradient will be maintained as long as water continues to evaporate from the cells at the surface. The osmotic gradient also provides the means by which water is drawn into the mesophyll cells from the vascular tissues.

The osmotic mechanism proposed above assumes that water moves through the leaf from one sap vacuole to the next. However, this is not the only route by which water molecules might move to the surface of the leaf. It is possible that they might diffuse along the cellulose cell walls of adjacent mesophyll cells. Evaporation from the walls of the surface cells would create a concentration gradient favouring the continued diffusion of water molecules towards the edge of the leaf.

There is some uncertainty as to which of these two routes is correct. The second would offer less resistance to the movement of water than the first but of course both may occur to varying extents.

STRUCTURE OF THE ROOT

To understand how roots take up water we must first examine their structure. The internal organization of a typical primary root is summarized in Fig 12.5. Basically there is a surface layer of epidermal-type cells, and a central core of **vascular tissues**. The two are separated by turgid **parenchyma cells** tightly packed but separated in places by small water-filled intercellular spaces. The vascular tissues are enclosed within a layer of somewhat specialized cells, the **endodermis**. On account of the fact that these cells contain numerous starch grains (easily detected by staining sections of root with iodine), the endodermis is referred to as the **starch sheath**.

The cells at the surface form the **piliferous layer**, so-called because it is covered with **root hairs**. When fully formed the root hairs are about 4 mm long, each being a slender outgrowth from a single cell. The piliferous layer is only present in the younger parts of a root, some 2 or 3 mm behind the apex. As the root gets older and increases in girth, the piliferous layer ruptures and sloughs off leaving the outermost layer of parenchyma cells, the **epiblem**, to become the functional outer layer.

UPTAKE OF WATER BY THE ROOTS

The main function of the root hairs is to anchor the plant to the soil, but they also participate in the uptake of water. Slender and flexible, they penetrate between the soil particles and absorb water from the intervening spaces. To aid this they lack a cuticle and are numerous: their great number increases the surface area over which absorption can take place. A considerable amount of water is also absorbed beyond the root hair zone towards the apex of the root. Here the surface cells have no cuticle, and water is absorbed directly from the surrounding soil water.

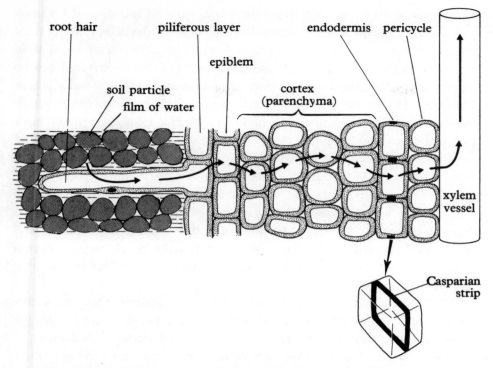

root hair   piliferous layer   endodermis   pericycle

epiblem

soil particle
film of water

cortex
(parenchyma)

xylem
vessel

Casparian
strip

Water is drawn into the root hair mainly by osmosis. On account of the presence of sugars and other metabolites, the concentration of solutes in the sap vacuole is greater than that of the surrounding soil water. Water molecules are therefore drawn across the cellulose wall and the semipermeable protoplast into the vacuole. Osmotic uptake may be aided by active transport involving the expenditure of energy, but little is known about this.

From the root hair water passes to the vascular tissues in the centre of the root via the intervening parenchyma cells. There are three possible pathways through which the water might flow:

(1) Through the sap vacuoles, water being drawn from one vacuole to the next by osmosis.

(2) Through the cytoplasm, water diffusing from cell to cell via the plasmodesma strands.

(3) Along and between the cell walls, water diffusing through the cellulose of adjacent cells and through the small intercellular spaces between them.

It is not possible to say categorically which of these three pathways is the correct one. The continual removal of water from the innermost parenchyma cells into the vascular tissues would provide the necessary osmotic gradient for the first pathway, and would also provide the diffusion gradient necessary for the second and third mechanisms. It is possible that all three operate to some extent. As with uptake into the root hairs, the movement of water from cell to cell may be aided by active transport.

INTO THE VASCULAR TISSUES OF THE ROOT

We come now to the question of how water passes from the parenchyma cells into the vascular tissues of the root. A possible hypothesis is that the high osmotic pressures developed in the leaf cells as a result of transpiration result in water being drawn into and along the vascular tissues. In other words the prime force moving water into the vascular tissues of the root is a 'pull' exerted

Fig 12.5 Structure of a dicotyledonous root showing one of the routes by which water and mineral salts may be drawn from the soil into the vascular tissues. The forces involved, and possible alternative pathways, are discussed in the text.

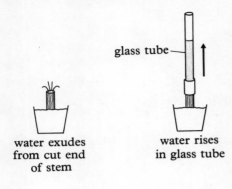

water exudes
from cut end
of stem

water rises
in glass tube

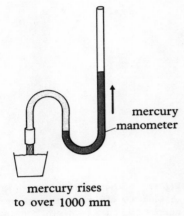

mercury rises
to over 1000 mm

Fig 12.6 Demonstration of root pressure in a potted plant. Explanation in text.

by the leaves. However, this is not the whole story. If the stem of a plant is severed, the cut end of the stump will exude copious quantities of water for a considerable time, suggesting that there is a force pushing water up the stem from the roots. This is called **root pressure**. The forces involved can be measured by attaching a suitable mercury manometer to the cut end of a stem (Fig 12.6). In this way it has been found that sizeable pressures can be set up by the roots of some plants: a stump of *Fuschia*, for example, can develop a pressure of over 90 kN/m^2.

In the face of these observations it would be difficult to deny the existence of root pressure, but what is it due to? The answer is not yet known for certain. One possibility is that water is actively secreted by the living root cells into the vascular tissues, a process involving the expenditure of energy. Active transport of water is not unknown in plants (see p. 231 for example), and it could account for the passage of water from the innermost parenchyma cells of the root into the vascular tissues, a phenomenon which is otherwise difficult to explain.

Which particular cells in the root might be responsible for this active movement of water? Here again we cannot be sure, but there are reasons for believing that it might be the **endodermis**. The endodermis cells are unique in possessing a **Casparian strip**, a band of corky material which runs round each cell in its radial and horizontal walls (Fig 12.5). If endodermis cells are plasmolysed the cytoplasm pulls away from the cellulose but remains stuck to the Casparian strip. Unlike the rest of the cell wall (cellulose), the Casparian strip is impermeable to water. What are the implications of these observations? It means that when water reaches the endodermis, it cannot enter the vascular cylinder via the radial and horizontal walls of the endodermis cells. It must pass through the cytoplasm. The endodermis forms a continuous cylindrical-shaped screen of living cytoplasm through which water must pass on its way to the vascular tissues. It is possible that water molecules are actively moved through this cytoplasmic screen into the vascular area. The starch grains, which are so numerous in the endodermis, may provide the raw material for the necessary energy. There is as yet no proof that this happens, nor do we know how important such a process might be in the normal uptake and transport of water. Until direct evidence is produced the endodermis must remain something of a mystery.

FROM ROOT TO LEAF

So far we have considered how water enters the vascular tissues in the root and is drawn out of them in the leaf. We turn now to the transportation of water from the roots to the leaves via the stem, the **transpiration stream**.

In order to understand this we must first look at the general organization of the **vascular tissues**. Evidence is presented later that the part of the vascular tissue responsible for water transport is the **xylem** so we will start with this.

STRUCTURE OF THE STEM

The way the xylem fits into the general organization of a dicotyledonous stem is shown in Fig 12.7. The xylem is contained in a series of **vascular bundles**, which also include the **phloem** (to be discussed later). **Collenchyma** and **sclerenchyma** tissue contribute to the strength of the stem, maintaining its erect form and preventing it from drooping. The vascular bundles are continuous with those in the roots and leaves. The stems of other vascular plants, e.g. ferns and conifers, are different in the arrangement of their tissues, but they all have vascular bundles containing xylem and phloem.

Fig 12.7 Stereogram of part of a dicotyledonous stem. The conducting tissues are confined to the vascular bundles: the xylem contains vessels and tracheids and the phloem contains sieve elements.

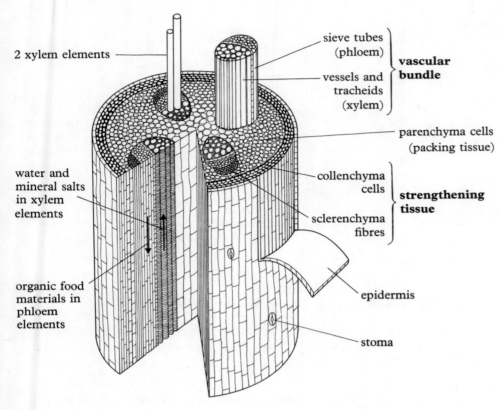

2 xylem elements

sieve tubes (phloem) — **vascular bundle**

vessels and tracheids (xylem)

parenchyma cells (packing tissue)

water and mineral salts in xylem elements

collenchyma cells — **strengthening tissue**

sclerenchyma fibres

organic food materials in phloem elements

epidermis

stoma

XYLEM TISSUE

The xylem contains two types of conducting element: **vessels** and **tracheids**. A vessel is formed from a chain of elongated cylindrical cells placed end to end. In the course of development the horizontal end walls break down partially or completely so that the cells are in open communication with each other. At the same time the cellulose side walls become impregnated with **lignin** (wood) rendering them impermeable to water and solutes. Unable to absorb nutrients, the protoplasmic contents of the cells die, leaving a hollow tube, the vessel. Its lignified walls are perforated by numerous **pits** where lignin fails to be deposited and only the primary cell wall remains. These pits permit the passage of water in and out of the lumen. In conifers the pits are characteristically bordered by a lignified rim. Such **bordered pits** contain a **torus**, a kind of plug which may play some part in controlling the passage of water. As the cells develop, lignified ribs of various kinds are laid down on the immediate inside of the walls. These give the vessel added strength and prevent its wall from caving in. **Spiral** and **annular thickenings** are particularly common and can often be seen beautifully in longitudinal stem sections soaked in acidified phloroglucin which stains the lignin red. Tracheids are similar to vessels except that they are five-sided in cross-section, and instead of being open at each end their tapering end walls are perforated by pits. Xylem tissue is illustrated in Figs 12.8, 12.9 and 12.10.

Fig 12.8 The xylem elements of higher plants. Each element starts as a single elongated cell that becomes lignified and loses its protoplasmic contents. The cells are lined up end to end so that subsequent perforation of the end walls results in the formation of greatly elongated tubes running from roots to leaves. The lignified thickenings prevent the walls caving in.

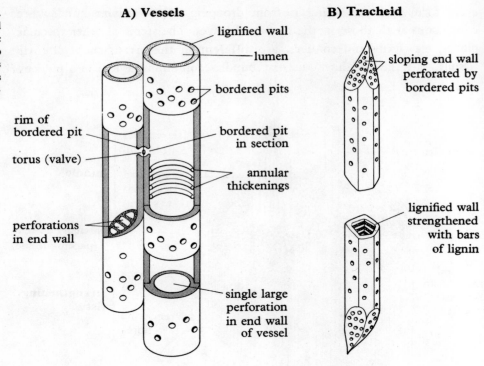

A) Vessels

lignified wall

lumen

bordered pits

rim of bordered pit

bordered pit in section

torus (valve)

annular thickenings

perforations in end wall

single large perforation in end wall of vessel

B) Tracheid

sloping end wall perforated by bordered pits

lignified wall strengthened with bars of lignin

C) Different types of thickening found in vessels

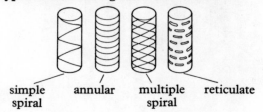

simple spiral   annular   multiple spiral   reticulate

Xylem tissue, together with phloem, is found in all parts of the plant. The vessels and tracheids of the roots, stem and leaves connect with one another to form a continuous system of water-conducting channels serving all parts of the plant.

THE XYLEM AND WATER TRANSPORT

That the xylem is responsible for water transport can be shown by various experiments. One of the earliest was done by the famous nineteenth century botanist, Eduard Strasburger, who demonstrated that a tree will continue to transpire even after its trunk has been encased in a steam jacket, thereby killing all the living cells. Similarly, allowing a cut stem to draw up picric acid or some other soluble poison has no immediate effect on the plant's ability to take up water, indicating that water transport occurs in dead cells.

More precise information has come from experiments in which cut stems are stood in a coloured dye solution such as eosin. Subsequent sectioning of the stem and examination under the microscope shows that only the xylem elements contain the dye. But this still leaves unanswered the question of whether water travels through the walls or the lumen of the vessels and tracheids. This question was tackled by the late H. H. Dixon, a leading investigator in plant water relations. He allowed plants to draw up fatty solutions which had the effect of blocking the lumina of the xylem elements

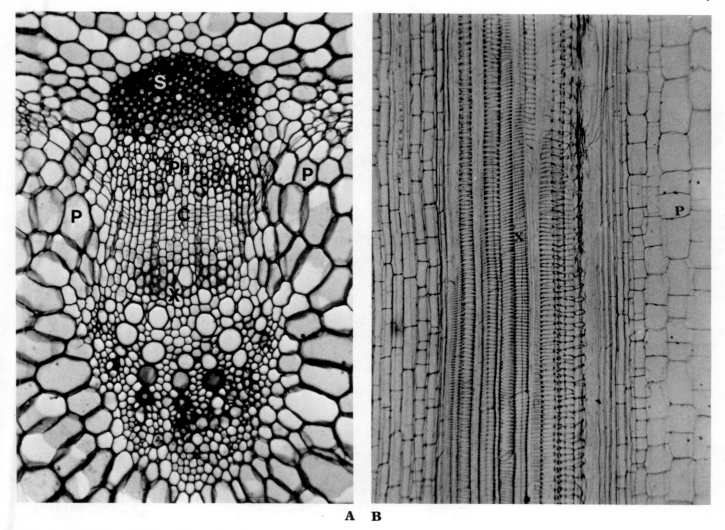

A   B

whilst leaving the walls unaffected. This was found to prevent further uptake of water, causing the plant to wilt.

So we conclude that water transport takes place through the lumina of the dead vessels and tracheids. For this function they are admirably suited. The pits allow horizontal movement of water in and out of them, and their hollow lumina permit the unimpeded flow of water along the tubes. Vessels, with their open ends, offer less resistance to the passage of water than tracheids whose end walls are merely perforated by pits.

THE ASCENT OF WATER UP THE STEM

If you cause a column of water to rise up a tube by means of a vacuum pump, the maximum height to which the water can be raised is 11 metres, the normal barometric height of water. How, then, can a continuous flow of water be maintained through the 20 m of an average oak tree, or the 100 m of a redwood?

To answer this question look at the model shown in Fig 12.11. A porous pot is attached to a glass tube which is dipped into a beaker of mercury. The porous pot and tube are filled with water, care being taken to ensure that there is no air in the system. The set-up is now placed in a dry atmosphere.

As water evaporates from the surface of the porous pot water rises up the tube drawing the mercury with it. If a vacuum pump was used the mercury

Fig 12.9 Microscopical structure of conducting tissues. A) Transverse section of entire vascular bundle of sunflower *Helianthus*. **S**, sclerenchyma fibres of pericycle; **Ph**, phloem; **P**, parenchyma; **C**, cambium; **X**, xylem. (*J. F. Crane*). B) Longitudinal section of xylem showing the tubular nature of the vessels and the different forms of thickening. (*J. F. Crane*)

Fig. 12.10 Electron micrograph of a carbon replica of a bordered pit in the spruce. The disc-shaped torus in the centre (**T**) is held in position by radial strands (**R**) which run from the sides of the pit (**S**). The strands represent the original primary cell wall. (*R. Schmid, Botanical Institute, Munich*)

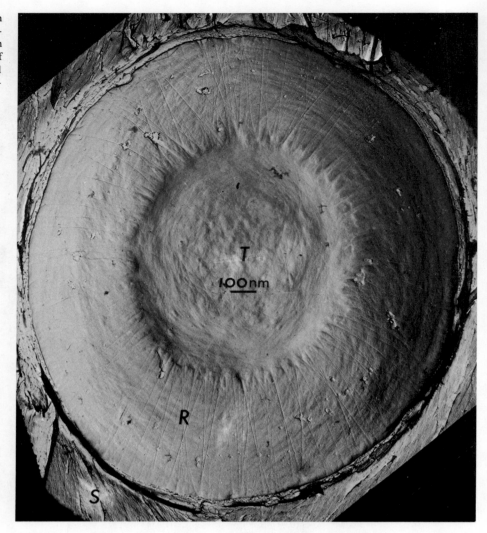

column would rise to a maximum height of only 760 mm, the normal barometric height of mercury. Yet when drawn by evaporation of water the mercury rises to over 1000 mm. It has been estimated that if there was no mercury in the system, i.e. if the porous pot, tube and beaker contained only water, the latter could be pulled up to a height of over one kilometre.

What is it about the porous pot system which makes this possible? The answer depends on the fact that its wall is perforated by numerous fine pores into which water passes before evaporating from the surface. As you know, water tends to adhere to the walls of a vessel containing it. This is owing to forces of **adhesion** between the water molecules and the molecules comprising the material lining the vessel. (Adhesion is defined as the force of attraction between unlike molecules). Now the narrower the container the greater will be the proportion of water molecules in contact with its walls. This results from the fact that the smaller the container the greater the area of its wall in comparison with the volume of water enclosed, i.e. the surface:volume ratio is greater. Being very narrow, considerable adhesive forces will be expected to develop in the pores of the porous pot, and these forces must be sufficient to support a considerable weight of water.

But the problem is not only to hold up the column of water but also to prevent it breaking in the middle. What is responsible for this? The answer is the force of **cohesion**. Cohesion is defined as the force of attraction between like molecules. In this case cohesion between the water molecules both in the walls of the porous pot and the glass tube prevents the water column from breaking.

The forces of adhesion and cohesion are demonstrated by the phenomenon

Fig 12.11 Mercury will rise much further in a glass tube if drawn by a column of water that is evaporating from the surface of a porous pot or leafy plant than it will if caused to rise by a vacuum pump.

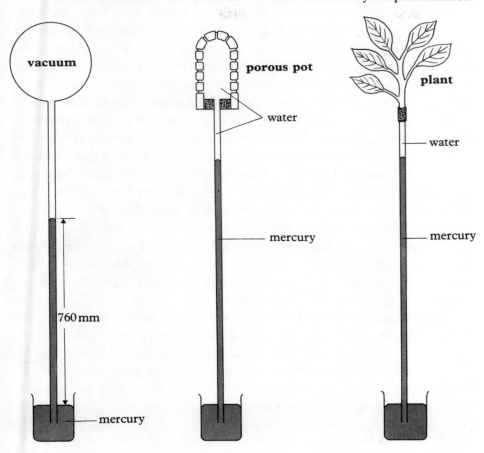

of capillarity, the rise of liquid in a capillary tube. If a series of open capillary tubes are dipped in water, the latter rises seemingly unaided to a height which depends on, amongst other things, the diameter of the tube: the narrower the tube the greater the height reached by the water.

That water rises at all can be tentatively explained as follows: the wall of the tube just above the surface of the water attracts water molecules upwards by adhesion. The latter attract other water molecules by cohesion. Thus the water is drawn upwards. The water continues to rise until the upward forces are equalized by the weight of water in the tube.

But why should the height reached by the water depend on the diameter of the tube? We can explain this by postulating that, because of the surface: volume ratios, the difference between the total adhesive force and any opposing forces, including the weight of the water, will be greatest in the narrowest tube, and lowest in the widest one.

Although it serves as a useful demonstration of adhesion and cohesion, capillarity, as such, is unlikely to play a significant part in the uptake of water. A plant does not draw up water into empty vessels and tracheids from scratch. Water fills the vessels and tracheids from the moment the plant starts to develop, and the water columns grow as the plant grows. The problem which a tall plant has to solve is to support the very long columns of water already present and to maintain their steady flow through the plant.

We must now return to our plant and see to what extent it complies with the porous pot analogy. You will remember that the effective evaporating surface of a plant consists of the cellulose walls of the leaf mesophyll cells. Studies on the fine structure of the cellulose wall have shown that it is perforated by numerous channels of molecular size which open at the surface as minute pores. As in the pores of the porous pot, considerable adhesive forces will be expected to develop in these narrow channels. So long as the total adhesive force is equal to the weight of the water it will prevent it from falling back. Within the vessels and tracheids in the leaves and stem, cohesive forces between the water molecules prevent the water column breaking in the middle.

The forces of cohesion and adhesion, then, provide a possible explanation of how continuous water columns are maintained in a tall plant, but how is the water kept moving from roots to leaves? One can only suppose that continual evaporation from the exposed surfaces results in water molecules replacing those that have been lost to the atmosphere. As a result of cohesion these will tend to pull other water molecules towards the surface, and so on.

It must be stressed that the mechanisms proposed here are tentative. There is much that is still not understood about the flow of water through plants and there is disconcertingly little experimental data available. Can you yourself spot any flaws in the argument presented here? What experiments might be carried out to settle the matter?

We have accounted for the ascent of sap without postulating the necessity for living tissues as such. Indeed Strasburger's demonstration that water can be drawn up stems whose cells have been killed seems to indicate unequivocally that living cells are not required, the process depending on purely physical forces. And yet there are some curious anomalies. For example, breaking the water columns in a tree trunk should put an end to further uptake. But it is claimed that this is not necessarily true. Again, changing the temperature should have no effect on water uptake except in so far as it affects the physical forces already discussed. And yet certain temperature changes that are known not to disturb the physical forces do affect the rate of water uptake. There is thus a school of thought which believes that physical forces may be aided by active processes occurring in the living cells. Certainly root pressure mentioned earlier will contribute towards the ascent of water.

THE FUNCTIONS OF TRANSPIRATION

Transpiration is the inevitable result of the necessity for the inside of the leaf to be open to the atmosphere. As such it might be regarded as more of an embarrassment than a help. But it does have a positive function, namely that it cools the leaves, an important effect particularly in hot conditions. The transpiration stream, as we have seen, is responsible for making good the water lost through transpiration. In addition it provides the pathway through which mineral elements are transported in the plant, to which subject we now turn.

Uptake of Mineral Salts

In Chapter 5 we saw that plants require in addition to carbon dioxide and water a variety of mineral elements. These are absorbed as the appropriate ions from the surrounding water in the case of aquatic plants, and from the soil in the case of terrestrial plants.

The data given in Fig 12.12 for the aquatic alga *Nitella* indicates that the concentration of certain ions may be many times greater in the cells than in the surrounding water, indicating that they enter the plant against a concentration gradient. Moreover certain ions are more concentrated than others. These observations suggest that ions are selectively absorbed by **active transport** involving the expenditure of energy.

There is good reason to believe that in terrestrial plants mineral ions are absorbed by similar means. It has been found, for instance, that the uptake of certain ions by young barley roots is increased by raising the temperature, and decreased by oxygen deprivation or treatment with a metabolic poison, all strong indications that active transport is involved (Fig 12.13).

Like water, mineral ions are taken up into the root hairs and other surface cells in the young parts of the root. After absorption they move towards the vascular tissues in the centre of the root by a process which is as yet unknown. Even the pathway through which they travel is uncertain. They may move along the cellulose walls or through the cytoplasm, or both. In the latter case the plasmodesmata (p. 27) would provide a route by which they might move from cell to cell. There is some evidence that the endodermis may concentrate the ions before they pass on to the vascular tissues, but the relative importance of active and passive mechanisms in the overall process is not known.

Once inside the vessels and tracheids, the ions are carried up the stem along with water in the transpiration stream. This has been shown by **ringing experiments**. All the living tissues are removed in a ring from around the central core of vessels and tracheids in a woody stem, and the plant is then placed in a solution containing radioactive phosphorus ^{32}P. Removal of the living cells in no way impedes the upward movement of the radioactive phosphorus which can subsequently be detected in the leaves by means of a Geiger counter.

From the xylem elements the ions are conveyed, probably by a combination of diffusion and active transport, to their two main destinations, the photosynthetic cells of the leaf and the various growing points in the plant. Here they are put to their sundry uses, for example the building up of amino acids and proteins. The ultimate fate of the various mineral ions is discussed in Chapter 10.

SPECIAL METHODS OF OBTAINING ESSENTIAL ELEMENTS

Some soils are deficient in mineral salts and plants living in them have special means of obtaining nitrogen and other essential elements. For example, the roots of many moorland and woodland plants, particularly coniferous trees, which live in soil rich in humus but deficient in mineral salts possess a **mycorrhiza**, an association between their roots and a fungus. In some cases the fungus is located on the surface of the root, in others it is internal; either way, it has the ability, not shared by the host plant, to break down the proteins of the humus into soluble amino acids, some of which are absorbed and utilized by the host. The majority of plants with mycorrhizal roots only obtain phosphorus, and perhaps nitrogen, compounds from the fungus. They can

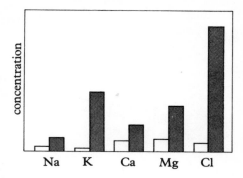

Fig 12.12 Diagram showing the relative concentrations of different ions in pond water (clear boxes) and in the cell sap of the green alga *Nitella* (shaded boxes). The much greater concentration of ions in the cells suggest that their absorption involves active transport. (*data by Hoagland*)

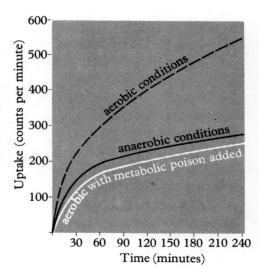

Fig 12.13 The influence of oxygen deprivation and metabolic poison on the uptake of sulphate ions by intact barley plants. The plants were provided with sulphate labelled with radioactive ^{35}S and the amount taken up by the plants was estimated by means of a Geiger counter. The much reduced uptake in anaerobic conditions and with addition of a metabolic poison suggests that active transport is involved. (*data supplied by R. Gliddon, Clifton College.*)

photosynthesize normally and can therefore manufacture their own carbohydrates, some of which are absorbed by the fungus. The association is therefore one in which both partners benefit, an example of **symbiosis**. Although most plants with mycorrhizal roots can photosynthesize, there are some that lack chlorophyll and cannot do so: a good example is the bird's nest orchid, a sickly looking inhabitant of dark woods. Such plants depend upon their mycorrhiza for carbohydrates as well as protein.

Another special adaptation is seen in leguminous plants – peas, beans, gorse, clover etc. Their roots contain bacteria of the genus *Rhizobium* which are able to absorb nitrogen from the air and build it up into amino acids. The bacteria gain entrance to the plant through its root hairs. Once inside they cause the parenchyma cells to divide into a **root nodule**. A vascular strand connects the nodule with the vascular tissues of the main root. The bacteria live and multiply in the parenchyma cells of the root nodule, absorbing carbohydrates from the host and nitrogen from the atmosphere, and synthesizing amino acids, some of which are passed on to the host. As with mycorrhiza the association is symbiotic, both partners gaining from it. A number of other micro-organisms can also fix atmospheric nitrogen. Some are free-living in the soil, but many form symbiotic associations with higher plants, and not only leguminous ones. Nitrogen fixation is much more widespread than was originally supposed, and it plays an important part in enriching the soil in available nitrogen compounds. Its beneficial effects are discussed on p. 161.

Plants living in nitrogen-deficient soil sometimes resort to feeding on animals as a means of obtaining nitrogen (see p. 136). Small animals, mainly insects, are trapped in a modified leaf and there subjected to the action of proteolytic enzymes. The products of digestion are absorbed into the leaf and transported away to wherever they are needed.

Transport of Organic Substances

So far we have considered the uptake and transport of the raw materials for photosynthesis and other metabolic processes in the plant. Now we must consider how the products of photosynthesis (sugars, amino acids and the like) are transported from the leaves where they are formed to the parts of the plant where they are needed. Obviously all the cells that are themselves unable to photosynthesize need a share in these materials, but the apices of the stem and side branches where cell division and growth are taking place are particularly needy. Transport of food materials to these **growing points** is greatest in the spring and summer when growth is most prolific. Later in the year many plants form **perennating organs** (tubers, bulbs, corms, etc.) to which food materials are transported for storage until the following season. When the next season arrives the stored food is transported in soluble form to the growing points of the new plant.

PHLOEM TISSUE

Transport of the soluble products of photosynthesis is called **translocation**, and a number of experiments have been done showing that it occurs in the part of the vascular tissue known as the **phloem**. In flowering plants phloem tissue consists mainly of elongated **sieve elements** placed end to end so as to form long **sieve tubes** running parallel with the long axis of the plant (Fig 12.7). Each sieve element is derived from a cell whose nucleus disintegrates during development. Its end walls, known as **sieve plates**, are perforated by numerous pores which allow the passage of materials from one sieve element to the next (Fig 12.14). The inside of the sieve element is filled with fine **cytoplasmic filaments** which are continuous, via the pores in the sieve plate, with similar filaments in the next sieve element. The cytoplasm of these filaments is structurally very simple: it contains no endoplasmic reticulum, mitochondria, plastids or other organelles. All these disintegrate during develop-

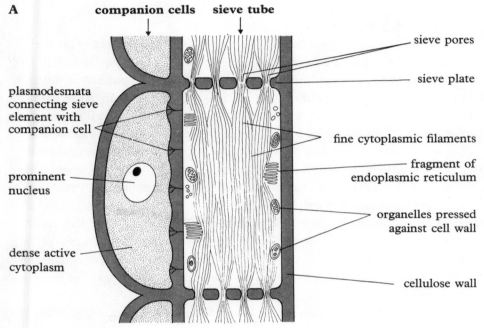

A

companion cells   **sieve tube**

sieve pores

sieve plate

plasmodesmata connecting sieve element with companion cell

fine cytoplasmic filaments

prominent nucleus

fragment of endoplasmic reticulum

organelles pressed against cell wall

dense active cytoplasm

cellulose wall

Fig 12.14 The structure of the phloem. A) Diagram of a sieve element and its companion cell. The sieve element has no nucleus and not many organelles, what few there are being pushed to the side. In contrast the companion cell has a prominent nucleus and dense cytoplasm containing numerous mitochondria, abundant endoplasmic reticulum and ribosomes. (*Modified after Clowes and Juniper*). B) Photomicrographs of sieve tubes from the phloem of flowering plants. C) Electron micrograph of part of a sieve tube and neighbouring companion cell. Notice the pores in the sieve plate and the plasmodesmata (**pd**) piercing the cellulose wall between the sieve tube and companion cell (**CC**). (B and C: *K. Esau and V. I. Cheadle, University of California*)

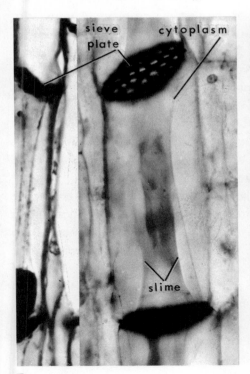

B

sieve plate   cytoplasm

slime

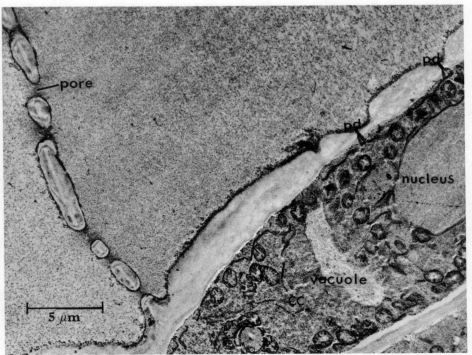

C

pore

pd

pd

nucleus

vacuole

cc

5 μm

ment. A few such organelles persist immediately adjacent to the cellulose wall, but elsewhere they are totally absent.

Closely applied to the side of each sieve element are one or more **companion cells,** which possess a nucleus, dense endoplasmic reticulum, ribosomes and numerous mitochondria. Plasmodesmata connect each sieve element with its adjacent companion cells. The latter are the site of intense metabolic activity, and one is tempted to assume that they provide the more inert sieve element with its metabolic needs. On purely structural grounds it would seem that translocation of food materials takes place along the sieve tubes. Assuming this is an active process, we may postulate that the energy is derived from the adjacent companion cells. However, this is pure speculation.

ESTABLISHING THE SITE OF TRANSLOCATION

How do we know that translocation occurs in the phloem? Some of the earliest investigations involved **ringing experiments** of the type described on p. 193. It so happens that in a mature tree trunk the phloem is confined to the bark, indeed the living component of bark is almost entirely phloem tissue. Now if a ring of bark is stripped off a tree trunk it can be shown that the sugar concentration increases immediately above the ring and decreases below it, indicating that the downward movement of sugars is blocked at that point.

More critical investigations have been carried out with **radioactive tracers**. If a plant is exposed to carbon dioxide labelled with radioactive ^{14}C, the ^{14}C becomes incorporated into the end products of photosynthesis which are subsequently detected in the stem. That these substances are confined to the phloem can be shown by cutting sections of the stem, placing the sections in contact with photographic film and making autoradiographs. It is found that the sites of radioactivity correspond precisely to the positions of the phloem.

These experiments indicate that organic food materials are transported in the phloem. In fact one can imagine that the sieve tubes carry a constant stream of materials from the leaves to the rest of the plant. A number of **parasitic organisms** make use of this and plug into the phloem of certain plants in order to gain nourishment. Thus several parasitic plants send out sucker-like processes that pierce the roots or stem of the host plant and link up with the sieve tubes. Dodder, a climbing plant belonging to the convolvulus family, attaches itself to the stems of various host plants in this way; broomrapes and toothwort (Fig 12.15) go for the roots. In all these cases the fact that the parasite acquires ready-made organic food means that it does not require chlorophyll and accordingly it lacks the green colour characteristic of most plants and has a yellow appearance.

A striking example of an animal parasite that feeds on the contents of the phloem are aphids (green flies). Aphids suck plant juices by means of a proboscis, a sharp tube rather like a hypodermic needle which is inserted into the stem or one of the leaf veins (see p. 138). The plant juices then pass up the proboscis to the aphid's stomach. Use has been made of this by T.E. Mittler of the University of California to study translocation. His method is beautiful for its simplicity and ingenuity. An aphid alights on a stem or leaf and inserts its proboscis into the tissue. A stream of carbon dioxide is then passed over the animal in order to immobilize it. The proboscis is then cut as close to the head as possible, and left sticking into the plant. It is found that fluid exudes

Fig 12.15 The great toothwort, *Lathraea squamaria*, in flower. A parasite on the roots of beech and hazel, its leaves are reduced to small scales entirely devoid of chlorophyll. (*B. J. W. Heath, Marlborough College*)

from the cut end of the proboscis and may continue to do so for days, providing a perfect technique for tapping the contents of the phloem with minimum injury to the plant. The fluid is collected with a micro-pipette, and chemical analysis shows that it contains concentrated sugars and amino acids. That these substances really come from the phloem can be confirmed by cutting thin sections of the part of the stem or leaf which has the proboscis in it. In this way it has been demonstrated that the tip of the proboscis pierces a single sieve element.

The rate at which fluid exudes from an aphid proboscis may exceed 5 mm^3/h, which must delight the average aphid as it means that it does not even have to suck in order to get a plentiful supply of food. For a botanist this provides a means of calculating the speed of translocation. An exudation rate of 5 mm^3/h means that an individual sieve element must be emptied and refilled between 3 and 10 times per second. From this it can be calculated that the speed of translocation in an individual sieve tube is of the order of 1000 mm/h, a figure which is about twice as great as those obtained from work with radioactive tracers. (Why the difference, do you think?).

THE MECHANISM OF TRANSLOCATION

We come now to the most ticklish question of all: what causes the movement of materials in the sieve tubes? On this there is as yet no general agreement. Although sugars and amino acids tend to move along concentration gradients, the speed at which they travel makes it impossible to account for the process simply by diffusion.

This has led to the suggestion that there may be a **mass flow** of materials through the phloem as a result of a turgor pressure gradient: highest in the leaves and lowest in the roots. The high turgor pressure of the leaf cells could be caused by their high osmotic pressure already mentioned in connection with transpiration. However if this mass flow occurs one would expect different substances to move at the same speed, but it is known that sugars and amino acids, for example, can move at different speeds, and even in opposite directions, within the same sieve tube.

Independent movement of different substances strongly suggests some form of **active transport**. This is supported by other lines of evidence. For example there is a close correlation between the speed of translocation and metabolic rate of the phloem. Again, lowering the temperature, decreasing the oxygen content and treatment with metabolic poisons all reduce the speed of translocation, suggesting that it is an active energy-requiring process which cannot be explained by physical forces alone.

The conditions in which life processes can take place are narrowly defined, and fluctuations in the immediate environment of the cells can play havoc with their normal functioning. The principles underlying the mechanisms by which the body adjusts to changing conditions are outlined in Chapter 13, the control of blood sugar being taken by way of illustration. This chapter also includes an account of the liver, one of the most important organs controlling internal conditions in the mammalian body. In subsequent chapters other aspects of control are discussed. Chapter 14 deals with elimination of nitrogenous waste matter and the control of water balance, Chapter 15 with the regulation of body temperature. The control processes discussed in these two chapters have been of paramount importance in the evolution of living things.

The human body must be able to adjust quickly to sudden changes in conditions. Nowhere can this be better seen than in the control of carbon dioxide, particularly during muscular exertion. This is the subject of Chapter 16. The last chapter in this section discusses the body's defences against disease as an aspect of internal control.

Edward White, Gemini IV, floats in space.
(NASA.)

Part III
Adjustment and Control

13 The Principles of Homeostasis

One of the generalizations to emerge from physiological studies in the present century is that the conditions in which cells can function properly are narrow. Even quite small fluctuations in osmotic pressure, temperature or the amounts of certain chemical substances, can disrupt biochemical activities and in extreme cases may kill the cells altogether.

This basic tenet of biology was first recognized by the French physiologist Claude Bernard. In 1857 he wrote *La fixité du milieu interieur est la condition de la vie libre* ('the constancy of the internal environment is the condition for free life'). Later we will examine the meaning of 'free life'. For the moment let us concentrate on the **internal environment**.

THE MEANING OF INTERNAL ENVIRONMENT

By internal environment is meant the immediate surroundings of the cells. In mammalian tissues the cells are generally surrounded by tiny channels and spaces filled with fluid. The latter is called **intercellular, interstitial,** or **tissue fluid**. It provides the cells with the medium in which they have to live, and represents the organism's internal environment. It is this that must be kept constant if the cells are to continue their vital functions. To see how this is achieved we must first understand how intercellular fluid is formed.

THE FORMATION OF INTERCELLULAR FLUID

Intercellular fluid is formed from blood by a process of **ultra-filtration** in which small molecules and ions are separated from the larger molecules and cells (Fig 13.2). When blood reaches the arterial end of a capillary it is under high pressure because of the pumping action of the heart and the fine bore of the capillaries. This pressure forces the fluid part of the blood through the walls of the capillaries into the intercellular spaces. Analysis of intercellular fluid shows that it consists of all the constituents of blood plasma, less the proteins. The walls of the capillaries act as a filter holding back the plasma protein molecules together with the cellular components of the blood, but allowing everything else to go through.

Once formed, the intercellular fluid circulates amongst the cells and eventually returns to the blood vascular system by one of two routes. At the venous end of the capillary systems the hydrostatic pressure of the blood is comparatively low, and is exceeded by the osmotic pressure of the plasma proteins which, owing to the removal of the other constituents of the plasma, are now much more concentrated than they were at the arterial end of the system. This causes intercellular fluid to be drawn back into the capillaries,

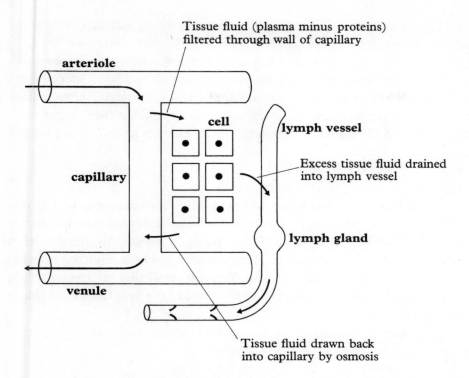

Tissue fluid (plasma minus proteins) filtered through wall of capillary

arteriole

cell

lymph vessel

capillary

Excess tissue fluid drained into lymph vessel

lymph gland

venule

Tissue fluid drawn back into capillary by osmosis

Fig 13.2   The formation of tissue fluid. Excess tissue fluid drained into the lymph vessel returns, as lymph, to the venous system. The lymph gland is the site of formation of lymphocytes which are important in defence against disease.

and so returned to the circulation. Excess intercellular fluid, however, is drained into the **lymph vessels**, whence it eventually passes back into the veins.

Intercellular fluid is the medium in which the cells are bathed. Through it respiratory gases diffuse, from it the cells extract all their metabolites, and into it they shed unwanted substances. This is Claude Bernard's *milieu interieur*, and since it is formed from the blood, it is the blood which must be kept constant. Much of an animal's physiology is concerned with doing just this.

WHAT FACTORS MUST BE KEPT CONSTANT?

The most important features of the internal environment that must be kept constant are:

(1) its chemical constituents, for example, glucose, ions etc;

(2) its osmotic pressure, determined by the relative amounts of water and solutes;

(3) the level of carbon dioxide;

(4) its temperature.

In addition certain chemical ingredients must be virtually eliminated from it altogether. The most important of these are nitrogenous waste products arising from protein metabolism, and toxic substances liberated by pathogenic micro-organisms.

The importance of a constant internal environment to the well-being of cells can be shown by removing tissues from the body. If they are subjected to conditions markedly different from those prevailing in the body they will die; but if maintained under the correct conditions they will survive. This was appreciated by the nineteenth century physiologist Sidney Ringer who perfected the art of keeping tissues and organs alive outside the body. He found, for example, that an excised heart of a frog or mammal would continue beating for a long time in a mixture of sodium, potassium and calcium ions. It is now known that virtually all tissues can be kept alive in a suitable 'cocktail' of ions similar to the tissue fluids. Such solutions, which vary according to the species, are known as **physiological salines** or **Ringer's solutions**.

Maintenance of a constant internal environment is known as **homeostasis**, a Greek word meaning 'staying the same'. Many physiological processes are homeostatic in that they are responsible, directly or indirectly, for regulating the internal environment. It is impossible to exaggerate their importance. Without them life would be impossible. As an example let us take the control of sugar.

THE HOMEOSTATIC CONTROL OF SUGAR

The normal value of sugar in the human bloodstream is approximately 90 mg/100 cm³, and even after the heaviest carbohydrate meal rarely exceeds 150 mg/100 cm³. What keeps it constant? Before trying to answer this question let us briefly consider what happens to glucose after it has been absorbed from the small intestine into the bloodstream. You will recall that after entering the hepatic portal vein, it is conveyed to the liver. In the liver three main things may happen to it:

(1) it may be broken down into carbon dioxide and water (cell respiration);

(2) it may be built up into glycogen and stored;

(3) it may be converted into fat and sent to the body's fat depôts for storage;

(4) instead of being metabolized or stored, it may pass on from the liver to the general circulation. In fact under certain circumstances the glycogen stores in the liver may be broken down so as to add to the level of glucose in the body.

The level of glucose in the blood and tissue fluids at any given moment is mainly determined by the relative extent to which these different processes occur in the liver. If there is too much glucose, as for example after a heavy meal rich in carbohydrate, the liver metabolizes what it can, and stores the rest as glycogen. If there is a deficiency of glucose, the liver breaks down glycogen into glucose, thereby raising the glucose level in the body. In cases of prolonged deficiency glucose may be formed from non-carbohydrate sources, even protein. The wasting away of the tissues, which occurs in extreme starvation, is because the body resorts to converting its protein into carbohydrate.

THE ROLE OF INSULIN

The liver cannot perform this homeostatic function unaided. It has to receive information instructing it what to do. This is provided in the form of the hormone **insulin**, which is secreted into the bloodstream by special groups of cells, the **islets of Langerhans**, in the **pancreas** (Fig 13.3). In Chapter 9 we discussed one function of the pancreas, namely the production of digestive enzymes. We see now that it also functions as an **endocrine organ**, secreting a hormone into the blood stream. On reaching the liver, insulin exerts its effects by increasing the oxidative breakdown of glucose, and facilitating the conversion of glucose to glycogen and fat. At the same time it inhibits the formation of glucose from glycogen and non-carbohydrate sources. Insulin thus achieves the overall effect of lowering the level of glucose in the body. In the absence of insulin the reverse takes place: oxidative breakdown and storage

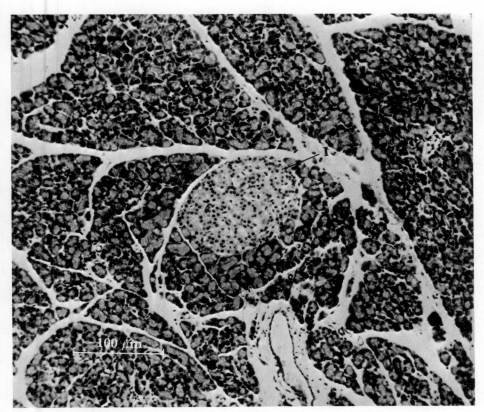

Fig 13.3 Photomicrograph of section of pancreas showing an islet of Langerhans (**I**) in the centre. The pancreas is both an endocrine (ductless) and an exocrine (ducted) gland. The islets secrete the hormone insulin into the bloodstream, while the surrounding cells secrete digestive enzymes into the pancreatic duct and thence to the duodenum. (*J. F. Crane*)

of glucose is inhibited, and additional glucose is formed from storage compounds. As a result the glucose level rises.

Clearly insulin plays a vital rôle in the regulation of glucose. Without it the liver cannot respond appropriately to the needs of the body. This can be illustrated by considering what happens if the pancreas is surgically removed from an animal. The result is a drastic increase in the general level of glucose in the blood, accompanied by a decrease in the glycogen content of the liver and muscles. The blood sugar level exceeds its normal value (a condition known as **hyperglycaemia**); when it reaches a critical level, glucose starts to be excreted in the urine, a condition called **glycosuria**. These conditions can be quickly reversed if insulin is injected into the bloodstream.

DIABETES

Certain individuals have islets of Langerhans which, for one reason or another, are unable to produce as much insulin as they should. The result is a condition known as **diabetes melitus**, the symptoms of which are similar to those seen in an animal deprived of its pancreas. There is an increase in the blood sugar level (hyperglycaemia), and sugar appears in the urine (glycosuria). If untreated, the condition is fatal. Diabetes can be prevented by regular injections of insulin. Unfortunately the hormone cannot be taken by mouth as it is a protein and is digested in the alimentary canal. A diabetic must therefore inject himself with insulin at regular intervals. The existence of insulin, and its rôle in regulating sugar and preventing diabetes, was discovered by the Canadian physiologists Frederick Banting and Charles Best in the early 1920s. Their discovery stands as a landmark in the history of physiology and clinical medicine.

WHAT CONTROLS THE SECRETION OF INSULIN?

We have seen that the liver, under instructions from the pancreas, regulates the body's sugar level. What regulates the production of insulin by the islets of Langerhans? There is good reason to believe that it is the amount of glucose in the blood itself which is the effective agent. If the blood sugar level is abnormally high, this stimulates the islet cells to produce correspondingly more insulin. On the other hand if the blood sugar level is low, less insulin is secreted. In other words the sugar itself switches on the mechanism by which it is itself regulated, an excess of sugar setting into motion the physiological processes which return the sugar level to its normal value. The sequence of events is summarized in Fig 13.4.

Fig 13.4 Homeostatic scheme for the control of glucose in the mammalian body. For further explanation see text.

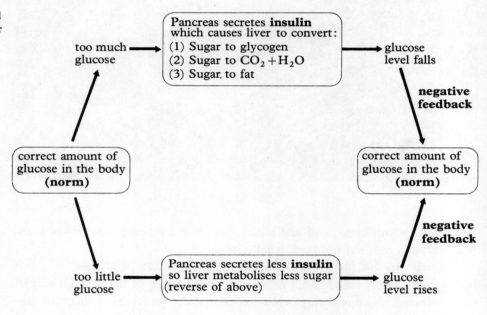

SOME GENERALIZATIONS ABOUT HOMEOSTASIS

Regulation of sugar illustrates an important principle of homeostasis, namely that the corrective mechanism is triggered by the very entity which is to be regulated. In other words, homeostasis involves a **self-adjusting mechanism,** the control process being built into the system.

The sugar story illustrates another principle of homeostasis: **negative feedback.** This is the term given to the fact that, in the case of sugar-regulation, an increase in the amount of sugar sets into motion processes which decrease it. Conversely a decrease in the sugar level sets into motion processes which increase it. In other words a change in the sugar content of the blood automatically brings about the opposite effect. In both cases the result is that the level of sugar is kept reasonably constant.

Another generalization emerging from this is that homeostasis must necessarily involve fluctuations, small though these may be. The control of sugar can obviously only work if there is a periodical increase, or decrease, in the sugar level. The blood sugar level does not, and cannot, remain absolutely constant, but wavers within narrow limits on either side of an optimum value which we can call the **norm** or **set-point.** Only by deviating from this norm can the homeostatic mechanism be brought into play which restores the sugar to its normal value. What invariably happens is that when the blood sugar falls, it overshoots the mark and drops below the norm, thus triggering the corrective processes which cause it to rise again, and so on.

Clearly, for any homeostatic mechanism to work there must be receptors capable of detecting the change, a control mechanism capable of initiating the appropriate corrective measures, and effectors that can carry out these corrective measures.

These basic principles underlie all homeostatic mechanisms and are summarized in Fig 13.5. Their effectiveness depends on negative feedback. Sometimes the corrective mechanism leading to negative feedback breaks down with the result that a deviation from the norm initiates further deviation. This is described as **positive feedback**, examples of which will be considered in later chapters.

We have seen how homeostasis works in the regulation of sugar. Later we shall apply it to the regulation of osmotic pressure, temperature and respiratory gases. It is by no means confined to physiological situations. The size of a population of animals or plants is kept constant by a homeostatic process, and essentially the same mechanism maintains the constancy of species over long periods of time (see Chapter 35).

Nor are homeostatic mechanisms confined to biology. To take a very obvious example, the **thermostat** operates on a homeostatic basis, switching itself on or off according to the temperature. Many machines involve similar principles, and even industrial processes, economic systems, and stabilization of prices can be analysed in similar terms.

The term now used to embrace all these concepts is **cybernetics**, the science of communication and control. A Greek word meaning 'steersman', the term was first coined by the French physicist and mathematician A. M. Ampère over a hundred years ago, but it was applied more widely by Norbert Wiener in 1948. The term is now used as a unifying concept to include all forms of self-adjusting mechanism both in machines and organisms.

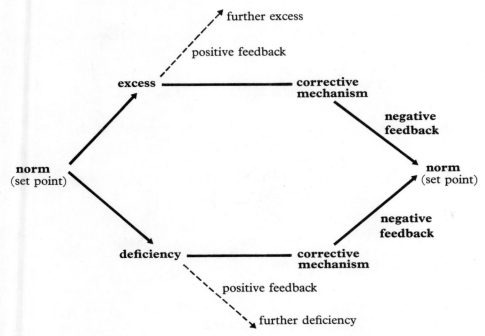

Fig 13.5 General scheme summarizing any homeostatic control process. Any deviations from the norm (set point) set into motion the appropriate corrective mechanisms which restore the norm (negative feedback). This contrasts with positive feedback where a deviation from the norm leads to further deviation.

The Mammalian Liver

Earlier in this chapter we saw that the liver plays a central rôle in the regulation of sugar. This is only one of several important homeostatic functions performed by this vital organ. In studying the liver we see a very close relationship between structure and function. We will therefore start by reviewing its functions, and then relate these to its structure.

FUNCTIONS OF THE LIVER

It has been estimated that the liver performs over 500 functions. For our purposes these can be conveniently reduced to twelve:

(1) Regulation of sugar. Through the intermediacy of insulin, the liver removes glucose from, or adds glucose to, the blood, depending on circumstances. Inside the liver cells the glucose is either built up into glycogen for storage, or broken down into carbon dioxide and water with the liberation of energy. If in considerable excess it is converted into lipids.

(2) Regulation of lipids. The liver cells remove lipids from the blood, and either break them down or modify them and send them to the fat depots.

(3) Regulation of amino acids, and hence proteins. The body is unable to store proteins or amino acids, and any surplus is destroyed in the liver. Excess amino acids, brought to the liver in the hepatic portal vein, are **deaminated** by the liver cells. In this process the amino (NH_2) group is removed from the amino acid, with the formation of ammonia. The amino acid residue is then fed into carbohydrate metabolism and oxidized. Meanwhile the ammonia must not be allowed to accumulate for it is highly toxic even in small quantities. Under the influence of specific enzymes in the liver cells, the ammonia enters a cyclical series of reactions (the **ornithine cycle**) in which it reacts with carbon dioxide to form the less toxic nitrogenous compound **urea** (Fig 13.6). The urea is then shed from the liver cells into the bloodstream,

Fig 13.6 The fate of excess amino acids in the body. In the process of deamination (A) the amino group is removed from the amino acid resulting in the formation of ammonia. In most animals this highly toxic substance reacts with carbon dioxide to form the less toxic compound urea, as summarized in (B). In fact urea-formation takes place via a complex cycle of reactions, the ornithine cycle, shown in outline in (C).

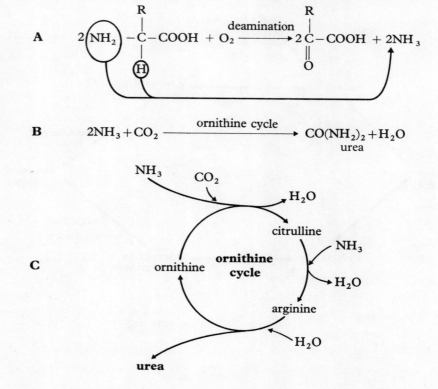

and taken to the **kidney** which eliminates it from the body. How this is done is explained in Chapter 14.

(4) Production of heat. The liver is therefore involved in temperature regulation. Its continuously high metabolic rate, coupled with its large size and excellent blood supply, make it ideal for the steady production and distribution of heat.

(5) Production of bile. This is synthesized by the liver cells and stored in the **gall bladder**. From the latter it passes down the bile duct to the duodenum. The **bile salts** play an important part in digestion by emulsifying fats in the small intestine. In addition the bile contains several important excretory products.

(6) The formation of cholesterol. This is a fat derivative, some of which is required as an important constituent of cell membranes, particularly of nerve cells. Excess cholesterol is excreted in the bile. If there is a considerable surplus, it may precipitate in the gall bladder, or bile duct, as **gall stones**. These sometimes block the bile duct, leading to obstructive jaundice in which the skin acquires a characteristic yellow appearance due to retention of bilirubin (see below) in the blood. The amount of cholesterol and other lipids in the blood is largely determined by dietary intake in conjunction with the activities of the liver. If there is a considerable excess in the blood, some of it may be deposited in the walls of certain arteries, obstructing the smooth passage of blood and often leading eventually to an intravascular clot. If this occurs in the coronary vessels of the heart, the result is a **coronary thrombosis** or 'heart attack'. The elimination of excess cholesterol is thus an important function of the liver.

(7) Elimination of sex hormones after they have performed their functions. Some are modified chemically by the liver cells, some are sent to the kidney for renal excretion, and others are expelled in the bile.

(8) Formation of red blood cells. In the foetus the liver is directly responsible for the formation of red blood cells, but in the adult this function is taken over by the bone marrow. However, the adult liver still plays an important part in the production of red blood cells for it produces a chemical substance called the **haematinic principle**, necessary for the formation of red blood cells in the bone marrow. For the liver cells to synthesize this principle, **vitamin B12**, a complex cobalt-containing ring compound, is required. If the liver's haematinic principle is lacking the result is **pernicious anaemia**, a fatal condition characterized by a drastic reduction in the amount of haemoglobin in the blood.

(9) Elimination of haemoglobin from used red blood cells. The cells are destroyed by phagocytic cells lining the blood vessels of the liver, spleen and bone marrow. The haemoglobin is broken down by the liver cells into the green pigment **biliverdin**, which is then reduced to the brown pigment **bilirubin**. This is eliminated in the bile, giving it its characteristic colour.

(10) Storing blood. The veins in the liver have great powers of expansion and contraction, to such an extent that the total volume of blood in the human liver can vary from 300 cm³ to 1500 cm³. This enables the liver to serve as a **blood reservoir**. Along with the spleen, it can regulate the amount of blood in the general circulation.

(11) Synthesis of plasma proteins. These include **fibrinogen**, the protein responsible for the clotting of blood. The other two plasma proteins,

albumen and **globulin**, have a number of functions connected with homeostasis, some of which have already been discussed.

(12) **Storage of vitamins** A, D, and B12, and a number of **minerals** such as potassium, iron and copper.

Reviewing these functions of the liver, one is struck by the extent to which it regulates the physical and chemical composition of the internal environment. Each of the twelve functions listed above is in some way connected with homeostasis. We can summarize the rôle of the liver by saying that it synthesizes certain vital substances required by the body, stores compounds which are of no immediate use, purifies the blood that passes through it, and generally regulates the internal environment.

STRUCTURE OF THE LIVER

The structure of the liver is directly related to these functions. It is the largest organ in the body, weighing about 1.5 kg, 3—4 per cent of the total body weight. It has an excellent blood supply, receiving more blood per unit time than any other organ. In fact it has been calculated that the blood-flow through the liver is well over a litre per minute.

The liver might well be described as the body's metabolic centre. As such, it must have a good blood supply from which it can draw its raw materials, and into which it can shed its products. Its blood supply is derived from two sources:

(1) the **hepatic artery** which brings oxygenated blood from the dorsal aorta, and

(2) the **hepatic portal vein** which brings blood rich in food materials from the gut. As much as three-quarters of the blood reaching the liver does so via the hepatic portal vein. The liver sheds its products into the **hepatic vein**: glucose, amino acids, lipids, plasma proteins, urea, cholesterol and of course carbon dioxide from the respiration of its cells. Bile is shed into the bile duct.

One of the remarkable things about the liver is that nearly all its functions are carried out by one type of cell, the **liver cells**. These are structurally undifferentiated, and all identical in appearance. They have no special structural features except numerous mitochondria and a prominent Golgi apparatus.

Microscopical examination of the liver shows an intimate association between the liver cells, blood vessels and bile channels. This is shown in Fig 13.7. The liver is composed of numerous **lobules** roughly cylindrical in shape and approximately 1 mm in diameter. Each lobule is filled with liver cells arranged in rows radiating from the centre towards the periphery (Fig 13.8). Running alongside each lobule are branches of the hepatic artery, hepatic portal vein and bile duct. The first two are sometimes referred to as the **interlobular blood vessels** since they are located between adjacent lobules. In the centre of each lobule is a branch of the hepatic vein, referred to as the **central** or **intralobular vein**. The interlobular vessels are connected with the central vein by a system of capillary-like **sinusoids** that run parallel to, and come into close contact with, the chains of liver cells. The latter are surrounded by fine channels called **canaliculi** which connect up with the bile ducts at the edge of the lobule. So the liver cells bear an intimate association with the sinusoids on the one hand, and the canaliculi on the other.

The whole of this lobule is filled with radiating rows of liver cells, canaliculi and sinusoids as shown in B

liver lobule

A

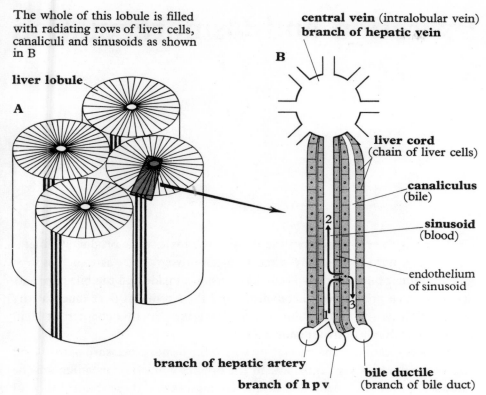

central vein (intralobular vein)
branch of hepatic vein

B

liver cord
(chain of liver cells)

canaliculus
(bile)

sinusoid
(blood)

endothelium
of sinusoid

branch of hepatic artery

branch of h p v

bile ductile
(branch of bile duct)

Fig 13.7   Structure of a liver lobule. The liver is made up of numerous lobules, cylindrical in shape and approximately 1 mm in diameter. There is an intimate relationship between the liver cells, sinusoids and canaliculi. In (B) the arrows indicate the flow of materials to and from the liver cells. What precisely do arrows 1, 2 and 3 represent? **hpv** stands for hepatic portal vein.

Blood reaches each lobule via the interlobular vessels: oxygenated blood arrives in the branches of the hepatic artery, blood rich in food materials in the branches of the hepatic portal vein. The blood then flows along the sinusoids towards the central vein. As it does so, the liver cells take up from the blood what they require, and shed their products into it. The only exception is the bile which is secreted, not into the sinusoids, but into the canaliculi whence it trickles to the bile duct.

Attached to the walls of the sinusoids are specialized **Kupffer cells**, part of the **reticulo-endothelial system**. These cells are phagocytic, and their function is to destroy old red blood cells. All other functions of the liver are carried out by the liver cells.

One of the many substances brought to the liver in its extensive blood supply is insulin which, as we have seen, controls its activities in relation to sugar. The pancreas and liver co-operate with one another in the homeostatic regulation of sugar. Such interaction between organs is fundamental to the homeostatic mechanisms of complex animals, as we shall see in subsequent chapters.

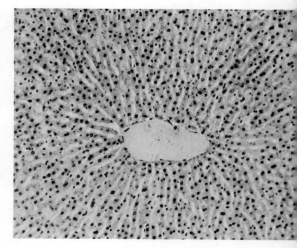

Fig 13.8   Photomicrograph of a single lobule of human liver in transverse section. Notice the large central vein and the radiating rows of liver cells and sinusoids. (*J. F. Crane*)

14 Excretion and Osmoregulation

Excretion is the removal from the body of the toxic waste products of metabolism. The term is generally taken to mean nitrogenous waste such as urea and ammonia, but other materials like carbon dioxide and the bile pigments are also waste products of metabolism, and their removal is as much a part of excretion as the elimination of urea. However, in this chapter we shall confine ourselves to **nitrogenous excretion**.

Osmoregulation is the process by which the osmotic pressure of the blood and tissue fluids is kept constant. In the present context its meaning may be extended to include regulation of the composition of these fluids. You will recall that the blood and tissue fluids contain a variety of solutes (glucose, salts and the like) dissolved in water. Regulation is necessary not only with respect to water content but also with respect to the concentrations, relative and absolute, of the various solutes normally present in the body fluids. The osmotic pressure is determined by the relative amounts of these solutes and water, and maintaining the correct balance between these two is the business of osmoregulation.

It may strike you as odd that these two physiological processes, excretion and osmoregulation, should be considered together. However, both are concerned with homeostasis and the physiological mechanisms involved are intimately bound up with each other. Indeed in the mammal the same organ performs both functions. This is the **kidney** to which we must now direct our attention.

The Mammalian Kidney

There are two kidneys, one on each side of the abdomen, and each contains between one and two million **nephrons**, loosely embedded in connective tissue and amply supplied with blood. The nephron can be regarded as the functional unit of the kidney, performing both functions of excretion and osmoregulation. To understand how it does this we must first look at its structure (Fig 14.1).

STRUCTURE OF THE NEPHRON

At one end of the nephron is a cup-shaped **Bowman's capsule**, about $200\mu m$ in diameter, located in the outer region (cortex) of the kidney. It is like a ball

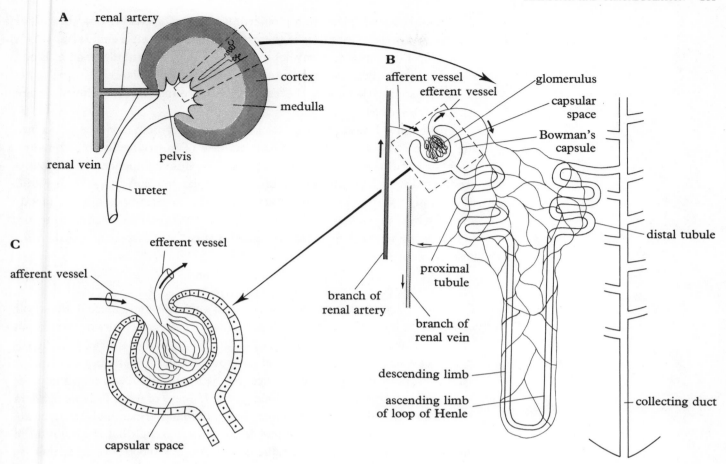

A) renal artery · cortex · medulla · renal vein · pelvis · ureter

B) afferent vessel · efferent vessel · glomerulus · capsular space · Bowman's capsule · distal tubule · proximal tubule · branch of renal artery · branch of renal vein · descending limb · ascending limb of loop of Henle · collecting duct

C) efferent vessel · afferent vessel · capsular space

Fig 14.1 Structure of the mammalian kidney. A) The kidney in horizontal section to show the position of nephrons relative to the whole kidney. B) Single nephron with its blood supply. C) Bowman's capsule and glomerulus on large scale. D) Photomicrograph of section through cortex showing Bowman's capsules, glomeruli and convoluted tubules. (*D: Flatters and Garnett Ltd., Manchester*)

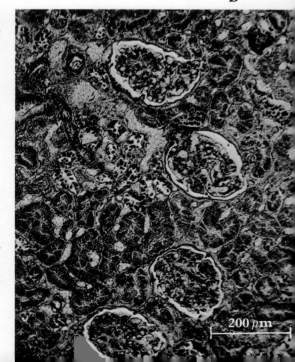

D

200 μm

which has been pressed in on one side. The invagination contains a dense network of capillaries called the **glomerulus**. Between the inner and outer linings of the capsule is a cavity, the **capsular space**, whose inner lining is closely applied to the walls of the glomerular capillaries. The barrier between the capsular space and the lumen of the capillaries is composed of only two layers of cells: the epithelium of the capsule and the endothelium of the capillaries. The significance of this will become clear later.

Leading from Bowman's capsule is a tubule, about 60 μm in outer diameter, whose lumen is continuous with the capsular space. The first part of the tubule, known as the **proximal convoluted tubule**, is contained in the cortex and is highly coiled. This leads to a U-shaped **loop of Henle** which plunges down into the deeper part of the kidney (the medulla) where it does a hair-pin bend and returns to the cortex. Here it leads on to another coiled section, the **distal convoluted tubule**, which opens into a **collecting duct**, along with several other nephrons. The ducts converge at the **pelvis** of the kidney, shedding their contents into the **ureter** which conveys the urine to the **bladder** for temporary storage.

The essential job of the kidney is to purify the blood which flows through it, extracting and eliminating harmful substances. To this end it has an extensive blood supply. It is true that the total flow of blood through the kidney per unit time is not as great as through the liver, but the flow per weight of tissue is far greater than for any other organ (over 700 cm³/min/100 g tissue). It has been estimated that the total length of blood vessels in both kidneys is of the order of 160 kilometres.

Blood is conveyed to each glomerulus by an **afferent vessel**, a branch of the **renal artery**. Blood leaves the glomerulus by an **efferent vessel** which breaks up into capillaries enveloping the whole of the tubule system. Blood is drained from these capillaries into the **renal vein**. So blood flows first to the glomerulus and thence to the capillaries surrounding the tubules before leaving the kidney.

Although a few substances are known to be secreted from the blood into the tubules, the kidney accomplishes its task of purifying the blood mainly by the use of **filtration** and **reabsorption**. The fluid components of the blood are filtered from the glomerulus into the capsular space. As the resulting fluid flows along the tubules useful substances are reabsorbed back into the bloodstream whilst unwanted substances flow on to the ureter and bladder as urine. Let us examine each of these processes in turn, first filtration and then reabsorption.

FILTRATION IN BOWMAN'S CAPSULE

Our knowledge of what happens in Bowman's capsule, and indeed in the whole nephron, is largely founded on a series of elegant experiments carried out by A. N. Richards at the University of Pennsylvania in the late 1930s. Richards inserted a micro-pipette into individual capsules and drew off very small quantities of fluid from the capsular space (Fig 14.2). The fluid was then analyzed. Most of these experiments were done on the kidneys of frogs and salamanders in which glomeruli can be seen more clearly than in the mammal, but later experiments on mammals gave similar results. It was found that the fluid in the capsular space has exactly the same composition as blood plasma minus the plasma proteins. It seems that the capsular fluid is formed by a process of **ultra-filtration** from the glomerular capillaries. The filtrate (**glomerular filtrate**) contains all the constituents of blood except for the plasma proteins and the blood cells: i.e. those components of blood which are too large to pass through the barrier between the glomerular capillaries and the capsular space.

In order for ultra-filtration to occur two things are necessary. First, the barrier must be so constructed as to retain molecules above a certain size whilst allowing smaller ones to pass through. Second, there must be a head of pressure to force fluid through the filter. Research on the kidney has shown that both these conditions are satisfied in the glomerulus.

The electron microscope has shown that the endothelial lining of the glomerular capillaries is pierced by numerous **pores**, approximately $0.1\,\mu m$ in diameter, quite large enough to let through all the constituents of plasma but too small to let the cells through. The endothelial cells rest on a basement membrane on the other side of which are the epithelial cells of the capsule (Fig 14.3). The electron microscope has shown the latter to be quite unlike ordinary epithelial cells. Instead of adhering together to form a continuous sheet, they are diffusely arranged into an irregular network. Each cell, known as a **podocyte**, bears numerous **major processes** from which much finer **minor processes** extend to the basement membrane of the capillary endothelium. Narrow spaces lie between adjacent epithelial cells and between the minor processes which attach them to the basement membrane. These spaces, like the pores in the capillary endothelium, are large enough to allow the passage of blood plasma.

If the pores and spaces described above are large enough to let through

Fig 14.2 Diagram to show how the fluid in Bowman's capsule can be collected for analysis. The proximal tubule is blocked by pressing on it with a fine glass rod. A micro-pippette filled with mercury is pushed through the wall of Bowman's capsule into the cavity beneath. As the mercury is withdrawn from the pipette, fluid is pulled out of the capsule. The glass rod and micro-pipette are both operated by micro-manipulators, mechanical devices which reduce the movements of the experimenter.

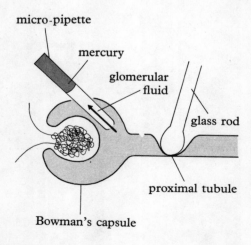

micro-pipette

mercury

glomerular fluid

glass rod

proximal tubule

Bowman's capsule

A

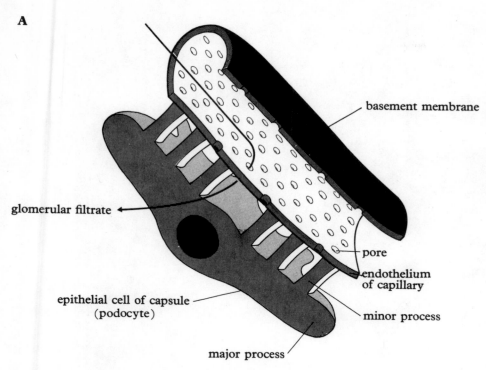

basement membrane

glomerular filtrate

pore

endothelium
of capillary

epithelial cell of capsule
(podocyte)

minor process

major process

Fig 14.3 The structural basis of ultra-filtration in the mammalian kidney. A) Diagram showing the relationship between the endothelium of a glomerular capillary and the epithelium of Bowman's capsule. The two together form a filter across which the fluid constituents of plasma pass from the blood to the cavity of the capsule. Explanation in text. B (*p. 214 overleaf*) shows two electron micrographs of sections through the wall of a glomerular capillary and capsule epithelial cells (podocytes). Can you say in what plane each section has been cut relative to the diagram in A? Can you identify the various structures seen in each electron micrograph? (B: *D. Pease, University of California*)

all constituents of blood plasma, which component of the barrier serves as the dialyzing membrane separating the proteins from the rest of the plasma? The answer is that it must be the **basement membrane**. This is the only continuous membrane between the blood and the capsular space, and it is the only constituent of the barrier which has not got holes in it large enough to let through the plasma proteins. It is exceedingly thin and as such offers minimum resistance to the passage of fluid.

The second requirement for ultra-filtration is pressure. It has long been known that the blood pressure in the kidneys is considerably higher than in other organs. The reason for this is that the efferent vessel in which blood is removed from the glomerulus is considerably narrower than the afferent vessel (see Fig 14.1C). The resulting high blood pressure in the glomerular capillaries tends to force the fluid constituents of the blood into the capsular space. This hydrostatic pressure is opposed by the osmotic pressure of the plasma proteins which tends to hold the plasma back in the capillaries. However, in normal circumstances the osmotic pressure of the plasma proteins is only about half as great as the hydrostatic pressure, so there will be a gradient ensuring the formation of glomerular filtrate. If the blood pressure is gradually reduced by replacing the heart with a mechanical pump there comes a point, round about 8.0 kN/m^2, when the pressure falls below the osmotic pressure of the plasma proteins, and the production of urine ceases.

REABSORPTION IN THE TUBULES

After the glomerular filtrate has been formed it flows along the tubules, and eventually emerges from a collecting duct as **urine**. It has been found that the chemical composition of the urine is very different from that of the glomerular filtrate, and it is therefore clear that the fluid is somehow modified as it flows along the tubule.

What happens to it has also been investigated by A. N. Richards. Using amphibian kidneys, he extracted small quantities of fluid, the so-called **renal**

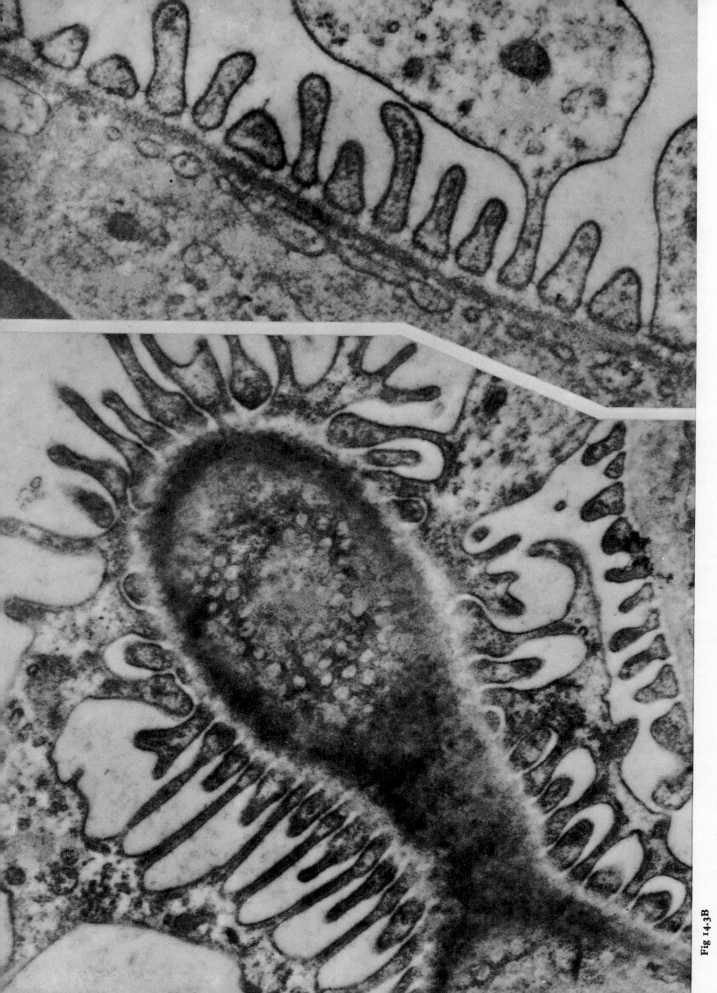

Fig 14·3B

fluid, from the proximal and distal tubules and then analyzed them chemically (Fig 14.4A). This is an exceedingly difficult task, bearing in mind that the tubules are a mere 20μm wide, and his success lay in the use of extremely fine micro-pipettes mounted in micro-manipulators. Many of his findings were confirmed by isolating a short section of the tubule and perfusing it with a fluid of known composition (Fig 14.4B). The fluid was introduced at one end of the isolated stretch and withdrawn from the other. It was then analyzed. As with the analysis of glomerular filtrate these experiments were first done on amphibian kidneys, but more recent investigations on the mammal have given basically similar results.

The composition of renal fluid may be conveniently expressed in terms of the **renal:plasma ratio**. This is the concentration of a particular constituent, say glucose, in the renal fluid divided by the concentration of the same constituent in the plasma. The renal:plasma ratio for glucose, chloride and urea at different points along the nephron is shown in Fig 14.5. Before reading further examine this figure carefully and interpret it as fully as you can.

Briefly, the graphs suggest that glucose is reabsorbed in the proximal tubule and chloride ions, accompanied by sodium ions, in the distal tubule. Urea is not reabsorbed at all; indeed there is evidence that it may be actively secreted into the tubules from the blood. Water is reabsorbed in both the proximal and distal tubules, indicated by the fact that the concentration of glucose rises in tubules which have been treated with **phlorizin**. Phlorizin renders the cells incapable of reabsorbing glucose so that any increase in its concentration must be because of the removal of water. Water is also reabsorbed in the collecting ducts.

The reabsorption of water can be explained, at least in part, by **osmosis**. The blood in the capillaries enveloping the proximal tubule is derived from

Fig 14.4 Investigating the functioning of the kidney tubules. A) Diagram showing how renal fluid can be obtained for analysis. A tiny drop of coloured oil is injected into a tubule. A fine micro-pipette is inserted into the tubule in front of the oil droplet. Fluid is then withdrawn at a rate that keeps the oil drop stationary. B) Technique for determining what happens to the constituents of renal fluid as it flows along the kidney tubule. Fluid of known composition is injected into one end of an isolated length of tubule and withdrawn from the other. It is then analysed for any changes that may have occurred in its composition.

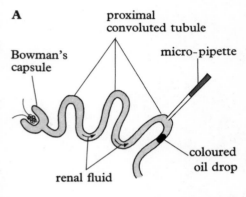

A

B

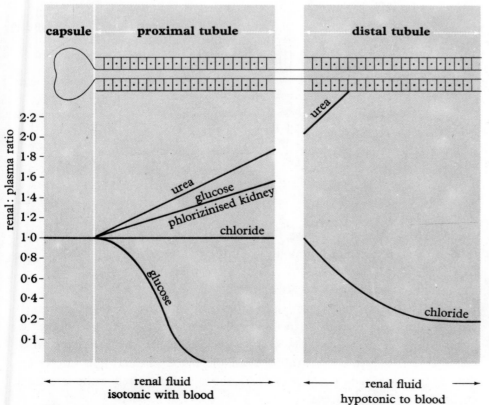

Fig 14.5 Changes in the renal:plasma ratio of urea, glucose and chloride in the kidney of the frog. A renal:plasma ratio of 1.0 means that the concentrations are the same in the renal fluid and the plasma. If the ratio is more than 1.0 the concentration is greater in the renal fluid than in the plasma; if less than 1.0 the concentration is lower in the renal fluid than in the plasma. This data tells us whether or not a substance is reabsorbed as it flows along the tubule. (*after A. N. Richards*)

the efferent vessel of the glomerulus. It is therefore rich in proteins which exert an appreciable osmotic pressure. Moreover, it is under a low hydrostatic pressure. Both these factors favour the passage of water molecules from the tubules into the capillaries.

There is good reason to believe that the reabsorption of glucose and salts involves **active transport**. Diffusion alone cannot account for the transfer of these materials for it occurs against a concentration gradient. That the process is an active one involving the expenditure of energy is supported by the fact that it can be slowed down or stopped by treating the tubules with a metabolic poison such as cyanide or dinitrophenol.

FINE STRUCTURE OF THE TUBULE CELLS

The fine structure of the cells lining the tubules correlates well with their function of reabsorption (Fig 14.6). The inner surface of the epithelial cells of the proximal tubule (and to a lesser extent the distal tubule) bears numerous **microvilli** about 1 μm in length. These are so numerous that in man the total surface area achieved by the proximal tubules of the two kidneys is of the order of 50 square metres. This helps to explain the fact that over two thirds of the water in the renal fluid is reabsorbed in the proximal tubule.

The relationship between the epithelial cells and the enveloping capillaries is extremely intimate. The surface area on the capillary side of the epithelial cells is increased by infoldings of the cell membrane. Between these infoldings lie numerous **mitochondria** which presumably supply energy for the active

Fig 14.6 The cellular structure of the kidney's proximal convoluted tubule can be related to its function of reabsorption. Intimately associated with the thin wall of adjacent capillaries, the cells bear numerous microvilli and contain densely-packed mitochondria.

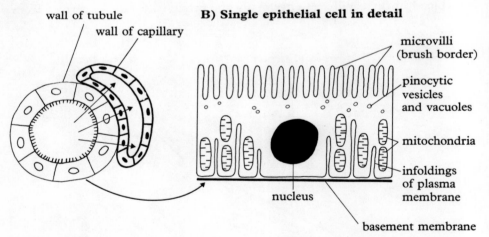

A) Cross section of proximal convoluted tubule

wall of tubule
wall of capillary

B) Single epithelial cell in detail

microvilli (brush border)
pinocytic vesicles and vacuoles
mitochondria
infoldings of plasma membrane
nucleus
basement membrane

absorption of salts and glucose. This is seen in both the proximal and distal tubules though it is more pronounced in the latter.

THE ROLE OF THE LOOP OF HENLE

It is in the interests of a terrestrial animal to conserve water. In mammals this function is performed by the loop of Henle. Mammals are the only vertebrates which can produce a markedly hypertonic urine, that is a urine which has a greater solute concentration than the blood plasma. It is interesting that desert mammals produce a particularly concentrated urine and have an extra long loop of Henle.

What exactly does the loop of Henle do? At one time it was thought that it directly reabsorbed water back into the bloodstream, but its rôle is now thought to be more subtle than that. It seems that it concentrates sodium chloride in the medulla of the kidney. This then causes vigorous osmotic extraction of water from the collecting ducts thereby concentrating the urine and making it hypertonic to the blood.

In achieving this the loop of Henle employs the principle of a **hair-pin countercurrent multiplier** (Fig 14.7). As the renal fluid flows along the

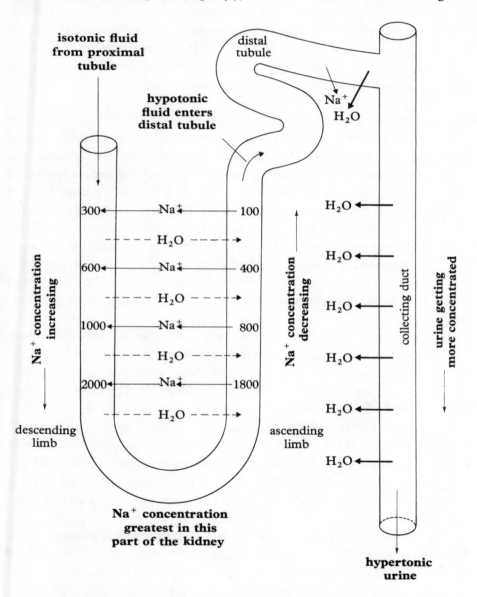

Fig 14.7 Diagram to show the function of the loop of Henle in the reabsorption of water. The figures in the descending and ascending limb refer to the Na^+ concentration in milliosmoles per kilogramme of water.

ascending limb, sodium chloride is actively removed from it and deposited in the descending limb. This active transfer of salt takes place at all levels of the loop of Henle. At any given level the effect of this is to raise the concentration in the descending limb very slightly above that in the ascending limb, the so-called **unit effect**. The effect at any one level is slight, but the overall result is multiplied by the length of the hair-pin. As the renal fluid flows down the descending limb towards the apex of the loop it becomes more and more concentrated; as it flows up the ascending limb it becomes more and more dilute. The effect of the system is to produce a region of particularly high salt concentration in the deep part of the medulla. The collecting duct passes through this region before opening into the pelvis. As the renal fluid passes down the collecting duct water is extracted from it by osmosis. This raises its concentration, resulting in the production of a markedly hypertonic urine.

What the loop of Henle is really doing is to increase the solute concentration of the kidney tissue. Its characteristic U-shape provides a counter-current system which establishes and maintains this high concentration. The effectiveness of the kidney in retaining water will be realized from the fact that of the 120 cm^3 of fluid filtered per minute only 1 cm^3 reaches the ureter. Over 99 per cent of fluid is reabsorbed in the kidney tubules. It has been estimated that if the kidney stopped reabsorbing water, complete dehydration of the body would occur in less than three minutes.

THE KIDNEY AS A REGULATOR

This brief account of the kidney illustrates its two principal functions, **excretion** and **regulation**. In that it totally removes urea from the blood, it is functioning as an excretory organ, clearing the body of toxic nitrogenous waste. Other toxic materials, poisons etc. are dealt with in a similar way. On the other hand glucose is completely reabsorbed, none being present in the urine. Only if the amount of glucose in the blood exceeds a certain threshold value does it begin to appear in the urine. In man this threshold value is about 180 mg/100cm^3 and is rarely reached even after the richest carbohydrate meal. Normally the liver—insulin mechanism discussed in the last chapter keeps the glucose level constant, the kidney being brought in only if fluctuations get out of hand.

It is in its treatment of salts and water that the effectiveness of the kidney as a regulator can be seen most clearly. The relative amount of water and salts reabsorbed is strictly geared to the body's needs, and it is upon this that the osmotic pressure of the blood and tissue fluids depends. The question is: how does the kidney know how much water and/or salts to reabsorb?

The answer is that the composition of the blood itself determines the reabsorptive activities of the kidney. The osmotic pressure of the blood cannot influence the kidney tubules directly but does so via hormones. Both water and salt reabsorption are under hormonal control but changes in the reabsorption of water play the most direct part in correcting fluctuations in osmotic pressure. Briefly the mechanism is as follows:

In the brain there are groups of **osmoreceptor cells** sensitive to the osmotic pressure of the blood, located in the **hypothalamus** at the base of the **pituitary gland** (see Chapter 18). These receptors are stimulated by a rise in the osmotic pressure of the blood, such as might occur if the body becomes dehydrated or a large quantity of salt is consumed. When the receptors are stimulated a hormone is released from the posterior lobe of the pituitary gland, whence it is conveyed via the bloodstream to the kidney where it causes the nephron to increase its reabsorption of water. This will obviously lead to the production of a more concentrated urine and the osmotic pressure of the blood will be lowered.

Now consider the converse situation. If the osmotic pressure of the blood falls, as it does after a large quantity of water has been drunk, the osmoreceptors will be less stimulated than before, less hormone will be produced, less water will be reabsorbed in the kidney and a more dilute urine will be produced. Result? The osmotic pressure of the blood and tissue fluids will rise again. The production of a copious flow of watery urine is known as **diuresis** and clearly the action of the hormone is to counteract this condition. It is therefore known as the **anti-diuretic hormone,** or ADH for short. Its importance can be realized by considering what happens to a person who has a faulty pituitary which fails to produce ADH. The patient permanently produces vast quantities of dilute urine and has to make good the loss by drinking a lot of water. If he fails to do this his body will become dehydrated very quickly. The clinical condition is called **diabetes insipidus.** The hormonal control of osmoregulation is summarized in Fig. 14.8.

Exactly how ADH exerts its effect on the kidney is not known for certain, but it appears to make the cells lining the distal convoluted tubule and collecting duct more permeable to water, thus facilitating its osmotic withdrawal into the surrounding blood vessels.

Whatever the mechanism, the regulation of water retention by ADH provides us with an admirable example of a **homeostatic feedback process.** Fluctuations in osmotic pressure are quickly detected and the corrective mechanism brought into action.

Fig 14.8 Diagram summarizing the sequence of events which takes place if the osmotic pressure of the blood and tissue fluids rises. In this case the rise in osmotic pressure is brought about by ingesting an excessive quantity of salt, but it might equally well result from dehydration of the tissues caused by excessive sweating or failure to drink any water. Stimulation of the osmoreceptors in the hypothalamus results in secretion of ADH by the posterior lobe of the pituitary. It used to be thought that nerve impulses were transmitted along the nerve tract to the posterior lobe, but it has been shown that ADH is secreted by specialized nerve cells in the hypothalamus, whence it flows down their axons to the pituitary. This is an example of neurosecretion, a quite common phenomenon amongst animals.

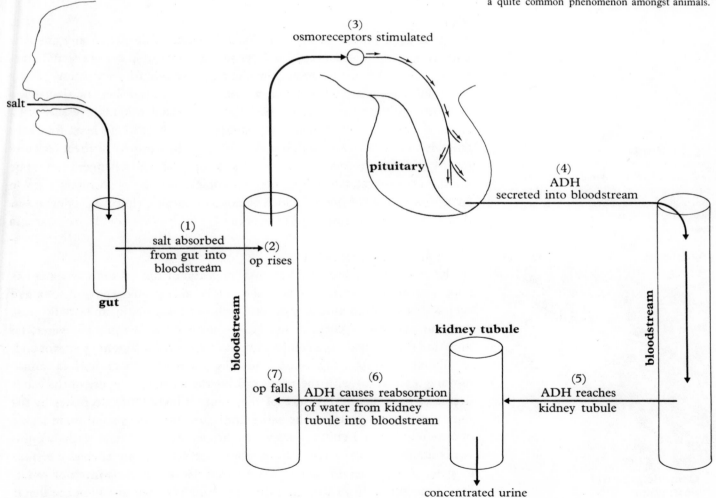

(3)
osmoreceptors stimulated

pituitary

(4)
ADH
secreted into bloodstream

salt

(1)
salt absorbed
from gut into
bloodstream

(2)
op rises

gut

bloodstream

bloodstream

kidney tubule

(7)
op falls

(6)
ADH causes reabsorption
of water from kidney
tubule into bloodstream

(5)
ADH reaches
kidney tubule

concentrated urine

Excretion and Osmoregulation in other Animals

If a fresh water fish such as a stickleback is placed in sea water it will die. The reason is that sea water contains about twice as much salt as the blood of the fish with the result that water is drawn out by osmosis. Although the skin is relatively impermeable, the lining of the mouth cavity and gills acts as a semipermeable membrane so that the fish behaves like an osmometer, losing water when placed in a hypertonic solution, i.e. one which has a higher osmotic pressure than the body fluids, and gaining water if it is returned to a hypotonic solution.

The result of transferring a fresh water fish to sea water illustrates an important generalization, namely that most aquatic animals cannot tolerate appreciable fluctuations in the salinity of their external medium. This is not to say that the osmotic pressure of their body fluids, the internal OP, is necessarily the same as the external OP. On the contrary, many aquatic animals can maintain a difference between the internal and external OP, but their ability to do so breaks down if the OP of the external medium deviates too far from its normal value.

From the point of view of osmoregulation animals fall into two groups: those that cannot regulate their osmotic pressure at all, and those that can, at least to some degree. The former are all marine, and as they are the more primitive we will start with them.

MARINE INVERTEBRATES

There is good reason to suppose that life began in the sea. Many animals, notably marine invertebrates such as sea anemones, spider crabs and starfishes, have remained in the sea throughout their evolutionary history. Their body fluids are isotonic with sea water; indeed in starfishes the tissues are perfused with sea water itself. Since their internal osmotic pressure (OPi) is equal to the external osmotic pressure (OPe) there is no need for these animals to osmoregulate and, not surprisingly, the majority of them lack any means of doing so. Any changes in OPe result in similar changes in OPi. For marine invertebrates this presents no problem for the open sea is a stable environment not subject to sudden changes in salinity. However, it does mean that their habitat is restricted to the sea.

FROM SEA TO FRESH WATER

In the evolutionary history of animals migration from sea to fresh water has happened on more than one occasion. To be able to undergo and complete such a migration an animal must be capable of maintaining an osmotic pressure independent of fluctuations in the salinity of the surrounding water. To illustrate this compare the results of immersing three different species of crab in diluted sea water (Fig 14.9). The fully marine spider crab *Maia* cannot osmoregulate at all with the result that as the osmotic pressure of the water decreases the osmotic pressure of the animal's body fluids decreases by the same amount. In contrast, the shore crab *Carcinus* is capable of some degree of osmoregulation. In diluted sea water the osmotic pressure of its body fluids is maintained above that of the surrounding water. This enables *Carcinus* to live in the brackish water of estuaries. However, its powers of osmoregulation break down if the external medium becomes too dilute, so that it

Fig 14.9 Graphs showing the effect of changing the solute concentration (**OP**) of the external medium on the internal osmotic pressure of three different genera of crabs. Note that *Maia* cannot osmoregulate at all, *Carcinus* can do so provided that the external medium is not too dilute, and *Eriocheir* can osmoregulate even in fresh water.

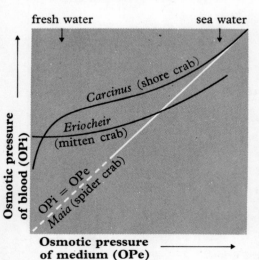

cannot migrate very far up rivers. The third species, the mitten crab *Eriocheir*, can tolerate much greater dilutions than *Carcinus*, enabling it to penetrate upstream into completely fresh water.

Although osmoregulation is necessary for permanent migration from sea to fresh water it is not the only means by which a marine animal can withstand dilution of its surrounding medium. There are two other possibilities.

(1) It may possess tissues that can tolerate a wide range of salinities. This is true, for example, of the lugworm *Arenicola* which survives in the comparatively dilute waters of the Baltic Sea, and the ragworm *Nereis diversicolor* which flourishes in the Gulf of Finland.

(2) It may avoid the effects of dilution by behavioural means. This is the method used by the estuarine snail *Hydrobia* which burrows into the mud when the tide is going out, thus escaping the periodical dilution of its external medium. This kind of trick enables animals to penetrate for variable distances upstream, but for full exploitation of fresh water the ability to osmoregulate is essential.

The problem facing a fresh water animal (or an estuarine animal for that matter) is that OPi is greater than OPe. The danger here is dilution of the tissues resulting from the osmotic influx of water across the exposed semipermeable surfaces of the body.

On purely theoretical grounds there are two possible solutions to this problem:

(1) Water might be eliminated as fast as it enters by means of a kidney or some equivalent structure, salts being reabsorbed from the water before it leaves the body resulting in the production of a hypotonic urine.

(2) Salts might be actively taken up from the external medium thereby offsetting the diluting effect of the inflowing water.

Both these solutions would involve the movement of ions against a concentration gradient and so would require the expenditure of energy (active transport). With these general principles in mind let us take a brief look at the osmoregulatory devices of several fresh water animals.

CONTRACTILE VACUOLE

This simple device is found in *Amoeba* and other fresh water protozoons. It is a small sac, lined with unit membrane, lying freely in the cytoplasm (Fig. 14.10). The cell membrane surrounding the animal is semipermeable and since OPi is greater than OPe, water enters the cell by osmosis. To counter this, water is secreted into the contractile vacuole as fast as it enters the body. As this happens the contractile vacuole expands and eventually discharges its contents to the exterior through a small pore in the cell membrane, after which the whole process is repeated.

Simple though this may seem, there are many unresolved questions about the contractile vacuole. However, it is reasonably well established that its function is osmoregulatory. If a fresh water protozoon is placed in water to which a small quantity of salt has been added the contractile vacuole discharges its contents less frequently than before. In fact the frequency of discharge decreases as the salinity of the medium increases.

That the mechanism by which the contractile vacuole collects water involves the expenditure of energy is suggested by numerous mitochondria

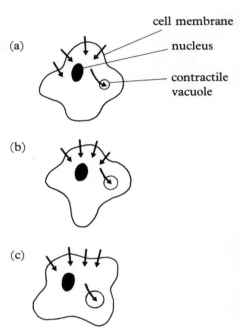

cell membrane
nucleus
contractile vacuole

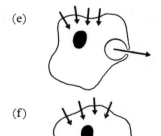

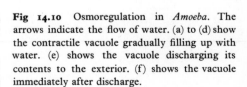

Fig 14.10 Osmoregulation in *Amoeba*. The arrows indicate the flow of water. (a) to (d) show the contractile vacuole gradually filling up with water. (e) shows the vacuole discharging its contents to the exterior. (f) shows the vacuole immediately after discharge.

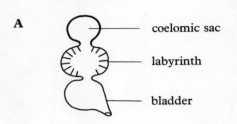

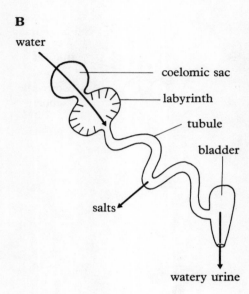

Fig 14.11 The antennal gland of (A) the shore crab, and (B) the crayfish. Notice the salt-re-absorbing tubule in the crayfish.

which lie close to the vacuole membrane. This view is supported by treating *Amoeba* with a metabolic poison. This puts the contractile vacuole out of action with the result that the animal takes up water, swells and dies. It is interesting that marine species of *Amoeba*, whose internal fluids are isotonic with the surrounding sea water, have no contractile vacuoles. If placed in a hypotonic medium a contractile vacuole develops.

ANTENNAL GLANDS

In the shore crab *Carcinus*, mentioned earlier, water is eliminated by a pair of **antennal glands**, so-called because of their position at the base of the antennae in the head. Each gland (Fig 14.11A) consists of a small **end sac** derived from the coelom, which is connected to a larger sponge-like cavity, the **labyrinth**. This in turn connects with a **bladder** which opens to the exterior by a small **pore** at the base of the antenna. The antennal glands are responsible for nitrogenous excretion: fluid rich in nitrogenous waste is filtered into the end sac and labyrinth from the surrounding blood. At first sight the antennal gland might also appear to perform an osmoregulatory function. However, it has been found that the urine is isotonic with the blood. This is because the antennal gland is incapable of holding back salts: it eliminates water and salts alike. However, the loss of salts in the urine is compensated by the fact that the gills absorb salts from the surrounding medium and secrete them into the blood against a concentration gradient. In this way OPi can be maintained at a higher level than OPe. The same thing happens in the mitten crab *Eriocheir*, but here the inward secretion of salts is sufficiently vigorous to enable the animal to flourish in fresh water.

In the fresh water crayfish we find that a considerable step forward has been taken. Like *Carcinus*, the crayfish has a pair of antennal glands but, unlike *Carcinus*, it is capable of producing a hypotonic urine. This is because the antennal glands in this case are able to reabsorb salts, a process which takes place in a coiled **tubule** linking the labyrinth with the bladder (Fig 14.11B). Analysis of fluid extracted from the antennal gland at various points along its length has shown that the contents of the end sac and labyrinth are isotonic with the blood, but as the urine flows along the coiled tubule salts are removed from it. Thus the antennal gland of the crayfish gets rid of excess water and holds back salts. The result is that OPi is considerably higher than OPe.

FRESH WATER FISHES

The best adapted fresh water animals are the bony fishes (teleosts). Fresh water fishes are liable to osmotic influx of water across the gills and lining of the mouth cavity and pharynx. Fresh water teleosts, such as the stickleback overcome this problem in three ways (Fig 14.12A):

(1) The rate at which glomerular filtrate is produced in the kidney is very high, much more so than in marine fishes. The cortex of the fish kidney is structurally similar to that of the mammal and the high filtration rate is achieved by the kidney possessing numerous large glomeruli.

(2) As the renal fluid flows along the kidney tubules salts are extensively reabsorbed back into the bloodstream with the result that the urine is marked-ly hypotonic to the blood.

(3) The small amount of salt lost in the urine is offset by the active uptake

A) Fresh water bony fish
(OPi > OPe)

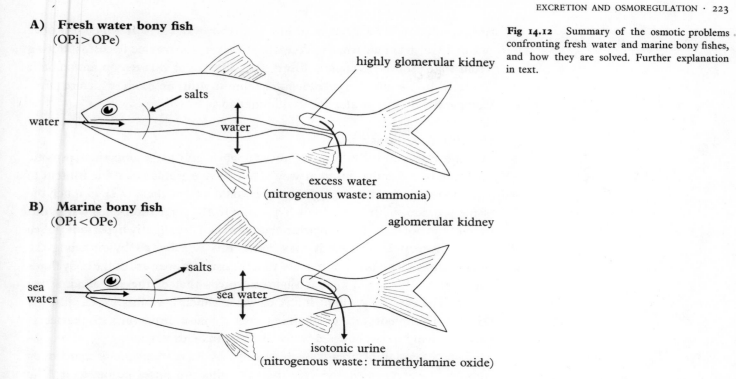

Fig 14.12 Summary of the osmotic problems confronting fresh water and marine bony fishes, and how they are solved. Further explanation in text.

salts

water

highly glomerular kidney

water

excess water
(nitrogenous waste: ammonia)

B) Marine bony fish
(OPi < OPe)

aglomerular kidney

salts

sea water

sea water

isotonic urine
(nitrogenous waste: trimethylamine oxide)

of salts by special **chloride secretory cells** in the gills. These take up salts from the external medium and move them against a concentration gradient into the bloodstream. The result is that chloride is some 800 times more concentrated in the animal's blood than in the surrounding fresh water. Needless to say this inward movement of ions requires the expenditure of energy (active transport). Thus, to summarize, fresh water fishes solve their osmotic problem by combining the production of a hypotonic urine with the active uptake of salts.

FROM FRESH WATER BACK TO THE SEA

There is good reason to believe that in the course of evolutionary history certain fishes have returned from fresh water to the sea. The problems attendant on this can be appreciated by examining the osmotic situation in present-day marine teleosts such as cod, whiting etc.

Unlike marine invertebrates, whose body fluids are isotonic with sea water, marine teleosts have body fluids which are hypotonic to their surroundings, i.e. OPi is less than OPe. The result is osmotic extraction of water from the body leading to dehydration of the tissues. This is overcome by a combination of three processes (Fig 14.12B).

(1) By swallowing sea water and having a kidney with a relatively low filtration rate. The kidney of marine teleosts, in contrast to their fresh-water relatives, contains relatively few glomeruli, small in size.

(2) By actively extruding salts by means of **chloride secretory cells** in the gills. In fresh water teleosts these cells are responsible for moving salt inwards. Here they do precisely the reverse, moving them against a concentration gradient from the blood to the surrounding sea water.

(3) By eliminating nitrogenous waste in the form of a compound which is soluble but non-toxic. Fresh water teleosts excrete their nitrogenous waste as ammonia, a highly toxic substance which requires a large volume of water

for its dilution. This is no hardship to a fresh water teleost which has more water than it knows what to do with. However, the marine teleost, suffering from water shortage, cannot afford such a loss. For this reason ammonia is replaced by the non-toxic compound **trimethylamine oxide** which requires comparatively little water for its elimination.

MARINE ELASMOBRANCHS

Marine elasmobranchs (sharks, rays, dogfish etc) get round their osmotic difficulties in a quite different way. They have a similar chloride content to marine teleosts and at first glance one would expect them to be in much the same position, with body fluids hypotonic to the surrounding sea water. But in fact the body fluids of a marine elasmobranch are slightly hypertonic to sea water. Surprising though it may seem, this is achieved by retaining the nitrogenous waste product **urea** so that the osmotic pressure of the body fluids is raised to the point that it slightly exceeds that of the surrounding sea water. The result is a slight influx of water which is readily expelled by the kidney. The importance of retaining urea is that it dispenses with the necessity of having to swallow sea water and eliminate excess salts.

The retention of urea is made possible by its reabsorption in the kidney tubules, together with the fact that the gills are impermeable to it. The remarkable thing is that the tissues can stand it. Normally a high concentration of urea alters the shape of proteins by breaking the hydrogen bonds linking adjacent polypeptide chains. This can completely disrupt the smooth functioning of cells, particularly when enzymes are affected. Elasmobranch proteins are immune to these effects. Indeed the tissues appear to thrive on urea. If the heart of an elasmobranch is perfused with a balanced salt solution lacking the usual high concentration of urea, it will stop beating!

MIGRATORY FISHES

In all the cases considered so far osmotic independence can be maintained only if the salinity of the external medium remains constant. If the latter fluctuates unduly then the animal's ability to osmoregulate breaks down. However, a notable exception to this are the **migratory fishes** which can continue to osmoregulate even when the external medium fluctuates drastically.

Consider the case of the salmon. This extraordinary fish spends the first three years of its life in fresh water after which it moves down to the sea. When full-grown, some two years later, it makes its way upstream to spawn, after which it usually returns to the sea again. Three times in the course of its life the salmon undertakes a dramatic journey from one extreme of osmotic environment to another. How it manages to do this is not known, but presumably it involves changes in the filtration rate of the kidney, and a reversal of the direction in which the chloride secretory cells transfer salt. From moving salts inwards in fresh water they must change to moving them outwards in sea water.

Similar events occur in eels. Here spawning takes place in the sea. After about three years the young eels, or elvers, swim upstream to fresh water where they complete their growth. When mature the adults migrate downstream to the sea again. As with the salmon these migrations necessitate osmoregulatory devices which can adjust to changing conditions.

FROM FRESH WATER TO LAND

The emergence of fresh water animals onto dry land presents a whole series of structural and physiological problems which can only be solved by the development of special adaptations. The adaptations that concern us in the present context are those which solve the osmotic problem. What is the osmotic problem facing a terrestrial animal?

Like marine teleosts terrestrial animals are liable to water loss, but whereas in fishes this is caused by osmotic removal of water, in terrestrial animals it is because of evaporation from permeable surfaces exposed to the atmosphere. Terrestrial animals endeavour to overcome this problem in the following ways:

(1) By having a waterproof integument. This obvious way of reducing loss of water has been successfully developed by a number of groups, notably insects, reptiles, birds and mammals. The **keratinous scales** of reptiles and the **cornified epithelium** of mammals both provide not only physical protection but also insulation against water loss, and this is one of the reasons why these animals can live successfully in hot, dry places. Of all terrestrial vertebrates amphibians are the least well adapted in this respect. Their skin is thin and moist and offers little resistance to evaporation. Indeed, the skin is used as a supplementary respiratory surface for which purpose it must be permanently moist. This is why amphibians are generally restricted to damp places.

Amongst the invertebrates insects have developed waterproofing to a degree unattained by any other invertebrate group, but they have achieved this by a technique quite different from that used by the vertebrates. On the surface of the insect cuticle is a microscopically thin layer of **wax**. This is impermeable to water and confers upon the cuticle its waterproofing properties. If the wax is removed the rate of water loss is greatly increased. This has been demonstrated by gradually warming up a cockroach nymph and measuring the rate of water loss at different temperatures (Fig 14.13). Below a certain **critical temperature** there is very little increase in water loss as the temperature is raised. But as soon as the critical temperature is reached the rate suddenly increases dramatically. It has been found that at the critical temperature, about 30°C for the cockroach, the arrangement of the wax molecules changes in such a way that the cuticle becomes much more permeable to water. Though waterproofing plays an important part in preventing excessive water loss, it is only a passive way of conserving water. Nor can it ever be complete, for the animal has got to breathe and there will always be a certain amount of evaporation from the respiratory surfaces however protected they may be.

(2) By reducing the glomerular filtration rate. You will recall that marine teleosts reduce their filtration rate by reducing the number and size of glomeruli in the kidney. Exactly the same is found in many terrestrial vertebrates particularly those that inhabit hot, dry places like the desert. Thus the desert frog, *Chiroleptes*, one of the few amphibians to flourish in hot deserts, has fewer and smaller glomeruli than its relatives living in moist temperate regions.

(3) By producing a non-toxic nitrogenous waste. You will remember that marine teleosts excrete nitrogenous waste in the form of trimethylamine oxide, an adaptation for conserving water. Similar trends are found in terrestrial animals. For example, amphibians and mammals excrete **urea** which,

Fig 14.13 Graph showing the rate of water loss from a cockroach nymph at constant humidity with increasing temperature. Notice the sharp increase in water loss at 30°C. (*after Beament*).

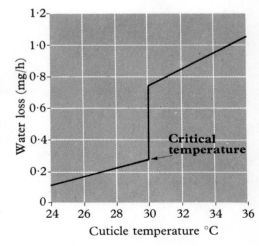

though toxic in high concentration, is less toxic than ammonia and therefore requires less water for its removal.

Reptiles, birds and insects have taken this a step further and excrete nitrogenous waste as **uric acid**, which, unlike urea, is insoluble in water. Water can therefore be extensively removed from it, the uric acid being excreted in semi-solid form.

(4) By reabsorbing water. One of the principal functions of the kidney tubules is to reabsorb water from the glomerular filtrate. As we saw earlier the loop of Henle, which is present in birds and mammals, is a special adaptation for this purpose. Desert mammals have an extra long loop of Henle and produce a more concentrated urine in consequence. The kangaroo rat, for example, produces urine which is four times as concentrated as man's, a feat which is made possible not only by its long loop of Henle but also by the fact that the level of ADH in its blood is unusually high.

As we saw in the last section, the extent to which water can be retained depends on the nature of the excretory product. Ammonia is wasteful of water, urea is better. But uric acid, being insoluble, is ideal. In birds the ureters and rectum open into a common cavity, the **cloaca**. Here water is extensively reabsorbed from the faeces and excretory waste. What passes out is a semi-solid sludge, well known to anyone who has tried to walk through a city square.

Insects have evolved a system which is essentially similar to that of reptiles and birds. This has been investigated by Sir Vincent Wigglesworth in the blood-sucking bug *Rhodnius* and there is good reason to believe that the process is basically the same in other insects. Between the midgut and rectum is a bunch of blind ducts called **Malpighian tubules**. These open into the gut at one end whilst their free ends float in the blood-filled body cavity (Fig 14.14). The insect's tissues produce nitrogenous waste in the form of soluble **potassium urate** which is liberated into the blood and taken up by the cells lining the Malpighian tubules. The tubules are muscular and their writhing movements may facilitate the absorption of urate by stirring up the blood. In the cells of the tubule the potassium urate reacts with water and carbon dioxide (from respiration) to form potassium bicarbonate and **uric acid**. The former is absorbed back into the blood, but the latter is deposited in the lumen of the tubule. As the uric acid moves towards the gut, water is vigorously reabsorbed back into the blood, to such an extent that the proximal end of the Malpighian tubule becomes filled with solid crystals of uric acid. Electron micrographs show that the epithelial cells bear numerous **microvilli** which increase the surface area for water reabsorption. Water is further reabsorbed by the folded walls of the **rectal gland** so that by the time the urine leaves the body it is very much more concentrated than the blood.

The remarkable ability of insects to conserve water has undoubtedly contributed towards their success, and this is in large measure due to the action of the Malpighian tubules and the rectal glands. Insects conserve water more effectively than any other group and this is why we find them in the hottest and driest of places.

(5) By means of behaviour. Many animals avoid, or at least reduce, the problem of water loss by modifying their behaviour or changing their habitat. The kangaroo rat, for example, cuts down evaporation from its

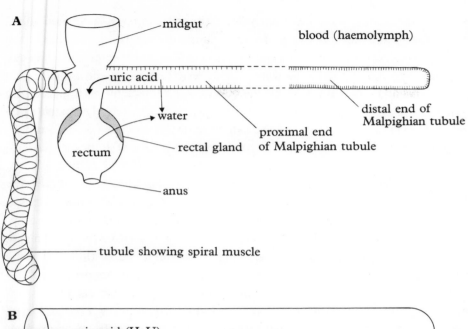

A

midgut

blood (haemolymph)

uric acid

water

distal end of
Malpighian tubule

proximal end
of Malpighian tubule

rectal gland

rectum

anus

tubule showing spiral muscle

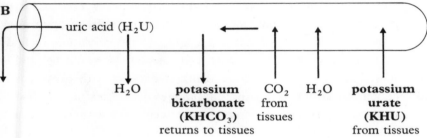

B

uric acid (H_2U)

H_2O

potassium
bicarbonate
($KHCO_3$)
returns to tissues

CO_2
from
tissues

H_2O

potassium
urate
(KHU)
from tissues

overall equation:

$$KHU + H_2O + CO_2 \longrightarrow KHCO_3 + H_2U \downarrow$$

lungs by as much as 25 per cent by remaining in its burrow during the heat of the day.

Many animals respond to drought by seeking out a damper environment. This is true of the earthworm which burrows deeper when the surface soil is dry. In normal circumstances the earthworm lives in soil which is so moist that it can in effect be regarded as a fresh water animal: the job of its excretory organ, the nephridium, is to get rid of surplus water and conserve salts, the reverse of most terrestrial animals. Certain animals go into a state of dormancy during the summer or dry season, a phenomenon called **aestivation**. The African and South American lungfishes, for instance, can survive the complete drying up of the swamps in which they live by burrowing into the bottom and encasing themselves in a cocoon of hard mud lined with mucus. Here they go into a state of suspended animation until the arrival of the next rainy season six or seven months later.

(6) By using metabolic water. One of the products of oxidative metabolism (cell respiration) is water. In some desert animals this is an important source of water. The amount of water yielded by oxidative metabolism depends on the food substance being used. When a gramme of carbohydrate is oxidized approximately 0.56 g of water are produced, but almost twice as much is formed when a gramme of fat is oxidized. For

Fig 14.14 Structure and functioning of the Malpighian tubules of an insect. A) Diagram showing relationship between Malpighian tubules and gut in the blood-sucking bug *Rhodnius*. B) Diagram to show how the nitrogenous waste, uric acid, is formed in *Rhodnius*. Work on other insects shows the process to be substantially the same though the details vary considerably. C) Electron micrograph showing microvilli projecting into the lumen of a Malpighian tubule. Slender extensions of the lining epithelial cells, the microvilli increase the surface area for reabsorption of water. (*David Chapman, Zoology Department, University of Cambridge, courtesy of Prof. T. Weis-Fogh*).

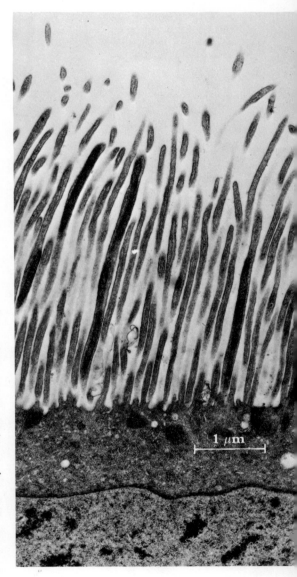

1 μm

C

Fig 14.15 The kangaroo rat, *Dipodomys spectabilis*, a desert rodent, shows many adaptations to life in a hot dry environment. (*K. Schmidt-Nielsen, Duke University*).

this reason a desert animal such as the kangaroo rat tends to metabolize fat rather than carbohydrate. In most animals breathing to obtain the necessary oxygen for the oxidation of fat involves more water loss than can be gained from the metabolic water. However, the kangaroo rat manages to lose less water than it gains by metabolism. This it achieves by reducing evaporation from its lungs, not sweating, and producing a highly concentrated urine and very dry faeces. In fact the kangaroo rat, the doyen of desert animals, produces and retains metabolic water so effectively that it never needs to drink.

(7) **By having tissues tolerant to water loss.** The camel, unlike the kangaroo rat, does sweat and yet it can go for as much as two months without drinking. It is popularly believed that water is stored in the hump, but in fact fat is stored here and water is obtained from its metabolism. However, the camel cannot produce more metabolic water than is lost by sweating and expiration. How then does it survive without drinking? The answer, discovered by Knut Schmidt-Nielsen, is that the camel's tissues are extraordinarily tolerant to dehydration. As the days go by more and more water is lost and the body fluids become progressively more concentrated, and yet it survives. In fact it will survive a water loss that reduces its body weight by as much as 30 per cent, 10 per cent more than would be fatal to man. Little wonder that when it does drink it does so with great gusto. Schmidt-Nielsen, an authority on the physiology of desert animals, reports that a camel which was given water after sixteen days without it, drank 40 litres in 10 minutes!

Osmoregulation in Plants

For convenience we shall discuss plants separately but it should be realized that they face the same basic problems as animals — particularly that of losing water. As we saw in Chapter 12 land plants lose water by transpiration, evaporation from the leaves. To a plant growing in well-watered soil this presents no problem as the water transpired can be replaced by uptake from the soil. But a plant living in dry conditions runs the risk of drastic dehydration and must take measures to avoid this contingency. Such plants are called **xerophytes** and are adapted to surviving drought in various ways.

In the first place some plants have tissues that are tolerant to dessication. Plants generally are much more capable of withstanding fluctuations in the water content of their tissues than animals. Indeed some are so adaptable that their protoplasm can, to all intents and purposes, be completely dried out and yet resume normal functioning later when water becomes available. This is seen in certain desert plants and is also true of many mosses and ferns.

Many plants survive dry periods as **seeds** or **spores**, a method of evading drought which is in some senses equivalent to aestivation in animals. The protoplasmic contents of spores and seeds are generally in a highly dessicated state and protected within a hard case. In this condition they may remain viable for long periods, germinating into a new plant when water becomes available and other conditions are suitable. The use of seeds for surviving dry conditions is well illustrated in many small desert plants. These germinate, grow to full size and flower during the short rainy season in the spring, turning the arid desert landscape into a spectacular carpet of colour within a matter of a few days. After the seeds have been dispersed the parent plants die and the seeds remain dormant in the dry soil until the following year.

Some plants manage to live in dry places by having extremely **deep roots** which absorb subterranean water deep down in the soil. Many Mediterranean trees and shrubs such as *Acacia* and *Oleander* do this. Other plants have **superficial roots** which grow out horizontally keeping close to the surface. This puts them in the best position to absorb the maximum amount of water after a short shower on a hot day.

Another method used by some plants is to **store water** in large parenchyma cells contained within swollen stems and leaves. This makes the tissues wet and juicy, for which reason such plants are called **succulents**. There are many stem succulents in the desert of which perhaps the best known is the giant saguaro cactus of Arizona (Fig 14.16). Water is stored in the thick stem and side branches, the leaves being reduced to slender protective spikes.

A

Fig 14.16 Three desert succulents. A) Prickly pear showing detail of the spines. (*B. J. W. Heath, Marlborough College*). B) Saguaro and barrel cacti in Saguaro National Monument, Arizona. (*US National Park Service*). In all three cases water is stored in thick fleshy stems, the leaves being reduced to protective spines. Excessive transpiration is prevented by thick leathery cuticle.

B

There is no point in being a succulent without having means of preventing the stored water from being lost. One of the most obvious methods is to **reduce the number of stomata,** a strategy used by many desert plants including the prickly pear (Fig 14.16). The stomata of some plants, for example the evergreen shrub *Hakea* of the Australian desert, are sunk down into **pits** in which humid air tends to accumulate, thus reducing the rate of evaporation from the leaves. Sinking of the stomata is accompanied in some plants by **folding of the leaves,** an adaptation seen in certain grasses which roll their leaves up in dry weather (Fig 14.17).

The adaptations mentioned so far are all structural ones, but some plants solve the problem physiologically by **reversing the normal stomatal rhythm**: opening the stomata at night and closing them by day. As a means of preventing excessive water loss this would seem an admirable idea, but obviously the stomata must be open for at least part of the daylight hours in order for photosynthesis to take place.

Transpiration is not confined to the stomata. You will remember from Chapter 12 that in ordinary plants a certain amount of water evaporates through the cuticle. Such cuticular transpiration is reduced in xerophytes by thickening of the **waxy cuticle** which is thereby rendered impermeable to water, an adaptation reminiscent of that found in insects. This is parti-

Fig 14.17 Photomicrograph of a leaf of marram grass, *Ammophila arenaria*, in transverse section. A common inhabitant of dry coastal sands, this xerophytic grass reduces water loss by folding its slender leaves in dry weather. Can you spot any other xerophytic adaptations? (*J. F. Crane*)

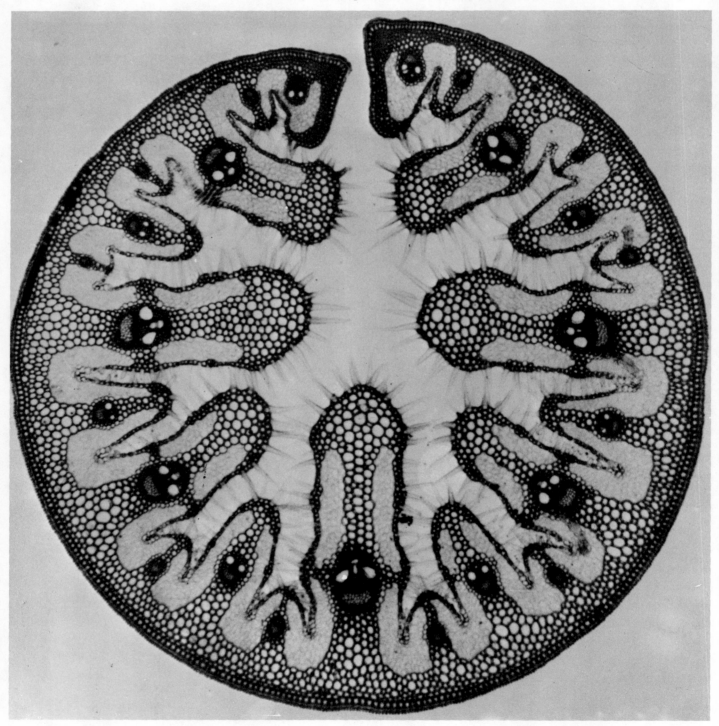

cularly well seen in desert cacti like the prickly pear, but it is also shown by many evergreen trees and shrubs of temperate regions. These plants suffer from water shortage in the winter not so much because of evaporation from the leaves but because frosts and snow decrease its availability from the soil, causing what is called **physiological drought**. One way of circumventing this problem is to **shed the leaves** before winter sets in, which is precisely what deciduous trees do. Getting rid of the leaves altogether is obviously one of the most effective ways of cutting down transpiration, and one of the most spectacular examples of it is shown by *Ocotillo*, the 'vine cactus' of the North American desert. This shrub comes out in leaf every time it rains and sheds the leaves immediately afterwards: this may happen as many as half a dozen times in a year.

You will recall that an aquatic animal may suffer from water loss because its surrounding medium has a higher osmotic pressure than its tissue fluids. Much the same applies to certain plants, namely those that live on mud flats and salt marshes. Known to botanists as **halophytes,** they are faced with the problem of having to absorb water which is at a considerably higher osmotic pressure than ordinary soil water. They get round this problem in two ways. First, their root cells exert a much greater osmotic pressure than those of ordinary plants, more than twice as great in some cases, which enables them to take in water by osmosis in the normal way. Second, many of them store water so that they will not run short in the event of the external osmotic pressure exceeding that of the root cells. This sometimes happens when water evaporates from mud flats at low tide. There is still plenty of water but its osmotic pressure is so high that it is unobtainable to the plants, another example of physiological drought.

Finally we must briefly mention those plants that have special methods of getting rid of excess water. On cool humid evenings in summer the leaves of some trees can be seen to exude water. This is known as **guttation** and is caused by more water being taken up into the plant than can be removed by evaporation. In many plants, particularly those of tropical rain forests where the humidity is high, water is actively secreted by special structures called **hydathodes** situated at the terminal ends of the veins round the edges of the leaves.

15 Temperature Regulation

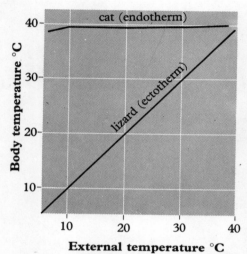

Fig 15.1 Relationship between the internal (body) and external temperatures for cat and lizard. The cat maintains a more or less constant body temperature irrespective of changes in the external temperature. The lizard's internal temperature depends on the temperature of its immediate surroundings. *(after Martin)*

It is a well known fact that irrespective of fluctuations in the environmental temperature, the body temperature of man remains at approximately 36.9°C. Many of the body's structures and physiological processes contribute towards the maintenance of this constant temperature. The importance of this aspect of homeostasis is appreciated when one realizes that the ill-effects accompanying a disease are due not so much to the presence of pathogenic micro-organisms as such, but to the increased temperature resulting from their activities. In fact if man's temperature rises much above 42°C death occurs.

A constant body temperature of about 36.9°C is necessary because it is the optimum temperature for the action of **enzymes**, upon which the organized functioning of the cells depends. Enzymes are proteins. If the temperature rises much above 40°C the proteins are denatured, and enzyme activity ceases. One of the first organs to be affected is the brain. Since this contains centres controlling respiration and the circulation, the rise in temperature disrupts the normal functioning of these important systems. An abnormally low body temperature can be equally serious: it has the effect of slowing metabolic activities, notably cell respiration and muscular contraction.

As far as temperature control is concerned, animals can be divided into two groups: those that regulate their body temperature by physiological means, and those that cannot do so (Fig 15.1). For many years the terms **homoiothermic** and **poikilothermic** have been given to these two groups respectively. Homoiothermic is a Greek word meaning 'having the same temperature'. Homoiothermic animals are popularly described as 'warm-blooded'; their body temperature is independent of environmental temperature so that in cold conditions their blood is at a temperature higher than that of their surroundings. Poikilothermic means 'having a variable temperature'. Poikilothermic animals are described as 'cold-blooded', their body temperature changing with fluctuations in the environmental temperature. If the environment is cold, so is their blood.

It is true that homoiothermic animals maintain a constant body temperature independent of the environment. However, it is not true that the temperature of poikilothermic animals is necessarily variable; nor is it true that their blood is cold. Many so-called cold-blooded animals maintain a surprisingly constant body temperature, well above that of the environment.

More useful terms are **endothermic** and **ectothermic**. Endothermic animals, as the word implies, generate heat from within the body, and keep it there. Ectothermic animals gain heat from the environment, i.e. from outside

232

the body. Both maintain reasonably constant temperatures, but by totally different means: endotherms by physiological means, ectotherms by behavioural means. The word poikilothermic is best reserved for the comparatively small number of animals whose body temperature really does fluctuate between wide limits. We shall return to this later.

HOW HEAT IS LOST AND GAINED

An animal in a situation where the environmental temperature is lower than the body temperature may lose heat by four physical processes: radiation, evaporation, conduction and convection.

   (1) **Radiation,** the diffusion of heat from a warm body to relatively colder objects via the air, can be a major source of heat loss. A man sitting in a room whose temperature is $21°C$ can lose as much as 60 per cent of his heat this way.

   (2) **Evaporation,** the change of a liquid to a vapour, is always accompanied by cooling. Evaporation of water from the surface of the body depends on various factors such as temperature, humidity and air currents, but it can account for a substantial loss of heat. A man at $21°C$ can lose 25 per cent of his heat by evaporation.

   (3) **Conduction** is the transfer of heat from the hotter to the cooler of two objects in contact with each other. A dog lying on a floor whose temperature is lower than that of the body, will lose heat to the floor by conduction. Similarly a man sitting in a cold room will lose heat to the floor and chair.

   (4) **Convection** is the movement of air resulting from local pockets of warm air being replaced by cooler air, and vice-versa. These air movements speed up loss of heat by radiation and evaporation.

   Of course these processes can work both ways. If the environmental temperature is higher than the body temperature, heat can be gained by radiation, conduction and convection.

   The problem in temperature regulation is to overcome, control or make use of these physical processes. Let us see how this is done in an endothermic animal.

RESPONSE TO COLD BY AN ENDOTHERMIC ANIMAL

Although many different systems co-operate in the process, the key structure involved in the temperature regulation of any endotherm is the **skin**. A vertical section of mammalian skin is shown in Fig 15.2. Literally all the structures included in the diagram play some part in temperature regulation.

   To understand the basic processes let us consider what happens when a mammal is subjected to severe cold. The tendency will be for heat to be lost from the body by the physical processes discussed earlier, but this will be offset by the following:

   (1) **The sub-cutaneous fat** in the dermis of the skin serves as an insulator and reduces heat loss from the body. It is interesting that animals living in very cold habitats, the polar bear and seal for example, have a particularly thick layer of sub-cutaneous fat.

   (2) **The hair is raised** and brought up into a more-or-less vertical position by contraction of the **erector-pili muscles**. The advantage of this is that air gets trapped in the spaces between the hairs. This air is warmed by the body and being a poor conductor of heat it serves as an insulatory layer around

A

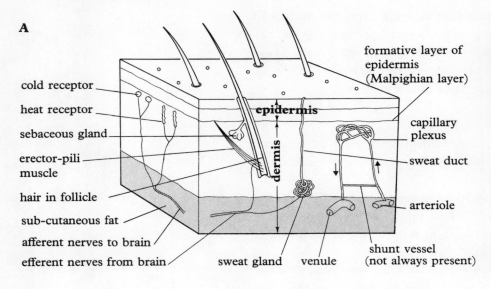

cold receptor

heat receptor

sebaceous gland

erector-pili
muscle

hair in follicle

sub-cutaneous fat

afferent nerves to brain

efferent nerves from brain

formative layer of
epidermis
(Malpighian layer)

epidermis

dermis

capillary
plexus

sweat duct

arteriole

sweat gland   venule

shunt vessel
(not always present)

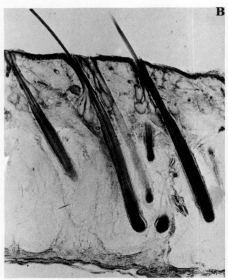

B

Fig 15.2 Structure of mammalian skin. A) Stereogram showing structures involved in temperature regulation. B) Vertical section of skin of scalp showing hairs, sebaceous gland and an erector pili muscle. (*Gene Cox, Micro-Colour International*)

the animal. In man the body hair is much reduced, its place being taken by clothes. In birds the same function is performed by the feathers.

(3) The superficial blood vessels in the skin constrict so that blood is diverted from the surface to the deeper layers. This reduces loss of heat from the blood to the surrounding atmosphere. In exposed structures such as the ears there are special **shunt vessels** interconnecting the arterioles and venules that take blood to and from the superficial capillaries of the skin. In cold conditions these shunt vessels dilate so that blood by-passes the surface of the skin. This is aided by a general reduction in the total volume of circulating blood, achieved by some of the blood being taken up into reservoirs such as the spleen. In prolonged conditions of extreme cold the blood may be diverted from the surface to such an extent and for so long that the cells die, resulting in frost-bite.

(4) Extra heat is produced by an increase in the **metabolic rate**, particularly of the liver and muscles. This starts as a general increase in muscular tone but may be followed by rhythmical involuntary contractions of the skeletal muscles (shivering) and contraction of the smooth muscles in the skin (goose flesh). Man frequently augments these processes by taking vigorous exercise. In winter as much as 40 per cent of the food we eat may be used in generating heat in resting conditions.

From this brief summary it can be seen that an endothermic animal's four techniques for coping with environmental cold fall into two types. The first three in our list are **physical mechanisms** designed to conserve heat by increasing the body's powers of insulation; the fourth is a **chemical mechanism** in which the body actively generates heat.

RESPONSE TO HEAT BY AN ENDOTHERMIC ANIMAL

The response to excessive environmental heat involves the reverse of the above processes. Heat production is cut down, and heat loss encouraged. The following is a summary of the situation.

(1) Animals living in hot climates have comparatively little subcutaneous fat. The fat deposits tend to be localized so as not to impede the loss of heat. Thus the camel's fat is stored in its hump, and in buffalo and bison it is located on top of the neck.

(2) **The hair is lowered** by relaxation of the erector-pili muscles so that it lies flat against the surface of the body. With no spaces between the hairs no air can be trapped against the skin. Insulation is reduced and heat can be more readily lost by radiation and convection.

However, it is important to realize that when the environmental temperature is *higher* than the body temperature, heat cannot be lost in this way. Under these circumstances the hair becomes important in insulating the body against excessive heat uptake. This is important in large animals like the camel which find it difficult to escape the heat of the sun.

(3) **The superficial blood vessels are dilated** so that blood is brought up near the surface from which it can lose heat to the surrounding atmosphere. The shunt vessels are constricted and the total blood volume is raised, thereby further increasing the flow of blood to the surface.

(4) **Sweating or panting occur.** Sweating involves the secretion of a watery fluid from **sweat glands** in the skin (Fig 15.2). Evaporation of the sweat from the surface of the body cools the skin and the blood flowing through it. As a means of dissipating heat, sweating is extremely important. Indeed when the environmental temperature exceeds the body temperature there is no other way of getting rid of heat. The cooling effect of sweating depends not only on the temperature of the surrounding air but also on its relative humidity, i.e. the degree to which the atmosphere is saturated with moisture. When the relative humidity is low, evaporation and hence cooling are rapid. When the humidity is high, evaporation and cooling are slow. This is why a temperature of 32°C in the desert with a relative humidity of only about 20 per cent is more comfortable than the same temperature in a tropical swamp where the relative humidity may be over 90 per cent. At very high humidities a large proportion of the sweat does not evaporate at all but drips from the skin, or sinks into the clothes, creating conditions which many people find intolerable.

In normal circumstances the evaporating power of the atmosphere is greatly enhanced by air movements. A gentle breeze will disperse the layer of high humidity that builds up round the body after a long period of sweating. This will encourage further evaporation to take place. The use of electric fans in hot weather is based on this principle.

In the dog and cat families there are no sweat glands (except in the pads of the paws) and heat is lost by panting. This greatly speeds up evaporation from the lungs with concomitant cooling of the blood. It also facilitates loss of heat from the blood as it flows through the pulmonary capillaries.

(5) **The metabolic rate falls** in hot conditions so that less heat is generated by the body. This is why animals are generally less active in hot weather than in colder conditions.

So we see that when the temperature of the environment changes, the body takes the necessary steps to maintain its own temperature at a constant level. But how does the body know when to 'turn on' its heating or cooling devices?

THE ROLE OF THE BRAIN IN TEMPERATURE REGULATION

For a long time, since about 1912 in fact, it has been suspected that there is a centre for controlling body temperature in the brain, and recent experiments have narrowed the site down to the **hypothalamus**. Electrical stimulation of this area brings about thermo-regulatory responses, and action potentials can be recorded from it as temperature changes occur. We can therefore describe this region of the hypothalamus as a **thermo-regulatory centre**.

The body also possesses millions of microscopically small receptors sensitive to temperature. These **thermoreceptors** are located in the skin (Fig 15.2) and are connected with the central nervous system by afferent nerves through which they signal changes in the environmental temperature to the brain (Fig 15.3).

Fig 15.3 Summary of the structures involved in the reflex control of body temperature in a mammal. Changes in the general metabolic rate are brought about by the hormones adrenaline and thyroxine which are secreted in extra large amounts in cold conditions.

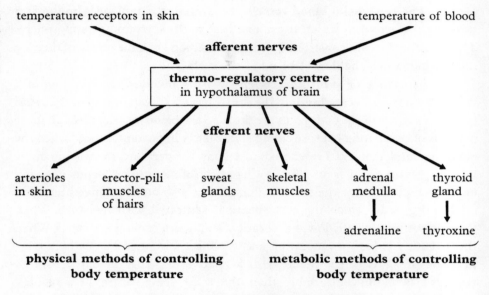

You will remember from Chapter 13 that for any homeostatic process to work, there must be a receptor, a control mechanism and an effector. We must now ask whether the thermo-regulatory centre in the hypothalamus serves only as a control device, or whether it functions as a receptor as well. In other words, does it simply relay information from the thermoreceptors to the effectors (erector-pili muscles, sweat glands etc.), or is the centre itself sensitive to changes in body temperature?

In order to answer this question ingenious experiments have been carried out in the USA by T. H. Benzinger's team at the Naval Medical Research Institute in Maryland. Their technique involves putting a man in a special form of **calorimeter** which enables simultaneous measurement of hypothalamic and skin temperature, and loss of heat by radiation, convection and sweating (Fig 15.4). The temperature of the hypothalamus is recorded by inserting a thermocouple into the outer ear and pressing it against the tympanic membrane (ear drum). Previous tests had shown that the temperature measured at this point was the same as that of the hypothalamus itself.

The subject, lying in the calorimeter at a constant temperature well above that of his body, was asked to consume a large quantity of iced sherbet at half-hour intervals. Immediately after taking the ice, three changes occurred:

(1) the temperature of the hypothalamus dropped, owing to withdrawal of heat from the blood by the ice in the gut;

(2) heat loss from the skin dropped, because of decreased sweating;

(3) the skin temperature rose owing to reduced heat loss from it.

The important thing to emerge was a perfect correlation between the fall in temperature of the hypothalamus and the decrease in the rate of sweating.

Fig 15.4 The calorimeter used by Benzinger in his experiments on human temperature control. The chamber is lined with a thin foil of material interlaced with thousands of thermoelectric junctions that measure heat loss from the skin. (*T. H. Benzinger, Naval Medical Research Institute, Bethesda, Maryland*)

The inescapable conclusion is that the decreased temperature of the blood is in some way detected by the thermo-regulatory centre, which then causes a decrease in the rate of sweating. That the skin receptors play little or no part in the response, is indicated by the skin temperature rising during this period. This would be expected to stimulate the heat receptors which would switch on the body's cooling processes, exactly the reverse of what actually happens.

It seems, then, that the hypothalamic centre functions as a **thermostat**.

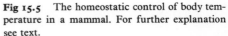

Fig 15.5 The homeostatic control of body temperature in a mammal. For further explanation see text.

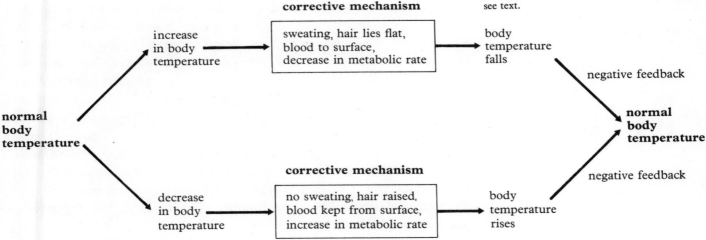

It is sensitive to temperature changes of the blood flowing through it, and responds by sending nerve impulses to the appropriate effectors. If the temperature of the blood is slightly higher than it should be, the thermo-regulatory centre detects this and sets into motion processes that collectively encourage heat loss. On the other hand, if the temperature falls below the optimum level, the centre initiates processes that produce and conserve heat. These events are summarized in Fig 15.5.

Does this mean that the thermoreceptors in the skin play no part in temperature regulation? On the contrary, while the hypothalamus detects temperature fluctuations inside the body, the skin receptors detect temperature changes at the surface. They enable the animal to feel whether the external environment is hot or cold. This information, acting via the thermo-regulatory centre, initiates voluntary activities such as taking muscular exercise in severe cold, or moving into the shade if it is very hot.

THE EFFECT OF LOWERING AND RAISING THE ENVIRONMENTAL
TEMPERATURE

We have seen that body temperature is regulated by a combination of physical (insulatory) and chemical (metabolic) means. To what relative extent does an animal depend on each of these processes in its life? To answer this let us consider what happens to a naked man if the environmental temperature is gradually lowered from a pleasant 29°C to freezing point. To begin with the subject relies entirely on purely physical mechanisms to maintain a constant temperature, his metabolic rate remaining unchanged. However, at about 27°C (the exact temperature is variable) the physical mechanisms are no longer capable of maintaining a constant body temperature on their own, and the metabolic rate starts to go up. This is called the **low critical temperature**, the lowest temperature at which physical mechanisms alone can regulate the body temperature. As the outside temperature is further lowered the metabolic rate continues to increase, until eventually the chemical mechanisms break down. At this point the **lower lethal temperature** is reached and the subject dies. (The experiment is not normally taken to this extreme!)

Now consider the result of increasing the environmental temperature. To start with, the various mechanisms promoting heat loss succeed in keeping the body's temperature constant, but there comes a point when the ability of the body to regulate its temperature breaks down. This is known as the **high critical temperature** and it varies according to humidity etc. The metabolic rate now starts to go up, and continues to do so for as long as the environmental temperature rises. No longer protected by the body's cooling processes, the chemical reactions in the cells become subject to the temperature rule, doubling their rate every time the temperature increases by 10°C. Moreover, once the temperature-regulating mechanisms fail, the metabolic rate goes on climbing even if the environmental temperature is no longer increased. The reason is that every time the metabolic rate increases, it generates more heat which raises the metabolic rate a bit more, and so on: an excellent example of **positive feedback** (see p. 205). As the body temperature continues to rise, **heat exhaustion**, characterized by cramps and dizziness, sets in. This is followed by death at a body temperature of about 42°C, the **upper lethal temperature** for most human beings.

TEMPERATURE REGULATION AND THE ENVIRONMENT

So we see that in man physical mechanisms alone regulate the body temperature between a high and low critical temperature. This is called the body's **efficiency range**. Above the high critical temperature the body's cooling mechanisms fail, the metabolic rate increasing as the temperature rises. Below the low critical temperature physical mechanisms are augmented by chemical means, the metabolic rate increasing as the temperature falls (Fig 15.6).

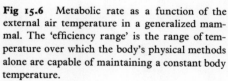

Fig 15.6 Metabolic rate as a function of the external air temperature in a generalized mammal. The 'efficiency range' is the range of temperature over which the body's physical methods alone are capable of maintaining a constant body temperature.

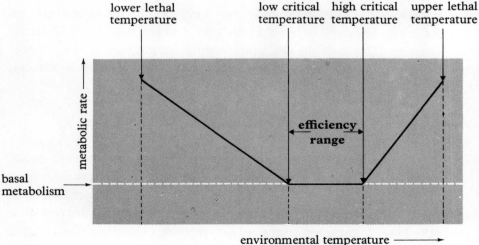

The low critical temperature has been determined for a number of different species, and it has been found that animals living in cold places have a lower critical temperature than those which live in warm places. Thus the kangaroo rat *Dipodomys* has a low critical temperature of 31°C, whereas for the arctic fox it may be as low as −40°C. Moreover, below the low critical temperature the curve of metabolic rate against environmental temperature is much less steep, and the lower lethal temperature much lower, for cold-dwellers than for warm-dwellers (Fig 15.7). These findings reflect the fact that animals in cold environments have better insulation mechanisms than those living in warmer places, an important aspect of survival.

As an endothermic animal, man is in rather an embarrassing position. As we have seen his low critical temperature is about 27°C, not much below that of the kangaroo rat. This is because of the scantiness of his hair, a deficiency for which he compensates by wearing clothes. Had he not developed the ability to fashion and wear clothes, man would never have been able to exploit the temperate regions of the world, let alone explore the poles. The wearing of clothes is an example of **behavioural control** of body temperature, to which we must now turn.

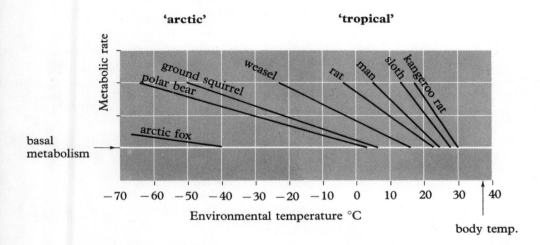

Fig 15.7 Metabolic rate of various mammals as a function of the external air temperature. Notice that the critical temperature at which the metabolic rate starts to rise is much lower for arctic than for tropical animals. (*after Scholander*)

BEHAVIOURAL CONTROL OF BODY TEMPERATURE

When an animal finds itself in a particularly hot environment, it may modify its behaviour in such a way as to cool itself. When cold it behaves so as to warm itself up. Such adaptive behaviour is seen in man when he puts on extra clothes or stamps his feet to keep warm.

Behavioural control is particularly important to ectothermic animals in which it is substantially the only method of temperature regulation. Even the lowliest of animals, protozoa, will actively seek out a region in their environment that provides the optimum temperature. Similar responses are shown by many other ectothermic animals such as insects and fish. But it is in the reptiles that behavioural control of temperature is most fully developed. Many lizards and snakes gain heat by lying in the sun (radiation etc.), or absorbing it from rocks and sand (conduction) (Fig 15.8). Desert lizards tend to be most active in the early morning and evening when it is neither too cold nor too hot. These are the times when one sees them scurrying about for food. In the heat of the day they retire to a shady place; and at night, when it can be extremely cold, they may burrow or seek out a crevice in which to build up a warm atmosphere from the heat generated by their metabolism. These animals are not insulated as a mammal is, but by shifting from place to place they do the next best thing, which is to make sure that the temperature of their immediate surroundings is always agreeable.

Fig 15.8 Heat gains and losses by an ecto-thermic animal. By absorbing heat from, or losing it to, its surroundings the animal maintains its body temperature at approximately 36°C. The photograph shows a lizard basking on a rock: it gains heat by radiation from the sun, convection currents in the air, reflection from nearby surfaces, and conduction from the rock. (*Bayard H. Brattstrom, California State College*)

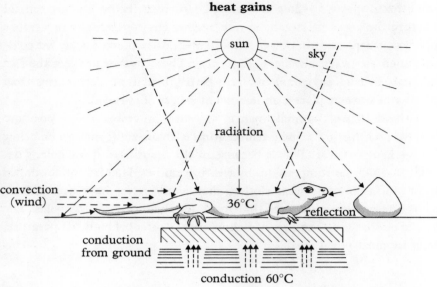

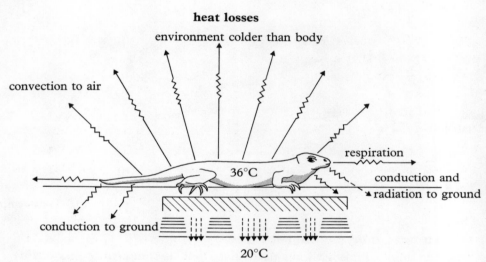

The same principle is involved in **migration**. Triggered by the cooler and shorter days of autumn, many birds that spend the summer months in northern regions, migrate southwards to warmer latitudes in winter. Thus swallows migrate from northern Europe to South Africa, golden plovers from Canada to the southern-most tip of South America, and so on. In forsaking their summer territories, such birds avoid the vicissitudes of a northern winter, including the lower temperatures that it inevitably entails.

TEMPERATURE AND HIBERNATION

We have seen that below the low critical temperature, the metabolic rate increases. This involves expenditure of energy and requires a plentiful supply of food to provide the fuel. In winter when the environmental temperature is low and food is short, this may be difficult and the animal may **hibernate**. Hibernation occurs in a variety of temperate and arctic animals, and is usually stimulated by cold. The animal responds by going into a deep sleep: the metabolic rate falls to the minimum required for maintaining the vital activities of the cells. The body temperature also falls, and is maintained at a much lower level than normal, typically about 18°C but as low as 2°C in the case of the hamster. Generally temperature regulation does not stop altogether

16 Control of Respiratory Gases

One of the impressive things about the human body is that, although conditions may change radically, it still manages to function adequately. These changes are often environmental, but sometimes the body imposes changes upon itself, for example, during heavy muscular exercise. Such activities may bring about profound alterations in the internal environment, and the body must make necessary adjustments if it is to continue operating efficiently.

It is important to realize that conditions change all the time, particularly in an active person. At the moment you are probably sitting in a chair reading this book; if you walk across the room, changes immediately occur in your body, to which adjustment must be made. However slight the exertion, there is bound to be a momentary increase in the metabolic rate. This will result in an increase in the amount of oxygen used and carbon dioxide produced. The oxygen content of the blood will, therefore, fall and carbon dioxide will rise. In this chapter we shall be concerned with how the body responds to this particular change.

THE ROLE OF BREATHING AND THE CIRCULATION

The respiratory and circulatory systems determine how much oxygen and carbon dioxide are present in the body at a given moment. If the amount of oxygen in the blood is low, and carbon dioxide high, the body responds by increasing (1) the rate and depth of breathing (**ventilation rate**), and (2) the rate at which the heart beats (**cardiac frequency**). At the same time the diameter of the arterioles serving those structures that are running short of oxygen and accumulating carbon dioxide, is increased. This local **vasodilatation** encourages the flow of blood through those structures that need it most. The overall effect of these responses is to keep the general level of oxygen and carbon dioxide constant. Reciprocal events occur if the oxygen content of the blood is abnormally high, and carbon dioxide low.

THE EFFECTS OF FLUCTUATIONS IN OXYGEN AND CARBON DIOXIDE

The importance of maintaining the respiratory gases at a constant level can be appreciated by considering the effects that follow if they fluctuate badly. To take **oxygen** first, a deficiency of oxygen (**anoxia**) deprives the tissues of a vital requirement for metabolism. The special senses, particularly vision, are impaired, as is the brain. This results in the adoption of an irresponsible slap-happy state of mind. This sometimes happens to aircraft pilots whose oxygen apparatus has broken down, and can result in gross misjudgement of situations. The trouble is that the person may be quite unaware that anything is wrong, and so he does nothing about it. Unconsciousness occurs suddenly, followed by paralysis (caused by irreparable damage to nerve cells) and death.

Excess oxygen can be equally dangerous. In an effort to use up the extra oxygen, the tissues metabolize at a frantic rate, using protein as well as carbohydrate and fat to keep pace with the oxygen supply. The cells 'burn themselves up', breaking down their own structures in the process. Once under way the situation is made worse by the inevitable accumulation of carbon dioxide. This results in an increase in ventilation rate and cardiac frequency, leading to even more oxygen being delivered to the tissues. This is another example of **positive feedback** setting in once the body's normal homeostatic mechanisms break down.

The cells are even more susceptible to changes in the level of **carbon dioxide**. Accumulation of this gas increases the acidity of the blood and tissue fluids, which affects the functioning of **enzymes**. The latter can only operate over a narrow range of pH, for mammals between 7.35 and 7.45. Even slight fluctuations above or below this range may inhibit them, thereby stopping essential metabolic processes. This is why breathing air rich in carbon dioxide is so dangerous. Rebreathing one's own expired air, by breathing in and out of a polythene bag for example, can be fatal in less than a minute.

CARBON DIOXIDE AS THE STIMULUS IN THE CONTROL PROCESS

The mechanism by which the level of respiratory gases in the body is kept constant provides another example of **negative feedback**. As long ago as 1905, Haldane and Priestley showed that in man the ventilation rate can be doubled by increasing the carbon dioxide in the air from its usual 0.04 per cent to 3.0 per cent. We now know that a change in the level of carbon dioxide in the blood is the effective stimulus initiating a change in the rate of respiration and circulation. For bringing about such changes, a small alteration in the amount of carbon dioxide is more effective than even a large change in the amount of oxygen. For this reason, the partial pressure of oxygen in the blood may vary considerably, but the partial pressure of carbon dioxide rarely deviates by more than a mere 13.3 N/m^2.

These facts can be demonstrated by recording a person's respirations with a spirometer (see p. 97). The effect of increasing the carbon dioxide level in the body can be shown by breathing in and out of the spirometer chamber with the carbon dioxide absorber removed. Since expired air is being rebreathed, the air taken into the lungs will get richer and richer in carbon dioxide. To compensate, the ventilation rate increases, as the body tries to get rid of the accumulating carbon dioxide (Fig 16.1, Curve A). After the subject has had time to recover the experiment is repeated, this time with the carbon dioxide absorber inserted between the subject and the spirometer chamber. Under these circumstances there is no such increase in the ventilation rate (Fig 16.1, Curve B).

A further experiment can be done to show the effect of decreasing the amount of carbon dioxide in the body. In this case the subject engages in **forced breathing** (extremely deep and rapid respirations) for several minutes. This has the effect of 'washing out' carbon dioxide from the lungs, so that its partial pressure is reduced. The subject then endeavours to breathe as naturally as possible. The result is that the ventilation rate is severely reduced,

Fig 16.1 Graphs showing the effect of rebreathing expired air on ventilation rate. As carbon dioxide accumulates the ventilation rate increases (graph **A**), but when pure oxygen is breathed there is no such increase (graph **B**).

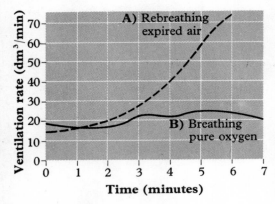

indeed breathing may stop altogether for a short period of time, a condition known as **apnoea**. This temporary inhibition of respiration provides an opportunity for carbon dioxide to build up, so that its partial pressure returns to the correct value.

THE ROLE OF THE BRAIN

In the **medulla** of the brain, there are groups of nerve cells that control the frequency of respiration and heart-beat. These are called the **respiratory** and **cardio-vascular centres** respectively. Efferent nerves convey impulses from the respiratory centre to the diaphragm and intercostal muscles, and from the cardio-vascular centre to the pacemaker of the heart. The two centres are directly responsible for increasing or decreasing the ventilation rate and cardiac frequency, according to circumstances. They do so by responding appropriately to the level of carbon dioxide, and to a lesser extent oxygen, in the bloodstream. If the partial pressure of carbon dioxide rises, the centres respond by increasing the ventilation rate and cardiac frequency. If it falls, they respond by decreasing the ventilation rate and cardiac frequency.

How are the medullary centres informed of the level of carbon dioxide in the blood? In the walls of certain arteries are sensory cells sensitive to changes in the partial pressure of carbon dioxide. As these sensory cells respond to a chemical, they are called **chemoreceptors**. They are grouped together at the base of the internal **carotid arteries** on each side of the neck, where they form the **carotid bodies** (Fig 16.2). The carotid bodies are sensitive to the smallest fluctuations in the partial pressure of carbon dioxide in the blood flowing past them. If the partial pressure of carbon dioxide rises, the sensory cells are stimulated, and impulses are conveyed via afferent nerves to the respiratory and cardio-vascular centres in the brain. The latter respond by sending impulses to the breathing apparatus and heart respectively, thereby bringing about an increase in the ventilation rate and cardiac frequency. This has been confirmed by experiments. In one experiment the carotid body was isolated by tying a ligature on each side of it, and then perfused with blood containing different amounts of carbon dioxide. It was found that increasing the carbon dioxide content of the blood had the immediate effect of speeding up the rate of breathing and heart beat. Cutting the carotid nerves abolished this response, clearly indicating the reflex nature of the response. Similar experiments can be carried out in which action potentials are recorded from the carotid nerves. Increasing the carbon dioxide in the blood raises the frequency of impulses in these nerves, thereby increasing the excitation of the respiratory and cardio-vascular centres. Similar responses can be produced by lowering the pH (i.e. raising the acidity) of the blood, or lowering the partial pressure of oxygen.

Responses to carbon dioxide are not confined to the heart and respiratory apparatus. The peripheral blood vessels also respond. When the carbon dioxide level rises, impulses are sent out from the cardio-vascular centre to the arterioles, which constrict. This **vasoconstriction** has the effect of raising the general blood pressure. At the same time accumulation of carbon dioxide in particular organs (for example the muscles during exercise) has the direct effect of causing the arterioles to dilate. This **local vasodilatation**, accompanied by a general rise in blood pressure, has the effect of greatly increasing the flow of blood through those parts of the body where carbon dioxide is

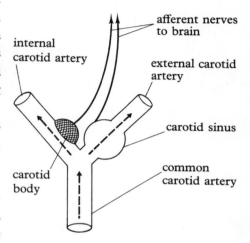

Fig 16.2 The carotid body and sinus are located in the neck where the common carotid artery splits into internal and external carotids. The carotid body is sensitive to carbon dioxide, the sinus to pressure. The broken arrows indicate the direction of flow of blood.

tending to build up. The control of carbon dioxide is summarized in Fig 16.3.

Fig 16.3 Summary of the reflex pathway by which the amount of carbon dioxide in the body is kept constant. An increase in the level of carbon dioxide in the blood results in an increase in cardiac frequency and the rate of ventilation of the lungs.

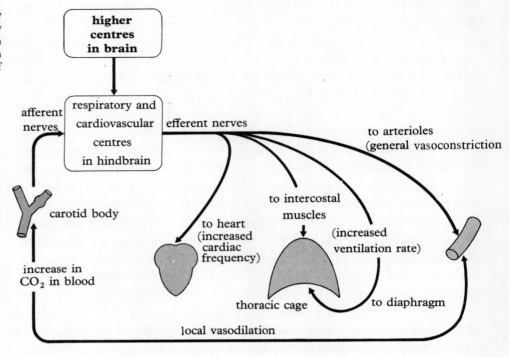

THE CONTROL OF BLOOD PRESSURE

From the previous section it will be clear that blood pressure plays an important part in the regulation of respiratory gases. It is also important in excretion and osmoregulation, and in temperature regulation. The mechanism by which blood pressure is controlled is closely related to the mechanism controlling carbon dioxide. At the base of the external carotid artery, on each side of the neck, is a small bulbous swelling, the **carotid sinus**, in whose walls are sensory cells sensitive to stretching (Fig 16.2). If the blood pressure rises, the walls of the carotid sinus are extended, and the **stretch receptors** stimulated. Impulses are conveyed via afferent nerves to the cardio-vascular centre in the medulla, which responds by slowing the heart beat and dilating the peripheral blood vessels, thereby lowering the arterial pressure. The carotid sinus is thus a sensitive pressure gauge, detecting changes in arterial pressure and signalling these to the cardio-vascular centre. The latter then brings about appropriate adjustments.

CEREBRAL CONTROL OF RESPIRATION AND CIRCULATION

It would be misleading to give the impression that the respiratory and cardio-vascular centres can only be influenced through the carotid and aortic receptors. To some extent they can be stimulated direct by carbon dioxide in the blood. Furthermore they are connected by nervous pathways to higher centres in the brain, and can be influenced by impulses reaching them from the **cerebral cortex**. This is most obvious in the case of respiration. Anyone can speed up or slow down his rate of breathing by a conscious act of the will. In these cases the higher centres exert either an excitatory or inhibitory influence on the respiratory centre.

Control over the circulation is less direct. No one can change the rate of his heart beat at will, but most people have experienced the increased pulse rate that accompanies excitement, shock or emotional experience. In this case impulses are conveyed from the higher centres to the **adrenal glands**, which respond by secreting the hormone **adrenaline** into the bloodstream. Adrenaline prepares the body for action in emergencies, and its effect on the heart is to increase the cardiac frequency. It also causes general constriction of arterioles, resulting in an increase in blood pressure. So the effects of adrenaline are identical with those produced by the nervous system, and its function is really to prepare the body for coping with a rise in carbon dioxide before it actually happens.

ADJUSTMENT TO HIGH ALTITUDES

In responding to changes in the partial pressure of carbon dioxide in the blood, the body not only prevents accumulation of this poisonous gas but also ensures that sufficient oxygen is delivered to the tissues at all times. The efficiency of these homeostatic processes can be tested by subjecting a person to an atmosphere containing an abnormally low partial pressure of oxygen. This is precisely what happens at high altitudes where the atmospheric pressure, and hence the partial pressure of oxygen, are considerably lower than at sea level. How does a person adjust to such conditions?

The answer depends on how high the altitude is, and how quickly he gets there. An aircraft pilot flying straight up to a great height without oxygen apparatus develops symptoms of anoxia at about 4,000 metres, and becomes unconscious at about 8,000 metres. On the other hand a mountaineer who ascends slowly over a period of days or weeks, has time to get used to the progressively rarefied atmosphere. At about 4,000 metres he begins to develop signs of oxygen lack – headache, nausea and fatigue (**mountain sickness**). But these unpleasant symptoms wear off as he becomes **acclimatized**. Adjustments occur in his respiratory and circulatory systems as the homeostatic responses to oxygen lack get pushed to their limit. At the same time there is an increase in the total number of red blood cells and the haemoglobin content of the body. The red cell count may increase from about 5.5 million per mm^3 to 8 million per mm^3 by the time the climber reaches 4,300 metres. Thus the oxygen-carrying power of his blood goes up, as the ventilation rate and cardiac frequency increase. In the Himalayan expedition of 1953, Hillary and Tensing spent three hours at a height of 9,300 metres without oxygen apparatus, levelling snow and pitching a tent. It was not easy and they had to rest every few minutes, but the fact that they managed to do it at all indicates the importance of acclimatization as a physiological process. An unacclimatized person at such a height would be unconscious within five minutes.

ADJUSTMENTS DURING EXERCISE

Acclimatization to high altitudes illustrates the way the body can adjust to a changing environmental situation. Let us now briefly consider another kind of adjustment, namely that which occurs during the performance of a sprint. This should be of interest to any biologist who happens to be an athlete, and will enable us to apply our knowledge of respiratory and circulatory physiology to a familiar situation. Try to list the various processes that will occur before reading further.

A)   Before and during the early stages of the sprint **adrenaline is secreted**

from the adrenal glands into the bloodstream. Secreted in response to impulses received from the cerebral cortex of the brain, it causes

(1) an increase in cardiac frequency;

(2) general constriction of arterioles except for those serving vital organs like the heart, lungs and brain;

(3) contraction of the spleen, the body's main blood reservoir.

B) During the sprint the **metabolic rate increases.** This is caused by the shortage of ATP that inevitably results from muscular exertion. There is never much ATP available as such: it is manufactured on the go. It has been found that a decrease in the amount of ATP activates an enzyme which initiates the further phosphorylation of sugar, another example of negative feedback.

C) The increased metabolic rate results in carbon dioxide building up in the muscle tissues. This causes **local dilatation of the arterioles**, leading to an increased blood flow through the muscles. It has been found that the increase in body temperature during the sprint renders the tissues more sensitive to carbon dioxide, thereby accentuating this mechanism.

D) Local accumulation of carbon dioxide in the muscles is quickly followed by a general rise in the concentration of carbon dioxide throughout the circulatory system. This **stimulates chemoreceptors** in the aortic and carotid bodies leading reflexly to

(1) increased ventilation rate,

(2) increased cardiac frequency, and

(3) constriction of arterioles.

The events listed so far result in a rise in arterial pressure, and a greatly increased flow of blood through the active muscles.

E) The increased arterial pressure results in **stimulation of stretch receptors** in the carotid sinuses. This results reflexly in a decrease in cardiac frequency which will tend to decrease the speed of the circulation. This, together with the dilatation of the muscle arterioles, safeguards the body from developing too high a blood pressure.

F) During the race the metabolic rate of the active muscles increases greatly and the demand for oxygen rises accordingly. Despite the mechanisms listed above insufficient oxygen is delivered to the muscles to keep pace with their demands. As a result the **muscles start respiring anaerobically**, with the formation of **lactic acid**. This accumulates during the race, but afterwards is circulated to the liver where it is converted back to glycogen. For this to happen oxygen is required. Constituting the so-called **oxygen debt**, it accounts for the heavy panting that ensues after the race.

G) While all this has been happening, the greatly increased metabolic rate results in a **rise in body temperature**. This is offset by the body's various cooling processes described in Chapter 15.

The same kind of adjustments occur in longer races. One difference, however, is that oxygen debt may be paid on the run, an equilibrium being reached between oxygen supply and oxygen usage. When this point is reached **second wind** is said to be acquired. It is clear, however, that for the short period involved in a sprint, the muscles can function perfectly efficiently under anaerobic conditions, provided of course that the oxygen debt is paid immediately afterwards. This enables an athlete to hold his breath during a short sprint, the hundred metres for example. Under these circumstances circulatory and metabolic changes occur as described above, but changes in breathing are temporarily suspended until after the race.

It follows that the performance of vigorous muscular activity at high altitudes imposes particularly heavy strains on the body, and requires long periods of acclimatization and training before it can be achieved efficiently. This is why the 1968 Olympic Games and the 1970 World Cup, held in Mexico City at the gruelling height of 2,460 metres, presented a special challenge to those teams coming from countries at lower altitudes.

Surprising though it may seem, much is still not understood about the detailed physiological changes occurring during vigorous muscular activity. For example the relationship between lactic acid and oxygen debt is by no means simple. What this brief and simplified account shows is that several different systems (respiratory, circulatory, nervous and endocrine) all co-operate in bringing about appropriate adjustments, which together maintain the continued efficiency of the body.

RESPONSE TO TOTAL OXYGEN DEPRIVATION

We have mainly discussed how the human body adjusts to a gradual lowering of the oxygen tension. What happens if an animal is suddenly deprived of oxygen altogether?

For most species the result is disastrous, death occurring within a matter of minutes. But certain animals, notably those capable of diving, can survive much longer. For example, a seal can remain under water for over 20 minutes, and certain species of whales for over an hour. How is this achieved?

By recording the heart beat, blood pressure and other variables in a variety of diving animals it has been found that very soon after the dive commences the cardiac frequency decreases dramatically (**bradycardia**) and the arterioles of all but the vital organs constrict. This rapid reflex response results in the body's oxygen store, derived from its haemoglobin and myoglobin, being sent to those organs that are least able to endure oxygen deprivation, namely the heart and brain.

It is now known that this response is not restricted to diving animals like seals and whales but is shown by many other animals confronted with sudden oxygen deprivation. It occurs in man – a human diver will develop bradycardia within 30 seconds after the beginning of a dive – and in fishes it occurs when they are taken out of water. It therefore seems to be a life-saving response of general importance.

17 Defence against Disease

The body of an animal is constantly being invaded by **micro-organisms**, particularly **bacteria** and **viruses**, which enter by the body's orifices, especially the mouth and nose, or through wounds. Many of these micro-organisms either feed on the tissues or liberate toxic substances into the bloodstream, thereby bringing about disease. Thus viruses cause such diseases as influenza, poliomyelitis and the common cold, whereas bacteria cause diseases like typhoid, diphtheria and tuberculosis. The activities of these pathogenic microbes, (or **germs**) are not, of course, confined to man. All animals and plants are susceptible, though the particular micro-organisms that plague them may be different.

In destroying the tissues and liberating toxic substances these micro-organisms change the internal environment and upset the smooth running of the body. Their control by the body's natural defence mechanisms is thus an aspect of homeostasis. Since we know more about these defence mechanisms in mammals than in other organisms we will confine our discussion to them.

The body's defences against disease can be divided into **passive** and **active mechanisms**. The passive processes are concerned with preventing the entry of micro-organisms, whereas the active processes destroy them once they manage to get in. We will deal with each in turn.

PREVENTING ENTRY

The **skin** with its hard, keratinized outer layer, serves as an effective barrier to most micro-organisms. However, much as an animal might like to envelop itself in an impenetrable barrier, this is patently impossible. Thus the alimentary and respiratory tracts are major pathways through which micro-organisms can get into the body. To some extent entry via the respiratory tract is prevented by **cilia** lining the trachea and bronchi. Micro-organisms and other undesirable particles get caught up in **mucus** secreted by numerous goblet cells and are carried by the beating cilia towards the glottis. **Coughing, sneezing** and **vomiting** all aid in expelling foreign bodies as Fig 17.1 makes only too clear, and the **acid** in the stomach serves as an effective sterilizing agent killing many bacteria that come in with the food.

The entry of micro-organisms through wounds is, as everyone knows, a major cause of infection. To some extent this is prevented, or at least cut down, by the **clotting of blood** when it is exposed to air. A necessary prerequisite for the healing of a wound, blood-clotting not only prevents excessive bleeding but also blocks the entry of germs. It is a fascinating process in which no less than 12 different chemical factors present in the

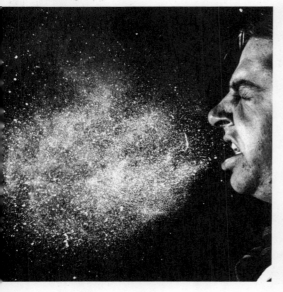

Fig 17.1 Short-duration flash photograph of a sneeze. Notice the vast number of droplets emitted from the mouth and respiratory tracts. This is one way in which pathogenic micro-organisms are transferred from one person to another. (*photo by Dr M. W. Jennison, Syracuse University, New York. Courtesy of the Society of American Bacteriologists*)

blood bring about the conversion of the soluble plasma protein **fibrinogen** into a meshwork of fine fibres called **fibrin**. The process is triggered by the disintegration of the blood **platelets**, tiny cell fragments, when they are exposed to air (Fig 17.2).

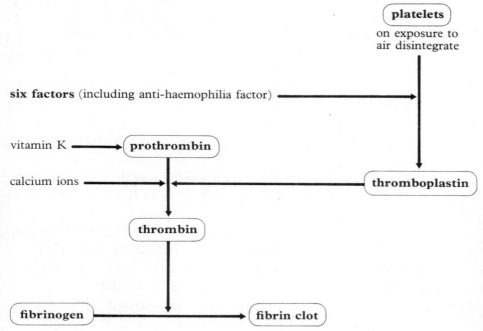

Fig 17.2 The clotting of blood plays an important part in preventing the entry of microorganisms through wounds. Exposure of the blood platelets to air sets into motion a complex sequence of chemical reactions which terminate in the conversion of the plasma protein fibrinogen into a meshwork of fine fibres called fibrin. Essential for the process is vitamin K, a group of quinone compounds found in abundance in vegetables. Vitamin K is required for the formation of prothrombin, the precursor of thrombin. Recent research has shown that it is also needed for the synthesis of two of the six factors required for the formation of thromboplastin.

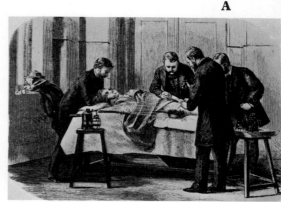

An enzyme has been discovered in tears, nasal secretions and saliva which is capable of destroying certain bacteria. On account of the fact that it dissolves the cell wall and therefore bursts the bacteria, this enzyme is called **lysozyme**. It is a powerful enzyme, effective even in low concentrations, and undoubtedly plays some part in keeping bacteria out of the body.

In considering temperature regulation we saw that man is unique amongst animals in supplementing his natural homeostatic mechanisms with artificial devices. This is also true of disease control. **Public health** and **hygiene, antiseptics** and **aseptic surgery** are all aimed at preventing the entry of germs into the body. The foundations of these preventive measures were laid a century ago by the French microbiologist Louis Pasteur, who established the germ theory of disease and showed that micro-organisms are not spontaneously generated from non-living matter as had been generally supposed, but come from pre-existing microbes. He discovered that micro-organisms could be destroyed by heat treatment, thus providing the basis for sterilization and preservation of food: sterilization of milk by heat treatment is still known as pasteurization.

Pasteur's work on micro-organisms and sterilization inspired the English surgeon Joseph Lister to look for other means of destroying bacteria: Lister's use of carbolic acid in the operating theatre marked the beginning of antiseptic surgery. Today antisepsis is gradually being supplanted by aseptic techniques, the ultimate aim of which is to eliminate germs from the environment altogether. It would be unrealistic to think that this can ever be achieved in the general environment, but it is aimed at in modern operating theatres and intensive care units in hospitals (Fig 17.3).

Despite the body's passive methods germs do get into the body, and,

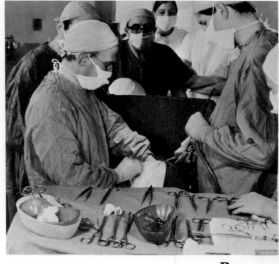

Fig 17.3 The crude conditions in which surgical operations were carried out a hundred years ago contrast sharply with the aseptic atmosphere in the modern operating theatre. A) shows the use made in the nineteenth century of the Lister carbolic spray for creating antiseptic conditions in the immediate vicinity of the wound. B) Today hospitals attempt to create a totally germ-free atmosphere in their operating theatres. (A: *Wellcome Historical Medical Museum and Library;* B: *University College Hospital, London*)

Fig 17.4 Defence against disease by phagocytosis.   A) The two types of leucocyte most closely concerned with phagocytosis: neutrophil (top) and monocyte. The neutrophils, making up about 70 per cent of the white cells, originate in the bone marrow along with the red cells. The monocytes constitute 4–8 per cent of the white cells and are formed in the lymph glands. During infection the lymph glands manufacture additional phagocytes.   B) A neutrophil attacks a clump of bacteria. The micro-organisms are taken up into a vacuole where they are digested by lysosome enzymes as described on p. 56.

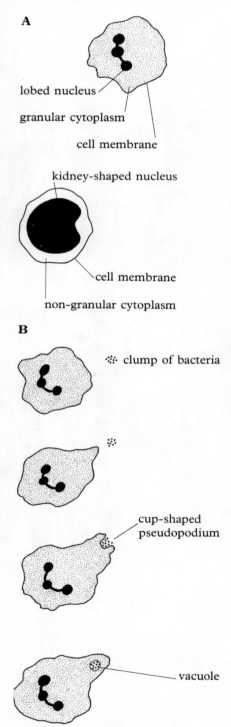

A

lobed nucleus

granular cytoplasm

cell membrane

kidney-shaped nucleus

cell membrane

non-granular cytoplasm

B

∴ clump of bacteria

cup-shaped pseudopodium

vacuole

once inside, they multiply prodigiously. How are they dealt with? The body's active defence mechanisms fall under two headings: **phagocytosis** and the **immune response**. These two processes are brought about mainly by the **white cells** (**leucocytes**) which occur in large numbers in the blood; about 7,000 per mm³ in human blood. There are five types of leucocyte which can be distinguished by examining a smear of blood stained with Leishman's or Wright's stain (see. p. 163). Some of these leucocytes are formed in the **bone marrow**, along with the red blood cells, others in the **lymph glands.**

PHAGOCYTOSIS

Phagocytosis is carried out mainly by the **neutrophils**. Manufactured in the bone marrow, these cells have a granular cytoplasm and a curious lobed nucleus rather like a string of sausages. They are **amoeboid**, moving about in contact with the endothelium of the blood vessels where they ingest bacteria (Fig 17.4). Once taken up, the bacteria are digested by lysosome enzymes as described on p. 56. These phagocytes are not confined to the blood vessels for they can squeeze between the endothelial cells and migrate into the tissue spaces, a process called **diapedesis**. Thus in the early stages of a local infection hundreds of neutrophils migrate from the capillaries to the infected area, where they undergo phagocytosis on a large scale. Fully engorged, a single neutrophil may contain as many as 20 visible bacteria, many of them still alive and moving.

In addition to the white blood cells many tissues contain their own phagocytes, called **macrophages** (Fig 17.5). Together they make up the body's **reticulo-endothelial system.** They are particularly numerous in the liver and lymph glands and are able to take up and concentrate various foreign particles and poisons as well as micro-organisms. The phagocytes, white blood cells and macrophages alike, have been aptly described as the policemen of the body, moving actively from place to place devouring bacteria as soon as they enter. Phagocytes are by no means confined to mammals; they are found in all vertebrates and a wide range of invertebrates as well.

THE IMMUNE RESPONSE

Each kind of micro-organism, it might be the diphtheria bacillus or the polio virus, contains a wide range of macromolecules which act as **antigens**. If and when a particular antigen gets into the body it stimulates certain cells, derived from the **lymphocytes**, to produce a corresponding protein called an **antibody**. The antibody combines with and in some way neutralizes the antigen, destroying the micro-organism in the process. This can take place in several different ways but generally the antibodies either adhere to the surface of the micro-organisms, making them clump together (**agglutination**), or they may cause them to burst (**lysis**). Sometimes agglutination is followed by lysis and the remains of the micro-organisms are then ingested by phagocytes.

The production of antibodies in response to antigen is called the **immune response**. Once specific antibodies have been produced against a particular microbe, defence against the disease is set up, at least for the time being, and **immunity** is established.

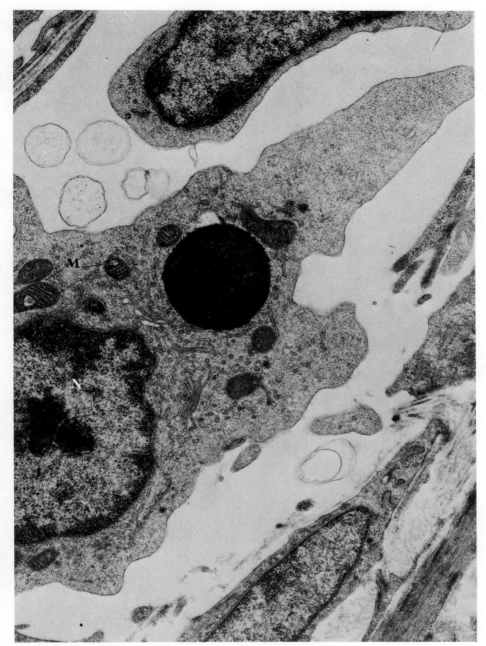

Fig 17.5 Electron micrograph showing a phagocytic cell in a blood vessel of the kidney. Notice the prominent nucleus (N) and mitochondria (M). What do you think the large black object might be? (*A. R. Akester, University of Cambridge*)

The kind of immunity just described is called **active immunity** because the body makes its own antibodies in response to the arrival of an antigen. However, during the development of a mammal a certain number of antibodies may 'leak' across the placenta from the mother to the foetus. This confers **passive immunity** on the young animal, at any rate for a short time after birth. The human infant, for example, may be protected from such diseases as measles and poliomyelitis as a result of passive immunity. However, the number of antibodies transferred in this way is, limited, and since it does not involve antibody-production by the young animal itself such immunity is generally short-lived.

In immunity, as in so many other aspects of homeostasis, man has augmented nature's methods with his own. Active artificial immunity can be established by injecting a small quantity of antigens, the **vaccine**, into the body (**immunization**). This induces the production of the appropriate antibodies which are then present at the ready if and when that particular micro-organism gets into the body. For example, protection against poliomyelitis may be secured by injection of a small quantity of killed-virus vaccine or oral administration of a live-virus vaccine. Immunization techniques were first developed by the eighteenth century physician Edward Jenner. He discovered that defence against smallpox could be established by injecting into the body a small quantity of plasma obtained from a dairymaid who had contracted cowpox. We now know that the cowpox virus causes a relatively mild disease but triggers production of the same antibodies as the smallpox virus. Injecting the cowpox virus into a human will therefore not cause him undue suffering but stimulates him to manufacture antibodies against the much more virulent smallpox. Jenner's was a momentous discovery in the history of medical science, and immunization was later extended to many other diseases.

A quite different technique is to give a person antibodies from another individual. In this passive form of immunity the recipient is not induced to produce his own antibodies but is supplied with them, ready-made as it were, from an outside source. Protection against diphtheria is given in this way. The antibodies are prepared by injecting toxin obtained from the diphtheria bacillus into a horse. The toxin is first rendered harmless by chemical treatment so the horse itself does not suffer from the disease, but responds by producing large quantities of the appropriate antibody. The horse serum, suitably prepared, can then be used against the disease.

The trouble with this procedure is that the protection, though immediate, is relatively short-lived, lasting only as long as the antibodies persist. As antibodies are proteins, and proteins are continually being broken down and replaced, they may last for only several weeks. On the other hand active immunity, where the individual is induced to manufacture his own antibodies, is longer-lived. This difference can be appreciated by comparing the two kinds of treatment that can be given against tetanus. Anti-tetanus serum (containing ready-made antibodies) gives immediate but short-lived protection, whereas anti-tetanus vaccine gives long-lived protection. In general, serum is given only in cases where an immediate dose of antibodies are needed in numbers that the individual cannot be expected to produce for himself.

BLOOD GROUPS

An unfortunate aspect of the immune response, as far as man is concerned, is that any living material introduced into the body is treated by the recipient as 'foreign', and antibodies react against it. This is the outcome, for example, of transfusing blood from one person to another. If the blood of the two individuals is not compatible the donor's red blood cells clump together in groups (agglutination) which may result in blockage of the recipient's blood vessels. The reason for this reaction is that the donor's red cells contain antigens which are complementary to antibodies present in the recipient's plasma. Unlike normal immune responses the recipient does not actually produce antibodies in response to the donor's blood. They are present all the time, agglutination occurring if the donor's blood happens to contain the corresponding antigens.

The entire human population can be divided into four groups on the basis of the reaction between the blood of different individuals when mixed together. These groups are called **A**, **B**, **AB** and **O**. The capital letters stand for the type of antigens present in the person's red blood cells. The corresponding antibodies are carried in the plasma and are represented by small letters: **a**, **b**, **ab** and **o**. Obviously if an individual has a particular antigen in his red cells he cannot have the corresponding antibody in his plasma, otherwise agglutination will occur. Thus a person belonging to group **A** has red blood cells containing type **A** antigens; his plasma will not contain type **a** antibodies, but it does contain type **b** antibodies. A person of group **B** has red blood cells containing type **B** antigens and type **a** antibodies in the plasma. Group **AB** contains both antigens **A** and **B** and neither antibodies. Group **O** has neither antigens but both antibodies.

Fig. 17.6 summarizes what happens when blood of different groups are mixed together in a transfusion. All is well provided that the recipient's blood does not contain antibodies corresponding to the donor's antigens. If it does agglutination results. Normally it does not matter if the donor's antibodies are incompatible with the recipient's antigens because, as the recipient generally receives a relatively small amount of blood from the donor, the dilution effect will minimize agglutination. Furthermore, red blood cells being comparatively short-lived, there is no permanent supply of antibody-producing cells being transferred in the transfusion.

In addition to the ABO system many other blood groups are now recognized. These include the **Rhesus system**, so-called because it was first discovered by injecting rabbits with red cells obtained from the Rhesus monkey. The majority of white people possess red blood cells containing an antigen called the **Rhesus factor**. Their blood is described as **Rhesus positive**. The remainder lack the Rhesus antigen and are called **Rhesus negative**. Unlike the ABO system, Rhesus negative blood does not automatically contain the Rhesus antibody. However, if Rhesus-positive blood finds its way into a Rhesus-negative recipient, the latter responds by producing the corresponding Rhesus antibody. Nothing further happens, but if the Rhesus-negative recipient subsequently receives another dose of Rhesus-positive blood, his Rhesus antibodies will cause agglutination of the donor's red cells, often with fatal results. It may seem unlikely to you that this would ever happen in practice, but in fact it can occur during pregnancy. The trouble occurs when a Rhesus-negative mother bears a Rhesus-positive child. Sometimes, particularly in the last month of pregnancy, fragments of the foetus' red blood cells, containing the Rhesus antigen, pass across the placenta into the mother's bloodstream. The mother responds by producing Rhesus antibodies which pass back across the placenta into the foetal circulation, destroying the child's red cells both before and immediately after birth. Generally the antibodies are not formed sufficiently quickly to affect the first child, but subsequent children, if also Rhesus-positive, suffer from massive destruction of their red blood cells. The condition is known as **haemolytic disease of the new-born (erythroblastosis foetalis)** and may be fatal unless the child's blood is replaced by transfusion with Rhesus-negative blood. Nowadays such transfusions are common practice and can be carried out while the foetus is still in the womb. It has recently been found that haemolytic disease can be prevented by treating the mother with an anti-Rhesus globulin that coats the foetal cells, thus blocking the Rhesus factor.

universal donor   universal recipient

		recipient		
	O ab	**A** b	**B** a	**AB** o
O ab	−	−	−	−
A b	+	−	+	−
B a	+	+	−	−
AB o	+	+	+	−

Fig 17.6 Summary of reactions that occur when bloods of different groups are mixed. Plus: agglutination; minus: no agglutination. The capital letters refer to antigens in the corpuscles; small letters to antibodies in the plasma. A universal donor (group **O**) can give blood to a recipient of any group without causing agglutination. A universal recipient (group **AB**) can receive blood from a donor of any group without agglutination.

A

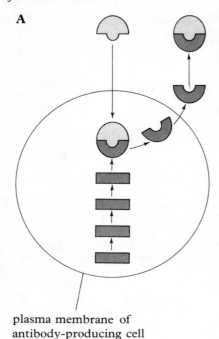

plasma membrane of
antibody-producing cell

B

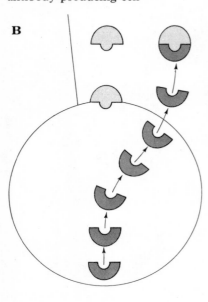

antigen molecules in light tone
antibody molecules in dark tone

Fig 17.7 How are antibodies produced? According to the 'instructive' hypothesis (A) the antigen enters the cell and there serves as a template for the assembly of appropriate antibodies which then leave the cell. In the 'clonal selection theory' (B) the antigen sticks to the surface of the cell at a receptor site into which it fits. It is postulated that the cell already has all the potential biochemical equipment for producing a particular antibody. The arrival of the antigen triggers it to proliferate and then start production. (*based on C. J. V. Nossal*, Scientific American, *Dec. 1964*)

[1] The clonal selection theory has been likened to buying a ready-made suit 'off the peg'. Why is this a misleading analogy?

HOW ARE ANTIBODIES PRODUCED?

Any explanation of how antibodies are formed must account for the fact that specific antibodies are produced on demand by the arrival of a particular antigen. We will briefly consider two possible explanations.

The first, known as the **instructive hypothesis**, is based on the supposition that the antigen enters the antibody-forming cell and there acts as a template for the production of appropriate antibodies. Since the construction of the antibodies is directed by the antigens themselves it is supposed that their shapes will be complementary to each other. The antibodies then leave the cell and combine with antigen molecules as depicted in Fig 17.7A.

The trouble with the instructive hypothesis is that, although it accounts for the specificity of the immune response, it fails to explain why antibodies are not formed to the individual's *own* antigens. This difficulty has led to the development of the **clonal selection theory** first proposed in 1960 by Sir Macfarlane Burnet, Director of the Walter and Eliza Hall Institute of Medical Research in Melbourne, Australia.

The clonal selection theory is based on the idea that a wide range of **lymphocytes**, each potentially capable of producing specific antibodies, is present in the body before birth. During development of the foetus these lymphocytes are thought to proliferate into **clones** of simple undifferentiated **plasma cells**. (A clone is a population of cells all derived from one original cell.) When an antigen gets into the body it 'selects' a plasma cell of the appropriate clone, adheres to its surface and causes it to proliferate into a further mass of cells, all of which proceed to manufacture the correct antibodies. Now it is thought that if a plasma cell makes contact with a complementary antigen during embryonic life it is in some way rendered incapable of producing antibodies. Thus any clones that might produce antibodies against the individual's own antigens are inactivated. However, if a foreign antigen gets into the body after birth it adheres to the correct plasma cell and initiates antibody-production. The plasma cell already has the necessary biochemical machinery for making the right antibodies. The antigen merely triggers it into action, probably by switching on the appropriate part of its DNA. The theory depends, of course, on the plasma cell 'recognising' its particular antigen. There is evidence that this is achieved by the antigen fitting into a receptor site on the cell surface, antibody production commencing only if the fit is exact (Fig. 17.7B).

To understand the basic difference between the instructive and clonal selection theories an analogy from everyday life might be useful. If you want to purchase a suit the best procedure is to go to a tailor, get fitted, and have one made to your particular specifications. This is analogous to the instructive hypothesis where antibodies are, as it were, tailor-made to fit a particular antigen. The 'selective' theory involves a rather different situation. Here we must imagine that the customer selects the right shop and rings the doorbell. This particular tailor already knows exactly what sort of suit to make, and he has all the necessary wherewithal to do so. Without admitting the customer he makes the suit and hands it to him through the door.[1] Like all analogies this must not be carried too far but it does illustrate the two contrasting mechanisms by which antibody-formation might be initiated.

The clonal selection theory can be extended to explain the enhanced immunological response given when the body receives a second dose of antigen. At one time this was thought to be due to the presence of antibodies left over from the first response. However, the results of recent research are more consistent with the idea that the first exposure to a specific antigen results in the formation of plasma cells which lie dormant until the next arrival of the same antigen, upon which they immediately produce appropriate antibodies. These 'memory cells', as it were, 'remember' their previous experience of antigen and are ready to respond when a second dose is received. This is why we normally only get certain infectious diseases, mumps for example, once.

How do the instructive and clonal selection theories stand today? The whole subject is very confused but the evidence tends to support a selective process. It has been established from experiments on isolated lymph glands that antibodies are not produced from lymphocytes themselves but only from the plasma cells. By using antigens labelled with radioactive isotopes it has been shown that the antigen never gets inside the plasma cell and therefore does not come into direct contact with the antibody. Contact with the surface of the cell is sufficient to start antibody production. This would seem to refute the idea that the antigens provide a template for the formation of anti-bodies. But does it necessarily mean that an antigen exerts no influence at all on the kind of antibodies a plasma cell produces?

REJECTION OF GRAFTS

An unfortunate by-product of the immune response is that if living tissue is **transplanted** from one individual to another, the recipient recognizes the foreign antigens in the transplanted tissue and destroys it. Destruction of a graft in this way is described as **rejection**. Though similar in principle, the rejection of grafts involves a somewhat different type of immune response from that described earlier. In this case the antibodies are cell-bound and not free in the blood plasma. Transplanted organs such as kidneys and hearts are rejected unless the two individuals share exactly the same complement of antibodies. This will only be the case if they are genetically identical, i.e. identical twins.

Will the recipient not accept foreign tissue under any circumstances? The answer is that it will, provided that the formation of antibodies against it is first prevented. One way of achieving this is to introduce some of the foreign tissue into the recipient before, or very soon after, birth i.e. at the time when, according to Burnet's clonal selection theory, antibody-production against the individual's own antigens is being abolished. This was first dis-covered by Sir Peter Medawar who injected cells obtained from an adult mouse into a foetus. He found that after birth the recipient would accept, permanent-ly, grafts of skin and other tissues from the donor. The recipient had been made **immunologically tolerant** to the donor's tissues by receiving a pre-natal injection of the donor's cells. Logically enough, such induced tolerance was found to be abolished by re-equipping the recipient with lymphocytes from another individual of the same strain. For this important achievement Medawar, together with Burnet, were awarded the Nobel Prize for medicine in 1960.

Since then a number of agents capable of inducing immunological toler-ance have been developed. These include a preparation made by injecting lymphocytes into a horse and collecting, then purifying, the antibodies pro-

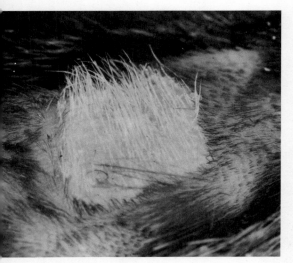

Fig 17.8 Photograph of a successful graft of rat tail skin (white hairs) about four months after transplantation to an adult mouse. Normally such a graft would be rejected but in this case the adult mouse was rendered immunologically tolerant by means of anti-lymphocytic serum. (*Sir Peter Medawar, National Institute for Medical Research*).

duced. The resulting **anti-lymphocytic serum** is capable of destroying the original lymphocytes to such an extent that an animal treated with the serum will subsequently accept grafts from another individual (Fig 17.8).

TRANSPLANTATION SURGERY

Transplantation surgery is an important field of medical endeavour in which great advances have been made in recent years. Kidneys have been transplanted for a number of years with varying degrees of success. The first heart transplant was carried out in 1967 by Christiaan Barnard in Cape Town, and since then attempts have been made to graft lungs and livers. For success the tissues of the donor and recipient must be as genetically similar as possible. How close they are can be determined by **tissue-typing**, a procedure which is a necessary preliminary to surgery. Subsequent rejection of the transplanted organ, or graft, is prevented, or at least slowed down, by **cobalt irradiation** and **immuno-suppressive drugs** which inhibit the body's normal immune response. The trouble is that in so doing they inhibit the body's defences against micro-organisms and therefore lay the patient wide open to disease. The high failure rate in transplantation operations is due to the patient succumbing to a disease, commonly pneumonia, or to rejection of the organ. In the future success will depend on using drugs which abolish the graft-rejecting response but leave the disease-responses unimpaired. If such drugs can be developed there is no reason why organs should not be transplanted freely between different individuals of the same species (**homografts**) or indeed between individuals of different species (**heterografts**) with reasonable success. Advances have already been made in this area at the research level and some optimistic people believe that in time it may become common medical practice.

However, it is as well to remember that tissue grafting and organ transplantation are highly unnatural events which never occur in nature. The fact that transplanted organs are rejected is merely the result of individual diversity, a reflection of the fact that every individual is genetically unique. Rejection is part and parcel of the body's normal defence against invading microbes. One can hardly expect the body to distinguish between pathogenic micro-organisms and useful tissue introduced by a benevolent surgeon.

INTERFERON

This brief account of the body's disease-defence mechanisms would be incomplete without mentioning **interferon**. This is a protein extracted from cells which has been found to prevent multiplication of viruses. Unlike antibodies it is non-specific, i.e. it is effective against a wide range of viruses, not just one. These include the influenza and polio viruses. However, it seems to be more effective against viruses in the same species of animal in which it is manufactured.

How effective is it? Interferon has been administered to chick embryos and rabbits and this has been found to give them protection against subsequent infection with vaccinia virus. This kind of discovery obviously raises the possibility of using interferon to protect man. It would, certainly, have great advantages: it is non-toxic, does not itself stimulate antibody-production, and its lack of specificity would confer general immunity against a wide range of virus diseases. But there are obvious difficulties: for example it would be necessary to prepare very large quantities and it would have to come from human cells, or those of a closely-related species like the chimpanzee. But these difficulties are by no means insuperable. Recently it has been found that double-stranded RNA extracted from the mold *Penicillium* induces interferon production when injected into rabbits. This discovery may open up a possible technique for getting large enough quantities for human use.

CHEMOTHERAPY AND ANTIBIOTICS

Man's endeavours to combat disease include the use of **chemotherapy**. This is the administration of chemical substances, natural or synthetic, that kill or prevent the reproduction of micro-organisms. The term is now extended to include the inhibition of dividing malignant cells as in cancer. The chemical substances used are called **chemotherapeutic agents**. Some of these substances are themselves produced by micro-organisms, in which case they are called **antibiotics**.

One of the best known antibiotics is **penicillin** which was discovered by the British bacteriologist Alexander Fleming in 1928 as the result of a happy accident. Fleming had been working on *Staphylococcus* and it happened that some spores of a mold floated into his laboratory through an open window and landed on one of his *Staphylococcus* colonies. To his surprise the bacteria were quickly destroyed. The mold was subsequently identified as *Penicillium notatum* for which reason the active substance killing the bacteria was called penicillin. In 1940 Howard Florey and Ernst Chain succeeded in isolating and purifying the active substance, thereby enabling it to be used in injections. Since then semi-synthetic penicillins have been developed which, unlike the natural product, resist the action of gastric juice and can therefore be taken by mouth. Since the early 1940s penicillin has saved countless millions of lives and has proved successful against numerous bacterial infections including pneumonia, meningitis, gonorrhoea, syphilis and anthrax.

Penicillin acts on growing bacteria, killing them and preventing their growth. However, its precise mode of action is unknown, as is the case with the majority of antibiotics. One of the few chemotherapeutic agents whose action is understood are the **sulphonamides**, complex organic ring compounds with a powerful anti-bacterial action. The sulphonamides are similar in their chemical structure to para-aminobenzoic acid, an essential metabolite in the reproduction of certain bacteria. They are believed to compete with para-aminobenzoic acid for the active site of an enzyme (see p. 91). In this way, though they do not actually kill the bacteria, they stop them reproducing.

Although chemotherapeutic agents have had an enormous influence on the control of disease, they have one serious drawback. Every time a new one is used resistant strains of micro-organisms arise against which further drugs have to be developed. New drugs should therefore be used with restraint and discrimination.

To survive an organism must react to changes in its environment. This necessitates mechanisms for detecting such changes and putting the appropriate responses into practice. Since the structures responsible for detecting the changes may be far removed from those responsible for performing the response, a means of rapid internal communication is required. In animals this generally takes the form of a nervous system, aided by hormones, the main topics discussed in Chapter 18. How environmental stimuli are monitored and the information fed into the nervous system is the subject of Chapter 19. How animals respond to information transmitted through their nervous systems is discussed in Chapter 20, with reference mainly to the contraction of muscle.

An animal's response to environmental stimuli may involve complex locomotory movements, and behaviour patterns. The principles underlying these important activities are discussed in Chapters 21 and 22.

The importance of responding appropriately to environmental changes applies no less to plants than to animals. However animals, with their need to find food and defend themselves from predators, respond more rapidly than plants. Plants generally respond to stimuli by means of growth movements which are discussed in Part V.

Photomicrograph of muscle fibres and bipolar nerve cells in a mesentery of the sea anemone *Metridium senile*. (*Elaine A. Robson, University of Reading.*)

Part IV Response and
        Coordination

18 · Nervous and Hormonal Communication

The **nervous system** provides the quickest means of communication within the body. In most animals the nervous system is divided into the **central nervous system** (CNS), a concentrated mass of interconnected nerve cells, and the **peripheral nerves** which link the CNS with the body's receptors and effectors. In general terms, the peripheral nerves signal changes in the environment, registered by the **receptors,** to the CNS which integrates the information it receives and sends appropriate messages to the **effectors** which respond accordingly. The receptors include the various sensory cells and sense organs whose collective function is to respond to different types of stimulation. As we shall see later most of these receptors are located near the surface of the body where they are in an ideal situation for detecting changes in the immediate external environment. Some, however, lie deep inside the body and register changes occurring within the body itself. The effectors include all those structures which respond to information received from the nervous system. Muscles and glands are the most obvious examples. In this chapter we shall be concerned with the nervous system, receptors and effectors being dealt with separately in later chapters.

The Nerve Cell and its Impulse

The nervous system is made up of many millions of **nerve cells (neurones)** together with various forms of supporting tissue in which they are embedded. The function of the nerve cell is to transmit messages **(impulses)**. As such it is the basic functional unit of the nervous system and we shall begin by discussing its structure and the way it works.

STRUCTURE OF NERVE CELLS
Fig 18.1 illustrates the structure of a vertebrate **motor neurone,** so-called because it transmits impulses from the CNS to a muscle. Although peculiar in shape, it possesses the same basic structures found in other cells: nucleus, mitochondria, cell membrane and so on. The cytoplasm contains prominent granules called **Nissl's granules** which have been found to be rich in RNA and are concerned with protein synthesis. The nucleated part of the cell, the **cell body** or **centron,** is located in the CNS and is connected with neighbouring neurones by arm-like processes called **dendrons** which terminate in the form

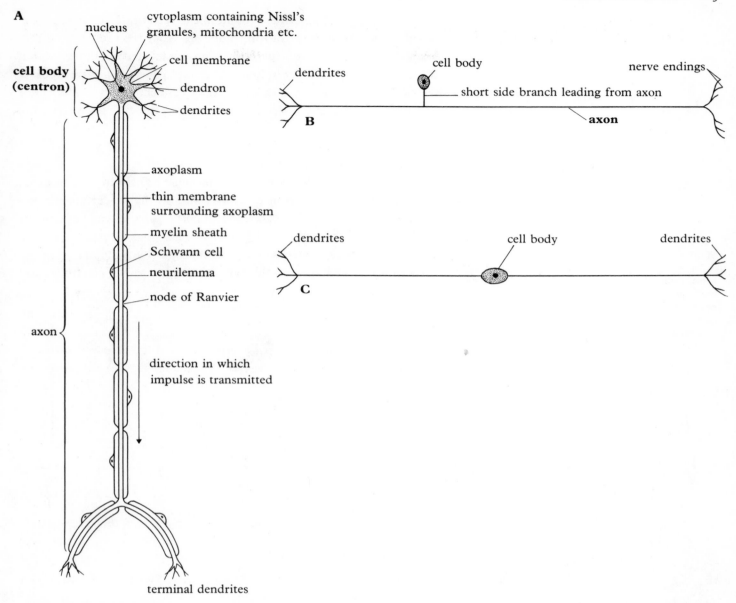

A

cell body (centron)

nucleus

cytoplasm containing Nissl's granules, mitochondria etc.

cell membrane

dendron

dendrites

axoplasm

thin membrane surrounding axoplasm

myelin sheath

Schwann cell

neurilemma

node of Ranvier

axon

direction in which impulse is transmitted

terminal dendrites

B

dendrites

cell body

nerve endings

short side branch leading from axon

axon

C

dendrites

cell body

dendrites

Fig 18.1 Three types of nerve cell found in vertebrates. Nerve cells are described as uni-polar, bipolar, multipolar etc. according to how many dendrons project from the cell body.   A) Diagram of a vertebrate motor nerve cell. The dendrites make contact with neighbouring neurones in the CNS. The terminal dendrites at the end of the axon connect up with muscle fibres.   B) Sensory nerve cell. The myelin sheath is not shown. The nerve endings at the right hand end connect with sensory cells (receptors) and the dendrites at the left hand end connect with nerve cells in the CNS.   C) Intermediate nerve cell. The dendrites connect up with sensory and motor neurones in the CNS.

of slender **dendrites**. One of the processes is drawn out to form an **axon** which enters a peripheral nerve and terminates in a muscle.

The entire neurone is able to transmit impulses but the axon, being an enormously elongated extension of the cell, is a special adaptation for this purpose. An individual peripheral nerve may contain several thousand axons some of which may be over a metre long. Each axon is filled with **axoplasm** continuous with the cytoplasm in the cell body and bounded by a thin membrane continuous with the plasma membrane of the cell body. The axon is enclosed within a fatty **myelin sheath**. This is surrounded by a thin **neurilemma** which, strictly speaking, is not part of the neurone but the membrane of another cell, the **Schwann cell**, which lies in intimate contact with the axon. The myelin sheath is interrupted at approximately one millimetre intervals by constrictions called **nodes of Ranvier**. Some vertebrate axons are non-myelinated, that is they lack a myelin sheath, but the majority are myelinated. Apart from its purely protective function, the importance of the myelin sheath is that it insulates the axon and also speeds up the transmission of impulses, a function we will return to later.

Two other types of nerve cell are found in the vertebrate nervous system. **Sensory (afferent) neurones** transmit impulses from receptors to the CNS. Their cell bodies, situated to one side of the axon, are located not in the CNS but in the dorsal root ganglia of the spinal nerves (see page 277). **Intermediate neurones** are confined to the CNS where they connect sensory and motor neurones with each other and with other nerve cells in the CNS. Their structure shows considerable variation.

INVESTIGATING THE NERVE IMPULSE

It has been realized for many years that the message transmitted by an axon, the **nerve impulse,** is an electrical phenomenon and that in certain respects the axon behaves like an electrical cable. In fact nerve impulses can be recorded and measured using an apparatus which is sensitive to small electrical changes. Such an instrument is the **cathode ray oscilloscope** (CRO) illustrated in Fig 18.2. Impulses are picked up from the nerve through a pair of electrodes and

Fig 18.2 A) How a nerve impulse can be recorded. A suitable stimulus is applied to one end of a nerve, either direct or via receptors. The electrical impulses are picked up by a pair of recording electrodes placed in contact with the nerve at the other end. The impulse is passed into an amplifier which magnifies it and thence to a cathode ray oscilloscope (CRO). When the impulse arrives the beam on the fluorescent screen of the CRO is momentarily deflected to form a spike. B) shows a student recording nerve impulses from the nerve cord of an earthworm in response to tactile stimulation of the head. C) shows the nerve cord hooked over the recording electrodes. (*B. J. W. Heath, Marlborough College*).

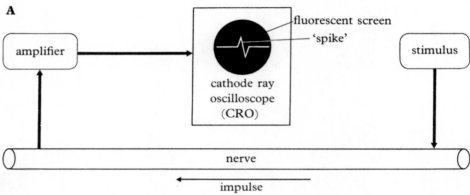

A

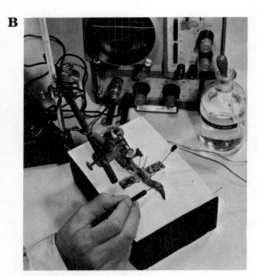

B

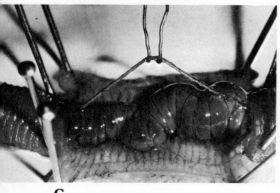

C

fed into an amplifier and thence to the CRO, on whose screen they are registered. This apparatus has been used extensively by neurophysiologists to analyse the pattern of impulses generated in various parts of the nervous system of different animals. But, useful though this information is, it tells us little about the fundamental nature of the nerve impulse.

What we really need to know is what happens inside an axon when an impulse is conveyed along it. This has proved a difficult question to answer mainly because the majority of axons are extremely fine, rarely exceeding 20 μm in diameter.

But fortunately certain animals possess unusually large axons. One such animal is the squid whose peripheral nervous system contains a series of **giant axons** serving the mantle muscles. The largest of these axons reaches a millimetre in diameter – still small, but large enough for a neurophysiologist to do something with.

The last twenty years have seen major advances in our understanding of the nerve impulse. These have been caused in large part by the remarkable experiments pioneered by A. L. Hodgkin and A. F. Huxley who have analysed the events occurring inside the giant axons of the squid. What they did was to insert a very fine electrode (a **micro-electrode**) into the inside of the axon and place a second electrode on the surface of the membrane surrounding it (Fig 18.3). The electrodes were connected to an oscilloscope for recording electrical events.

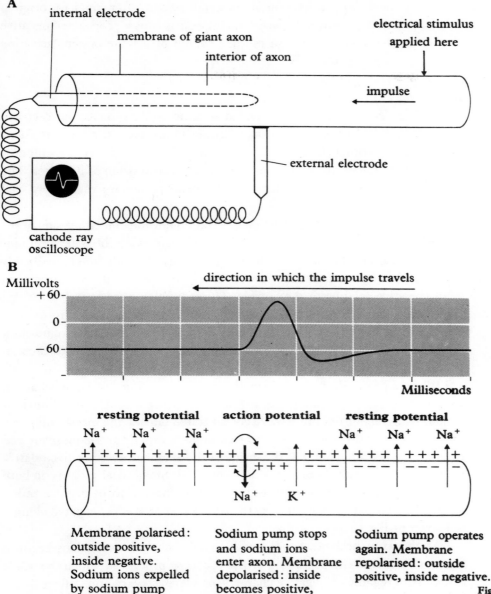

A

internal electrode

membrane of giant axon

interior of axon

electrical stimulus applied here

impulse

external electrode

cathode ray oscilloscope

B

Millivolts

direction in which the impulse travels

+60

0

−60

Milliseconds

resting potential   action potential   resting potential

Na⁺ Na⁺ Na⁺   Na⁺ Na⁺ Na⁺

Na⁺   K⁺

Membrane polarised: outside positive, inside negative. Sodium ions expelled by sodium pump mechanism.

Sodium pump stops and sodium ions enter axon. Membrane depolarised: inside becomes positive, outside negative.

Sodium pump operates again. Membrane repolarised: outside positive, inside negative.

Fig 18.3 The nature of the nerve impulse. A) shows the technique used to investigate the electrical events occurring in the giant axon of the squid. One micro-electrode is inserted into the interior of the axon and the other is placed in contact with the membrane surrounding it. The electrodes are connected to an oscilloscope. Electrical activity is recorded when the axon is resting and also when it transmits an impulse generated by applying an electrical stimulus at the right hand end of the axon. B) is a summary of the events occurring in the giant axon. The tracing shows the resting and action potentials as recorded by an oscilloscope. Below is shown the electrical charges inside and outside the axon and the movement of ions across the membrane at rest and during passage of the impulse. The local circuits in front of the action potential cause depolarization of the next section of the membrane so that the impulse is propagated along the axon from right to left. Notice that the entry of sodium ions into the axon is followed by loss of potassium ions, which initiates restoring of the resting potential.

THE ELECTRICAL NATURE OF THE NERVE IMPULSE

It has been found that when the axon is in the resting state, that is when it is not transmitting an impulse, a potential difference is maintained between the inside and outside of the axon. The inside of the axon is negatively charged to the extent of approximately − 60 millivolts, and the outside is positively charged. As this is the situation which prevails when the axon is at rest it is known as the **resting potential**. It shows that the membrane surrounding the axon is **polarized,** that is it is capable of maintaining a different electrical charge on its two sides.

When an impulse passes the electrodes this situation is momentarily reversed, the inside of the axon suddenly becoming positive to the extent of almost + 60 millivolts, and the outside negative. This sudden reversal of the resting potential which accompanies the impulse is called the **action potential,** and clearly it involves the sudden **depolarization** of the nerve membrane. As you will see from the time relations in Fig 18.3 the action potential is extremely short-lived. It lasts for about a millisecond, after which the original resting potential is restored again. The impulse is thus a propagated negative charge on the outside of the membrane caused by a wave of depolarization which passes along the axon.

IONIC BASIS OF THE NERVE IMPULSE

More recent experiments have provided an ionic explanation for the electrical events described in the last section. The ions in the axon have been replaced by radioactive isotopes and their movements across the nerve membrane studied both in resting conditions and during transmission of impulses. The explanation is far from complete but the main points are reasonably well established.

When the axon is at rest the concentrations of ions on either side of the membrane are very different. Briefly, there is an excess of potassium ions and organic anions inside the axon, and an excess of sodium ions immediately outside. This unequal distribution of ions results in the inside of the axon being negative relative to the outside. The sodium ions are actively extruded across the membrane whose permeability properties prevent them from entering the axon once they have been expelled from it. This active removal of sodium is called the **sodium pump mechanism** and it is the major cause of the resting potential.

When an impulse passes along the axon the sodium pump mechanism suddenly breaks down and the membrane becomes permeable to sodium ions which, being about ten times more concentrated outside the membrane, rush into the axon by diffusion. This event takes place rapidly and explosively and reverses the resting potential: the inside of the axon becomes positive relative to the outside, resulting in the action potential. Small local circuits on both sides of the membrane at the leading end of the region of depolarization excite the next part of the axon, so the action potential is propagated along it as shown in Fig 18.4.

As the axon becomes flooded with sodium ions potassium ions begin to leave. This marks the beginning of the recovery process in which the inside of the axon regains its negative charge. Complete repolarization of the membrane is achieved by the sodium pump mechanism starting up again, the sodium ions being vigorously extruded from the axon once more. Depolarization of the membrane corresponds to the rising phase of the action potential, repolarization to the falling phase. The relationship between these various events is summarized in Figs 18.3 and 18.4. As can be seen from Fig 18.3 the whole process is complete in approximately 2 ms.

So we see that the nerve impulse can be explained in terms of changes occurring in the properties of the membrane surrounding the axon. The question naturally arises: how is the membrane able to maintain an ionic difference in the first place? In particular, how is it able to remove sodium ions and deposit them outside the axon? The complete answer to this question is not known but experiments have been done showing that the expulsion of sodium ions is an active process requiring energy from respiration. For example, treating an axon with a metabolic poison such as dinitrophenol prevents the extrusion of sodium ions which will, however, be started up again if the poison is washed away. That the energy comes from the splitting of ATP is indicated by the fact that, when treated with a metabolic poison, the axon

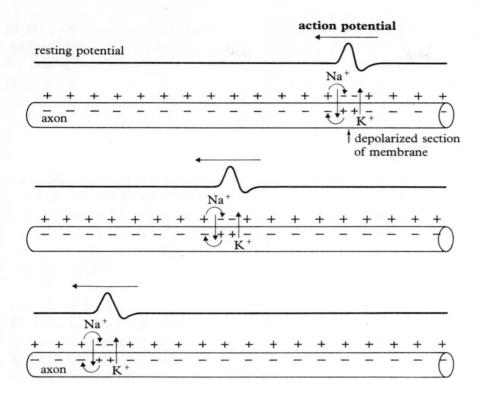

action potential

resting potential

Na⁺

K⁺

↑ depolarized section
of membrane

axon

Na⁺

K⁺

Na⁺

axon   K⁺

Fig 18.4 Diagram showing three stages in the propagation of an impulse along an axon. The action potential is transmitted as a wave of depolarization which spreads rapidly along the axon. Local circuits on both sides of the membrane cause depolarization of the next section of the membrane so the action potential advances from right to left in the diagram.

loses ATP at about the same rate that the sodium pump mechanism runs down. Another interesting discovery is that when a poisoned axon is treated with ATP the sodium pump mechanism starts up again and the resting potential is at least partially restored. Maintenance of the resting potential is thus an energy-requiring process involving the active transport of sodium ions across the membrane surrounding the axon. During transmission of an impulse active transport ceases and the membrane becomes permeable to sodium ions which stream in by diffusion.

The account of the nerve impulse given here is based on the giant axons of the squid and it is reasonable to ask if other axons, including those of vertebrates, behave similarly. The answer is that there is no reason to believe that the latter differ in any fundamental respects. In fact experiments carried out during the last few years suggest that the ionic explanation of the resting and action potentials applies in general to all nerves.

Properties of Nerves and Nerve Impulses

Now that we understand the fundamental nature of the nerve impulses we can go on to consider some of the more important properties of nerves and the impulses they transmit.

STIMULATION

What sort of stimuli generate nerve impulses? In normal circumstances impulses are set up in nerve cells as a result of excitation of receptors. But an axon can also be excited by direct application of any appropriate **stimulus** that

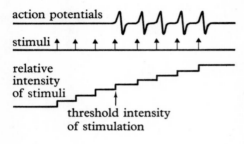

Fig 18.5 The all-or-nothing law. One end of a nerve is stimulated with eight electrical shocks of gradually increasing intensity and action potentials are recorded from the other end. The results, shown below, indicate that a stimulus either evokes a full response or no response at all.

action potentials

stimuli

relative
intensity
of stimuli

threshold intensity
of stimulation

Fig 18.6 Graph demonstrating the excitability of a nerve following stimulation. For about 0.5 ms after an impulse has been transmitted the excitability of the nerve falls to zero (absolute refractory period), after which it gradually returns to normal (relative refractory period). How would you explain the brief supernormal phase following recovery?

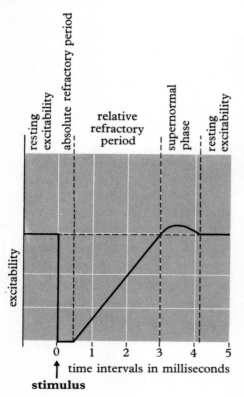

resting
excitability

absolute refractory period

relative
refractory
period

supernormal
phase

resting
excitability

excitability

0   1   2   3   4   5

time intervals in milliseconds

stimulus

causes local depolarization of the membrane. In general nerves can be stimulated by mechanical, osmotic, chemical, thermal and electrical stimuli. Such stimuli will hardly be expected to influence axons directly in the body but they are important experimentally. Electrical stimulation is particularly useful because the strength, duration and frequency of electrical stimuli can be controlled and the axons are not damaged by this form of stimulation.

THE ALL-OR-NOTHING LAW

Fig 18.5 shows what happens when one end of an axon is stimulated with a series of electrical shocks of gradually increasing intensity, and action potentials are recorded from the other end. You will notice that if the strength of the stimulus is below a certain **threshold intensity** no action potential is evoked. If, however, the stimulus is above the threshold a full-sized potential is evoked. Further increase in the intensity of the stimulus, however great, does not give a larger potential. This is the **all-or-nothing law** which states that the response of an excitable unit (in this case an axon) is independent of the intensity of the stimulus. In other words, the size of the impulse is independent of the size of the stimulus. All that is necessary for a full action potential to be produced is that the stimulus is above the threshold required to excite the axon. This is an important concept on which much of the functioning of the nervous system depends.

REFRACTORY PERIOD

After an axon has transmitted an impulse it is impossible for it to transmit another one for a short period. The reason is that after it has been active the axon has to recover, ionic movements have to occur and the membrane has to be repolarized, before another action potential can be transmitted. The period of inexcitability that accompanies the recovery phase of the axon is called the **refractory period** and typically it lasts about 3 milliseconds. It can be divided into an **absolute refractory period** during which the axon is completely incapable of transmitting an impulse, followed by a **relative refractory period** during which it is possible to generate an impulse provided that the stimulus is stronger than usual (Fig 18.6).

The importance of the refractory period is that, together with transmission speed, it determines the frequency at which an axon can transmit impulses. For most axons this ranges from 500 to 1000 per second. This, in turn, determines the pattern of muscular responses as we shall see in Chapter 20.

TRANSMISSION SPEED

It is obvious that if the nervous system is to be efficient as a means of communication its neurones must transmit impulses as rapidly as possible. In fact **transmission speeds** vary enormously, depending on the type of neurone and the animal in question. Transmission speeds of over 100 m/s are found in certain mammalian axons. On the other hand the neurones of many invertebrate animals transmit impulses at speeds of 0.5 m/s or less.

What determines the speed at which an axon transmits impulses? In general two factors are important: the possession or otherwise of a **myelin sheath**, and the **diameter of the axon**.

MYELIN SHEATH

You will recall that the axon is surrounded by a **myelin sheath,** composed of fatty material interrupted at approximately one millimetre intervals by nodes of Ranvier. At these nodes the myelin sheath is absent and the axon is surrounded only by its thin membrane and the neurilemma. Experiments have shown that when a myelinated axon transmits an impulse the membrane at the nodes of Ranvier undergoes depolarization, and ionic exchanges occur as they do along the whole length of the squid axon. What happens is that the myelin sheath causes the action potential to leap from node to node thereby speeding up its transmission along the axon. The majority of vertebrate axons are myelinated to some extent, and there is no doubt that the development of myelin sheaths in these animals has been important in increasing the efficiency of their nervous systems in the course of evolution.

AXON DIAMETER

The other factor influencing transmission speed is the diameter, or cross-sectional area, of the axon. In general the thicker the axon the faster it will transmit impulses. Why this should be so is uncertain, though it would seem to be connected with the greater area of membrane over which ionic exchange can take place. The development of axons with unusually large diameters is an adaptation found mainly amongst the invertebrate animals. Many annelids, including the earthworm and numerous marine worms, possess these **giant axons.** So do prawns, crayfish, and various other crustaceans, and we have already met them in the squid, a cephalopod mollusc. In all these cases the giant axons are associated with rapid escape responses whose effectiveness depends on quick transmission of impulses from receptors to the muscles. Fig 18.7 shows the giant axons in the nerve cord of the earthworm: the median axon is 160 μm in diameter and it transmits impulses at up to 45 m/s, about a hundred times faster than the ordinary neurones in the nerve cord.

We are left then with the general picture that rapid transmission has been achieved in the vertebrates by myelination, but in the invertebrates by the

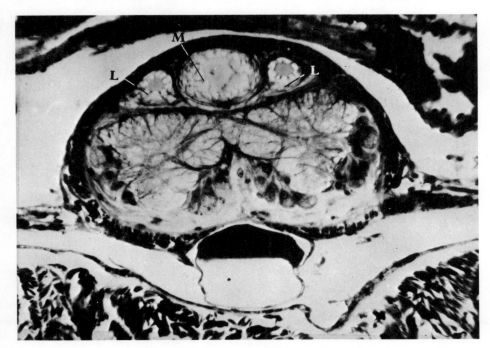

Fig 18.7 Photomicrograph of the ventral nerve cord of the earthworm showing the giant axons. The median giant axon (M) with a diameter of 160 μm can transmit impulses at 45 m/s. The lateral giant axons (L), 60 μm in diameter, transmit at 10 to 15 m/s. Ordinary neurones transmit at less than 0.5 m/s. *(M. B. V. Roberts).*

development of giant axons. The pattern of giant axons varies considerably from one group of invertebrates to another. There is considerable variation even within the different families of marine worms. This leads us to suppose that they have evolved independently in many different invertebrate groups, an indication of the adaptive importance of rapid transmission speed in the lives of these animals.

It might well be asked whether giant axons or myelination provide the best solution to the problem of speed. All we can say is that even the largest giant axons fail to transmit impulses as rapidly as certain myelinated axons, and they take up more space into the bargain. The giant axon of the fanworm *Myxicola*, is well over a millimetre in diameter in its widest part and more or less fills the entire nerve cord. Despite its size it transmits impulses at a maximum speed of only 20 m/s. This is fast enough to give the worm a rapid escape response but it is only a fifth of the transmission speed achieved by some of the faster myelinated axons. Thus myelination provides maximum transmission speed with the greatest economy of space.

B

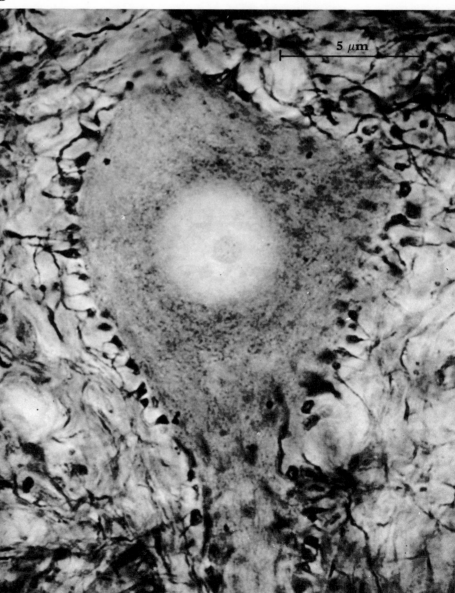

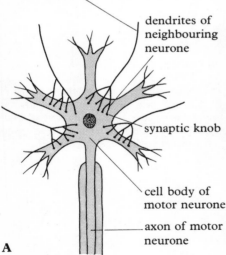

process from neighbouring neurone

dendrites of neighbouring neurone

synaptic knob

cell body of motor neurone

axon of motor neurone

A

Fig 18.8 Nerve cells connect with one another via synapses. A) Synaptic knobs on the surface of a vertebrate motor nerve cell. A single neurone may be covered by as many as 50,000 synaptic knobs derived from many different neighbouring neurones. B) Photomicrograph showing synaptic knobs on the surface of the cell body of a nerve cell in the spinal cord of a cat. (*J. Armstrong, University College, London*)

The Synapse

The precise point where one nerve cell connects with another is called a **synapse** and a typical motor nerve cell in the spinal cord of a vertebrate is covered with hundreds of **synaptic knobs** derived from numerous adjacent nerve cells (Fig 18.8). From pharmacological evidence it has long been thought that transmission across most synapses occurs not by electrical transmission but by a chemical process and this theory had received spectacular support in recent years from electron-microscopy and neurophysiology.

STRUCTURE OF THE SYNAPSE

Analysis of nervous tissue using the electron microscope has shown that the synaptic knob has a highly organized internal structure (Fig 18.9). Within the knob are numerous **mitochondria** and sac-like **vesicles**. The vesicles are thought to contain molecules of a **transmitter substance** which play an important part in synaptic transmission. Between the end of the knob and the membrane of the adjoining nerve cell is a small but definite gap, the **synaptic cleft**. This is bounded on one side by a **pre-synaptic membrane** (which belongs to the knob) and on the other side by a **post-synaptic membrane** (which belongs to the adjoining nerve cell). The space between these two membranes is about 20 nm across and represents a distinct break in continuity between the two nerve cells.

Fig 18.9 Structure of the synapse. A) Diagram of a synaptic knob. The synaptic cleft separating the pre- and post-synaptic membranes is approximately 20 nm wide. B) Electron micrograph of a synaptic knob in the spinal cord of a fish. How many of the structures shown in the diagram can you see in the electron micrograph? *(E. G. Gray, University College, London).* C) Schematic representation of transmission across the synapse. After the vesicle has discharged its contents into the synaptic cleft it moves back towards the centre of the knob and is recharged with transmitter substance.

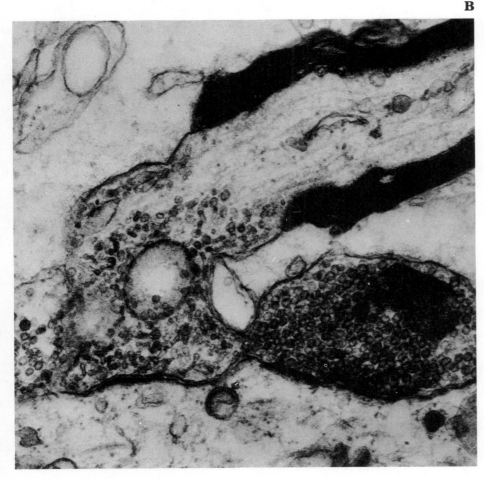

A

end part of dendrite

mitochondria

synaptic vesicles

synaptic knob

pre-synaptic membrane

synaptic cleft

post-synaptic membrane

interior of post-synaptic nerve cell

B

C

impulse

mitochondrion

synaptic vesicle filled with transmitter substance

synaptic cleft

1 8
2 7
3 4 6
5

pre-synaptic membrane

post-synaptic membrane

depolarized region of post-synaptic membrane

TRANSMISSION ACROSS THE SYNAPSE

Modern electro-physiological techniques have made it possible to insert extremely fine micro-electrodes into a nerve cell body and record the potential changes that occur when impulses arrive at one or more synaptic knobs. These investigations, together with parallel experiments on the chemical events occurring at synapses, have led to the following theory of **synaptic transmission** (Fig 18.9C). It is thought that when an impulse arrives at a synaptic knob it causes a synaptic vesicle to move towards the pre-synaptic membrane. The vesicle attaches itself to the membrane and discharges its contents, the transmitter substance, through the membrane into the synaptic cleft. The molecules of transmitter substance then diffuse across the cleft and attach themselves to specific sites on the post-synaptic membrane which is consequently depolarized locally. This local depolarization causes an influx of sodium ions into the post-synaptic nerve cell so that a positive charge develops in that part of the cell. This charge, which can be recorded with an oscilloscope, is known as an **excitatory post-synaptic potential (EPSP)**. If the EPSP's become sufficiently extensive, that is if the positive charge inside the nerve cell builds up to a critical level, an action potential is generated in the nerve cell.

It is clear, therefore, that transmission across the synapse occurs by chemical means. The arrival of an impulse at the synapse causes the release of a chemical transmitter substance which diffuses across the gap and brings about excitation of the post-synaptic nerve cell. Not surprisingly this method of transmission results in an appreciable delay (as much as one millisecond in some cases) when an impulse reaches a synapse, and there is no doubt that synapses have the cumulative effect of slowing down transmission in the nervous system. They also prevent impulses going in the wrong direction. An impulse can pass along an axon in either direction but it can cross a synapse in only one direction. This is because the synaptic vesicles are found on only one side of the synaptic cleft, namely the pre-synaptic side.

THE TRANSMITTER SUBSTANCE

The transmitter substance at the majority of synapses is **acetylcholine** (usually abbreviated **ACh**), an ammonium base which is known to have a profound effect on the permeability of nerve membranes. Obviously, if it is to be effective as a transmitter, ACh must not be allowed to linger at a synapse after it has depolarized the post-synaptic membrane. This is prevented by an enzyme **cholinesterase** which is present in high concentration at the synapse. The moment that acetylcholine has done its job it is hydrolyzed by cholinesterase and thereby rendered inactive.

THE NERVE-MUSCLE JUNCTION

A special kind of synapse is the **nerve-muscle (neuromuscular) junction**, the point where the dendrite of a motor nerve cell makes contact with a muscle fibre. At the nerve-muscle junction the membrane of the muscle fibre is modified to form an **end plate** to which the dendrite is attached. Electron microscope studies have shown that the internal structure of the dendrite at this point is very similar to a synaptic knob (Fig 18.10), and there is good reason to believe that transmission at the nerve-muscle junction takes place in the same way as at synapses between contiguous nerve cells. When an impulse arrives at the nerve-muscle junction acetylcholine is discharged from synaptic vesicles into the synaptic cleft. The acetylcholine diffuses across the gap and depolarises the muscle end plate. **End plate potentials** can be recorded and it has been shown that if these build up sufficiently an action potential is fired off in the muscle fibre.

THE ACTION OF DRUGS AND POISONS

The discovery that transmission at synapses occurs by chemical means provides us with an explanation of the action of various drugs and poisons on the nervous system.

Clearly any chemical that destroys acetylcholine, inhibits its formation, or prevents its action, will stop synaptic transmission. **Atropine**, for example, does not prevent acetylcholine being formed but it stops it depolarizing the post-synaptic membrane, and therefore causes synaptic block. **Curare**, the poison used on the tips of arrows by South American Indians, has a similar effect specifically on nerve-muscle junctions.

On the other hand chemicals that destroy or prevent the formation or action of cholinesterase will be expected to enhance and prolong the effects of acetylcholine, and thus facilitate transmission across the synapse. **Eserine** has this effect since it prevents cholinesterase from destroying acetylcholine. **Strychnine** also enhances synaptic transmission to such an extent that a person suffering from strychnine poisoning will give convulsive contractions of all his muscles upon the slightest stimulation. Some of the poisonous **nerve gases** used for military purposes work in a similar way.

NORADRENALINE

The occurrence of acetylcholine is very widespread in the animal kingdom but it is not the only transmitter known. Certain nerves belonging to the sympathetic nervous system of vertebrates produce **noradrenaline** at their terminals. Nerves that produce acetylcholine are called **cholinergic nerves** and those producing noradrenaline are called **adrenergic**. Transmission at the endings of adrenergic nerves is probably fundamentally similar to cholinergic transmission.

Particularly interesting is that noradrenaline is chemically similar to the hormone **adrenaline** secreted by the **adrenal glands,** and in fact the effects produced in the body by the sympathetic nervous system are generally similar to those brought about by adrenaline. Here, then, is an interesting functional connection between the nervous and endocrine systems, both of which are concerned with communication in the body (see p. 295).

Like acetylcholine, the action of noradrenaline can be inhibited or enhanced by various drugs though these, of course, are different from those which affect the cholinergic system. There is evidence that some of the synapses in the human brain involve the use of noradrenaline and it is thought that certain psychoactive drugs exert their effects by action on the synapses. For example, **mescaline** and LSD (**lysergic acid diethylamide**) may produce their hallucinatory effects by interfering with noradrenaline to which they are closely related chemically.

SUMMATION AND FACILITATION

The way a nerve cell responds to impulses reaching it is very complex and

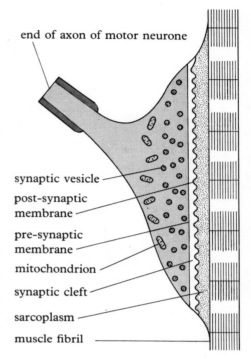

Fig 18.10 Diagram of a nerve-muscle junction showing its essential similarity to an ordinary nerve-to-nerve synapse. The post-synaptic membrane, formed from the sarcolemma surrounding the muscle fibre, is greatly folded to form the muscle end plate.

end of axon of motor neurone

synaptic vesicle
post-synaptic membrane
pre-synaptic membrane
mitochondrion
synaptic cleft
sarcoplasm
muscle fibril

depends on various conditions. A nerve cell may fire off an impulse as a result of excitation through a single synaptic knob. In such cases it is thought that the arrival of an impulse at a single knob releases enough transmitter substance to depolarize the post-synaptic membrane and generate an action potential in the nerve cell. Other nerve cells only fire when excited through two or more synaptic knobs simultaneously. In this case a single synaptic knob fails to produce enough transmitter substance to fire an impulse in the post-synaptic nerve cell but sufficient transmitter is produced by two knobs acting simultaneously. The effects produced by each knob add together, or summate, to excite the nerve cell, a phenomenon called **spatial summation**. Again, it has been found that some nerve cells which fail to generate an impulse when excited through a single synaptic knob will send out an impulse when they receive two or more impulses from a single knob in quick succession. Here again a process of summation is taking place: the amount of transmitter released when the knob receives a single impulse is insufficient to trigger off an impulse in the post-synaptic nerve cell but a second impulse, provided that it arrives sufficiently soon after the first one, releases a further dose of transmitter which, together with the first, is sufficient to generate an impulse in the nerve cell. This is known as **temporal summation** and it involves a process known as **facilitation**. The first impulse fails to cross the synapse but it leaves an effect which facilitates the passage of the second impulse. We shall meet this again in connection with more primitive nervous systems (p. 289).

INHIBITION

Synaptic transmission is further complicated by the fact that certain synaptic knobs inhibit the nerve cells to which they are attached. Little is known about how they achieve this effect but it seems that the arrival of an impulse at such synaptic knobs makes the inside of the post-synaptic nerve cell more negative than usual. This negative charge can be recorded on an oscilloscope and is called the **inhibitory post-synaptic potential (IPSP)**. Clearly the IPSP will make it more difficult for excitatory post-synaptic potentials (EPSP's) to build up and generate an impulse in the nerve cell.

ACCOMMODATION

One further property of synapses must be mentioned. If a synapse is bombarded with impulses at high frequency over a long period there comes a time when the post-synaptic nerve cell fails to respond and impulses are no longer generated in it. It is as if the synapse gets tired and the transmission mechanism breaks down. What in fact happens is that the supply of transmitter substance gets exhausted and its resynthesis cannot keep pace with the rate at which impulses reach the synapse. The synapse is said to **fatigue** or **accommodate**. The rapidity with which different synapses fail in this way is very variable; some do not accommodate at all whilst others may do so after they have transmitted a comparatively small number of impulses. An accommodated synapse must be given time to recover, to regenerate a new supply of transmitter substance, before it can transmit successfully again.

THE ROLE OF SYNAPSES IN THE NERVOUS SYSTEM

At this point you may well be wondering what the function of synapses is.

After all, in so far as they slow down the rate of transmission and are highly susceptible to drugs and fatigue, they would seem to be more of a hindrance than a help. But the fact is that they play an extremely important part in the functioning of the nervous system as a whole.

The vast number of synaptic connections allows for great flexibility in the integrative functions of the nervous system. They are equivalent to the switchboard in an elaborate telephone exchange, enabling messages to be diverted from one line to another, and so on. But there is a further point. We have seen that synapses offer a certain degree of resistance to the passage of impulses. An impulse reaching a synapse may facilitate the passage of subsequent ones or it may inhibit it. Sometimes an impulse may cross the synapse unimpeded. The point is that this pattern of resistance determines the flow of impulses within the whole nervous system. One example will serve to illustrate this. When you inadvertently put your finger on something hot the response may not necessarily be confined to the muscles of the arm. In addition to pulling your hand away you may jump back and let out a cry. Numerous effectors and nervous pathways are involved in this response, not to mention the organs of special sense and equilibrium. It is the nervous system that coordinates this response, the condition of the synapses determining which muscles should contract and which should not. For example, in this particular response the **flexor (biceps)** muscle of the arm contracts powerfully but the **extensor (triceps)** does not. What ensures that this happens is the condition of the synapses in the CNS. Those that connect up with axons supplying the flexor muscles are in a state of excitation, whereas those connected with axons that supply the extensor muscles are temporarily inhibited. In short, the synapses are like gates allowing impulses to flow along those nervous pathways appropriate to the requirements of the organism at a particular moment.

It is also believed that synapses in the brain may play an important part in **learning** and **memory**. Modifications in the pattern of synaptic transmission may provide a means by which information derived from different sense organs is associated and stored within the nervous system.

Reflex Action

So far we have considered the functioning of individual nerve cells and synapses but little has been said about the way these units are organized to form an effective system of communication. For this to be accomplished the nerve cells must receive impulses from **receptors** and in turn hand them on to **effectors**. This is the basis of **reflex action**.

A reflex action is an automatic response mediated through the nervous system. In man the knee-jerk and withdrawal of the hand from a hot object are examples. Amongst the lower animals the escape responses mentioned earlier in connection with giant axons provide good examples of comparatively simple reflexes.

Fig 18.11 Schematic diagram to show the essential components of a reflex arc. There is much variation in the number of neurones comprising a reflex pathway from receptor to effector. In the simplest reflexes a sensory neurone connects directly with a motor neurone but in most cases sensory and motor neurones are linked by at least one intermediate neurone.

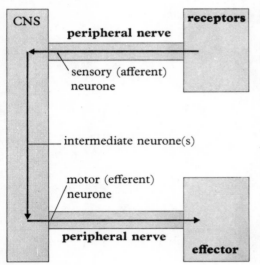

STRUCTURE OF A GENERALIZED REFLEX ARC

The structural basis of reflex action is the **reflex arc** which represents the series of units through which impulses have to pass in order to bring about a reflex response (Fig 18.11). The **receptors** may be scattered sensory cells in the skin, or an organ of special sense like the eye. Their stimulation results in impulses being generated in **sensory** (**afferent**) **neurones** located in the peripheral nerves. The afferent neurones convey impulses to the CNS where they make synaptic connection with **intermediate** (or **internuncial**) neurones. These in turn connect with **motor** (**efferent**) neurones which extend into the peripheral nerves and eventually reach the **effectors,** usually glands or muscles. It is important to realize that this general description of a reflex arc applies to any animal which has a CNS. It applies to worms, insects and squids just as much as it does to man.

THE VERTEBRATE REFLEX ARC

A vertebrate reflex arc involving the **spinal cord** is illustrated in Fig 18.12A. Each **spinal nerve** is attached to the spinal cord by separate **dorsal** and **ventral roots** which contain the sensory and motor nerve cells respectively. The **grey matter** of the cord contains the intermediate nerve cells together with the cell bodies of the motor nerve cells whose axons run out into the ventral roots.

With this anatomical picture in mind we can trace what happens in a spinal reflex response such as the withdrawal of one's hand from a hot object. Stimulation of pain receptors in the skin fires off impulses in sensory neurones contained in the brachial (arm) nerve. These impulses enter the spinal cord via dorsal roots, traverse the intermediate neurones in the grey matter and leave the spinal cord via the appropriate efferent neurones in the ventral roots. Eventually they reach the flexor muscles in the arm which contract accordingly.

In the vertebrate nervous system there are numerous reflex arcs corresponding to the series of spinal nerves. It is important to realize that these reflex arcs are not separate but are interconnected by **longitudinal neurones** located in the **white matter** of the cord. This is illustrated in a simplified manner in Fig 18.12B. Not only do these longitudinal tracts connect adjacent reflex arcs together but they also connect them with the higher centres in the brain. The functional importance of these longitudinal tracts can be illustrated by returning to our example. In this case the afferent pathway is simple: impulses are transmitted into the spinal cord via sensory neurones in the brachial nerve. However, the response may involve contraction of leg muscles as well as arm muscles, necessitating the spread of impulses to reflex arcs other than those that deal with the arm. We can usefully take this example a step further. If one were foolish (or perhaps a masochist) one could keep one's hand in contact with the hot object and leave it there. By a conscious effort of the will the normal reflex is suppressed. This illustrates the inhibitory influence which the brain can have on reflex action. In both these examples the neurophysiological processes involved require that successive reflex arcs are joined together and with the brain, and once again we see that the critical factor determining which reflex arcs are operative at a given time is the state of the synaptic connections.

A

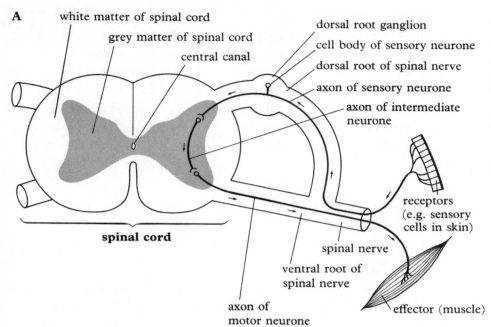

white matter of spinal cord
grey matter of spinal cord
central canal

dorsal root ganglion
cell body of sensory neurone
dorsal root of spinal nerve
axon of sensory neurone
axon of intermediate neurone

spinal cord

receptors (e.g. sensory cells in skin)

spinal nerve
ventral root of spinal nerve

axon of motor neurone

effector (muscle)

Fig 18.12 Vertebrate reflex arc. A) Diagrammatic cross-section of the spinal cord to illustrate a typical vertebrate reflex arc. There are three neurones, the sensory and motor neurones being linked by an intermediate neurone. In a simple reflex like the knee-jerk, the sensory neurone connects directly with the motor neurone but this is exceptional. The arrows indicate the direction in which impulses are transmitted. B) Stereogram of the spinal cord showing the longitudinal tracts of neurones in the white matter. The arrows indicate the direction of transmission of impulses. Ascending tracts in the dorsal and dorso-lateral regions of the cord conduct impulses towards the brain. Descending tracts in the ventral and ventro-lateral regions conduct impulses from the brain.

B

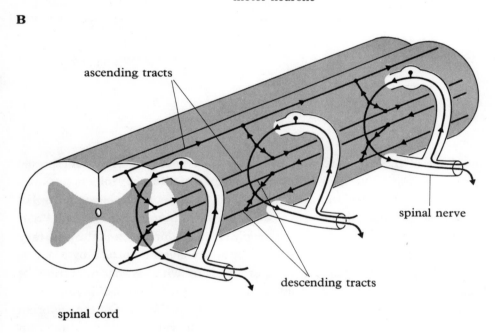

ascending tracts

spinal nerve

descending tracts

spinal cord

Organization of the Nervous System

Now that we have discussed the basic principles on which the functioning of the nervous system depends, we can consider its general organization in the body. The **CNS,** with its thousands of interconnected neurones is like a central telephone exchange, the **peripheral nerves** corresponding to the lines that link the exchange with the subscribers. This principle applies to all animals with a CNS, but the detailed organization of the nervous system depends on the animal in question.

THE VERTEBRATE CNS

In all vertebrates the CNS develops as a longitudinal tube of nervous tissue towards the dorsal side of the embryo (see p. 417). The anterior end of this **neural tube,** the portion of it contained in the head, swells up to form the **primary brain vesicle,** which later becomes subdivided into three regions: the **forebrain, midbrain** and **hindbrain** (Fig 18.13B). At a later stage the forebrain becomes further subdivided into a greatly expanded **endbrain** and, behind this, the **'tweenbrain.**

Fig 18.13 Diagram illustrating how the primary brain vesicle of a primitive vertebrate like the dogfish becomes differentiated during development. A) Shows the division of the primary brain vesicle into four main regions. B) Shows the structures associated with each region.

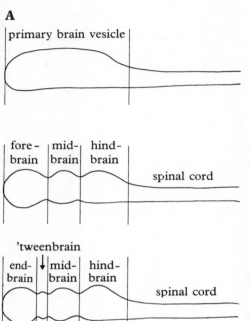

A

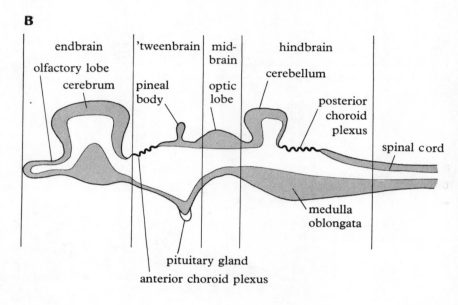

B

Each of these main regions of the brain is associated with certain specific structures (Fig 18.13A). The endbrain has a pair of **olfactory lobes** anteriorly. Further back it forms the **cerebrum** which, in vertebrates above the level of fishes, expands dorsally to form a pair of **cerebral hemispheres.** The 'tweenbrain gives rise to two outgrowths: the **pineal body** dorsally, and the **pituitary gland** ventrally. In development, the pituitary gland is formed by a downgrowth from the floor of the 'tweenbrain fusing with an upgrowth from the roof of the mouth cavity. In front of the pituitary is the **optic chiasma** which carries nerve fibres from the eyes to the midbrain. The roof of the midbrain expands to form a pair of **optic lobes.** The anterior part of the hindbrain expands to form the **cerebellum** dorsally, the ventral and posterior part becoming the **medulla oblongata.** The latter is continuous with the **spinal cord,** a more or less uniform tube extending the full length of the body. The entire CNS is hollow, the cavities (**ventricles**) of the brain being continuous with one another, and with the **central canal** of the spinal cord. These cavities are filled with **cerebrospinal fluid,** whose origin is explained below.

It is obviously essential that the delicate tissues of the CNS should be adequately nourished and protected. To this end the entire CNS is enveloped by a system of membranes or **meninges.** The innermost membrane, called the **pia mater,** is very delicate. The outermost membrane, the **dura mater,** is tougher. These two membranes are separated by a narrow space, the **arachnoid layer,** containing a network of delicate fibres. Filled with cerebrospinal fluid, the arachnoid layer cushions the CNS and facilitates diffusion of oxygen and nutrients into the nervous tissue beneath.

The CNS receives oxygen and nutrients from two vascular membranes, the **anterior** and **posterior choroid plexuses**, which form the roof of the 'tweenbrain and hindbrain respectively. Continuous with the pia mater, the choroid plexuses keep the arachnoid space and the cavities of the brain and spinal cord, charged with cerebrospinal fluid. The whole of the CNS, and its meninges, are enclosed within the axial skeleton, the **cranium** protecting the brain and the **vertebral column** the spinal cord. These skeletal structures are pierced by holes, or **foramina**, which allow the entry and exit of the peripheral nerves.

THE PERIPHERAL NERVES

The peripheral nerves are divided into two groups: the **spinal nerves** connected to the spinal cord, and the **cranial nerves** connected to the brain. The cranial nerves are associated mainly with receptors and effectors in the head, whilst the spinal nerves serve receptors and effectors in the rest of the body.

In the mammalian system the **spinal nerves** form a regular series, one pair in each body segment, each being connected to the spinal cord by a dorsal and ventral root (p. 276). We have already seen that the spinal nerves contain both sensory and motor fibres, for which reason they are described as **mixed nerves**. Peripherally each splits into numerous branches, carrying groups of axons to the receptors and effectors in that area of the body. In fishes the segmental arrangement is very regular, each spinal nerve corresponding to a **myotome muscle block**. In other vertebrates the regular series is somewhat modified, mainly as a result of the development of paired limbs. In the trunk the spinal nerves are gathered together to form large compound nerves or plexuses, the **brachial plexus** serving the arms, and the **sciatic plexus** the legs.

The pattern of **cranial nerves** is more irregular. There are two main reasons for this. Firstly, the dorsal and ventral roots are not joined and indeed may be widely separated. Secondly, the development of the brain and other specialized structures in the head, such as sense organs and jaws, has led to distortions in the pattern of cranial nerves, so that at first sight their origin and distribution seem to be crazy. However, by studying the embryological development of the head, it is possible to see that fundamentally it is divided into a regular series of segments, each served by a pair of nerves, one on each side. This fascinating exercise in structural embryology brings order out of chaos, and shows that the arrangement of the cranial nerves is not as illogical as it appears at first sight.

There are ten cranial nerves in primitive vertebrates, twelve in mammals. Some contain nothing but sensory fibres, others nothing but motor fibres; some of the larger ones are mixed nerves. The pattern of cranial nerves is best seen in a primitive vertebrate like the dogfish. This animal has the advantage of having a cartilaginous skeleton which, being softer than bone, facilitates dissection. The cranial nerves of the dogfish are summarized in Fig 18.14 and Table 18.1.

The basic arrangement in mammals, including man, is similar, though there are certain differences mainly connected with the fact that in mammals the gills are replaced by lungs and various elaborations have been added to the anterior end of the respiratory and alimentary tracts. This has led to modifications in the distribution of nerves VII, IX and X. Instead of sending

No.	Name	Type	Branches		Structures Served
I	Olfactory	S			Olfactory organ
II	Optic	S			Eye (Retina)
III	Oculomotor	M	(four)		Superior rectus eye muscle
					Anterior rectus eye muscle
					Inferior rectus eye muscle
					Inferior oblique eye muscle
IV	Pathetic (Trochlear)	M			Superior oblique eye muscle
V	Trigeminal	S&M	Ophthalmic	(1)	Snout
			Maxillary	(3)	Upper jaw
			Mandibular	(4)	Lower jaw (mandible)
VI	Abducens	M			Posterior rectus eye muscle
VII	Facial	S&M	Ophthalmic	(2)	Snout
			Buccal	(5)	Buccal cavity
			Palatine	(6)	Palate
			(Hyomandibular)	(7)	
			Prespiracular	(8)	Spiracle muscle
			(Postspiracular)	(9)	
			External mandibular	(10)	Lower jaw
			Hyoidean	(11)	Hyoid arch
VIII	Auditory	S			Ear (semi-circular canals)
IX	Glossopharyngeal	S&M			1st gill pouch
X	Vagus	S&M	Branchials	(12)	2nd–5th gill pouches
			Visceral	(13)	Heart and viscera
			Lateral line	(14)	Lateral line sense organs

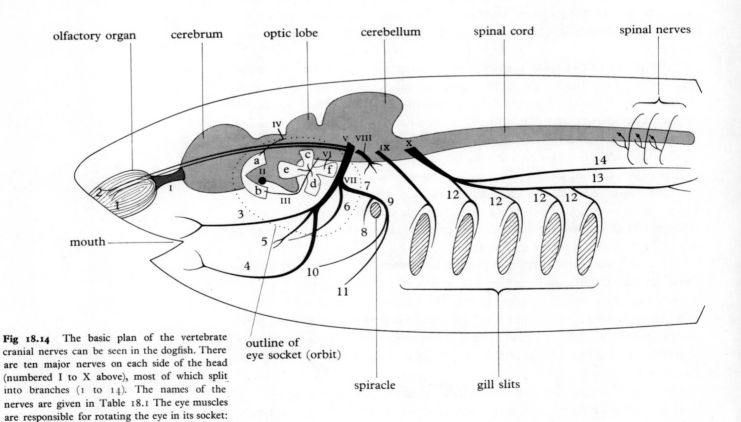

Fig 18.14 The basic plan of the vertebrate cranial nerves can be seen in the dogfish. There are ten major nerves on each side of the head (numbered I to X above), most of which split into branches (1 to 14). The names of the nerves are given in Table 18.1 The eye muscles are responsible for rotating the eye in its socket: a, superior oblique; b, inferior oblique; c, superior rectus; d, inferior rectus; e, anterior rectus; f, posterior rectus. The eye muscles are innervated by nerves III, IV and VI.

branches to the spiracles and gill pouches, these nerves have branches serving such structures as the salivary glands, tongue and larynx.

In all vertebrates the distribution of cranial nerves is confined to the head and neck. The only exception to this is the vagus, literally the 'wandering' nerve, which passes down the neck, through the thorax, into the abdominal cavity, giving off branches to such organs as the heart and gut. The vagus forms part of the **autonomic nervous system** to which we now turn.

THE AUTONOMIC NERVOUS SYSTEM

This part of the nervous system is concerned with controlling the body's involuntary activities, such as the beating of the heart, movements of the gut, and secretion of sweat. It is a complex system, and we can only touch on its basic organization here.

The autonomic system is divided into two parts: the **sympathetic** and **parasympathetic** systems. Both contain nerve fibres serving structures over which the body has little or no voluntary control, and which are therefore not innervated by the voluntary part of the nervous system. In both cases nerve fibres emerge from the brain or spinal cord, and pass to the organs concerned. There are many such pathways in each of the two systems. Along the course of each pathway there is a complex set of synapses constituting a **ganglion**. The nerve fibres on the proximal side of the ganglion are called **pre-ganglionic fibres,** those on the distal side **post-ganglionic fibres**.

The main structural difference between the sympathetic and parasympathetic systems relates to the position of the ganglia (Fig 18.15). In the

Table 18.1 (*opposite, far left*) Summary of the cranial nerves of the dogfish. The numbers in brackets after the names of the branches refer to the numbers in Fig. 18.14. The branches whose names are in brackets are non-terminal. **S**: sensory, i.e. nerve contains sensory (afferent) fibres only. **M**: motor, i.e. nerve contains motor (efferent) fibres only. The eye muscles are shown in Fig. 18.14.

Fig 18.15 General organization of the vertebrate autonomic nervous system. Notice the relative positions of the ganglia in the sympathetic and parasympathetic divisions. In the sympathetic system the pre-ganglionic fibres are short, post-ganglionic long. In the parasympathetic system the pre-ganglionic fibres are long, post-ganglionic short.

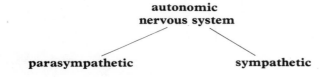

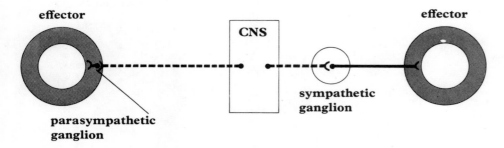

sympathetic system the ganglia lie alongside the vertebrae, close to the spinal cord, with the result that the pre-ganglionic fibres are short, the post-ganglionic fibres long. In the parasympathetic system the ganglia are embedded in the wall of the effector itself, so the pre-ganglionic fibres are long and the post-ganglionic fibres short. The main functional difference between the two systems concerns the transmitter substance produced in each case: **noradrenaline** is produced at the ends of the sympathetic nerve fibres, **acetylcholine** at the ends of the parasympathetic fibres. The effects produced by the two systems generally oppose one another: if the sympathetic causes a certain muscle to contract, the parasympathetic relaxes it.

The general plan of the mammalian autonomic system is shown in Fig 18.16 to which reference should be made while reading the following description. In the trunk the ganglia of the **sympathetic system** form a segmental series on each side of the vertebral column. Each ganglion is connected to the appropriate spinal nerve (ventral root) by a pair of slender **rami communicantes** which carry nerve fibres to it from the spinal cord. Successive ganglia are linked together by a **sympathetic nerve** which arises from the brain: the resulting series of interconnected ganglia is called the **sympathetic chain**. From each sympathetic ganglion, nerves pass to the appropriate effectors, either direct or via further ganglia. The latter, also located close to the vertebrae, contain synapses linking the pre- and post-ganglionic fibres. The complex neuronal interconnections in these ganglia ensure the rapid spread of excitation to all the appropriate effectors.

The **parasympathetic system** consists of the whole of the vagus nerve and its branches, plus certain other cranial and spinal nerves (see Fig 18.16).

Fig 18.16 Diagram summarizing the main features of the mammalian autonomic nervous system. See text for further details. Pre-ganglionic fibres are represented by fine arrows, post-ganglionic fibres by bold arrows.

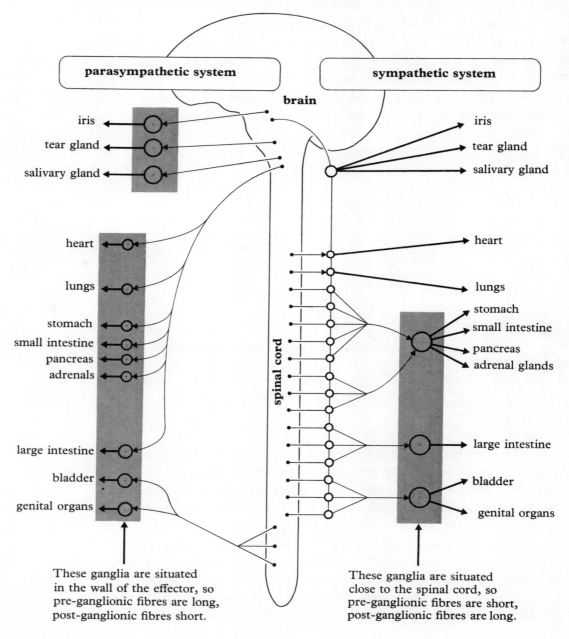

parasympathetic system

sympathetic system

brain

iris
tear gland
salivary gland

iris
tear gland
salivary gland

heart

lungs

stomach
small intestine
pancreas
adrenals

spinal cord

heart

lungs

stomach
small intestine
pancreas
adrenal glands

large intestine

bladder

genital organs

large intestine

bladder

genital organs

These ganglia are situated in the wall of the effector, so pre-ganglionic fibres are long, post-ganglionic fibres short.

These ganglia are situated close to the spinal cord, so pre-ganglionic fibres are short, post-ganglionic fibres are long.

Parasympathetic System	Sympathetic System
Slows heart	Accelerates heart
Dilates arteries	Constricts arteries
Constricts bronchioles	Dilates bronchioles
Constricts iris (pupil)	Dilates iris
Stimulates tear gland	
Causes flow of saliva	
Speeds up gut movements	Slows gut movements
Relaxes bladder and anal sphincters	Contracts bladder and anal sphincters
Causes contraction of bladder	Causes relaxation of bladder
	Contracts erector pili muscles
	Increases sweat secretion

Table 18.2 Summary of the responses evoked by the two parts of the mammalian autonomic nervous system. Note that the effects produced by the sympathetic oppose those produced by the parasympathetic. The precise rôle of the parasympathetic system in the dilatation of arteries is controversial. It should also be noted that the sympathetic system, though causing general vasoconstriction, dilates those arteries serving vital organs such as skeletal muscle. The functions of the sympathetic supply to the salivary and tear glands are uncertain and complex. The sympathetic nerves to the sweat glands produce the chemical transmitter acetylcholine, characteristic of the parasympathetic system, instead of noradrenaline. In addition to the functions listed, the sympathetic system causes secretion of adrenaline from the adrenal medulla (see p. 295). The actions of this hormone are the same as those initiated by the sympathetic system.

Since, in this case, the synapses between pre- and post-ganglionic fibres are located within the walls of the effectors, there are no ganglia in the course of the nerves themselves.

The functions of the sympathetic and parasympathetic systems are summarized in Table 18.2. Notice the opposing effects achieved by the two systems. Also notice that most of the effects produced by the sympathetic system have the general result of preparing the body for an emergency, for example, dilating the pupil, speeding up the heart, tightening the anal and bladder sphincters, causing the hair to stand on end and sweat to be secreted. These are all reactions we associate with a sudden shock.

Bearing in mind that the autonomic system is responsible for controlling involuntary activities, it may surprise you to see that it controls the emptying of the bladder and the opening of the anal sphincter. After all, these are activities over which man has very definite voluntary control. However, other animals cannot normally control these functions, and even man has to learn to do so. What this illustrates is that the neuronal interconnections in the nervous system are such that voluntary control *can* be imposed on the autonomic system. Does this mean that a person could, by a stupendous act of perseverence, voluntarily bring about peristalsis of his gut or slow down the beating of his heart? The answer is yes, provided that he is prepared to enter upon a long programme of mental and physical 'training'. This is essentially what mystics endeavour to do when practising the art of yoga. It is interesting to speculate what is going on in the nervous system during these strange activities.

FUNCTIONS OF THE VERTEBRATE BRAIN IN GENERAL

The general function of the brain is to coordinate the body's activities. To accomplish this it contains numerous centres for coordinating specific functions such as locomotion, respiration, and so on. These **ganglia** or **nuclei**[1] are composed of numerous nerve cells and synapses, and are connected with the peripheral nerves and spinal cord. Collectively they perform three functions: they receive impulses from receptors, they integrate these impulses, and they send out impulses to the appropriate effectors. By **integration** we mean that if impulses arrive simultaneously from several different receptors, the centre interprets and correlates the incoming information

[1]Not to be confused with the nuclei of cells.

before sending out impulses to the appropriate effectors. In addition certain centres, particularly those in the forebrain, are responsible for **memory**, i.e. they store information so that the behaviour of the animal may be modified as a result of past experience.

Various techniques are used to determine the function of a particular region of the brain. One method is to transect the brain at a certain level, or remove a specific structure such as the cerebellum, and observe the effects on the animal's behaviour. Similar information can be obtained in humans by observing the behaviour of people suffering from brain lesions, inflicted for example by gunshot wounds, in which certain identifiable regions of the brain have been damaged or destroyed. Much work has been done along these lines during war time. Another technique is to stimulate various regions of the brain with carefully controlled electrical shocks, and observe the result. Alternatively it is possible to stimulate individual receptors such as the eye, and record electrical activity from particular parts of the brain. The above techniques are attempts at direct **physiological analysis** of the brain. A quite different approach is to study the **behaviour patterns** of intact animals, both in their natural surroundings and in the laboratory, and then attempt to explain them in terms of cerebral function. This is the approach used by psychologists and students of animal behaviour, about which we shall have more to say in Chapter 22.

With this in mind we will summarize the functions of different parts of the brain in two vertebrates: a fish, which demonstrates the situation in a comparatively simple vertebrate, and the mammal with its infinitely more complex arrangement.

FUNCTIONS OF THE MAIN PARTS OF THE BRAIN IN A FISH

In fishes, functional localization in the brain is comparatively clear-cut (Fig 18.13). The forebrain receives and relays impulses from the olfactory organs, the midbrain does the same for the eyes, and the hindbrain is concerned with movement, taste and pressure sense (the fish's equivalent of hearing) along with various visceral functions such as respiration and circulation of the blood. Certain integrative functions are carried out in the 'tweenbrain and midbrain, and the cerebellum is responsible for balance and equilibrium. What is so remarkable about the fish's brain is that the forebrain does very little integrating, this function being performed by the more posterior parts of the brain. Removal of the forebrain abolishes the animal's sense of smell, but it has no observable effect on its general behaviour: certainly it does not affect its visible movements. Indeed, removal of the entire brain anterior to the hindbrain has little or no observable effect on locomotion. If, however, the hindbrain is removed the animal goes flabby, and movement becomes completely uncoordinated.

We therefore deduce that in fishes there are fairly rigid functional divisions in the brain. The forebrain is associated with the sense of smell, the midbrain with sight, and the hindbrain with movement, hearing, and various visceral functions. It is interesting that in certain fishes great enlargement of particular regions of the brain can be correlated with the development of particular senses. For example, the dogfish has an unusually large forebrain (for a fish) with particularly large olfactory lobes, this being correlated with its comparatively good sense of smell. Salmon and pike have a large midbrain, correlated with an excellent visual sense. Carp have a well developed hindbrain associated with their particularly delicate sense of taste. In all fishes coordination of movement is carried out mainly by the hindbrain.

FUNCTIONS OF THE MAIN PARTS OF THE BRAIN IN MAMMALS

If the situation in fishes is comparatively simple this is certainly not true of mammals. The human brain is shown in Fig 18.17, from which you will notice, that although the same regions are present, their relative sizes are very different from the fish. In particular the forebrain is enormously enlarged, mainly by great development of the **cerebral hemispheres**, and this is associated with profound changes in its function.

If the brain of a mammal is sectioned immediately in front of the **medulla oblongata** all voluntary movement ceases. However, certain visceral functions, such as controlled respiration and circulation, continue unaffected. In fact the medulla contains centres controlling, not only respiration and circulation, but also swallowing, salivation and vomiting. Thus electrical stimulation applied to the right region in the medulla will elicit the vomiting reflex, and so on. In Chapter 16 we discussed the rôle of the respiratory and vascular centres in the medulla in controlling rhythmical breathing and circulation, and this needs no further amplification here.

Just in front of the medulla is the **cerebellum**, a greatly folded dorsal expansion of the hindbrain. Removal of it makes the animal unbalanced, and renders it incapable of making accurate voluntary movements. The cerebellum is responsible for maintenance of posture and equilibrium, and the fine adjustment of movement. Man owes his manual dexterity to its efficient functioning.

At first sight, the midbrain appears to be one of the least important parts of the mammalian brain. Removal of the **corpora quadrigemina** (equivalent to the optic lobes of lower vertebrates) has little effect on the animal's behaviour, except with regard to eye movements and certain auditory reflexes. However, the floor of the midbrain contains a ganglionic centre, the **red nucleus**, which is of considerable interest. If the brain is transected immediately behind the red nucleus, or if the nucleus is destroyed, a condition known as **decerebrate rigidity** results. The extensor muscles of the limbs go into a state of tonic contraction, so that the limbs are held out rigidly from the body. If the brain is transected anterior to the red nucleus, decerebrate rigidity does not occur. It is therefore apparent that this nucleus plays an important part in the control of movement and posture, and specifically inhibits excessive contraction of the postural muscles. In addition to this specific function, the midbrain contains nerve tracts connecting the anterior and posterior parts of the brain.

The 'tweenbrain contains numerous ganglia and synaptic connections from which impulses are relayed to the more posterior parts of the brain. The great **ascending** and **descending tracts**, connecting the forebrain with the spinal cord, are contained in the dorsal and lateral parts of the 'tweenbrain, the so-called **thalamus**. The ventral part, the **hypothalamus**, contains centres controlling such functions as sleep and wakefulness, feeding and drinking, speech, body temperature and osmoregulation. Electrical stimulation of the hypothalamus has shown that there is a fair localization of function in this important part of the brain. For example, electrical stimulation applied to the

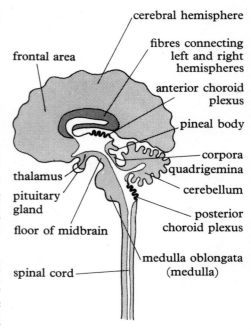

Fig 18.17 Structure of the human brain. In primitive vertebrates like the dogfish the fundamental division of the brain into fore-, mid- and hindbrain, can be clearly seen. In mammals these divisions are obscured by the great expansion of the forebrain to form the cerebral hemispheres.

frontal area

cerebral hemisphere

fibres connecting left and right hemispheres

anterior choroid plexus

pineal body

corpora quadrigemina

thalamus

pituitary gland

cerebellum

floor of midbrain

posterior choroid plexus

spinal cord

medulla oblongata (medulla)

appropriate part of the hypothalamus of a cat will cause it to feed, drink or go to sleep.

One of the least understood parts of the 'tweenbrain is the **pineal body**. There is evidence that this functioned as an eye in certain fossil vertebrates. The only present-day animal to have a third eye is the Tuatara lizard, *Spheno-don*, found on the offshore islands of New Zealand. In this primitive reptile the pineal eye, situated on top of the head, has a lens, a pigmented retina, and is connected to the brain by a nerve which passes through a hole in the cranium. There is evidence that it is light-sensitive. Little is known of its function in mammals, though the presence in it of secretory cells suggests an endocrine function.

Much more is known about the **pituitary gland**. This is certainly an endocrine organ, secreting a variety of hormones controlling such functions as water and salt balance (i.e. osmoregulation), growth, metabolism, and sexual development. In that it regulates the activities of other endocrine organs, it can be looked upon as the master gland of the endocrine system. It has long been suspected that the hypothalamus is associated with sexual behaviour and many of its effects are exerted through the intermediacy of the pituitary gland. There is thus a close functional connection between the CNS and the pituitary gland. This is further demonstrated by the fact that some of the pituitary hormones are secreted by modified nerve cells. Known as **neurosecretory cells**, the hormone flows down their axons before being liberated into the bloodstream.

The evolution of the vertebrate brain centres on the great development of the forebrain. In fishes we saw that the function of the forebrain is merely to receive impulses from the olfactory organs, and relay them to the lower parts of the brain. In mammals the forebrain, enormously enlarged by expansion of the **cerebral hemispheres**, does much more than this. The olfactory function still remains, but it is confined to the deeper, more primitive parts of the forebrain. The great expansion of the cerebral hemispheres is mainly caused by development of a new area, the **neopallium**, which first makes its appearance in amphibians and reptiles, but only reaches full development in mammals. In the latter the neopallium forms the entire roof and sides of the forebrain. It is divided into several distinct histological layers of which the most superficial, the **cerebral cortex**, consists of millions of densely-packed nerve cells interconnected in so complex a manner as to defeat the imagination (Fig 18.18). This is the most elusive part of the brain, and very little is known about the way it works.

What, then, can be said about its functions? A certain amount can be learned from animals that have had it removed, so-called **decorticate animals**. A decorticate dog, for example, loses its sense of smell and is blind, but can still move in a coordinated manner. It feeds and sleeps regularly, and will growl or bark when disturbed. By far the most striking change is its complete lack of interest in its surroundings. It fails to recognize people, stops answering to its name, and does not respond to other dogs, even those of the opposite sex. It is impossible to train it to acquire simple habits, and in fact it loses all ability to learn. To quote D. Whitteridge, a leading brain physiologist, it loses all those characteristics that justify a description of the dog as a relatively intelligent animal. Here is the clue to the function of the cortex: it is the seat of intelligence and higher activities of the brain.

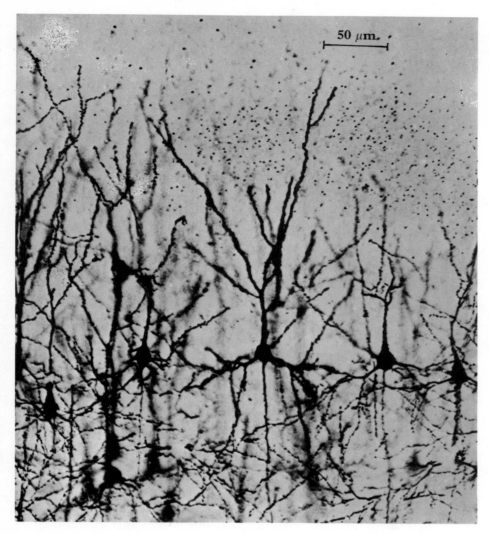

50 μm.

Fig 18.18 Nerve cells and processes in the cerebral cortex of a cat, stained by Golgi's silver technique. Only a proportion of the nerve cells take up the stain so that there are really many more than are shown. At the top of the photomicrograph are numerous small dendrites in cross section. (*J. Armstrong, University College, London*)

In more advanced mammals such as apes, and presumably in man too, the effects of decortication are more profound. The changes in behaviour described above are accompanied by sensory impairment and inability to perform coordinated movements. This turns the animal into a moribund vegetable — alive, but insensitive to stimulation, and incapable of performing any positive voluntary action. The animal is deaf, blind and unconscious, a condition which contrasts sharply with that seen in lower vertebrates where removal of the forebrain has little or no observable effect on movement and general behaviour. This illustrates that in the course of vertebrate evolution the forebrain has become increasingly important as a centre for integration and coordination. In fishes it is merely a relay centre serving the sense of smell; in mammals it controls nearly all the voluntary activities of which the animal is capable.

LOCALIZATION IN THE CORTEX

Many experiments have been carried out to discover which regions of the cortex are responsible for particular functions. On the basis of these investigations, the cortex can be divided into **sensory, motor** and **association areas**. Sensory areas receive impulses from receptors, and can be further subdivided into regions associated with particular senses. Stimulation of a receptor leads to electrical activity in the appropriate sensory region of the

Fig 18.19 Electrical stimulation of the brain enables one to assign specific functions to different parts of the cerebral cortex. Stimulation of the motor areas gives discrete movements of the appropriate parts of the body. Stimulation of the sensory areas evokes appropriate sensations. Note the visual sensory area at the posterior end of the occipital lobe: stimulation of this area evokes the sensation of light. (*modified after Penfield & Forster*)

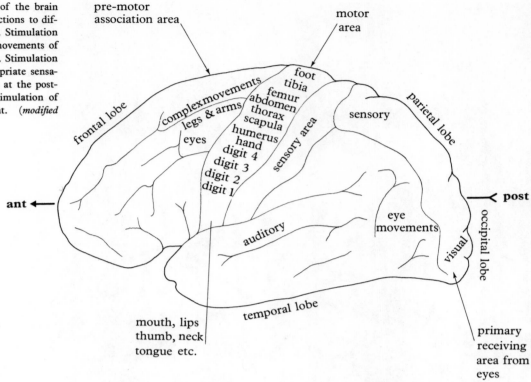

cortex. Removal of the latter abolishes that particular sense, at least temporarily. Motor areas send out impulses to the voluntary muscles via descending tracts that plunge down into the spinal cord. The motor areas show considerable localization of function, electrical stimulation of a particular region initiating a very precise response such as flexion of the little finger. These regions are shown in Fig 18.19. The association areas are more mysterious. Collectively their function is to sort out, integrate and store information before it is sent to the appropriate motor areas, but little is known about what goes on in them.

Still less is known about the **frontal lobes**. These are particularly well developed in man, and appear to be the seat of memory, imagination, thought and intelligence. They are responsible for those subtle and highly individual aspects of behaviour that we usually label personality. Surgical removal of the frontal lobes in monkeys and chimpanzees abolishes their ability to learn and solve simple problems with the use of tools (see p. 363). Such experiments obviously cannot be done on human subjects, but observations on patients whose frontal lobes have been damaged, have confirmed that they are the seat of intelligence and higher mental activities. Sometimes surgical cutting of some of the nerve tracts in the frontal lobes is used to relieve extreme mental anxiety, and not infrequently this operation leads to a lowering of the patient's mental state, and impairment of his finer judgments.

It is possible to record electrical activity from the cerebral cortex using a machine called an **electroencephalograph**. Electrodes, held in contact with the scalp by wax or tape, pick up electrical activity which is amplified and recorded. The recordings fall into three distinct groups: **alpha, beta** and **delta waves**. Each wave has a characteristic frequency and size (Fig 18.20). The alpha waves are associated with the visual cortex: they occur regularly when the eyes are closed, but are replaced by bursts of electrical activity correspond-

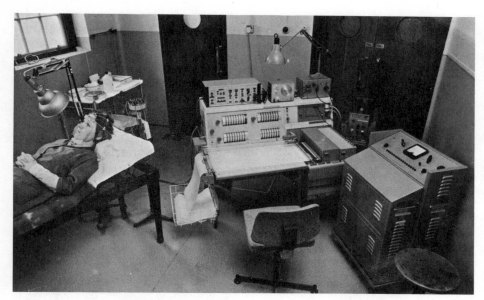

A

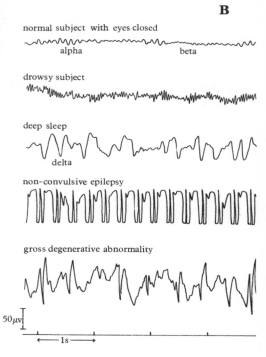

Fig 18.20 Electrical activity in the cortex of the brain can be recorded by means of an electro-encephalograph machine (A). Electrical potentials picked up by electrodes placed in contact with the scalp are amplified and recorded to give an electroencephalogram (EEG). Some sample EEG recordings are shown in B. Notice the different wave forms recorded. If a normal subject opens his eyes the alpha waves disappear and only beta waves are recorded. Note the delta waves characteristic of deep sleep, and the large abnormal waves accompanying the two mental disorders shown. (A: *Frenchay Hospital Bristol*).

B

normal subject with eyes closed

alpha   beta

drowsy subject

deep sleep

delta

non-convulsive epilepsy

gross degenerative abnormality

50μv|

←—— 1s ——→

ing to specific visual stimuli when the eyes are opened. The beta waves occur all the time, and are generated at a higher frequency than the alpha waves. During deep sleep the alpha and beta waves are replaced by the delta waves, which are larger and slower than the other two. These brain waves tell us little about how the brain works, but they are important in the diagnosis of psychological disorders and brain diseases. For example, **epilepsy** is characterized by abnormally large waves, accompanied by periodical spikes: these abnormal waves occur spontaneously all the time, not just when an epileptic fit is in progress. In fact they may occur in a person who has never had an epileptic fit in his life, but might have one under certain conditions. It is now possible to analyze the frequency of abnormal waves by computer, and this provides a much surer way of diagnosing disorders of the brain.

Superimposed on this background of brain waves are bursts of electrical activity corresponding to specific mental processes. In recent years attempts have been made to associate the occurrence of electrical activity with simple learning processes. For example, it has been found that there is a close correlation between the development of electrical potentials, and expectancy based on past experience. Experiments along these lines may ultimately lead us to a fuller understanding of the brain. Certainly the unravelling of brain function offers one of the most exciting challenges in biology.

Fig 18.21 Dahlia anemones in the open and closed conditions. The large specimen is nearly 300 mm across. Notice the mouth in the centre of the oral disc, and the slender tentacles round the edge (*Douglas P. Wilson*)

The Primitive Nervous System

In this chapter we have concentrated mainly on the nervous systems of advanced forms such as mammals, where there is a distinct CNS. For comparison we turn briefly to animals which lack a CNS, namely the coelenterates, the phylum that includes *Hydra* and its rather more complex relatives the sea anemones (Fig 18.21). These lowly animals have a diffuse **nerve net** which ramifies throughout the body wall (Fig 18.22A). The dendrites of mainly quadripolar nerve cells make synaptic contact with one another, as well as with sensory cells and muscle fibres. Although there is no concentration of

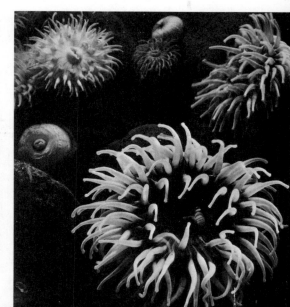

Fig 18.22 The primitive nervous system of the sea anemone.   A) The nerve net found in the tentacles and mouth region.   B) Through-conduction tract running in sheet-like mesenteries that project into the enteron cavity. For functional explanation see text. C) Photomicrograph of part of the nerve net and through-conduction tracts in a mesentery of the sea anemone *Metridium*. *(E. J. Batham, University of Otago, New Zealand)*.

A

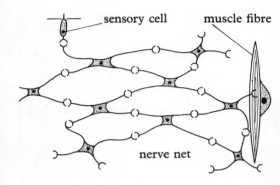

B

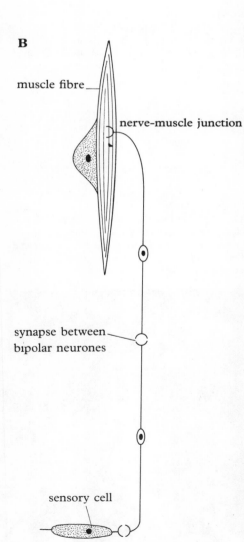

nervous tissue to form anything remotely describable as a CNS, we have here the necessary ingredients for reflex action, namely receptors, neurones and effectors. When, for example, a tentacle is prodded with a needle, the animal responds by closing up.

INTERNEURAL AND NEUROMUSCULAR FACILITATION

However, the story is not quite as straightforward as this, as has been shown by simple experiments on sea anemones. If a tentacle is touched, the tentacle pulls away but, unless the stimulus is violent, the response is confined to that particular tentacle. If, however, it is stimulated again, the response spreads to neighbouring tentacles. With continued stimulation, the response spreads to more and more tentacles, and eventually the whole animal closes up. Why is a full response not given to a single stimulus? The reason is that, when a receptor is stimulated, an impulse is generated in the nearest neurone, but

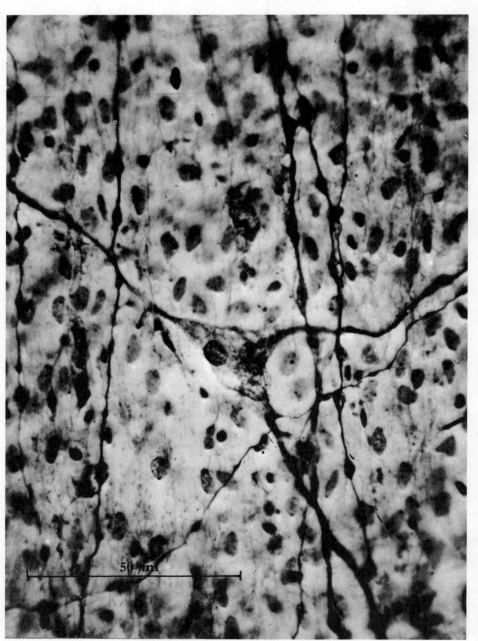

C

when the impulse reaches the first synapse it fails to cross it. However, it leaves a facilitating effect which enables a second impulse to get through. This impulse reaches the next synapse where it leaves a facilitating effect enabling a third impulse to get through, and so on. Since this type of facilitation takes place at synapses between contiguous neurones, it is known as **interneural facilitation**. It enables the slow but progressive spread of excitation throughout the nerve net. To elicit a total response, stimulation must therefore be maintained for long enough to bring in the whole nerve net.

Interneural facilitation is characteristic of all coelenterates including *Hydra*, but an additional phenomenon is seen in sea anemones. If the base of the animal is stimulated with a single electrical shock, no response is given. If, however, a second shock follows the first one within a sufficiently short space of time, a full response is given, the animal closing up. Instead of a slow spread of excitation, a comparatively rapid transmission of impulses takes place from one end of the animal to the other.

On this basis, we can predict that in sea anemones there exists, in addition to the nerve net, **through-conduction tracts** which can transmit a single impulse the full length of the body (Fig 18.22B). It is envisaged that the first of two impulses gets as far as a nerve-muscle junction, but fails to cross it. However, it leaves a facilitating effect enabling a second impulse to reach the muscle and cause a contraction. Since this kind of facilitation occurs between a nerve and its muscle, it is known as **neuromuscular facilitation**.

Anatomically the body cavity of a sea anemone is divided up by a series of radiating partitions called **mesenteries**. These sheet-like structures bear, amongst other things, well developed **longitudinal musculature** which, on contraction, pulls the tentacle end downwards and closes the animal up. Staining the mesenteries with dyes specific to nervous tissue, shows them to bear vertical **nerve tracts** made up of bipolar neurones in series (Fig 18.22 B,C). There is reason to suppose that these are the through-conduction pathways.

THROUGH-CONDUCTION

The function of the through-conduction pathways in sea anemones is to transmit impulses from one end of the body to the other with the minimum of delay. We see this process carried further in higher animals, with the development of the CNS. We have already seen that rapid transmission is one of the chief functions of a CNS.

The first glimmerings of a CNS are seen in animals like flatworms where longitudinal nerves run backwards from a nerve centre in the head. Greater centralization is seen in annelids and arthropods, whose nervous system is based on the general plan shown in Fig 18.23. A pair of **cerebral ganglia**, representing the brain, lie immediately above the pharynx at the anterior end of the body. From the cerebral ganglia a pair of **circum-pharyngeal connectives** encircle the pharynx, uniting ventrally to form a **sub-pharyngeal ganglion**. From this a **ventral nerve cord** extends to the posterior end of the body. The nerve cord is swollen in each body segment to form a **ganglion** from which **peripheral nerves** pass to the receptors and muscles of that segment.

The ventral nerve cord is a most important development as far as through-conduction is concerned. Impulses entering it from the various **receptors** are transmitted through it to the appropriate **effectors**. In some animals the

Fig 18.23 Ground plan of the nervous system of an invertebrate such as an annelid or arthropod.

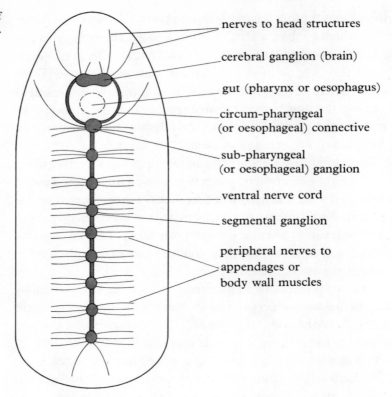

nerves to head structures

cerebral ganglion (brain)

gut (pharynx or oesophagus)

circum-pharyngeal
(or oesophageal) connective

sub-pharyngeal
(or oesophageal) ganglion

ventral nerve cord

segmental ganglion

peripheral nerves to
appendages or
body wall muscles

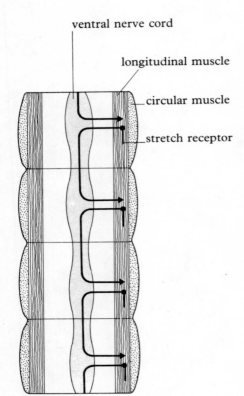

ventral nerve cord

longitudinal muscle

circular muscle

stretch receptor

Fig 18.24 Movement in the earthworm is not controlled by the brain but is effected by a series of intersegmental reflexes. When tension is applied to the body wall in segment 1, stretch receptors in the longitudinal muscle are stimulated and impulses are transmitted via the nerve cord to the longitudinal muscle in segment 2, causing it to contract. This initiates a similar reflex in the next segment, and so on. Propagation of locomotory waves will continue even after the brain has been removed.

speed of through-conduction is greatly increased by the nerve cord containing one or more **giant axons** (see p. 269). The reflex arcs, of which the giant axons form a part, can be traversed by a single impulse, and therefore require no facilitation. From primitive beginnings we therefore see the development of rapid reflex action.

THE ROLE OF THE BRAIN IN LOWER ANIMALS

If the head is removed from an earthworm, the latter will still move in a perfectly coordinated manner: in fact quite small lengths of a worm are capable of fully coordinated movements. Much the same applies to many other invertebrates: even a decapitated insect (a locust for example) is capable of hopping.

This tells us that in these creatures the brain plays little or no part in coordinating locomotion. Instead, this function is spread throughout the whole nervous system. To illustrate this, let us return to the earthworm. If you watch this animal on a rough surface, you will see that it moves forward by means of muscular bulges that pass along the body from front to rear. The brain is not required for initiating or maintaining these waves of contracture: this is done by reflexes arising within each body segment (Fig 18.24). One of the most dramatic instances of this diffuse control of movement is seen in planarian flatworms. These animals have an eversible pharynx located on the ventral side of the body. By powerful muscular contractions the pharynx sucks up food into the gut. Now if the pharynx is separated from the rest of the body, it will continue to suck in isolation, its movements being just as coordinated as in the intact animal.

What, then, is the function of the brain in these lowly animals? Its chief function is to serve as a nerve centre receiving impulses from the various receptors in the head and relaying them, via the nerve cord, to effectors in the more posterior parts of the body. It seems to do little by way of integration, and so plays little part in coordination as such.

CEPHALIZATION

As an evolutionary development, the brain is an integral part of the **head**. The head itself appears to have developed as a result of directional locomotion. If we go back for a moment to the coelenterates we find radially symmetrical animals that do not move in a set direction (i.e. they have no leading end). However, most other animals are bilaterally symmetrical with an antero-posterior axis, the anterior end being the leading end. Since this is the end to come into contact with new environmental situations first, it is not surprising to find that in the course of evolution it has become particularly well equipped with highly sensitive receptors for receiving stimuli. In higher animals these receptors take the form of eyes, nose, ears, etc., but in more primitive animals like flatworms the sensory cells are not, for the most part, concentrated into elaborate sense organs. It is simply that at the anterior end they are more numerous and concentrated than elsewhere. In flatworms the only receptors that can remotely be described as sense organs are the 'eyes', and these are merely small groups of light-sensitive cells quite incapable of forming any kind of image.

With the development of the head as the main region of stimulus-reception, the flow of impulses into the nervous system in this part of the body increases. The brain develops primarily as a means of accommodating all this 'sensory information', and relaying the appropriate signals to the rest of the body. In animals like flatworms and earthworms, the brain seems to have little else to do. Only in later stages of evolution do we see it becoming important as an integration centre for coordinating movement and controlling the more advanced types of behaviour.

The development of a head in evolution is known as **cephalization**. Different animals show different degrees of cephalization according to how primitive or advanced they happen to be. The structural complexity of the head in higher forms results mainly from its neuro-sensory function, but also from the fact that, logically enough, it contains the mouth and associated **feeding structures**. The necessity for the latter to be properly coordinated results in further elaboration of the nervous system in this part of the body.

The complexity of an animal's head also depends on the type of life it leads. Predictably, the 'best' heads are found in highly active, free-living animals whose survival depends on responding quickly to sudden changes in the environment. On the other hand, the head (particularly its sensory appendages) is reduced in animals that live in highly sheltered environments. This is well seen in parasites and burrowers. For example, the earthworm, a burrowing vegetarian, has a very much simpler head than its much more active, carnivorous relative, the ragworm (*Nereis*). The sensory tentacles, palps, cirri and eyes of *Nereis* are totally absent in the earthworm.

Hormonal Communication

Hormones are organic compounds produced in one part of the body, from which they are transported to other parts where they produce a response. A minute quantity may exert a profound effect on the organism's development, structure or behaviour. Here we shall only be concerned with the general principles involved.

Hormones are secreted by **endocrine organs** directly into the bloodstream. The word endocrine means 'internal secretion' and the endocrine organs are therefore **glands of internal secretion**. Since they shed their secretion into the bloodstream, they have no ducts and are hence known as **ductless glands**. Once in the bloodstream, the hormones are carried round the body, bringing about responses in various places. Structures that respond to them are called **target organs**.

Though they may be widely separated from one another spatially, endocrine organs do not exist in functional isolation. They influence one another and, through their interactions, are integrated into a highly coordinated system, the **endocrine system**.

HORMONAL COMPARED WITH NERVOUS COMMUNICATION

A basic similarity between the endocrine and nervous systems is that both provide means of communication within the body of an organism. Both involve transmission of a message which is triggered by a stimulus and produces a response. The target organs of a hormone are equivalent to a nerve's effectors. The main difference between the two systems concerns the nature of the message. In the endocrine system the message takes the form of a **chemical substance** conveyed through the blood system. In the nervous system the message is a discrete, all-or-nothing **action potential** transmitted along a nerve fibre. All other differences spring from this fundamental one. They can be listed as follows.

(1) Because of the comparatively high speed at which impulses are transmitted along nerves, nervous responses are generally evoked more rapidly than hormonal ones.

(2) Since it is shed into the bloodstream, there is nothing to stop a hormone being carried to every part of the body. Nervous impulses, however, are transmitted by particular neurones to specific destinations.

(3) As a result of (2), hormonal responses are often widespread, sometimes involving the participation of numerous target organs far removed from one another. In contrast, nervous responses may be very localized, involving, perhaps, the contraction of only one muscle.

(4) Hormonal responses frequently continue over a long period of time: obvious examples of such long-term responses are growth and metabolism. Nervous responses, on the other hand, are usually rapid and short-lived, such as the contraction of a muscle.

Despite these obvious differences, there is one fundamental similarity between the two systems: both involve **chemical transmission**. We saw earlier that in the nervous system transmission of the message across the neuromuscular junctions is achieved by a chemical substance. The latter is equivalent to a hormone in the endocrine system. The principal difference between them is that the neuromuscular transmitter has to travel a mere fraction of a micrometre, whereas a hormone may have to travel the full length of the body to achieve its full effect.

This may seem a rather academic point of comparison, but in fact it provides a basis for linking the two systems. This is best illustrated by the **adrenal glands**. The middle part of these glands, the **adrenal medulla**, secretes the hormone **adrenaline** which is chemically almost identical to the transmitter substance **noradrenaline** produced at the ends of the sympathetic nerves. It is interesting that adrenaline evokes the same responses as impulses in the sympathetic nerves: acceleration of the heart, constriction of arterioles, dilatation of the pupils, etc. In addition adrenaline induces a marked increase in metabolic rate, so that the combined effect of the endocrine and nervous systems is to prepare the body for emergency.

We see, then, that there is a close connection between the endocrine and nervous systems. In the case of the adrenal medulla and sympathetic nerves, the connection is so close that one suspects that the two share a common evolutionary origin. Innervated by the sympathetic nervous system, the adrenal medulla can be looked upon as an enormous conglomeration of modified nerve cells which, being far removed from any effectors, shed their transmitter substance into the bloodstream.

PRINCIPLES OF HORMONE ACTION ILLUSTRATED BY THE THYROID GLAND

The **thyroid gland** can be taken to illustrate the basic principles of hormonal communication and action. Situated in the neck, the thyroid gland secretes **thyroxine**, a complex organic compound containing iodine. The iodine required for its synthesis is obtained from the diet. The structure of the thyroid gland (Fig 18.25) demonstrates the basic requirement of any endocrine organ, namely a close association between the secretory cells and the bloodstream. The secretory cells are arranged round a series of hollow **follicles**. Iodine is taken up into the cells from the bloodstream by active transport. The follicles contain an inactive precursor of the hormone, **thyroglobulin**, which is thyroxine conjugated with a protein. A proteolytic enzyme secreted by the follicle epithelium separates thyroxine from the protein. The free hormone then passes through the wall of the follicle into the bloodstream.

Thyroxine is responsible for controlling the **basal metabolic rate**, and is therefore particularly important in growth. Under-secretion of it during development (**hypothyroidism**) causes arrested physical and mental development, a condition called **cretinism**. A cretin aged 14 or 15 is stunted and pot-bellied, and so mentally retarded that he may not even be able to feed himself. In adults the situation is not quite so serious since growth by this time is complete. The condition is called **myxoedema,** the symptoms being a decreased metabolic rate, increase in the amount of subcutaneous fat, coarsening of the skin and general physical and mental sluggishness. Over-production of thyroxine (**hyperthyroidism**) leads to a condition called **exophthalmic goitre,** so-called because of the characteristic swelling of the thyroid and protrusion of the eyeballs that accompanies it. Other symptoms include a greatly increased metabolic rate, loss of weight, and accelerated heart beat, all this being accompanied by a general physical and mental restlessness. Plainly the thyroid is important in determining its owner's general mental and emotional state, and it is interesting to speculate to what extent our own

A

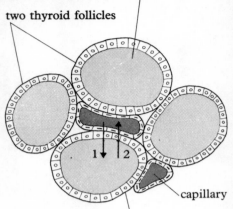

lumen of follicle
containing thyroglobulin

two thyroid follicles

capillary

cuboidal epithelial cells

Arrow 1: iodide I⁻

Arrow 2: thyroxine

B

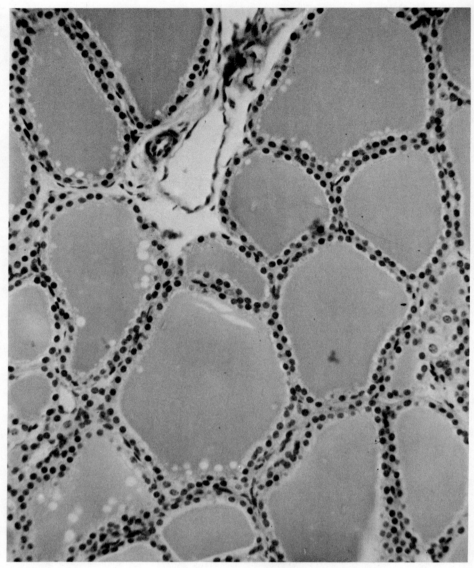

Fig 18.25 (A) The thyroid gland consists of numerous follicles in close association with thin-walled blood capillaries. Iodine, taken up into the follicles by active transport across the epithelial cells, is combined with tyrosine to form thyroxine. The latter is stored in the vesicle in combination with a protein as thyroglobulin. It is liberated into the bloodstream as thyroxine. In the photomicrograph (B) notice the thin-walled capillaries lying between the follicles. (B: *J. F. Crane.*)

personalities may be altered by minor fluctuation in thyroid activity.

A deficiency of thyroxine in the body may be caused either by the thyroid gland failing to function properly, or to deficiency of iodine in the diet. The latter is common in certain iodine-poor regions of the world, for example, parts of Switzerland and the Great Lakes in the USA. In such areas the problem is overcome by adding iodine to the drinking water or supplying the inhabitants with iodized salt. The remedy for myxoedema used to be to take thyroid orally; the old prescription was one fried sheep's thyroid weekly with black-currant jelly. Nowadays carefully regulated quantities of thyroxine are administered to the patient.

An over-abundance of thyroxine is generally caused by an over-active, often excessively large, thyroid gland. In the old days the remedy was to remove a chunk of the thyroid, and hope for the best. If severe myxoedema resulted, small pieces of the thyroid were grafted back. Nowadays surgical removal of thyroid tissue is carried out with much greater precision. Sometimes surgery is avoided altogether by injecting controlled doses of **radioactive iodine** into the patient's bloodstream: this is taken up by the thyroid cells, killing those in which it accummulates above a certain level.

CONTROL OF THYROXINE-PRODUCTION

How is the secretion of thyroxine kept to the requirements of the body? This is achieved by a **negative feedback process** of the kind discussed in Chapter 13. The shedding of thyroxine into the bloodstream is triggered by a hormone secreted by the anterior lobe of the **pituitary gland** (Fig 18.26). This is called **thyroid-stimulating** or **thyrotrophic hormone.** Now the production of thyrotrophic hormone is regulated by thyroxine itself: a slight excess of thyroxine inhibits the anterior lobe of the pituitary which responds by secreting less thyrotrophic hormone. This in turn reduces the activity of the thyroid gland, leading to a drop in the amount of thyroxine produced. This then removes the inhibitory influence on the pituitary so that more thyrotrophic hormone will be produced again, and so on. This is a good example of homeostasis and negative feedback, the principles of which are discussed in Chapter 13.

THE ROLE OF THE PITUITARY GLAND

The production of thyrotrophic hormone illustrates one of the most important functions of the pituitary gland, namely to regulate the activity of other endocrine organs. The pituitary secretes other trophic hormones which activate such endocrine organs as the adrenal cortex and gonads. The pituitary is in turn inhibited by the hormones secreted by these target organs. The pituitary gland is closely influenced by the brain, and, through the brain, by the receptors. In this way environmental changes as well as the animal's general mental state, can influence hormonal activity. We shall return to this in Chapter 22 .

NEUROSECRETION

It was suggested earlier that the adrenal medulla can be regarded as modified nervous tissue which secretes the hormone adrenaline. In development the secretory cells of the medulla are derived from the same group of cells that elsewhere give rise to sympathetic ganglion cells. However, as they differentiate they become rounded and thoroughly unlike nerve cells in appearance, so their final form belies their origin. However, there are many instances of nerve cells, recognizable as such, producing a hormone. This process is known as **neurosecretion** and the cells that do it are called **neurosecretory cells.** For example, the hormones shed from the posterior lobe of the pituitary gland, (including the **anti-diuretic hormone** discussed in Chapter 14) are secreted not by the pituitary itself but by neurosecretory cells in the hypothalamus. These cells look like ordinary nerve cells. The secretion is produced in the cell bodies whence it flows down their axons to the pituitary where it passes into the bloodstream.

Neurosecretion is of widespread occurrence in the animal kingdom and it further emphasises the close connection between the nervous and endocrine systems. We shall come across it again in Chapter 27. Other aspects of the endocrine system, including the rôle of hormones in regulating the growth of insects and plants, are also discussed in Chapter 27.

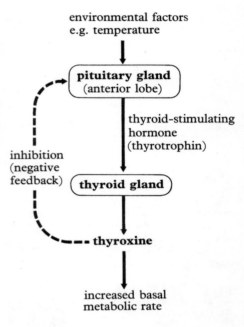

Fig 18.26 The principles of endocrine action illustrated by the thyroid gland. The secretion of thyroxine is regulated by the pituitary gland which in turn is subject to environmental influences acting via the brain.

environmental factors
e.g. temperature

pituitary gland (anterior lobe)

thyroid-stimulating hormone (thyrotrophin)

inhibition (negative feedback)

thyroid gland

thyroxine

increased basal metabolic rate

19 Reception of Stimuli

Fig 19.1 Scanning electron micrograph of sensory tufts projecting from the olfactory epithelial cells in the nose. The slender processes that make up each tuft are sensitive to airborne molecules that impinge upon them. (*A. Boyde, University College, London*).

It is obviously important to an organism, particularly an active one, that it should be aware of conditions around it so as to respond appropriately to any changes that may occur. These changes, or **stimuli,** may be local, such as a prick with a pin, or more general, such as the level of illumination. In either case the stimulus is registered by **receptors** from which impulses are relayed to the CNS. The receptors consist of **sensory cells** which may be single, scattered more or less uniformly over the whole body, or concentrated to form a **sense organ**. In some animals the sense organs reach a high degree of elaboration, as in the mammalian eye and ear for example.

Receptors can be classified according to the type of stimulation they respond to:

(1) Chemoreceptors are stimulated by chemicals, for example receptors mediating the senses of smell and taste.

(2) Mechanoreceptors are sensitive to mechanical deformation, for example, receptors sensitive to touch, pressure, tension (stretch), sounds (ear) and displacement of the body (balance).

(3) Photoreceptors are stimulated by light, for example, eyes and isolated light-sensitive cells.

(4) Thermoreceptors are stimulated by temperature, for example, receptors located in the skin, sensitive to warmth and cold.

Although receptors are commonly associated with the integument where they detect changes in the external environment, receptors also occur inside the body where they register internal changes such as the concentration of carbon dioxide in the blood, the degree of tension in the muscles, or the position of the head relative to the force of gravity. These two types of receptor, distinguished by whether they are stimulated by external or internal stimuli, are known as **exteroceptors** and **interoceptors** respectively. Those specifically concerned with giving details of position and movement are termed **proprioceptors,** and they are of the utmost importance to an animal in achieving equilibrium and coordinated locomotion. The detailed structure of these different receptors and their distribution in the body depend on the species in question.

In this chapter we shall start by considering the structure and functioning of individual receptor cells, and then explore how these are integrated to form complex sense organs.

Sensory Cells

Sensory cells possess the characteristics of cells in general plus certain unique features of their own depending on the kind of stimulation they are adapted to receive. Commonly the distal end of the cell is drawn out into one or more slender processes which are adapted to receive a specific stimulus (Fig 19.1). The cell responds by generating an action potential in an axon attached to the sensitive part of the cell.

Structurally a distinction is made between two kinds of sensory cell (Fig 19.2). In the simplest case the sensitive device consists of the fine non-myelinated termination of an axon. This **sensory terminal** represents the specialized peripheral end of an afferent neurone. It is continued inwards as the conducting part of the axon proper which terminates in the CNS as fine dendrites. These make synaptic connection with the next nervous element to which the message is transmitted. The cell body of this sensory unit may be found in a variety of different positions: close to the sensory terminal, on a short side branch of the main axon, or in the CNS.

In the second kind of sensory cell an epithelial cell is modified to receive the stimulus and to transmit the response to an associated afferent neurone with which it is in synaptic contact. The two types of sensory cell just described are known as **primary** and **secondary receptors** respectively. The main difference between them is that in primary receptors there is no synapse between the sensory device and the axon, whereas in secondary receptors the sensory device and axon are separated by a synapse.

Each type of sensory cell is specialized in position and structure to respond to one kind of stimulus. For example, all the photoreceptor cells of a vertebrate are located in the retina of the eye which is specially adapted to collect and focus light. In this position the receptor cells are protected from other forms of stimulation. In addition they possess a specialized ultra-structure which enables them to absorb light energy and transform it into nerve impulses. The same principle applies to other receptors: ear, balance organs and so on. Nevertheless some receptors may respond to more than one kind of stimulus. Thus in certain circumstances mechanoreceptors respond to temperature, photoreceptors to pressure, etc. An everyday example of this is the sensation of light experienced on pressing one's eyeballs.

FUNCTIONING OF A SENSORY CELL

The job of a sensory cell is to respond to a specific stimulus by initiating an action potential in the axon to which it is attached. This can be demonstrated in a wide variety of animals by exposing a short length of **cutaneous nerve** and placing it in contact with a pair of recording electrodes connected to an amplifier and oscilloscope (Fig 19.3). When the **tactile receptors** in the epidermis are stimulated by touching the skin, bursts of action potentials appear on the screen of the oscilloscope. The sensitivity of the receptors is most striking: in many animals the slightest touch with a fine needle is sufficient to evoke a response.

Although it is easy enough to demonstrate what receptors do, it is less easy to show how they do it. Small and often inaccessible, most sensory cells defy investigation by conventional physiological techniques. There is one receptor, however, which has been less reluctant than others to reveal its

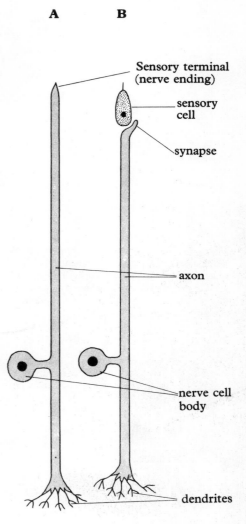

Fig 19.2 The two types of sensory cell found in vertebrates: A) a primary receptor in which the terminal end of an afferent nerve fibre (axon) is modified to receive stimuli. B) a secondary receptor which is a modified epithelial cell that makes synaptic connection with an afferent nerve fibre.

A   B

Sensory terminal (nerve ending)

sensory cell

synapse

axon

nerve cell body

dendrites

Fig 19.3 The properties of a receptor, in this case the tactile receptors in the skin, can be studied by recording action potentials from its afferent nerve. B) shows how, with continued stimulation of the receptors, the discharge of action potentials falls off and eventually ceases altogether (adaptation).

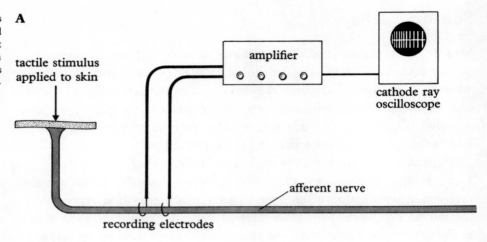

A

tactile stimulus applied to skin

amplifier

cathode ray oscilloscope

afferent nerve

recording electrodes

B

stimulus commences

Fig 19.4 Highly schematic diagram to show the development of the generator potential and firing of a nerve impulse when a sensory cell is stimulated. In this case stimulation is achieved by deflection of a sensory hair. This is typical of many mechanoreceptors. Explanation in text.

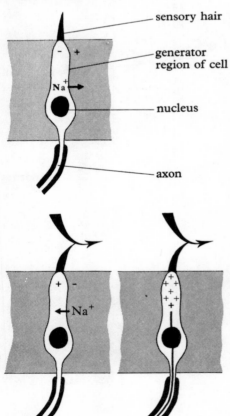

sensory hair

generator region of cell

nucleus

axon

secrets. This is the **muscle spindle** found in vertebrates, groups of nerve endings which respond to stretching of the muscle fibres surrounding them. Physiologists have inserted micro-electrodes into muscle spindles and have recorded and measured the electrical changes that take place when they are stimulated by stretching. The following account of how receptor cells work is largely based on such investigations.

HOW DOES A RECEPTOR CELL WORK?

On stimulation, the sensitive part of the receptor cell, known as the **generator region**, develops a local, non-conducted electrical change called the **generator potential**. The generator potential is caused by depolarization of the membrane surrounding this part of the cell, brought about (probably) by exchange of ions similar to that which takes place in the transmission of nerve impulses (see p. 266). The magnitude of the generator potential depends on the intensity of the stimulus. If the stimulus is weak, only a slight potential develops. However, if the stimulus is strong enough the generator potential may build up to such an extent that it reaches the necessary level, or **threshold**, to fire off an **action potential** in the axon (Fig 19.4). If the generator potential is maintained after an action potential has been conducted away, a second one will be fired off. In fact action potentials will generally go on being fired as long as the generator potential persists.

FREQUENCY OF DISCHARGE

The ability of a receptor to generate impulses in an afferent nerve is thus dependent on the development of a local potential change. Since this discovery a number of interesting facts about the generator potential have emerged. In the first place it has been found that the larger the generator potential, the shorter the interval between successive action potentials. In other words the size of the generator potential (which itself is determined by the intensity of stimulation) determines the frequency of action potentials discharged from the receptor. Thus a weak stimulus brings about a comparatively small generator potential and a low-frequency discharge of impulses in the axon; stronger stimuli produce larger generator potentials and a higher-frequency discharge of impulses. This is the basis of graded reactions to stimulation.

ADAPTATION

If a steady stimulus is maintained, the generator potential gradually declines and the frequency of action potentials decreases. Eventually the generator potential falls below the firing threshold and no further action potentials are discharged (Fig 19.5). When this state is reached the receptor is said to be **adapted**. The complete explanation of adaptation is not known but it may be due, at least in part, to the fact that with prolonged stimulation the membrane surrounding the generator region of the cell becomes progressively less permeable to the inward migration of sodium ions. The speed at which a receptor adapts depends on the size and duration of its generator potential in relation to the firing threshold. These in turn depend on the properties of its membrane. Some receptors adapt rapidly, others slowly or not at all.

The importance of adaptation is that it protects the organism from excessive discharge of impulses in its afferent nerves. To illustrate the point of this, think what happens if you put on a coarse shirt. At first the tickling sensation is almost intolerable but within half an hour or so the unpleasant sensation disappears and there is no further discomfort. This is because the tactile receptors in the skin stop firing impulses (see Fig 19.3). Of course it is not only receptors that adapt in this way. Synapses do so too (see p. 274), and it is sometimes difficult to decide whether an animal's failure to respond to repeated stimulation is due to adaptation of receptors or failure of synapses in the nervous system.

FUSION OF STIMULI

Watching a film, one is not normally conscious of the flicker on the screen. If a sensory cell is stimulated repetitively the stimuli can be detected separately only if the frequency is not too great. If the stimuli exceed a certain frequency, they appear to fuse into one continuous stimulus. To be detectable the frequency must be sufficiently low for the generator potential produced by each stimulus to die down before the next stimulus is received. If the frequency of stimulation is increased, there comes a point when the generator potentials fuse, resulting in a continuous stream of action potentials being discharged from the receptor. When this point is reached the separate stimuli can no longer be detected.

The principle of stimulus fusion applies to receptors in general but it has been particularly investigated in the eye. In man flicker fusion occurs typically at about 50/s. However, the exact figure depends on the intensity of the light and the duration of each flash. Can you explain this variation in terms of the generator potential?

THE INITIAL EVENTS IN THE RECEPTION OF STIMULI

We have seen that sensory cells do their job by developing a generator poten-

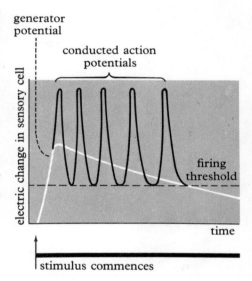

Fig 19.5 The electrical changes that occur when a sensory cell is stimulated. The black bar represents the duration of the stimulus. Action potentials are discharged when the generator potential is above the firing threshold.

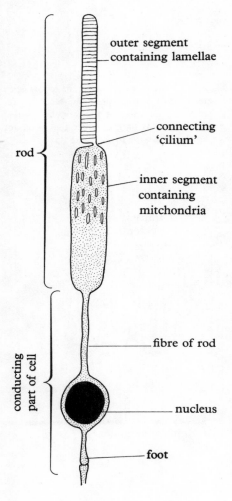

Fig 19.6 Diagram of a rod from the human retina. The molecules of photochemical pigment, rhodopsin, are thought to be located on the lamellae. The connecting 'cilium' is so-called because it contains an internal array of filaments similar to that of cilia. The foot makes synaptic contact with a nerve cell which connects with a fibre of the optic nerve.

outer segment containing lamellae

connecting 'cilium'

rod

inner segment containing mitchondria

fibre of rod

conducting part of cell

nucleus

foot

tial which fires off action potentials. But how does a stimulus evoke a generator potential in the first place? In most cases the answer to this question is not known, but there is one receptor, the mammalian eye, where we are beginning to gain an insight into the initial events that take place in the receptor process.

In the light-sensitive **retina** of the eye are two types of sensory cells, **rods and cones**. Both are stimulated by light, but the rods are more sensitive than the cones and are therefore concerned with vision in conditions of low illumination, i.e. at night. It is the rods that we shall be concerned with here.

The structure of a rod is shown in Fig 19.6. The outer segment of the cell contains a reddish pigment called **rhodopsin** (old name: **visual purple**) whose absorption spectrum corresponds closely to the curve of light-sensitivity obtained in conditions of low illumination when only the rods are functioning. Briefly, when light strikes the rod this photochemical pigment is broken down, and it is thought that this depolarizes the membrane of the cell and evokes the generator potential. The pigment is then rapidly resynthesized and can be used again. This resynthesis is experienced when we enter a dimly lit room from bright light. At first nothing can be seen, but gradually we begin to make out our surroundings. Vision becomes possible when the photochemical pigment, previously broken down by the bright light, has been regenerated. This is known as **dark adaptation**.

Rhodopsin itself is a complex protein (**opsin**) conjugated with a comparatively simple light-absorbing component called **retinine**. Retinine, an aldehyde of **vitamin A** (**carotene**), can exist in two different isomeric forms known as '**cis**' and '**trans**' **isomers**. When a quantum of light strikes a molecule of rhodopsin, the retinine changes from the 'cis' to the 'trans' form. This initiates the splitting of the rhodopsin into its protein opsin and free retinine. The resynthesis of rhodopsin from opsin and retinine takes place by a complex series of endergonic reactions. The rod is well adapted to perform its functions: numerous **mitochondria** for furnishing the necessary energy occur close to the receptive part of the cell, and stacks of sheet-like **lamellae** increase the surface area for holding the pigment molecules.

Though rather less is known about them, cones operate on the same principle as rods. The initial process in the receptor mechanism involves the splitting of a photochemical pigment similar to, but not identical with, rhodopsin. The cone pigment is less readily broken down than rhodopsin and it regenerates more slowly. Thus the cones only function in conditions of high light intensity, i.e. they are for day vision. We shall return to this later.

What does all this tell us about sensory cells in general? Photoreceptors, with their special task of converting light energy into nerve impulses, are highly specialized with many features peculiar to themselves. Nevertheless it is possible that in all sensory cells the stimulus triggers off a chain of chemical reactions which depolarize the generator part of the cell. In other words the link between the stimulus and generator potential is a chemical process.

MUTUAL INHIBITION

A sensory cell often has synaptic contacts with adjacent sensory units. These synapses may be **inhibitory**, making it more difficult for the membrane of the sensory cell to be depolarized, in much the same way that inhibitory synapses lower the excitability of post-synaptic nerve cells (p. 274).

This phenomenon has been investigated in the compound eye of *Limulus*, the horseshoe crab. This animal, a native of the east coast of North America, is not a crab at all but a marine arachnid related to spiders. The compound eye of *Limulus* lends itself particularly well to research on reception of stimuli because the individual receptor units, the **ommatidia**, are sufficiently large for a single one to be stimulated with a fine beam of light. Impulses so generated can be recorded by micro-electrodes placed in contact with individual fibres of the optic nerve.

Research on *Limulus* has shown that if two ommatidia are illuminated simultaneously, they fire impulses at a lower frequency than when each is stimulated on its own. It seems that the two ommatidia exert an inhibitory influence on each other, so the frequency of impulses fired by both sensory cells is lowered.

What is the significance of this **mutual inhibition**? To understand this look at Fig 19.7. Suppose that regions A and B of the compound eye are illuminated simultaneously, A with a strong beam of light, B with weaker light. Each group of ommatidia exerts an inhibitory effect on the other. However, A exerts a stronger inhibitory effect on B than B does on A. The result is that the beam of light illuminating region A is not registered, in terms of impulse frequency, as brightly as it really is, but the difference between the intensities of the two lights appears greater than it really is. In other words the contrast between well illuminated and dimly illuminated objects is exaggerated. This reciprocal inhibition is greatest between immediately adjacent regions of the eye. The further apart they are, the smaller is their inhibitory influence over each other.

The importance of this from the visual point of view is that it heightens the contrast at light-dark boundaries and therefore enhances contours, an effect which is well known in human vision and is often accentuated by artists in their drawings and paintings. A similar process of mutual inhibition may lead to a sharpening of the sense of pitch in the human ear.

INHIBITION THROUGH EFFERENT NERVES

At first sight the idea that receptors are inhibited through efferent nerves may seem to defeat their object. However, such inhibition may be of value in cases where the activities of an animal result in stimulation of its own receptors. To take an example, the **lateral line canals** of fishes and amphibians contain groups of receptor cells sensitive to local movements in the surrounding water. In this way the animal is warned of the presence of other animals in its immediate vicinity. Now it has been shown that these receptor cells have efferent nerve fibres going to them. Impulses in these fibres inhibit the receptors, thereby preventing them from responding to stimulation. This only happens when the animal is actively swimming, so its lateral line receptors are not excited by its own movements.

SENSITIVITY

One of the most noticeable features of receptors is their extreme sensitivity. For example, it has been estimated that in the human eye only a single quantum of light is required to excite a single rod. The tactile receptors of certain insects will respond to a deflection of as little as 3.6 nm, a property which makes them highly sensitive to airborne sounds.

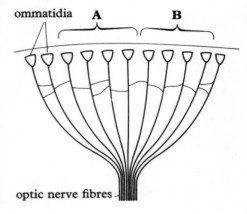

Fig 19.7 Mutual inhibition in the compound eye of *Limulus*. The compound eye is made up of numerous densely packed receptor units called ommatidia (see p. 309). When two or more ommatidia are stimulated simultaneously they exert an inhibitory influence over each other, this being made possible by the cross-connections between adjacent nerve fibres. The significance of this is explained in the text.

Fig 19.8 The principles of sensitivity and precision illustrated by the rods and cones in the retina of the eye. Groups of rods converge onto a single optic nerve fibre, thereby increasing sensitivity. In contrast each cone has its own optic nerve fibre, thereby increasing precision. Further details in text.

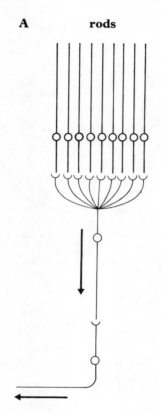

A   rods

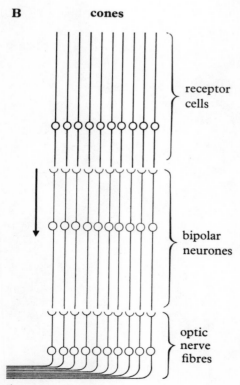

B   cones

receptor cells

bipolar neurones

optic nerve fibres

Such high sensitivity is achieved partly by the chemical and electrical events within a single sensory cell being triggered by the minutest of stimuli. However, for some receptors at least, there is more to it than this. In the human eye, for example, there are about 150 million sensory cells (about 7 million cones, the rest being rods) but only one million fibres in the optic nerve. This means that a group of rods must share, or converge onto, a single optic nerve fibre. Microscopical examination of the retina shows that this is indeed the case (Fig 19.8A). Numerous rods make synaptic contact with a single bipolar neurone which in turn connects with the cell body of a single optic nerve fibre. This is known as **retinal convergence**.

The importance of retinal convergence is that it enables the eye to be more sensitive than it would otherwise be. Look at it this way. A quantum of light hitting a single rod may be sufficient to excite that rod but insufficient to generate an action potential in the bipolar neurone attached to it. However, several rods, stimulated simultaneously, might be able to excite the bipolar neurone. On the basis of extremely careful experiments it has been estimated that six rods stimulated simultaneously are enough to fire an impulse in an optic nerve fibre. We have met this process of **summation** before in connection with synaptic transmission (p. 274), and now we see how it can effectively increase the sensitivity of a receptor. It is interesting that the cones, which are considerably less sensitive than the rods, show little or no convergence (Fig 19.8B).

PRECISION

Another property of receptors that we must explain is the precision of the information they transmit. Again we see this most clearly demonstrated in the eye. Imagine yourself looking at two black dots on a piece of paper which is gradually moved further and further away. There comes a point when the two dots can no longer be distinguished, or resolved, as separate entities, and they appear as one. The ability of the eye to resolve two or more stimuli separated spatially is known as its **visual acuity**. In comparison with lower animals the vertebrate eye has particularly high visual acuity, a property that it owes entirely to the cones. What is it about the cones that makes this possible?

To answer this question consider what must happen if two objects are to be distinguished (Fig 19.8B). If their images fall on the same cone they will obviously not be perceived separately. Even if they fall on next-door cones they cannot be distinguished, for the result will be exactly the same as a single dot falling on both cones. To be seen as two separate objects their images must fall on two different cones separated by at least one in between. From this it follows that the closer together the cones, the higher will be the acuity of the eye. The structure of the retina fits in exactly with this idea. Most of the 7 million cones are concentrated in a special area, the **fovea**, in the centre of the retina. The fovea is not much more than a millimetre across and the cones in it are thinner than elsewhere, enabling more of them to be packed into this comparatively small area.

If the cones are to signal two separate images to the CNS they must obviously connect with different optic nerve fibres. Plainly they would lose their identity if they were to share the same afferent pathway. For this reason cones show little or no convergence, at least in the central part of the fovea. In general each cone has its own bipolar neurone which connects with a single optic nerve fibre. This, coupled with their close proximity to each other, enables the eye to discriminate between objects very close together.

The same principle applies to the sense of touch. Imagine two tactile stimuli (needle pricks, for example) being applied simultaneously to your skin, or, better still, try it yourself. If the distance between the two stimuli is sufficiently great, the two will be perceived separately. If, however, this distance is reduced, there comes a point when it is no longer possible to discriminate between the two stimuli, and the sensation is that of a single prick. This critical distance varies with the part of the body stimulated. At the tip of the tongue and on the finger-tips it is as low as 1.0 mm, whereas on the thigh it may exceed 60.0 mm. In most parts of the skin large numbers of tactile receptors converge onto single afferent nerve fibres. For discrimination the two stimuli must fall on the receptive fields of two distinct nerve fibres separated by at least one in between (Fig 19.9).

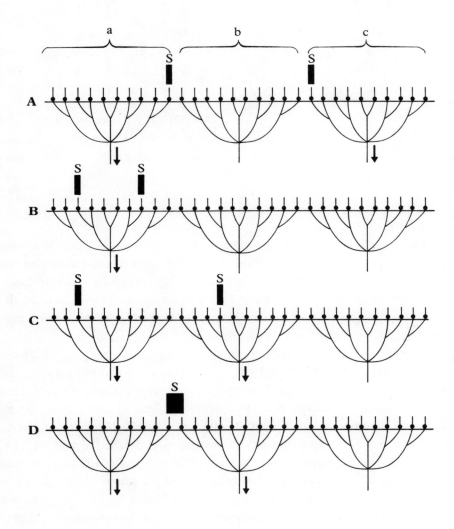

Fig 19.9 The principle of stimulus discrimination. For two simultaneously applied stimuli to be distinguished they must activate two separate afferent neurones separated by one in between (A). If they fall within the field subtended by a single afferent neurone they cannot be distinguished as separate stimuli (B). The same applies if they fall on next-door fields (C) for this is no different from a single stimulus exciting receptor cells at the adjacent edges of the two fields (D). The letters **a**, **b** and **c** indicate the receptive fields subtended by three adjacent afferent neurones. **S**, stimuli. Arrows indicate discharge of impulses in afferent fibres.

Sense Organs

So far we have confined our attention to the working of individual sensory cells and their interactions. However, in the course of evolution sensory cells have tended to become massed together to form localized sense organs. A sense organ includes many supplementary structures whose function is to

Fig 19.10 Structure of the human eye. Notice the extensive ancillary equipment (lens, cornea, iris, etc.) which protects the eye and ensures that the retina receives the right kind of stimulation.

protect the sensory cells, ensuring that they receive the kind of stimulation to which they are adapted to respond. To illustrate this let us look briefly at the eye and ear.

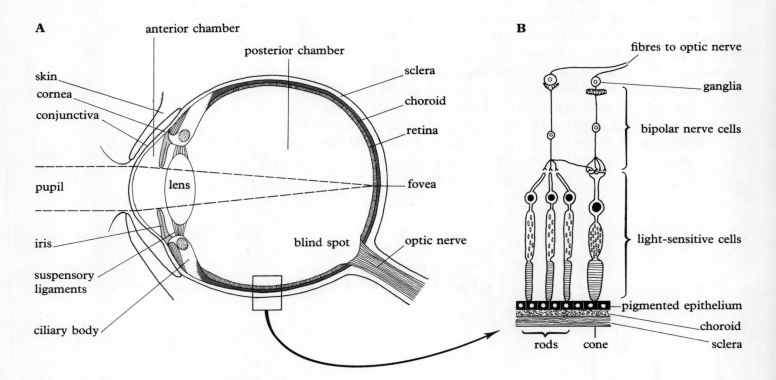

THE MAMMALIAN EYE

The structure of the mammalian eye is shown in Fig 19.10. The photoreceptor cells are concentrated in the **retina** which lines most of the interior of the fluid-filled eye ball. The retina is nourished by a vascular **choroid layer** immediately beneath it and protected by a thick connective tissue coat, the **sclera**. Heavy pigmentation in the choroid layer shields the retina and prevents light passing straight through the back of the eye.

The front of the eye contains an elaborate complex of structures which are mainly concerned with catching light rays and bringing them to focus on the retina. The structure most directly involved in this is the **lens**. Rather like a transparent rubber balloon filled with jelly, the lens is held in position by **suspensory ligaments** attached to the **ciliary body** which encircles the lens. The ciliary body contains two sets of smooth muscles, radial and circular, controlled by the autonomic nervous system. When the radial muscles contract the perimeter of the lens is pulled outwards giving it a flattened shape; when the circulars contract the tension on the lens is released so that it returns to a more spherical shape. By altering its shape the lens can **accommodate** for near and far objects, becoming flatter for distant objects and rounder for closer objects (Fig 19.11).

The front of the eye is protected by the thick transparent **cornea** behind which lies the **iris**. The latter is continuous with the ciliary body and also contains circular and radial muscles. Differential contraction of these two sets of muscles has the effect of varying the size of the aperture, or **pupil**, like the opening and closing of an iris diaphragm. In this way excessive light is

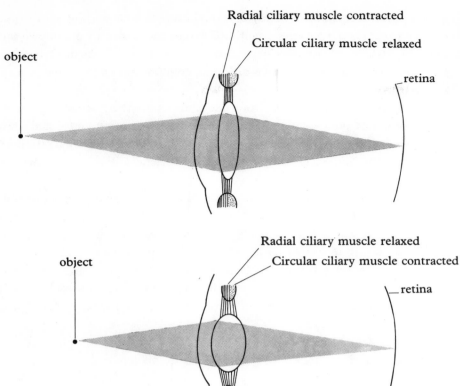

Radial ciliary muscle contracted

Circular ciliary muscle relaxed

object

retina

Radial ciliary muscle relaxed

Circular ciliary muscle contracted

object

retina

Fig 19.11 Accommodation of the eye for a near object is achieved by increasing the curvature of the lens. This is brought about by contraction of the circular ciliary muscle accompanied by relaxation of the radial muscles, thereby releasing the tension on the lens. For far objects the lens is flattened by the reverse process: the radial ciliary muscles contract and the circulars relax.

prevented from falling on the retina. The iris muscles, like those of the ciliary body, are controlled by the **autonomic nervous system,** and **atropine** (which inhibits acetylcholine) prevents the circular muscles from contracting. To dilate the pupils opticians generally put a few drops of atropine into their patients' eyes before examining the retina with an ophthalmoscope. Female film stars do the same thing to keep their pupils open when filming in bright light.

The important point to note in this brief description of the eye is that the various supplementary structures, lens, ciliary body, iris etc., play an essential part in ensuring that the sensory cells in the retina operate to maximum advantage.

THE RETINA

The eye's light-sensitive cells are located in the retina (Fig 19.10). These are the rods and cones mentioned earlier. The **cones** are packed together in the **fovea,** where their function is to perceive the surrounding environment as accurately as possible in conditions of good illumination; in a nutshell, **daylight vision.** They contain a photochemical pigment, **iodopsin,** which is not readily bleached even by light of high intensity. The cones are capable of **colour perception** (see below) and have high **visual acuity.** The latter is because they are densely packed, each one having its own connection with an optic nerve fibre (see p. 304).

The **rods** lie outside the fovea in the more peripheral parts of the retina. Their function is **night vision,** i.e., to perceive as much as possible of the environment in conditions of low illumination. To this end they contain **rhodopsin,** a pigment that is readily bleached by only a small quantity of light and is rapidly regenerated. They also show considerable **retinal convergence** (see p. 304). Their resulting sensitivity makes them able to operate in semi-darkness, which is why nocturnal animals have mainly rods in their retinas. However, they cannot perceive different colours and their visual acuity is relatively poor.

So rods and cones differ from one another in their sensitivity, rate of regeneration of photochemical pigment, colour perception and visual acuity, all of which are related to their separate functions. The functional differences between them can be related to everyday experiences. For example why is it that:

(1) When reading a book the words you are looking at directly are clear and sharp, whereas surrounding words are blurred?

(2) If you are trying to make out a particularly faint star in the sky, it is better not to look directly at it but slightly to one side of it?

(3) The flicker on a cinema screen can be seen when one looks at the screen out of the corner of one's eye, but not when one looks directly at it?

(4) In dim illumination brightly coloured objects appear as black, white or various shades of grey?

(5) When one enters a dimly lit room from bright sunlight, the room seems pitch dark to begin with, but gradually objects become visible (dark adaptation)?

COLOUR VISION

The now generally accepted theory of colour vision, the **trichromatic theory**, depends on the principle of colour mixing, i.e., that all colours can be produced by mixing the primary colours blue, green and red, in various proportions. The theory proposes that in the retina there are three functionally distinct types of cone sensitive to these three wavelengths. If any one of these cone types is stimulated on its own, it will give the appropriate sensation because of its connections in the brain. It is reasoned that a particular colour is perceived by its wavelengths stimulating one, two, or all three cone types to a varying degree. In other words the sensation, the colour 'seen', is determined by the relative excitation of the three receptors.

Can it be shown that there are in fact three types of cone in the retina? The answer is yes. To mention one experiment, the particular wavelengths of light absorbed by single cones have been determined by projecting fine beams of light of known wavelength through an isolated retina placed on a slide on a microscope stage. After passing through the eyepiece the intensity of light is measured by means of an extremely sensitive photomultiplier. The results are compared with light passing through a control tissue lacking photoreceptors. These experiments, carried out by George Wald and his collaborators at Harvard University, have shown that in the retina of man and monkeys there are **three types of cone** with maximum absorption at 450, 525 and 550 nm respectively, i.e., in the blue, green and red parts of the spectrum.

Colour vision is therefore the result of differential stimulation of three different types of cone. The intensity of colour is thought to be determined by the absolute frequency at which impulses are discharged from the three receptors; the quality of colour by the relative frequency of discharge from each.

A

ommatidia

fibres connecting
ommatidia with
optic ganglion

optic ganglion

optic nerve

B

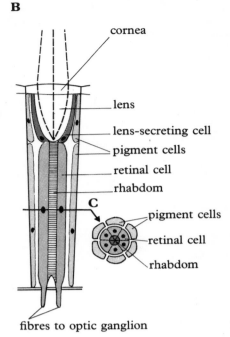

cornea

lens

lens-secreting cell

pigment cells

retinal cell

rhabdom

C

pigment cells

retinal cell

rhabdom

fibres to optic ganglion

Fig 19.12 The compound eye of an insect. The eye is made up of numerous facets or ommatidia, each complete with its own lens and composed of a group of retinal cells surrounded by pigment cells. The light-sensitive part of the ommatidium is the rhabdom, an elongated structure formed by the fusion of densely packed microvilli on the inner side of each retinal cell. It is believed that the rhabdom represents the sensory part of the ommatidium where, on receiving the stimulation of light, a photochemical reaction takes place. This leads to depolarization of the membranes of the retinal cells and the firing of impulses in the optic nerve. The field of vision of each ommatidium is such that there is considerable overlap between the fields of adjacent ommatidia. A) Diagrammatic vertical section of compound eye and optic ganglion. B) Single ommatidium in longitudinal section. C) Cross-section of ommatidium.

THE COMPOUND EYE

As a separate evolutionary development, the compound eye of arthropods (insects etc.) contrasts interestingly with the vertebrate eye. The **ommatidia** of the compound eye are much larger than the rods and cones, with the result that comparatively few can be packed into an equal area (Figs 19.12 and 19.13). The result is that arthropods have a much lower visual acuity than vertebrates. The honey bee, for example, has an acuity about a hundredth that of man, and most arthropods are worse off than that. On the other hand, the time taken for an individual ommatidium to receive a stimulus, fire an impulse and recover, is considerably shorter than for a rod. The result is that flicker fusion (see p. 301) occurs at higher frequencies in insects than in man. In the honey bee, for example, flicker fusion occurs at approximately 200/s whereas in man the figure is more like 50/s. This, coupled with the fact that the compound eyes cover a very substantial area of the head, means that the insect eye, though incapable of high visual acuity, is very good at detecting movement over a wide field (Fig 19.14). This, aided by rapid transmission of impulses through the nervous system, means that reaction times are much reduced, as is well known to anyone who has tried to swat a fly.

THE MAMMALIAN EAR

Fig 19.15 shows a simplified diagram of the mammalian ear. It consists of three chambers: an air-filled **outer** and **middle ear**, and a fluid-filled **inner ear**. The middle ear is separated from the outer ear by the **tympanic membrane**, or **ear drum**. It is separated from the inner ear cavity by an **oval window** (**fenestra ovalis**) and **round window** (**fenestra rotunda**) both of which are guarded by membranes. Spanning the middle ear chamber from the tympanic membrane to the oval window are three **ear ossicles**, the **malleus**, **incus** and **stapes**, which are held in position by muscles. The **Eustachian tube,** connecting the middle ear with the pharynx, ensures that the air pressures on the two sides of the tympanic membrane and equal. The inner ear

Fig 19.13 Vertical section through the compound eye of the prawn *Leander aspersus*. The pigment which normally lies between the ommatidia has been dissolved so that the details of the individual ommatidia can be seen. (*Sir Francis Knowles, King's College, London*)

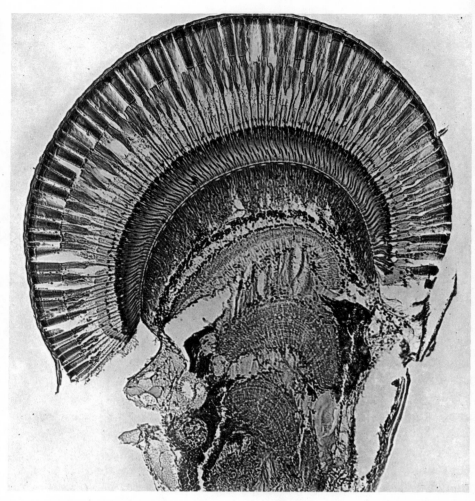

Fig 19.14 The head of a midge (*Chironomus* species) showing the great area of the head occupied by the compound eyes. (*J. F. Crane*)

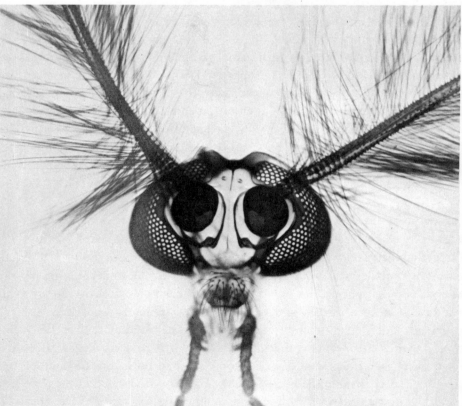

contains the **membranous labyrinth** which is made up of the **vestibular apparatus** and **cochlea**.

The ear performs two basic functions, **balance** and **hearing**. Balance is dealt with by the vestibular apparatus, hearing by the cochlea. Let us look at hearing first.

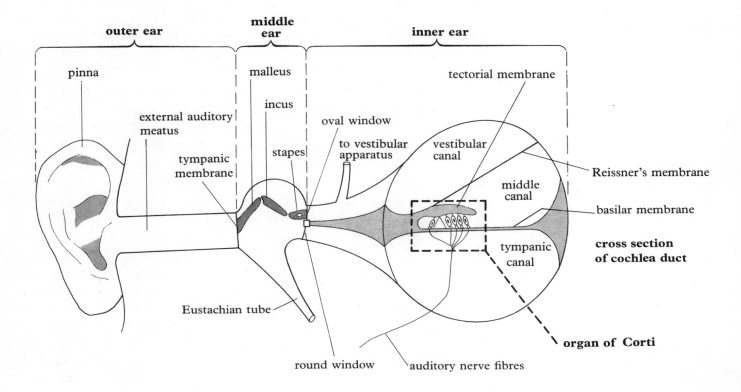

HEARING

The cochlea is a coiled tube consisting of three parallel canals: the **vestibular canal** (connecting with the oval window), the **middle canal**, and the **tympanic canal** (connecting with the round window). These three canals are separated from each other by membranes: **Reissner's membrane** is situated between the vestibular and middle canals, the **basilar membrane** between the middle and tympanic canals. The relationship between the three canals and the membranes separating them is shown in cross-sectional view on the right hand side of Fig 19.15.

This figure also shows the sensory cells that respond to sound. Into the middle chamber projects a shelf, the **tectorial membrane**, which runs parallel with the basilar membrane for the full length of the cochlea. **Sensory cells** span the gap between the basilar and tectorial membranes. Their bases are rooted in the basilar membrane where they connect with nerve fibres that join the **auditory nerve**. At the other end they bear fine hairs which just reach the tectorial membrane. This part of the cochlea, the part that actually responds to sound, is called the **organ of Corti**.

So much for the structure. How does it work? In simple terms what happens is this. Sound waves enter the outer ear and impinge on the tympanic membrane which vibrates accordingly. These vibrations are transmitted by the three ear ossicles to the oval window. This results in displacement of fluid in the vestibular canal, which in turn causes movement of Reissner's mem-

Fig 19.15 Diagram to illustrate the basic structure and mode of action of the mammalian ear. The sensory cells are located in the cochlea duct. As with the eye, the purpose of other structures is primarily to provide these sensory cells with appropriate stimulation. The three ossicles in the middle ear increase the force with which movements of the tympanic membrane are transmitted to the oval window. The ossicles are held in place by ligaments and muscles which are not shown in the diagram. The muscles contract reflexly in response to very loud sounds, thus dampening the ear and protecting the cochlea from damage caused by excessive vibrations. Further explanation in text.

brane. This displaces fluid in the middle canal which moves the basilar membrane, thereby displacing fluid in the tympanic canal. Displacement of this latter fluid is taken up by stretching of the membrane covering the round window.

Thus vibrations of the tympanic membrane, set up by sound waves hitting it, are transmitted via a series of ossicles, membranes and fluid-filled cavities, to the basilar membrane. Here, deep in the inner ear, sensory stimulation takes place. Movement of the basilar membrane distorts the sensory cells of the organ of Corti, resulting in impulses being fired in the auditory nerve.

A complete account of hearing must of course explain how the ear discriminates between sounds of different **intensity** and **pitch**. The intensity of a sound is determined by the amplitude of the sound waves impinging upon the tympanic membrane. This in turn determines the amplitude with which the basilar membrane vibrates. Loud sounds, that is sound waves of large amplitude, bring about greater displacement of the basilar membrane than softer sounds, with the result that impulses are fired at higher frequency in the auditory nerve fibres.

What about pitch? The pitch of a sound is set by the frequency of the sound waves, i.e. the **wavelength**. High tones are caused by sound waves of high frequency (short wavelength), while low tones are caused by sound waves of lower frequency (long wavelength). The pitch of a sound thus determines the frequency at which the basilar membrane vibrates. Careful experiments on the inner ear have shown that the sensory cells in different regions of the cochlea respond to different frequencies. Those towards the apex respond to low tones (down to about 65 hertz in man) while those towards the base respond to high tones (up to about 16,000 hertz in man). It has been estimated that these high frequency sounds vibrate the basilar membrane by a mere one billionth of a centimetre.

There is still controversy over many aspects of how the cochlea responds to sound. For instance, what is the precise nature of the stimulus that excites the sensory cells? Are they pulled, pushed, or what? The hairs that project from these cells (see Fig. 19.15) are stiff, and current opinion is that at rest they just touch the tectorial membrane. It is believed that movement of the basilar membrane subjects them to shearing forces, possibly accentuated by the rods that run from the basilar to the tectorial membrane. The stresses so created are then transmitted to the sensitive part of the cell.

With regard to pitch discrimination, how is localization of response in the cochlea accomplished? The mechanism seems to depend on the stiffness of the basilar membrane gradually decreasing from the base to the apex of the cochlea. It is thought that movement of the ear ossicles sets up a travelling wave which passes along the basilar membrane for a certain distance, and then dies out. How far it gets depends on the frequency. High frequency waves travel only a short distance, low frequency waves much further. This, coupled with differential sensitivity of the receptor cells, could result in different parts of the cochlea responding to different frequencies. Under certain circumstances the whole of the cochlea may be activated by a particular sound, a very loud, low pitch sound for example. The sensation received is determined by the pattern of impulses generated in the many fibres of the auditory nerve.

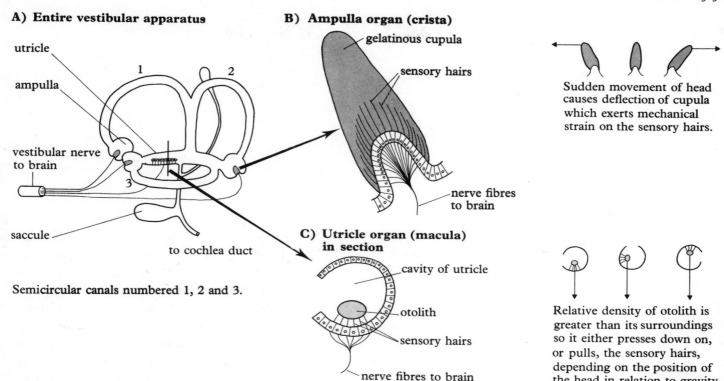

A) Entire vestibular apparatus

utricle

ampulla

vestibular nerve to brain

saccule

to cochlea duct

Semicircular canals numbered 1, 2 and 3.

B) Ampulla organ (crista)

gelatinous cupula

sensory hairs

nerve fibres to brain

Sudden movement of head causes deflection of cupula which exerts mechanical strain on the sensory hairs.

C) Utricle organ (macula) in section

cavity of utricle

otolith

sensory hairs

nerve fibres to brain

Relative density of otolith is greater than its surroundings so it either presses down on, or pulls, the sensory hairs, depending on the position of the head in relation to gravity.

BALANCE

The vestibular component of the inner ear consists of an assembly of fluid-filled sacs and canals (Fig 19.16). These include three **semicircular canals** in planes at right angles to each other. At one end of each semicircular canal is a swelling, the **ampulla**, which contains a receptor. The semicircular canals open into the **utricle** which in turn connects with the **saccule**. The utricle and saccule both contain receptors.

The ampulla receptors consist of groups of sensory cells whose hairs are embedded in a dome-shaped gelatinous cap, the **cupula**. The semicircular canals are sensitive to movements of the head, and the fact that they are in three different planes ensures sensitivity to movement in any plane. Thus if you shake your head the horizontal canal is activated; if you nod your head one of the vertical canals is brought into play, and so on. Because of the inertia of the fluid in the semicircular canals, the cupula gets deflected in a direction opposite to that in which the head is moving. This puts a mechanical strain on the sensory cells, causing them to fire impulses in the afferent nerve fibres.

The utricle and saccule give information on the position of the head. These receptors consist of a patch of sensory cells, the free ends of which are embedded in a concretion of calcium carbonate called an **otolith**. According to the position of the head, the pull of gravity on the otoliths will vary. The different distortions of the sensory cells resulting from this affects the pattern of impulses discharged in the afferent nerve fibres.

This account of the eye and ear has necessarily been brief but it should convey something of the complexity reached by sense organs in the course of evolution. Again it must be stressed that in fundamental respects the individual sensory cells behave like those of other receptors. The complexity of the eye and ear lies mainly in the extensiveness of the ancillary apparatus involved in activating and protecting the receptor cells.

Fig 19.16 Diagram of the mammalian vestibular apparatus in the inner ear, showing the structure and mode of action of the two main types of balance receptor. In both cases stimulation is achieved by mechanical distortion of sensory hairs. (*based on Davson and Eggleton*)

20 Effectors

An effector is a structure which responds, directly or indirectly, to a stimulus. Most effectors are controlled by the nervous system and respond when they receive impulses from efferent nerves. However, certain effectors have no nerves going to them and respond to direct stimulation.

The most common effectors are **muscles** and **glands**. Less widespread, but no less important to the animals possessing them, are **pigment cells**, **light-producing organs** and **electric organs**. **Cilia** and **flagella** are also effectors, though they may not necessarily be under the control of a nervous system. The **stinging cells** of coelenterates provide an example of effectors which are completely independent of nervous control.

In this chapter we shall concentrate on muscle, since more is known about this than most other effectors. Other specialized effectors will be briefly discussed at the end of the chapter.

DIFFERENT TYPES OF MUSCLE

It is customary to divide vertebrate muscles into three types according to their location in the body:

(1) Skeletal muscle. This is attached to the skeleton and is therefore concerned in locomotion. As it is innervated by the voluntary part of the nervous system, it is also known as **voluntary muscle**. When viewed under the microscope its fibres are seen to have stripes running across them, for which reason it is called **striated** or **striped muscle**. Characteristically it contracts, and fatigues, rapidly.

(2) Visceral muscle. This is found lining the gut, blood vessels and various cavities, and is innervated by the autonomic (involuntary) part of the nervous system, for which reason it is also known as **involuntary muscle**. Characteristically unstriated, it contracts, and fatigues, comparatively slowly.

(3) Cardiac muscle. This has already been described (p. 171). Situated in the wall of the heart, its fibres are striated and joined by cross-connections. It contracts rhythmically, without fatigue, and does not have to be initiated by the nervous system. Its contractions are therefore **myogenic**, i.e. generated from within the muscle tissue itself.

The fundamental question is: how does muscle contract? In attempting to answer this we will focus our attention, as physiologists have done, on skeletal muscle. But first let us look briefly at what skeletal muscle is capable of doing.

Properties of Skeletal Muscle

When a muscle is activated it either shortens or, if it is attached to a rigid skeleton, develops tension. Its properties can be investigated by attaching it to a movable lever that writes on a revolving drum (**kymograph**). The muscle can be activated by stimulating it, or its nerve, with brief electric shocks. A preparation commonly used in this kind of work is the **calf** (**gastrocnemius**) **muscle** of the frog. The muscle is removed from the leg and set up as shown in Fig 20.1. The muscle is stimulated with single or repetitive shocks, either direct or through its nerve, and the resulting contractions are recorded on the drum.

A **B**

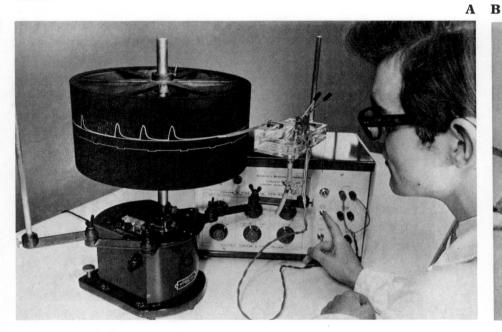

Fig 20.1 Investigating the physiological properties of skeletal muscle.   A) When the operator presses the key of the stimulator an electrical shock is sent into the muscle, or its nerve, through a pair of electrodes. The muscle responds by contraction, pulling on the lever which makes a tracing on the revolving smoked drum (kymograph).   B) shows how the muscle is attached to the lever. The electrodes are placed in contact with the muscle or its nerve. In this case they are in contact with the nerve. (*B. J. W. Heath, Marlborough College*)

THE SINGLE TWITCH

If a single shock, strong enough to evoke an electrical response, is sent into the muscle the latter responds by giving a quick, sharp contraction or **twitch** (Fig 20.2A). It takes about 0.1 s for the contraction to reach its height, and a further 0.2 s for relaxation to be complete. The entire response is therefore over in approximately 0.3 s. The short delay, or **latent period**, between the application of the stimulus and the onset of the muscular contraction is caused partly by the inertia of the apparatus but also by the time taken for the electrical response to be translated into a contraction. Typically this lasts about 0.02 s.

SUMMATION

Now consider what happens if two shocks are delivered in succession (Fig 20.2 B and C). If the interval between the shocks is sufficiently long, two distinguishable twitches are given; but if the interval is gradually shortened there comes a point when the two twitches fuse, or **summate**, to give one smooth contraction. Each impulse initiates a contraction, but the second contraction starts before the muscle begins to relax, so that the summated response is larger and of longer duration than a single twitch.

Fig 20.2 Kymograph tracings obtained by stimulating the gastrocnemius muscle of the frog.   A) shows a simple 'twitch' in response to a single stimulus.   B) and   C) show the effect of two stimuli in succession, the interval between the two stimuli being sufficiently short in C for the two twitches to fuse into a single contraction (summation).   D), E) and F) show how a smooth tetanus can be produced by increasing the frequency of a train of stimuli. In all cases the stimuli are dotted and a time scale is shown below each tracing.   E) and   F) appear at the top of p. 317.

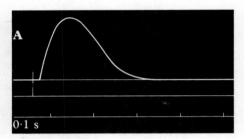

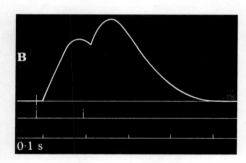

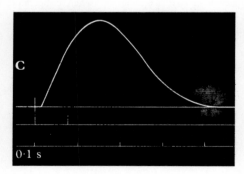

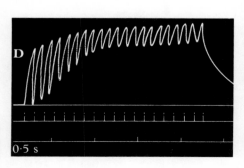

TETANUS

We can take the above experiment a stage further. Instead of stimulating with only two shocks, we send a whole train of shocks into the muscle. If they are of sufficiently high frequency, the muscle goes into a maintained contraction or **tetanus**, relaxing only when the stimuli are switched off (Fig 20.2F). Numerous twitches have fused, as it were, to give a single prolonged response. If the frequency of stimulation is not great enough, separate twitches can be discerned. (Fig 20.2D and E). To produce a smooth tetanus a minimum frequency of between 15 and 30 shocks per second is required.

What is the significance of this? When the muscles of an intact animal contract under natural conditions they do so tetanically and not in simple twitches. For example, if you bend your arm, the flexor muscle undergoes a tetanic contraction as a result of trains of high-frequency impulses reaching it from the brachial nerve. How long the contraction goes on for depends on the duration of the train of impulses, and that depends on you. Just as an experimenter can stop tetanizing a muscle by switching off the stimuli, so you (using your brain) can voluntarily stop your flexor muscle contracting by 'turning off' the impulses which are streaming down to it.

FATIGUE

A tetanic contraction cannot go on indefinitely, and if stimulation is continued the muscular responses gradually decline, and eventually disappear altogether. This **fatigue** is brought about by various factors operating within the muscle. If, instead of being stimulated direct, the muscle is excited through its nerve, the responses decline even more quickly. This is owing to exhaustion of the transmitter substance (acetylcholine) at the neuromuscular junctions.

ELECTRICAL ACTIVITY IN MUSCLE

The experiments described above involve using a whole muscle. More detailed investigations are carried out on single muscle fibres which can be teased out of the whole muscle, isolated and experimented on individually. Micro-electrodes can be inserted into the muscle fibre so that its electrical as well as mechanical properties may be demonstrated. What emerges from these experiments is that muscle fibres are remarkably similar to nerve cells. When not active they have a **resting potential**, and when activated this is momentarily reversed to give an **action potential**. This is generally of longer duration, and is transmitted more slowly, than that of a nerve cell, but otherwise the two are similar and have the same ionic basis.

Like nerve, muscle obeys an **all-or-nothing law**. By stimulating an isolated muscle fibre with shocks of varying intensity, it can be shown that the size of the action potential, and the resulting contraction, are independent of the intensity of stimulation. If the strength of a stimulus is below the threshold required to excite the fibre, no action potential and no contraction are given. If the stimulus is above this threshold an action potential and contraction are given, and further increase in the intensity of stimulus will not enhance this response.

Muscle also shows the phenomenon of **refractory period**. After contraction there is a brief period of complete, followed by partial, inexcitability, the **absolute** and **relative refractory periods** respectively. These are generally slightly longer than for nerve.

From an electrical standpoint nerve and muscle are therefore very similar. The main difference is that in muscle the action potential is accompanied by contraction. In other words the electrical energy of the action potential is somehow converted into the mechanical energy of contraction.

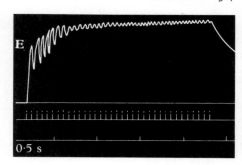

How Muscle Contracts

To understand how a muscle contracts it is first necessary to look at its structure. We will start with what you yourself can see under a light microscope, and then go on to its fine structure as revealed by the electron microscope.

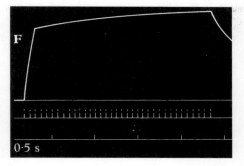

THE LIGHT MICROSCOPE STRUCTURE OF MUSCLE

A whole muscle, such as the frog's calf muscle, is made up of many hundreds of **muscle fibres** varying in length from 1 to 40 millimetres. If an individual fibre is sectioned longitudinally, stained, and examined under the microscope, the structures shown in Fig 20.3C can be seen. The fibre is filled with a specialized

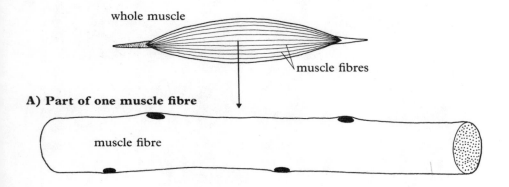

whole muscle

muscle fibres

A) Part of one muscle fibre

muscle fibre

B) Muscle fibre stained to show striations

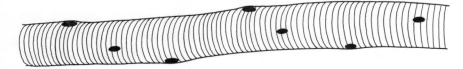

C) Muscle fibre in longitudinal section

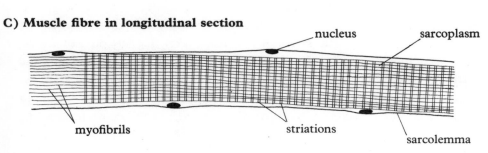

nucleus   sarcoplasm

myofibrils   striations   sarcolemma

Fig 20.3 Structure of skeletal muscle. A muscle is composed of numerous muscle fibres bound together by connective tissue. When viewed under the light microscope each fibre can be seen to contain numerous longitudinal fibrils (myofibrils) and to display a characteristic pattern of striations.

cytoplasm called **sarcoplasm** in which about 100 nuclei are spaced out evenly just beneath the bounding membrane or **sarcolemma**. These nuclei are not separated from one another by cell membranes, so the fibre is multinucleate.

Numerous parallel **striations** traverse the fibre from one side to the other. Little else can be seen, though in good sections it is sometimes possible to make out slender threads running along the length of the fibre. These are called **myofibrils** (or just **fibrils**), and we shall see presently that they play a crucial rôle in the contraction process.

Until the early 1950s this was all that was known about the structure of muscle; and a frustrating picture it was, for it gave no clue as to how contraction takes place. However, since then the electron microscope has revealed a wealth of information that goes a long way towards explaining how it works.

THE FINE STRUCTURE OF MUSCLE

Under the light microscope little detail can be discerned in the striations, and the myofibrils appear as extremely thin lines. However, in the electron microscope the internal structure of individual myofibrils shows up clearly, and the reason for the striations becomes apparent. This is shown in Fig 20.4.

Fig 20.4 Low magnification electron micrograph of part of a skeletal muscle fibre showing six fibrils with their characteristic pattern of dark and light bands. (*H. E. Huxley, MRC Laboratory of Molecular Biology, Cambridge*).

1 μm

Notice that each myofibril is divided up into alternating **light** and **dark bands**. As the light and dark bands of adjacent myofibrils lie alongside each other, one gets the impression of continuous striations running right across the entire fibre.

Fig 20.5 will help you to interpret the electron micrograph in Fig 20.4. You can see that each dark band has a comparatively light region in the middle. This is called the **H zone,** and has darker regions on either side. Running across the middle of the H zone is a dark line (the **M membrane**), and traversing the middle of the light band is an even darker line, the **Z membrane**. The region of a myofibril from one Z membrane to the next is called a **sarcomere.** For descriptive purposes the sarcomere can be regarded as the basic unit of the myofibril, the whole myofibril consisting of a long chain of such units placed end to end.

The explanation of the banding pattern is made clear in Fig 20.5B. The myofibril is composed of numerous longitudinal filaments of two types: thick and thin. The **thick filaments** are confined to the dark band. The **thin fila-**

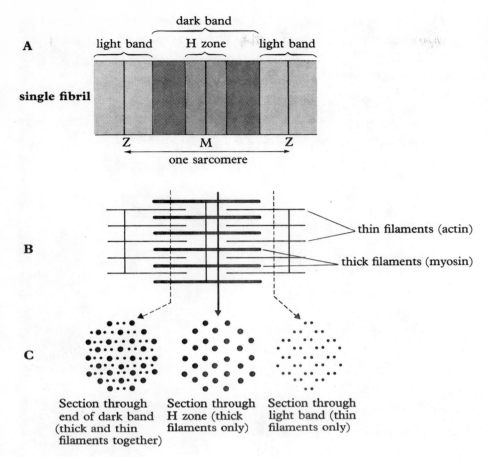

A

dark band

light band | H zone | light band

single fibril

Z   M   Z

one sarcomere

B

thin filaments (actin)

thick filaments (myosin)

C

Section through end of dark band (thick and thin filaments together)

Section through H zone (thick filaments only)

Section through light band (thin filaments only)

Fig 20.5 The detailed structure of skeletal muscle has been worked out with the electron microscope. The light and dark bands seen in low magnification electron micrographs (A) are due to each fibril consisting of alternating groups of thick and thin filaments which interdigitate between one another, as shown in B. The groups of filaments are held in alignment by transverse membranes (**Z** and **M** in the diagrams). The interlocking arrangement of filaments has been confirmed by examining cross-sections of fibrils in the electron microscope (C).

ments occur in the light band, but extend in between the thick filaments into the dark band. The segments on either side of the H zone are therefore particularly dark because they contain both thick and thin filaments. The H zone consists of thick filaments only, which is why it is slightly lighter than the two ends of the dark band. This interpretation of the banding pattern is not just a hypothesis. It is confirmed by high resolution electron micrographs such as the one shown in Fig 20.6. Furthermore, cross sections of individual myofibrils give the appearance shown in Fig 20.5 C.

THE CHEMISTRY OF MUSCLE

So far this description has been purely anatomical. But while electron-microscopists were tackling the fine structure of muscle, biochemists were investigating its chemistry. It had been found some years before that muscle is composed of two proteins: **actin** and **myosin**. When these proteins are extracted from muscle fibres and placed in a solution of ATP, contraction occurs. This means that from a purely biochemical standpoint all that is required for contraction is a mixture of actin, myosin and ATP.

Where is the actin and myosin in relation to the fine structure of the myofibril? The answer has been found by treating muscle with chemicals that selectively dissolve one or other of these proteins, and then examining the myofibrils under a phase-contrast microscope. It was found that if a muscle is treated with a solution that selectively dissolves myosin, all the dark bands disappear. If however, it is treated with a solution that removes actin, all the light bands disappear. The conclusion is that the thick filaments are composed of myosin, the thin filaments of actin.

Fig 20.6 One complete sarcomere in a skeletal muscle fibril. Notice the thick and thin filaments in between one another, and the bridges connecting them together. (*H. E. Huxley, MRC Laboratory of Molecular Biology, Cambridge*).

So the myofibrils consist of alternating sets of thick myosin and thin actin filaments which overlap in the dark band. What then happens when the muscle contracts?

THE SLIDING HYPOTHESIS

From the structure of the myofibril it would seem a reasonable proposition that contraction occurs by the thick and thin filaments sliding between each other. The thin filaments in each light band are held together by the Z membrane and the thick filaments in each dark band by the M membrane, so they would be expected to slide as a unit. The orderly arrangement of the filaments would therefore be maintained during the contraction process.

This sliding hypothesis was first put forward by Hugh Huxley and Jean Hanson of London University. By a combination of different kinds of microscopy and logical deduction, they secured good evidence to support it. They argued as follows. If it is true that the filaments slide, the banding pattern in the myofibril should change as contraction occurs. When the muscle is fully relaxed (or stretched) the light bands, and H zones, should be comparatively long. The dark ends of the dark band, however, should be relatively short. When contracted, the light bands, and the H zones, would be expected to get much shorter, and the dark ends of the dark bands longer. All this should be accompanied by a change in the overall length of the sarcomeres; from being long in the relaxed muscle they should become comparatively short.

This is precisely what Huxley and Hanson found. They studied the banding pattern of living muscle fibres and of isolated myofibrils contracting in a solution of ATP and found the change in the banding pattern to be exactly as they had predicted. Convincing confirmation of the sliding hypothesis has been obtained by comparing the electron microscopic appearance of stretched myofibrils with those at normal resting length. As living materials cannot be viewed in the electron microscope it is obviously impossible to see the sliding process actually taking place, but examining electron micrographs of muscle in different states of contraction is the next best thing. The diagram in Fig 20.7 is based on such electron micrographs. In severely contracted

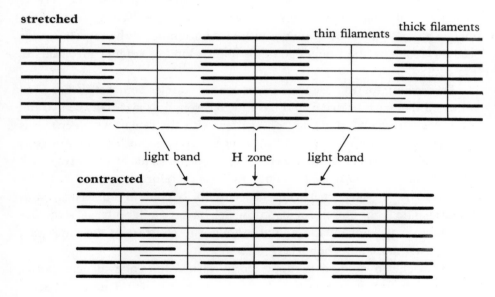

Fig 20.7 The sliding hypothesis proposes that when a muscle shortens the thick and thin filaments slide in between one another as shown here. Although this has never been seen taking place, it is supported by electron microscopy and other evidence. Note that when contracted the dark bands should remain the same length but the light bands and H zones should shorten. This is in fact the case.

muscles the ends of the filaments actually meet, the thin ones first and then the thick ones. This has the effect of creating new bands at the positions occupied by the M and Z membranes, which can be interpreted by postulating that where they meet the filaments crumple or overlap. In the case of the thin filaments, this has been confirmed by the discovery that in transverse sections of myofibrils cut through this region there are twice the normal number of thin filaments. When a muscle is contracted to this extent the light bands and H zones disappear altogether, both types of filaments occurring together throughout the length of the fibril.

WHAT PROPELS THE FILAMENTS?

We are now faced with the ticklish question: how are the two sets of filaments

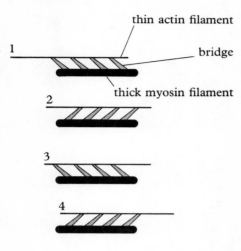

Fig 20.8 Diagram summarizing the proposed ratchet mechanism by which the thin filaments may be pulled towards the thick ones when a muscle contracts.

moved in between each other? Here again the electron microscope comes to the rescue. In high-magnification electron micrographs, such as the one shown in Fig 20.6, **bridges** can be seen connecting the thick and thin filaments where they overlap. These bridges project from the thick filaments at roughly 6.0 nm intervals, describing a spiral pathway round the edge of the filament. In the region of overlap each thick filament is surrounded by a hexagonal array of six thin filaments. Successive bridges projecting from the thick filament are attached to each of these thin filaments in turn so the spiral pattern of bridges repeats itself once every six bridges, a distance of about 40 nm.

Recent research on muscle has been concerned with showing how these bridges might cause contraction. It has been known for a long time that actin and myosin will interact to form another protein, **actomyosin**. Evidence suggests that during shortening of the muscle each bridge attaches itself to a thin filament, actomyosin being formed at the point of contact, and then contracts. The concerted contraction of many bridges has the effect of pulling the thin filaments past the thick ones. After it has contracted, each bridge detaches itself from the thin filament and re-attaches itself at another site further along. The cycle is then repeated. Shortening of the muscle is thus brought about by the bridges going through a kind of **ratchet mechanism**. (Fig 20.8).

To account for the known rate at which muscles contract, it has been estimated that each bridge would have to go through its cycle between 50 and 100 times per second. This figure, together with the rate at which energy is liberated from contracting muscle, leads us to suppose that for each bridge to go through a complete cycle the splitting of one molecule of ATP is required. The ATP consumption of contracting muscle is therefore considerable, which explains why the gaps between adjacent fibrils contain numerous mitochondria. The flight muscles of insects, which have a similar internal structure to vertebrate skeletal muscle and are thought to contract in the same way, contain even more mitochondria (Fig 20.9).

The ratchet hypothesis is supported by various lines of evidence. Special techniques in electron microscopy have shown the detailed structure of actin and myosin, and X-ray analysis supports the view that the bridges undergo substantial movement during contraction. The ratchet idea also has the merit of explaining many facts about muscle which have been known for a long time. You should consult recent issues of scientific periodicals for details of recent developments.

Although much is known about the electrical properties of muscle and the way it contracts, we know little about how the electrical and mechanical processes are linked. However, exciting information has come from studies on the endoplasmic reticulum of striated muscle, in this case known as the **sarcoplasmic reticulum**. Highly organized, the sarcoplasmic reticulum includes a system of **transverse tubules** which run from the sarcolemma to the Z membranes. Connected with these tubules are vesicles containing calcium ions. Calcium ions are necessary for the splitting of ATP, and there is evidence that when an action potential passes along the surface of the muscle fibre, an electrical signal is transmitted via the transverse tubules to the vesicles which then pump their calcium ions into the sarcoplasm. This raises the calcium ion concentration sufficiently for ATP-splitting, and con-

Fig 20.9 Highly active muscles require a particularly large supply of energy. In this longitudinal section of insect flight muscle numerous densely-packed mitochondria (**M**) can be seen between adjacent fibrils. Within each mitochondrion the cristae are so close together that they can barely be distinguished. (*M. J. Cullen, University of Oxford*)

traction occurs. After the muscle has contracted it is thought that the vesicles take up the calcium ions again, thereby lowering their concentration to a level below that at which ATP-splitting can occur.

This is an interesting lead but it still leaves many questions unanswered. The general problem of how an electrical impulse derived from the nervous system is translated into action by an effector is obviously fundamental to understanding how effectors work.

Other Effectors

Although muscles are amongst the most widespread effectors, they are not the only ones, and this chapter would be incomplete without a brief mention of other specialized examples.

CHROMATOPHORES

Chromatophores are specialized **pigment cells** found in the skin of fishes, amphibians, reptiles, octopuses and squids, etc. Many such animals can change their colour, or shade, by concentrating or expanding the pigment in their chromatophores. This is achieved either by muscle fibres pulling the flexible wall of the cell outwards, as in octopuses and squids, or by migration of pigment within the chromatophore, as in most other animals. These movements are controlled partly by nerves, partly by hormones. In vertebrate chromatophores with a nerve supply, the sympathetic system causes concentration, the parasympathetic dispersal, of the pigment.

Chromatophores are important in adaptive coloration, particularly where colour change is possible. Darkening or lightening of the body so that it merges with the background is an important aspect of camouflage, and a sudden colour change may also serve as a warning signal. So far as colour change is concerned, the doyen of animals is the chameleon whose chromatophores contain a wide range of pigments including green, yellow, red and black. By varying the relative concentration of these different pigments, it can adopt a wide variety of colours depending on the background. These subtle responses are initiated through reflexes involving the eyes.

ELECTRIC ORGANS

Found in certain fishes, notably the South American electric eel and the Mediterranean electric ray, these organs consist of stacks of sheet-like **electroplates**, modified muscle fibres innervated by the autonomic nervous system. Instead of contracting they send out electric pulses. The total discharge of all the plates simultaneously can exceed 700 volts, which is more than enough to electrocute quite sizeable prey.

It is now known that many fishes produce much lower voltages, emitting a pattern of electric pulses which are not used for killing prey but for locating objects in their immediate vicinity. The fish is sensitive to disturbance of the **electric field** surrounding it, a kind of radar system in fact.

LIGHT-PRODUCING ORGANS

Luminescence is common in marine organisms, particularly deep sea fishes. Such animals have light-producing cells which are under nervous control. Luminescence is brought about by oxidation of an organic compound, **luciferin**, catalysed by the enzyme **luciferase**. For this reaction, ATP is required.

Light-producing organs are sometimes quite elaborate. That of the shrimp has a reflecting layer behind the light-producing cells, and a special lens for concentrating the light. The function of luminescence is not always clear, though in many species it probably brings the sexes together for spawning.

THE NEMATOBLAST: AN INDEPENDENT EFFECTOR

The coelenterates (*Hydra*, sea anemones, jelly fish) possess specialized cells called **nematoblasts**. There are many different types, of which that illustrated in Fig 20.10 is but one. In all cases the cell contains a small vesicle, the **nematocyst**, containing a **coiled thread** which can be everted. Its function depends on the form of the everted thread, but in the one illustrated in

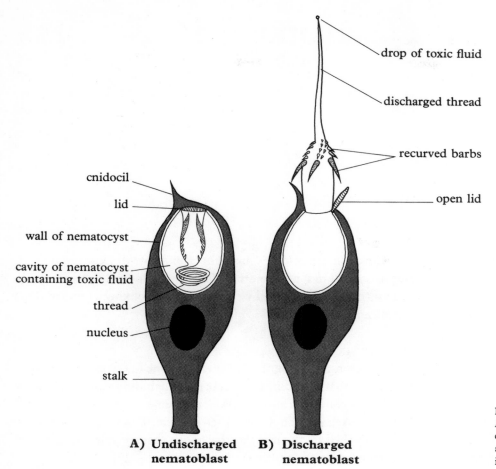

drop of toxic fluid

discharged thread

recurved barbs

open lid

cnidocil

lid

wall of nematocyst

cavity of nematocyst
containing toxic fluid

thread

nucleus

stalk

**A) Undischarged
nematoblast**

**B) Discharged
nematoblast**

Fig 20.10 The nematoblast (stinging cell) of *Hydra* is an independent effector. Its explosive discharge is triggered by bending of the cnidocil, aided by chemical stimulation. The response is independent of the nervous system.

Fig 20.10 it is to pierce and paralyse prey. It is believed that a small quantity of poison is injected into the victim from the distal end of the everted thread.

If a tentacle of a sea anemone is stimulated electrically through micro-electrodes, nematocysts are discharged only in the immediate vicinity of the electrodes. There is no evidence of their being excited as a result of impulses reaching them through the nervous system. Nematoblasts are therefore **independent effectors**: each contains within itself both the sensory and motor apparatus required to bring about discharge. It is thought that the hair-like **cnidocil** functions as the sensory trigger.

What sort of stimulus excites the nematoblast? **Mechanical stimulation** alone is ineffective: if a tentacle is stroked with a smooth glass rod only a small proportion of the nematoblasts discharge, though the number may be increased if the stimulating surface is roughened. **Chemical stimulation** by itself is also ineffective: saliva mixed with the sea water induces little or no discharge. However, mechanical accompanied by chemical stimulation evokes a massive response: thus a glass rod that has been dipped in saliva causes the discharge of far more nematoblasts than can be excited by mechanical or chemical stimuli alone. This is an experiment you can easily do for yourself in the laboratory.

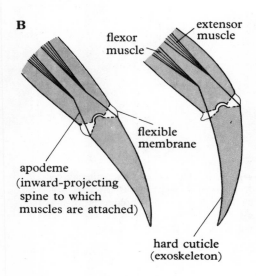

A

pectoral girdle

flexor muscle
(biceps)

humerus

extensor muscle
(triceps)

radius & ulna

B

extensor
muscle

flexor
muscle

flexible
membrane

apodeme
(inward-projecting
spine to which
muscles are attached)

hard cuticle
(exoskeleton)

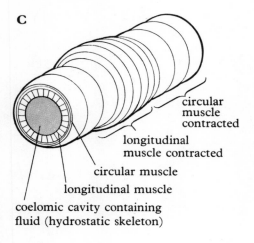

C

circular
muscle
contracted

longitudinal
muscle contracted

circular muscle

longitudinal muscle

coelomic cavity containing
fluid (hydrostatic skeleton)

Fig 21.1 Diagrams comparing the arrangement of the muscles and skeleton in a vertebrate, an arthropod, and a soft-bodied invertebrate. A) the human arm; B) the distal joint of an insect leg; C) an earthworm.

21 Locomotion

With few exceptions locomotion is brought about by the contraction of **muscles** against some kind of **skeleton**. The **musculo-skeletal system** is therefore the basis of locomotion and we must start by looking at its arrangement in different animals.

SKELETON AND MUSCLES

Three types of skeleton are recognized (Fig 21.1). The one you are probably most familiar with is that found in all vertebrates, including yourself. This is the **endoskeleton,** so-called because the skeletal elements, bone or cartilage, are internal to the muscles which are attached to them. Thus in our own limbs the bones are ensheathed by the muscles that move them (e.g. in the arm, the biceps and triceps muscles are external to the humerus – Fig 21.1A).

A rather different arrangement is found in arthropods (insects, lobsters, etc). Here the hard cuticle performs the function of a skeleton which, as it lies outside the muscles, is called an **exoskeleton.** For example, the flexor and extensor muscles of an insect's leg are enclosed within the box-like exoskeleton to which they are attached (Fig 21.1B). Despite this structural distinction, both endo- and exoskeletons are jointed and operate on the same mechanical principles.

A totally different system is found in soft-bodied invertebrates like sea anemones and earthworms (Fig 21.1C). In these animals there is no hard skeleton at all, its place being taken by a fluid under pressure. This fluid functions as a **hydrostatic skeleton,** being surrounded by muscles which contract against it. In the earthworm circular and longitudinal muscles in the body wall contract against the coelomic fluid in the body cavity. Normally the coelomic fluid is kept under pressure, the shape and form of the body being maintained in much the same way that the shape of a balloon is maintained when blown up with air.

In the human arm the biceps and triceps muscles produce opposing effects, the biceps flexing the arm and the triceps extending it. For this reason these muscles are said to be **antagonistic** to one another. It is obviously necessary that when the biceps contracts, the triceps should relax, and vice versa. This is achieved by the synapses in the nervous system (see p. 275). These ensure that when nerve impulses are transmitted to the biceps muscle they are not transmitted to the triceps, and vice versa. This principle also applies to the flexor and extensor muscles in the arthropod limb, and the circular and longitudinal muscles in the body wall of the earthworm.

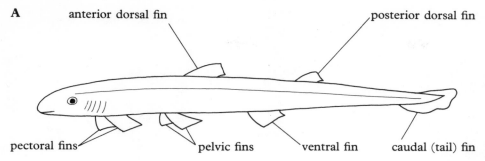

A

anterior dorsal fin   posterior dorsal fin

pectoral fins   pelvic fins   ventral fin   caudal (tail) fin

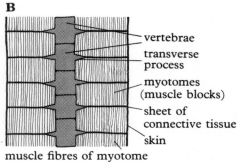

Fig 21.2 Structures involved in locomotion in the dogfish. A) external view of fish showing fins: the dorsal, ventral and caudal fins are all median, the pectoral and pelvic fins are paired. B) diagrams to show the operation of the myotome muscles in relation to the vertebral column in the swimming of a fish. The muscles contract alternately on each side of the body, bending the vertebral column from side to side.

B

vertebrae

transverse process

myotomes (muscle blocks)

sheet of connective tissue

skin

muscle fibres of myotome

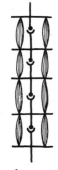

muscles equally contracted on both sides of vertebral column

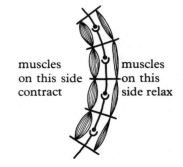

muscles on this side contract   muscles on this side relax

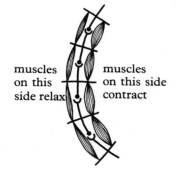

muscles on this side relax   muscles on this side contract

In considering the locomotion of any animal three things must be taken into account:

(1) **Propulsion:** the animal must be propelled in the appropriate direction.

(2) **Support:** the animal must be supported by its body acting against the particular medium in which it lives.

(3) **Stability:** the animal may become temporarily unstable while moving but eventually stability must be restored.

How these three problems have been solved depends on whether the animal moves in water, on land or in the air. We will therefore illustrate the basic principles by considering locomotion in fishes, land-living vertebrates and birds.

Movement in Water

Compared with air, water, particularly sea water, has a high relative density i.e. it is a comparatively viscous medium offering considerable resistance to the movement of objects through it. In the larger aquatic animals (fishes, porpoises etc.), this disadvantage has been overcome by streamlining. On the credit side water provides substantial support and is a tolerably thick medium on which propulsive devices can gain a purchase.

PROPULSION IN FISHES

In the dogfish (Fig 21.2A) propulsion comes from the side-to-side lashing of the tail which is equipped with a broad **caudal fin** for increasing the surface area. These movements are brought about by contraction of segmentally-arranged **myotome muscles** which stretch from the transverse processes of one vertebra, and the sheet of connective tissue continuous with it, to the next (Fig 21.2B). The tail sweeps from side to side by alternate contractions of the myotomes on each side of the body. Clearly the myotomes on the left and right sides are antagonistic, and reciprocal innervation is required to coordinate them.

When a dogfish is swimming, forces are set up as shown in Fig 21.3A. As the tail sweeps across towards the right it pushes against the water, as a result of which it experiences a force which can be resolved into a forward and sideways component. The forward component is responsible for driving the fish through the water. The sideways component will tend to swing the tail towards the left and the head towards the right. This **lateral drag** is counteracted by the pressure of water against the head and anterior dorsal fin. The

A) Carangiform locomotion e.g. dogfish

B) Anguilliform locomotion e.g. eel

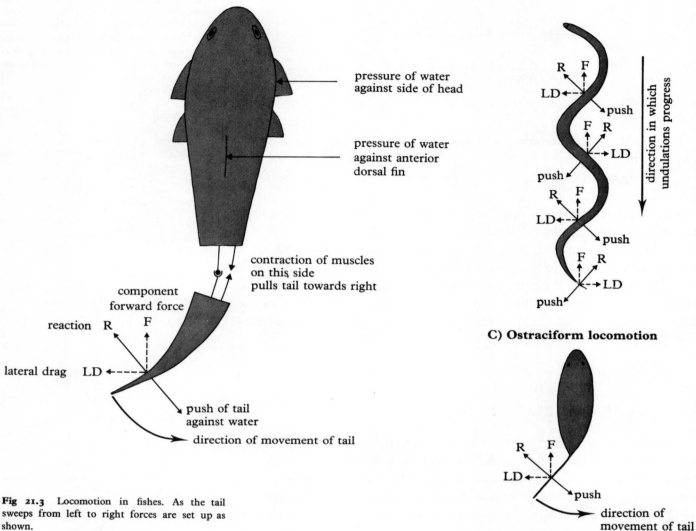

pressure of water
against side of head

pressure of water
against anterior
dorsal fin

contraction of muscles
on this side
pulls tail towards right

component
forward force

reaction R F

lateral drag LD

push of tail
against water

direction of movement of tail

direction in which
undulations progress

C) Ostraciform locomotion

direction of
movement of tail

Fig 21.3 Locomotion in fishes. As the tail sweeps from left to right forces are set up as shown.

result is that the fish moves forward without swinging from side to side to any great extent.

In the dogfish the posterior half of the body lashes from side to side (**carangiform locomotion**). Most other fishes use basically the same method of propulsion though different amounts of the body may be involved. In eels the entire body is thrown into lateral undulations which, progressing from front to rear, exert forces similar to those exerted by the dogfish's tail. This is called **anguilliform locomotion** (Fig 21.3B). On the other hand there are a few fishes, such as the trunk fish *Ostracion*, in which propulsion is achieved by a very small tail flapping vigorously from side to side (**ostraciform locomotion**). In all these cases the mechanical principles are the same. The same principles also apply to invertebrates such as roundworms which move by lashing their bodies from side to side, and also to spermatozoa and flagella.

This method of locomotion is by no means confined to aquatic animals. Snakes move by lateral undulations which pass along the body from front to rear as in eels. So long as there are solid objects like stones or tufts of grass that the body can push against, the animal can move forward. A striking demonstration of this is seen in eels. An eel placed on a flat slippery surface

A) Dogfish

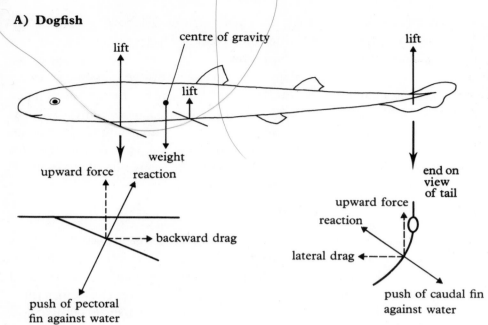

Fig 21.4 How lift is produced in fishes. In the dogfish lift is provided by the action of the paired fins, particularly the pectorals, and the caudal fin during active swimming. In bony fishes the air-filled swim bladder provides buoyancy, so lift does not have to be supplied by the fins. (*modified after Harris*)

B) Bony fish (teleost)

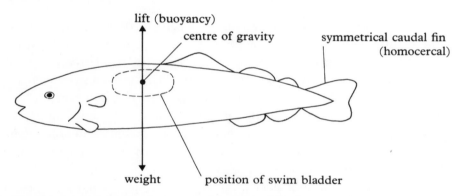

makes little or no progress, but if placed on a peg-board it progresses quickly and efficiently by gaining a purchase on the pegs. This enables eels to move overland in their journeys up rivers.

SUPPORT IN FISHES

A dogfish is heavier than water and sinks if it stops actively swimming, so support comes from the process of locomotion itself. There are two mechanisms, one dealing with the anterior end, the other with the posterior end. (Fig 21.4A). Anteriorly, support is derived from the large **pectoral fins**, and to a lesser extent the pelvics, which are held at a slight angle to the body. So long as the fish is moving forward these fins experience a force which can be resolved into upward and backward components. The upward component provides support (**lift**), keeping the anterior end up in the water. The backward component, (**backward drag**) though a nuisance, is normally far less than the forward force exerted by the tail and therefore does not seriously impede the fish's progress through the water.

Support at the posterior end depends on the structure of the tail. If you look at the tail in Fig 21.4A you will notice that there is more caudal fin ventral to the body then dorsal to it. This kind of fin is described as **heterocercal**, and it generates an upthrust as well as a forward force as it sweeps from side to side. As the tail moves through the water the ventral lobe of the caudal fin, being flexible, trails behind. The force resulting from the pressure of the trailing fin on the water can be resolved into upward and sideways components. The latter (lateral drag) is counteracted by the pressure of water against the head and dorsal fin. Clearly the lift achieved by the caudal fin is dependent on the flexible ventral lobe: if this lobe is removed from the tail of a dogfish in an aquarium, the fish cannot get off the bottom of the tank. In large cartilaginous fishes like sharks, the upward force is increased by the tail plunging downwards at the end of each sweep. This can be seen clearly in slow-motion ciné films of sharks swimming under water.

BUOYANCY IN BONY FISHES

Bony fishes like the trout have an air-filled **swim bladder** to keep them up in the water. This has far-reaching effects on their external structure. Not having to provide lift, the caudal fin becomes symmetrical above and below the body, the so-called **homocercal** condition (Fig 21.4B). The pectoral fins are also different: freed from having to provide lift, they become small fan-like appendages situated on either side of the body. They can be pulled into the side or stuck out at will, and are used for steering and braking.

There are two types of swim bladder, open and closed. The **open type** is connected to the pharynx by a duct, air being taken in or expelled through the mouth. Thus by 'blowing bubbles' a goldfish can raise its relative density and sink to a lower level in the water. On the other hand, to occupy a higher level it must first swim to the surface and gulp air into its swim bladder. The **closed swim bladder** is more sophisticated. Gas, mainly oxygen, is secreted into it, or withdrawn from it, by special gas glands in its lining. In this way the density of the fish can be adjusted with the minimum disturbance to the animal, enabling it to keep at the required depth. The majority of advanced teleosts have swim bladders of this type.

How did the swim bladder originate? It is thought that it evolved in the fresh-water ancestors of modern bony fishes as an air-breathing device, a kind of primitive lung. Evidence suggests that it arose as an outgrowth of the pharynx, just as lungs do, and that it served as a supplementary respiratory surface enabling these primitive fishes to breathe air at times when the swamps and rivers in which they dwelt dried up. Some modern fishes still use their air sacs for respiration – the tropical lung fishes for example – but in most fishes it has migrated from the ventral to the dorsal side of the pharynx to become the hydrostatic swim bladder.

STABILITY IN FISHES

A fish is liable to three kinds of instability: yawing, pitching and rolling (Fig 21.5A). In the dogfish, **yawing**, the lateral deflection of the anterior part of the body resulting from the propulsive action of the tail, is counteracted by the general massiveness of the head and the pressure of water against the side of the body and the vertical fins. In many bony fishes these corrective forces are enhanced by lateral flattening of the body. **Pitching**, the tendency for the nose to plunge vertically downwards, is counteracted in the dogfish by dorso-ventral flattening of the body and the large flap-like horizontal fins. **Rolling**, rotation of the body about the longitudinal axis, is counteracted by both the vertical and horizontal fins acting like the stablilizers on ships.

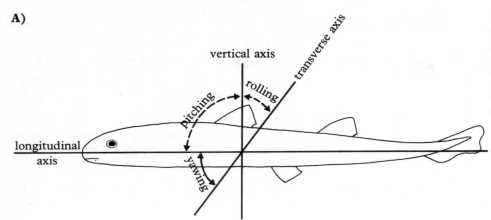

A)

Fig 21.5 Stability in fishes. A) illustrates the three types of instability to which the dogfish is liable: yawing, rotation about the vertical axis; pitching, rotation about the transverse axis; and rolling; rotation about the longitudinal axis. B) all three types of instability are counteracted by the combined actions of the paired and unpaired fins. (B *modified after Marshall and Hughes*)

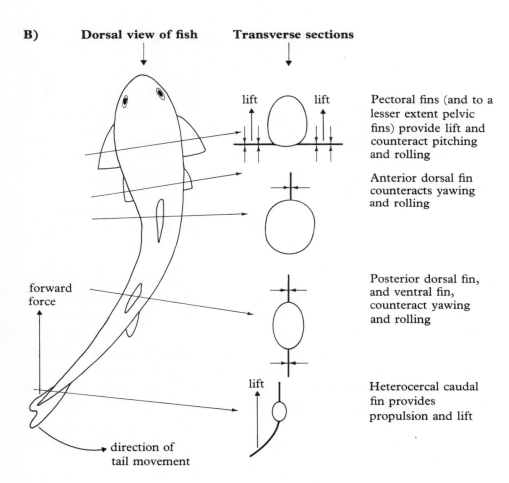

B)

Dorsal view of fish   **Transverse sections**

lift   lift

Pectoral fins (and to a lesser extent pelvic fins) provide lift and counteract pitching and rolling

Anterior dorsal fin counteracts yawing and rolling

forward force

Posterior dorsal fin, and ventral fin, counteract yawing and rolling

lift

Heterocercal caudal fin provides propulsion and lift

direction of tail movement

THE ACTION OF CILIA AND FLAGELLA

Flagella, and flagella-like structures such as sperm tails, achieve their propulsive action using principles similar to those involved in anguilliform locomotion: undulations pass along the flagellum from base to tip driving the organism in the opposite direction. Cilia, such as those of *Paramecium*, employ a different principle: each cilium, held out rigidly from the surface of the body, performs an effective backstroke, propelling the organism forward rather like the breaststroke of a swimmer. On completing its backstroke the cilium returns to its starting position, bending as it does so in order to minimise backward drag. Then, held out rigidly once more, it repeats its backstroke.

Movement on Land

The basic problems facing fishes also face tetrapods, but they have been solved differently. This is because tetrapods live in air, a much rarer medium than water. The only tetrapods that successfully use air to support their bodies and provide a basis for propulsion are birds and bats. All others use the ground for support and propulsion, the limbs playing a vital rôle in both.

THE MUSCULO-SKELETAL BASIS OF LOCOMOTION IN TETRAPODS

The locomotory machinery of the limbs consists of a series of **bones** which articulate smoothly at well lubricated **joints**. (Fig 21.6). The bones are con-

Fig 21.6 Two kinds of joint found in tetrapods: A) Ball and socket joint, in this case between femur and pelvic girdle, permits movement in any plane, including rotation; B) Hinge joint, eg. between upper and lower sections of a limb, permits movement in one plane only. Note that the basic structure of the joint is the same in both cases. The joint is enclosed within a fibrous capsule lined with a synovial membrane which secretes synovial fluid. The latter acts as a lubricant preventing friction at the joint when the bones move relative to each other. The articular cartilage covering the articulating surfaces, being relatively soft, serves as a cushion preventing jarring.

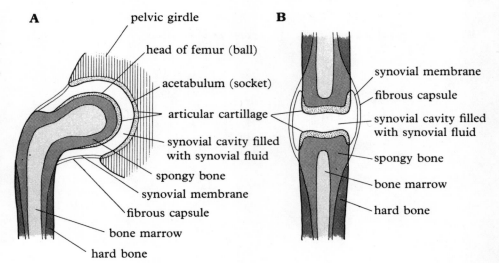

nected to one another by tough connective tissue **ligaments** which, being elastic, are admirably suited to bearing sudden stresses. They are composed of tightly packed bundles of **elastic fibres**. The bones are moved relative to one another by an elaborate system of **muscles**. The muscles are attached to the bones by **tendons**, tough connective tissue consisting almost entirely of **collagen fibres**[1]. The latter, though non-elastic, are capable of bearing sudden stresses because the muscles to which they are attached have considerable elasticity. All the muscles have their **origin** on one bone and their **insertion** on another situated further away from the centre of the body. Thus the muscles span one, or sometimes two, joints and when they contract they cause the bones to move relative to each other.

The action of the different muscles can be investigated in a pithed frog, i.e. one whose brain has been destroyed. The skin is stripped off one of the hind legs, and individual muscles are stimulated by a weak electric current delivered through a pair of needle electrodes. When the actions of the superficial muscles have been established these can be carefully dissected away and the ones underneath tested. In this way a full picture of the actions of the different muscles can be built up. Although the muscles are numerous and complex, they can be divided into seven groups according to the effects they produce (Fig 21.7).

(1) **Protractor muscles** pull the base of the limb forward.
(2) **Retractor muscles** pull the base of the limb backwards.
(3) **Adductors (depressors)** pull the limb inwards towards the body.

[1] The microscopical structure of connective tissue and bone is described on pp. 36–8.

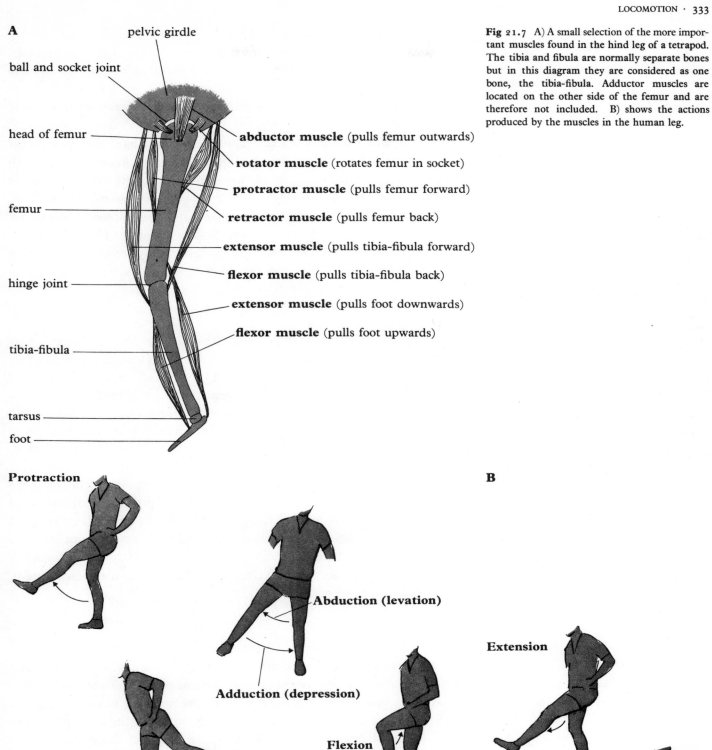

A

pelvic girdle

ball and socket joint

head of femur

femur

hinge joint

tibia-fibula

tarsus

foot

abductor muscle (pulls femur outwards)

rotator muscle (rotates femur in socket)

protractor muscle (pulls femur forward)

retractor muscle (pulls femur back)

extensor muscle (pulls tibia-fibula forward)

flexor muscle (pulls tibia-fibula back)

extensor muscle (pulls foot downwards)

flexor muscle (pulls foot upwards)

Fig 21.7 A) A small selection of the more important muscles found in the hind leg of a tetrapod. The tibia and fibula are normally separate bones but in this diagram they are considered as one bone, the tibia-fibula. Adductor muscles are located on the other side of the femur and are therefore not included. B) shows the actions produced by the muscles in the human leg.

Protraction

Retraction

Abduction (levation)

Adduction (depression)

Flexion

B

Extension

Rotation

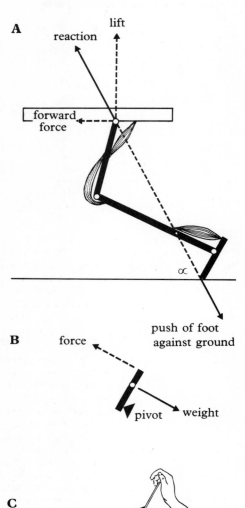

Fig 21.8 Diagrams to illustrate the action of the hind limb of a tetrapod in propelling the body forward. A) Model of the limb, showing the retractor and extensor muscles; B) the foot acting as a second class lever, C) an analogy for the lever action of the foot: it is like prizing open the lid of a box.

(4) Abductors (levators) pull the limb outwards away from the body.

(5) Rotators swivel the whole, or part, of the limb at one of its joints.

(6) Flexors pull two parts of the limb towards each other, i.e. they close joints.

(7) Extensors pull two parts of the limb away from each other, i.e. they open joints.

Some of the above muscles are **extrinsic**, that is they have their origin outside the limb itself, i.e. on the limb girdles. The rest are **intrinsic**, having both their origin and insertion within the limb. Clearly the different sets of muscles oppose one another: thus protractors are **antagonistic** to retractors, adductors to abductors, and flexors to extensors. The different muscles are supplied by axons from the limb nerve, the brachial nerve in the case of the forelimb, sciatic nerve in the case of the hind limb. By channelling impulses down the appropriate axons, the central nervous system ensures that each set of muscles contracts at the right time.

PROPULSION IN TETRAPODS

In propelling the body forward the most important muscles are the retractors and extensors. When they contract the limb acts as a **lever** (Fig 21.8). The foot presses downwards and backwards against the ground, resulting in an equal and opposite force which is transmitted along the length of the limb against the body. This force can be resolved into a vertical and horizontal component. Both are obviously useful, the former for lifting the body off the ground, the latter for propelling it forward. The relative magnitude of these two forces will depend on the angle between the ground and the main axis of the limb (the angle α in Fig 21.8A). If this angle is 90°, and the point of contact of the foot with the ground is immediately beneath the centre of gravity, there will be an upward force but no forward force, with the result that the body is thrust vertically upwards. On the other hand if the angle is a small one, and the foot is a long way behind the centre of gravity, the forward force will be considerable but the upward force relatively small. This is what a sprinter aims to achieve as he takes his position at the starting post.

SUPPORT IN TETRAPODS

In tetrapods little or no support is afforded by the medium (air) so the limbs must hold the body off the ground. In so doing they act as **struts**. For maximum efficiency the struts should be as straight as possible. To appreciate this we have only to compare a mammal with a more primitive tetrapod such as a lizard (Fig 21.9). In the latter the limbs splay out from the body giving the animal a bow-legged appearance. Holding the body off the ground necessitates the contraction of powerful adductor muscles running from the ventral surface of the limb girdle to the proximal limb bone. One cannot help thinking that walking for a heavy reptile like a crocodile must be a great strain, like 'push-ups' for a human. Little wonder that such animals spend much of their time lying on their bellies. In mammals this problem has been overcome by bringing the limbs into a straight line immediately beneath the body so that the weight is transmitted along what are essentially four straight struts. Predictably, the adductor muscles of the forelimbs are considerably reduced, as is the ventral part of the limb girdle.

In most tetrapods it is necessary for the **vertebral column** (Fig 21.10) to

A) Primitive stance

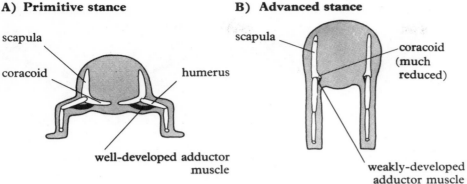

scapula
coracoid
humerus
well-developed adductor muscle

B) Advanced stance

scapula
coracoid (much reduced)
weakly-developed adductor muscle

C

bridge the gap between the fore and hind limbs. The tetrapod skeleton can, in fact, be looked on as a bridge in which the limbs represent the piers and the vertebral column the span. What kind of bridge is it? This question was investigated some years ago by the great zoologist D'Arcy Thompson. On the basis of mathematical considerations he came to the conclusion that the backbone is comparable to a **cantilever bridge**, like the Forth railway bridge in Scotland. His argument was based on the fact that both are made up of interconnected **compression** and **tension members** (Fig 21.11). In the backbone the chain of **vertebrae** pushing against each other is equivalent to the horizontal compression member of the bridge. The **neural spines** projecting from each vertebra represent the oblique compression members. In the backbone the vertebrae are bound together by a complicated set of **ligaments**. Some of these run between the neural spines and others connect the various

Fig 21.9 Diagrammatic end-on views of a primitive and an advanced tetrapod to show their respective stances. Note that the primitive condition requires the development of well-developed adductor muscles to keep the body off the ground. C) is a photograph of a rock lizard showing the primitive splayed-out position of the limbs. (*Zoological Society of London*)

A

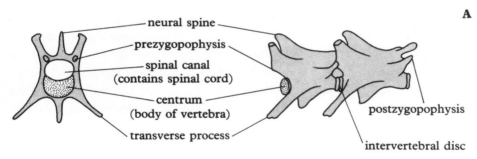

neural spine
prezygopophysis
spinal canal (contains spinal cord)
centrum (body of vertebra)
transverse process
postzygopophysis
intervertebral disc

Fig 21.10 Construction of the mammalian vertebral column. A) Diagram of lumbar vertebrae. The prezygopophyses point upwards and inwards and fit under the postzygopophyses which point downwards and outwards. Between successive centra are pads of fibro-cartilage which prevent jarring. The articulation between vertebrae is such that there is little movement between them. The various processes projecting from the vertebrae (e.g. neural spines and transverse processes) are for the attachment of muscles and ligaments. B) Photograph of lumbar region of vertebral column of rabbit showing how the vertebrae fit together. (*B. J. W. Heath, Marlborough College*)

dorsal projections with transverse processes. These two sets of ligaments represent the horizontal and oblique tension members respectively.

An important property of a cantilever is that if it is severed in the centre of one of its arches it will not collapse. This is because the unit of construction is not two piers connected by a span, but a single pier with half a span on each side of it. (This is why a cantilever bridge is also known as a bracket bridge.) The importance of this to a tetrapod is that the vertebral column in the neck can support a heavy head without collapsing.

Although certain details of D'Arcy Thompson's analysis are open to question the basic idea is sound, and it helps us to understand the functional reasons underlying the construction of the tetrapod skeleton.

STABILITY IN TETRAPODS

For a tetrapod stability is a major problem. At rest the problem is comparatively slight for the four legs are planted, fairly and squarely, on the ground. The animal is therefore like a four-legged table, the centre of gravity falling some-

B

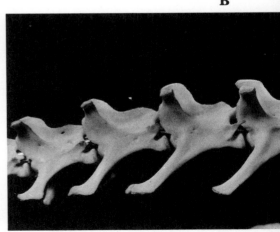

Fig 21.11 The tetrapod skeleton as a cantilever bridge. The centra and neural spines of the vertebral column are equivalent to the compression members in the bridge (thick, dark grey lines), the binding ligaments to the tension members (thin black lines). Can you detect any flaws in this analogy?

cantilever bridge

tetrapod skeleton

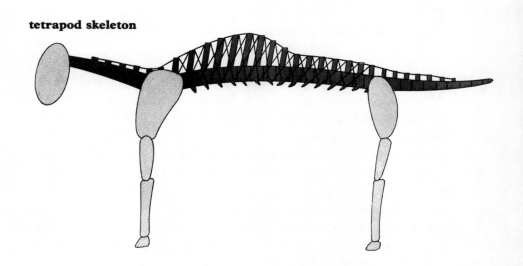

where inside the area delineated by its four legs (Fig 21.12A). But when it is in motion problems arise because periodically at least one leg must be taken off the ground. When this happens the tetrapod changes to being a **tripod**. If a tetrapod is to remain stable when it takes a foot off the ground, it must first shift its centre of gravity into the triangle delineated by the three legs that remain in contact with the ground. If this fails to happen the animal will topple over.

The sequence of events that occurs during tetrapod locomotion is based on this simple fact of mechanics. Imagine a slow-moving tetrapod with its legs in the position shown in Fig 21.12B-1. In order to progress forward the first thing it must do is to move its weight forward so that its centre of gravity falls within the triangle delineated by the right fore, right hind, and left fore limbs (Fig 21.12B-2). It then lifts the left hind limb and brings it up behind the left fore limb (Fig 21.12B-3). It then lifts its left fore limb and places it out in front (Fig 21.12B-4). The centre of gravity is now shifted into the triangle delineated by the left fore, left hind, and right fore limbs (Fig. 21.12B-5). The right hind limb can now be raised and brought up behind the right fore limb. The animal thus progresses in a **diagonal pattern** in which the order of leg-raising is left hind, left fore, right hind, right fore, and so on.

This diagonal sequence is very clearly seen in primitive slow-moving tetrapods like salamanders and newts. In mammals mechanical stability is augmented by reflexes arising from the **vestibular apparatus** in the inner ear (see p. 313). This enables such animals to maintain equilibrium when two, three, or even all four legs are taken off the ground simultaneously. But even in a galloping horse, as slow-motion cinéphotography shows, the diagonal pattern

A) Stationary

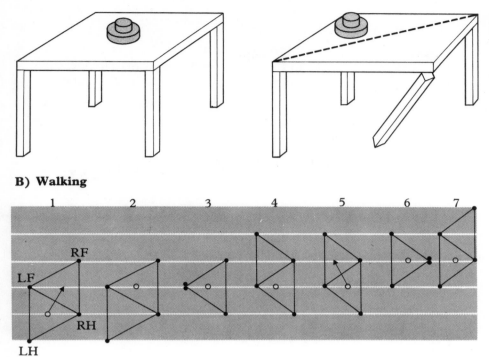

B) Walking

Fig 21.12 The problem of stability in a tetrapod. A) A stationary tetrapod can be looked upon as a table. If all four legs are in contact with the ground, the centre of gravity can fall anywhere within the area delineated by the four legs. If one leg is taken off the ground the table will topple over unless the centre of gravity is shifted so that it falls inside the triangle delineated by the three legs that remain in contact with the ground.   B) Successive stages in the diagonal locomotory pattern of a walking tetrapod. The order in which the feet are moved is: left hind, left fore, right hind, right fore. Note that a foot can only be raised if the centre of gravity (represented by the circle) is first shifted into the triangle delineated by the three feet which remain in contact with the ground.

can still be detected, the legs being slightly out of phase with each other during the locomotory cycle.

Plainly the position of the centre of gravity plays an important part in an animal's life. In an animal like a horse the centre of gravity lies towards the front of the body so that the animal can raise one of its hind legs with no risk of instability. But if it is to raise one of its fore legs it must first move its weight back so that its centre of gravity is nearer the hind legs. In rabbits, squirrels, bears and kangaroos the reverse is true. Here the centre of gravity lies towards the rear so that the fore legs can be lifted off the ground without loss of stability. This enables such animals to sit on their haunches, a useful thing to be able to do, for it enables the animal to survey its surroundings from an elevated vantage point. In the kangaroo the centre of gravity is so far back that it would fall over backwards were it not for its strong muscular tail acting as a support. In this way the kangaroo can rest on a tripod whilst surveying its surroundings (Fig 21.13). The ability to stand permanently on the hind legs (bipedalism) has been an important development in the evolution of man.

Fig 21.13 The Red Kangaroo (*Megaleia rufa*) uses its strong muscular tail to provide support when in the upright position. (*Australian News and Information Bureau*)

Movement in Air

As any aeronautical engineer would admit, the technical problems connected with movement in air are considerable. This is largely due to the rarity of the medium which provides little support and no purchase for the propulsive devices. For these reasons the principles involved in movement through air are rather different from those involved in aquatic locomotion.

Active flight has been successfully developed in only three groups of animals: birds, bats and insects. In all three cases the flight mechanism depends on the possession of **wings** which act as **aerofoils**.

FLIGHT IN BIRDS

When an airstream flows over an aerofoil the latter experiences a force at approximately right angles to the undisturbed airstream. To illustrate this, consider a bird such as a gull which has been actively flying in still air, and then **glides**. Although the air is still, the forward motion of the bird creates a flow of air over the wings, as a result of which the forces shown in Fig 21.14

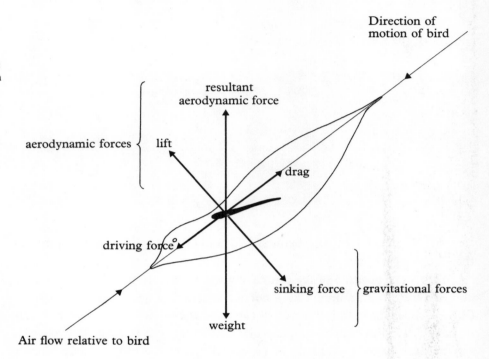

Fig 21.14 Diagram illustrating the forces operating in a bird gliding in still air. Explanation in text. *(modified after Gray)*

are developed. As a result of these forces the bird does not drop like a stone but glides along an inclined pathway. The weight of the bird acting downwards can be resolved into two components: a **sinking force** and, at right angles to this, a **driving force** propelling the body obliquely downwards. Assuming that the bird is moving at constant speed, these two forces are opposed by equal forces acting in the opposite direction. The sinking force is opposed by a **lift force** acting obliquely upwards, and the driving force by a **drag force**. The resultant of these two is the **total aerodynamic force**, and is equal and opposite to the weight.

The speed of gliding depends on the weight of the bird and the size, and shape, of the wings. A heavy bird with small wings obviously glides faster than a light bird with large wings. The distance a bird can glide in still air depends on the height it starts gliding from and the angle of glide, i.e. the angle between its downward pathway and the horizontal. For an expert glider such as the albatross this angle is small and the bird can glide almost horizontally without losing much height. This is achieved by holding the wings at such an angle to the main axis (angle of attack) that drag is minimized and maximum lift is obtained.

Even the best gliders cannot maintain an absolutely horizontal pathway in still air, but if the air is rising the bird can maintain its level or even climb.

A

To maintain a horizontal level, the speed at which the air rises relative to the ground must equal the speed at which the bird drops relative to the air. If the air speed is greater than this the bird will be given so much lift that it gains height as well as moving forward.

Birds are constantly using upward air currents for gaining height. These upcurrents arise in several ways, for example when air, warmed by the earth's surface, rises and is replaced by cooler air from above (**thermal upcurrents**), or when horizontal wind hits a vertical obstruction such as a cliff (**obstructional upcurrents**). When gulls are gliding on the windward side of a cliff they are making use of such obstructional upcurrents.

In circumstances when little or no support can be gained from upward air currents, the same effect can be achieved by flapping the wings: **active flight**, as opposed to passive gliding. In this case the flapping of the wings creates an airflow over them which produces much the same system of forces as in gliding flight. At the completion of the downstroke the wings are returned to their original position, anterior edge first so as to minimize downward drag.

The wings are operated by powerful **depressor** and **levator muscles** rich in **myoglobin** (see p. 165). The way these flight muscles are attached to the skeleton is shown in Fig 21.15. The depressor muscle, responsible for the powerful downstroke runs from the underside of the **humerus** to the **sternum**: clearly when it contracts it will pull the wing downwards. The tendon of the

Fig 21.15 The structural basis of flight in birds. A) shows the skeleton of a pigeon with its deep keel on the lower side of the sternum   B) shows how the great flight muscles run from the humerus to the sternal keel. (A: *B. J. W. Heath, Marlborough College*)

B

foramen triosseum
through which tendon of levator
muscle passes to its insertion on
upper surface of humerus

scapula

humerus

coracoid

clavicle ('wish bone')

keel of sternum

levator muscle
(pulls humerus up)

depressor muscle
(pulls humerus down)

levator muscle has its insertion on the upper surface of the humerus whence it passes through the **foramen triosseum** (bounded by the scapula, coracoid and clavicle) and so to the sternum. In order to accommodate these two very large muscles the sternum of birds is greatly expanded and has a deep **keel** to increase its surface area.

FLIGHT IN INSECTS

Insect flight obeys much the same aerodynamic principles as bird flight, but its musculo-skeletal basis is quite different. Instead of being attached to the wings, the flight muscles are attached to the walls of the **thorax**. On contrac-

A

B

tion, these muscles pull the walls of the thorax, thereby altering its shape. How this brings about the flapping of the wings is shown in Fig 21.17. The base of each wing is attached both to the roof (**tergum**) and the side (**pleuron**) of the thorax. The wing's attachment to the roof (**tergal attachment**) is median to the attachment to the side (**pleural attachment**).

The thorax contains two sets of flight muscles: **dorso-ventral** muscles run from the roof to the floor of the thorax, and **longitudinal muscles** run from the anterior surface of the dome-like roof to the posterior surface. Now when the dorso-ventral muscles contract the tergal attachment of the wing is pulled downwards relative to the pleural attachment, with the result that the wing goes up. When the longitudinal muscles contract the concavity of the roof is increased and the tergal attachment rises relative to the pleural attachment, with the result that the wing goes down. Since the muscles responsible for these movements are not directly attached to the wings, they are known as **indirect flight muscles**. There is in addition a set of **direct flight muscles** attached to the base of the wing itself, and these are responsible for making adjustments to the wing stroke and folding the wings at rest. Although the indirect flight muscles might appear to be working at considerable mechanical disadvantage, a tiny contraction is sufficient to produce a sizeable movement of the wing tip. In the wasp, for example, the muscular contractions are miniscule but the wings move through a sector of 150°.

One of the remarkable things about insect flight is the frequency with which the wings can beat. Large insects like butterflies, locusts and dragonflies beat their wings comparatively slowly, i.e. at rates ranging from approximately 10 to 50 per second. In these cases **nerve impulses** are sent to the two sets of antagonistic muscles at the same frequency as the wing-beat. But many insects beat their wings much faster than this: the housefly at about 200/s, *Drosophila* at 250/s, and mosquitoes at up to 600/s. Certain midges have been shown to have a wing-beat frequency of over 1000/s!

This presents a problem, for in no known animal can impulses be transmitted to muscles causing them to contract and relax at such frequencies. How do insects manage this? The answer lies in the mechanical properties of the thorax. When one set of muscles contracts the resistance offered by the elasticity of the thorax gradually increases. However, there comes a point when the resistance of the thorax suddenly ceases and the wings click into their new position. This has the effect of releasing tension on the contracting muscle, and stretching its antagonist. As a result of this sudden stretching the latter is stimulated to contract. The result is rapid alternate contractions of the two opposing muscles, the contraction of one stimulating the other. The rate of oscillation is determined mainly by the mechanical properties of the thoracic wall.

Since the muscles stimulate each other there would seem to be no need for nerve impulses to be sent to them. In actual fact it has been found that in the blowfly impulses are delivered to the flight muscles at a frequency of about 3/s when the wings are beating at 120/s. In other words the muscles receive one impulse every 40 wing beats. Nerve impulses obviously initiate flight, after which occasional impulses are required to maintain the muscle fibres in an active state.

The ability to contract repetitively is also an intrinsic property of the muscle itself. This can be shown by adding ATP to glycerin-extracted insect

A) Side view of thorax showing wing attachment

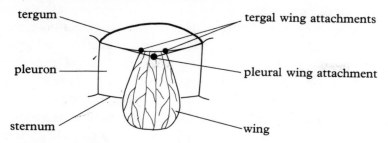

tergum

tergal wing attachments

pleuron

pleural wing attachment

sternum

wing

Fig 21.17 Flight in insects. When the dorso-ventral muscles contract, the wings go up; when the longitudinal muscles contract, they go down. This is because of the way the wing bases are attached to the tergum and pleuron. Further explanation on p. 340. *(Based on Ramsay)*

B) Side view of thorax showing flight muscles

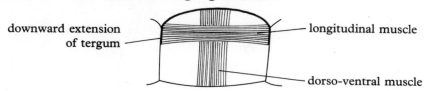

downward extension of tergum

longitudinal muscle

dorso-ventral muscle

C) Cross-section of thorax

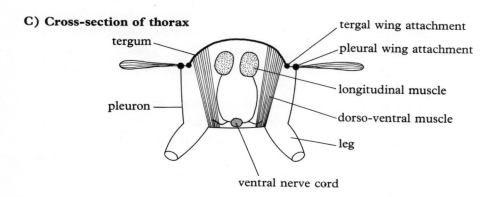

tergum

tergal wing attachment

pleural wing attachment

longitudinal muscle

pleuron

dorso-ventral muscle

leg

ventral nerve cord

flight muscle and ordinary muscle. The former gives repeated cycles of contraction, the latter only one contraction. Insects are therefore adapted for flight not only in the mechanics of their musculo-skeletal system but also in the organization of the nervous system and the physiology of their muscle.

Amoeboid Movement

Finally we turn to a type of locomotion which is quite different from the other methods discussed in this chapter, namely that displayed by amoeboid animals – *Amoeba* itself, white blood cells, phagocytes etc.

Amoeba consists of an elastic plasma membrane containing cytoplasm. The latter is differentiated into two regions, a fluid **endoplasm** in the centre surrounded by a stiffer region towards the periphery called the **ectoplasm**. When *Amoeba* moves, the fluid endoplasm flows forward inside the 'wall' of ectoplasm to form a temporary projection of the cell called a **pseudopodium**. When it reaches the leading end of the advancing pseudopodium, the endoplasm everts, rather like the folding-back of a cuff, and is somehow converted into the stiffer ectoplasm. At the other end of the cell the reverse happens. The ectoplasm *inverts* and is converted into fluid endoplasm. So *Amoeba* moves by a fluid core flowing forward through a tube of its own making.

How does this process take place? There are two theories to account for it. The older theory, dating back to the 1920s, proposes that the change of endoplasm to ectoplasm, and of ectoplasm to endoplasm, is simply the result of the cytoplasm changing from one colloidal state to another. At the leading end of a pseudopodium the cytoplasm changes from a fluid sol (endoplasm) to a more solid gel (ectoplasm). The endoplasm flows forward as a result of the ectoplasm at the rear end contracting, the whole thing functioning rather like the hydrostatic skeleton described on page 326. No one has ever proved this theory, but it was thought that the conversion of sol to gel was caused by protein molecules changing their shape: folding up in the sol condition and opening out in the gel condition. The actual folding process at the rear could provide the pressure forcing the endoplasm forward.

In 1961 Robert D. Allen of Princeton University put forward a totally different theory to account for amoeboid movement. On the basis of careful observations on the behaviour of the endoplasm both in intact amoebae and ruptured cells, he suggested that the protein molecules are extended in the endoplasm and folded in the ectoplasm, the reverse of the situation claimed by the old theory. Allen claims that forward movement is not caused by squeezing at the rear end, but by a pulling force at the front. This, it is believed, is achieved by the protein molecules of the endoplasm contracting at the front end before they pass into the ectoplasm.

It is not known for certain which of the two theories is correct, but the results of studies on the properties of the cytoplasm are more consistent with the second. The story illustrates the general point that at the cellular level we are a long way from understanding how movement takes place. Much the same applies to the movement of cilia and flagella.

22 Behaviour

When we talk of an animal's behaviour we mean all its varied activities; its movements, reactions, changes in posture, and so forth. In this chapter we shall look at some examples of behaviour and see to what extent they can be explained in terms of receptors, nerves and effectors.

A WORD ABOUT TECHNIQUES

In describing an animal's behaviour one must try to be as objective and accurate as possible. Difficulties may be encountered at once because behaviour patterns are often either slow, or else fast or repetitive. One must therefore resort to various recording techniques. **Kymographs** (see p. 315) may be used to make a graphic record of the muscular responses of simple animals like worms and sea anemones; **slow-motion cinématography** is used to record the activities of fast-moving animals like birds; and **time-lapse photography** is used for slow-moving animals such as sea anemones. For extremely rapid movements, such as jumping in salticid spiders, **multiple-flash photography** may be used.

To gain a full picture of an animal's behaviour it is often necessary to record its activities, which are frequently repetitive, over a long period of time. To do this by direct observation would drive all but the most persistent investigator insane, and therefore a variety of **activity recorders** are used. Ranging from simple mechanical gadgets to complex electronic devices, each is tailor-made to fit the particular animal under investigation. This is not to imply that direct observation is valueless; where feasible some of the best work still depends on it.

In studying behaviour one tries to keep the animal in as natural conditions as possible. The trouble is that in its normal environment the animal may be inaccessible and its behaviour obscured. On the other hand if it is brought into the laboratory its normal behaviour patterns may be hopelessly interfered with by captivity or by subjecting it to artificial experimental situations. This problem is immediately multiplied if the animal is operated on in any way, such as having recording electrodes stuck into its brain. A student of behaviour must therefore compromise between the inconvenience of studying the animal in its natural surroundings and the artificiality of subjecting it to the unnatural conditions of the laboratory. Frequently he does first the one and then the other.

The ultimate aim is to explain the activities of an animal in terms of its underlying neuro-sensory mechanisms. Unfortunately the complexity of the sensory and nervous systems, coupled with limitations in physiological techniques, makes it impossible to do this except in a few simple cases. It is often necessary, therefore, to resort to an indirect approach in which observed

Fig 22.1 Courtship dance of the wandering albatross, South Georgia. At this stage in the mutual performance excitement mounts and the male raises his wings, points his head vertically upwards, and shrieks. The female (left) looks on. Later she, too, raises her wings, and both birds approach one another slowly until their bills almost touch. (*Alfred Saunders, courtesy of Frank Lane*)

behaviour patterns are interpreted in terms of hypothetical processes taking place in the nervous system. In the last analysis only the tools of the neurophysiologist can show whether or not these hypothetical processes are correct.

TYPES OF BEHAVIOUR

Behaviour can be divided into that which is shown by all members of a species (**species-characteristic behaviour**) and that which varies from one individual to another (**individual-characteristic behaviour**). The former includes the stereotyped behaviour patterns distinctive of particular species, for example courtship and copulation of many animals (Fig 22.1). The latter include the behaviour learned by an animal during its lifetime; for example the 'tricks' performed by individual dogs. This does not mean, however, that all learned behaviour is individual-characteristic, or that all so-called instinctive behaviour is species-characteristic.

It must be made clear at the outset that rigid classifications have no place in behaviour. An act of behaviour, however simple or complex, must be seen as the result of an interaction between an individual's genetic constitution and its environment. We take it for granted that structural development results from these dual forces (see Chapter 31); the same is no less true of the development of behaviour.

Certain fundamental processes enter into an animal's behaviour. These include **reflex action, orientation** and **learning** which, for purposes of analysis, will be examined separately in the course of the chapter.

Reflex Action

A reflex action is a simple act of behaviour in which some kind of stimulus evokes a specific, short-lived response. We have already discussed the physiological basis of reflexes in Chapter 18. Here we are more concerned with the way they fit into the behavioural repertoire of animals in their natural environment.

ESCAPE RESPONSES

As an example of reflex action let us take the escape reactions found in many invertebrate animals. These responses bring about rapid withdrawal of the animal from a harmful, or potentially harmful, stimulus. Thus a squid and a crayfish both dart backwards, the crayfish by rapidly flexing its abdomen, the squid by squirting a jet of water through its funnel; an earthworm quickly withdraws into its burrow and a fanworm jerks back into its tube, and so on. In all these examples the escape response is mediated by a reflex arc in which stimuli are received by receptors, impulses are transmitted through the nervous system and a contraction occurs in the appropriate muscles. The rapidity of the response is caused partly by the speed at which the muscles contract, but also by the high speed at which the impulses are transmitted through the nervous system. This high transmission velocity shortens the delay between stimulation and the onset of the muscular response, and it results in a synchronous contraction of all the muscle fibres.

ANALYSIS OF THE EARTHWORM'S ESCAPE RESPONSE

To appreciate the rôle of such an escape response in the general behaviour of the animal consider the earthworm, an animal which can be easily observed in its natural environment and in the laboratory. We will analyse the earthworm's escape response in some detail because it provides a good example of behaviour that can be explained in terms of receptors, nerves and effectors.

If you go into the garden on a warm, wet night you will see hundreds of worms lying on the surface of the ground (Fig 22.2). However, if you touch one it quickly disappears into its burrow. Even slight vibration of the ground is enough to cause this to happen. The tail of the worm normally remains in the opening of the burrow. On stimulation, the powerful **longitudinal muscle** in the body wall (Fig 22.3) contracts rapidly and, at the same time, the bristle-

Fig 22.2 An earthworm lying on surface of ground at night with its tail tucked into its burrow. (*flash photograph by B. J. W. Heath, Marlborough College*)

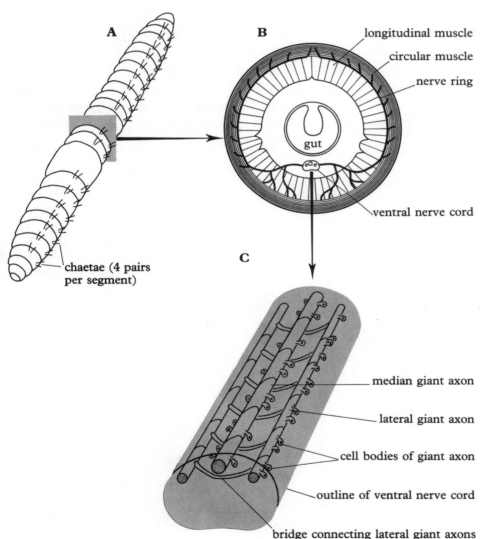

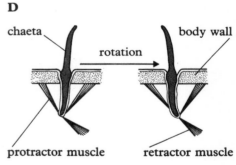

Fig 22.3 Structures involved in the escape response of the earthworm. A) Whole worm showing bristle-like chaetae. (In reality there are many more segments than are shown.) B) Transverse section showing ventral nerve cord and peripheral nerves serving the body wall muscles. C) Part of ventral nerve cord showing the giant axons which transmit impulses rapidly to the longitudinal muscle and chaetal muscles. D) Chaetae before and after rotation.

like **chaetae** are protracted from muscular sacs in the body wall. Contraction of the longitudinal muscle causes the whole body to shorten, protraction of the chaetae gives the tail a firm grip on the wall of the burrow into which the worm retracts. This grip is enhanced by dorso-ventral flattening of the extreme posterior tip of the body which has the effect of pushing the sharp chaetae into the sides of the burrow. But this is not all. You will see from Fig 22.3D that the chaetae are hooked. When the anterior end of the worm is touched the tips of the chaetae all point towards the head. This gives the worm a better grip and prevents it being pulled out of its burrow as it might well be when pecked at by a bird. If the tail end is stimulated the chaetae rotate so that they point backwards. This gives the head end a better grip if the worm is burrowing and the tail is seized by an aggressor. These movements are brought about by contraction of an elaborate set of muscles attached to the inner end of the chaetae.

The escape response of the earthworm is clearly adaptive to the circumstances in which it lives. Moreover, in this case, as in many other simple reflexes, it is possible to explain the response in terms of its underlying neural mechanism. It is mediated by a system of **giant axons** in the nerve cord which transmit impulses at high speed from one end of the body to the other (see p. 269). There are three such axons, a median one and a pair of laterals. The latter are interconnected at intervals by 'bridges' which pass beneath the median axon (Fig 22.3C). That these giant axons transmit impulses which evoke the rapid response has been shown by transecting them with a very fine blade. When this is done the rapid contraction is blocked at that point, as are the impulses associated with it. If the median giant axon alone is cut the rapid response evoked by stimulating the head end is confined to the part of the body in front

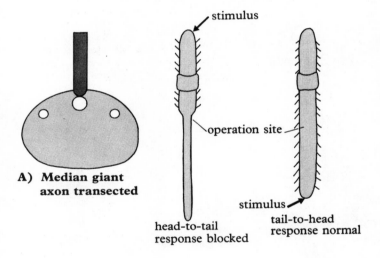

stimulus

operation site

A) Median giant axon transected

head-to-tail response blocked

stimulus

tail-to-head response normal

Fig 22.4 Establishing the neural basis of the earthworm's escape response involves transecting the giant axons.   A) If the median axon is transected the head-to-tail response is blocked.   B) If the lateral axons are transected the tail-to-head response is blocked.   C) Since all three giant axons transmit impulses equally well in either direction, the explanation must lie in the fact that the median axon is connected to touch receptors at the anterior end of the body, whereas the lateral axons are connected to touch receptors at the posterior end.

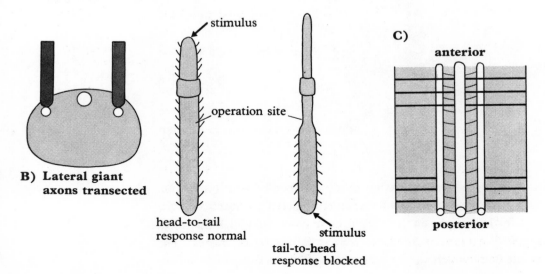

stimulus

operation site

B) Lateral giant axons transected

head-to-tail response normal

stimulus

tail-to-head response blocked

C)

anterior

posterior

of the operation site. However, the response evoked by stimulating the tail end still occurs uninterrupted throughout the length of the body. If the lateral giant axons alone are cut, leaving the median intact, the 'tail-to-head' response becomes blocked at the operation site while the 'head-to-tail' response remains unaffected (Fig 22.4). An obvious explanation of these results is that the giant axons transmit impulses in one direction only, the median transmitting impulses from head to tail and the laterals from tail to head. However, stimulating one end of an isolated nerve cord with electrical shocks and recording action potentials from the other end shows that the giant axons can transmit impulses in either direction. We must therefore turn to an alternative explanation, namely that the median giant axon is connected to receptors towards the head whilst the lateral giant axons have their sensory connections at the tail end of the body. Thus a touch applied to the head will activate the median axon, whereas touching the tail end will activate the laterals. There is now ample evidence that this is in fact the correct explanation, though no-one has verified it histologically.

ESCAPE RESPONSE OF THE SQUID

The effectiveness of giant axons in mediating escape responses is that, owing to their rapid transmission, all the muscle fibres contract synchronously. In the squid this synchronization has been achieved in an ingenious way. In this animal's response the circular muscles of the body wall (mantle) undergo a sudden powerful contraction, resulting in water being ejected through the funnel with great force so that the animal is propelled backwards. The nerves supplying the body wall muscles radiate from a pair of ganglia situated at the anterior end of the mantle cavity (Fig 22.5). Each nerve contains a giant axon which innervates the circular muscle. Now it has been shown that the diameter of each axon varies with its length, i.e. with the distance the impulses have to travel before they reach the muscle. The shorter axons supplying the anterior end of the mantle are relatively thin; the longer axons supplying the more posterior parts of the mantle are thicker and have correspondingly higher transmission speeds. The result is that all the impulses arrive at the muscles simultaneously despite their having to travel different distances before doing so. All the mantle muscles are thus made to contract synchronously.

In this section escape responses have been discussed at some length because they illustrate how relatively simple acts of behaviour can be analyzed and explained in terms of their underlying neural mechanisms. Sadly, this is not so easy to achieve with more complex forms of behaviour, as we shall see.

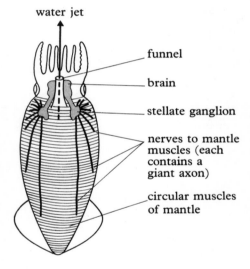

Fig 22.5 Neural basis of the squid's escape response. Giant axons, supplying the circular muscles of the mantle, radiate from a pair of stellate ganglia. When impulses reach the ends of the giant axons they initiate a synchronous contraction of the circular muscle which forces a jet of water through the funnel, propelling the animal backwards.

Orientation

In so far as it involves responding to stimulation, orientation has much in common with reflex action, and there is no sharp distinction between the two. Indeed many orientation responses are reflexes. However, the response is generally more complex; a behaviour pattern rather than the simple, short-lived response typical of a reflex. In an orientation response the organism takes up a particular position in relation to a stimulus. This kind of behaviour is obviously important in the natural environment for it enables organisms to move towards desirable stimuli and away from harmful ones. In this way green flagellates are guided towards light, animals towards food, parasites towards their hosts, and spermatozoa towards the female's eggs.

KINESIS

The simplest kind of orientation behaviour, known as **kinesis,** is that which occurs when an animal is subjected to an unpleasant stimulus acting from no particular direction. Under these circumstances the speed of random locomotion increases. This results in the organism quickly getting back to a region where the noxious stimulus is less intense. The speed of movement then decreases so that it tends to remain in the more agreeable situation. In the total absence of noxious stimulation the organism may stop moving altogether. This kind of behaviour can be seen very clearly in woodlice (Fig. 22.6.) If woodlice are placed in a 'choice-chamber', half of which has a humid atmosphere, and the other half a drier atmosphere, the animals move much faster in the dry half than the humid half. As a result they eventually congregate on the humid side. In general the speed of movement is related to the intensity of the noxious stimulus.

Sometimes the rate of turning also plays an important part, though the precise way it operates varies from species to species. In general what may happen is this. If the organism crosses a sharp dividing line between a favourable and unfavourable environment, its rate of turning increases, thus raising its chances of a rapid return to the favourable area. However, if it penetrates far into the unfavourable zone its rate of turning may gradually decrease so that it moves in long straight lines before turning sharply. This increases its chances of re-entering the good area. When it does get back the organism generally stops turning altogether. However, in some cases the rate of turning may increase dramatically so that the organism moves this way and that, trapped, as it were, in a tight circle. This will tend to keep it in the favourable area. The importance of turning in orientation behaviour is well shown by free-living flatworms. These animals will glide about on the bottom of an enamel dish leaving a trail of sticky mucus wherever they go. If half the dish is covered and the other half lit, the worms eventually come to rest in the dark half. This is because the rate of turning increases in the illuminated half. The actual tracks of an animal can be revealed by removing the water from the dish and sprinkling fine charcoal on the bottom. On washing gently the charcoal sticks to the mucus trail.

TAXIS

If a stimulus comes from a specific direction the animal may orientate itself with respect to the source of the stimulus. Such a response is called a **taxis.** Well-known examples are the swimming of *Euglena* and other green flagellates towards light (**phototaxis**), and the movement of spermatozoa towards chemical substances secreted by eggs (**chemotaxis**).

To see the more complicated aspects of taxic responses we return to flatworms. If a small piece of liver is placed 100 mm from a hungry flatworm in a dish of water the animal glides towards the food. As it moves it waves its head from side to side and appears to compare the intensity of stimulation on the two sides of the head. Once it is within twenty or thirty mm of the food it

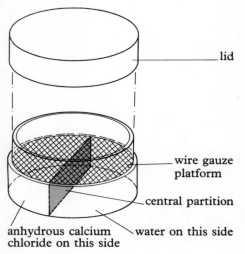

Fig 22.6 A choice-chamber suitable for experiments on orientation behaviour of woodlice. The water on the right of the central partition gives a humid atmosphere in that half of the chamber. The anhydrous calcium chloride gives a drier atmosphere on the left. The woodlice are placed on the wire gauze.

lid

wire gauze platform

central partition

anhydrous calcium chloride on this side

water on this side

approaches it in a straight path without turning its head. Under these circumstances it seems to be able to compare the intensities of stimulation on both sides simultaneously, and presumably moves in a direction which gives equal stimulation to the two sides.

In this kind of behaviour we can only surmise what is going on in the nervous system, but its adaptive significance is clear enough. It is in fact what a man would do if he had to find his way about using only his sense of smell. Imagine yourself, blindfolded, ears plugged, hands tied behind your back, in an empty room. A plate of roast beef is put down on a table at one end. How will you find it? Surely you will move towards that part of the room where the smell is strongest, and as you do so you will turn your head and sniff first on one side and then on the other, comparing the intensity on either side of you as you progress towards the food. When you get sufficiently close you find that you no longer need to move your head in order to locate the source of the smell.

Elaborate orientation responses are found in organisms as low in the evolutionary scale as the protists. In *Euglena* (Fig 2.19, p. 31) there is a light-sensitive swelling, **a photoreceptor**, at the base of the flagellum at the anterior end of the body. To one side of the swelling is an opaque pigment spot. *Euglena* rotates about its long axis as it moves forward. If light is shining from the side, the pigment spot periodically casts a shadow on the light-sensitive swelling. The flagellate responds by changing direction in such a way that the light is directed at it from in front so the pigment spot is uniformly illuminated all the time. By this means the organism is guided towards light.

Taxic responses of the sort just described are associated with lowly animals. With the development of a more complex CNS and more elaborate receptors, particularly eyes, environmental signs and landmarks can be used in guiding animals in their natural environment. Here orientation overlaps with learning for these signs can only be used if the animal has learned to associate them with a particular situation.

Fig 22.7 Sound spectrograms of chaffinch songs. A) and B) Normal full songs of two wild birds in an aviary, each showing individual characteristics which have been learned from other birds. C) The basic innate song pattern given by a bird which has been reared in isolation. Notice its simplicity compared with the learned patterns. (*from W. H. Thorpe:* Learning and Instinct in Animals, *Methuen, and* Bird Song, *Cambridge University Press*)

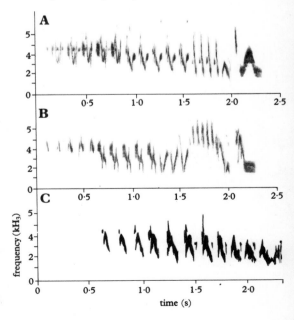

Analysis of Species-characteristic Behaviour

It has already been pointed out that species-characteristic behaviour is controlled by genes interacting with the environment. Some idea of the relative importance of genetic and environmental influences may be gained by studying the behaviour patterns of an animal reared in isolation. The animal is separated from others, including its mother, as soon as possible after birth. On reaching maturity its responses to stimuli, and its behaviour generally, are compared with those of animals reared normally.

The results show that certain acts of behaviour, particularly those associated with feeding, sex and protection, can still be performed in a reasonably normal manner. However, the behaviour is often less elaborate and more stereotyped than in normal circumstances. For example chaffinches reared in isolation sing a simpler song than those brought up with other adult birds (Fig 22.7). This simple song is the same in all members of the species, and in normal circumstances the refinements are learned from other birds, added on as it were. This kind of investigation shows that, although the rudiments of

Fig 22.8 Results of experiments showing that red colour is the most effective stimulus eliciting the begging response in young herring gulls. The strength of the response, as measured by the frequency of pecking, is represented by the length of the bar beneath each model. Note that an entirely red beak evokes an even greater response than the red spot. *(after Tinbergen and Perdeck)*

Model presented to chick

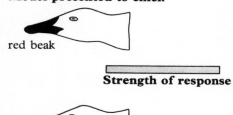
red beak

Strength of response

red spot

black spot

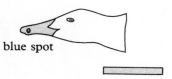
blue spot

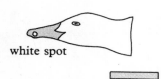
white spot

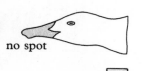
no spot

behaviour can develop in isolation, the full expression of a behaviour pattern may require contact with other individuals.

At this point a word should be said about **instinct**. This word is popularly taken to mean species-specific behaviour which is inborn, i.e. under genetic control, developing independently of the environment. But in fact no behaviour can be entirely free of environmental influences. It is not even valid to say that certain parts of a behaviour pattern are instinctive. We come back to the point made on p. 344, namely that a behaviour pattern is a single entity which must be looked upon as the result of an interaction between genes and environment. Even the basic song of the chaffinch, though plainly not learned from other chaffinches, cannot be regarded as independent of the environment, for it is impossible to deprive any animal reared in isolation of all external influences. To do so would require total sensory deprivation. For this reason the word instinct is of limited value in behaviour studies.

One of the simplest examples of species-characteristic behaviour is seen in feeding, which we can use to introduce the basic concepts. If a predatory animal which has not fed for some time, a tiger for example, senses the presence of food in its environment it responds by searching for it, and on discovering its whereabouts, it devours it. But the animal will only respond if it is hungry: i.e. it must be **motivated**. Indeed, if sufficiently motivated it may search before it senses the presence of food. The seeking out of the food involves orientation and various other behavioural processes; it is generally adaptable and may be modified by past experience, as can be abundantly seen in the hunting activities of many a predatory beast. The final consuming of the food, the 'goal' in the overall sequence of events, tends to be the least flexible part of the proceedings. It is usually stereotyped and relatively independent of learning processes.

The behaviour pattern just described depends on stimuli and motivation. We shall now examine the rôle of each of these in behaviour generally.

THE ROLE OF STIMULI

For convenience we may distinguish between three kinds of stimuli: those which determine the animal's state of responsiveness (called **motivational stimuli**), and those which elicit particular responses when the animal encounters them (called **releasing stimuli** or just **releasers**), and those which bring an act of behaviour to an end (**terminating stimuli**). In feeding behaviour the smell of food may raise the animal's state of responsiveness, may make it more 'conscious' of being hungry as it were, and therefore serve as a motivational stimulus. On the other hand the sight of the food may act as a releasing stimulus unleashing feeding behaviour. Finally a full stomach may act as a terminating stimulus bringing feeding behaviour to an end.

RELEASERS

In general a releaser is any feature of the environment which can be positively shown to evoke a behavioural response. In some cases, as in courtship, the releaser may be a feature possessed by another individual: its colour, shape or a particular marking.

One of the techniques used to investigate releasers is to make models of a stimulus object and present them to the animal in question. The models are modified in various ways so that, by elimination, those features which elicit a response can be determined. In this way it has been found that a stimulus object may possess a wide range of characteristics (shape, colour and so on) only some of which are effective as releasers. This has been investigated by Niko Tinbergen and his colleagues in the herring gull. Herring gull chicks are fed on fish regurgitated by their parents. To make the parent regurgitate, the chick pecks at its parent's bill. What is it about the bill that elicits this behaviour? The herring gull's bill is yellow with a red spot towards the tip of the mandible. Tinbergen made a series of cardboard models in which he varied the shape and colour of the head and bill, and the colour and contrast of the mandibular spot. Each model was then presented to the chicks in as natural a manner as possible, and their responses were measured in terms of the number of pecks made at it. The results showed that the shape of the head behind the bill plays no part at all in eliciting the response. The most important feature turned out to be the red colour of the mandibular spot. In fact the maximum response was evoked by a model whose entire bill was painted red (Fig 22.8).

The same kind of thing applies in **territorial behaviour**. In the male robin the red breast is the effective stimulus eliciting threat in other males. In the male stickleback the red belly serves much the same purpose and it has been shown that other features of the fish, such as its shape, play only a small part in fetching a response (Fig 22.9).

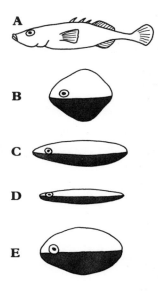

Fig 22.9 Models used by Tinbergen to elicit aggression in male sticklebacks. In each case black represents red. Models (B) to (E), though crudely shaped, provoked more attacks than the realistic model (A) which lacked the red belly. (*after Tinbergen*)

STIMULUS SELECTION

To respond to a specific stimulus there must be a mechanism in the receptors or brain which filters out the relevant features of the stimulus object from irrelevant ones. Because of this selection process the herring gull chicks respond to the colour of the mandibular spot but not, for example, to the shape of the head. The neuro-sensory process which does this selecting is called the **stimulus filtering mechanism** and, although it is clear that it must exist, we have only a rudimentary idea of what it entails in physiological terms.

THE FUNCTION OF RELEASERS

The value of releasers, however, is clear enough. They serve as signals initiating appropriate behaviour and coordinating interactions between individuals. This latter is of great importance because an animal's environment is necessarily composed of other animals, and it is desirable that the right kind of behaviour should take place between different individuals. Above all, open conflict should be avoided.

The importance of this can be seen in aggressive behaviour. Aggression often involves a **threat display** in which the animal adopts a certain posture, or reveals certain markings, which intimidate a rival who may have encroached upon the animal's territory or made advances towards its mate. Such threat displays are particularly common in fishes and birds. For example cichlid fish, which are very easy to keep and observe in a heated aquarium, open their mouths or puff out their gill region (Fig 22.10). The importance of these threat displays is that they often prevent actual fighting from breaking out. Frightened by the threatening posture of the aggressor, the rival accepts defeat and departs. Sometimes the defeated animal, instead of withdrawing, performs an **appeasement display**, as can sometimes be seen in dogs. An aggressive dog threatens another by standing erect, baring its teeth and snarling; the weaker rival cowers and exposes the side of its neck. This submissive attitude evokes

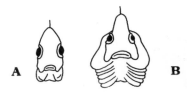

Fig 22.10 Threat display of a cichlid fish. A) normal posture. B) threat posture showing the puffing out of the gill region. (*after Baerends*)

non-aggressive behaviour in the other dog. The result is that the two part company without actually fighting. It is interesting to speculate to what extent open combat between human beings is avoided by facial expressions and other signals.

In sexual reproduction releasers may be very important in changing an animal's normally aggressive behaviour to sexual behaviour. An example of this is provided by the courtship of certain spiders. The female's normal tendency is to attack and eat any other spider that approaches her. However, when a male advances towards the female he signals to her in a characteristic way which in one species involves waving his palps. This acts as a stimulus, changing the female's behaviour from predatory to sexual. The story has a sad ending, however, for after copulation her behaviour reverts to aggressive and she devours her mate.

As was mentioned earlier, stimuli often bring a piece of behaviour to an

Fig 22.11 Courtship and mating in the three-spined stickleback. The stimuli which guide this sequence of events are discussed in the text. (*based on Tinbergen*)

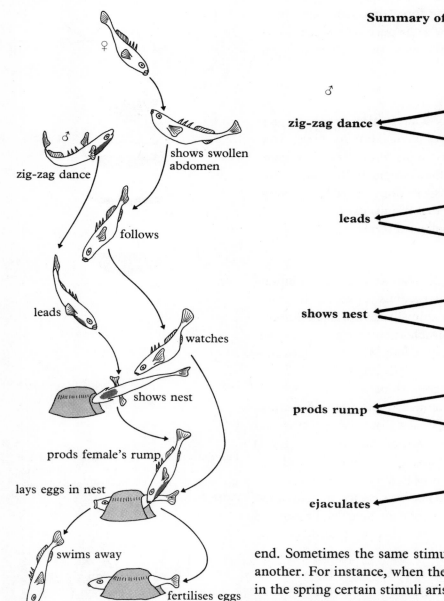

Summary of events

♂		♀
		appears
zig-zag dance		
		shows swollen abdomen
leads		
		follows
shows nest		
		enters nest
prods rump		
		lays eggs
ejaculates		

end. Sometimes the same stimulus may end one piece of behaviour and start another. For instance, when the male stickleback reaches its breeding grounds in the spring certain stimuli arising from its immediate environment bring its migratory behaviour to an end and initiate territorial behaviour in which it becomes aggressive towards other males. Can you think of other examples of stimuli which terminate one act of behaviour and start another?

It must not be imagined that stimuli are always external. Sometimes they arise from within the animal itself. This is seen nicely in drinking behaviour. Drinking is controlled by specialized osmoreceptors in the brain (see p. 218). The chloride concentration of the blood, the osmotic pressure of the body fluids, and the volume of the cells in the body may all influence drinking. Experimentally it can be initiated by injecting minute amounts of saline into the brain, thus demonstrating the importance of an internal stimulus in starting a piece of behaviour.

The importance of internal stimuli in ending behaviour is convincingly demonstrated by *Rhodnius*, a blood-sucking bug. Blood is sucked up through its proboscis by means of a muscular abdominal pump. It has been shown that the pump stops operating, and sucking ceases, when the pressure of blood in the abdomen reaches a certain critical level. Basically similar mechanisms operate in mammals where, for instance, chemical stimulation of the stomach wall, together with mechanical stretching, bring feeding to an end. Chemical changes in the blood may also be instrumental in bringing about the cessation of feeding.

SYNCHRONIZATION IN SEXUAL BEHAVIOUR

In sexual reproduction stimuli between the two participating individuals ensures synchronization of reproductive activity. This has been investigated by Tinbergen in a series of classical observations and experiments on the three-spined stickleback. Since Tinbergen's work clarifies many aspects of species-characteristic behaviour it is worthwhile considering it in detail.

The males migrate upstream to their breeding grounds in the spring. Each acquires a small piece of territory from which he chases away any intruders of either sex. Then he builds a small tunnel-like nest of weeds, open at both ends. During this time his belly region takes on a bright red coloration. His nest complete, he now becomes interested in females which at this season of the year are laden with eggs. If and when a female appears he swims towards her in a curious zig-zag fashion (zig-zag dance; Fig 22.11). When she sees him the female responds by swimming towards the male with her head and tail turned upwards, thereby displaying her swollen abdomen. The male then turns abruptly and swims quickly to the nest, the female following. He then shows her the nest entrance by poking it with his snout. At the same time he turns over on his side and raises his dorsal spines towards her. She enters the nest, her head sticking out of one end and her tail out of the other. The male now gives the female's rump several prods with a trembling motion, and this stimulates her to lay her eggs. When she has discharged all her eggs she leaves the nest and the male enters it, ejaculating over the eggs. He then chases the female away and goes off in search of another mate. Altogether he may escort as many as five different females to his nest.

After mating is over the male guards the nest and fans the eggs with his pectoral fins. Water is thus kept moving over them and this keeps them well supplied with oxygen. When the eggs hatch the male keeps the youngsters together and defends them until they are capable of looking after themselves.

By means of models Tinbergen has been able to demonstrate the stimuli involved in courtship. The stimulus which releases the male's zig-zag dance is the female's swollen abdomen; the size of the abdomen is all that matters: a stuffed female or even a bloated male will do equally well. The signal which causes the female to follow the male is the male's red belly; she will follow a red model equally persistently. The female is finally made to lay her eggs by the poking of her rump. It is not even necessary for a fish to do this; a glass rod is good enough. Finally the fanning of the clutch by the male is induced by the eggs themselves: take them away and fanning ceases.

So we see that the courtship behaviour of the stickleback consists of a series of stereotyped behaviour acts performed alternately by the male and female. Each act presents certain **sign stimuli** which elicit the next phase of activity in the other sex. The result of this **chain behaviour** is complete coordination and synchronization of reproductive activity between the two sexes. However, it must be stressed that this is an idealized sequence; most reproductive behaviour is much more complicated than this simple sequence suggests.

MOTIVATION

For a releaser to elicit a response the animal must be in the right kind of 'mood'. For example, food will not elicit feeding behaviour if the animal has just had a good meal. Similarly in sexual behaviour the arrival of a potential mate will not lead to reproductive behaviour if the animal is not ready for it. The term used to describe the internal state which must precede a specific act of behaviour is **motivation**.

Motivation results from the animal's internal physiological state, particularly the level of different hormones in the body. This in turn is determined by **motivational stimuli** such as temperature and light. For example, in birds sexual behaviour in spring is brought on by increasing day-length resulting in enlargement of the gonads. The eyes receive the stimulus of light which is transmitted to the brain and thence to the pituitary gland which secretes gonadotrophic hormones. These activate the gonads which in turn produce sex hormones inducing reproductive behaviour. Similarly courtship and mating in the stickleback will only take place in the spring when, for various environmental and physiological reasons, the female is ripe and the sex urge reaches its height.

The stickleback story illustrates how an animal's motivation can change in the course of a complex behavioural sequence. In the spring the male migrates upstream to its breeding grounds, this migratory behaviour being brought on by hormonal changes involving the pituitary and thyroid glands. On reaching the breeding grounds its behaviour changes to territorial, this being due to motivational stimuli derived from the territory itself. Its behaviour next becomes directed towards nest-building, then courtship and mating, and finally fanning the eggs and protecting the young. In each of these phases the appropriate motivation reaches a peak and then declines, making way for the next phase. The waxing and waning of motivation is seen particularly clearly in the fanning of the eggs by the male. The amount of time spent fanning increases each day, reaching a maximum on the eighth day when the eggs are about to hatch. It has been shown that fanning activity is proportional to the amount of carbon dioxide in the water round the eggs, and inversely proportional to the amount of oxygen. If a batch of eggs is removed after six days and replaced by a new clutch, the male continues to

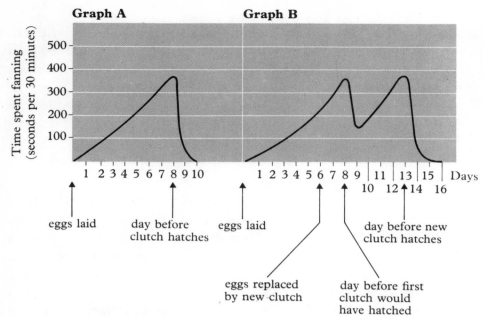

Graph A **Graph B**

Time spent fanning (seconds per 30 minutes)

eggs laid   day before clutch hatches   eggs laid   day before new clutch hatches

eggs replaced by new clutch

day before first clutch would have hatched

Fig 22.12 Fanning of eggs by male stickleback gradually increases over an eight day period and then rapidly declines just before the eggs hatch (Graph A). If the eggs are replaced by a new batch on the sixth day (Graph B) fanning still drops on the eighth day, but not completely. (*based on data by Tinbergen*)

fan the eggs as if they belonged to the original clutch. But the day before the original clutch would have hatched the fanning time drops, but remains at a level higher than on the first day (Fig 22.12). This suggests that fanning is controlled by internal as well as external factors, stimuli from the eggs having the immediate effect of inducing fanning and the longer-term effect of gradually raising the urge to fan.

PHEROMONES

A pheromone is a chemical substance produced by one animal which influences the behaviour of another. It may achieve this either by acting as a releaser or by building up motivation towards a particular type of behaviour. Perhaps the best known examples of pheromones are the secretions of mammals which are used for marking out territory or attracting a mate. Similarly the chemical attractants responsible for bringing the sexes together in insects, marine worms and many other species are pheromones. For example, unmated female cockroaches secrete a chemical substance from the surface of their body which stimulates males to court them. Males are aroused even if they touch a piece of filter paper which has been in contact with a female, and therefore contains some of this pheromone. In social animals like bees, pheromones play an important part in directing the development and behaviour of the different castes in the colony.

THE ROLE OF HORMONES

Hormones are certainly involved in the building up of motivation. This has been confirmed in many cases by investigating the behaviour of animals which have either had hormones injected into them, or endocrine organs removed.

How do hormones influence behaviour? The answer to this question is not fully known but the following possibilities are all supported by evidence.

(1) During development, and possibly even in the adult, hormones may affect the growth of nervous connections in the brain.

(2) Hormones may alter the sensitivity of peripheral receptors. It is known, for instance, that in rats the male hormones enhance sexual behaviour by raising the sensitivity of the penis.

(3) Hormones may directly influence the performance of effectors, either enhancing or suppressing them as the case may be. In extreme cases a hormone may actually cause degeneration of a muscle, thereby abolishing a particular response altogether.

(4) Hormones may directly affect nerve cells and synapses within the CNS. In this way they may block inhibitory pathways or open up excitatory ones along the lines described in Chapter 18.

THE HYPOTHALAMUS AND BEHAVIOUR

In certain instances hormones have been shown to influence specific regions of the brain, notably the **hypothalamus**. Many aspects of reproductive behaviour induced by injecting sex hormones into the body can be abolished by removing the hypothalamus. For example, mature specimens of the African clawed toad *Xenopus laevis* can be made to mate in the laboratory at any time of the year by injecting them with gonadotrophic hormone. However, this response can be abolished, or modified, by removing certain parts of the brain, particularly the hypothalamus.

More and more evidence is coming to hand implicating the hypothalamus in behaviour. Electrical stimulation of particular parts of it produce specific responses which are abolished if that part of the brain is removed or damaged. It has been found that certain regions of the hypothalamus initiate behaviour whereas other regions inhibit it. In very general terms stimulation evokes feeding and drinking, sleep and rage, attack and escape, urination and defaecation, and responses to heat. But we are a long way from understanding the precise neurophysiological processes involved and it would be a mistake to place undue emphasis on any one part of the brain. Undoubtedly the neural circuits governing all types of behaviour are exceedingly complex, involving elaborate interconnections between different regions of the brain.

DISPLACEMENT ACTIVITY

When an animal is in a state of stress it will sometimes engage in behaviour which is out of context or irrelevant. This is called **displacement activity**. For example, two birds that are fighting may suddenly preen themselves, or peck at the ground, or get into roosting positions. In the middle of a boundary dispute a male stickleback may suddenly adopt a vertical position with its head pointing downwards and start digging in the sand.

Such displacement activity in the examples just quoted seems to be the result of two opposing forces: fighting and escape. As such it may sometimes serve the useful purpose of preventing or diminishing open conflict.

This kind of thing also happens in man. When in a tense situation we frequently perform displacement activities such as stroking the forehead, scratching an ear, or walking up and down. Even smoking, particularly when carried out at a cocktail party, can be regarded as a displacement activity.

At first sight displacement activities may seem to be unimportant side-effects of behaviour. However, some zoologists believe them to be the basis of many normal behaviour patterns. Much of courtship behaviour, for instance, may have evolved from displacement activities arising from frustration. The male's sexual motivation builds up but cannot be released until the appropriate signal is given by the female. So some of his sexual motivation is channelled into other forms of behaviour which constitute courtship. If this view is correct it is clear that displacement activities have come to perform a very important rôle as releasers eliciting appropriate responses in other individuals.

VACUUM ACTIVITY

When an animal is frustrated it will sometimes perform an act of behaviour in the absence of the normal releaser. Its motivation builds up but no sign stimulus is provided to release the appropriate behaviour. As a result it performs in the wrong situation. Such behaviour is known as **vacuum activity**. For example, a cock deprived of a mate will display to an inanimate object such as a bucket, a bird may go through the motions of building a nest even if there is no nest material available, and so on. These phenomena, too, have their counterparts in human behaviour. Can you think of any?

Learning

Learning can be defined as an adaptive change in behaviour resulting from past experience. Learned behaviour is therefore acquired during the lifetime of an individual as a result of constant experience. Learning is characterized by flexibility: what is learned may vary from one individual to another, and the resulting behaviour can be modified if the environment changes. Clearly a learned behaviour pattern cannot be inherited though the ability to learn is almost certainly inherited. Learning may be classified into five categories: **habituation, associative learning, imprinting, exploratory learning**, and **insight learning**. These divisions are ones of convenience and do not reflect fundamentally different mechanisms.

HABITUATION

If an animal is subjected to repeated stimulation it may gradually cease to respond. If the stimuli are not harmful the animal learns not to react to them. Sometimes it is difficult to decide whether the decline in an animal's response is caused by synaptic accommodation (p. 274) or adaptation of the receptors (p. 301). Technically the latter is not regarded as habituation though it may appear remarkably similar to it. Generally habituation, once established, lasts longer than sensory adaptation.

As an example of habituation, consider the escape response of the peacock worm *Sabella* which can be easily observed in a sea-water aquarium (Fig 22.13). As has been mentioned before, fanworms jerk back into their tubes when touched, a rapid response mediated by a giant axon in the ventral nerve cord. Now if a tentacle is repeatedly stimulated the worm quickly stops reacting though it may continue to give slow movements of the body. Eventually these also cease. That adaptation of the touch receptors is not involved can be shown by stimulating another region of the tentacle. If waning of the response was due to sensory adaptation, shifting the point of stimulation should renew the response, but it does not. We therefore conclude that a **synaptic block** has occurred somewhere in the reflex, possibly between the afferent nerve(s) in the tentacle and the giant fibre in the nerve cord.

Fig 22.13 Peacock worms, *Sabella pavonina*. If the crown of tentacles is touched the worm jerks back into its tube. This escape response shows rapid habituation on repetition. (*Douglas P. Wilson*)

The development of such synaptic blocks has been investigated in some detail in the earthworm and helps us to understand more precisely the neural basis of habituation. If the head of a worm is prodded repeatedly the rapid contractions of the longitudinal muscle (see p. 345) gradually weaken and finally disappear altogether. This does not take long: twelve good pokes is usually enough to abolish the response. Now it is known that the touch receptors in the skin are slow to adapt and that the longitudinal muscle, if stimulated directly, will go on contracting for a very long time. This means that waning of the response must be caused by transmission failure at synapses somewhere in the giant axon reflex (Fig 22.14).

Fig 22.14 Establishing the neural basis of habituation involves stimulating the nerve on one side of a synapse and recording action potentials from the nerve on the other side. In this way it has been shown that habituation in the earthworm's escape response is due to accommodation of synapses situated between A) afferent neurones and giant axon, and B) giant axon and efferent neurones.

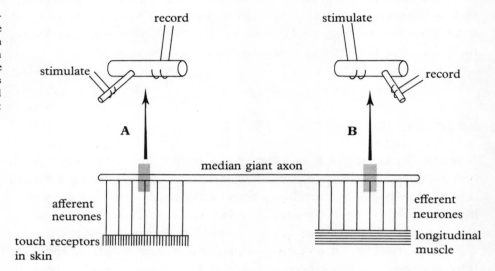

By systematically stimulating on one side of each possible synapse, and recording action potentials from the other, it has been shown that rapid failure occurs at synapses situated between the afferent neurones and the median giant axon, and at synapses between the median giant axon and the efferent neurones. On repeated stimulation these synapses quickly accommodate and must be given time to recover before they will transmit again.

What is the value of this kind of habituation? Put simply, it prevents the animal performing its escape response so frequently that it has no opportunity to do anything else. In more complex situations habituation may prevent an animal repeatedly responding to releasers that lead nowhere. Can you think of examples of this?

ASSOCIATIVE LEARNING

In this process an animal learns to associate a particular response with a **reward** or **punishment**. The words reward and punishment may seem to be appropriate only in the artificial conditions of the laboratory. However, they have their counterparts in the natural environment: finding food or a mate is equivalent to a reward, and being attacked by another animal is equivalent to a punishment. In associative learning the animal remembers its past experiences and modifies its behaviour accordingly. This is not to suggest that there is necessarily any conscious effort involved; indeed in the majority of animals the associative processes are carried out in the nervous system below the level of the 'conscious' part of the brain. Many animals, such as flatworms and earthworms, have no conscious existence at all but are still capable of learning.

THE CONDITIONED REFLEX

A basic type of associative learning is the conditioned reflex which was investigated in the classical experiments of the great Russian physiologist Ivan Pavlov, at the turn of the century. Pavlov worked on the production of saliva by dogs in response to food. As we saw in Chapter 9 the smell, sight and taste of food induces saliva to flow into the mouth cavity. Pavlov mounted his dogs in a harness and collected their saliva by means of a tube leading from the salivary duct. What Pavlov did was to teach his dogs to associate the arrival of food with some other stimulus so that the latter alone would cause saliva to flow. In one experiment he rang a bell immediately before producing the food. This was repeated many times, and eventually the dog salivated as soon as it heard the bell: it had learned to associate the bell with the food. Pavlov called the new stimulus, in this case the bell, the **conditioned stimulus**, and the response a **conditioned reflex**. As everyone now knows, dogs are notoriously easy to condition. Having taught his dogs to associate food with a bell, Pavlov taught them to associate the bell with another stimulus such as a light. In fact they conditioned themselves to the laboratory situation more than he bargained for: a dog that was to be tested would run ahead of him into the laboratory and jump onto the stand, ready to be harnessed in for an experiment!

Since Pavlov did his classical experiments on dogs many other animals have been conditioned in the laboratory, vertebrates and invertebrates alike. Conditioning is also important in the wild. For example, predators learn to associate unpalatable animals with certain markings or coloration, and will thus avoid eating them. In general terms conditioning allows animals to modify their behaviour in such a way that maximum rewards are obtained and punishment is avoided.

TRIAL AND ERROR LEARNING

Suppose that a hungry dog is allowed to roam around a room. As soon as it jumps onto a certain chair we reward it with food. Quite quickly the dog learns to associate the chair with a reward and consequently goes straight to it as soon as it enters the room. In this case the dog has learned to associate a reward not with a particular stimulus but with its own behaviour. This is called **trial and error learning**. It is clearly a type of conditioning though it differs from Pavlov's classical conditioned reflex in the way it becomes established.

Trial and error learning is widespread amongst animals from flatworms to man. To investigate an animal's ability to learn by this process many different kinds of experiment can be done. The one feature all these experiments have in common is that the animal is confronted with a 'choice' between two or more courses of action. If it makes the 'wrong' choice it is punished, if it makes the 'right' choice it is rewarded. The experiment may require the animal to choose between turning left or right in a simple maze, or between attacking or not attacking an object which is presented to it. There are many possible variations on these themes. The animal is tested at intervals. Each test constitutes a trial, and a variable number of trials are carried

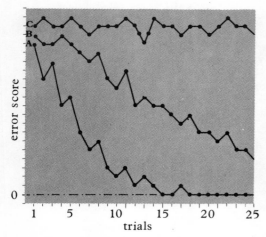

Fig 22.15 The kind of results obtained in learning experiments. Animal A learns quickly; B moderately quickly; C fails to learn at all.

Fig 22.16 Trial and error learning by an octopus. In the upper diagram an untrained octopus is seen leaving its shelter and advancing towards the crab at the other end of its tank. If it attacks the crab with white square it is given an electric shock (note the electrodes sticking out of the white square). If it attacks a crab alone it is allowed to take it back to its shelter and eat it. Six trials are carried out each day, three with crab only, and three with crab plus white square. The results are shown in the lower diagram: after a few days the octopus learns to attack the crab alone (Graph A), but *not* to attack the crab with white square (Graph B). (*based on Boycott and Young*)

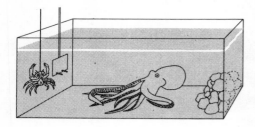

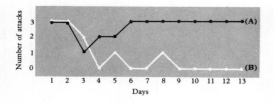

[1] How could trial and error experiments of the kind described here be extended to investigate the capabilities of an animal's receptors, in particular its powers of sensory discrimination?

out each day; how frequently depends on the animal in question. The number of errors which the animal makes is then plotted against the time. An alternative procedure is to test a large number of animals at intervals and plot the percentage which make errors against time. Either way, graphs of the sort shown in Fig 22.15 are obtained.

To put these generalizations into practical terms consider the situation shown in Fig 22.16. The octopus has a well-developed brain and good sense organs. It lives in a 'den' amongst rocks from which it emerges to attack moving prey such as crabs. If a crab is lowered into one end of an aquarium tank, the octopus will normally attack. In the experiment illustrated in Fig 22.16 an octopus is presented with a crab plus a white square. On attacking the crab it is given a brief electric shock. Three such trials are carried out each day, together with three in which the octopus is presented with a crab on its own (no electric shock). After about ten trials the animal no longer attacks a crab accompanied by a white square, though it will continue to attack crabs presented on their own. By trial and error the octopus has come to associate the white square with punishment[1].

Trial and error learning is by no means confined to animals with well-developed brains. It is shown, at an elementary level, even by lowly animals like the earthworm. Here is a simple experiment you could quite easily do for yourself. An earthworm is placed in the stem of a T-shaped tube. If it turns left it is given an electric shock; if it turns right it is returned to its box of leaf mould without punishment. It is claimed by people who have enough patience to complete this experiment that the worm learns by experience to associate turning left with punishment and eventually always turns right. Similar experiments have been done with many other animals including free-living flatworms.

A much more complicated situation is illustrated in Fig 22.17. In this case an ant is placed at the entrance to a complicated maze which has an empty food box at the other end. In the first trial the animal goes up numerous blind alleys as it traverses the maze. It is not punished for these mistakes but is rewarded with food when eventually it reaches the end. In subsequent trials it makes fewer and fewer mistakes, and gets through the maze progressively quicker, until eventually it makes no mistakes at all. This kind of experiment has been tried on a wide range of animals. Rats and ants are particularly adept at mastering mazes, rats rather more so than ants. Some animals such as amphibians, which are normally rather slow at solving even quite simple mazes, learn more quickly if they are punished for taking wrong turnings.

The ability of an animal to learn by trial and error is reflected in three things:

(1) the speed with which it ceases to make errors;

(2) the length of time it can remember without repeated trials; and

(3) the complexity of the situation to which it will respond.

For animals as a whole there is now a wealth of information on these three aspects of trial and error learning. As a broad generalization higher animals like rats and primates can learn complex situations quickly and retain the relevant information for a comparatively long time, whereas lower animals like worms and many insects can only learn simple problems rather slowly and the memory is retained for a relatively short time. The octopus is a special case. For a mollusc it has a remarkably large and well-developed brain whose learning capabilities are on a level with those of some of the vertebrates.

LEARNING AND THE OCTOPUS BRAIN

The large size of the octopus' brain, and its ability to withstand surgical operations, make it an ideal animal for studying the effect on memory of removing parts of the brain. Experiments of this sort were started by J. Z. Young and his colleagues just after the war and are now being continued by several groups of investigators, mainly at the Zoological Station in Naples. Their work is throwing considerable light on fundamental processes of perception, discrimination and learning. It would be impossible to go into the details of all the experiments so just one will be mentioned to illustrate the kind of thing which has been done. An octopus which has had the so-called vertical lobe of its brain removed cannot be trained to ignore a crab, accompanied by a white square, if the trials are separated by the kind of time intervals shown in Fig 22.16. But if the interval between successive trials is reduced to, say, five minutes the animal is capable of learning. However, when the trials are stopped it is found that the operated animal forgets more quickly than an unoperated one. An octopus with its vertical lobe intact will remember not to attack for more than two weeks. But an octopus which has lost its vertical lobes will only remember for an hour or two. From these observations it seems that the octopus has two interrelated memory systems, one short-term and the other long-term. There is good reason to believe that much the same applies to the human brain.

A NEURAL THEORY OF LEARNING

On the basis of research on the octopus brain J. Z. Young has put forward a theory accounting for trial and error learning. His theory is based on the idea that the brain contains numerous memory units made up of interconnected nerve cells. When a specific stimulus is registered by a receptor such as the eye a **classifying cell** in the brain is activated. This has connections with two **memory cells**, one of which is connected with the nervous pathway responsible for attack, the other with the pathway responsible for retreat. It is supposed that the memory cells are interconnected with each other and with other centres in the brain. When a specific stimulus is received by a receptor such as the eye, the classifying cell is activated and impulses are discharged into the attack pathway. If, however, the animal receives an unpleasant stimulus as a result of attacking, the retreat pathway is facilitated and the attack pathway is inhibited, presumably by the kind of synaptic processes described in Chapter 18. The result is that the next time the animal receives the same stimulus, impulses are prevented from entering the attack pathway and pass into the retreat pathway. On the other hand if attack results in a reward, the retreat pathway is inhibited and the attack pathway is facilitated. It is supposed that the memory cells are connected with centres in other parts of the brain, and these may be responsible for long-term memory.

A BIOCHEMICAL THEORY OF LEARNING

A rather different (but not incompatible) theory of learning has been put forward by American investigators working on free-living flatworms. As was

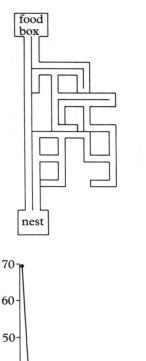

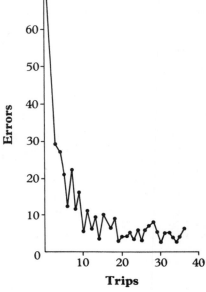

Fig 22.17 Maze-learning by the ant *Formica incerta*. A plan of the maze is shown in the upper diagram, and the results obtained with one representative individual are shown in the lower diagram. Notice how quickly the animal solves the maze. (*after Schneirla*)

Fig 22.18 Exploratory learning in rats. Group 1 were given a reward (food) at the end of each run. Groups 2 and 3 were given no reward until the third and seventh days respectively (the points marked X in the graph), after which they were rewarded each time. Notice the slow learning prior to X, and the very rapid reduction in errors immediately after X. This is attributable to previous exploratory learning. (*after Blodgett*)

Plan of maze

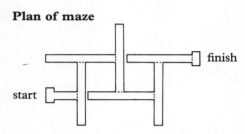

Results

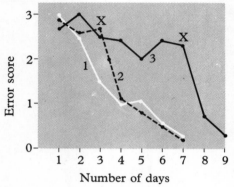

mentioned earlier free-living flatworms can be trained to turn one way or the other in a simple T maze. The theory of learning put forward by these workers is based on the remarkable discovery that if a previously trained flatworm is crushed up and fed to an untrained worm, the latter will learn more quickly. Moreover it is claimed that the active principle which induces this change is the nucleic acid RNA. Other workers claim that RNA is also implicated in mammalian learning. RNA extracted from the brain of a trained rat and injected into an untrained individual is said to cause the latter to learn faster. It is therefore claimed that at the biochemical level learning may involve a change in the animal's nucleic acid. However, it must be pointed out that these experiments, though immensely interesting, have not yet been confirmed and many biologists are extremely sceptical of them.

EXPLORATORY LEARNING

If a hungry rat is placed in a maze of the type illustrated in Fig 22.18 it will master the maze after about 12 trials. This is assuming that it is rewarded at the end of each trial and has never been in the maze before the experiments begin. However, if a rat which is not hungry or thirsty is allowed to spend some time in the maze, without any reward, before the trials begin, it is found that it subsequently requires fewer trials to gain complete mastery of the maze. It seems that in its preliminary experience of the maze it picked up clues about it which it subsequently used in finding its way from one end to the other. It is very striking how a well nourished rat which has no need to search for food or water will nevertheless run busily here and there sniffing into corners and investigating every nook and cranny. This is what is meant by **exploratory learning**. An example is given in Fig 22.18.

Exploratory behaviour is extremely important in the lives of many animals for it enables them to find their way about their environment. They learn all its characteristics and remember its landmarks.

IMPRINTING

In his charming book *King Solomon's Ring* the German zoologist Konrad Lorenz describes how young geese follow the first thing they see after they are born. Generally, of course, the first object they see is their mother, but Lorenz found that they would follow any moving object – a cardboard box, a tractor, or even himself. Lorenz' account of how he crawled round a field on his hands and knees followed by a group of quacking goslings is one of the classic stories of animal behaviour.

The story illustrates a basic principle, namely that many young animals tend to follow, or **imprint** on, their parents. This is obviously advantageous during the early stages of an animal's life when parental protection is important for survival. But there is an increasing body of evidence that imprinting also plays a profound rôle in determining the individual's behaviour later on. To quote an example, if a young mallard is reared by foster parents of another species it subsequently pairs with a member of that species rather than its own. Again, if a male mallard is reared by a foster father, instead of a foster mother, it subsequently pairs with a male rather than a female. Such homosexual pairings are by no means uncommon in the animal kingdom. Not surprisingly these sorts of findings have attracted the interest of psychiatrists.

INSIGHT LEARNING

This is the highest form of learning and in many ways the most difficult to interpret. It can be defined as the immediate comprehension, and response to, a new situation without trial and error. As such it would appear to involve some kind of mental reasoning or 'intelligence'. In human behaviour the concept of insight is perhaps applicable (there are moments when most of us have a sudden 'brainwave') but it is doubtful if it can be applied to animals. Even in man it is difficult to decide to what extent an apparent insight is based on past experience, trial and error, imprinting and so on. In fact intelligence (a notoriously difficult word to define) probably involves all the different processes discussed in this chapter.

Taking this broad view, an animal's intelligence may be assessed by the speed with which it solves a problem which it has not encountered before. A wide range of tests has been devized to measure the intelligence of different animals. But although they give us some idea of how quick an animal is on the uptake, it is difficult to compare the relative abilities of different species.

The classical studies on insight have been made on chimpanzees. For example, in order to get at food which has been placed out of reach a chimpanzee will resort to all manner of strategies such as fitting sticks together, piling boxes on top of each other or swinging a ball on the end of a chain (Fig 22.19). The solution is frequently arrived at quite suddenly, suggesting insight, but experiments show that chimpanzees benefit from previously handling the tools which they use for solving such problems. What they can do is to use the experience that they have gained in one situation to solve a problem in another context. And this, surely, is what we ourselves are trying to do all the time.

Fig 22.19 Insight learning? The ball has to be aimed at the grapes which, if hit, roll down the slope for the chimpanzee to eat. (*F.A.E. Hall, Zoological Society of London*)

To appreciate how organisms reproduce it is necessary to understand how cells divide. This is discussed in Chapter 23. Chapter 24 deals with reproduction of the whole organism, and the way this fits into the life cycle is explained in Chapter 25.

Reproduction must obviously be accompanied by growth and development. The purely structural aspects of these vital processes are presented in Chapter 26. How growth is regulated by internal and external influences is discussed in Chapter 27. Since growth is the basis of how plants respond to stimuli, this topic is also treated in Chapter 27.

Built into the process of reproduction are mechanisms ensuring that the offspring are basically similar, though not identical, to their parents. The basis of this is discussed in the remaining chapters. Chapters 28 and 29 deal mainly with the classical laws of genetics and their implications, whilst Chapter 30 presents the results of some of the more modern research into the structure and action of genes. The final chapter discusses the rôle of genes in controlling the development of organisms.

Catkin of male flowers of willow.
(*B. J. W. Heath, Marlborough College.*)

Part V   Reproduction, Development and Heredity

23 Cell Division

Reproduction involves the multiplication of cells. The formation of a bud in a simple animal like *Hydra*, the development of sex cells, and the growth of a young animal or plant into an adult all involve cell multiplication.

In order to multiply, cells undergo **cell division**: one divides into two, two into four, four into eight, and so on. The word division is misleading in some ways because it implies that the process always involves halving the cell and its contents. In fact we now know that cell division is accompanied or preceded by the formation of new cell components so that the products of cell division, the daughter cells, are essentially similar to the parent cell. Understanding cell division is largely a question of appreciating how this uniformity is preserved.

In any description of cell division the **chromosomes** occupy a central position. As the vehicles of heredity they determine the characteristics of the cell and its progeny, and it is essential that they should be correctly distributed between the daughter cells. It is known that a cell normally has a fixed number of chromosomes and these occur in pairs: the so-called **diploid** condition. Two types of cell division are recognized according to the behaviour of the chromosomes (Fig. 23.1). In the first of these, the daughter cells finish up containing exactly the same number of chromosomes as the parent cell. This is called **mitotic cell division** (or just **mitosis**) and is the type of cell division which takes place during an organism's growth. In the other type of division, known as **meiotic cell division** (or **meiosis**) the daughter cells finish up with half the total number of chromosomes present in the parent cell. This kind of division generally takes place in the formation of gametes, though in the more complex plants it occurs in the formation of spores. Its full implications will be discussed later.

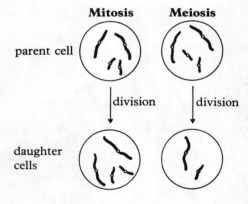

Fig 23.1 What do mitosis and meiosis achieve? A cell normally contains two of each type of chromosome, the diploid condition (two large and two small chromosomes in the parent cells depicted above). Mitosis preserves this condition. Meiosis, however, results in the daughter cells containing only one of each type of chromosome, the haploid condition.

Mitosis

For purposes of description mitosis is divided into four stages: **prophase**, **metaphase**, **anaphase** and **telophase**. At each of these stages certain crucial events take place, particularly in regard to the chromosomes. However, it is important to realize that mitosis is a continuous process and there are no sharp breaks between one stage and the next. Typically the entire process takes about an hour and is followed by a **resting stage** (called **interphase**) during which the daughter cells grow and prepare for the next division. This involves synthesis of new materials and replication of organelles. In actively dividing cells interphase lasts between 12 and 24 hours.

OBSERVING MITOSIS

One of the best places to see mitosis is the tip of a growing root, young bean roots for example. The root tip is cut off, sectioned or macerated, and treated with a dye such as acetic orcein which stains the chromosomes. In good preparations the various stages of mitosis can be clearly seen. The trouble is that the chromosomes are locked in a fixed position and nothing can be seen of their movements. It is like looking at still photographs of an athlete in action: useful for analysis but conveying nothing of the dynamic movements involved. In recent years, however, ciné films have been made of unstained living cells under the phase-contrast microscope.[1] As cell division is slow, time-lapse photography is used, successive frames being exposed at intervals of, say, half a minute. When projected, the movements of the chromosomes, speeded up many times, can be followed in detail. Needless to say, the technical difficulties are severe but nevertheless some good films have been made, particularly by the American (formerly Polish) biologist, A. Bajer. One of the biggest difficulties is finding a suitable tissue, one which contains actively dividing cells visible under the microscope without preliminary treatment which might damage the cells. Bajer found endosperm tissue, the nutritive tissue surrounding the embryo inside seeds, to be almost ideal. Its cells divide prolifically and, being a rather soft runny tissue, a thin smear can be made on a microscope slide without harming the dividing cells.

Let us now follow what happens when a diploid cell undergoes mitotic division. For convenience we will take a cell which happens to contain only four chromosomes: a pair of long ones and a pair of short ones. The events are summarized in Fig 23.2, which should be consulted with the text.

INTERPHASE

During interphase (Fig 23.2A) the cell has the same appearance as any non-dividing cell. The chromosomes are not visible as distinct bodies either under the light microscope or the electron microscope. At this stage they are strung out in the form of long **chromatin threads** swollen at intervals into visible chromatin granules. Not until prophase do the chromatin threads condense to form visible chromosomes.

To describe interphase as a resting stage is a complete misnomer. Far from being inactive the cell is growing and preparing for division. During interphase two things happen, both of which are essential if the cell is to divide. Firstly, the genetic material (DNA) replicates, i.e. doubles itself, so that sufficient DNA is made available for each of the two daughter cells. A cell never divides until this new genetic material has been formed. Secondly, the cell builds up a sufficiently large store of energy, a kind of 'energy reservoir', to carry the process through. That this accumulation of energy takes place during interphase rather than during division itself can be shown by inhibiting respiration at different stages in mitosis. If a cell is treated with a metabolic poison such as cyanide during interphase, mitosis fails to take place. On the other hand if the cell is treated with poison after mitosis has started, the process will still go through to completion.

[1] Why should the phase-contrast microscope be used for this?

Fig 23.2 Mitosis in a generalized animal cell. The sequence of events is almost exactly the same in plant cells. Two pairs of chromosomes are shown: a long pair and a short pair. In **A** the black dot in the nucleus is the nucleolus and the rods outside the nuclear membrane are the centrioles.

A) Interphase
Cell has normal appearance of non-dividing cell condition: chromosomes too threadlike for clear visibility.

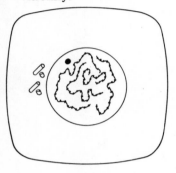

B) Early prophase
Chromosomes become visible as they contract, and nucleolus shrinks. Centrioles at opposite sides of the nucleus. Spindle fibres start to form.

C) Late prophase
Chromosomes become shorter and fatter—each seen to consist of a pair of chromatids joined at the centromere. Nucleolus disappears. Prophase ends with breakdown of nuclear membrane.

D) Early metaphase
Chromosomes arrange themselves on equator of spindle. Note that homologous chromosomes do not associate.

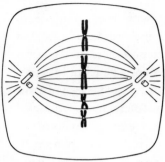

E) Late metaphase
Chromatids draw apart at the centromere region. Note that the chromatids of each chromosome are orientated toward opposite poles of the spindle.

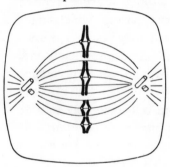

F) Early anaphase
Chromatids part company and migrate to opposite poles of cell, the centromeres leading.

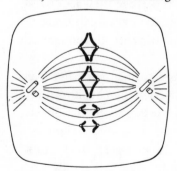

G) Late anaphase
Chromosomes reach their destination.

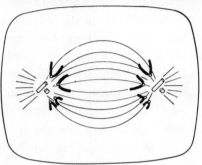

H) Early telophase
The cell starts to constrict across the middle.

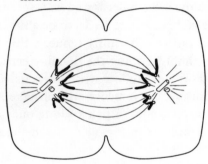

I) Late telophase
Constriction continues. Nuclear membrane and nucleolus reformed in each daughter cell. Spindle apparatus degenerates. Chromosomes eventually regain their threadlike form and the cells return to resting condition (interphase).
Note that the daughter cells have precisely the same chromosome constitution as the original parent cell.

A third important event that takes place during interphase is the formation of new cytoplasmic organelles – mitochondria, ribosomes, chloroplasts, and the like. If this did not happen successive cell divisions would result in a steady depletion of the cells' contents. Formation of new organelles takes place either by the production of new materials in the cytoplasm, or by the replication of existing organelles. Either way, the synthetic activity involved necessitates the expenditure of much energy, and a high metabolic rate is typical of cells preparing for mitosis.

One of the most prominent organelles to replicate during mitosis are the **centrioles**. Found in the cells of all animals and certain primitive plants, the centrioles consist of a pair of short rods lying just outside the nuclear membrane. The cells of higher plants lack centrioles and yet undergo normal mitosis, so they are not essential for the process. However, in cells where they are present their movements are an integral part of the mitotic sequence. By interphase they have already replicated, so in Fig 23.2A two pairs of centrioles are shown, one pair being destined to go into each daughter cell.

PROPHASE

If interphase is concerned with preparing the cell for division, prophase (Fig 23.2B, C) can be described as 'mobilization for action'. Certain clearly visible events can be seen under the microscope. The most obvious is the condensing of the chromatin threads to form distinct chromosomes. Long and thin at first, they gradually become shorter and fatter. As the chromosomes shorten it becomes increasingly clear that each consists of a pair of bodies lying close to each other. These are called **chromatids**. They tend to lie parallel along most of their length but are particularly closely associated in the vicinity of a specialized region called the **centromere**. We shall see later that this represents the centre of movement of the chromosome. The chromatids of one chromosome are usually referred to as **sister chromatids**.

It is essential to appreciate what these chromatids represent. Initially (i.e. in early interphase) each chromosome consists of a single, highly attenuated thread. During replication of the genetic material an exact duplicate of this single thread is produced. The two threads together (the original one and its newly-formed duplicate) are completely invisible until prophase when they make their appearance as the two chromatids. In a later chapter we shall consider the molecular basis of these events but for the moment it is important to realize that the two chromatids making up a chromosome are identical: they contain exactly the same genetic material point for point along their length.

Condensation of the chromosomes is one of the most important events in prophase. If it failed to happen it would be impossible for the chromosomes to move around within the cell without getting tangled up. While this is happening other changes take place. A series of protein fibres are formed in the cytoplasm. These span the cell from end to end, forming a structure which, because of its appearance, is called the **spindle** (Fig 23.3). The two ends of the spindle are known as its **poles** and the middle region as its **equator**. In cells where centrioles are present, these now lie one at each of the two spindle poles. This is achieved by one of them skirting the nuclear membrane so that it comes to occupy a position at the opposite side of the nucleus. Meanwhile the nucleolus disappears. Prophase ends with the breakdown of the nuclear membrane.

Fig 23.3 Phase contrast photomicrograph of isolated spindle apparatuses obtained from sea urchin eggs. Note the spindle fibres and the prominent asters. The asters are the fibres radiating from the poles and are present only in animal cells. The cells are at metaphase as can be seen by the positioning of the chromosomes on the equator of the spindle. (*Daniel Mazia, University of California*)

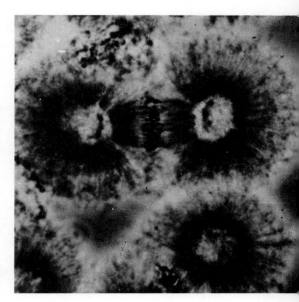

METAPHASE

When the nuclear membrane disrupts the chromosomes migrate to the central plane of the cell and arrange themselves round the equator of the spindle (Fig 23.2D, E). This is achieved through the organization of a series of fibres which run from the centromere of each chromosome to one of the two poles. At this moment the chromatids of each chromosome move slightly apart at the centromere region, the sister chromatids being orientated towards opposite poles as shown in Fig 23.2E.

Fig 23.2 makes it quite clear that homologous chromosomes behave entirely independently in the way they arrange themselves on the spindle. As far as is known they do not influence each other in any way, and they certainly do not associate with one another. Thus the two long chromosomes are attached to different fibre systems, as are the two short chromosomes. The significance of this will become clear later.

ANAPHASE

For anyone who has watched a ciné film of mitosis this is the most spectacular part of the whole process (Fig 23.2 F, G). Suddenly the chromatids belonging to each chromosome part company and move towards opposite poles of the spindle.

In recent years there has been much debate on the possible mechanism by which this migration of chromatids might take place. It used to be thought that the centromeres repel each other, causing the chromatids to move apart, but there is now evidence that the chromatids are pulled to the poles of the spindle. Observations on the spindle apparatus of live cells in the polarizing microscope have revealed two types of spindle fibre (Fig 23.4). One type runs right across the cell from one pole to the other; the other type runs from the poles to the centromere of each chromatid. It is possible that the latter type of fibre shortens (like a muscle fibre?) and pulls the chromatid, centromere first, towards the pole. With the centromere in the lead, the rest of the chromatid will tend to trail behind. Energy is certainly required for this unique process. Mitochondria have been shown to congregate round the spindle apparatus, and chemical analysis of isolated spindles has shown them to contain an active enzyme which splits ATP. But despite these interesting leads we are still a very long way from understanding what precisely moves the chromatids during anaphase.

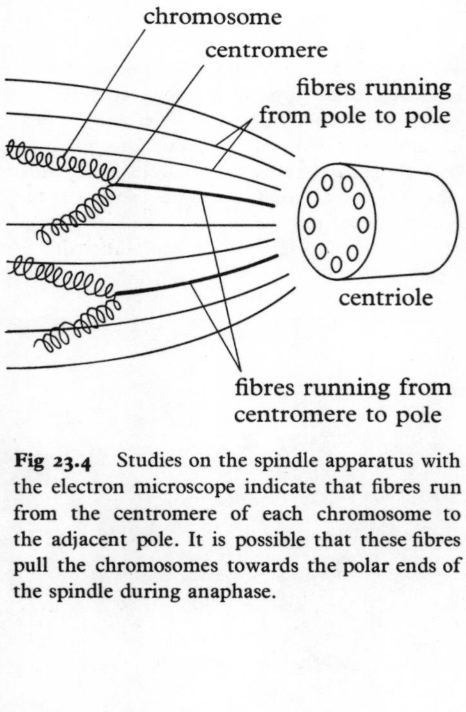

TELOPHASE

The chromatids are destined to become the chromosomes of the daughter cells. On reaching the polar ends of the spindle they become densely packed together. Meanwhile the cell divides into two (Fig 23.2 H, I). In animal cells this takes place by means of a constriction of the plasma membrane which cuts across the equator of the spindle. In plants a new wall, the **cell plate**, grows across the middle of the cell: in root tip cells, where the spindle is about the full width of the cell, the plate is deposited as scattered droplets and vesicles which coalesce; in larger cells this is followed by expansion of the cell plate outwards until it fuses with the wall. While this is happening the spindle apparatus breaks down and the nuclear membrane is re-formed. The nucleolus reappears and the chromosomes gradually uncoil and return to their original thread-like form. Mitosis is now complete; the daughter cells enter interphase and prepare for the next division.

Fig 23.4 Studies on the spindle apparatus with the electron microscope indicate that fibres run from the centromere of each chromosome to the adjacent pole. It is possible that these fibres pull the chromosomes towards the polar ends of the spindle during anaphase.

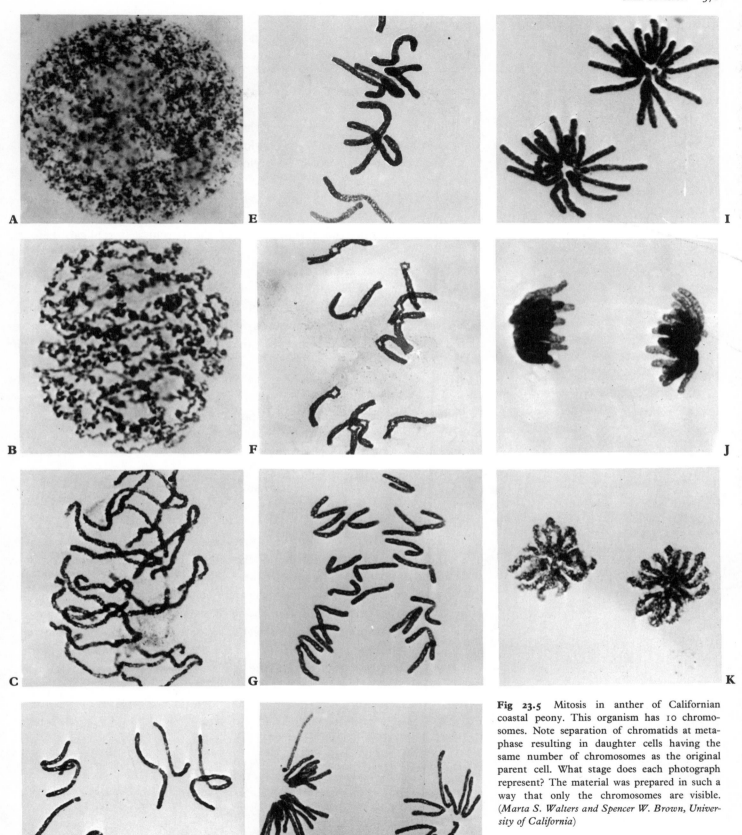

Fig 23.5 Mitosis in anther of Californian coastal peony. This organism has 10 chromosomes. Note separation of chromatids at metaphase resulting in daughter cells having the same number of chromosomes as the original parent cell. What stage does each photograph represent? The material was prepared in such a way that only the chromosomes are visible. (*Marta S. Walters and Spencer W. Brown, University of California*)

THE ESSENTIAL PRINCIPLE UNDERLYING MITOSIS

Mitosis is full of intriguing details and unexplained problems. For example, what triggers the whole process? What makes the chromosomes behave as they do? What causes the cell to divide down the middle? These are several questions that might be asked. But interesting though they are, these details are not our main concern. The really important thing about mitosis is that the daughter cells receive precisely the same number and types of chromosomes as the original parent cell: two long ones and two short ones in the case described here. In other words the diploid constitution is maintained from one generation of cells to the next.

Two salient features of mitosis ensure that the chromosome constitution is preserved:

(1) the fact that the chromosomes of the parent cell replicate before the cell divides, and

(2) the arrangement of the chromosomes on the spindle.

As a result of replication of the chromosomes the parent cell in effect contains twice its normal number of chromosomes before mitosis begins. The arrangement of the chromosomes on the spindle ensures that the chromosomes are distributed evenly between the two daughter cells.

THE ROLE OF MITOSIS

Mitosis is the type of cell division that takes place during the **growth** of an organism, for example in the development of a fertilized egg into an adult human being. The fact that these divisions are mitotic means that all the cells of the body have the diploid chromosome constitution, 23 pairs in the case of the human.

Mitosis is also the basis of **asexual reproduction**. It is the process that occurs when a protozoon undergoes binary or multiple fission: when a bud develops in *Hydra*, and when new plants develop from vegetative organs like corms and bulbs. Here, too, mitosis ensures that the chromosome complement, and hence the genetic constitution, of the offspring is the same as the parent's.

One of the places where mitosis can be observed particularly clearly is in the tapetum of certain plants such as peony. The tapetum is a single layer of large cells lining the pollen sacs in the anthers (see p. 395). Suitably stained squashes of developing anthers show the tapetal cells in various stages of mitosis. The photomicrographs shown in Fig 23.5 were obtained in this way. Can you relate these pictures to the stages depicted diagrammatically in Fig 23.2?

Meiosis

We have seen that a built-in feature of mitosis is the preservation of the parental chromosome condition. In meiosis, however, the number of chromosomes is halved, the daughter cells receiving only one of each type of chromosome instead of two. The daughter cells are said to be in the **haploid** state, haploid coming from the Greek word *haploos*, meaning single.

THE RÔLE OF MEIOSIS

In animals meiosis occurs in the formation of **gametes** (the sex cells) such as eggs and sperm. To understand its significance we must pause for a moment to consider the function of the gametes. By definition a gamete is a cell which cannot develop further until it fuses with another gamete. Thus a spermatozoon and an egg have no future unless they unite to form a **zygote** which then develops into an adult organism. In the process of fertilization the nuclei of the two gametes fuse to form the nucleus of the zygote. Now if the gametes were to have two of each type of chromosome, the zygote would obviously have four of each type, twice the normal number. Let us assume that this zygote develops into an adult which itself produces gametes with four of each type of chromosome. If two of these gametes were to unite, a zygote would be formed with eight of each type of chromosome, and so on. In other words if gametes were formed by mitosis rather than meiosis, the chromosome number would double with each succeeding generation. By halving the chromosome number meiosis ensures that this does not happen. When gametes with the haploid number of chromosomes unite, the normal diploid condition is restored.

THE ESSENTIAL PRINCIPLE UNDERLYING MEIOSIS

How does meiosis achieve this halving of the chromosome number? The answer lies in the behaviour of the chromosomes during prophase and metaphase. Meiosis consists of two successive divisions: the parent cell splits into two (**first meiotic division**) and the products then divide again (**second meiotic division**), giving a total of four daughter cells. In the first division homologous chromosomes get separated from each other and go into different cells. The second division is concerned with separating the chromatids. With these basic ideas in mind let us examine in detail the events that take place in meiosis (Fig 23.6). As in mitosis the process is divided for convenience into a series of stages. These are given the same names as in mitosis but each is followed by I or II indicating whether it belongs to the first or second meiotic division.

PROPHASE I

In many respects this is similar to the prophase of mitosis: the nucleolus disappears, the centrioles, if present, become arranged at opposite sides of the nucleus and the chromosomes condense (Fig 23.6B–D). However, there is one fundamental difference. In mitosis homologous chromosomes do not associate with each other in any way. But in meiosis they come to lie side by side, a process known as **synapsis**. In this condition each pair of homologous chromosomes constitutes a **bivalent**. As prophase proceeds the chromosomes may become intimately coiled round each other, and later, when they move slightly apart, the chromatids of the two homologous chromosomes remain in contact at certain points called **chiasmata** (singular **chiasma**). Extremely important genetic changes may occur in association with the chiasmata, but these are dealt with elsewhere (see p. 572). Although chiasma-formation is typical of most meiotically dividing cells, we shall start by considering a pattern of meiosis where it does not occur. Such **achiasmate divisions** are by no means uncommon and they are simpler to understand than the more complex situations involving the formation of numerous chiasmata.

A) Interphase

Cell in normal non-dividing condition with chromosomes long and threadlike.

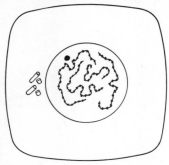

B) Early prophase I

Chromosomes contract, becoming more clearly visible. Nucleolus shrinks.

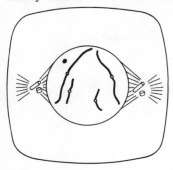

C) Mid prophase I

Homologous chromosomes come together (synapsis) forming a bivalent.

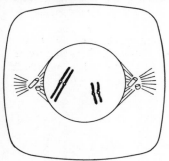

D) Late prophase I

Each chromosome seen to consist of a pair of chromatids.

METAPHASE I

As in mitosis the chromosomes move to the equator of the spindle (Fig 23.6E). The important difference from mitosis is that here homologous chromosomes do this together; in other words each bivalent behaves as a unit. Moreover they arrange themselves in such a way that sister chromatids orientate towards the same pole, whereas the two homologous chromosomes orientate towards opposite poles. So the positioning of the chromosomes at this stage is quite different from what it is in mitosis. The cell is now poised ready for the separation of the homologous chromosomes.

ANAPHASE I

The homologous chromosomes, each made up of a pair of chromatids joined at the centromere, move towards opposite poles of the spindle. The sister chromatids no longer remain in parallel alignment but diverge from one another (Fig 23.6F).

TELOPHASE I

When the chromosomes reach their respective poles the cell generally divides across the middle as in mitosis (Fig 23.6G). Usually the daughter cells go into a short resting state (interphase), but sometimes the chromosomes remain condensed and the daughter cells go straight into prophase of the second meiotic division.

THE SECOND MEIOTIC DIVISION

Separation of homologous chromosomes has already been achieved by the first meiotic division, and the purpose of the second division is merely to separate the chromatids from one another. In prophase II (Fig 23.6H) a new spindle apparatus is formed. In metaphase II (Fig 23.6I) the chromosomes move to the equator of the spindle, the chromatids orientating towards opposite poles as in mitosis. Anaphase II (Fig 23.6J) sees the chromatids separating and moving apart from each other. The chromatids become the chromosomes of the daughter cells. On reaching the poles the cell divides across the middle in the usual way (Telophase II: Fig 23.6K). Characteristic of telophase, the spindle apparatus disappears, and the nucleolus and nuclear membrane are reformed. The chromosomes uncoil and regain their threadlike form. Meiosis is complete: four cells (sometimes referred to as a **tetrad**) have been formed, and each cell has the haploid number of chromosomes.

From this brief description of meiosis it is clear that of the two successive divisions it is the first which is responsible for separating the homologous chromosomes and halving the chromosome number. For this reason the first meiotic division is described as a **reduction division**. The function of the second is to separate the chromatids from each other.

CHIASMATA

As was mentioned earlier meiosis generally involves the formation of chiasmata (**chiasmate meiosis**). In the course of prophase I the intimate association between homologous chromosomes weakens so that they move slightly apart. It can now be seen that the chromatids are attached to each other at certain points. These are the chiasmata.

Chiasma formation is very varied. Every bivalent forms at least one chiasma, but in fact as many as eight may be present, and these can be formed between any two of the four chromatids (Fig 23.7).

E) Metaphase I
Chromosomes arrange themselves on equator of spindle

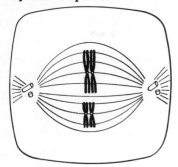

F) Anaphase I
Homologous chromosomes part company and migrate to opposite poles of the cell.

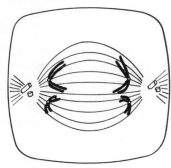

G) Telophase I
The chromosomes have reached their destination and the cell constricts across the middle as in mitosis.

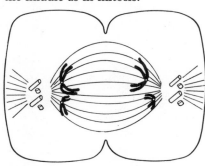

H) Prophase II
The two daughter cells prepare for the second meiotic division: centrioles have replicated and a new spindle is formed.

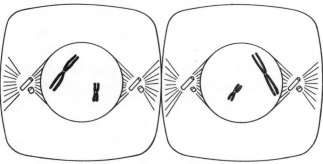

I) Metaphase II
Chromosomes arrange themselves on the spindle in the usual way.

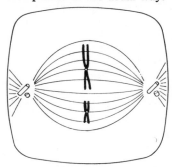

J) Anaphase II
Chromatids part company and migrate to opposite poles of the cell.

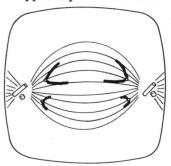

K) Telophase II
Cell constricts across the middle and the nuclear membranes and nucleoli are reformed as at the end of mitosis. Chromosomes regain their threadlike form and the cells go into the resting state (interphase).
Note that the daughter cells have half the number of chromosomes present in the original parent cell.

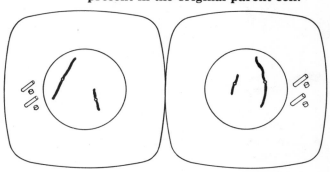

Fig 23.6 Meiosis in a generalized animal cell. As in the diagrams of mitosis only two pairs of chromosomes are shown: a long pair and a short pair. The sequence of diagrams starts at the top of p. 374: compare the sequence of events with mitosis (p. 368). In C–E homologous chromosomes are shown diagrammatically lying side by side; in reality the association is so intimate that at first the four chromatids cannot be distinguished from each other.

Fig 23.7 Chiasmata. A) is a photomicrograph of a bivalent at late prophase from testis of grasshopper. B) is an interpretive diagram. Chiasmata are numbered according to which chromatids are in contact. Note that chiasmata can be formed between any two chromatids. (*Bernard John, University of Southampton*)

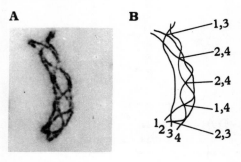

Except at the chiasmata the homologous chromosomes in a bivalent become increasingly separated from one another. This results in the bivalents adopting characteristic shapes, which vary according to the number and position of the chiasmata. A bivalent with only one chiasma assumes a cross-shaped configuration; the formation of two chiasmata results in a ring; and where three or more are present a series of interconnected loops results. However, not all the loops need necessarily be visible because consecutive ones may be at right angles to each other like links in a chain. Moreover, the chromosomes continue to contract at this stage and this, coincident with the separation of homologous chromosomes, has the effect of pushing the chiasmata towards the end of the bivalent, a process called **terminalization**. This may result in changes in shape of the bivalents which are sometimes difficult to interpret. Bearing all this in mind can you interpret the meiotic figures seen Fig 23.8?

Chiasmata have two functions; one mechanical, the other genetic. The mechanical function is to hold the two homologous chromosomes together while they manoeuvre themselves onto the spindle prior to segregation. The genetic function is discussed in detail in Chapters 29 and 35 and will only be touched on here. The chiasmata represent places where the chromatids may break and rejoin, a portion of one chromatid changing places with the equivalent portion of another. This enables exchange of genes to occur between homologous chromosomes, a process known as **crossing over**. Since the two homologous chromosomes contain different genes, crossing over promotes genetic variety, which in turn plays an important part in the process by which evolution has taken place.

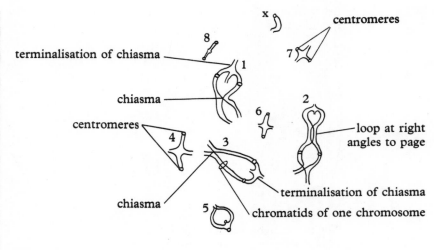

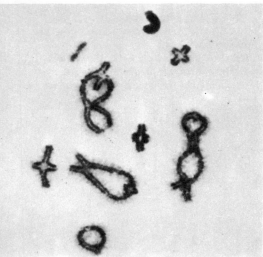

Fig 23.8 Chromosomes at late prophase in a meiotically dividing cell from grasshopper testis. A) photomicrograph; B) interpretive diagram. Note pairing of homologous chromosomes. In bivalents 1, 2, 3, and 4 each chromosome can be seen to consist of a pair of chromatids. Individual chromatids cannot be detected in bivalents 5, 7, and 8, and only with difficulty in 6. The **X** chromosome has no partner. (*Bernard John, University of Southampton*)

OBSERVING MEIOSIS

To observe stages in meiosis one must choose a reproductive tissue in which gametes or spores are being produced (see Chapter 24). In flowering plants meiosis can be observed in developing pollen grains or embryo sacs. This means examining the contents of anthers and ovules repectively. In animals meiosis can be observed in the testes or ovaries. In all these cases the tissues must be stained appropriately in order to show up the chromosomes, this being achieved either by sectioning or making squash preparations. The photomicrographs shown in Fig 23.9 were obtained from the testis of the grasshopper, a convenient choice because the chromosomes are large and

A **B** **C** **D**

E **F** **G** **H**

comparatively few in number, enabling individual chromosomes to be seen and followed clearly. Examine the photomicrographs carefully and compare them with the diagrams in Fig 23.6.

THE DIFFERENCES BETWEEN MITOSIS AND MEIOSIS

This brief account of cell division will have shown that there are many detailed differences between mitosis and meiosis. However the one salient difference is that in meiosis homologous chromosomes associate with one another, whereas in mitosis they do not. This is clearly seen in late prophase when the chromosomes make their appearance as distinct bodies, and again at metaphase when they arrange themselves on the spindle (Fig 23.10). If you think about it, all other differences, including dissimilarities in the genetic constitutions of the daughter cells, stem from this one fundamental difference in the behaviour of the chromosomes.

Fig 23.9 Meiosis in testis of grasshopper. This organism has 8 pairs of chromosomes plus, in the male, one **X** chromosome. Note halving of the chromosome number: ignoring the **X** chromosome, the parent cell has a total of 16 chromosomes (diploid), whereas the products have only 8 (haploid). The **X** chromosome has no partner: in the first meiotic division it goes to one of the two daughter cells, and its constituent chromatids separate in the second meiotic division. (*Bernard John, Australian National University*)

A) Metaphase of mitosis **B) Metaphase I of meiosis**

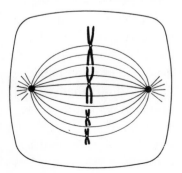

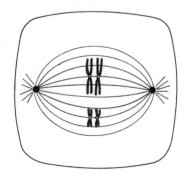

Fig 23.10 The essential difference between mitosis and meiosis lies in the behaviour of the chromosomes. This can be seen at metaphase when they arrange themselves on the spindle. In mitosis (A) homologous chromosomes do not associate with one another, whereas in meiosis (B) they come together and then segregate.

24 Reproduction

In Chapter 23 we saw how individual cells multiply. We shall now concern ourselves with reproduction of the whole organism.

Organisms reproduce in two different ways, sexually and asexually. **Sexual reproduction** generally involves the fusion of specialized sex cells, the **gametes**, derived from two different individuals. Gametes are always haploid, and they cannot develop further unless they fuse appropriately. In contrast, **asexual reproduction** does not involve gametes, nor the participation of two individuals.

It is important to realize from the outset that reproductive processes, as they have come to be evolved, entail more than just proliferation of the organism. In as much as it involves fusion of gametes, sexual reproduction brings about the mixing of genetic material from two different individuals. This is important in confering new characteristics on the offspring, a matter we shall return to in later chapters. We shall see also that reproduction is closely associated with protection against adverse conditions, and survival over unfavourable periods.

THE GENETIC IMPORTANCE OF SEX: BACTERIAL CONJUGATION

The genetic importance of sexual reproduction has been elegantly demonstrated by studies on bacteria. In 1946 Joshua Lederberg and Edward Tatum produced the first reliable evidence that bacteria have sex. They were working on the colon bacillus *Escherichia coli*, which inhabits the mammalian large intestine. *E. coli* can be grown on a nutrient medium containing only glucose and inorganic salts. All the amino acids required for the synthesis of its proteins, the bacteria can manufacture for themselves. Tatum and Lederberg found that radiation treatment produced mutant strains lacking the ability to synthesize one or more amino acids[1]. To enable such strains to survive, it was necessary to add these particular amino acids to the minimal medium. In one particular experiment two mutant strains were obtained, one incapable of synthesizing amino acids A, B and C, the other incapable of synthesizing D, E and F. The two strains were then grown together on a medium containing all six amino acids. After several hours, during which the bacteria reproduced, single cells were isolated from the culture, and tested for their nutritional requirements by growing them on different media. It turned out that some of the bacterial cells were capable of growing on the minimal medium, i.e. they were able to synthesize all the required amino acids.

It could be argued that these particular bacteria, or their immediate ancestors, had undergone a mutation back to the original type. However, this

[1] A mutant strain is one displaying heritable abnormalities, and it is descended from cells whose genes have undergone a spontaneous change known as a mutation (see p. 574).

can be discounted on the grounds that it is very unlikely that three genes in both mutants would have changed in such a short time: the probability of this happening would be about one in 10^{15}! It is, therefore, concluded that some kind of genetic exchange had taken place between the two strains, the genes responsible for the synthesis of A, B and C being recombined with those responsible for the synthesis of D, E and F.

Such genetic recombination presupposes some kind of **conjugation**. This has now been confirmed. When bacterial cells collide, a temporary bridge, visible in the electron microscope, may be set up between them (Fig 24.1). Through this bridge genetic material passes from one to the other. One of the two conjugants acts as the **donor**, or 'male', the other as a **recipient**, or 'female'. To be a donor, the bacterium must contain a **sex factor** which is absent from the other conjugant. The injection of this factor into a recipient turns the latter into a 'male' which henceforth behaves as a donor.

How do we know all this? From isotope labelling and special staining techniques, it has been established that *E. coli* has only one chromosome. Normally conjugation takes about two hours during which the entire chromosome may be injected into the 'female'. Now the conjugants can be separated at any moment after the beginning of pairing, by violently agitating them. In this process the bridges are broken, and the conjugants torn apart. Any change in the nutritional characteristics of the bacteria can then be determined by growing them on different media. New characteristics gained can be related to the genetic material injected by the donor.

To cut a very long story short, it has been found that the longer the period of conjugation, i.e. the longer the time before the conjugants are separated, the greater the number of new nutritional characteristics acquired by the recipient, i.e. the greater the amount of genetic material transferred.

Here we see sexual reproduction at its simplest: the direct transfer of genetic material from one organism to another. Though much more elaborate, this is essentially what happens in higher organisms, as we shall see.

Fig 24.1 The simplest form of sexual reproduction is illustrated by conjugation between two colon bacteria (*Escherichia coli*). Evidence is presented in the text suggesting that when a bridge is set up between two bacterial cells as seen in the electron micrograph (A), the chromosome belonging to one of the two bacteria is injected into the other as depicted in B. The conjugants may separate from one another at any time after the beginning of the injection process. The number of new transmissible characteristics acquired by the recipient depends on how much of the donor's chromosome is injected into it. (A: *François Jacob, Institut Pasteur, Paris*)

A

B

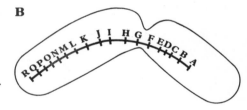

EVOLUTIONARY DEVELOPMENT OF GAMETES

In most organisms sexual reproduction involves the union of gametes, a process known as **syngamy**. Generally the gametes differ from each other in structure, size and behaviour, for which reason they are known as **heterogametes**. Eggs and sperm are heterogametes, syngamy taking place by a process of **fertilization** in which a sperm penetrates an egg and donates its nucleus to it. However, primitive organisms, such as the green algae and fungi, often produce gametes that are identical with each other (**isogametes**), or only very slightly different (**anisogametes**). In many unicellular organisms the gametes, as well as being identical with each other, are also structurally similar to the parent cells. At the simplest level there are no gametes at all, genetic material being transferred directly from one individual to another. This is essentially what happens in bacteria and, with certain elaborations, in some unicells.

To be of any genetic value syngamy must occur between gametes derived from different parents, and it is interesting that even in isogamous organisms fusion does not occur between gametes from the same parent. Although structurally identical, the gametes can recognize whether or not they come from the same parent, and react accordingly. In some algae and fungi it is

Fig 24.2 Structure of spermatozoa. A) Detailed structure of human spermatozoon based on electron micrographs. B) Spermatozoa of other organisms for comparison. (*based on various authors*)

possible to separate members of a single species into **plus** and **minus strains**: plus conjugates with minus, but plus will not conjugate with plus, nor minus with minus. This foreshadows the development of separate sexes in later evolution. Some primitive organisms also show a tendency for one gamete to be migratory and the other stationary, and this foreshadows the development of **sperm** and **eggs** in higher forms. Fully differentiated eggs and sperm are found in organisms as primitive as seaweeds and coelenterates, and the sperm produced by these lowly forms are basically similar to man's.

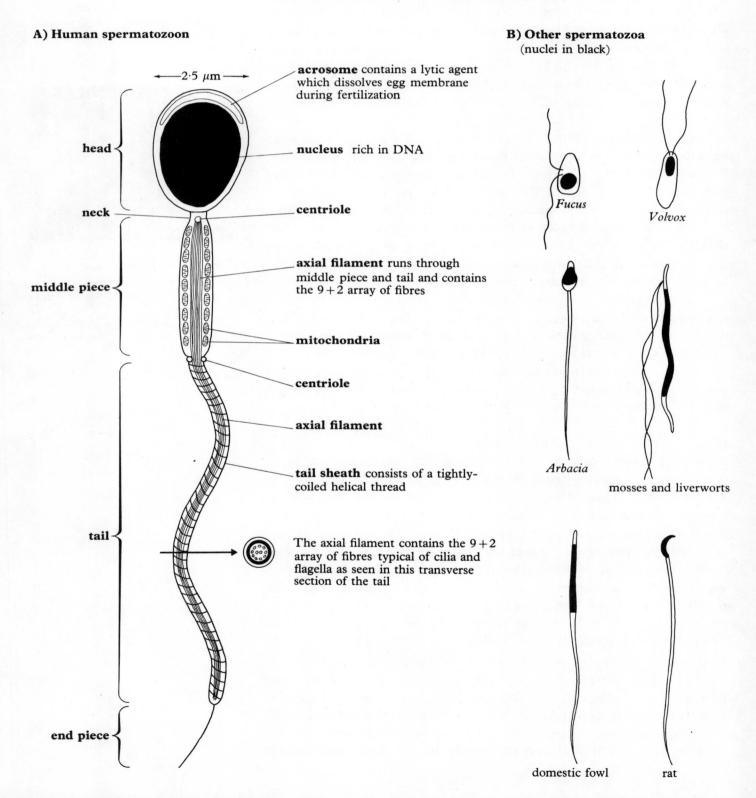

A) Human spermatozoon

2·5 μm

acrosome contains a lytic agent which dissolves egg membrane during fertilization

nucleus rich in DNA

head

neck

centriole

middle piece

axial filament runs through middle piece and tail and contains the 9 + 2 array of fibres

mitochondria

centriole

axial filament

tail sheath consists of a tightly-coiled helical thread

tail

The axial filament contains the 9 + 2 array of fibres typical of cilia and flagella as seen in this transverse section of the tail

end piece

B) Other spermatozoa
(nuclei in black)

Fucus

Volvox

Arbacia

mosses and liverworts

domestic fowl

rat

THE SPERMATOZOON

The structure of the human spermatozoon is now fairly well understood, thanks largely to the electron microscope (Fig 24.2A). It is divided into five regions: **head, neck, middle piece, tail** and **end piece**. The head carries genetic material in a large **nucleus** that very nearly fills it. The nucleus is rich in DNA, which accounts for more than half the dry weight of the head. The amount of DNA present in sperm nuclei is constant, and is half that found in somatic cells. This is because it is **haploid**, containing only half the full number of chromosomes. The nucleus is surmounted by a thin cap, the **acrosome**, which we shall see plays an important part in fertilization.

The rest of the spermatozoon is concerned with propulsion. Running down the centre of the middle piece and tail is an **axial filament** consisting of two central fibres surrounded by nine peripheral ones, the famous 9 + 2 **structure** typical of cilia and flagella. The sperm tail is thus a modified flagellum. By lashing from side to side, it propels the spermatozoon in the liquid medium through which it has to swim before reaching the egg. In the middle piece the axial core is surrounded by densely-packed **mitochondria**. Rich in respiratory enzymes, the middle piece is concerned with releasing energy for driving the spermatozoon.

The sperm of other organisms show many variations on the theme outlined above, but most of them have in common a prominent nucleus, and either cilia or flagella for motility (Fig 24.2B).

THE EGG CELL

Since eggs do not have to propel themselves they have a less complicated structure than spermatozoa and are very much larger. The human egg is just over 0.1 mm in diameter (120 μm), compared with the spermatozoon whose head is only 2.5 μm across in the widest part.

Fig 24.3 shows a generalized egg cell. A large haploid **nucleus** is surrounded by dense **cytoplasm** containing the usual organelles found in any living cell. In addition there are two specialized inclusions: **cortical granules** and **yolk droplets**. The cortical granules form a layer just beneath the surface. The yolk, a mixture of protein and fat, is concentrated towards the lower end of the egg, away from the nucleus. This gives the egg a definite polarity: the non-yolky end with the nucleus and active cytoplasm is called the **animal pole**. The more inert yolky end is the **vegetal pole**. The function of the yolk is to provide a source of nourishment for the embryo, at least in its early stages of development. Eggs differ markedly in the amount of yolk they contain (compare for example the vast quantity of yolk in a bird's egg with the much smaller amount in a frog's egg), and this can have a profound effect on later development. Some eggs have no yolk at all, in which case nourishment is supplied either by separate yolk-filled cells or by a nutritive tissue surrounding the embryo. The endosperm tissue in flowering plants is a good example of the latter (see p. 398).

The egg cell is surrounded by an envelope consisting of two unit membranes: the inner one is the **plasma membrane**, the outer one the so-called **vitelline membrane**. Beyond this is a **jelly coat** of variable thickness made of glycoprotein.

FERTILIZATION

For various reasons it is difficult to study fertilization in mammals, and our

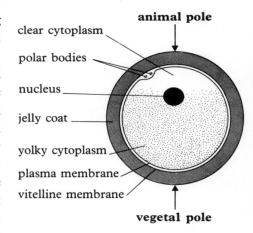

Fig 24.3 Diagram of a generalized animal egg cell. The polar bodies are small non-functional eggs formed during meiosis (see p. 384). The cytoplasm towards the animal pole contains the usual cell organelles, e.g. mitochondria, endoplasmic reticulum etc.

animal pole

clear cytoplasm

polar bodies

nucleus

jelly coat

yolky cytoplasm

plasma membrane

vitelline membrane

vegetal pole

Fig 24.4 Acrosome reaction in the sea urchin *Arbacia*. When the tip of the sperm head comes into contact with the vitelline membrane of the egg, the acrosome turns inside out forming the slender filament and releasing the lytic agent which softens the vitelline membrane. (*modified after Austin*)

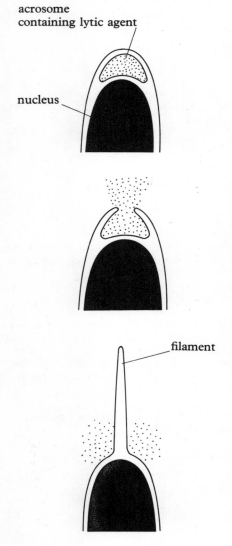

acrosome
containing lytic agent

nucleus

filament

knowledge of what happens is based largely on lower animals such as sea urchins and marine worms. The eggs of such animals are easily collected, and fertilization takes place externally in sea water where it can be observed under the microscope. The following description is based on the sea urchin *Arbacia*, but as far as is known the main sequence of events is the same in other organisms as well.

The spermatozoa come into contact with the eggs by random movement. There is no evidence that in animals the sperm are attracted to the egg, but in lower plants this certainly happens. For example, in bracken (bracken has sperm, believe it or not), the eggs secrete malic acid to which the sperm are attracted, an example of **chemotaxis** (see p. 348).

When the head of a spermatozoon hits the vitelline membrane, a remarkable process takes place (Fig 24.4). The acrosome at the tip of the head bursts open, releasing a substance which softens the vitelline membrane at the point of contact. Meanwhile the membrane lining the acrosome turns inside out, forming a thin filament which pierces the egg membranes. The whole process is called the **acrosome reaction**, and it enables the spermatozoon to penetrate into the cytoplasm of the egg. Further sperm are prevented from entering by a rapid chemical change at the surface of the egg, the visible sign of which is the thickening of the vitelline membrane. This results from the cortical granules migrating through the plasma membrane and applying themselves to the inner surface of the vitelline membrane. Fluid accumulates immediately beneath the now thickened vitelline membrane, with the result that the latter appears to lift off from the egg surface. Now known as the **fertilization membrane**, it serves as an effective barrier to the entry of further spermatozoa.

After a spermatozoon has penetrated the egg, its tail is usually discarded. The head and middle piece are drawn, by forces not yet understood, through the cytoplasm towards the nucleus. There is much variation in the way the two nuclei fuse. Commonly the nuclear membranes break down, a spindle is formed, and the now-visible sperm and egg chromosomes (the **paternal** and **maternal chromosomes** respectively) arrange themselves on the spindle as in mitosis. The diploid number of chromosomes is thus restored and the fertilized egg, or **zygote**, is ready for its first mitotic division. The sequence of events involved in fertilization is summarized in Fig 24.5.

EVOLUTION OF REPRODUCTIVE METHODS

Fundamental to sexual reproduction is the method by which eggs and sperm are brought together. At its simplest this takes place by the gametes of both sexes being liberated into the surrounding water (**external fertilization**). This method is obviously only possible for aquatic organisms, or for terrestrial animals such as amphibians that return to water for breeding. In most terrestrial animals, however, the sperm are introduced into the female, fertilization occurring inside her body (**internal fertilization**). Generally necessitating the use of some kind of **intromittent organ** (e.g. penis), internal fertilization has two great advantages:

(1) it is a surer method with better chances of sperm meeting eggs;

(2) it means that the fertilized egg can be enclosed within a protective covering before it leaves the female's body, or, better still, it can be retained within the body of the female and protected there during development. This last has been exploited by the placental mammals in which the embryo is

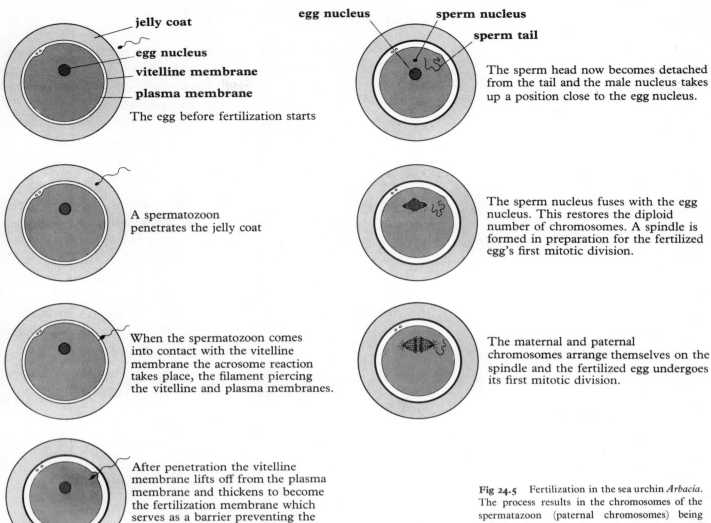

jelly coat
egg nucleus
vitelline membrane
plasma membrane

The egg before fertilization starts

A spermatozoon
penetrates the jelly coat

When the spermatozoon comes
into contact with the vitelline
membrane the acrosome reaction
takes place, the filament piercing
the vitelline and plasma membranes.

After penetration the vitelline
membrane lifts off from the plasma
membrane and thickens to become
the fertilization membrane which
serves as a barrier preventing the
entry of further spermatozoa.

fertilization membrane

egg nucleus   sperm nucleus
sperm tail

The sperm head now becomes detached
from the tail and the male nucleus takes
up a position close to the egg nucleus.

The sperm nucleus fuses with the egg
nucleus. This restores the diploid
number of chromosomes. A spindle is
formed in preparation for the fertilized
egg's first mitotic division.

The maternal and paternal
chromosomes arrange themselves on the
spindle and the fertilized egg undergoes
its first mitotic division.

Fig 24.5 Fertilization in the sea urchin *Arbacia*. The process results in the chromosomes of the spermatazoon (paternal chromosomes) being united in the same nucleus as the chromosomes of the egg (maternal chromosomes). Since the gametes are haploid, the fertilized egg (zygote) is diploid.

protected and nourished within the womb. This results in the offspring being born at a comparatively advanced stage of development, a phenomenon known as **viviparity**. Terrestrial plants are comparable: although they do not possess intromittent organs as such, special techniques have evolved for transferring the male gametes to the egg cells, early development of the embryo taking place within the body of the parent plant.

After birth the new-born mammal is nourished with milk produced by the mother's mammary glands. Milk-production (**lactation**) is but one manifestation of the fact that mammals care for their young, in which respect they are rivalled only by birds. Not only do they feed and protect them while they are unable to fend for themselves, but later they may teach them to cope with various situations in the environment, including hunting for their own food. Particularly well developed in the great cats and social mammals, training reaches its climax in man. Viviparity, coupled with care of the young, raises the chances of survival of the individual and has undoubtedly contributed to the success of the mammals as a group.

With these generalizations in mind let us look at the details of sexual reproduction in two terrestrial organisms, the mammal and flowering plant.

Sexual Reproduction in Man

In delving into the problem of how man reproduces his kind, three basic questions have to be answered:

(1) How and where are the eggs and sperm produced?

(2) How are they brought into contact with each other?

(3) What happens to the zygote?

Although we shall discuss these questions in relation to man, the basic principles apply to many other animals as well. Let us take each question in turn.

GAMETOGENESIS

Formation of gametes, or **gametogenesis**, takes place in the **gonads**: the **testis** produces sperm (**spermatogenesis**), and the **ovary** eggs (**oögenesis**). The overall sequence of events is essentially the same in both sexes. Cells in a particular region of the gonad, we shall see exactly where later, divide repeatedly. The daughter cells grow, divide again, and then differentiate into the appropriate gametes, eggs or sperm as the case may be. The original cells, called the **primordial germ cells**, are diploid, but the gametes are haploid. Gametogenesis thus involves meiosis. The full sequence of events is shown in Fig 24.6.

Fig 24.6 Summary of the sequence of events involved in the formation of sperm and eggs in a mammal. Explanation in text.

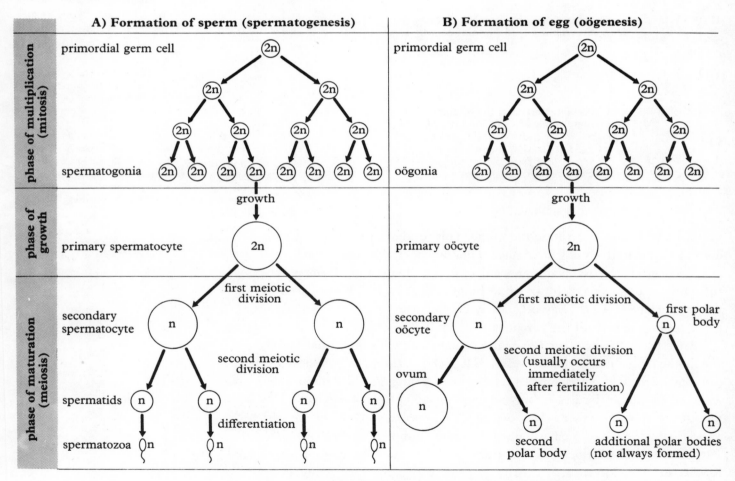

In spermatogenesis primordial germ cells divide repeatedly by mitosis to form diploid **spermatogonia**. There then follows a brief period of growth, during which each spermatogonium increases in size to form a primary **spermatocyte**. This then divides meiotically, the first meiotic division giving two **secondary spermatocytes,** and the second a group of four **spermatids.** Finally each spermatid, at this stage a seemingly unspecialized cell, differentiates into a **spermatozoon** with characteristic head and tail. Clearly a large number of spermatozoa are formed from a single primordial germ cell. As the latter themselves are very numerous the total number of spermatozoa that can be formed is very great indeed. We shall have more to say on this point presently.

Oögenesis is basically similar to spermatogenesis, but different in detail. A primordial germ cell proliferates mitotically to form **oögonia,** but only one of these grows into a **primary oöcyte.** The others degenerate. The amount of growth that takes place at this stage is much greater than in spermatogenesis: this is what makes the egg so much larger than the spermatozoa. The primary oöcyte now undergoes meiosis, but the divisions are unequal resulting in the products differing greatly in size. The first meiotic division produces a **secondary oöcyte** and a very much smaller **polar body** which may be seen adhering to it. The second meiotic division results in the production of another polar body. The secondary oöcyte now becomes the functional egg or **ovum,** with the haploid number of chromosomes. Meanwhile the first polar body may undergo a second meiotic division resulting in a total of three polar bodies, all haploid of course. The events described are essentially similar to the formation of spermatozoa, but whereas in spermatogenesis all four products of meiosis develop into functional gametes, in oögenesis only one of the four becomes a functional egg. The rest are extruded as non-functional polar bodies which eventually degenerate. The polar bodies are the inevitable result of the fact that meiosis involves two successive divisions; their function is simply to receive half the chromosomes, thereby making the ovum haploid.

HISTOLOGY OF THE TESTIS

If you look at a microscopical section of a mature testis, you can see sperm at different stages of development, as described above. The testis is composed of numerous **seminiferous tubules,** in whose walls spermatogenesis takes place (Fig 24.7). The primordial germ cells are formed from the **germinal epithelium** lining the outside of the tubule, and as cell divisions proceed the daughter cells move towards the lumen of the tubule. The final transformation of spermatids to spermatozoa takes place in the part of the wall immediately adjacent to the lumen. At this stage the heads of the developing sperm are enveloped by large **Sertoli cells,** from which they derive their nourishment. The sperm tails project into the fluid-filled lumen of the tubule. Eventually the mature spermatozoa become detached and are released into the lumen. Spermatogenesis is particularly sensitive to temperature. If the temperature is too high the sperm fail to develop, or develop abnormally. This is why the testes are held in scrotal sacs outside the abdominal cavity, and it is also why the scrotal sacs of hairy mammals are devoid of hair. If the testes are insulated, or moved back into the abdomen by surgical means, a low sperm-count results.

HISTOLOGY OF THE OVARY

Comparable events take place in the ovary (Fig 24.8). Primordial germ cells, derived from the germinal epithelium lining the outside of the ovary, divide to form oögonia which migrate inwards towards the centre. Each oögonium is enveloped by a layer of **follicle cells,** also formed from the germinal epi-

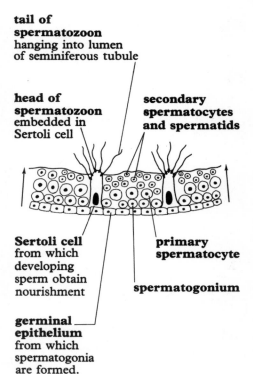

Fig 24.7 Section through the wall of a seminiferous tubule showing the formation of spermatozoa. Spermatogonia are formed by proliferation of the germinal epithelium. These then grow into spermatocytes which divide meiotically into spermatids. The spermatids attach themselves to giant Sertoli cells where they mature into spermatozoa.

tail of spermatozoon hanging into lumen of seminiferous tubule

head of spermatozoon embedded in Sertoli cell

secondary spermatocytes and spermatids

Sertoli cell from which developing sperm obtain nourishment

primary spermatocyte

spermatogonium

germinal epithelium from which spermatogonia are formed.

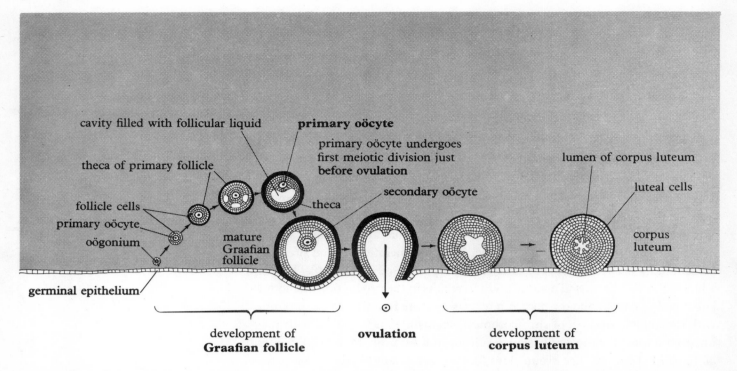

Fig 24.8 Formation of an egg in the mammalian ovary. An oögonium, formed from the germinal epithelium, grows into a primary oöcyte which becomes enclosed within a Graafian follicle. The primary oöcyte undergoes its first meiotic division into a secondary oöcyte shortly before being shed from the ovary. After ovulation the Graafian follicle develops into a corpus luteum.

thelium. The whole structure, about 50 μm in diameter, is called a **primary follicle**. At birth there are between 200,000 and 400,000 primary follicles in each ovary, but only 200 to 400 complete their development. The rest degenerate into **atretic follicles** which are visible in the ovary as small cyst-like bodies that never produce eggs.

Let us follow what happens to one of the viable follicles. As the oögonium grows into a primary oöcyte, the follicle cells surrounding it proliferate to form a wall many cells thick. As this occurs, a fluid, called **follicular liquid**, collects between the cells, forming little pools. As more and more fluid accumulates, the pools coalesce, eventually forming one large cavity filled with follicular liquid. The oöcyte is embedded in a little hillock of follicle cells projecting into this cavity. Meanwhile connective tissue inside the ovary forms a protective sheath surrounding the follicle. This is divided into two layers: a highly vascular **theca interna**, and a less vascular, more fibrous, **theca externa**. The whole thing is now called a **Graafian follicle**, after the seventeenth century physician Regner de Graaf who first described it. While all this is happening, the follicle moves slowly back towards the surface of the ovary, increasing in size as it does so. From a mere 50 μm in diameter, it eventually reaches a diameter of about 12 mm, and causes a distinct bulge on the surface of the ovary. When the time is ripe (see p. 390), this bulge ruptures and the oöcyte, pinched off from its attachment to the wall of the follicle, is extruded from the ovary. This is known as **ovulation**, and in a human female who has reached puberty, it occurs in one of her two ovaries approximately once every 28 days. Shortly before ovulation, the egg, which you will remember is a primary oöcyte at this stage, undergoes its first meiotic division. The second meiotic division does not normally take place until fertilization. After ovulation the rôle of the follicle in protecting and nourishing the egg comes to an end. However, this does not mean that it has no further part to play. What happens to it will be discussed later, but in the meantime we must see how the sperms and eggs are brought together.

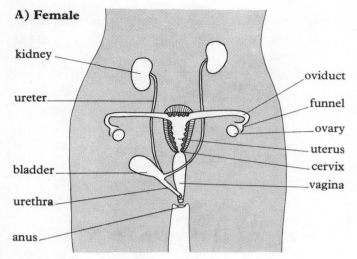

A) **Female**

kidney

ureter

bladder

urethra

anus

oviduct

funnel

ovary

uterus

cervix

vagina

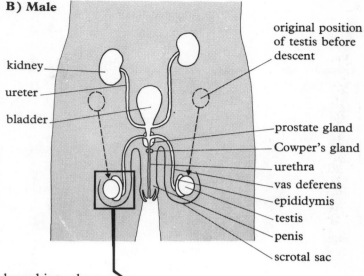

B) **Male**

kidney

ureter

bladder

original position
of testis before
descent

prostate gland

Cowper's gland

urethra

vas deferens

epididymis

testis

penis

scrotal sac

WHAT BRINGS SPERM AND EGGS TOGETHER?

In the majority of aquatic animals eggs and sperm are released into the surrounding water, and it is largely fortuitous as to whether or not they come into contact. Man, in common with other terrestrial animals, has developed a more certain means of bringing the gametes together. In the male a system of tubes conveys sperm from the testes to an intromittent organ, the **penis**, from which they are introduced into the **vagina** of a female. In the female a plumbing system comparable to that of the male conveys the sperm to the upper end of the oviducts, where fertilization takes place.

The anatomy of the male and female reproductive systems is shown in Fig 24.9. For descriptive purposes the reproductive and excretory tracts are considered together as the **urinogenital system**. This is because the two systems are closely associated, particularly in the male where the duct of the penis, the **urethra**, is a common passage for both sperms and urine. Study the diagrams carefully, and notice that although they may look rather complicated, the reproductive tract is really nothing more than a rather complex tube running from the gonad to the exterior.

There are about 1,000 seminiferous tubules in each testis, giving a total length of over 500 metres. Here the sperm are formed, after which they pass, via the **vasa efferentia** and **epididymis**, into the **vas deferens**. For the spermatozoa to be introduced into a female, it is necessary for the penis to be inserted into the vagina in an act of **copulation** or **coitus**. Under conditions of erotic excitement, the arteries in the penis dilate and the veins constrict, and the resulting high blood pressure causes the penis to become erect. This enables copulation to proceed. Repeated tactile stimulation of sensory cells towards the tip of the penis, triggers off a reflex causing contraction of the vas deferens. This sweeps the sperm down into the urethra where they are mixed with secretions from the seminal vesicles and prostate glands. These secretions are essential for maintaining the sperm in a viable and motile state. The resulting suspension, known as **semen**, is expelled from the penis by powerful contractions of the urethra, a process called **ejaculation**. The reflex mechanism bringing about erection and ejaculation also inhibits urination, so there is no possibility of sperm being contaminated with urine. Ejaculation comes as the climax of copulation, the so-called **orgasm**, and it occurs with enough force to project the sperm into the top of the vagina and even into the uterus.

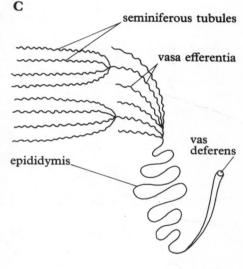

C

seminiferous tubules

vasa efferentia

vas
deferens

epididymis

Fig 24.9 Human urinogenital system in ventral view. In the male, the bladder as well as the vasa deferentia open into the urethra which is therefore a common urinogenital duct carrying both urine and sperm to the outside. In the female, however, the urinary and genital systems open separately to the exterior. C) illustrates diagramatically the organization of the tubules in the testis. Each testis contains approximately 1,000 seminiferous tubules arranged in bundles of about 100. Each bundle opens into a vas efferens, and all the vasa efferentia open into a single epididymis. The epididymis, a coiled tube approximately 6 metres long, leads into the vas deferens. (A *and* B *based on Knowles*)

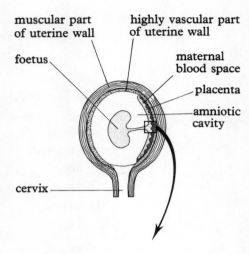

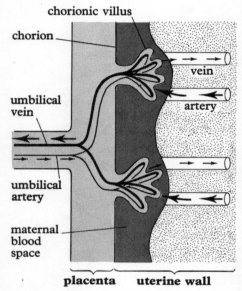

oxygenated blood ——————

deoxygenated blood ————

Fig 24.10 The mammalian foetus is protected within a fluid-filled amniotic cavity inside the uterus and receives its nourishment via the placenta. In the placenta numerous finger-like chorionic villi, containing capillary loops derived from the umbilical artery, project into the maternal blood space in the wall of the uterus. This space is kept charged with blood from branches of the mother's uterine artery. Across the thin barrier separating foetal from maternal blood, exchange of materials takes place: soluble food substances, oxygen, water and salts pass into the umbilical vein from the mother's blood; carbon dioxide and nitrogenous waste, brought to the placenta in the umbilical artery, pass into the mother's blood. The placenta is therefore the foetus' excretory organ as well as its respiratory surface and source of nourishment.

Up to this point the sperm have relied on muscular contraction of the male genital tract to propel them. Once in the female, however, they are at least partly dependent on their own powers of propulsion for further progress. By rhythmical undulations of the tail, they swim through the watery mucus lining the female's genital tract. The ability of the sperm to reach the egg is something to be marvelled at, not only because of the sheer motive power involved (it has been estimated that a spermatozoon swimming the full length of the female genital tract is equivalent to a man swimming the Atlantic – in treacle!), but also because of the seemingly determined way in which they do it. How the spermatozoa propel themselves is not completely understood, but presumably the impressive array of densely-packed mitochondria provides the necessary energy, and we know that they can resort to anaerobic respiration when oxygen is scarce. Nevertheless, it is doubtful if the sperm could ever complete the journey without the aid of muscular contractions of the uterus and oviduct. Perhaps surprisingly there is no evidence that the sperm are attracted to the egg by chemical means, although this is common in lower organisms. Be that as it may, some of the sperm eventually reach the egg in the upper reaches of the oviduct, where fertilization occurs.

Despite all the safeguards only a small proportion of the sperm ever reach the egg. To increase the chances of success, a vast number are produced. The semen of a fertile man generally contains more than 100 million spermatozoa per cm^3, and there can be as many as 300 million in a single ejaculation. Egg production is less prolific: the human female normally sheds only one egg every 28 days during her sexual life of about thirty years. So she only produces about 400 eggs in her entire life. However, it would be wrong to think that the females of all species are as economical as this. The fact that most mammals produce litters rather than single births bears witness to this, and the Icelandic cod can produce well over 50 million eggs in a single spawning!

DEVELOPMENT OF THE ZYGOTE

After fertilization the zygote is propelled down the oviduct by peristaltic contractions. The lining of the oviduct is ciliated, but the cilia are so small in comparison with the zygote that they probably play little or no part in propelling it. By the time the zygote reaches the uterus (in man the journey takes up to a week), it has already divided mitotically to form a hollow sphere of cells, the **blastocyst**. After several days the blastocyst becomes embedded in the lining of the uterus, a process called **implantation**, and there continues its development. The outermost layer of cells surrounding the blastocyst, the **trophoblast**, develops small finger-like outgrowths, **trophoblastic villi**, which project into the surrounding tissue of the uterine wall. Nutriments are absorbed across these villi. As development proceeds the embryo, now known as the **foetus**, becomes enveloped by protective **foetal membranes**. The innermost membrane, the **amnion**, encloses a fluid-filled **amniotic cavity** in which the foetus, protected by the cushioning effect of the fluid, is suspended. As the foetus grows, the amniotic cavity expands until eventually it fills the entire uterus. Meanwhile two other membranes, whose origin is discussed in Chapter 26, unite to form the **placenta**. This develops in association with the uterine wall, and is connected to the foetus by the **umbilical cord** (Fig 24.10). Within the placenta, foetal and maternal blood vessels come into intimate association with each other, and across their walls exchange of respiratory gases, food materials and nitrogenous waste takes place. The umbilical cord, with the placenta at the end of it, is in a literal sense, the foetus' lifeline.

The mammal provides us with an example of **internal development**. Instead of eggs being laid, the embryo is protected and nourished in the uterus of the mother, and the offspring are born at a reasonably advanced stage of development (**viviparity**). To accommodate the growing foetus during the period of pregnancy the uterus expands enormously, and this is accompanied by thickening of its wall and a great increase in vascularization. When the foetus reaches a certain size birth, or **parturition**, takes place.

In the process of parturition the foetus, and shortly afterwards the placenta, are expelled through the cervix and vagina by powerful contractions of the uterus. At this stage the **mammary glands** in the breasts, which have been developing greatly during pregnancy, start secreting milk (**lactation**). This provides the baby's main source of nourishment during its first few months of life.

The time from fertilization to birth is called the **gestation period**. In man it lasts approximately nine months, but in other mammals it ranges from as little as 18 days (mice) to as long as 18 months in the Indian elephant.

The events described above cannot take place throughout an individual's life. At first the gonads do not function properly and this continues until the onset of **puberty** at the age of about 12 in girls, 14 in boys, when eggs and sperm start to be produced. Men can go on producing sperm until the age of about 70, but a woman stops ovulating at about 45, when she reaches the so-called **menopause**.

THE SEXUAL CYCLE

The striking thing about the female's rôle in reproduction is that all the events are synchronized, so that each occurs at just the right moment. For example, long before ovulation takes place the uterine wall starts to prepare itself for implantation, so that by the time the fertilized egg reaches the uterus the lining is ready to receive it. Similarly, changes in the structure of the uterine wall during pregnancy, and the onset of lactation at birth, occur at the appropriate moments. What controls the timing of these events?

The answer is rooted in the fact that the female's reproductive behaviour occurs in a cycle. This is called the **sexual cycle**, or alternatively the **menstrual** or **oestrous** cycle. The events that occur in the course of the cycle follow a set pattern which is regulated by hormones. These are secreted from the pituitary gland and the ovary itself. If pregnancy occurs, the normal cyclical pattern is interrupted and a third source of hormones comes into play: the placenta.

In the human female the most obvious, and rather tiresome, manifestation of the sexual cycle is the monthly discharge of blood known as **menstruation**. But this merely marks the end of a whole series of changes that have occurred in her body during the previous 28 days. To understand what has happened we must return to the ovary.

During the first 14 days following the beginning of a menstrual discharge, the Graafian follicle develops as described earlier. After ovulation, the empty follicle undergoes certain changes (Fig 24.8). The follicle cells enlarge considerably, and a yellow pigment accumulates in them. Eventually

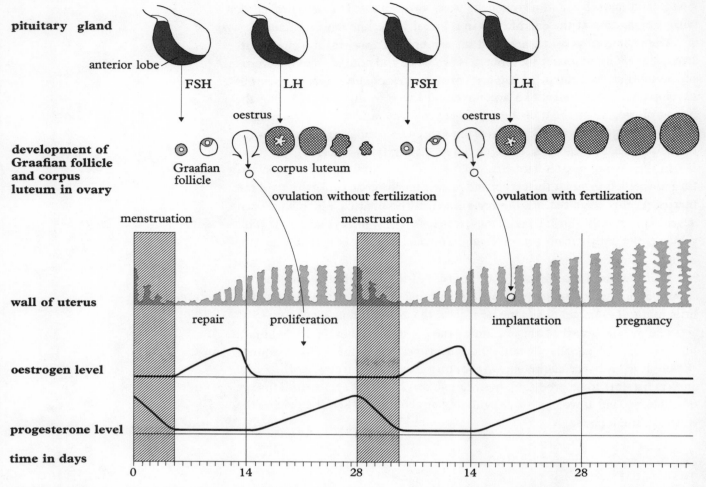

pituitary gland

anterior lobe

FSH   LH   FSH   LH

oestrus   oestrus

development of
Graafian follicle
and corpus
luteum in ovary

Graafian
follicle

corpus luteum

ovulation without fertilization   ovulation with fertilization

menstruation   menstruation

wall of uterus

repair   proliferation   implantation   pregnancy

oestrogen level

progesterone level

time in days

0   14   28   14   28

Fig 24.11 The sexual cycle of the human female showing the relationship between the secretions of the pituitary gland, the development of the Graafian follicle and corpus luteum, the degeneration and repair of the uterine wall, and the levels of oestrogen and progesterone in the blood. Further explanation in text.   (*based on Schröder*)

they fill the cavity of the original follicle, turning it into a solid **corpus luteum** ('yellow body'). Assuming that the egg is not fertilized, the corpus luteum persists in the ovary for about 14 days, and then degenerates. Now while the follicle and corpus luteum are developing, the wall of the uterus prepares itself for receiving a blastocyst. The inner layers, which together make up the **endometrium**, become thickened, folded and invaded with numerous blood vessels in preparation for implantation. If, however, fertilization does not occur, the unfertilized egg never becomes implanted, and passes straight out of the genital tract. Under these circumstances the corpus luteum degenerates, and the endometrium of the uterus breaks down and sloughs off. This is accompanied by the loss of a considerable quantity of blood and constitutes menstruation.

These changes, occurring in the ovary and uterus, are synchronized by hormones (Fig 24.11). Just after menstruation, the anterior lobe of the **pituitary gland** starts secreting **follicle stimulating hormone (FSH)**. FSH has two effects: it causes a Graafian follicle to develop in the ovary, and it stimulates the tissues of the ovary to secrete another hormone called **oestrogen**. The immediate effect of oestrogen is to bring about healing and repair of the uterine wall following menstruation. In the course of the next two weeks the amount of oestrogen in the body builds up until its concentration reaches a point when it stimulates the anterior pituitary to start producing a second hormone. This is called **luteinising hormone (LH)**.

LH brings about ovulation and causes the Graafian follicle to change into a corpus luteum. It then stimulates the corpus luteum to secrete yet another hormone, **progesterone**, whose principal function is to cause proliferation of the uterine wall in preparation for implantation. Progesterone also inhibits FSH–production by the anterior pituitary, so no further follicles develop and oestrogen production is reduced. During the following two weeks the concentration of progesterone gradually rises and, as it does so, it inhibits the production of LH from the anterior pituitary. This in turn causes the corpus luteum to stop secreting progesterone, at which point menstruation takes place. With the sudden drop in the level of progesterone, the anterior pituitary starts secreting FSH again, and the cycle is repeated.

Our knowledge of the control mechanism described above is based mainly on experiments performed on small mammals like hamsters and guinea pigs. Space precludes a detailed account of these experiments, but one will be mentioned to illustrate the kind of investigation that has been carried out. At one time there was some uncertainty as to whether LH or FSH causes ovulation. That it is LH and not FSH was shown by administering FSH to young animals whose ovaries contained developing follicles. It was found that a dose sufficient to bring about ovulation in an intact animal, failed to do so in an animal whose pituitary gland, the source of LH, had been removed.

The interaction between the different hormones illustrates the principle of **negative feedback** developed in Chapter 13. Because they stimulate the gonad, the two pituitary secretions, FSH and LH, are termed **gonadotrophic hormones**. These, in turn, are regulated by the ovarian hormones secreted by the gonads themselves. The system is summarized in Fig 24.12.

IN THE EVENT OF PREGNANCY

The sequence of events described above is what occurs if fertilization does not take place. If, however, the egg is fertilized it becomes implanted in the uterine wall, and **pregnancy** results (Fig 24.11). Under these circumstances

Fig 24.12 Scheme summarizing the interaction of the hormones controlling the female sexual cycle. Black arrows signify stimulation; broken arrows inhibition.

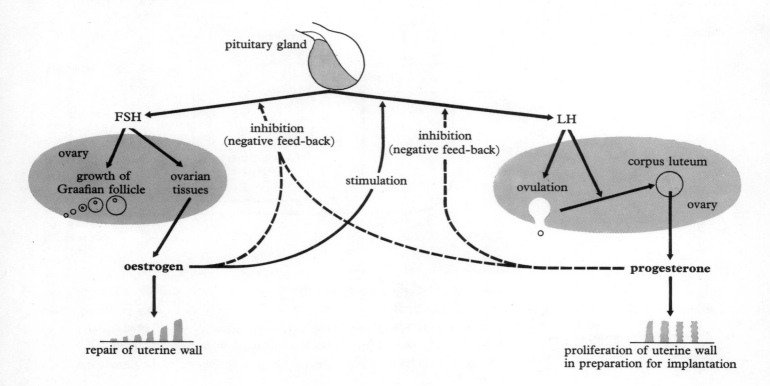

the corpus luteum, instead of degenerating, persists and continues to secrete progesterone. This, coupled with a small but steady secretion of oestrogen from the ovary, maintains the continued development of the uterus, and of course prevents menstruation. By inhibiting the anterior pituitary from producing FSH, the progesterone also prevents further follicles from developing during pregnancy[1]. After the first three or four months of pregnancy, the corpus luteum begins to regress, and the placenta takes over the job of secreting progesterone and oestrogen.

In the course of pregnancy the level of oestrogen in the blood rises and progesterone falls, and it has been suggested that this plays some part in bringing about parturition. Certainly oestrogen promotes uterine contraction and progesterone inhibits it; miscarriages caused by premature birth are sometimes caused by insufficient progesterone, which can be made good by injection. However, a more direct cause of parturition is another hormone, **oxytocin,** secreted by the posterior lobe of the pituitary. This causes the uterine muscle to contract. Oestrogen achieves its effect by making the uterine muscle more sensitive to oxytocin. What kind of mechanism might be responsible for initiating the production of oxytocin from the pituitary?

Progesterone and oestrogen are also responsible for the growth, during pregnancy, of the mammary glands, in readiness for milk-production. After parturition the flow of milk is initiated by yet another pituitary hormone – **prolactin,** secreted from the anterior lobe. Surgical removal of the pituitary during lactation stops the flow of milk immediately. Before birth prolactin secretion is inhibited by oestrogen and progesterone. The sudden drop in these two hormones at the end of pregnancy permits the onset of lactation. The action of the various hormones during pregnancy is summarized in Fig 24.13.

Fig 24.13 Action of the female hormones during pregnancy. Black arrows signify stimulation, broken arrows inhibition.

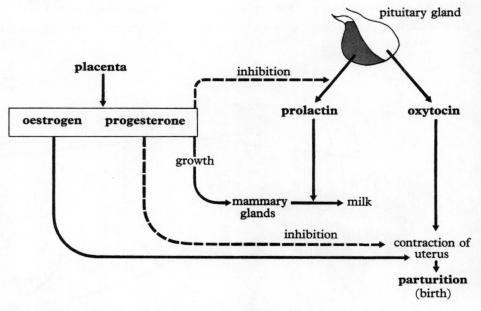

THE SEXUAL CYCLE IN OTHER MAMMALS

The period of time during which the human female can be successfully impregnated is limited to around the time of ovulation. This period is called **oestrus.** In most mammals oestrus is accompanied by heightened sexual excitement (oestrus comes from the Greek word *oistros* meaning 'mad desire') and the animal is described as being 'on heat'. At this stage the female may pro-

[1] This fact has been made use of in developing the contraceptive pill. The pill contains progesterone plus varying amounts of oestrogen. It works by suppressing the development of follicles in the ovary.

duce various secretions which evoke sexual activity in the male. This ensures that the female is impregnated at the right time. In some species, rabbits for example, this is made even more sure by the fact that ovulation is delayed until mating occurs. In man, however, no such safeguards exist: there is no evidence of increased sexual awareness at the time of ovulation, and it is largely fortuitous as to whether or not copulation takes place at the right time.

In the human female oestrus occurs approximately midway between one menstrual period and the next, and as the sexual cycle lasts 28 days a woman will obviously come into oestrus some 12 or 13 times in the course of a year. In few other mammals is the frequency so low. Even in large mammals like cows the oestrus cycle only lasts about three weeks, and in small mammals like rats and mice oestrus occurs at approximately four to five day intervals.

Rats, mice, cows and women can reproduce at any time of the year, but in many other mammals there are definite **breeding seasons**. During the breeding season the female may come into oestrus once, as in dogs and cats, or many times as in hamsters and horses. The number and timing of the breeding seasons are also variable: commonly there is only one, occurring in the spring or summer, but some mammals, dogs and cats for example, have two.

There has been much speculation as to what regulates the reproductive activities of animals. Some animals are strongly affected by temperature and daylight. For example, it has been shown in several species of birds that maturation of the gonads can be induced by artificially increasing the day length. In some other animals the moon has been implicated. The Pacific palolo worm spawns at dawn exactly one week after the November full moon, much to the delight of the local natives who regard the palolos as a great delicacy.

SEXUAL ACTIVITY IN THE HUMAN MALE

In the human male there is no sexual cycle as such, but the gonads are regulated by gonadotrophic hormones identical with those secreted in the female. The anterior pituitary secretes FSH and LH (generally called **interstitial cell stimulating hormone**, ICSH, in the male). The detailed action of these hormones is not known, but FSH appears to promote spermatogenesis, while LH causes the interstitial cells between the seminiferous tubules to secrete **androgens** or **male hormones**. Chemically different from the ovarian hormones, the function of the androgens is to stimulate development of the male's secondary sexual characteristics, such as body hair and deep voice.

Sexual Reproduction in the Flowering Plant

Having discussed human reproduction at some length, we turn for comparison to the flowering plant. We shall see that in many basic respects they are similar, but the flowering plant has special problems connected with the fact that it is stationary. This necessitates devices for bringing the gametes together and for distributing the reproductive products.

Fig 24.14 Structure of a generalized flower. The sepals and petals on the near side have been removed so as to expose the stamens and carpel. The latter has been sectioned vertically.

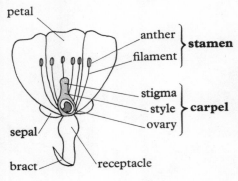

STRUCTURE OF THE FLOWER

The function of sexual reproduction is performed by the **flowers**. The flower develops from special flower buds. As the bud opens it is seen to consist of a series of circlets or whorls of structures (Fig. 24.14). The outermost whorl is composed of leaf-like **sepals** which together constitute the **calyx**; then come the **petals** constituting the **corolla**; then the **stamens** constituting the **androecium**; and finally, in the centre, one or more **carpels** constituting the **gynoecium**. The stamens and carpels bear the male and female structures respectively, for which reason they are known as the **essential** parts of the flower. The sepals and petals are known as the **accessory** or **non-essential** parts. The whole flower is generally borne in the angle between the main stem and a small leaf-like **bract**.

The flowers of different plants show great diversity, particularly in the number of units within each whorl. For example, there may be a single carpel in the centre (e.g. garden pea), or numerous carpels (e.g. buttercup); between these two extremes are many intermediate conditions and degrees of fusion. Sometimes a particular whorl is absent altogether: if the carpels are missing so that only stamens are present, the flower is described as **staminate**; if the stamens are missing so that only carpels are present, it is described as **carpellate**. The significance of such **unisexual flowers** will be discussed later. Sometimes sepals and petals may be indistinguishable from one another, forming what is called a **perianth**. There is also much variation in the shape and symmetry of the flower: for example, it may be radially symmetrical (**actinomorphic**) as in the buttercup family, or bilaterally symmetrical (**zygomorphic**) as in the pea family. Many of these variations are related to the way the male and female gametes are brought together. We will ignore them for the moment, and concentrate on the events that occur inside the male and female parts of a 'typical' flower.

INSIDE THE STAMEN

Each stamen consists of an **anther** borne at the end of a slender flexible **filament**. The anther contains four **pollen sacs** in which the **pollen grains** develop (Fig 24.15A). The latter are the bearers of the male gametes. Initially each pollen sac contains a densely-packed mass of large **pollen mother cells** all of which are diploid. Each of these cells undergoes meiosis to give a tetrad of four haploid pollen grains (Fig 24.15B,C). Within each pollen grain, the haploid nucleus divides mitotically into two nuclei: one of these is called the **generative nucleus**, the other the **tube nucleus**. Later we shall see what happens to these two nuclei. The above events are accompanied by thickening of the wall of the pollen grain which frequently becomes elaborately sculptured.

INSIDE THE CARPEL

Each carpel consists of an **ovary** surmounted by a slender **style** which terminates as a small swelling called the **stigma**. The ovary is hollow, its cavity containing one or more **ovules**. Each ovule starts as a small protuberance projecting into the cavity. As development proceeds, the ovule grows and generally bends over as shown in Fig 24.16A. Its stalk or **funicle** is attached to the wall of the ovary by a cushion of specialized tissue called the **placenta**. Initially the ovule consists of a uniform mass of cells called the **nucellus**. As

A) Section of anther

pollen sac containing
pollen grains

wall of anther
tapetal cells
stomium
vascular bundle

dehisced anther

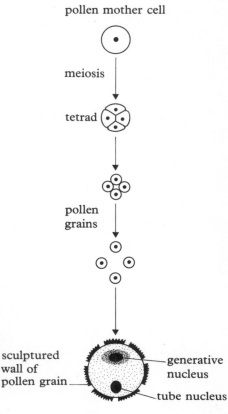

B) Development of pollen grain

pollen mother cell

meiosis

tetrad

pollen
grains

sculptured
wall of
pollen grain

generative
nucleus

tube nucleus

single pollen grain

C

Fig 24.15 The male structures of a flowering plant. The pollen grains, formed by meiotic cell division, are haploid. C) is a photomicrograph of a pollen sac containing developing pollen grains. Notice the tetrads which will later separate as depicted in B. The large cells surrounding the pollen sac form the tapetal layer which provides the developing pollen with nourishment. (C: *J. F. Crane*)

it develops, the wall of the ovule differentiates into protective inner and outer integuments which envelop the softer nucellus tissue beneath (Fig 24.16B). The integuments do not completely enclose the nucellus but leave a small opening at one end, the **micropyle**. The other end of the ovule is called

A) Development of ovule

B) Mature carpel

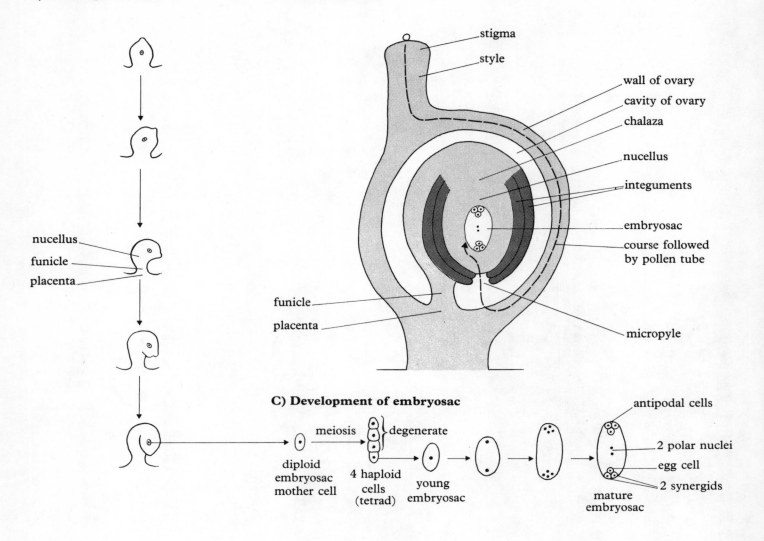

C) Development of embryosac

Fig 24.16 The female apparatus of a flowering plant is located in the ovary at the base of the carpel. The ovary shown in B contains a single ovule, at the centre of which is the embryosac. How the ovule develops is shown in A. The eight nuclei in the embryosac are all haploid since the embryosac develops from a haploid cell that has been formed by meiosis (C).

the **chalaza**. In the centre of the ovule a structure called the **embryosac** develops: this contains the female gamete (**egg cell**). The egg cell is haploid, and how it comes to be so necessitates understanding how the embryosac develops.

The embryosac arises from a single cell, the **embryosac mother cell** (Fig 24.16C). This undergoes meiosis to form a row of four haploid cells. Three of these cells degenerate, leaving the remaining one to develop into the embryosac. This becomes vacuolated and enlarges rapidly. Meanwhile its nucleus undergoes three successive mitotic divisions, giving a total of 8 daughter nuclei, four at each end of the developing embryosac. One nucleus from each group now moves to the centre of the embryosac: these are called the **polar nuclei,** and we shall return to them later. The remaining six nuclei become separated by cell walls and remain at the ends of the embryosac. So the mature embryosac finally contains three cells at each end, and two free nuclei in the centre. All are haploid. One of the cells at the micropyle end becomes the functional **egg cell**. The remaining cells eventually disintegrate: the two at the micropyle end are **synergids** (non-functional eggs); the three at the other end are called the **antipodal cells** and play little or no part in subsequent events.

TRANSFER OF POLLEN

Drying out of the wall of the anther causes shrinkage of the cells, thereby creating tensions that eventually cause the anther to split open down the sides (Fig 24.15). The line of splitting, called the **stomium**, represents a line of weakness in the wall of the anther. The result is that the pollen grains are released.

The pollen grains must now be transferred from the open anthers to the stigmas, a process called **pollination**. Sometimes the pollen grains fall onto the stigmas of the same flower (**self-pollination**) but more often they are conveyed either by wind, insects or some other agent, to another flower (**cross-pollination**). Insect pollination is promoted by a variety of devices which serve to attract the insect: these include brightly coloured petals and/or sepals, scent-production, and secretion of **nectar**. Nectar, a sugary fluid, is secreted by glandular **nectaries** at the base of each petal, and serves as a bait.

GROWTH OF THE POLLEN TUBE AND FERTILIZATION

Because the egg cell is embedded within the tissues of the carpel, fertilization involves an extraordinary process in which an outgrowth from the pollen grain, called the **pollen tube**, grows down to the embryosac, taking the male gametes with it.

The pollen grains adhere to the stigma as a result of the stigma cells secreting a sticky substance. The latter also stimulates the pollen grain to germinate, sending out its pollen tube. The pollen tube grows into the stigma, down the style, to the ovary, pushing its way between the cells and deriving nourishment from the surrounding tissues. Exactly what causes it to do this is still not fully understood: there is evidence that the pollen tube is negatively aerotropic, i.e. it grows away from air, and it may also be attracted towards chemical substances present in the carpellary tissues. In many species the process is remarkably rapid, taking place in a matter of minutes.

When the pollen grain germinates, the tube nucleus occupies a position at the tip of the growing pollen tube (Fig 24.17). Meanwhile the generative nucleus divides mitotically into a pair of **male gamete nuclei**. These follow behind the tube nucleus as the pollen tube grows down the style. On reaching the ovary, the pollen tube enters an ovule either via the chalaza, or, more usually, via the micropyle (Fig 24.16). When it reaches the centre of the ovule, it penetrates the wall of the embryosac and bursts open. At about this time the tube nucleus disintegrates, leaving a clear passage for the entry of the male nuclei.

We come now to the crux of the reproductive process: fertilization itself. We have already noted that there are two male nuclei in the pollen tube. One of these fuses with the functional egg cell to form a diploid **zygote**; the other fuses with both polar nuclei to form a triploid nucleus. This latter is known as the **primary endosperm nucleus**, and we shall see what happens to it in a moment. This **double fertilization**, entailing the fusion of two male gametes with female cells, is unique to flowering plants.

AFTER FERTILIZATION

As a result of fertilization certain further changes take place in the floral parts of the plant. Briefly, these are as follows:

Fig 24.17 Union of male and female nuclei in a flowering plant. Having been transferred to a stigma the pollen grain germinates (A): the pollen tube grows out of the pollen grain through one of the germ pores where the hard coat is absent. When the pollen tube reaches the embryosac fertilization occurs (B): one male nucleus fuses with the two polar nuclei to form the triploid primary endosperm nucleus, the other male nucleus fuses with the egg cell to form the diploid zygote.

A) Germination of pollen grain and growth of pollen tube

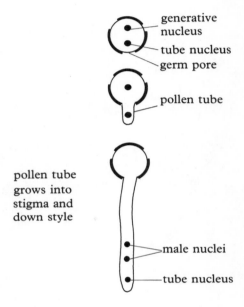

generative nucleus
tube nucleus
germ pore
pollen tube

pollen tube grows into stigma and down style

male nuclei
tube nucleus

B) Fertilization

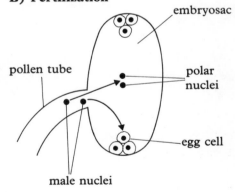

embryosac
pollen tube
polar nuclei
egg cell
male nuclei

Fig 24.18 After fertilization in a flowering plant the zygote develops into the embryo which is here seen embedded in endosperm tissue in the now greatly expanded embryosac. The ovule is destined to form the seed, and the ovary the fruit. The integuments become the seed coat.

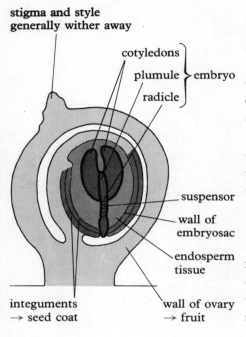

stigma and style generally wither away

cotyledons
plumule ⎱ embryo
radicle ⎰

suspensor

wall of embryosac

endosperm tissue

integuments → seed coat

wall of ovary → fruit

(1) The zygote divides mitotically, growing and developing into the **embryo** (Fig 24.18). The embryo becomes differentiated into **radicle** (young root), **plumule** (young shoot) and either one or two **cotyledons** ('seed leaves'). The embryo is attached to the wall of the now expanding embryosac by a **suspensor** through which it may, in some cases, derive nourishment.

(2) The triploid primary endosperm nucleus divides mitotically into a mass of nuclei which become separated from one another by thin cell walls. This becomes a semi-fluid food-storage tissue called the **endosperm** which surrounds the developing embryo and provides it with nourishment.

(3) To accommodate the developing endosperm tissue and the growing embryo, the embryosac expands. As a result, the nucellus becomes crushed out of existence, so the embryo and endosperm come to fill the whole of the area inside the integuments.

(4) The ovule develops into the **seed**, the integuments becoming the **seed coat**. The outer integument becomes the **testa**, the inner integument the **tegmen**. Both layers are tough and protective, often lignified. The final transformation of the ovule into the seed involves the removal of water from an initial 90 per cent to a mere 15 per cent, thus forming a dormant, resistant structure that can withstand adverse conditions.

(5) While the ovule develops into the seed, the ovary develops into the **fruit**. Fruits come in a wide range of shapes and forms but they all have a common function, namely to protect the seeds, and aid in their dispersal. Other structures may also contribute towards this function, including the sepals, styles, receptacle and even the bracts (see p. 405).

THE FLOWERING PLANT COMPARED WITH OTHER ORGANISMS
Sexual reproduction of the flowering plant may seem very different from the mammal, but in basic respects they are similar. Thus both involve the production and union of haploid nuclei, and the protection and nourishment of the developing embryo.

However, a word must be said about the flower. In common parlance it is usual to refer to the flower as the sexual apparatus of the plant, the stamens as the male organs, the carpels as the female organs. Strictly speaking this is incorrect. The flower is really an asexual spore-forming device, the spores being represented by the pollen grains and embryosac respectively. In Chapter 25 we shall see that in lower plants such as mosses and ferns, such spores germinate to form a separate individual with gamete-producing sex organs. No such thing happens in flowering plants where the gametes develop within the spores. Therefore, in the strict sense, the flowering plant has no sex organs at all.

Parthenogenesis

Parthenogenesis is the development of a new individual from an unfertilized egg. Although many instances of artificial parthenogenesis are claimed, as a natural occurrence it is restricted to certain groups of invertebrates, notably insects. In some species it is the only known method of reproduction.

There are two kinds of parthenogenesis: diploid and haploid. In **diploid parthenogenesis** the eggs, instead of being formed by meiosis, are formed by mitosis, with the result that they are diploid instead of haploid. The resulting adult will therefore have the normal diploid constitution. This is what happens at certain stages in the life cycle of aphids. In the summer months wingless females produce further generations of mainly wingless females by diploid parthenogenesis, a rapid and efficient way of increasing numbers without necessitating the presence of males.

In **haploid parthenogenesis**, eggs are produced by meiosis in the usual way, and are therefore haploid. These develop, without being fertilized, into a new animal whose cells are therefore haploid. A well known example of this type of parthenogenesis is seen in the life cycle of the honeybee. The bee colony contains three distinct types of individual: the queen, workers and drones. The drones, which are all fertile males, develop from unfertilized haploid eggs laid by the queen.

That an adult animal with a haploid chromosome constitution should exist at all is surprising enough; that it should develop without the stimulus of fertilization is even more remarkable. For many years biologists have tried to discover what induces parthenogenesis by subjecting the unfertilized eggs of many different animal species to various environmental influences. This has included pricking, violent shaking, chemical treatment, high and low temperatures, and altering the density or osmotic pressure of the surrounding medium. In some cases such treatment has resulted in cleavage, and on occasions the eggs have given rise to reasonably advanced embryos, even larvae. What about adults? There are several instances of adult frogs, and at least one authenticated case of an adult mammal, being formed by artificial parthenogenesis. In the latter case an unfertilized egg was removed from a female rabbit, activated by pricking, and then popped back into the uterus. Hormone treatment had been previously given to the female so that her uterine mucosa was prepared for implantation. Normal development ensued, and a visibly normal offspring was produced.

The whole subject is shrouded in mystery. Why, for example, should it be so difficult to induce artificial parthenogenesis? What is the natural stimulus which allows it to happen in some organisms but not in others? What enables an unfertilized egg to develop into a certain kind of individual, for example, a drone bee or a wingless female aphid? We have to admit that we do not know the answer to these questions.

Self versus Cross Fertilization

The majority of animals have separate sexes, i.e. they are **dioecious**. However most plants (and some animals) are **hermaphroditic** (**monoecious**), i.e. each individual has both male and female organs. The advantage of this is that, as every individual is a potential producer of fertilized eggs, the fecundity of the species is increased. The disadvantage is that the organism runs the risk of self fertilization. As was mentioned earlier, self fertilization precludes the possibility of genetic mixing, and if carried out on a large scale over many

Fig 24.19 Two devices for encouraging cross pollination. A) The massive inflorescence of the desert succulent *Agave* provides a bright splash of colour attracting insects. B) The flower of yellow toadflax is adapted for cross pollination by bees. When probing into the flower for nectar, the bee may deposit pollen from another flower onto the stigma. The anthers are below the stigma thus reducing the chance of their pollen falling onto the stigma. (A: *M. B. V. Roberts*; B: *after L. J. F. Brimble*, Intermediate Botany, *Macmillan*)

A

B

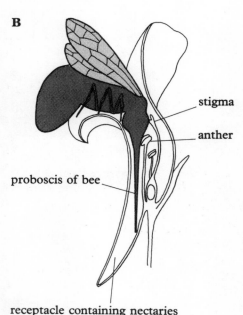

stigma

anther

proboscis of bee

receptacle containing nectaries

generations it may lead to the decline of the species. In general, offspring resulting from self fertilization are less vigorous and productive than those resulting from cross fertilization. It is therefore not surprising to find that monoecious organisms generally have anatomical and physiological mechanisms that prevent, or at least reduce, the chances of self fertilization.

In flowering plants this means having mechanisms that encourage **cross pollination** and prevent **self pollination**. The different devices that have evolved include the possession of brightly coloured petals and/or sepals, the secretion of nectar, and the emission of a potent scent for attracting insects which convey the pollen from one flower to another. Commonly a large number of small flowers are clustered together at the end of a floral shoot. The resulting **inflorescence** provides a bright splash of colour, and a concentrated source of scent and nectar, for attracting insects (Fig 24.19A). Self fertilization may be prevented by the stamens and carpels being located in different flowers, or maturing at different times (stamens first, carpels later, or vice versa). This latter strategy is also used by certain hermaphroditic animals in which the testes develop before the ovaries. Some plants are self-sterile, i.e. the plant's pollen fails to germinate on its own stigmas.

Amongst insect-pollinated plants there are sometimes elaborate mechanisms for promoting cross pollination. In many flowers the stigma is higher than the surrounding stamens, thus making it impossible for pollen to fall onto the stigma of the same flower. Moreover if a large insect, such as a bee, visits the flower, its body will brush against the stigma before it reaches the anthers. Any pollen the bee has picked up from another flower will, therefore, be deposited on the stigma (Fig 24.19B). In such flowers self pollination is by no means impossible, but cross pollination is more likely.

A highly specialized mechanism is found in the arum lily, *Arum maculatum*, otherwise known as 'lords and ladies' or 'parson in the pulpit'. The floral shoot of this extraordinary plant bears separate male and female flowers, and terminates in the form of a colourful bulbous swelling which produces a pungent smell (Fig 24.20). At the base of the shoot, or **spadix** as it is called, are female flowers, each consisting of a single carpel. Immediately above these is a ring of hair-like sterile flowers. Then comes a mass of male flowers each consisting of two to four stamens, and above these is yet another ring of hair-like sterile flowers. The entire inflorescence is enclosed in a sheath-like bract called the **spathe** which remains closed below the uppermost ring of hairs, but unfolds above this level to reveal the end of the spadix.

Attracted by its disgusting smell, small dung flies crawl to the base of the spadix where nectar is secreted. If the insect happens to be covered with pollen from another arum lily, the carpels are pollinated. The insects are prevented from climbing out until all the carpels have been pollinated. In the course of the ensuing twelve hours or so, the stamens ripen. When this has happened changes occur which make it possible for the flies to crawl up to the male part of the inflorescence, where they become covered with pollen. They then escape, possibly to visit another inflorescence. What kind of changes might occur inside the spathe enabling the flies to escape?

One of the many intriguing things about the arum lily is the swollen tip of the spadix. Club-shaped and fleshy in texture, its colour ranges from pink to purple. Furthermore it is claimed that its temperature is significantly higher than that of the surroundings. This, together with its elevated position and odious smell, suggests that it may be responsible for attracting insects. But which of its features – colour, smell or temperature – serves as the attractant? To answer this question, models of the arum lily have been con-constructed, and their ability to attract insects tested. These models were equipped with spadixes of different colours, and some were fitted with electric heating coils to raise their temperature. To cut a long story short, colour and temperature were found to play no part in attracting insects; a model would only attract insects if it was scented either with chemicals extracted from the tip of a real spadix, or with decaying animal matter. There is still much that is not understood about this remarkable plant, but it serves to illustrate the extent to which an organism may be adapted to ensure cross fertilization.

Amongst animals, devices for preventing self fertilization are no less elaborate. In general, copulation of monoecious animals involves mutual exchange of spermatozoa in which the positioning of the genital apparatus is such that self fertilization is virtually impossible. In certain free-living flat-worms, for example, the insertion of the penis of each individual into the sperm receptacle of the other, automatically prevents sperm and eggs of the same individual coming together (Fig 24.21). Such mechanical devices can be quite effective in minimizing self fertilization.

Despite these safeguards, self fertilization does occur in many monoecious organisms, particularly in plants where preventive devices are generally less foolproof than in animals. This need not necessarily be a bad thing provided that it is not the only method of reproduction. In the absence of a pairing partner, self fertilization is better than nothing, and it does at least ensure an increase in numbers, even though genetic variety may be curtailed. In fact some plants, particularly those in which cross pollination is uncertain, have mechanisms which positively favour self fertilization.

Asexual Reproduction

Since it does not involve the participation of two individuals, asexual reproduction has the disadvantage of not allowing genetic mixing to occur. However in many organisms it is useful for rapid multiplication. For convenience asexual methods of reproduction can be divided into five types: **fission, spore-formation, budding, fragmentation** and **vegetative propagation.** Let us look at each in turn.

FISSION

In this process the organism divides into two or more equal-sized parts. **Binary fission,** the division of the organism into two daughter cells, is characteristic of bacteria and protists (Fig 24.22). The rate of reproduction achieved this way can be prodigious: for example, in favourable conditions bacterial cells divide once every 20 minutes, which means that in 24 hours a single cell could give rise to a population exceeding 4,000 million million million (4×10^{21}). This astronomical figure is achieved because the increase is **exponential,** one cell dividing into two, two into four, four into eight, eight into sixteen, and so on. In its early stages such exponential growth seems rather unim-

Fig 24.20 The arum lily, *Arum maculatum*, provides an example of a highly specialized pollination mechanism. A) shows the bulbous end of the spadix projecting from the spathe. In B) the spathe has been cut away to reveal the specialized inflorescence. Explanation on p. 400. (*B. J. W. Heath, Marlborough College*)

Fig 24.21 Copulation in the free-living flat-worm *Planaria lugubris* involves mutual exchange of sperm. The erect penis of each partner is inserted into the sperm receptacle of the other. Since erection is necessary for ejaculation, it is virtually impossible for spermatozoa to get into the oviducts of the same individual. (*modified after Ullyott and Beauchamp*)

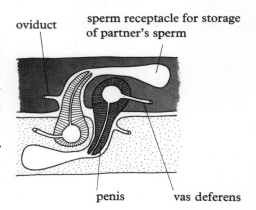

oviduct   sperm receptacle for storage of partner's sperm

penis   vas deferens

Fig 24.22 Electron micrographs showing three successive stages in the asexual division of a cell of the bacterium *Staphylococcus aureus*. (*Audrey M. Glauert, Strangeways Research Laboratory, Cambridge*)

A

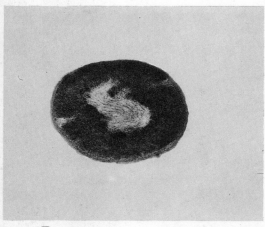

B

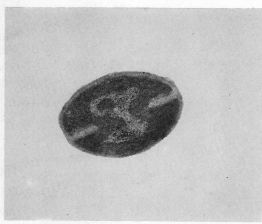

C

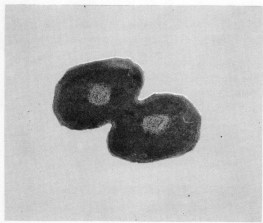

pressive, but it gains momentum as the numbers increase. To revert to bacteria, only eight are formed in the first hour, and only 512 by the end of the third hour. But by the end of a day. . . . !

Even more fantastic reproductive rates are achieved by **multiple fission** which is shown by various protists, particularly parasitic ones. In this case the nucleus divides repeatedly and each daughter nucleus breaks away together with a small portion of the cytoplasm (see p. 533). This splitting process is termed **schizogony**, and a cell that does it is called a **schizont**. The detailed mechanism varies, as does the number of offspring produced. In the malarial parasite multiple fission produces as many as 1,000 merozoites from a single schizont during the parasite's asexual cycle when it invades the liver cells. Multiple fission occurs again later when the merozoites invade the red blood cells. In this case multiplication is rather less prolific, not more than 24 daughter cells being formed from a single parent.

SPORE FORMATION

Spores are unicellular bodies formed by cell divisions in a parent organism. Having become detached from the parent, and if conditions are suitable, they germinate, directly or indirectly, into a new individual. Spores are produced by bacteria, protists and many plants. They come in a wide range of forms, and are produced and dispersed in many different ways. They are generally very small and light, thereby aiding distribution by wind, water and animals. Some have thick resistant walls which enable them to survive unfavourable conditions including drought. They are usually numerous. The ability of the fungi to produce vast numbers of airborne spores explains why members of this important group of saprophytes and parasites manage to spread so quickly. Examples of spore formation, some with specialized dispersal mechanisms, are discussed in Chapters 25 and 33.

BUDDING

In this method of reproduction an organism develops an outgrowth which, on detachment from the parent, becomes a self-supporting individual. Budding is characteristic of yeast cells and a wide variety of lower animals including *Hydra*, certain flatworms and several annelid groups. In multicellular animals like *Hydra* budding takes place by proliferation of undifferentiated cells which then differentiate into the appropriate structures.

FRAGMENTATION

Sometimes an organism may be broken into two or more pieces each of which grows into a new individual. As a means of reproduction, fragmentation depends on the organism having good powers of **regeneration**, a property which for the most part is confined to the lower animals. Sponges and hydroid coelenterates, for example, have astonishing powers of regeneration: if a sponge is macerated by pressing it through fine gauze, the separated cells come together in groups and grow into new individuals. Given the right conditions, very small fragments of free-living flatworms will regenerate into new individuals, and in the marine nemertine worm *Lineus* literally hundreds of simultaneously regenerating individuals may be formed from a single worm.

There are two questions here: what causes regeneration, and can this be

regarded as a natural method of reproduction? The answer to the first question is not completely known. Successful regeneration depends on the animal possessing large numbers of undifferentiated cells which, when required to do so, can proliferate and develop into any kind of tissue. The nervous system, or rather a chemical substance present in nervous tissue, seems to play an important part in triggering regeneration, a fact which has been established from scores of experiments involving the transplantation of nervous tissue from one part of the body to another. For example, if a small piece of ventral nerve cord is grafted into the body wall of an earthworm (and if conditions are favourable!), a head will grow out of the side. But, irritatingly, we have no idea as to the identity of the active chemical ingredient in the nervous tissue.

Whether or not regeneration should be regarded as a natural method of reproduction depends on what causes fragmentation to occur. If an experimenter cuts the animal up with a knife, it can hardly be regarded as natural. If, on the other hand, it can be shown that fragmentation occurs spontaneously, then this qualifies as a *bona fide* method of reproduction. To revert to the nemertine *Lineus*: at certain times of the year (spring and summer) these worms develop rings of constriction which cut the body up into short fragments. The rate of fragmentation is increased by raising the temperature, and decreased by lowering it. The evidence suggests that, at any rate in this animal, fragmentation and regeneration is a natural reproductive process.

VEGETATIVE PROPAGATION

This is the name given to the process of asexual reproduction in plants where part of the body becomes detached and develops into a new self-supporting plant. Scores of examples could be given. Detached stems, and even leaves, often take root and grow into new individuals, as is well known to any gardener who propagates his plants by taking cuttings. An example of a self-propagating leaf is shown in Fig 24.23. Some plants reproduce vegetatively by sending out a side branch, known as a **runner**, which gives rise to new plants. The runner grows out from one of the lower axillary buds of the parent plant. At its nodes small axillary buds develop into new plants which later become separated from each other by decay of the internodal portions of the runner. Creeping buttercup and strawberry are examples.

Often vegetative propagation involves the formation of some sort of storage organ, which lies in the soil over the winter and develops into one or more plants the following year. For this reason such devices are known as **perennating organs**. They may be formed from a modified stem, root, leaves or bud, depending on the plant in question. Thus the potato **tuber** is a swollen stem, as is the **corm** of a plant like the crocus. Like any normal stem, these structures have apical buds and nodes with leaves and axillary buds (p. 425). It is from the latter that new plants (or, in some cases, new storage organs) are formed. The basic principles involved in the formation and subsequent growth of a perennating organ are illustrated in Fig 24.24. In the case of root structures, the storage organ may be formed from the main taproot, as in carrot and parsnip, or from lateral (adventitious) roots, as in dahlia and lesser celandine. **Bulbs** are really swollen buds in which food is stored in thick fleshy leaves projecting from a much shortened vertical stem. Apical and axillary buds, situated amongst the leaves, develop into new plants either directly or via further

Fig 24.24 The general principle of perennation illustrated by a potato plant. Towards the end of the growing season, the plant forms swollen tubers at the ends of horizontally growing underground stems. The tubers lie dormant in the soil until the following year when they sprout new plants from their axillary buds. In the diagram only one new plant is shown growing out of the old tuber; but in fact all the axillary buds are potentially capable of producing new plants. The old tuber withers and dies.

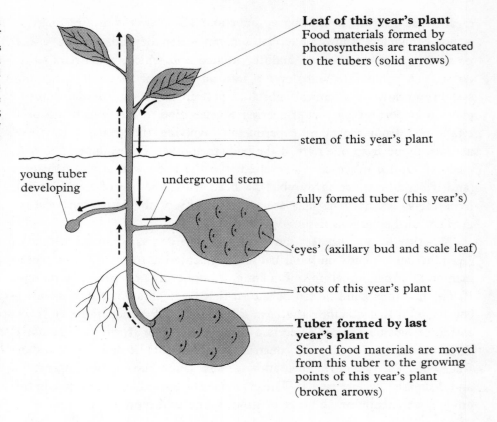

Leaf of this year's plant
Food materials formed by photosynthesis are translocated to the tubers (solid arrows)

stem of this year's plant

young tuber developing

underground stem

fully formed tuber (this year's)

'eyes' (axillary bud and scale leaf)

roots of this year's plant

Tuber formed by last year's plant
Stored food materials are moved from this tuber to the growing points of this year's plant (broken arrows)

bulbs. These are but a few examples of vegetative propagation; the main point to note is that many different parts of a plant can contribute towards this process.

Dispersal

Built into the reproductive processes of many organisms are mechanisms ensuring wide dispersal of the progeny. This is particularly important to sessile or slow-moving organisms that remain in one place throughout adult life. Wide dispersal is important, otherwise the offspring may compete with one another, and indeed with the parents, thus lowering the survival rate. Moreover wide dispersal enables the species to gain a foothold in new un-exploited localities.

An organism may be dispersed in a variety of forms: as **spores, seeds, fruits** or **larvae**. Larvae are themselves motile (see p. 422) and are important in the dispersal of many sessile and slow-moving aquatic animals. Spores, seeds and fruits, however, depend on natural agents such as wind, water and animals. To this end they are adapted appropriately. For example, many spores and seeds are extremely small and light, aiding airborne dispersal. Heavier seeds and fruits are equipped with floatation devices such as wings and hairy parachutes (Fig 24.25A). Dispersal is further aided by spores or seeds being discharged from the parent plant by an 'explosive' mechanism involving the sudden splitting open (**dehiscence**) of the spore or seed-containing body. Alternatively, they may be scattered from a receptacle borne at the end of a long

A

B

flexible stalk that sways in the wind or is knocked by passing animals. Such 'pepper pot' mechanisms are seen in the dispersal of poppy seeds and of spores in mosses.

The rôle of animals in dispersal is equally varied. **Hooked fruits** such as cleavers and burdock (Fig 24.25B) may cling to their bodies as they brush past. Other fruits are edible, the hard seed being indigestible so that it passes out with the faeces, unharmed and still capable of germination. In parasitic animals dispersal is often achieved by an intermediate host or **vector**. This is beautifully illustrated by the mosquito which carries the malarial parasite from one main host to another (see p. 533).

Water is important in the distribution of aquatic larvae and certain plants, indeed any organisms that can float or survive immersion. The latter include certain palms (coconut palm for instance) whose fruits have thick fibrous walls containing air pockets, thus enabling them to float. Undoubtedly this has been important in the colonization of previously unoccupied land masses during evolution. The ancestors of many of those curious organisms that today inhabit the Galapagos Islands must have got there via the ocean, either floating or attached to driftwood.

Fig 24.25 Two devices aiding dispersal. A) Hairy parachute of dandelion. The 'pappus' of hairs, derived from the sepals, is borne at the end of a slender stalk. At the lower end of the stalk is the fruit developed from the ovary. B) The hooked 'fruit' of burdock. In this plant the small flowers are massed together into a compact inflorescence. This is surrounded by overlapping bracts which, after fertilization, become woody and hooked, thus enabling the inflorescence to cling to rough surfaces – clothes, hairs, etc. (*B. J. W. Heath, Marlborough College*)

25 The Life Cycle

Fig 25.1 The life cycle of man and most other organisms follows the plan outlined below. There is much variation in different organisms as to the way the zygote develops into the adult, but the fundamental division of the life cycle into diploid and adult phases is the same throughout. n = haploid; 2n = diploid.

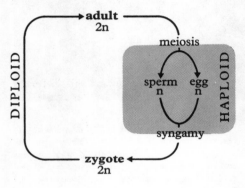

Fig 25.2 The life cycle of many plants, notably mosses and ferns, shows alternation of generations in which a haploid gamete-producing gametophyte plant alternates with a diploid, spore-producing sporophyte. Notice that meiosis and syngamy divide the life cycle into haploid and diploid phases.

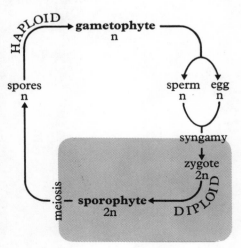

The progressive sequence of changes which an organism goes through from the moment of fertilization to death constitutes its life cycle. In the course of its life cycle the organism normally produces a new generation of individuals which repeat the process. New generations are produced by **reproduction**, which we have seen may be **sexual** or **asexual**. In the course of the life cycle **mitosis** and **meiosis** occur. Let us start by looking briefly at the life cycle of man which, despite the structural complexity of the adult, is comparatively simple (Fig 25.1).

THE LIFE CYCLE OF MAN

The adult human produces eggs or sperm. If these gametes fuse successfully, a **zygote** is formed which develops into a new adult. The details of this complicated business are discussed elsewhere. At present we are only concerned with the overall picture. The cells of an adult human are diploid. Gametes, however, are formed by meiosis and have the haploid constitution. The general term for the union of gametes is **syngamy**, which literally means 'joining together'. Since it is formed by the union of two haploid gametes, the zygote is diploid; and as the zygote divides mitotically, the adult will also be diploid.

Meiosis and syngamy divide the life cycle into two distinct phases: the **diploid phase** which embraces the zygote and the adult, and the **haploid phase** containing only the gametes. The majority of animals, and certain plants, have life cycles which conform to this general plan. There are, of course, many variations on the theme. For example, one or more **larval stages** may be interpolated between the zygote and the adult. Whilst this may increase the general complexity of the life cycle, it in no way affects the overall pattern.

ALTERNATION OF GENERATIONS

A radical departure from the basic life cycle described above is seen in certain seaweeds and more particularly in mosses and ferns and their relatives. The life cycle of these plants, summarized in Fig 25.2, starts with a sexually mature adult plant. Since it produces gametes, it is called the **gametophyte** (literally 'gamete plant'). The gametophyte is haploid, as are the eggs and sperm it produces. The latter fuse in the usual way to produce a diploid zygote, but this does not develop directly into a new gametophyte. Instead it grows (by mitotic cell divisions) into another plant which is quite distinct from the gametophyte though it may in some cases be dependent on it for nourishment. The function of this plant is to produce **spores**, for which reason it is referred to as the **sporophyte** (literally 'spore plant'). Formed by meiosis and therefore haploid, the spores are small, light, and readily dispersed by wind. When they alight on a suitable surface, they germinate into a gametophyte, which then repeats the sequence of events.

From this brief account, it is clear that a moss or a fern consists of two distinct plants, a **haploid gametophyte** and a **diploid sporophyte**, which alternate with each other within the life cycle. This phenomenon is known as **alternation of generations**.

THE LIFE CYCLES OF MOSS AND FERN

The description just given is a general one and takes no account of the variations and complexities of individual plant species. Figs 25.3 and 25.4 show how the detailed life cycles of the common moss and fern conform to this general pattern. In both cases the gametophyte bears special gamete-forming organs: **antheridia** produce sperm, and **archegonia** eggs. Sperm are released from the antheridia and brought into contact with the eggs by

Fig 25.3 Life cycle of a common moss such as *Funaria*. In spring, egg-containing archegonia and sperm-producing antheridia are located within leaf-lined rosettes in the haploid gametophyte. For successful transference of sperm the archegonial and antheridial rosettes must contain water. Hair-like paraphyses may help to hold water in the rosettes. Sperm are released by rupture of the antheridia. At the same time the necks of the archegonia open, creating an open passage through which the flagellated sperm can swim to the eggs. Transference of sperm from antheridial to archegonial rosette is thought to be aided by rain-splash. After fertilization, the zygote grows into the sporophyte which remains attached to, and dependent on, the gametophyte. Spores develop by meiosis within spore sac inside sporangium (capsule). In dry weather the operculum falls off and the flexible teeth bend outwards. The tissues at the distal end of the capsule dry out and the haploid spores are released, to be dispersed by wind and air currents. On landing on moist ground, each spore germinates into a green filamentous protonema which produces buds that grow into the gametophyte, thus completing the life cycle.

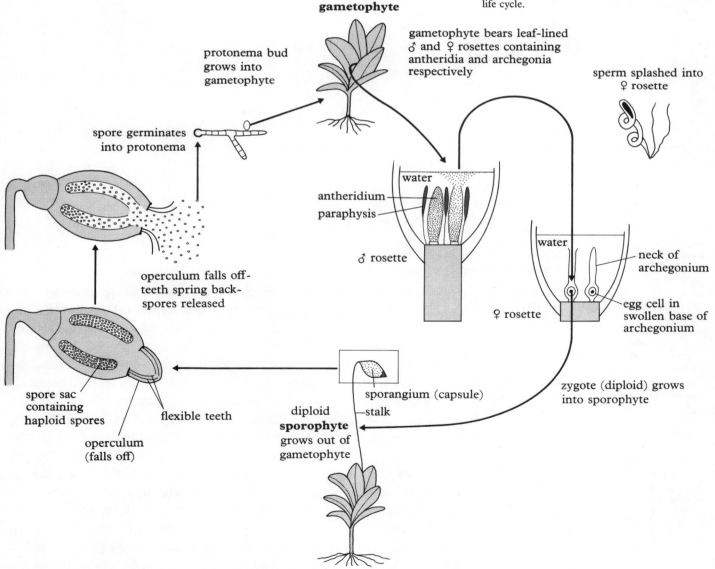

haploid **gametophyte**

protonema bud grows into gametophyte

gametophyte bears leaf-lined ♂ and ♀ rosettes containing antheridia and archegonia respectively

sperm splashed into ♀ rosette

spore germinates into protonema

water

antheridium

paraphysis

♂ rosette

water

neck of archegonium

♀ rosette

egg cell in swollen base of archegonium

operculum falls off- teeth spring back- spores released

spore sac containing haploid spores

flexible teeth

operculum (falls off)

diploid **sporophyte** grows out of gametophyte

sporangium (capsule)

stalk

zygote (diploid) grows into sporophyte

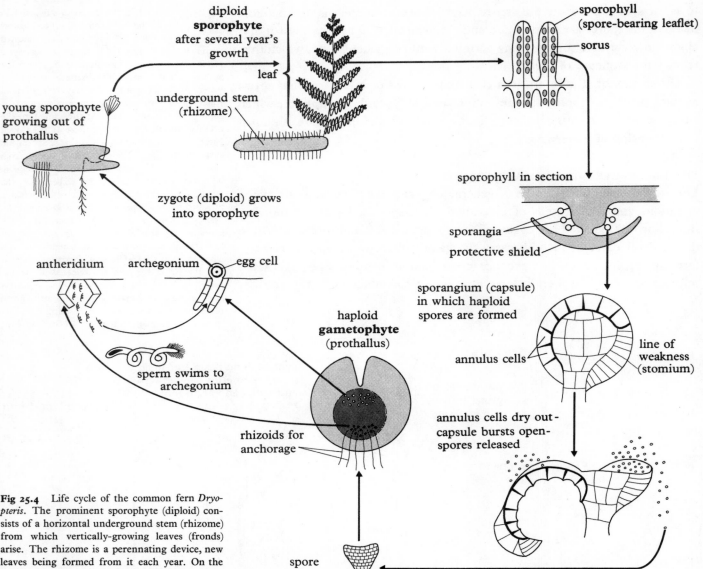

Fig 25.4 Life cycle of the common fern *Dryopteris*. The prominent sporophyte (diploid) consists of a horizontal underground stem (rhizome) from which vertically-growing leaves (fronds) arise. The rhizome is a perennating device, new leaves being formed from it each year. On the undersides of spore-bearing leaflets (sporophylls) groups of sporangia develop. Each group, protected by an umbrella-like shield, is called a sorus, and inside the sporangia haploid spores are formed by meiosis. The mature sporangium (capsule) is topped by a row of annulus cells which in dry weather readily lose water by evaporation. The resulting tension ruptures the capsule at the stomium, thereby releasing the spores. If moisture is present, each spore germinates into a simple heart-shaped prothallus, a flat plate of photosynthetic cells anchored to the soil by filamentous rhizoids. The prothallus represents the gametophyte: antheridia and archegonia, located on its underside, produce sperm and eggs respectively. The gametophyte is absolutely dependent on having a damp environment, not only to prevent its drying out, but also for transference of sperm. After rupture of the antheridium, the ciliated sperms swim to an egg cell at the base of an archegonium. The zygote grows into a young sporophyte which, once it has established roots and leaves, becomes self supporting, thus completing the cycle. Note that in contrast with the moss, the sporophyte is the dominant generation, the gametophyte being much reduced.

specialized mechanisms that differ in the two cases. The zygote grows into a sporophyte. In mosses this is attached to, and dependent upon, the gametophyte; in ferns it is a separate self-supporting plant. In both cases the spores are formed in specialized spore-bearing structures called **sporangia** from which they are released by a dispersal mechanism which ensures that they are scattered over as wide an area as possible. The spores then germinate into the gametophyte. This is direct in ferns, but in mosses an extra stage is interpolated into the cycle: each spore germinates into a thread-like **protonema** which in turn produces **buds**. These then germinate into the gametophyte.

The two life cycles are therefore basically similar but different in detail. The most noticeable difference is the relative emphasis which each places on the gametophyte and sporophyte (Fig 25.5). In mosses the gametophyte is the 'dominant' generation, the sporophyte being comparatively simple, short-lived and dependent, almost parasitic, on the gametophyte. In ferns it is the other way round. The sporophyte is the 'dominant' generation:

it is large (some of the tree ferns are over 7 metres high), differentiated into leaf, stem and roots, and has a complex internal organization on a level with that of flowering plants. In comparison the gametophyte (or **prothallus** as it is called) is extremely simple. A mere millimetre or two in diameter, it is a flat plate of photosynthetic cells anchored by simple hair-like rhizoids to the surface of the soil.

REDUCTION OF THE SPOROPHYTE OR GAMETOPHYTE

Alternation of generations is basic to the life cycle of almost all plant groups. It is most clearly seen in mosses and ferns, where both generations can be easily recognized. In other groups it may be obscured by one or other generation being reduced or absent. For example, in the life cycle of a number of primitive plants, certain green algae for instance, the zygote is diploid but instead of giving rise to the adult by mitosis, it first undergoes meiosis and then develops into the adult. The haploid adult may be regarded as the gametophyte generation, the sporophyte and spores being non-existent. In some forms the zygote divides meiotically to form motile flagellated spores called **zoospores**, which in turn develop into the adult. In such cases there is no sporophyte plant as such, but the zygote functions as a sporophyte in that it gives rise to spores (Fig 25.6).

In the kind of plant just mentioned the sporophyte is absent altogether. In the more advanced forms, such as conifers and flowering plants, it is the other way round: the sporophyte is the dominant generation and the gametophyte is much reduced. In fact the latter, far from being an independent plant, is incorporated into the body of the sporophyte. But before we can really understand how this comes about we must return to ferns.

Fig 25.5 In these two photographs the gametophyte of a moss and fern can be compared. A) the moss *Polytrichum* showing the leafy gametophyte with sporophytes attached.   B) the fern *Dryopteris* showing the simple thalloid gametophyte (prothallus) from which a young sporophyte is beginning to grow. (*B. J. W. Heath, Marlborough College*)

A

B

Fig 25.6 In primitive plants there may be no sporophyte as such, the zygote undergoing meiosis into haploid zoospores which then give rise to the adult. The life cycle illustrated here is based on the filamentous alga *Oedogonium*. Other primitive plants show variations on this theme: for example, the unicellular *Chlamydomonas* usually produces isogametes rather than differentiated eggs and sperm; and the zygote of the filamentous *Spirogyra* develops into the adult direct rather than via motile zoospores. For further details a botanical textbook should be consulted.

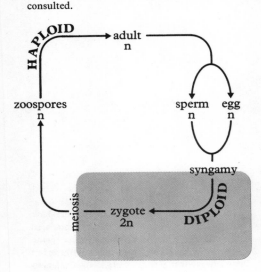

CLUB MOSSES AND HORSETAILS

In the common fern (and mosses too for that matter) archegonia and antheridia are formed on one and the same gametophyte. In other words the gametophyte is hermaphroditic, producing both eggs and sperm (Fig 25.7A). However, a number of plants closely related to ferns have separate male and female gametophytes, bearing antheridia and archegonia respectively. In some cases these are formed from two distinct types of spore, shed from two distinct types of sporangium. A plant that shows this is *Selaginella*, one of the club mosses (not a moss at all really, but a primitive fern). The sporangia are formed on special leaves, the **sporophylls**, which are densely packed to form cone-like structures projecting from the main stem. The sporophylls bear two types of sporangia. Small **microsporangia** produce a large number of tiny **microspores**. These germinate to form a sperm-producing male gametophyte. Larger **megasporangia** produce a single large **megaspore**. This germinates into a female gametophyte. The gametophytes, both male and female, are small and simple but self-supporting.

The life cycle of *Selaginella* is summarized in Fig 25.7B. The distinction between the two sexes makes it seem more complex than the fern, but in fact it is fundamentally the same: meiosis occurs in the formation of the spores, exactly as it does in other plants with alternation of generations. A situation intermediate between the club mosses and more advanced ferns is found in another fern-like group, the horsetails. In these plants there are separate male and female gametophytes, but these develop from just one kind of spore which is formed in a single type of sporangium. With this background let us examine the situation in flowering plants.

FLOWERING PLANTS AND CONIFERS

The actual plant itself, herb, shrub or tree as the case may be, is the sporophyte, and the gametophyte is contained within it. We have seen that the function of the sporophyte is to produce spores, and this is precisely what a flowering plant does. But what, and where, are the spores? As in the club mosses there are two kinds: the microspores are the pollen grains, and the megaspore is the embryosac. There is no gametophyte plant as such: the male gametophyte is represented by the protoplasmic contents of the pollen grain, and the female gametophyte by the protoplasmic contents of the embryosac. No antheridia or archegonia are present. The sperm are represented by the male nuclei in the pollen tube, and the egg by the egg cell in the embryosac.

If they really do represent the spores, we would expect the pollen grains and embryosac to be formed by meiosis, and therefore haploid. This is precisely what we find (see Chapter 24). The whole of the rest of the plant is diploid. Moreover we can usefully extend these homologies to other parts of the flower. The spore-containing microsporangia are represented by the anthers which contain the pollen grains, and the megasporangium by the ovule which surrounds the embryosac. Where, then, are the equivalents of the sporophylls? The general definition of a sporophyll is a spore-bearing leaf. In ferns the sporophylls are recognizable as ordinary leaves, but in flowering plants they are modified to form the stamens and carpels respectively.

In the club mosses, the sporophylls are grouped together to form cone-like structures. Precisely the same happens in flowering plants, the equivalent of the 'cone' being the flower itself. This is more easily seen in conifers (firs,

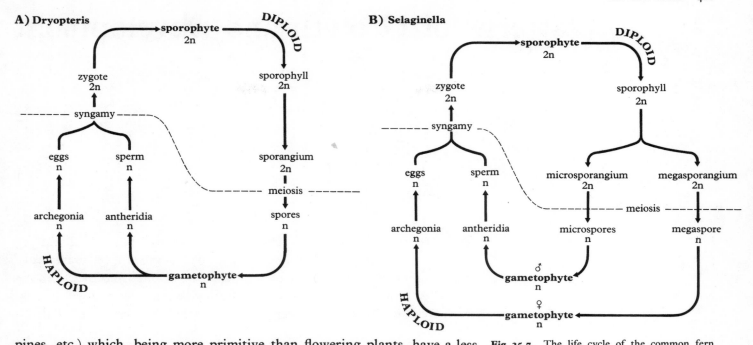

Fig 25.7 The life cycle of the common fern *Dryopteris* and the club moss *Selaginella* compared. The life cycles are basically similar. The principal difference is that *Selaginella*, unlike *Dryopteris*, produces two different kinds of spore: microspores, formed in microsporangia, germinate into sperm-producing male gametophytes; megaspores, formed in megasporangia germinate into egg-producing female gametophytes.

pines, etc.) which, being more primitive than flowering plants, have a less modified structure. Each cone is a collection of closely-packed sporophylls. The male cones are composed of **microsporophylls** each of which bears a pair of pollen sacs (microsporangia). The female cones are made up of **megasporophylls**, each bearing two ovules (megasporangia). In other basic respects they resemble flowering plants. There is no free-living gametophyte stage, the male and female gametophytes being represented by the contents of the pollen grain and embryosac respectively, just as in flowering plants.

The flowering plant represents the end product of a line of evolution in which the gametophyte, a prominent self-supporting plant in the more primitive groups such as mosses, gradually degenerates until eventually it becomes incorporated into the body of the sporophyte. Meanwhile, the sporophyte gains in structural importance, being very much the dominant generation in both ferns and flowering plants. We shall return to this in Chapter 36.

ALTERNATION OF GENERATIONS IN ANIMALS?

Do animals show alternation of generations? At first sight the answer might appear to be yes. After all, don't butterflies and moths show an alternation in their life cycle between the imago and the caterpillar? They certainly do, but for this to be regarded as true alternation of generations, in the sense that we have been using the term here, it would have to be shown that the caterpillar, like the sporophyte, is capable of asexual reproduction and that the imago, like the gametophyte, is haploid. In fact neither is true.

A more likely case is to be found in the coelenterates. The colonial hydroids like *Obelia* show an alternation between a gamete-producing jelly fish-like individual, the **medusa**, and an asexually reproducing hydra-like **polyp**. On grounds of reproduction this would seem to be a genuine alternation of generations. However, it does not qualify from the genetical point of view, for the medusa and polyp are both diploid, meiosis occurring in the formation of the gametes. There is in fact no authenticated case of alternation of generations in the animal kingdom.

26 Patterns of Growth and Development

In the course of its life cycle an organism changes from a fertilized egg into an adult. This process of development involves growth, cell movements, and differentiation. In this chapter we shall be mainly concerned with the overall pattern of changes that occurs during development and will not say much about the causative mechanisms responsible for initiating and controlling them.

Growth

Growth is the permanent increase in size acquired by an organism in the course of its development. Three distinct processes contribute to growth: **cell division, assimilation**, and **cell expansion**. Cell division is the basis of growth in all multicellular organisms, but plainly to grow to the size of the parent cell, the daughter cells must be able to synthesize new structures from raw materials which they absorb from their surroundings. This is what is meant by assimilation, and it results in cell expansion. In many plants, growth is aided by vacuolated cells taking up water by osmosis and expanding like balloons. Because of differential thickening of their walls, some of the cells acquire a greatly attenuated form, resulting in rapid elongation of the stem. We shall return to this presently (p. 425).

MEASURING GROWTH

Growth can be estimated by measuring some parameter of the organism, such as height or weight, at suitable intervals over a known period of time. This is not always easy, because of difficulties in selecting the right parameter. Commonly a linear dimension such as height or length is measured, and to aid such estimations a slowly revolving **kymograph** or **time-lapse photography** may be used. The drawback in measuring a single linear dimension such as height, is that it takes no account of growth in other directions, which in some cases may be considerable. One can sometimes get round this by measuring changes in volume, though this may call for considerable mathematical ingenuity if the organism has an irregular shape. A parameter often used is weight, but this may not necessarily be a measure of growth as such, and can be particularly influenced by variations in the fluid content of the body. To overcome this one may resort to estimating **dry weight**, i.e. the weight after all moisture has been driven off by heating. The trouble here is that it kills the organism, so meaningful measurements must be obtained by taking random samples at intervals from a large population of individuals. The latter must be the same age and growing under identical conditions. This technique is commonly adopted with plants.

One's difficulties do not end even after the best parameter has been chosen, for growth shows many irregularities as a result of fluctuations in the environment or diet. Moreover the organism's shape and form may change as it grows. It is often very difficult to eliminate these factors, try as one may.

A misleading aspect of measuring growth by changes in the size or weight of the whole organism, is that it fails to recognize that different parts of the organism may grow at different rates and stop growing at different times. In man, for example, the brain grows rather more slowly than other organs, and stops completely at about the age of five, though the rest of the body continues to grow for another 20 years or so. The growth of different parts of the body at rates peculiar to themselves, higher or lower than the growth rate of the body as a whole, is known as **allometric growth**. Most organisms show it to some extent. A full description of an organism's growth must obviously take this into account.

THE GROWTH CURVE

If an organism's measurements (height, weight or whatever it is) are plotted against time, a **growth curve** is obtained (Fig 26.1A). Despite difficulties in measuring growth, the general pattern turns out to be the same for most organisms. Growth tends to be slow at first, then it speeds up, and finally it slows down as adult size is reached, giving an S-shaped curve. In man and certain other vertebrates growth stops altogether when adult size is reached (in the early twenties in most men), but in the majority of organisms growth continues in adult life, though slowly.

RATE OF GROWTH

The growth curve enables us to re-express the growth of an organism in terms of **growth rate**. This can be done by estimating the increase in size that takes place during successive intervals of time, i.e. **growth increments**, and plotting these against time. In most organisms the growth rate increases steadily until it reaches a maximum, after which it gradually falls, giving a bell-shaped curve such as the one shown in Fig 26.1B.

PERCENTAGE GROWTH

We have seen that growth of a multicellular organism takes place by cell division, each generation of daughter cells undergoing assimilation and expansion. In other words the products of the growth process are themselves capable of growing. This is quite different from, say, the growth of a crystal, in which new material is added to the surface of the existing crystal from the outside. Different, too, from the building of a house in which new bricks, obtained from an outside source, are added one by one to the existing structure.

In contrast, an organism's growth is essentially an internal process. This means that the amount of growth which takes place at any stage is dependent on the bulk of tissue already present, i.e. on how much growth has

Fig 26.1 Growth curves constructed from data on lupins. (*after Simpson and Bede*, Life: An Introduction to Biology, *Harcourt Brace Jovanovich, Inc. 1965*)

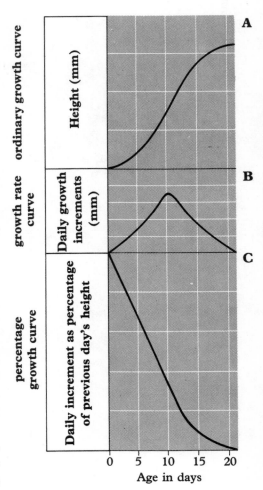

already occurred. More meaningful, therefore, than absolute growth is **percentage growth**, in which increase in growth over a period of time is expressed as a percentage of the growth that has already taken place. For example, between the ages of 1 and 2 a baby boy's weight might increase from 10 to 12 kilogrammes. His absolute increase is therefore 2 kg, but his percentage increase is $\frac{2}{10} \times 100 = 20$ per cent. Now, in the same period of time the weight of a teenage boy might go up from 50 to 55 kg, giving an absolute increase of 5 kg. However, his percentage increase is only $\frac{5}{50} \times 100 = 10$ per cent, i.e. less than that of the baby. So the baby, though it puts on less weight in a year, has a higher percentage growth rate.

If percentage increase is plotted against time, we get a curve of the sort shown in Fig. 26.1C. Represented this way, the growth rate does not rise and then fall. On the contrary, growth is fastest at the beginning of life (in a human baby while it is still in the mother's uterus), after which it gradually slows down. This is true of most organisms, both animal and plant.

INTERMITTENT GROWTH IN ARTHROPODS

The smooth growth curve shown in Fig 26.1A is typical of most animals, but there is one notable exception, arthropods. If the weight changes of, say, an insect are plotted against time, a curve of the sort shown in Fig 26.2 is obtained. Instead of increasing smoothly, growth takes place in a series of steps. These correspond to the sequence of stages, or **instars**, in the insect's development (see p. 423).

This intermittent growth is made necessary by the hard **cuticle** (**exoskeleton**) which prevents overall growth of the whole body. Periodically the cuticle is shed, and only then can growth take place, while the new cuticle underneath is still soft enough to allow the body to expand. In some cases rapid expansion is achieved by the insect swallowing air or water. The distension of the gut pushes the soft integument outwards. The new cuticle then hardens, after which further growth is impossible until the cuticle is shed again. Rapid expansion of an insect's body, including the wings, following the shedding of the cuticle, is illustrated in Fig 26.3.

The shedding process is known as **moulting** or **ecdysis**, and it takes place by the secretion of a **moulting fluid** immediately beneath the cuticle. This dissolves the soft inner part of the cuticle, leaving only the hard outer part. Meanwhile the new cuticle, soft at first, is secreted by the epidermis. Protected from the enzymatic action of the moulting fluid by its protective surface, it is destined to become the hard cuticle of the next instar. The cuticle is composed of **chitin**, a complex nitrogen-containing polysaccharide. Hardening of the outer part is achieved by the chitin being impregnated with tanned (hardened) proteins. Waterproofing of the cuticle is achieved by the deposition of a thin layer of wax at the surface.

During the early stages of growth, radical changes take place in the body: old structures are reorganized and new ones formed. These progressive changes, which must be undergone before the organism acquires its adult characteristics, constitute **embryology**; while they are taking place the organism is referred to as an **embryo**. Studying the embryology of any animal or plant entails asking two basic questions: what changes take place, and why do they occur? In this chapter we shall be concerned with the first of these questions.

Fig 26.2 Intermittent growth in an arthropod. The graph shows the increase in weight of the waterboatman *Notonecta glauca* over a period of nearly 70 days. The sudden weight increases (marked by arrows) correspond to moulting of the cuticle. (*data from Teissier*)

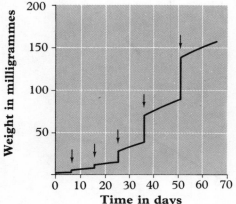

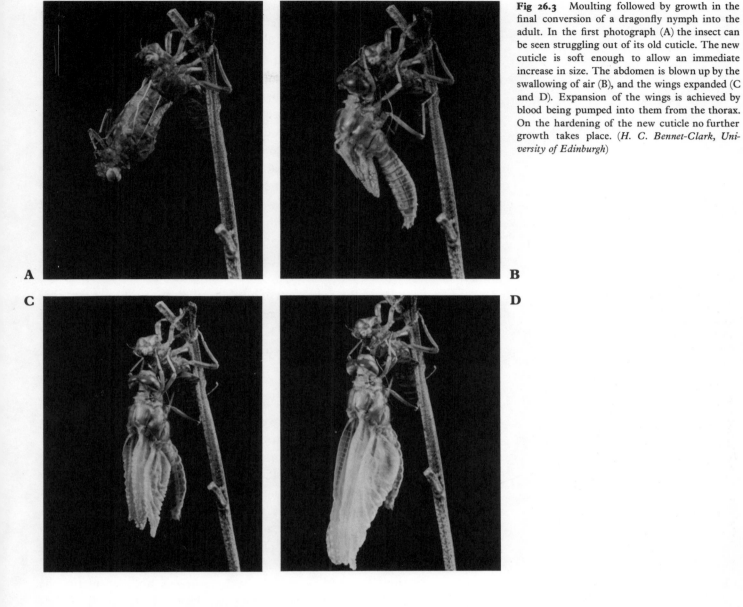

A

B

C

D

Fig 26.3 Moulting followed by growth in the final conversion of a dragonfly nymph into the adult. In the first photograph (A) the insect can be seen struggling out of its old cuticle. The new cuticle is soft enough to allow an immediate increase in size. The abdomen is blown up by the swallowing of air (B), and the wings expanded (C and D). Expansion of the wings is achieved by blood being pumped into them from the thorax. On the hardening of the new cuticle no further growth takes place. (*H. C. Bennet-Clark, University of Edinburgh*)

Animal Development

Embryological development is normally triggered by the act of **fertilization**. In the majority of animals the subsequent changes that take place fall into three stages:

(1) **Cleavage**: division of the zygote into daughter cells.

(2) **Gastrulation**: arrangement of these cells into distinct layers.

(3) **Organogeny**: formation of organs and organ systems.

The following account is based mainly on the invertebrate chordate *Amphioxus* and amphibians such as the frog or salamander. These animals show the main sequence of events in chordate development particularly clearly.

CLEAVAGE

Development starts with cleavage, the zygote dividing repeatedly by mitosis

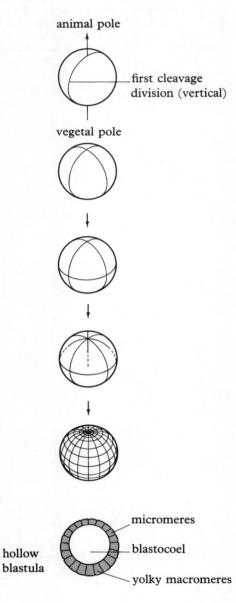

Fig 26.4 Cleavage in an amphibian. Successive vertical and horizontal mitotic divisions of the fertilized egg result in the formation of a hollow blastula.

Fig 26.5 Gastrulation in a primitive chordate like *Amphioxus* takes place by a simple process of invagination. It results in the formation of a two-layered gastrula.

into progressively smaller cells or **blastomeres**. In animals generally, there are many different types of cleavage, determined mainly by the amount of yolk present in the egg. If there is little or no yolk, the cleavage divisions are symmetrical, resulting in blastomeres of equal size. More often, however, a moderate amount of yolk is concentrated towards the vegetal pole: this causes cleavage to be unequal, the cells at the animal pole being smaller than the ones beneath. The small cells are called **micromeres**, the larger ones **macromeres**. Sometimes there is so much yolk that the vegetal end of the egg does not divide at all, so cleavage is confined to the animal pole.

Cleavage is not a haphazard affair: it is a highly organized process in which a series of successive divisions take place in planes at right angle to each other. In amphibians, for example, two vertical divisions at right angles to one another are followed by a horizontal division; thereafter alternate vertical and horizontal divisions take place (Fig 26.4). As the micromeres divide faster than the larger yolky macromeres, they come to be smaller as well as more numerous.

Cleavage results in the formation of a spherical mass of cells. At first this is solid, but as cleavage continues the cells draw away from the centre, leaving a fluid-filled cavity in the middle. The mass of cells is now known as the **blastula**, and its cavity as the **blastocoel**.

GASTRULATION

The word gastrulation literally means 'formation of a gut', and this is precisely what results from the various cell movements that take place. The exact way gastrulation occurs varies from one species to another, but at its simplest it involves a process of invagination at one end of the blastula (Fig 26.5). Rather like pushing in a tennis ball on one side, it results in the formation of a two-layered cup-shaped **gastrula**. The outer layer of cells, the **ectoderm**, is destined to give rise to the skin and associated structures; the inner layer ultimately forms the lining of the gut and various other structures. Gastrulation results in the virtual obliteration of the blastocoel and its replacement by a new cavity, the **archenteron** or 'primitive gut'. Its opening to the exterior, where invagination occurred, is called the **blastopore**, and represents the future posterior end of the embryo. During gastrulation the lips of the blastopore represent the point where cells that originally lined the blastula, turn inwards to form the lining of the archenteron. Some of these cells are destined to give rise to particular structures in the embryo; in fact their fate is sealed as far back as the blastula. These cells include the **chorda-mesoderm**: the area destined to give rise to the **notochord** and **mesoderm** respectively. Originally located on the dorsal side of the blastula, these cells are drawn inwards at the

Blastula

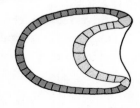

Gastrulation

Completed gastrula

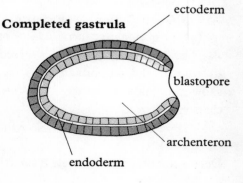

dorsal and lateral lips of the blastopore to form the roof of the archenteron. The remaining cells, forming the ventral surface and sides of the archenteron, constitute the **endoderm** which is destined to become the lining of the alimentary canal.

The above account is based on *Amphioxus*. In few chordates is gastrulation as simple as this. In amphibians, for example, invagination of the blastula is accompanied by mass migration of cells in various directions, including downward movement of micromeres over the surface of the larger macromeres below (**epiboly**). In birds and mammals there is no invagination at all, endoderm-formation taking place entirely by cell migration. These modifications are mainly caused by the presence of large quantities of yolk. In amphibians gastrulation takes place above the large yolky macromeres, the blastopore being a narrow slit to begin with. As gastrulation continues this slit becomes crescent-shaped and eventually a complete circle, the yolky cells within the circle forming the so-called **yolk plug** (see Fig 26.8C).

During and immediately after gastrulation the embryo increases in length so that, in effect, it becomes a tube within a tube. Eventually the anterior end of the archenteron acquires an opening to the exterior, which becomes the mouth. Perhaps surprisingly, the anus is not formed from the blastopore: the latter closes up and a new connection with the exterior, formed just beneath it, becomes the anus.

FORMATION OF NEURAL TUBE AND NOTOCHORD

Meanwhile three important structures are laid down in what remains of the blastocoel: these are the central nervous system, notochord, and mesoderm. The formation of the CNS is much the same in all chordates. The ectodermal roof of the gastrula, known as the **neural plate**, sinks downwards and a pair of neural folds grow over the top of it as shown in Fig 26.6. The resulting invagination, the neural groove, eventually separates from the rest of the ectoderm as a result of the neural folds fusing in the mid-dorsal line. Now known as the **neural tube**, it eventually develops into the whole of the CNS: the anterior end expands to form the brain, whilst the rest becomes the spinal cord (see p. 278). The formation of the neural tube is termed **neurulation** and the embryo at this stage is called the **neurula**.

The formation of the notochord and mesoderm varies in different chordates. In *Amphioxus*, which again can be taken to illustrate the basic pattern, these two structures are derived from the lining of the archenteron. As you can see from Fig 26.6, the notochord is derived from a strip of cells on the mid-dorsal side of the archenteron. It comes to lie immediately beneath the neural tube where it forms a skeletal rod, stiffening the dorsal side of the embryo. In *Amphioxus* and other invertebrate chordates it persists in the adult. However, in the vertebrates it is replaced by the vertebral column. We shall return to this later.

FORMATION OF MESODERM

In *Amphioxus* the mesoderm arises as two pouch-like evaginations of the archenteron on either side of the body (Fig 26.6). The rest of the archenteron wall then closes over to become the endodermal lining of the gut. The origin of the mesoderm from the wall of the archenteron can be clearly seen in an invertebrate chordate like *Amphioxus*, but is obscured in higher chordates. In

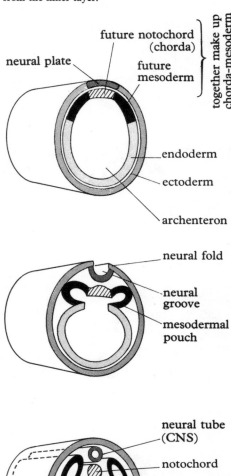

Fig 26.6 In *Amphioxus* the CNS, notochord and mesoderm are formed by folding of the two layered wall of the gastrula. The CNS is derived from the outer layer, the notochord and mesoderm from the inner layer.

Fig. 26.7 Development of the mesoderm in a typical chordate. Explanation in text.

A

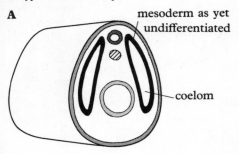

mesoderm as yet undifferentiated

coelom

B

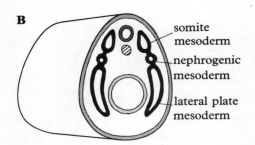

somite mesoderm

nephrogenic mesoderm

lateral plate mesoderm

C

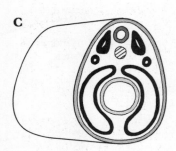

D

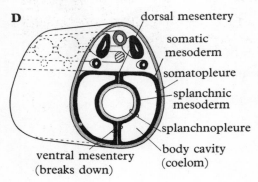

dorsal mesentery

somatic mesoderm

somatopleure

splanchnic mesoderm

splanchnopleure

ventral mesentery (breaks down)

body cavity (coelom)

amphibians the mesoderm is formed at an earlier stage by inward migration of cells into the blastocoelic space from the lateral lips of the blastopore. The same applies, with certain modifications, to other chordates like birds and mammals. In the bird this inward migration of mesoderm occurs on either side of a deep invagination, the **primitive streak**, which extends posteriorly from the blastopore.

Whatever its mode of formation, the mesoderm comes to lie between the ectoderm and endoderm. The embryo is therefore composed of three layers of cells: **ectoderm, mesoderm** and **endoderm**. These are called the **germ layers** and, although they may arise differently in the various groups, they are fundamental features of the ground plan of most animals.

ORIGIN OF THE COELOM

Also fundamental to most animals is the possession of a **coelom**. This is a cavity enclosed by the mesoderm. It is clear from Fig 26.6 that in *Amphioxus* this cavity is derived from the archenteron; however in other animals it may be formed by an initially solid mass of mesodermal cells drawing away from one another so as to create a cavity in the middle. However formed, the coelom becomes the main body cavity in the adult.

FURTHER DEVELOPMENT OF MESODERM AND COELOM

The formation of organs results largely from differentiation of the mesoderm. This is best seen in amphibians (Fig 26.7). If you look at a transverse section of a frog embryo shortly after gastrulation, you will see that the mesoderm consists of a uniform sheet of tissue, containing a coelomic cavity, on either side of the body (Fig 26.7A). This soon becomes split up, by the development of horizontal constrictions, into three regions:

(1) the **somite mesoderm** on either side of the notochord;
(2) the **lateral plate mesoderm** on either side of the gut;
(3) the **nephrogenic mesoderm** in between the other two.

These three masses of mesodermal tissue are further distinguished from one another by the fact that the somite and nephrogenic mesoderms are **metamerically segmented**, i.e. they are divided up into a series of identical units recurring along the length of the body. The lateral plate mesoderm, on the other hand, is unsegmented.

What happens to these three portions of the mesoderm? The coelomic cavities enclosed by the lateral plate mesoderm expand to form the general body cavity. As a result, the mesoderm itself comes to lie against the ectoderm and endoderm, where it forms the **somatic** and **splanchnic mesoderm** respectively. These become the peritoneal lining of the body cavity. The layer of mesoderm and ectoderm forming the body wall is called the **somatopleure**, and the layer of mesoderm and endoderm lining the gut is called the **splanchnopleure**. With the expansion of the left and right coelomic cavities, the mesodermal tissues of the two sides meet in the midline above and below the gut. The mesoderm above the gut persists as the **dorsal mesentery** by which the gut is suspended in the body cavity. Below the gut, in the abdominal region, the mesoderm breaks down and disappears, so that the body cavity is continuous between the two sides. Anteriorly, this part of the mesoderm does not break down but forms the tubular **heart**, the section of the coelom surrounding it becoming the **pericardial cavity**.

Meanwhile, the somite mesoderm contributes to the formation of three important structures. Cells on its outer side take up a position immediately beneath the ectoderm, forming the **dermis** of the skin, the overlying ectoderm becoming the **epidermis**. Cells from its inner side migrate inwards, surrounding, and eventually replacing, the notochord where they become the **axial skeleton** (vertebrae etc). The remainder of the somite, called the **myotome**, becomes the **skeletal musculature**.

While all this is going on, the nephrogenic mesoderm in each segment sends out tubules which unite and grow back to the cloaca, the common cavity that eventually receives the anus and the external openings of the urino-genital system. Further elaboration of the nephrogenic mesoderm results in the formation of the **kidney**.

Fig 26.8 Five stages in the development of the salamander *Amblystoma opacum*. The embryos were kept at a constant temperature of 20°C. A) Cleavage looking towards the animal pole: 4 cell stage ($11\frac{1}{2}$ hours). B) 8 cell stage ($13\frac{1}{2}$ hours): note the four large macromeres lying beneath the four smaller micromeres. C) Late gastrula (79 hours): blastopore and yolk plug clearly visible. D) Formation of neural tube (105 hours): the neural folds have met in the midline in the middle region of the body. E) Embryo at 209 hours: outline of segmental somites just visible towards dorsal side. (*John A. Moore, University of California*)

A

B

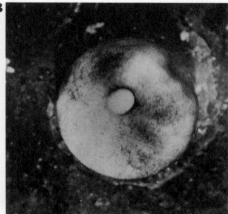

C

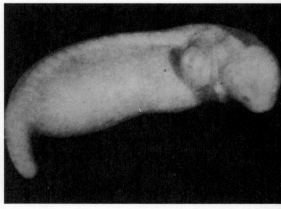

D

Thus is the basic ground plan of a chordate laid down. Subsequent development involves the further differentiation of the three germ layers, particularly the mesoderm, into specific organs. The **gut** increases in length and becomes coiled, and various accessory organs such as the **liver** and **pancreas** develop as outgrowths from it. The neural tube differentiates into **brain** and **spinal cord**, and the **peripheral nerves** develop. The **lungs** arise as an outgrowth from the back of the **pharynx**, and **blood vessels** are moulded out of the mesoderm. **Gills**, if present, develop as pouches of the pharynx which break through to the exterior on either side of the body. For details of these various events a textbook of embryology should be consulted.

A series of photographs illustrating the development of an amphibian is given in Fig 26.8. To what extent can you relate these photographs to the events described here?

DEVELOPMENT OF BIRDS AND MAMMALS: THE EXTRA-EMBRYONIC MEMBRANES

In this account of animal development we have concentrated mainly on lower chordates such as *Amphioxus* and amphibians. To what extent do higher chordates such as birds and mammals depart from this basic pattern?

They depart from it in two main ways. Firstly, the eggs of birds, for instance, contain much more yolk, with the result that cleavage is confined to the animal pole. This results in the embryo being situated on top of a **yolk sac** from which it derives nourishment via a system of **vitelline blood vessels**. Although there are many detailed differences in the way it is formed, the embryo itself is similar to the amphibian embryo illustrated in Fig 26.7D.

E

The second difference is that the embryos of birds and mammals are surrounded by a system of protective membranes which develop from tissues outside the embryo itself. Because of their origin they are known as **extra-embryonic membranes**. How these membranes are formed is illustrated in Fig 26.9. You will notice in Fig 26.9A that the endodermal lining of the yolk sac is continuous with the lining of the archenteron, and that the yolk sac is surrounded by somatopleure and splanchnopleure continuous with those layers in the embryo itself. Now the somatopleure below the embryo on the upper side of the yolk sac becomes thrown into a circular fold which grows upwards, finally enveloping the embryo and enclosing it in a fluid-filled **amniotic cavity** (Fig 26.9B, C). The inner lining of this cavity is called the **amnion,** the outer one the **chorion.** They are separated by a coelomic cavity (**extra-embryonic coelom**) continuous with the coelom in the embryo.

While the amniotic cavity is being formed, a further cavity develops (Fig 26.9B, C). This arises as an outgrowth from the hindgut which pushes its way into the extra-embryonic coelom. Known as the **allantois**, it expands distally into a balloon-like sac whose lining fuses with the mesodermal component of the chorion to form the **allanto-chorion.**

What exactly are the functions of all these cavities and membranes? In both birds and mammals the function of the fluid-filled amniotic cavity is to cushion the delicate embryo and protect it from physical disturbance. The function of the allantois differs in the two groups. In birds it becomes a depository for nitrogenous waste in the form of uric acid: being surrounded by the shell there is no way of getting rid of nitrogenous waste, so it has to be stored out of harms way. The allanto-chorion becomes a highly vascular respiratory surface connected to the embryo by blood vessels. Located just beneath the shell, it allows gaseous exchange to occur between the blood of the embryo and surrounding atmosphere.

In eutherian mammals the allanto-chorion develops into the **placenta**. The chorion develops finger-like outgrowths, the **chorionic villi,** which project into blood spaces in the wall of the mother's uterus. The villi become invaded by capillary loops through which blood circulates from and to the embryo. Across the thin walls of the chorionic villi, exchange of materials takes place between embryonic and maternal blood. As the placenta expands, the yolk sac gradually diminishes. At the same time the amniotic cavity expands, pushing the amnion and chorion outwards until finally the amnion fuses with the chorion, and the chorion with the lining of the uterus. The net result of these changes is that the embryo (now known as the **foetus**) comes to lie in the middle of a large fluid-filled amniotic cavity which completely fills the uterus. Thus cushioned, it is protected from jarring and asymmetrical pressure changes that might adversely affect its development. The only connection between the embryo itself and the wall of the uterus is the stalk of the allantois which becomes the **umbilical cord**. This contains the **umbilical artery** and **vein** which convey foetal blood to and from the placenta (see p. 388). The extra-embryonic membranes of the mammal are referred to as the **foetal membranes**.

As was explained on page 388, the placenta provides the means by which the foetus obtains oxygen. At birth important changes take place in the circulation, associated with the fact that the respiratory function of the placenta is taken over by the lungs. In the foetus the umbilical vein conveys

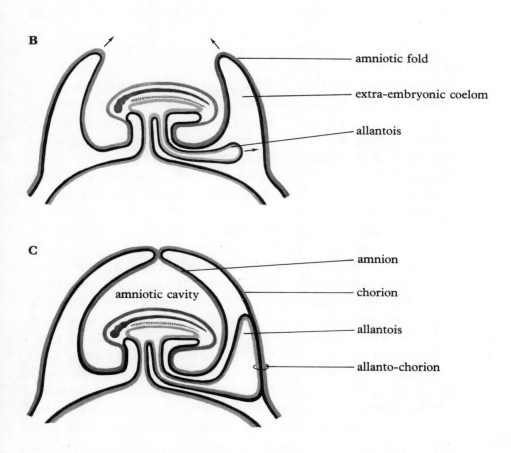

Fig 26.9 Formation of the extra-embryonic membranes in birds and mammals. The functions performed by these membranes in the two groups are discussed in text. Mesoderm in dark tone, ectoderm in medium tone, endoderm in light tone.

oxygenated blood to the posterior vena cava whence it enters the right atrium of the heart. The lungs are functionless and most of the blood bypasses them by flowing through the **foramen ovale,** a hole connecting the right and left atria, and the **ductus arteriosus,** a vessel linking the pulmonary artery with the aorta (Fig 11.9). As a result of various pressure changes and nervous reflexes triggered principally by the rise in the partial pressure of oxygen in the blood when the baby takes its first breath, the foramen ovale and ductus arteriosus close soon after birth. The result is that from now on all the blood returning to the right atrium is sent to the lungs. Failure of the foramen ovale and/or ductus arteriosus to close results in a 'blue baby': a proportion of the blood continues to bypass the lungs, resulting in inadequate oxygenation of the tissues.

LARVAL FORMS

From the foregoing account it will be clear that development involves a progressive increase in complexity as new structures are gradually formed. Sometimes, however, the egg develops first into a **larva** which then changes into the adult. Usually very different in form and habit, the larva changes into the adult by a sudden dramatic transformation known as **metamorphosis**. The best known larvae are the tadpole of the frog and the caterpillar of butterflies, but many other animals, mainly invertebrates, also have larvae. These include the 'tadpole' larva of sea squirts, the nauplius and zoea of crustaceans, and the ciliated trochophore of marine annelids and molluscs. Ciliated larvae are produced by a number of other invertebrates and include the planula of coelenterates, the miracidium of parasitic flukes, and the dipleurula of echinoderms. Many of these larvae are discussed in other parts of the book.

Though varied in structure, all these larval forms have three basic features in common:

(1) they are markedly different from the adult in their structure;

(2) they can lead an independent life, fending for themselves in ways which are generally different from those adopted by the adult;

(3) they are incapable of sexual reproduction, and only in specialized instances do they undergo asexual reproduction.

The fact that they are so different from the adult in structure and habits, enables larvae and adults to exploit different habitats, never coming into direct competition with one another.

What is the significance of larval forms? Looking at animals generally, larvae have three basic functions:

(1) If small and motile, they may be responsible for distribution of the species. This is particularly important in animals that have restricted mobility because they are slow-moving, sessile, or parasitic. Thus the ciliated dipleurula larva enables the spread of slow-moving echinoderms like the starfish; the ciliated planula larva does the same for sessile coelenterates like sea anemones; and the ciliated miracidium enables parasitic flukes to get from host to host.

(2) In some cases larvae are primarily responsible for feeding and growth, prior to the formation of the adult. This applies to insects, particularly butterflies and moths. In some cases the larva has well-developed feeding devices, whereas the adult may be totally incapable of feeding. Certain adult moths, for example, have vestigial mouthparts and live for only 24 hours, during which time their sole function is to reproduce. The longer-lived caterpillar, however, has well developed mandibles for chewing plant food. Larval and adult insects feed on different food so that they do not compete with each other.

(3) In certain specialized instances, notably the parasitic flukes, the larvae may be capable of asexual reproduction, thereby greatly increasing the number of offspring produced. In the liver fluke *Fasciola hepatica* the intermediate host, a snail, harbours several successive larval stages, each responsible for producing large numbers of new larvae by a process of internal propagation (see p. 534).

METAMORPHOSIS

Metamorphosis, the change from larva to adult, generally involves a profound reorganization of the body, often involving considerable breakdown of larval tissues. Nowhere can this be better seen than in insects. As you may know, insects have two different types of life history. **Hemimetabolous insects** like locusts, cockroaches, grasshoppers, termites and earwigs, show what is rather misleadingly called **incomplete metamorphosis**: the egg develops into the adult via a series of **nymphs** which are essentially miniature adults lacking wings. Moulting and growth take place between each nymphal stage. In contrast, **holometabolous insects** like butterflies, moths, beetles and flies, show **complete metamorphosis**: the egg develops into a **larva** which is strikingly different from the adult. After an active life of feeding and growth, typically with several moultings, the larva enters a seemingly dormant stage called the **pupa** or **chrysalis**. Although immobile, and to all outward appearances completely inactive, the pupa is the site of considerable internal activity. Apart from the CNS and a number of small groups of cells called **imaginal buds**, the larval organs are broken down by phagocytes into a fluid mass. From the imaginal buds adult organs are formed: the cells multiply and differentiate, the necessary nourishment coming from the now dissolved remains of the other larval tissues. The adult body complete, and environmental factors permitting, the insect hatches out of the chrysalis as the adult or **imago**.

Metamorphosis does not always entail as complete a dissolution of larval tissues as in holometabolous insects. For example, the metamorphosis of the tadpole into an adult frog involves more a modification of pre-existing structures than their total replacement. In amphibians these changes are associated with the move from water to dry land: the tail is lost (by phagocytosis), the gills are replaced by lungs, and the legs (already present in rudimentary form) develop fully. Meanwhile changes take place in the heart and blood vessels. As in insects these changes are controlled hormonally, in this case by thyroxine. Injecting a tadpole with thyroxine can induce precocious metamorphosis. But the control of metamorphosis is another story to which we shall return in the next chapter.

Development in the Flowering Plant

In flowering plants development commences with the growth of the zygote into a simple **embryo** within the seed: the embryo is differentiated into an embryonic shoot (**plumule**) and root (**radicle**) and either one or two seed leaves (**cotyledons**). The embryo is surrounded by the nutritive **endosperm tissue**, and the whole is enclosed and protected within the tough **seed coat**. The structure of a generalized seed is shown diagrammatically in Fig 26.10A. Some seeds, such as broad bean, have large fleshy cotyledons full of stored food, and little or no endosperm tissue. Others, such as sunflower, have a lot of endosperm tissue and relatively small scale-like cotyledons. We shall see that this is closely related to the way the seed germinates. Notice the **epicotyl** (the base of the plumule above the point of origin of the cotyledon stalks), and the **hypocotyl** (the base of the radicle below the point of origin of the cotyledon stalks). The significance of these two regions will become clear presently.

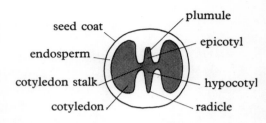

Fig 26.10 Structure and germination of seeds. A) Structure of a generalized seed. B) and C) show the two types of germination described on p. 424. In hypogeal germination the epicotyl elongates so the cotyledons remain below ground. In epigeal germination the hypocotyl elongates so the cotyledons are thrust up above ground.

A) Generalized seed

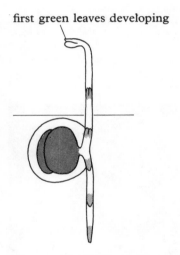

B) Hypogeal germination

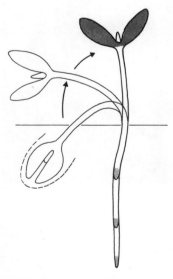

C) Epigeal germination

TYPES OF GERMINATION

When the seed germinates, the seed coat ruptures by growth of the plumule and radicle. On emerging, the plumule grows upwards and the radicle downwards. There are two types of germination: hypogeal and epigeal. **Hypogeal germination** (Fig 26.10B) takes place by elongation of the epicotyl, with the result that the plumule is thrust upwards out of the ground, leaving the cotyledons, still enclosed within the ruptured seed coat, below the ground (hypogeal means 'below the ground'). Seeds germinating in this way have much stored food in their large cotyledons. This provides the growing embryo with nourishment until the first green leaves develop at the tip of the plumule. Examples of seeds showing hypogeal germination are broad bean and wheat, the former having two cotyledons, the latter only one. In wheat the tip of the plumule is protected by a sheath, the **coleoptile**, which bursts open when the first leaves appear.

In **epigeal germination** (Fig 26.10C) the hypocotyl elongates, with the result that the plumule and cotyledons are thrust upwards out of the ground. Seeds germinating this way generally have small cotyledons which, once exposed to light, develop chlorophyll and start to photosynthesize. Before this happens, nourishment is provided by the endosperm tissue of which there is always a large amount in this kind of seed. Examples of seeds showing epigeal germination are sunflower and castor oil.

THE PHYSIOLOGY OF GERMINATION

Germination starts with the rapid uptake of water by the seed, resulting in a dramatic increase in weight. In the broad bean the amount of water absorbed is $1\frac{1}{2}$ times the original weight of the seed. The initial uptake of water is caused by colloidal particles in the seed coat, and later in the endosperm and embryo itself, imbibing it. With the subsequent increase in the number of solute molecules in the embryonic tissues, water is taken up by osmosis. The swelling of the embryonic tissues ruptures the seed coat, thereby allowing the growing plumule and radicle to emerge (Fig 26.11). The absorbed water activates the enzymes, and allows the hydrolysis of stored food materials (eg. starch) into soluble products capable of being translocated from the storage tissues to the growing points of the embryo. This rapid mobilization of food reserves is a necessary requirement for successful germination.

Apart from water, the other requirements are correct illumination, suitable temperature, and the presence of oxygen. Some seeds require light for germination, though the amount needed may be very little; some only germinate in darkness; others in light or dark. The rôle of light in triggering germination is discussed in Chapter 27. Temperature requirements are also variable, and depend to a large extent on what part of the world the seeds come from: the optimum temperature for germination of temperate seeds is generally lower than that for tropical and sub-tropical seeds.

GROWTH AND DEVELOPMENT OF THE SHOOT AND ROOT

It is characteristic of plants that growth and development are achieved by active cell division and differentiation in certain localized regions of the plant. Such regions are called **meristems**. This contrasts with animals where development goes on all over the body. In a growing plant the principal meristems are located at the tips of the stems and roots where they form the **apical meristems** or growing points.

Fig 26.11 The broad bean shows hypogeal germination, the epicotyl elongating so that the cotyledons and seed coat remain below ground. Notice that the plumule emerges in such a way that the delicate leaves at its tip are bent back and thereby protected from damage. Root hairs are evident on the radicle, and two lateral roots are just beginning to sprout. One of the cotyledons is just visible beneath the ruptured seed coat. During germination food reserves are mobilized in the cotyledons and sent to the growing points. (*B. J. W. Heath, Marlborough College*)

The cells at the extreme tip of a growing stem or root are in a state of active mitotic division. Further back the cells develop vacuoles and expand, later differentiating into epidermis, parenchyma, sclerenchyma or vascular tissue, depending on their topographical position in the plant. The process of expansion is illustrated in Fig 26.12. To begin with, several small vacuoles appear in the cytoplasm; these then coalesce to form the large sap vacuole typical of a mature plant cell. While this happens the vacuoles take in water by osmosis, pushing the as yet thin and stretchable cell wall outwards. In this way the cell expands. The final shape acquired by the cell depends on the thickening of its cellulose walls and this is the basis of differentiation. If the cellulose is laid down uniformly, a spherical cell results typical of parenchyma. If it is laid down unevenly a long, narrow cell may be formed. Differentiation involves further elaboration of the cell wall: excessive deposition of cellulose at the corners results in the formation of collenchyma cells, and impregnation of the cellulose with lignin results in the formation of woody vessels, tracheids and sclerenchyma fibres. The apex of a stem or root can therefore be distinguished into three zones: at the extreme tip is a **zone of cell division**; just behind this is a **zone of cell expansion**; and further back still is a **zone of differentiation** where the cells are developing into specialized types (Fig 26.13). This applies to stems and roots alike.

In the stem the apical meristem gives rise to **leaf primordia** on either side. These grow up and envelop the apex, forming the **apical bud**. In this way the delicate meristematic tissues are afforded protection. In the angle between the leaves and the main stem are formed **axillary** (**lateral**) **buds**. These have dividing cells at the tip potentially capable of forming side branches. Leaves and axillary buds occur at regular intervals called **nodes**; the region of the main stem in between is called the **internode**.

In the root the apical meristem is protected by the **root cap**, a mass of rather loose cells that are continually sloughed off and replaced by new cells from the apical meristem as the young root pushes down into the soil.

In both stem and root, the tissues behind the zone of differentiation are referred to as **permanent tissues**. They have been formed by **apical** or **primary growth** and constitute the primary structure of the stem or root.

SECONDARY GROWTH

Apical growth, as just described, increases the length of stems and roots. How, then, is their girth increased from year to year? This is achieved by a process of **secondary growth** in the permanent zone. Such growth takes place in woody perennials like trees and shrubs, but it does not occur in herbaceous annuals and biennials which only live for one or two years.

Secondary growth takes place in both the stem and root by the division of meristematic cells located between the xylem and phloem in the vascular part of the plant (Fig 26.14). These cells comprise the **cambium,** and unlike other cells in the primary stem and root, they retain the capacity to divide mitotically. Initially the cambium is restricted to a series of small groups of cells wedged between the xylem and phloem in each vascular bundle. The first step in secondary growth involves the linking of these groups to form a cylinder, or ring, of cambium tissue separating the xylem from the phloem. This is achieved by radial divisions of the cambium cells. The xylem is situated on the inside of the cambium ring, the phloem on the outside.

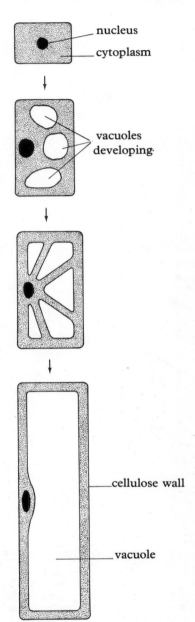

Fig 26.12 Expansion of a cell in the growing region of a flowering plant takes place by the osmotic uptake of water into the developing sap vacuole. In the case illustrated below, expansion results in the cell acquiring an elongated shape. When fully extended, further expansion is prevented by thickening of the cellulose wall.

nucleus

cytoplasm

vacuoles developing

cellulose wall

vacuole

A) Stem

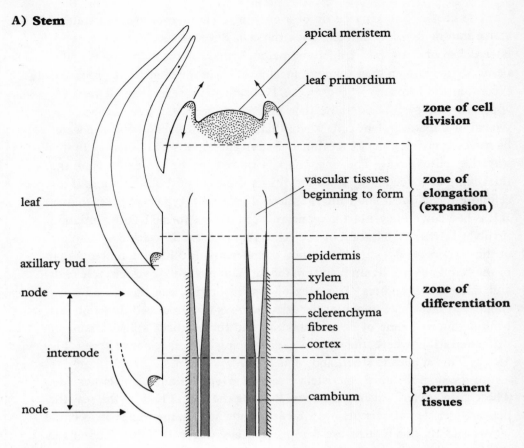

apical meristem

leaf primordium

zone of cell division

zone of elongation (expansion)

vascular tissues beginning to form

leaf

axillary bud

node

internode

node

epidermis

xylem

phloem

sclerenchyma fibres

cortex

zone of differentiation

cambium

permanent tissues

B) Root

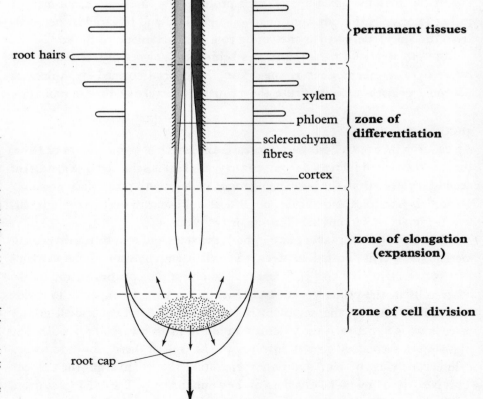

permanent tissues

root hairs

xylem

phloem

sclerenchyma fibres

cortex

zone of differentiation

zone of elongation (expansion)

zone of cell division

root cap

Fig 26.13   Growth of stem and root in a flowering plant. (A and B respectively). Cell division in the apical meristem produces daughter cells that, with the continued formation of new cells at the tip, get pushed back towards the zone of elongation. Increase in length of both stem and root is achieved mainly by cell elongation. There is no sharp distinction between the zones of elongation and differentiation, the cells beginning to differentiate while they are still expanding.   C (opposite page) is a photomicrograph of a longitudinal section of the stem apex of lilac: **A,** apical meristem; **L₁,** young leaves developing; **L₂,** older leaves; **B,** axillary bud in the angle between leaf and main stem; **V,** vascular strand consisting of elongated cells differentiating into xylem and phloem.   D) is a longitudinal section of the root tip of broad bean showing mitotically-dividing cells in the zone of cell division   (*C: J. F. Crane; D: I. D. Burton and G. H. Northridge, Marlborough College*)

C

D

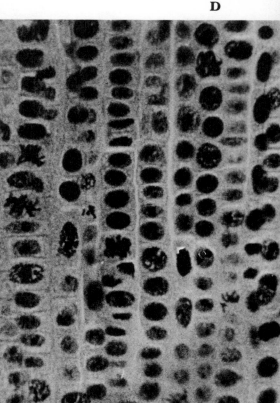

The cells of the **cambium ring** now divide tangentially to form **secondary xylem** tissue on the inside, and **secondary phloem** on the outside. In between adjacent vascular bundles they form **secondary parenchyma**, thereby increasing the girth of the medullary rays. In the stem, similar columns of parenchyma are formed within the secondary xylem and phloem, resulting in the formation of new medullary rays. These allow horizontal transport of water and solutes inside the thickened stem. Much more xylem is formed than phloem with the result that the phloem tissue, together with the cambium ring itself, gradually gets pushed outwards. Radial divisions of the cambium cells occur in order to keep pace with its ever increasing circumference. This is made clear in Fig 26.15.

Secondary growth is restricted to the middle months of the year. In the spring the first-formed xylem contains a high proportion of large vessels with

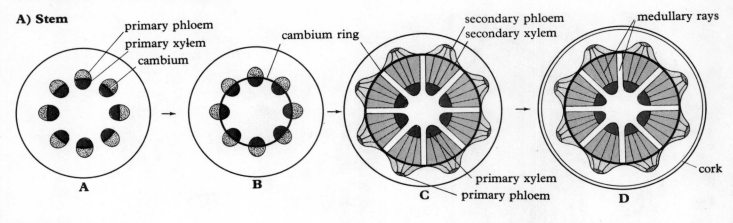

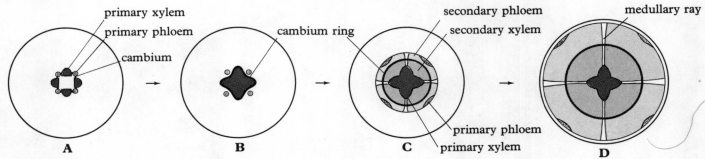

Fig 26.14 Secondary growth in stem and root. In both cases the cambium tissue (shown in black throughout) forms a complete ring which proliferates internally to form secondary xylem and externally to form secondary phloem.

relatively thin walls to carry the spring flow of the transpiration stream. As the summer progresses, the vessels become narrower and thicker-walled, and an increasing number of thick-walled sclerenchyma fibres are formed. This produces a harder and denser wood, providing a marked contrast to the whiter, soft spring wood (Fig 26.16). The result of this seasonal growth is the formation of a series of concentric **annual rings** which can be counted if the stem is severed. This provides an accurate method of estimating the age of trees: for example, counting these annual rings has shown that some of the giant redwoods in California are over 3,000 years old. From the plant's point of view the importance of secondary growth is that it greatly increases the thickness and strength of the stem, thereby enabling it to grow to a much greater height than would otherwise be possible. Fig 26.17 bears witness to this.

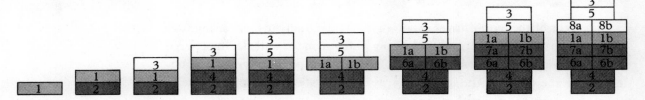

Fig 26.15 Diagrams illustrating how successive tangential divisions of a cambium cell result in the formation of secondary phloem and xylem in a dicotyledonous stem or root. Occasional radial divisions increase the number of cambium cells to allow for the expanding circumference of the cambium ring. The figures illustrate a hypothetical sequence in which new cells might be cut off: 1 represents the formative cambium cell; 2, 4, 6, and 7 are secondary xylem cells; 3, 5, and 8 are secondary phloem cells.

The great increase in girth resulting from the above procedure would inevitably rupture the surface tissues were it not for the fact that they too undergo a secondary growth process. Just beneath the epidermis a layer of cells, called the **cork cambium**, divides tangentially to form new surface tissues (Fig 26.18). Those cut off on the inside of the cork cambium form **secondary cortex**; those cut off to the outside form the so-called **periderm**. The walls of these cells become impregnated with **suberin**, a fatty material which renders them impermeable to water and respiratory gases. They form

Fig 26.16 Photomicrograph of secondary xylem of a redwood tree (*Sequoia sempervirens*) as seen in a transverse section of the trunk. Notice the marked contrast between the thin-walled spring wood and the much thicker-walled autumn wood. (*Forest Products Laboratory, United States Department of Agriculture.*)

Fig 26.17 The General Grant Tree (*Sequoia gigantea*) one of the largest redwoods in King's Canyon National Park, California. It is over 80 metres tall, and 11 metres wide at the base. At 66 metres above the ground the trunk is still 4 metres thick. This great tree is thought to be over 3,000 years old. (*Harrington Wells*)

the dead corky tissue characteristic of bark. Thus, by the meristematic activity of the cork cambium, the surface of the stem is kept intact and is afforded added protection. Bark, as such, is impermeable to water and respiratory gases, but periodically its corky cells, instead of being tightly packed, form a loose mass called a **lenticel**. In these regions gaseous exchange can take place freely between the inside and outside of the stem.

In this chapter we have examined the structural changes that occur during growth and development of animals and plants. We must now turn to the way they are controlled.

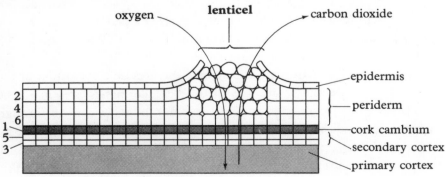

Fig 26.18 The formation of cork at the surface of a stem or root depends on the meristematic activity of the cork cambium beneath the epidermis. The order in which the cork cambium cuts off new cells is indicated by the numbering to the left. The cells at the base of the periderm, i.e. those closest to the cambium layer, are the youngest, have cellulose walls and are alive. The ones near the surface have impermeable walls impregnated with suberin, are dead, and form the cork.

27 The Control of Growth

In the last chapter we saw that growth is brought about by cell division, expansion and assimilation. Obviously any factor that directly or indirectly affects these processes will influence growth. For convenience we can make a distinction between **internal** and **external factors**.

External factors include a whole host of environmental influences: temperature, light, diet, oxygen, and so on. One of the most important is diet. The organism has got to be supplied with the raw materials from which to make its structures, and the metabolites for releasing the necessary energy. If the diet is deficient, either in a raw material such as protein, or an essential vitamin or trace element, poor growth will result.

Protein, as we saw in Chapter 5, forms the structural foundation of the body, and for a growing child laying down tissues at least 20 per cent of the food taken in must be protein. Even in a full grown man the figure is around 15 per cent. Because of the relatively low protein content of plant foods, a person on an exclusively vegetarian diet may have considerable difficulty getting the required amount. This is a particularly acute problem in those parts of the world where populations are dense and livestock scarce. Assuming his diet cannot be supplemented with animal foods, the only thing a person can do in these circumstances is to consume a vast amount of vegetable food. If this too, is in short supply he may find himself in dire straits. Severe protein deficiency results in a condition called **kwashiorkor**, an African word meaning 'the rejected one'. A child suffering from kwashiorkor has an emaciated appearance, is physically weak and shows retarded growth.

These principles apply no less to plants than animals. Unless they possess special adaptations, plants growing in nitrogen-deficient soils cannot manufacture enough protein, and show stunted growth and development. In addition to food supply, two other external factors influence plant growth, namely light and oxygen. Light is required for the synthesis and action of chlorophyll without which photosynthesis cannot take place. Oxygen does not generally affect the growth of above-ground parts of a plant, but it can have a profound influence on the growth of roots as Fig 27.1 clearly shows. Well aerated soil is essential for good root growth.

Internal factors affecting growth include the genetic constitution of the organism and the relative quantities of different hormones present in the body. These two are connected in that the genes influence growth through the intermediacy of hormones. But the hormones are also influenced by environmental factors: for example, lack of iodine in the diet results in cretinism, one of whose symptoms is stunted growth. Iodine deficiency has this

Fig 27.1 Effect of oxygen concentration on the growth of tomato roots. The plants were grown in separate culture solutions, the percentage concentration of oxygen in the air above each solution being, from left to right; 1, 3, 5, 10 and 20 per cent. Notice the pronounced retardation of growth brought about by lowered oxygen concentration. (*L. C. Erickson, University of California*)

effect because it is needed for the synthesis of **thyroxine**, the hormone that regulates the metabolic rate (see p. 295).

The general principle here is that growth, like so many other aspects of an organism's biology, depends on an interaction between external (environmental) and internal influences. Nowhere can this be better seen than in the way growth is controlled in flowering plants.

The Control of Plant Growth

Consider the basic problem: after a seed has germinated the young shoot grows straight up and the young root pushes down into the soil. What makes the shoot and root behave in this way? Over the years a large number of experiments have been done in an attempt to answer this fundamental question.

EVIDENCE FOR INVOLVEMENT OF A HORMONE

Consider experiment 1, Fig 27.2. The seeds of a suitable plant such as maize or broad bean are allowed to germinate and grow in the dark. Now if the tip of a growing shoot is cut off just in front of the zone of elongation, no further growth takes place. If, however, the tip is replaced growth is resumed. This simple experiment suggests that the tip in some way exerts an influence over the more posterior part of the shoot, causing growth to take place.

That this influence is exerted by a diffusible chemical substance is suggested by experiments 2 and 3. In experiment 2 a razor blade is stuck horizontally into the right hand side of the shoot just in front of the zone of elongation. The result is that growth stops on that side but continues on the left, so the shoot bends over to the right. Whatever it is that normally promotes growth is being blocked by the razor blade. In experiment 3 the shoot is decapitated and the tip placed on a small block of agar jelly. After about two hours the tip is discarded and the agar block placed on the cut end of the shoot. The result is that growth, temporarily inhibited by removal of the tip, is resumed. We conclude that a growth-stimulating substance, produced in the tip, has accumulated in the agar whence it diffuses back into the stump. Since this substance fulfils the definition of a hormone (see p. 294), and since it promotes growth, we can call it a **growth hormone**.

Experiment 4 lends further support to the hormone hypothesis. A shoot is decapitated and the tip placed in contact with an agar block. After a couple of hours the block is placed on top of a decapitated shoot in such a way that it covers half the cut end. The result is that more growth takes place on the side which has the block on it, so the shoot bends away from that side. This famous experiment was first performed in 1928 by the Dutch botanist Fritz Went. In subsequent experiments Went investigated the effect of varying the concentration of hormone in the agar block. He produced different concentrations by collecting the hormone for one hour and then placing the block in contact with one or more further blocks so that the hormone, by spreading through the larger volume of agar, became diluted. He found that, over low concentrations, the degree of curvature developed by a decapitated shoot is

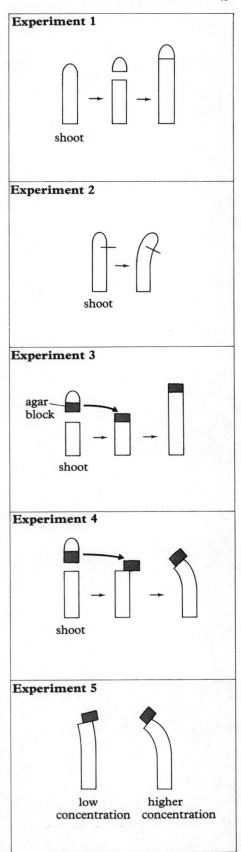

Fig 27.2 Experiments demonstrating that a hormone regulates growth in a young shoot and root. See adjacent text for full explanation. Agar blocks shaded. Experiments 6 and 7 are on p. 432.

Experiment 6

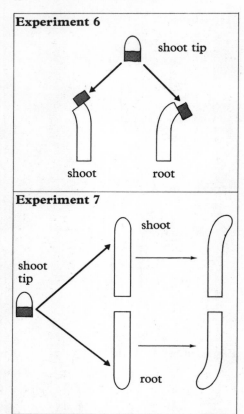

shoot tip

shoot

root

Experiment 7

shoot tip

shoot

root

Fig 27.3 *(below)* A) Structure of the naturally occurring auxin indoleacetic acid (IAA). B) Curvature of bean stems induced by treatment with IAA. The IAA was applied to the right hand leaf, whence it moved down the right hand side of the stem causing accelerated growth on that side resulting in curvature to the left. The degree of curvature is related to the concentration of IAA: the plant on the extreme right was treated with less IAA than the others. (*John W. Mitchell, Agricultural Research Service, Beltsville, Maryland*)

A

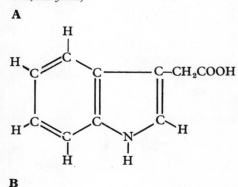

H—C—CH₂COOH

B

proportional to the concentration of hormone in the block (experiment 5). This is a convenient way of estimating the concentration of such a substance: the biological activity of the chemical can be assayed in terms of the curvature of the shoot.

The hormone demonstrated by these simple experiments appears to stimulate growth of the shoot. But what about the root? Experiment 6 throws light on this. The hormone is collected in an agar block as before. The block is then placed on the stump of a decapitated shoot exactly as in experiment 4. As we would expect, the shoot bends away from the side which has the block on it. If, however, the same is done with a decapitated root the reverse happens: the root bends towards the side with the block on it, suggesting that growth is inhibited on that side. We conclude that at the concentrations used in this experiment the hormone stimulates growth in the shoot, but inhibits it in the root.

This conclusion is confirmed by experiment 7. The hormone is collected in an agar block as before. It is then extracted from the block and the same amount painted on the left hand side of a shoot and a root. The result is that the shoot bends over to the right, the root to the left.

AUXINS

It is now known that the growth hormone, whose actions are described above, is not just one substance but several, known collectively as **auxins**, of which the most common is **indoleacetic acid (IAA)** (Fig 27.3). This compound has an extremely powerful effect on growth. A solution of 0.01 milligrammes in a litre of water applied to the side of a shoot is sufficient to cause bending. IAA has now been isolated from a wide variety of sources, and in addition many substances with similar effects have been synthesized in the laboratory. These are known as synthetic auxins, or just plant **growth substances** (one can hardly call them hormones as they are not naturally-occurring) and they are now used extensively in regulating the growth of plants of agricultural and horticultural importance. Some plant species respond very readily to them (Fig 27.4).

With a ready source of auxins at hand, quantitative experiments can be carried out to show the precise effects on growth of different amounts of auxin. The results of one such experiment are shown in Fig 27.5. You will notice that the effect of the hormone depends on its concentration. It is capable of stimulating growth in both the root and stem, but for the former a much lower concentration is needed, certainly much lower than that used in experiment 6, Fig 27.2. At the concentration required to achieve a maximum growth response in the stem, the root is strongly inhibited. And the concentration needed to cause a positive response in the root is too low to stimulate growth in the stem.

These results are consistent with what we know of the distribution of auxins in a plant. They are formed at the tip of the shoot and are transported downwards to the roots. Very small amounts are also produced in the root tip. The result is that there is a gradient along the main axis of the plant, the concentration being highest at the tip of the shoot and lowest in the root. Hence growth is promoted in both these parts of the plant.

RESPONSE TO GRAVITY

Unlike animals, plants often respond to an external stimulus acting from a

A   B

given direction by a movement which involves growth. Such growth responses are called **tropisms,** and auxins help us to explain them.

Consider experiment 1, Fig 27.6. A young broad bean seedling is placed horizontally in a dark chamber and left to continue its development. After a day or so, it is found that the shoot has bent upwards and the root downwards.

Fig 27.4 Some plants respond particularly quickly to growth substances, and are used for testing the effectiveness of different chemicals. Application of the substance in a suitable carrier such as lanolin to stems or leaves of the test plant (A) may induce a marked growth response within a matter of hours (B). (*United States Department of Agriculture*)

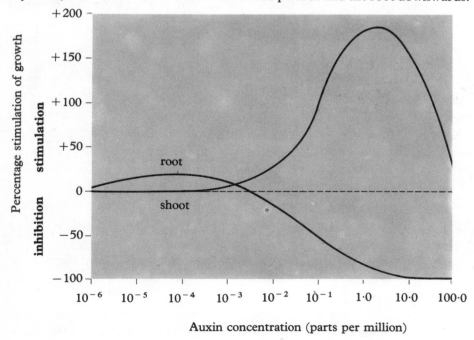

Auxin concentration (parts per million)

Fig 27.5 The effect on growth of applying different concentrations of auxin to the roots and shoots of oat seedlings. The results are expressed as percentage stimulation of growth compared with untreated controls. A positive result means relative stimulation compared with the control, a negative result means relative inhibition. Note that the concentration which produces maximum growth in the root (between 10^{-5} and 10^{-4} parts per million) has no effect whatsoever on the shoot; and the concentration which produces maximum growth in the shoot (about 1.0 part per million) inhibits growth in the root. (*after Audus*)

Fig 27.6 Experiments investigating the effect of gravity on the growth of the young shoot and root of broad bean. Agar blocks shaded.

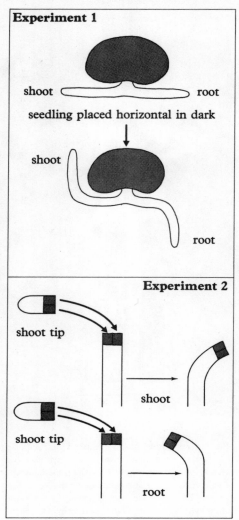

Experiment 1

shoot ⟶ ⟵ root

seedling placed horizontal in dark

shoot

root

Experiment 2

shoot tip

shoot

shoot tip

root

This is a plant's normal response to gravity and is called **geotropism**. Roots grow towards gravity and are described as **positively geotropic**, but shoots grow away from it and are therefore **negatively geotropic**.

We can explain geotropism by postulating that because of gravity the auxin tends to accumulate on the lower side of the shoot and root. In the shoot this causes more growth to take place on the lower side than the upper, resulting in an upward curvature. In the root the reverse happens: growth is inhibited on the lower side, resulting in a downward curvature.

This hypothesis is supported by experiment 2. The tip of a shoot is removed and placed horizontally in contact with an agar block which is divided down the middle by a thin metal strip. The object is to collect the hormone, but it is expected that owing to gravity more will accumulate in the lower half of the block than in the upper half. Two blocks are prepared in this way. After several hours one is placed on the cut end of a decapitated shoot, the other on a root. As would be expected, the shoot bends over towards the side of the agar block that was uppermost in the collection process, but the root bends in the opposite direction. So here again we have the seemingly paradoxical situation in which auxin promotes growth of the shoot, but inhibits growth of the root.

Geotropism can be demonstrated by means of a **klinostat**. The simple model illustrated in Fig 27.7 consists of a vertical wheel which can rotate at various speeds. A seedling can be attached to the wheel and the effect of rotation on the geotropic response investigated. It is found that if the speed of rotation is very slow, the shoot and root assume the shape of a corkscrew. On the other hand if rotation is sufficiently fast, they show no such twisting and make no observable response to gravity. In this case each side of the shoot and root is exposed to the gravitational stimulus for such a short time that they do not have time to respond. The response is thus governed by the duration of the stimulus.

Although roots are generally positively geotropic, and stems negatively so, there are some interesting exceptions to this. For example, the 'breathing' roots of mangroves grow vertically upwards and project from the surface of the water where they can absorb oxygen.

RESPONSE TO LIGHT

It is, of course, well known that stems grow towards light, i.e. they are **positively phototropic**. Roots are either **negatively phototropic** or do not respond at all. The response of a shoot to directional illumination is shown in Fig 27.8, experiment 1. Growth is slower on the illuminated side relative to the dark side, with the result that the shoot bends over towards the light. Quantitative experiments in which shoots are subjected to light of varying intensity and duration suggest that the response is generally proportional to the total amount of light energy received by the shoot. Thus strong light of short duration generally produces the same degree of curvature as weak light of long duration.

In explaining phototropism we have to consider two different hypotheses:

(1) light might cause an unequal distribution of the hormone in the shoot, more on the dark side than the light side;

(2) the cells responsible for growth in the zone of elongation might be prevented from responding to the hormone on the illuminated side of the shoot.

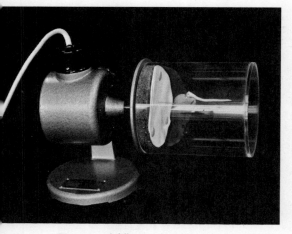

Fig 27.7 A klinostat in operation. A broad bean seedling is attached to the cork. The filter paper is soaked in water so as to keep the atmosphere in the perspex cylinder moist. Rotation of the wheel at the speed used in this particular experiment resulted in the radicle growing horizontally. (*B. J. W. Heath, Marlborough College*)

To find out which of these hypotheses is true experiment 2 can be done. A small tinfoil cap is placed over the tip of the shoot. It is found that when the capped shoot is illuminated from one side, it does not bend over towards the light. If, however, a tinfoil ring is placed round the shoot adjacent to the zone of elongation, the shoot bends over towards the light in the normal way. This simple test shows that the phototropic response is brought about by an unequal distribution of the hormone rather than a failure of the cells to respond to it.

This is confirmed by experiment 3. The tip of a decapitated shoot is placed on a divided agar block and illuminated from the right hand side. The block is then placed on the cut end of a shoot, and the latter bends over towards the right hand side. We conclude that the left half of the agar block contains more hormone than the right half, and this is confirmed by direct chemical estimation of the amount of hormone on the two sides of the agar block.

Further information on the effect of light on the distribution of auxin is provided by the experiment illustrated in Fig 27.9. In this case auxin is collected from intact and divided shoots placed in the dark and illuminated from one side. In both cases auxin is collected over a 3-hour period. The results of A and B show that the total amount of auxin collected is approximately the same whether the shoot is kept in the dark or illuminated, from which we draw the conclusion that, though it may influence its distribution, light does not destroy the auxin. The results of C indicate that when a directionally illuminated tip is not divided auxin accumulates on the dark side, but if it is divided there is no significant difference between the two sides. From this we conclude that, provided there is no obstruction, auxin is transported laterally to the darker side of the shoot.

OTHER RESPONSES

Plants respond to a wide range of directional stimuli besides light and gravity. Thus roots often grow towards moisture (**hydrotropism**) or towards certain chemical substances in the soil (**chemotropism**). A good example of chemotropism, encountered in Chapter 24, is the growth of the pollen tube towards the ovary in flowering plants.

Some plants give a growth response to touch (**thigmotropism**). The best example of this is provided by the tendrils of climbing plants which bend round an object with which they come into contact. In this case growth is slowed down on the side of the tendril experiencing the stimulus of touch (Fig 27.10). A rapid sense of touch, comparable with animals in its speed of action, is seen in the spectacular closing of the leaves of *Mimosa pudica* in response to touch, and the reactions of carnivorous plants such as sundew and Venus flytrap to contact with prey (p. 137).

Plants also respond to a variety of diffuse stimuli that do not come from any particular direction, for example temperature, humidity, and general level of illumination. The opening and closing of flowers and leaves provide a good example of such **nastic responses**. In all these cases the response involves either differential cell division in one part of the plant, or expansion of certain cells. Either way, the result will be a localized bending movement in the appropriate part of the plant. How the stimulus is transmitted to the part of the plant which responds is quite unknown.

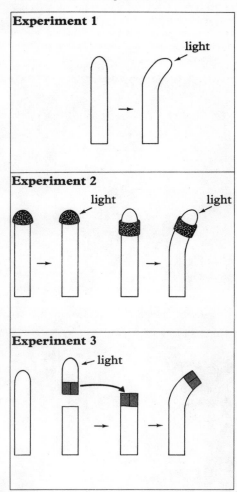

Fig 27.8 Experiments investigating the response of a young shoot to directional light. Agar blocks shaded. Full explanation in text.

Experiment 1

Experiment 2

Experiment 3

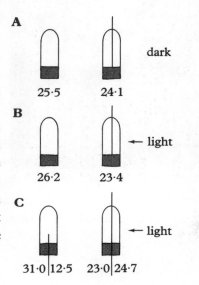

Fig 27.9 The effect of (A) darkness, and (B and C) directional illumination on the distribution of auxin in intact and divided maize coleoptile tips. Explanation in text. (*Briggs et al 1957*)

A   25·5   24·1   dark

B   26·2   23·4   light

C   31·0 | 12·5   23·0 | 24·7   light

Fig 27.10 Positive thigmotropism in young grapevine tendrils. A) Part of grapevine showing young tendrils. B) Immediately after photograph A was taken, a twig was placed in contact with one of the tendrils. C) The same tendril and twig one hour later. Notice that the tendril has curved round the twig. (*Patrick H. Wells, Occidental College, Los Angeles*)

A

B

C

MODE OF ACTION OF AUXINS

Since these experiments were done much progress has been made in our understanding of auxins. But we still do not understand how they facilitate cell expansion in the zone of elongation. One of several possibilities is that they enable the cellulose walls to be stretched more easily by the osmotic forces developed in the vacuoles. There is also evidence that auxins can promote cell division as well as elongation. A striking demonstration of this is seen if the tip of a stem is removed and the cut surface treated with IAA. The result is a tumorous growth of loosely-packed cells at the cut end of the stem.

OTHER AUXIN EFFECTS

IAA is now known to exert profound effects on many other aspects of plant growth and development besides stem and root elongation. For example, it inhibits the formation of side branches from lateral buds. This can be demonstrated by removing the apical bud of a plant, whereupon one or more lateral buds give rise to side branches. If, however, IAA is applied to the cut end of the main stem, no such branching occurs. Plainly the apex suppresses the lower parts of the plant, a phenomenon known as **apical dominance**, and only if this is removed will lateral growth take place. This, of course, is the theory behind pruning; cutting the main stem removes the source of auxin, thus encouraging the sprouting of side branches lower down.

On the other hand IAA stimulates the growth of adventitious roots, lateral roots which develop from the stem rather than the main root. Cuttings can be encouraged to 'take' in this way. The cut end of a stem dipped in IAA, or a commercial equivalent such as 'rootone', obligingly responds by sprouting roots. Rootone is prepared by mixing small quantities of IAA or a synthesized growth substance with an inert powder.

You will remember from Chapter 24 that fruits develop from the ovary. Normally this will only happen if fertilization has taken place but occasionally fruit-formation will occur in the absence of fertilization. This phenomenon, known as **parthenocarpy**, can be induced by treating unpollinated flowers with IAA, or a closely related synthesized chemical, and is commonly used by horticulturalists to bring about a good crop of fruit. It is thought that under natural conditions the developing embryo itself produces the hormone that stimulates the ovary to develop into the fruit.

In association with other plant hormones auxin is also involved in the initiation of secondary growth by division of cambium tissue, the ripening of fruit, and falling of leaves. Many of these effects can be investigated by introducing the growth substance into stems or branches, using techniques such as that illustrated in Fig. 27.11.

Natural and synthetic auxins are of great importance to man. The synthetic auxins are particularly valuable because some of them have special properties of their own. One of the best known is 2, 4-dichlorophenoxyacetic acid, mercifully shortened to 2,4-D. Even in extremely low concentration this induces distorted growth and excessive respiration leading to the death of the plant. Since it is much more effective on broad-leaved than narrow-leaved plants, it can be used as a selective weedkiller, ridding lawns of plantain, dandelions and other unsightly plants without destroying the grass.

GIBBERELLINS

Way back in the 1920s a Japanese farmer found that his rice seedlings suddenly started growing fantastically tall. He called these plants 'foolish seedlings'. It was discovered subsequently by Japanese research workers that this condition is caused by a fungus, *Gibberella fujikoroi*, which secretes a mixture of related compounds called the **gibberellins**. We now know that plants produce their own gibberellins in varying quantities and these have a very powerful effect on growth. The most important is gibberellic acid whose formula is shown in Fig 27.12A.

The most noticeable effect of gibberellins, and one which immediately distinguishes them from auxins, is the stimulation of rapid growth in dwarf varieties of certain plants. If treated with gibberellic acid a dwarf plant can be made to grow to the normal height. The dwarf condition is inherited and is thought to be caused by a shortage of gibberellin brought about by a genetic deficiency. The tall varieties on the other hand have enough gibberellin to make them grow to full height. As Fig 27.12B clearly shows, the effect of the gibberellin is to increase the length of the internodal regions.

Unlike auxins, gibberellins play no part in the bending of a shoot and have nothing to do with tropic movements. Their principal function is to cause elongation of the main stem. A spectacular demonstration of this is seen in

Fig 27.11 A growth-regulating substance being injected into a fruit tree through a hole bored into one of its branches. Its effect on growth, fruit set, fruit drop, longevity, etc., can then be studied. (*United States Department of Agriculture*)

A) GIBBERELLIN
Gibberellic acid (GA₃)

Fig 27.12 Structure and action of gibberellic acid (GA₃). B) compares the length of an internode in bean plants treated with (a) 250 microgrammes of lanolin, and (b) 250 microgrammes of lanolin containing 63 trillionths of a gramme of gibberellic acid. Photographs were taken 4 days after treatment. (*United States Department of Agriculture*)

B

Fig 27.13 Bolting in cabbages. The three plants on the right were treated with 0.1 milligrammes of gibberellic acid weekly. The control pair on the left received no treatment. Normally cabbages do not elongate and flower until the second year. However, first year plants treated with gibberellin show enormous elongatation and flower as shown below. (*S. H. Wittwer, Michigan State University*)

bolting, the precocious growth and flowering of plants that normally require exposure to cold or long days before they bloom. For example, the cabbage plant shown in Fig 27.13 was treated with less than one milligramme of gibberellin and developed a stem 5 metres long!

Like auxins, gibberellins exert their growth-promoting effects mainly by causing cell elongation, though they stimulate cell division as well. They tend to inhibit growth of the main root, and can cause parthenocarpy in certain plants, possibly because they initiate the formation of IAA. Unlike auxins they stimulate the growth of side branches from lateral buds, and inhibit the growth of adventitious roots. Thus in certain respects their actions are antagonistic to those of auxins. Gibberellic acid, like IAA, is much used in current research on plant growth. Amongst many recent advances is the amassing of evidence suggesting that it may exert its action by unlocking segments of DNA, thus enabling genes to work (see p. 512).

KININS

Another group of active growth substances were discovered in 1956. Known as **kinins**, they were first extracted from coconut milk but are now known to be of general occurrence. The formula of one of them, **kinetin**, is shown in Fig 27.14.

Kinins occur in very small quantities in plants but they are most abundant in tissues where rapid cell division is occurring. They promote cell division but, interestingly, will only do so in the presence of auxin. Indeed neither auxins nor kinins *alone* will stimulate division. This illustrates an important general point, namely that different plant hormones interact with one another in exerting their effects.

The interaction between auxins and kinins has been strikingly demonstrated by immersing undifferentiated callus tissue in a series of solutions containing different proportions of IAA and kinetin. Callus tissue is formed at the surface of woody plants in response to wounding. The callus can be induced to form buds or roots, to continue proliferating into callus tissue, or stop growing altogether, depending on the relative quantities of IAA and kinetin provided. This tells us that the interaction between different hormones can be of critical importance in determining the growth of specific organs in the plant.

LIGHT AND PLANT GROWTH

Light has a profound effect on the development of plants. It affects photosynthesis, the synthesis of chlorophyll, phototropic movements, stem and root elongation, the opening of stomata, and flowering. All these processes are directly or indirectly connected with growth. Growing a plant in darkness, or semi-darkness, results in **etiolation**; the plant is a pale yellow colour owing to lack of chlorophyll, the stem is frequently elongated and distorted, and the leaves fail to expand.

In Chapter 19 we saw that, before light can be responded to, it must be absorbed by some kind of photoreceptor substance. This applies just as much to plants as it does to animals. In recent years progress has been made in elucidating the characteristics of photoreceptor substances in plants.

It has been known for a long time that the seeds of many plants, certain varieties of lettuce for example, germinate only if they are exposed, at least briefly, to light. Sometimes only a brief flash of light is needed. In the early 1950s research workers in the United States Department of Agriculture carried out systematic tests to find out which particular wavelengths of light were effective in bringing about germination. They discovered that red light, in the range 580 to 660 nm was the most effective, whilst far-red light between 700 and 730 nm inhibited germination. Far-red light is at the end of the visible spectrum, almost at the beginning of the infra-red band. It is barely visible to us, and indeed some people cannot see it at all.

The interesting thing is that if a flash of red light is followed immediately by exposure to far-red, the stimulatory effect of the red light is cancelled out and germination is inhibited. In fact if seeds are exposed to alternating red and far-red flashes, the response is determined by the last flash in the series: the seeds germinate if the last flash is red; they fail to do so if it is far-red.

On the basis of these, and other, experiments it was postulated, and has since been confirmed, that light is absorbed by a single substance called **phytochrome**. This occurs in extremely small amounts, about 1 part in 10 million, in the tips of growing shoots. Despite the tiny quantities involved, it was successfully extracted and isolated around 1960 and has been shown to be a pale blue compound consisting of a pigment molecule attached to a protein.

Its absorption properties correspond nicely with the results of the germination experiments described above. It exists in two interconvertible forms; one absorbs red light and has its absorption peak at 665 nm; the other absorbs far-red light with its peak at 725 nm. These two forms of phytochrome are therefore designated **P665** and **P725** respectively (Fig 27.15).

Now when P665 absorbs red light it is rapidly converted into P725, and when P725 absorbs far-red light it is rapidly converted into P665. In the dark P725 is slowly converted into P665.

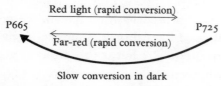

In natural sunlight P665 is converted into P725, and P725 into P665, but the former reaction predominates as sunlight contains more red than far-red light. So P725 tends to accumulate during daylight hours whilst at night it is converted slowly back into P665.

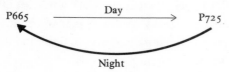

The significance of this is that P725 initiates enzymatic action, i.e. is biologically active, whereas P665 is inactive. This does not mean that P725 invariably stimulates growth. On the contrary, it sometimes inhibits it as we shall see later.

Returning now to the lettuce experiment, we can see why far-red light inhibits germination. Treatment with red light brings about conversion of inactive P665 to active P725, but if red treatment is followed immediately by far-red the P725 is converted back into P655 before it has had time to act.

The phytochrome system is now known to be important in many other

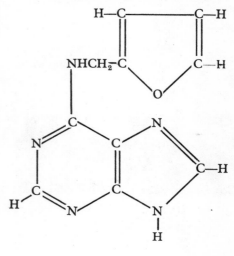

Fig 27.14 Structure of kinetin. This, along with other kinins, stimulates cell division in stems, roots, flowers, etc. Kinins also promote cell expansion in leaves, and stimulate growth of lateral buds by releasing them from the suppressive influence of apical dominance. They also release certain seeds from dormancy and are involved in the mobilization of food materials in leaves. Their involvement in cell differentiation is discussed in the text.

KININ

6-furfurylamino purine (kinetin)

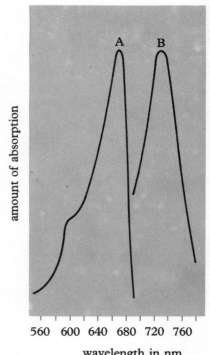

Fig 27.15 Absorption of light by phytochrome. A) red-absorbing form; B) far-red-absorbing form. A has its maximum absorption at 665 nm wavelength, whereas B has its maximum absorption at 725 nm. *(after Hendricks)*

Fig 27.16 The rôle of red and far-red light in stem elongation. All three bean plants received 8 hours of daylight each day. The centre plant was given an additional 5-minute exposure to elongation-promoting far-red light during the night. The plant on the right was given the 5 minutes of far-red light plus a 5-minute exposure to elongation-suppressing red light. Notice that the latter has counteracted the effect of the far-red. (*United States Department of Agriculture*).

growth processes. In some cases red light is the effective stimulus, in other cases far-red. Thus:

(1) Elongation of the stem is stimulated by far-red light but inhibited by red light (Fig 27.16). Far-red produces the same kind of effects as are seen in etiolation.

(2) Leaf expansion is stimulated by red light and inhibited by far-red. Brief exposure to very dim red light is enough to bring about leaf expansion but this effect is cancelled out by far-red treatment.

(3) The growth of lateral roots is stimulated by far-red, and inhibited by red.

(4) Flowering depends on alternating periods of light and dark. This has been shown to involve the phytochrome system and is discussed in detail in the next section.

How does phytochrome exert its action? The full answer is not known but it seems that after absorbing the appropriate wavelength of light the phytochrome is activated and causes production of a hormone, possibly from an inactive precursor, which then affects growth by influencing either cell division or cell expansion. That hormones are involved is suggested by several lines of evidence including the discovery that the amount of light required to initiate a response can sometimes be reduced by previous treatment with gibberellin.

PHOTOPERIODISM

In general terms this is the influence of the relative lengths of day and night on the activities of an organism. Although the best known example is flowering, many other responses in both animals and plants are regulated by day-length, i.e. the duration of the **photoperiod**.

The influence of the photoperiod on flowering can be demonstrated by exposing certain plants to a brief flash of light in the middle of the night. If this treatment is continued for a sufficient number of days, flowering will be delayed. This is made use of by horticulturalists in guaranteeing a supply of plants like chrysanthemums and poinsettias at Christmas. Such plants can be made to flower early by giving them extra darkness. In some cases one long night is all that is necessary.

With certain other plants, petunias for example, the reverse is true: light treatment induces early flowering whereas dark treatment delays it. By careful adjustment of the photoperiod, early and late varieties of plants can be made to flower simultaneously, thereby enabling plant breeders to cross them.

On the basis of their differing responses to light and dark, flowering plants can be divided into three groups:

(1) Those that require long days and short nights, **long-day plants**, for example, petunias, spinach, radishes and lettuce. Long-day plants flower only when the light period exceeds a certain critical length in each 24-hour cycle. This varies, but on average is about 10 hours.

(2) Those that require short days and long nights, **short-day plants**, for example, chrysanthemums, poinsettias, cocklebur. Short-day plants flower only when the light period is shorter than a critical length in each 24-hour cycle. For cocklebur this is $14\frac{1}{2}$ hours.

(3) Those that are indifferent to day-length, **day-neutral plants**, for example, tomato and cotton.

There is no hard and fast dividing line between long- and short-day plants: all gradations between the two exist. Not surprisingly, long-day plants tend to flower in the summer, or to inhabit temperate regions where days are long and nights short. Short-day plants, on the other hand, tend to flower in the winter and spring, or to live nearer the equator.

Many experiments, too numerous to describe in detail, indicate that the phytochrome system is involved in the photoperiodic control of flowering. To mention the results of one experiment, it has been found that only red light inhibits the flowering of short-day plants, and this inhibitory effect can be cancelled out by following the red treatment with far-red light (Fig 27.17). Bearing in mind that far-red light reconverts P725 back into P665, it seems that a short-day plant will only flower if a sufficient proportion of its phytochrome is in the P665 form. The trigger to flowering could be either a high enough accumulation of P665 or a low enough concentration of P725. Current opinion favours the latter view, i.e. P725 inhibits flowering and its conversion back to P665 removes the inhibition, thus allowing flowers to develop.

In long-day plants the reverse seems to be true: accumulation of P725, resulting from long exposure to light, stimulates flowering.

How does phytochrome exert its effects on flowering? In the first place the photoperiodic stimulus is perceived by the leaves. This has been shown by covering a whole plant with a light-proof cover except for one leaf which is then subjected to light/dark treatment. Under these conditions the flowering response still takes place.

From the leaves the message is transmitted to the buds where flower-formation is initiated. That the message takes the form of a hormone has been demonstrated by ingeniously grafting two short-day plants together. A plant which has been allowed to flower by exposure to short days is joined to another plant which has been prevented from flowering by being kept in long-day conditions. The result is that the latter blooms. A hypothetical scheme summarizing the photoperiodic control of flowering is given in Fig 27.18.

Fig 27.17 The influence of red and far-red light on the photoperiodic control of flowering. All three plants received 8 hours of daylight per day. The plant on the left with unopened buds was given an additional 8 hours of flower-suppressing red light. The tall centre plant with flowers was given an extra 8 hours of flower-stimulating far-red light. (*United States Department of Agriculture*)

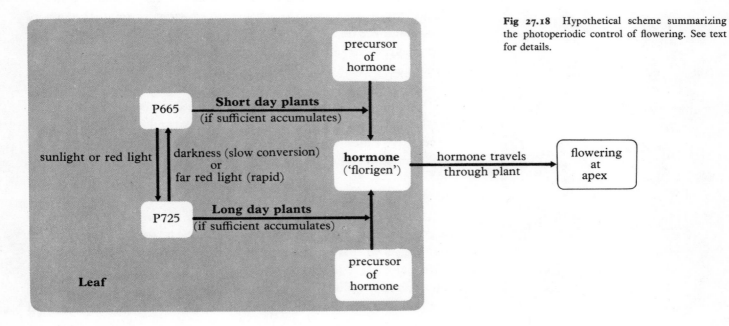

Fig 27.18 Hypothetical scheme summarizing the photoperiodic control of flowering. See text for details.

TEMPERATURE AND PLANT GROWTH

Temperature is no less important than light in influencing the growth of plants. It plays an important part in controlling germination (see p. 424) and directly influences the rate of metabolism and cell division. Flowering is generally favoured by an increase in temperature, the optimum temperature being related to the part of the world that the plant comes from. In general, tropical plants germinate, and subsequently flower, at higher temperatures than temperate plants. Indeed, the flowering of certain temperate plants may actually be brought forward by the germinating seeds being subjected to cold treatment, a phenomenon known as **vernalization**. Vernalization is an important means of inducing early flowering in crop plants but little is known of the underlying mechanism. Grafting experiments suggest that the stimulus, perceived by the apical meristem, causes the production of a hormone that, directly or indirectly, initiates flowering. In some plants vernalization is only effective if followed by long-day treatment, suggesting that cold treatment causes the production of a hormone precursor that is subsequently activated by light.

Control of Growth in Animals

Fig 27.19 Dwarfism (right) is caused by absence of certain cells in the anterior lobe of the pituitary gland which normally secretes a growth hormone. Gigantism (left) is caused by excessive secretion of the growth hormone. A pituitary dwarf may be less than 600 mm in height, a giant may top 2.7 metres. *(British Broadcasting Corporation)*

Growth in animals is also regulated by hormones. In mammals the hormone most directly involved is secreted by the anterior lobe of the pituitary gland, whence it is carried via the bloodstream to the epiphyses of the bones and other growing points in the body. This **growth hormone** has the general effect of increasing the metabolic rate, extra energy being diverted to cell division and protein synthesis.

Sometimes the pituitary secretes an over-abundance of growth hormone. If this happens during adolescence the rate of growth increases throughout the body resulting in a very large, but correctly proportioned, individual. This is known as **gigantism** and a similar condition can be induced in rats by injecting them with growth hormone. Occasionally over-secretion of growth hormone occurs in adult life, in which case new bone tissue is laid down in the body's extremities, particularly the hands and feet. These become greatly enlarged, a condition called **acromegaly** (from the Greek *akros*: end, *megas*: great).

If the pituitary produces too little growth hormone, **dwarfism** results. Growth takes place very slowly and the adult is short and stunted. Intelligence and reproductive functions are, however, unimpaired. Gigantism and dwarfism are illustrated together in Fig 27.19.

ACTION OF A GROWTH HORMONE ILLUSTRATED BY INSECTS.
Insects are particularly interesting in this respect because they enable us to see how growth hormones may exert their action. The control of growth and moulting has been particularly investigated by V. B. Wigglesworth at Cambridge and C. B. Williams at Harvard.

If a nymph of the blood-sucking bug *Rhodnius* is decapitated a day or two after gorging itself with blood, it will continue to live for several months, but moulting is prevented. If, however, decapitation is delayed for a week or so after a meal, moulting will still take place.

That a hormone is produced by the head during this period can be shown by the rather macabre experiment illustrated in Fig 27.20. A *Rhodnius* nymph decapitated the day after feeding (and thereby prevented from moulting), is connected by a capillary tube with another larva decapitated a week after feeding. The capillary tube makes it possible for blood to pass freely from one nymph to the other. The result is that the first larva moults as usual. How can we explain this?

We now know that distension of the gut following a meal causes a hormone to be produced by neurosecretory cells in the brain. This flows into the thorax where it stimulates a gland to secrete a second hormone. This does not happen at once: it is necessary for a certain amount of the brain hormone to accumulate before it can trigger the thoracic gland. This second secretion, a steroid, is known as the **moulting hormone,** and it brings about shedding of the cuticle and growth.

Little is known about how the moulting hormone exerts its effects, though evidence suggests that it activates the genes needed to produce the enzymes necessary for growth. One piece of evidence supporting this theory is the observation that certain regions of the chromosomes swell up to form **chromosome puffs** when exposed to the hormone. This can be seen clearly in the giant chromosomes of *Drosophila*. There is reason to believe that these puffs represent the site of RNA synthesis through which the genes exert their action (see Chapter 30). The moulting hormone certainly raises the metabolic rate, and increases the rate at which amino acids are built up into proteins in the growing tissues. It may well turn out that growth hormones generally exert their effects through the action of genes.

METAMORPHOSIS

You will remember that at successive moults an insect acquires, either gradually or suddenly, the features characteristic of the adult. What brings about this metamorphosis? A detailed answer cannot be given but we do know that it is suppressed by a **juvenile hormone** secreted by a gland called the **corpus allatum** in the brain. So long as this hormone is present in the blood the epidermal cells, under the influence of the moulting hormone, produce a cuticle characteristic of a juvenile stage, nymph or larva as the case may be. At metamorphosis the corpus allatum stops secreting its juvenile hormone, and the moulting hormone, in the absence of the juvenile hormone, causes the epidermal cells to lay down the adult type of cuticle (Fig 27.21).

The involvement of the corpus allatum has been shown by simple experiments: a larva whose corpus allatum has been removed develops adult characteristics precociously. Conversely, if the corpus allatum from a third or fourth stage larva is implanted into the abdomen of a fifth stage larva, the latter fails to pupate and undergo metamorphosis; instead it develops into a giant larva.

The juvenile hormone therefore inhibits metamorphosis and specifically causes the retention of larval characters. How it achieves this is uncertain but it is likely that it suppresses the genes responsible for producing adult structures. Once this inhibitory influence is removed these genes leap into action.

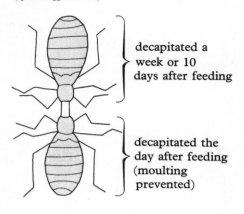

Fig 27.20 Experiment demonstrating that in *Rhodnius* moulting is triggered by a secretion from the head. Explanation in adjacent text. *(after Wigglesworth)*

decapitated a week or 10 days after feeding

decapitated the day after feeding (moulting prevented)

Fig 27.21 Scheme summarizing the hormones controlling growth in an insect. For shedding of the cuticle the moulting hormone is required. Moulting hormone accompanied by juvenile hormone causes the epidermis to produce a larval cuticle; moulting hormone alone causes it to produce an adult cuticle. *(based on Wigglesworth)*

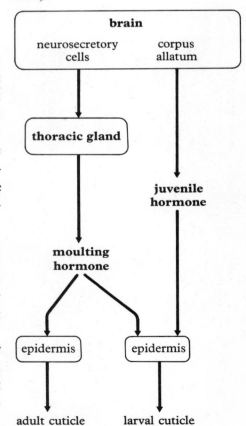

Another striking case of a hormone controlling metamorphosis is seen in amphibians. As is well known, development of the tadpole culminates in its undergoing a comparatively rapid metamorphosis into the adult, a process involving extensive changes in its structure. There is good evidence that in this case the process is controlled by **thyroxine** secreted by the **thyroid gland** (see p. 295). Injecting thyroxine into a tadpole results in precocious metamorphosis; conversely, removing the thyroid from a tadpole prevents metamorphosis. Quantitative information has been obtained by treating tadpoles with radioactive iodine. At metamorphosis there is a marked decrease in the radio-iodine content of the thyroid gland, this being due to discharge of the hormone from it. Evidence suggests that metamorphosis is controlled by variations in the amount of thyroxine secreted by the gland, coupled with changes in the ability of the different tissues to respond to it. Here, as in insects, the hormone may be exerting its effects by influencing the genes responsible for the production of adult structures.

Dormancy and Suspended Growth

Many organisms enter a state of **dormancy** either as adults or sometime during their life history. Growth and development cease, and the metabolic rate may fall to the point that it is only just sufficient to keep the cells alive. In this way the organism can survive for many months, even years, without exhausting its food reserves. Dormancy enables an organism to withstand unfavourable conditions, such as drought, food shortage and winter cold. It also allows time for distribution by natural agents such as wind and water currents, whilst at the same time allowing any necessary internal changes to take place.

Dormancy is found in seeds, buds, spores, eggs, and storage organs like tubers and bulbs. It also occurs in some adult organisms. The dormancy of seeds is usually associated with their hard resistant coat. The fertilized eggs of many lower plants, particularly the fungi, secrete a thick wall around themselves, thereby becoming resistant **zygospores** which may remain in a dormant state for long periods of time. Much the same applies to asexually produced spores.

Plant storage organs enable plants to survive the winter from one growing season to another, for which reason they are known as **perennating organs**. Buds provide a means whereby new foliage leaves and flowers can develop after a period of suspended growth during the winter months. This is important in deciduous trees and shrubs which shed their leaves before winter sets in.

SURVIVAL DURING DORMANCY

There is much variation in the length of time a dormant structure can survive and still remain viable. The oldest known viable seeds are those of the lotus found embedded in peat on the site of an ancient lake in Manchuria. Lotus seeds have an exceedingly hard coat which must either decay or be broken open before germination can take place. When the coats of these particular seeds were opened, every single one germinated successfully. Radioactive carbon dating showed them to be over 1,000 years old. Even greater longevity has been claimed for wheat seeds found in Egyptian tombs. Over 2,000 years old, these are said to have germinated successfully when moistened. However, considerable doubt has been thrown on these stories by the fact that the embryos of such seeds examined by professional botanists have invariably been found to be damaged. On the animal side, the record for longevity in the dormant state is probably held by the gall midge *Sitodiplosis mosellana* whose pupa is capable of surviving for at least 18 years.

There is also much variation in the severity of the conditions which can be endured by dormant tissues. Overwintering buds and plant storage organs can of course endure freezing, and spores and seeds can survive long periods of drought, but certain organisms do far better than this. Dry seeds have been shown to survive freezing in liquid air (-190°C), or liquid nitrogen (-250°C), and the cysts of the ciliate protozoon *Colpoda* can withstand boiling. Dormancy is therefore an efficient way of coping with adverse conditions.

THE MECHANISM OF DORMANCY

The phenomenon of dormancy raises two fundamental questions: what induces it, and what brings it to an end? The general answer is that dormancy is brought on, and subsequently ended, by environmental factors, particularly temperature and light, acting via hormones. As a general statement this would apply to animals as well as plants. However, when it comes to hard facts and details we are still very much in the dark, for little is known of the external and internal processes controlling dormancy.

Seed dormancy is thought to be caused by a variety of factors, some or all of which may occur in any one particular species. These factors include oxygen lack, drying out, or the presence of substances which inhibit germination. Alternatively (or perhaps in addition), it may be necessary for the embryo to undergo further development before it can germinate, or the seed coat may be so hard that germination simply cannot occur. One of the most promising developments in this field has been the discovery that the reluctance of certain seeds to germinate is caused by the presence of high concentrations of a **germination inhibitor**. The fact that dormancy is often broken by gibberellic acid suggests that this germination inhibitor may be inactivated by gibberellins.

The dormancy of buds during winter is well known to everyone. There is abundant experimental evidence that the cessation of growth, and the formation of dormant buds, is induced by the shorter days of autumn: in other words the stimulus is a photoperiodic one, akin to that which induces flowering. Generally a period of low temperature, typical of winter, is necessary before growth can be resumed the following spring. There are two possible causes of bud dormancy. It may be due to shortage of the normal growth-stimulating hormone, auxin or gibberellin. Only when this hormone builds up to a certain level, can dormancy be broken. Alternatively (or maybe in addition), dormancy may be brought on by accumulation of a growth inhibitor which must subsequently fall to a sufficiently low level for growth to be resumed. Both hypotheses are supported by evidence: the first by the fact that dormancy can often be arrested by treatment with IAA or gibberellic acid; the second by the fact that chemical substances extracted from buds have been shown to be active inhibitors of growth. Both hypotheses would therefore seem to be correct, but it is impossible to say how they relate to each other.

The growth-inhibiting substance extracted from buds is called **dormin** and it has been found to be chemically identical with a substance extracted from ageing leaves. The latter, known as **abscisin**, is responsible for bringing about the falling of leaves, a prerequisite to winter dormancy in deciduous trees. Leaf fall is brought about by development of a layer of cells, the **abscission layer**, at the base of the leaf stalk: in the course of time the calcium pectate making up the middle lamella of the abscission cells dissolves, with the result that the cells separate from one another and the leaf falls off. It has been found that abscisin stimulates this process but auxins inhibit it. It seems that during the life of a leaf auxin production decreases, whilst the amount of abscisin increases. Eventually there is so much abscisin relative to auxin that leaf fall occurs. How exactly the two hormones exert their action is unknown, but they may influence the activity of the enzyme responsible for the formation of insoluble calcium pectate. It is possible that auxin raises the concentration of this enzyme, and abscisin lowers it. In this way they might cause dissolution of the calcium pectate and separation of the abscission cells.

Dormancy is also seen in perennating organs. As explained in Chapter 24, a perennating organ may be a modified stem, root, leaves, or bud, depending on the plant in question. Whatever part of the plant the organ is developed from, the fundamental cycle of events is the same.

Soluble food materials are translocated from the photsynthesizing leaves of the parent plant to the developing organ, where they are built up into reserves, generally starch. Meanwhile the rest of the plant dies, the perennating organ remaining dormant in the soil during the winter. The following year the food reserves are mobilized and translocated to the growing regions of the new plant, after which the perennating organ withers and dies. Perennating organs include stem tubers, corms and rhizomes (all modified stems); root tubers (swollen taproot or adventitious roots as the case may be); and bulbs. A bulb is a swollen bud in which food is stored in thick fleshy leaves. The formation of swollen structures such as tubers and bulbs involves a suppression of the normal elongation process and its replacement by lateral expansion. At the same time there is a deterioration of the existing stem and root structures, so that the storage organ becomes isolated. The development of tubers and bulbs can be induced by a photoperiodic stimulus similar to that which induces bud formation, and there is evidence that such stimuli act via hormones produced in the leaves and subsequently translocated to the parts of the plant where the storage organs are formed.

Amongst animals, a type of dormancy akin to that seen in plants is shown by insects. Known as **diapause**, it can occur at any stage in the life history: egg, larva, pupa or adult. In the dormant state, the creature may survive and remain viable for months or even years. Diapause seems to share the same kind of causative mechanism as that underlying dormancy in plants, i.e. hormones regulated by environmental stimuli. An insect's growth and development can be arrested by removing its brain, the source of the hormone that normally promotes growth and development. In other words diapause seems to be caused by the normal growth-promoting hormone not being produced, at any rate in adequate quantities. What is more, these hormonal changes are related to day-length, as in plants. As soon as the hours of daylight fall below a certain critical level, diapause sets in. The critical amount of daily light varies from insect to insect. In the cabbage white butterfly, it is about 12 hours: when the light falls below this level, the pupae enter diapause and do not resume development until the following spring. Surprisingly, the light stimulus is not registered by the eyes: covering them does not abolish the diapause response. The light probably acts directly on the brain where, via the production of another substance or substances, it inhibits the growth-stimulating hormone. How is diapause ended? In some species the change to longer days may be the effective stimulus, but in many insects an obligatory period of cold has been found to be necessary before growth can be resumed. What exactly this chilling does is not known.

Superficially similar to diapause is **hibernation** which occurs on a seasonal basis in many animals. As in diapause and other forms of dormancy, metabolism drops to a low ebb so that the animal only draws slowly on its reserves. Hibernation is primarily a means of avoiding the necessity to maintain a high body temperature during the winter cold. To do so would necessitate keeping up a continuously high metabolic rate, a task which would require the consumption of more food than the animal could reasonably acquire during the winter. This is discussed in Chapter 15. Another kind of seasonal dormancy is **aestivation** which, in contrast with hibernation, occurs in summer in response to drought. It is illustrated by the African and South American lungfishes which can survive the complete drying-out of the swamps where they normally live (see Chapter 14).

In this chapter we have ranged over a wide variety of organisms, but it will be apparent that in many fundamental respects the part played by internal and external factors in controlling growth are the same.

28 Mendel and the Laws of Heredity

No one can help being struck by the often remarkable similarity between parents and their offspring, but it is no less noticeable that they also differ from each other in many respects. The science of heredity, or genetics, attempts to explain both the similarities and the differences between parents and offspring.

In Chapter 24 we saw that an individual animal or plant develops from a zygote which is formed by the union of gametes. To a considerable degree most of an individual's characteristics result from 'information' transmitted from the parents to the offspring via the gametes. Understanding the nature of this information is the central problem of heredity.

We now know that this information is contained in definite structures called **genes** which are located in the **chromosomes**. In recent years spectacular advances have been made in unravelling the chemical structure of the genes, putting us on the road towards understanding how inheritance works. But the first breakthrough in the study of heredity took place over a hundred years ago, long before genes or even chromosomes had been discovered. The man responsible for this was an Austrian monk, Gregor Mendel, and the story of heredity starts with him.

Fig 28.1 Gregor Mendel, who first put the study of heredity on a firm scientific basis. (*E. O. Dodson, University of Ottawa*)

MENDEL

Mendel was a member of the Augustinian monastery in Brünn, Austria (now Brno, Czechoslovakia). His ambition was to be a teacher but he repeatedly failed the necessary examinations and had to content himself with a job as substitute science teacher at the main school in Brünn. From all accounts he was a successful teacher, ever willing to spend time with his pupils and to show them his scientific experiments. He had always been interested in the problem of heredity and this led him to carry out breeding experiments on plants. As the subject for his research he chose the garden pea, which has a number of sharply contrasting and easily recognizable characteristics: for example, long and short stem, red and white flowers, smooth and wrinkled seeds, etc. With such clear-cut differences it is possible to cross or self-pollinate certain plants and examine the characteristics of their offspring. Starting in about 1856 Mendel carried out a vast number of such experiments in the garden of his monastery. With great perseverence and diligence he carefully isolated plants, transferred pollen from one to another, collected and sowed the seeds, and laboriously counted and recorded the different types of offspring. The conclusions he drew from his findings form the foundations on which the study of heredity is built.

Monohybrid Inheritance

In the early stages of his work Mendel studied the inheritance of just one pair of contrasting characteristics, what is now known to geneticists as **monohybrid inheritance**. In one such experiment he chose a pure-breeding tall pea plant and crossed it with a dwarf plant. This was done by taking pollen from one plant and dusting it on the stigma of the other, precautions being taken to ensure that the latter was not contaminated with its own pollen or that of any other plant. The seeds resulting from this cross were collected and sown. It was found that all the seeds gave rise to tall plants, no dwarf plants being produced at all. These plants belong to what is called the **first filial generation** (or **F1** for short). One of these tall F1 plants was then self-pollinated, precautions again being taken to prevent its being pollinated by any other kind of pollen. The resulting seeds were sown and the offspring (belonging to the **second filial** or **F2 generation**) were carefully examined. Mendel found that the F2 plants were a mixture: some were tall and some dwarf. In one particular cross he counted 787 tall plants and 277 dwarf ones, i.e. approximately three-quarters were tall and one-quarter dwarf. In other words the proportion of tall to dwarf plants approximates to a ratio of 3:1.

The results of this cross can be summarized as follows:

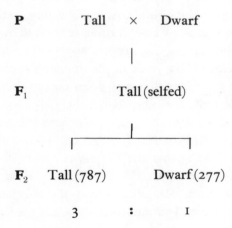

$$\textbf{P} \qquad \text{Tall} \quad \times \quad \text{Dwarf}$$

$$\textbf{F}_1 \qquad\qquad\qquad \text{Tall (selfed)}$$

$$\textbf{F}_2 \quad \text{Tall (787)} \qquad \text{Dwarf (277)}$$

$$3 \qquad : \qquad 1$$

Other monohybrid crosses produced similar results as can be seen from Table 28.1.

Table 28.1 Summary of Mendel's experiments on the inheritance of single pairs of characters in the garden pea *Pisum sativum*. Notice that for all the characters investigated a ratio approximating to 3:1 is obtained. (*after Mendel*)

Character Investigated	Cross	F₂ products	Ratio
Form of seed	Smooth × wrinkled	5474 smooth, 1850 wrinkled	2.96 : 1
Colour of cotyledons	Yellow × green	6022 yellow, 2001 green	3.01 : 1
Colour of seed coat	Grey-brown × white	705 grey-brown, 224 white	3.15 : 1
Form of pods	Inflated × constricted	882 inflated, 299 constricted	2.95 : 1
Colour of unripe pods	Green × yellow	428 green, 152 yellow	2.82 : 1
Position of flowers	Axial × terminal	651 axial, 207 terminal	3.14 : 1
Length of stem	Long × short	787 long, 277 short	2.84 : 1

CONCLUSIONS FROM THE MONOHYBRID CROSS

What conclusions can we draw from these results? The first striking fact to notice is that in neither the F_1 nor the F_2 generations are there any medium-sized plants, that is plants intermediate between the tall and dwarf conditions. From this we conclude that inheritance is not a process in which the features of the two parents are blended together to produce an intermediate result, like the mixing of black and white paints to produce grey. Rather it is a process in which definite structures, or particles, which may or may not show themselves in the outward appearance of the organism, are transmitted from parents to offspring . That such particles exist is borne out by the observation that they can be combined in one generation but separated in the next, as is witnessed by the recovery of both the original parental types in the F_2 generation. For these reasons inheritance may be described as **particulate**. We now call these particles genes but Mendel referred to them simply as factors. Although he never saw them, nor realized what form they might take, he appreciated their existence and is rightly credited with their discovery.

The second fact to notice is that there are no dwarf plants in the F_1 generation, despite the fact that one of the parent plants was dwarf. However, some dwarf plants appear in the F_2 generation. From this two conclusions can be drawn. First, although the F_1 plants are tall, they must receive from their dwarf parent a factor for dwarfness which remains 'hidden' in the F_1 plants and does not reveal its presence in the outward appearance of the plants until the F_2 generation. The second conclusion is that, as the factor for dwarfness fails to show itself in the F_1 generation, it must in some way be swamped by the factor for tallness. Only in the absence of this factor for tallness will the factor for dwarfness show itself in the outward appearance of the plant. In other words, the gene for tallness is **dominant** to the gene for dwarfness. The latter gene is described as **recessive**.

Although Mendel knew nothing of chromosomes and genes he postulated that his factors must be transmitted from parents to offspring via the gametes. If we are right in assuming that the F_1 plants contain factors for dwarfness as well as for tallness, it is reasonable to suppose that each F_1 plant receives via the gametes one factor for tallness from the tall parent and one factor for dwarfness from the dwarf parent. In other words the adult plants contain a pair of factors controlling their size, but the gametes contain only one of these two factors.

GENES AND THEIR TRANSMISSION

All these ideas are summarized in the genetic diagram shown in Fig 28.2. The gene for tallness is represented by **T**, and the gene for dwarfness by **t**. For reasons that will be explained later (p. 453) we will assume that each contains a pair of identical genes: **TT** in the case of the tall plant, **tt** in the case of the dwarf plant. When an organism contains identical genes like this, it is said to be **homozygous**. In making this statement we are describing the genetical constitution of the parent plants, or at least that part of it which determines their size. The genetical constitution of an organism is known as its **genotype**. The outward appearance of the organism, i.e. the way the genes express themselves in the structure of the organism, is known as the **phenotype**. So the genotype of the tall plant is **TT**, its phenotype being a long stem; the genotype of the dwarf plant is **tt**, its phenotype being a short stem.

Now each of the gametes produced by the tall parent will contain one **T** gene, those produced by the dwarf parent will contain one **t** gene. Fertilization brings the **T** and **t** genes together so that all the F1 offspring have the genotype **Tt**. They will, however, be phenotypically tall because the **T** gene is dominant to the **t** gene. When an organism contains two dissimilar genes in this way, it is said to be **heterozygous** (in contrast to the homozygous condition where the two genes are identical). In this case the **T** gene is dominant and expresses itself in the phenotype. The **t** gene, however, being swamped by the dominant **T** gene, is described as recessive. In general terms, a dominant gene, by definition, can express itself even when it is in the heterozygous condition. A recessive gene, however, can only express itself when in the homozygous condition. An individual homozygous for a recessive gene is known as the **double recessive**.

It is clear from what we have said so far, that the gene controlling size in the pea plant exists in two varieties. One functions normally and is responsible for producing a tall plant. The other, however, influences development in such a way that, if two are present together, a dwarf plant is produced. Such genes are known as **alleles**. Allele comes from the Greek work *allelon* meaning 'of one another'. This is an appropriate term for it implies that alleles, or allelic genes, are different versions of one another – which is what in essence they are.

To return to our example: each of the F1 plants, being heterozygous, produces two types of gamete: half will contain the **T** gene, the other half the **t** gene. On self-pollinating one of the F1 plants, these two types of gamete will fuse randomly to produce offspring possessing all three genotypes: **TT**, **Tt** and **tt**.

There are various ways of showing how this comes about; two are illustrated in Fig 28.2. A convenient graphic method is to employ a chequer board or **Punnett square**, so-called because it was first used by the Cambridge geneticist R. C. Punnett. In this device the gametes of one parent are written along the top of a series of boxes, and the gametes of the other parent are written down the side. The products of the various fusions are written in the appropriate boxes and their relative numbers can be estimated. In our example the **T** gene is possessed by half the male gametes and half the female gametes. The proportion of zygotes which receive **TT** will therefore be $\frac{1}{2} \times \frac{1}{2} = \frac{1}{4}$. The same reasoning applies to the **t** genes: half the male gametes and half the female gametes contain the **t** gene so the proportion of zygotes receiving **tt** will be $\frac{1}{4}$. In the case of the heterozygous (**Tt**) offspring there are two possibilities. Half the male gametes are **T** and half the female gametes are **t**, so the proportion of **Tt** zygotes will be $\frac{1}{2} \times \frac{1}{2} = \frac{1}{4}$. But, equally, half the male gametes are **t** and half the female gametes are **T**, and this will also give a proportion of **Tt** zygotes of $\frac{1}{4}$. Thus the total proportion of **Tt** zygotes will be $\frac{1}{4} + \frac{1}{4} = \frac{1}{2}$. To summarize: $\frac{1}{4}$ of the F2 offspring will be **TT**, $\frac{1}{2}$ will be **Tt**, and $\frac{1}{4}$ will be **tt**. As **T** is dominant to **t**, the plants with **TT** and those with **Tt** will both be tall. Thus a total of $\frac{3}{4}$ will be tall, and $\frac{1}{4}$ dwarf, which agrees with the 3:1 ratio found by Mendel.

IN TERMS OF PROBABILITY

It is important to appreciate what these figures mean in practice. To say that three quarters of the offspring will be tall and one quarter dwarf is only one

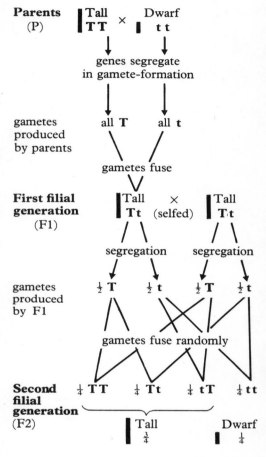

Fig 28.2 Interpretation of what happens when a pure-breeding tall pea plant is crossed with a pure-breeding dwarf plant. This is an example of monohybrid inheritance upon which Mendel's First Law, the Law of Segregation, is based. Symbols explained at foot of p. 450.

Punnett square to show fusion of F1 gametes

way of expressing the situation, and obviously only applies when there is a reasonable number of offspring produced. Another way of putting it is to say that if an F2 plant is selected at random, there is a 3 in 4 chance of its being tall, and a 1 in 4 chance of its being dwarf; in other words the **probability** of its being tall is $\frac{3}{4}$, and dwarf $\frac{1}{4}$. In the rather unlikely event of only one plant being produced in the F2 generation, the probability of its being tall is $\frac{3}{4}$, and dwarf $\frac{1}{4}$. This situation is hardly applicable to peas but it does apply to organisms which produce small numbers of offspring, man for example.

THE ROLE OF CHANCE

Returning to peas, when we say that three quarters of the F2 plants will be tall, what we really mean is that approximately three quarters of the plants are expected to be tall. It is unlikely that exactly three quarters will be tall because chance plays an important part in sexual reproduction. For example, although allelic genes are separated from each other during gametogenesis, it does not follow that they are distributed between the viable gametes in exactly equal numbers. After all, a good many gametes (or potential gametes) may die or be impotent. It must also be remembered that the union of gametes is a completely random affair. Whether or not a particular male gamete unites with a particular female gamete is a matter of sheer chance. For these reasons the actual ratios obtained in genetical experiments will only approximate to the expected ratios.

Not surprisingly observed ratios deviate from expected ones to the greatest extent when the sample one is dealing with is small. Think, for example, of what happens when a handful of coins is thrown onto the floor. One would expect approximately half the coins to land heads upwards and half tails upwards: in other words the expected ratio of heads to tails will be 1:1. In practice this 'ideal' situation is seldom realized. However, the greater the number of coins the closer the ratio is likely to be to the expected one. The difference between an expected ratio and an observed ratio is known as the **sampling error**, and it follows that the larger the sample, the smaller the error. Returning to Mendel's peas: the number of offspring produced in the F2 generations was considerable and the sampling errors comparatively small.

MENDEL'S FIRST LAW: THE LAW OF SEGREGATION

Mendel put his conclusions into a general statement known as **Mendel's First Law,** or the **Law of Segregation**. He said: an organism's characteristics are determined by internal factors which occur in pairs. Only one of a pair of such factors can be represented in a single gamete. Restated in modern terms: *the characteristics of an organism are controlled by genes occurring in pairs. Of a pair of such genes, only one can be carried in a single gamete.*

EXPLANATION OF MENDEL'S FIRST LAW

The explanation of Mendel's First Law is provided by **meiosis,** although of course Mendel himself knew nothing of this process. You will remember that in meiosis homologous chromosomes segregate from each other, as a result of which a gamete receives only one of each type of chromosome instead of the normal two. Genes also occur in pairs, each being located on one of two homologous chromosomes. When homologous chromosomes segregate in meiosis they take their genes with them, and thus the gametes receive only one of a pair of alleles — just as they receive only one of a pair of chromosomes. This is illustrated in Fig 28.3. It was the striking similarity between the segregation of Mendel's factors in inheritance and the segregation of homologous chromosomes in meiosis which provided one of the earliest pieces of evidence that the genes are located on the chromosomes.

BREEDING TRUE

One of the facts to emerge from Mendel's monohybrid cross is that a particular phenotype may be produced by more than one genotype. For example, a tall pea plant may be **TT** (homozygous) or **Tt** (heterozygous). Either way it will be tall, and there is no way of distinguishing between the two genotypes by their external appearance.

An obvious way of establishing whether a given tall plant is homozygous or heterozygous is to self-pollinate it. If the resulting offspring are all tall we conclude that the parent plant is **TT**. If, however, we get a mixture of tall and short plants we conclude that the parent plant is **Tt**.

The point is that a homozygous organism, when self-fertilized, produces offspring all of which are identical with the parent. Exactly the same result is given if a homozygous organism is crossed with another homozygote. This is known as **breeding true**. If the offspring are self-fertilized or crossed amongst themselves, another generation of identical organisms is produced. If this is continued a **pure line** is established. Mendel took great pains to use only true-breeding plants as the original parents in his experiments, a fortunate choice because had he not done so his results would have been inconsistent and much more difficult to interpret.

TEST CROSS

In the last section we saw that one possible way of establishing the genotype of a plant is to self-pollinate it, but what do we do when confronted with an organism that is incapable of self-fertilization, as most animals are?

By way of illustration let us take an animal which is much used in genetical experiments nowadays: the fruit fly, *Drosophila melanogaster*. *Drosophila* has a whole series of contrasting features, one of which involves the size of the wings. Most flies have long wings which extend beyond the end of the abdomen. This is the normal condition and flies possessing it are known as the **wild type**. In contrast, some have vestigial wings which are short, stumpy and functionless. The gene for the long-winged condition (**V**) is dominant to the gene for vestigial wing (**v**). If a homozygous long-winged fly is mated with a vestigial-winged fly, the F1 offspring are all long-winged, with the genotype **Vv**. If two of these F1 flies mate with each other (it is impossible to self-pollinate fruit flies!) a mixture of long-winged and vestigial-winged flies are produced in the proportions of 3:1

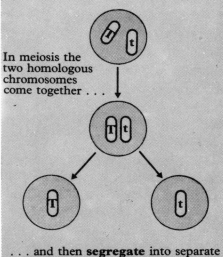

Fig 28.3 Meiosis provides the explanation of Mendel's First Law, the Law of Segregation. The segregation of genes in inheritance corresponds to the segregation of homologous chromosomes in meiosis. In the diagrams below the chromosomes are considered as single units and are not shown to consist of pairs of chromatids.

In describing the genotype of a plant as:

T   t

we mean that there is a pair of homologous chromosomes carrying the genes for tallness. One chromosome of the pair carries a **T** gene and the other a **t** gene, thus:

In meiosis the two homologous chromosomes come together . . .

. . . and then **segregate** into separate gametes

Thus each gamete contains only one of the original pair of genes.

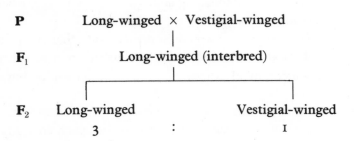

P	Long-winged × Vestigial-winged	
F₁	Long-winged (interbred)	
F₂	Long-winged	Vestigial-winged
	3	:	1

Fig 28.4 Monohybrid inheritance in the fruit fly *Drosophila melanogaster*. The Law of Segregation has been shown to apply to many organisms besides peas. The diagram on the right shows the result of crossing a pure-breeding long-winged fruit fly with a vestigial-winged fly. The result is similar to those obtained by Mendel for peas. Gene for long wing (**V**) dominant to gene for vestigial wing (**v**).

This is what we would expect and is comparable to Mendel's experiment with tall and short peas. A full interpretation of what happens is given in Fig 28.4.

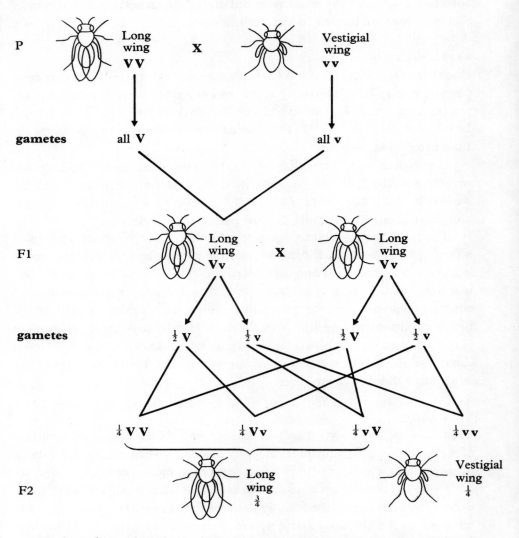

But how do we decide whether a given F2 long-winged fly is homozygous (**VV**) or heterozygous (**Vv**)? The simplest way is to cross it with a vestigial-winged fly. We know that a vestigial-winged fly must be **vv** (double recessive) — it cannot by anything else. If our unknown fly is **VV**, then crossing it with a vestigial-winged fly will give nothing but long-winged flies. If, however, the unknown fly is **Vv**, then the cross will give a mixture of long- and vestigial-winged flies in equal proportions. This is summarized in Fig 28.5. Because this experiment is carried out in order to determine the organism's genotype, it is called a **test cross**. Such test crosses with the double recessive are the routine method of establishing an organism's genotype, whether it is capable of self-fertilization or not.

The *Drosophila* case serves to demonstrate that the principles of monohybrid inheritance which Mendel established in peas apply equally well to other organisms; molds, flies, mice and men, to mention but a few.

MONOHYBRID INHERITANCE IN MAN

Man is a notoriously difficult organism in which to study heredity, but a number of conditions are known to be associated with a single pair of genes which are transmitted in a Mendelian fashion. One of these is **albinism**, a condition in which external pigment fails to develop, resulting in the person having a light skin, white hair and pink eyes. Albinism is caused by a recessive gene (**a**) so it will only exert its effects in the homozygous state (**aa**). The gene for normal skin (**A**) is dominant. The genotype of a normal person is therefore **AA** or **Aa**, and of an albino **aa**.

Fig 28.5 The principle underlying the test cross. How can we establish the genotype of a long-winged fruit fly? The answer is to cross it with the double recessive. If the 'unknown' fly is homozygous there will be nothing but long-winged flies in the progeny (left). If the 'unknown' fly is heterozygous there will be an equal proportion of long-winged and vestigial-winged flies in the progeny (right).

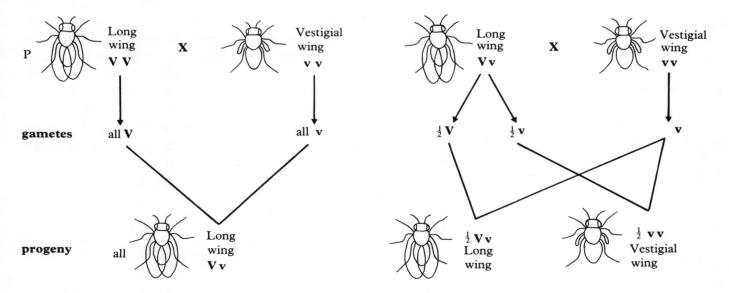

Let us imagine that a phenotypically normal couple have an albino child. For this to happen the child must obviously be **aa**, and both parents must be heterozygous (**Aa**). In other words the parents, though not themselves albinos, carry the albino gene, for which reason they are known as **carriers**. If this happened in practice the couple would almost certainly want to know the likelihood of their next child being an albino. The answer to this can be worked out quite easily (Fig 28.6). Both parents are heterozygous (**Aa**) so each

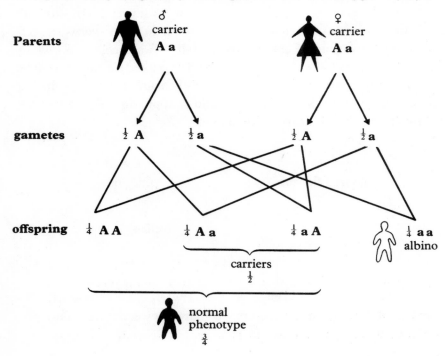

Fig 28.6 Albinism, an example of monohybrid inheritance in man. The diagram shows the genotypes and phenotypes of the children that might result from a union between two carriers of the albino gene. Note that there is a one in four chance of any one child being an albino. If a carrier marries an albino what is the probability of their producing an albino child?

Fig 28.7 Chondrodystrophic dwarfism (achondroplasia) is caused by a single dominant gene that occasionally arises as a mutation in the reproductive cells. Inherited in a Mendelian fashion, the gene causes abnormal growth of bones preformed in cartilage, resulting in a stunted body, bow legs, and large dome-like head. Approximately 80 per cent of chondrodystrophic dwarfs die within a year of birth but those that survive show normal mental development and can have children. The typical circus dwarf is a victim of this genetic abnormality. (*Wellcome Museum of Medical Science*)

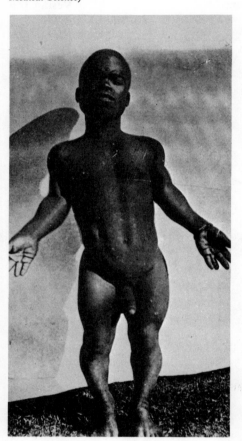

produces **A** and **a** gametes in roughly equal numbers. These fuse randomly to produce three types of genotype: **AA**, **Aa** and **aa**. A 3:1 ratio is obtained, just as it is whenever two heterozygous organisms are crossed. In practical terms this means that the probability of their next child being an albino is $\frac{1}{4}$, one chance in four.

Albinism, though distressing, is not particularly debilitating. However, a number of other inherited conditions are much more serious. These are known as **congenital diseases** and are of great medical importance. For example, **cystic fibrosis**, a disease in which connective tissue develops in various glands of the body, particularly the pancreas, is caused by a recessive gene which is transmitted like the albino gene. Only people homozygous for this gene will develop the condition which is usually fatal well before the patient reaches adult life. On the other hand, **chondrodystrophic dwarfism** (**achondroplasia**) (Fig 28.7), characterized by shortened and deformed legs and arms, is brought about by a dominant gene which therefore exerts its effects in either the homozygous or heterozygous conditions. In actual fact individuals homozygous for chrondrodystrophic dwarfism are likely to be rare: it is usually assumed (but established in only a few cases) that homozygotes for harmful dominants are lethal.

Dihybrid Inheritance

So far we have considered the inheritance of only one pair of contrasting characteristics. But Mendel did not stop at this. He went on to study the inheritance of two pairs of characteristics; what is now called a **dihybrid cross**. In one such experiment he crossed a pure-breeding tall pea plant possessing coloured flowers with a dwarf plant possessing white flowers. In the F1 generation he obtained nothing but tall plants with coloured flowers. These were then self-pollinated. In the F2 generation four different kinds of plant were produced: (1) tall with coloured flowers; (2) tall with white flowers; (3) dwarf with coloured flowers; and (4) dwarf with white flowers. In other words the offspring showed the two pairs of characteristics (tall – dwarf, coloured – white) combined in every possible way. As before, Mendel counted the different types of plant and in one particular case he got 96 tall coloured; 31 tall white; 34 dwarf coloured; 11 dwarf white, giving a ratio of approximately 9 : 3 : 3 : 1. The experiment can be summarized as follows:

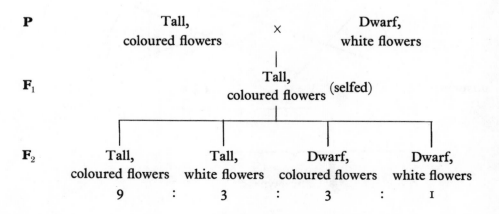

CONCLUSIONS FROM THE DIHYBRID CROSS

What conclusions can we draw from these results? To begin with, the fact that all the F1 plants are tall with coloured flowers tells us that tall is dominant to dwarf, and coloured flower to white flower. Fig 28.8 shows how the genes are transmitted. **T** represents the gene for tallness, **t** for dwarfness, **C** for coloured flower, and **c** for white flower. Being pure-breeding, the parent plants must be homozygous for both pairs of alleles. The genotype of the tall plant with coloured flowers must therefore be **TTCC**, and the dwarf plant with white flowers **ttcc**. Bearing in mind Mendel's First Law (the Law of Segregation), the gametes produced by the parent plants must be **TC** and **tc** respectively. All the F1 offspring will therefore be **TtCc**, heterozygous for both pairs of alleles.

The next step in the argument is crucial. If all four possible combinations of characteristics are to show up in the F2 generation, we must conclude (as Mendel did) that the F1 plants produce four kinds of gamete: **TC, Tc, tC** and **tc**. The Punnett square in Fig 28.8 shows the different ways these gametes can fuse, together with the genotypes of the F2 offspring. To be tall the genotype must contain at least one **T** gene; to be coloured it must contain at least one **C** gene. From the chart it can be seen that of 16 possible fusions, 9 will give tall coloured plants, 3 tall white, 3 dwarf coloured, and 1 dwarf white; and the observed ratio can be accounted for only if all the possible fusions occur with equal likelihood.

Fig 28.8 Interpretation of what happens when a pure-breeding tall pea plant with coloured flowers is crossed with a dwarf plant with white flowers. This is an example of dihybrid inheritance upon which Mendel's Second Law, the Law of Independent Assortment, is based. The symbols used are explained in the text at the top of this page.

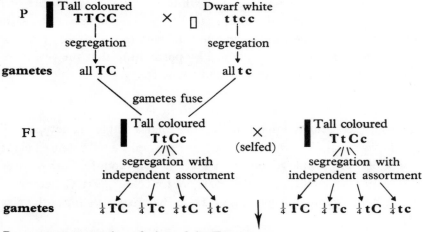

Punnett square to show fusion of the F1 gametes:

		male gametes (pollen nuclei)			
		$\frac{1}{4}$ **TC**	$\frac{1}{4}$ **Tc**	$\frac{1}{4}$ **tC**	$\frac{1}{4}$ **t c**
female gametes (egg cells)	$\frac{1}{4}$**TC**	$\frac{1}{16}$ **TTCC** Tall coloured	$\frac{1}{16}$ **TTCc** Tall coloured	$\frac{1}{16}$ **TtCC** Tall coloured	$\frac{1}{16}$ **TtCc** Tall coloured
	$\frac{1}{4}$**Tc**	$\frac{1}{16}$ **TTcC** Tall coloured	$\frac{1}{16}$ **TTcc** Tall white	$\frac{1}{16}$ **TtcC** Tall coloured	$\frac{1}{16}$ **Ttcc** Tall white
	$\frac{1}{4}$**tC**	$\frac{1}{16}$ **tTCC** Tall coloured	$\frac{1}{16}$ **tTCc** Tall coloured	$\frac{1}{16}$ **ttCC** Dwarf coloured	$\frac{1}{16}$ **ttCc** Dwarf coloured
	$\frac{1}{4}$**tc**	$\frac{1}{16}$ **tTcC** Tall coloured	$\frac{1}{16}$ **tTcc** Tall white	$\frac{1}{16}$ **ttcC** Dwarf coloured	$\frac{1}{16}$ **ttcc** Dwarf white

F2 $\frac{9}{16}$ Tall coloured $\frac{3}{16}$ Tall white $\frac{3}{16}$ Dwarf coloured $\frac{1}{16}$ Dwarf white

MENDEL'S SECOND LAW: THE LAW OF INDEPENDENT ASSORTMENT

What general conclusion emerges from this? The main conclusion, surely, is that the two pairs of genes are transmitted independently from parents to offspring and assort freely. This idea is embodied in **Mendel's Second Law**, the **Law of Independent Assortment**, which, as Mendel stated it, says that each of a pair of contrasted characters may be combined with either of another pair. In modern terms we would say that *each member of a pair of alleles may combine randomly with either of another pair.*

IN TERMS OF PROBABILITY

Expressed in terms of probability, transmission of the genes determining size and flower colour are **independent events**. If we consider the genes for size on their own, the probability of any one F2 plant being tall is $\frac{3}{4}$, and of its being dwarf $\frac{1}{4}$. Similarly if we consider the flower colour genes alone, there is a probability of $\frac{3}{4}$ that an F2 plant will be coloured, and of $\frac{1}{4}$ that it will be white. What, then, is the probability of an F2 plant being tall and coloured? Assuming that the genes are transmitted independently, the answer is $\frac{3}{4} \times \frac{3}{4} = \frac{9}{16}$. In practical terms this means that the chances of any one F2 plant, chosen at random, being tall and coloured are 9 out of 16. It also means that in a large random sample of F2 plants approximately 9 out of 16 of them will be expected to be tall and coloured.

We can apply similar reasoning to the other possible combinations. The probability of an F2 plant being tall and white is $\frac{3}{4} \times \frac{1}{4} = \frac{3}{16}$; dwarf and coloured $\frac{1}{4} \times \frac{3}{4} = \frac{3}{16}$; dwarf and white: $\frac{1}{4} \times \frac{1}{4} = \frac{1}{16}$. These figures agree with those obtained by the Punnett square method and with the results of Mendel's experiments. They can only be explained by postulating that the two pairs of genes are transmitted independently and assort freely.

RELATIONSHIP BETWEEN GENOTYPE AND PHENOTYPE

It is clear from the Punnett square in Fig 28.8 that a single phenotype may be produced by several different genotypes. For example, a tall coloured plant may have one of four possible genotypes: **TTCC** (homozygous for both pairs of genes), **TTCc** (homozygous for tall, heterozygous for colour), **TtCC** (heterozygous for tall, homozygous for colour), or **TtCc** (heterozygous for both pairs of alleles).

To ascertain the genotype of an F2 plant we could self-pollinate it and examine the kinds of offspring produced. Thus, to take the examples listed above, **TTCC** on selfing would be expected to breed true to both characters, **TTCc** should breed true to tall but not to coloured, **TtCC** should breed true to coloured but not to tall, and **TtCc** should not breed true to either character. The only genotype we can be certain of from its phenotype alone, is that belonging to the short white plant which, being double recessive, must be **ttcc**.

TEST CROSS

The best way of establishing the genotype of a tall coloured plant is to test cross it with the double recessive (**ttcc**). You will immediately appreciate that if the 'unknown' plant is homozygous for both characters, the offspring will all be tall and coloured (note why). Now consider the outcome if it happens to be heterozygous for both pairs of alleles. In this case it will produce four types of gamete: **TC, Tc, tC,** and **tc.** The double recessive, however, should

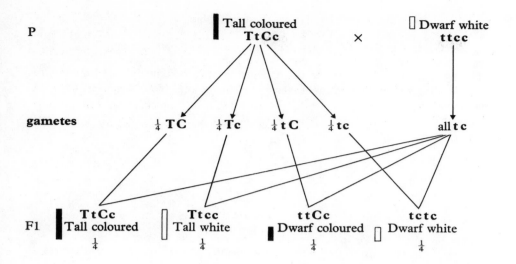

Punnett square to show fusion of the F1 gametes:

		gametes			
		TC	**Tc**	**tC**	**tc**
gametes	**tc**	$\frac{1}{4}$ **TtCc** Tall coloured	$\frac{1}{4}$ **Ttcc** Tall white	$\frac{1}{4}$ **ttcC** Dwarf coloured	$\frac{1}{4}$ **ttcc** Dwarf white

produce only one type of gamete: **tc**. The fusion of the gametes is shown in Fig 28.9 from which it is clear that four types of offspring should be produced in approximately equal proportions: tall coloured, tall white, dwarf coloured, and dwarf white. Mendel tried this experiment and this is precisely what he found. In one experiment, for example, he obtained 47 tall coloured, 40 tall white, 38 dwarf coloured, and 41 dwarf white. There are two other possible genotypes for a tall coloured plant: **TTCc** and **TtCC**. What should be the result of test-crossing each of these with the double recessive?

EXPLANATION OF MENDEL'S SECOND LAW

Independent assortment of genes, like segregation, is explained by **meiosis**. It is assumed that the genes concerned are carried on different chromosomes: that is the genes determining size are located on one pair of chromosomes, and the genes determining flower colour are located on another quite distinct pair of chromosomes (Fig 28.10). Now in metaphase of the first meiotic division homologous chromosomes line up side by side on the spindle prior to separating. In doing this, different pairs of homologous chromosomes behave independently of each other: the way one pair of homologous chromosomes arrange themselves on the spindle and subsequently separate has no effect whatsover on the behaviour of any other pair of chromosomes. In other words the arrangement of the different pairs of chromosomes relative to one another is completely random. The result of this is made clear in Fig 28.10 which shows exactly how the four different types of gametes (**TC**, **Tc**, **tC** and **tc**) can be formed from a heterozygous tall plant with coloured flowers (**TtCc**). We can summarize the situation by saying that the genes for size and flower colour segregate and assort independently because they are carried on separate chromosomes which themselves segregate and assort independently in meiosis.

Fig 28.9 Diagram showing the result of crossing a pea plant heterozygous for tallness and flower colour with the double recessive. Note that, as in the cross shown in Fig 28.8, all four possible types of offspring are produced.

Fig 28.10 Meiosis provides the explanation of Mendel's Second Law, the Law of Independent Assortment. The free assortment of genes in inheritance corresponds to the free assortment of chromosomes during meiosis. In the diagram (*right*) the chromosomes are considered as single units and are not shown to consist of pairs of chromatids.

In describing the genotype of a plant as:

T t C c

we mean that there are two pairs of chromosomes, one pair carrying the genes for tallness and the other pair carrying the genes for flower colour. The two chromosomes carrying the 'tallness' genes are homologous with one another, and the two chromosomes carrying the 'flower-colour' genes are homologous.

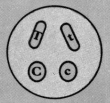

In meiosis the homologous chromosomes come together (**assort**) but they arrange themselves on the spindle **independently** of each other. Thus they may arrange themselves . . .

. . . like this: or like this:

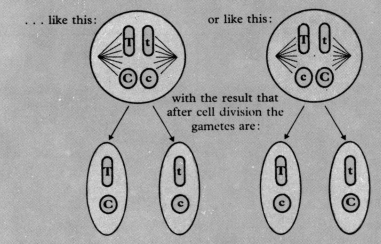

with the result that after cell division the gametes are:

Thus meiosis results in the formation of four types of gamete which, since the process is random, occur in approximately equal numbers.

This close parallel between the behaviour of genes as seen in genetical crosses and the behaviour of chromosomes in meiosis provides another piece of evidence that the genes are located on the chromosomes.

INDEPENDENT ASSORTMENT IN THE FRUIT FLY

As with segregation, independent assortment can be demonstrated in other organisms besides peas. Take *Drosophila* for example. As well as flies with long and vestigial wings, there are forms with grey and ebony-coloured bodies. The grey colour is the normal condition and is dominant to the ebony colour. If we cross a fly possessing long wings and a grey body with one possessing vestigial wings and an ebony body, the F1 offspring are all long-winged and grey-bodied. If two of these are crossed, the F2 generation yields four types of fly: long-winged, grey-bodied; long-winged, ebony-bodied; vestigial-winged, grey-bodied; and vestigial-winged, ebony-bodied. These occur in a ratio of 9:3:3:1 respectively.

From these results we conclude that the genes determining wing length and body colour are located on different pairs of chromosomes. The genes separate and assort freely, just as the genes for size and flower colour do in Mendel's experiments with peas.

Mendel in Retrospect

In interpreting monohybrid and dihybrid inheritance in terms of meiosis we have taken our discussion of heredity beyond the facts that were known to Mendel, but this has been necessary in order to provide a satisfactory explanation of the basic principles. Mendel's research reveals the mind of a genius. There are four main reasons for this. Firstly, he saw the importance of studying one phenomenon at a time. He did not attempt to follow the inheritance of all the many characteristics of pea plants simultaneously, but started by confining himself to one pair of characteristics. Secondly, he was not content with merely describing the different types of offspring produced, but counted them as well: in other words he realized the importance of expressing results quantitatively. Thirdly, and this is perhaps where his greatest insight lay, he interpreted his results and drew general conclusions from them. It is all the more to his credit that he managed to formulate his two laws without any knowledge of chromosomes, genes or meiosis.

In 1866 Mendel published his results and conclusions in the journal of the local scientific society. It received little publicity and made no impact on the scientific world. Disappointed, Mendel sent a copy to Karl von Nägeli, the distinguished Swiss botanist. But Nägeli failed to recognize its importance. He wrote a patronizing letter to Mendel telling him that his experiments were incomplete and suggesting that he should plant more peas, curious advice in view of the fact that Mendel's data was based on more than 21,000 plants!

In 1868 Mendel was appointed Abbot of his monastery and this put an end to his scientific research. He died in 1884, highly respected as a priest and teacher but unrecognized as a scientist. It was not until 1900 that the Dutch biologist Hugo de Vries came across Mendel's paper by accident and brought it to the attention of geneticists. By this time chromosomes had been discovered and meiosis described. Only then, 34 years after its publication, was the full significance of Mendel's work realized.

29 Chromosomes and Genes

In the last chapter we considered several examples of dihybrid inheritance in which the genes segregate and assort independently, giving all possible combinations in predictable proportions in the F2 generation. However, this does not always happen. Sometimes the genes fail to assort freely, or do so in unexpected proportions. Apart form their intrinsic interest these situations are important because they tell us much about the relationship between chromosomes and genes.

LINKAGE

One of the many inherited characteristics in *Drosophila* concerns the width of the abdomen. Some flies have a broad abdomen (the normal condition), others have a narrow abdomen (the recessive condition). The gene for broad abdomen is dominant to the gene for narrow abdomen, just as the gene for long wing is dominant to the gene for vestigial wing. If a long-winged, broad fly is crossed with a vestigial-winged, narrow fly, the F1 offspring all have long wings and broad abdomens, as we would expect. But if two of these flies are mated, the F2 generation yields mainly long-winged, broad flies and vestigial-winged, narrow flies in the proportion of 3:1, thus:

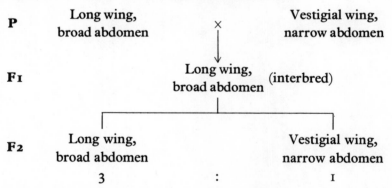

Before reading further can you suggest a hypothesis to account for this seemingly curious result?

The explanation in fact is that the genes determining the length of the wings and the width of the abdomen are located on the same chromosome, resulting in their being transmitted together (Fig 29.1). Such genes are said to be **linked,** and this general phenomenon is known as **linkage.** It does not, however, follow that linked genes can never be separated. In the example quoted above a small proportion of the offspring show free assortment, i.e. a few long-winged narrow flies and vestigial-winged broad flies turn up in

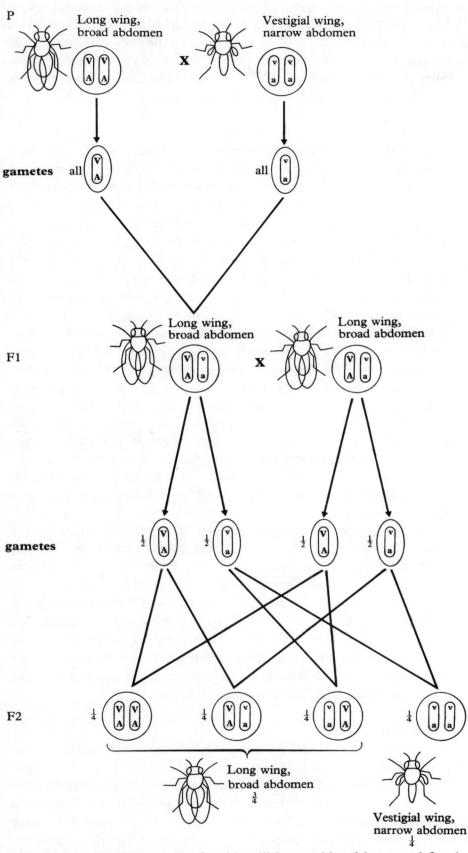

P — Long wing, broad abdomen   X   Vestigial wing, narrow abdomen

gametes   all   all

F1 — Long wing, broad abdomen   X   Long wing, broad abdomen

gametes   ½   ½   ½   ½

F2   ¼   ¼   ¼   ¼

Long wing, broad abdomen ¾   Vestigial wing, narrow abdomen ¼

Fig 29.1 Linkage in the fruit fly. The genes for wing length and width of abdomen are located on the same chromosome with the result that they are transmitted together and fail to show independent assortment. Gene symbols: vestigial wing, **v**; long wing, **V**; narrow abdomen, **a**; broad abdomen, **A**. **V** is dominant to **v**, and **A** to **a**.

the F2 generation. The reason for this will be considered later, and for the moment we will assume that linked genes normally stay together. If, in a breeding experiment, the transmission of two (or more) pairs of characters show independent assortment, we can be sure that the genes are carried on different chromosomes. If, however, little or no independent assortment is shown, we conclude that the genes are located on the same chromosome. We now know that an individual chromosome may contain a very large number of genes controlling a wide variety of characteristics. Genes linked together on the same chromosome constitute a **linkage group**.

LINKAGE GROUPS AND CHROMOSOMES

Our knowledge of the relationship between chromosomes and genes is based on experiments started in the early part of this century by T. H. Morgan and his co-workers in America. Morgan's influence on the science of genetics is hardly less great than Mendel's. It was Morgan who first realized the potential of *Drosophila* for breeding experiments, and indeed his research group was responsible for establishing many of the varieties (mutants) which form the basis of present-day breeding experiments. From a long series of experiments it has been established that *Drosophila melanogaster* has four linkage groups. The genes in any one linkage group are transmitted together but independently of the genes in the other linkage groups. The interesting thing is that there are also four pairs of chromosomes (Fig 29.2), a powerful piece of evidence that genes are located on the chromosomes. Similar results have been obtained in other organisms. In general the number of linkage groups found in a particular species corresponds to the number of different types of chromosome (the haploid number) characteristic of the species. Thus in the different species of *Drosophila* the haploid number varies from 3 to 6, and in each case the number of linkage groups is the same. In maize there are 10 pairs of chromosomes and 10 different linkage groups; in peas 7 pairs of chromosomes correspond to 7 linkage groups. For obvious reasons an extensive study of linkage in man is difficult to undertake but we would expect there to be 23 linkage groups, corresponding to the 23 pairs of chromosomes.

Studies of linkage groups enable us to predict not only the number of chromosomes in a particular species, but their relative sizes as well. If a linkage group contains a large number of genes, we conclude that the chromosome concerned is correspondingly large. On the other hand if a linkage group contains only a few genes, the chromosome is presumably comparatively small. This is borne out by *Drosophila melanogaster* in which two of the linkage groups are large (i.e. each contains a comparatively large number of genes), one is of medium size, and the remaining one is very small. The chromosomes bear a similar relationship to each other: there are two large chromosomes, a medium-sized one, and a very small dot-like one (see Fig 29.2).

SEX DETERMINATION

The genes carried by the medium-sized chromosomes in *Drosophila melanogaster* include those that determine the individual's sex, for which reason these chromosomes are referred to as the **sex chromosomes**. In the female the two sex chromosomes, both rod-shaped in appearance, are identical and are known as the **X chromosomes** (Fig 29.2). The genotype of the female is therefore **XX**. In the male, however, the two sex chromosomes differ from each other: one is a rod-shaped **X** chromosome, the other is hook-shaped, the so-called **Y chromosome**. The male genotype is therefore **XY**. The sex chromosomes are obviously an exception to the general rule that homologous chromosomes

Fig 29.2 The fruit fly *Drosophila melanogaster* has only four pairs of chromosomes which are readily distinguished from each other by their size. The black pair are the sex chromosomes, the **Y** chromosome being hooked.

female

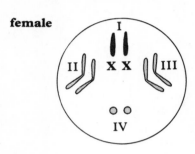

male

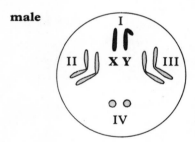

are identical in appearance. Being different, they are known as **heterosomes** (which literally means 'different bodies'), in contrast to all the other pairs of chromosomes which are identical and are called **autosomes**. But despite this structural anomaly the sex chromosomes are transmitted in a normal Mendelian manner as can be seen in Fig 29.3. Clearly, a female will produce

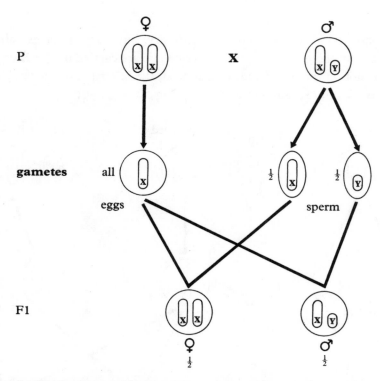

Fig 29.3 Sex determination in man. An individual's sex is determined by his or her genetical constitution, a female possessing two **X** chromosomes in her cells, a male one **X** and a **Y** chromosome. The sex chromosomes are transmitted in a normal Mendelian fashion as shown on the left.

only one kind of gamete as far as the sex chromosomes are concerned: all her eggs will contain an **X** chromosome. For this reason the female is termed **homogametic**. The male, however, will produce two kinds of gamete: half the sperms will contain the **X** chromosomes, the other half **Y**. The male is therefore the **heterogametic sex**. On fusing randomly, half the zygotes receive two **X** chromosomes and give rise to females; the other half receive an **X** and a **Y** chromosome and give rise to males.

The male is not invariably the heterogametic sex. For example, in birds the male is **XX** and the female is **XY**. In some insects the female is **XX** and the male is **XO**, the **Y** chromosome being absent.

SEX LINKAGE

It follows that all the genes carried on the sex chromosomes are transmitted along with those determining sex, that is they are **sex-linked**. Consider, for example, the inheritance of eye colour. In *Drosophila* there are red-eyed and white-eyed varieties, red eye being dominant to white. The result of crossing a red-eyed fly with a white one depends on which parent is red and which is white. If the father is white the F1 gives nothing but red-eyed flies, males and females being in equal proportions:

P White-eyed ♂ × Red-eyed ♀

↓

F1 Red-eyed♂'s and ♀ s

If, however, the father is red, the F1 yields equal numbers of red and white flies, all the red ones being females and the white ones males:

P   Red-eyed ♂ × White-eyed ♀

F1   Red-eyed ♀'s   White-eyed ♂'s
              $\frac{1}{2}$ $\frac{1}{2}$

At first sight these results seem extraordinary, but they can be explained by assuming that the genes controlling eye colour are carried on the **X** chromosome, but not on the **Y** chromosome. Indeed, the latter proves to carry few, if any, genes. A full explanation is given in Fig 29.4.

Fig 29.4 Sex linkage in the fruit fly. Genes controlling eye colour are located on the **X** chromosomes and are therefore sex-linked. The result of crossing red- and white-eyed flies depends on which parent is red-eyed and which white. This can be seen by comparing the two genetic diagrams below. Gene for red eye, **R** (dominant); gene for white eye, **r** (recessive).

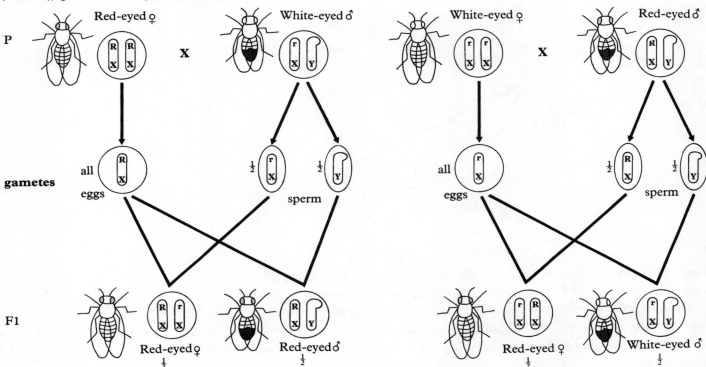

The principles of sex linkage seen in *Drosophila* apply equally well to other organisms including man. In most animals the male is heterogametic (**XY**) and the female homogametic (**XX**) and, as in *Drosophila*, the **Y** chromosome is usually genetically empty. As a male always receives his **Y** chromosome from his father he cannot inherit any of his father's sex-linked traits. A female, however, receives an **X** chromosome from her father which she may subsequently transmit to her children. The latter may therefore show their grandfather's sex-linked traits. To put it in general terms: a male transmits his sex-linked traits to his grandchildren via his daughters; he cannot transmit them to or through his sons.

The way this works in man can be seen clearly in the inheritance of red-green **colourblindness** (Fig 29.5). The gene for colourblindness is recessive and sex-linked. A colourblind man, married to a normal woman, transmits his 'colourblind gene' to his daughters who, since they are heterozygous, will not be colourblind but are carriers of the defective gene. His sons will not have the defective gene because they receive their **X** chromosome from the mother. But consider what happens if one of the daughters marries a normal

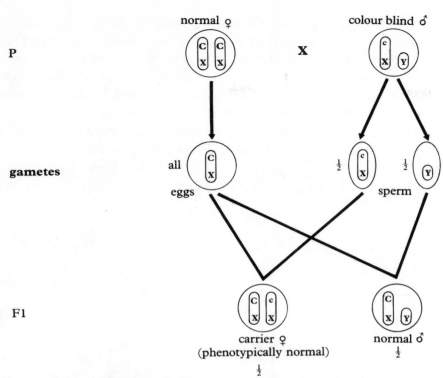

Fig 29.5 Colourblindness, an example of sex linkage in man. The diagram shows the possible result of a union between a normal woman and a colourblind man. None of the children will be colourblind but a daughter will certainly be a carrier of the defective gene. Gene for colour blindness, **c** (recessive); gene for normal vision, **C** (dominant).

man. If they produce a son there is a probability of ½ that he will be colour-blind. A daughter will be phenotypically normal but there is a probability of ½ that she will be a carrier. As would be expected, red-green colourblindness is more common in men than in women.

Another sex-linked trait, much more serious than colourblindness, is **haemophilia** or 'bleeders disease'. In this distressing condition the blood takes an abnormally long time to clot, resulting in profuse and prolonged bleeding from even the slightest wound. Haemophilia, like colourblindness, is caused by a recessive gene carried on the **X** chromosome. As can be seen in Fig. 29.6,

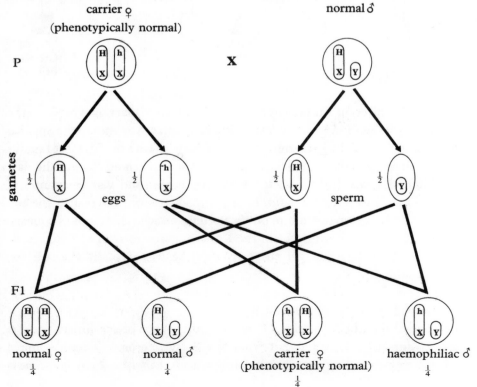

Fig 29.6 Haemophilia, another example of sex linkage in man. If a normal man marries a woman who carries the haemophilia gene, there is a probability of ¼ that they will produce a haemophiliac son. Haemophilia gene, **h** (recessive); normal gene, **H** (dominant).

if a normal man is married to a woman who happens to carry the gene for haemophilia, there is a probability of $\frac{1}{2}$ that if they produce a son he will be a haemophiliac, and if they produce a daughter she will be a carrier. Both children may then transmit their defective genes to the grandchildren, and so on. This is precisely what has happened in the Royal families of Europe during the last hundred years.

It seems that a haemophilia gene arose by mutation in one of the gametes (it might have been the egg or the sperm) which was responsible for producing Queen Victoria. Of her nine children one was a haemophiliac (Leopold) and two were carriers (Beatrice and Alice). Fig 29.7 shows how the defective genes

Fig 29.7 Genetic chart showing the spreading of the haemophilia gene through the Royal Houses of Europe. The gene probably arose as a mutation in Queen Victoria or her immediate ancestors, since she bore one afflicted son and two carrier daughters, and there was no record of haemophilia in her ancestry. Males are represented by squares, females by circles in the chart. (*from* Winchester, Genetics, *Houghton Mifflin Co*)

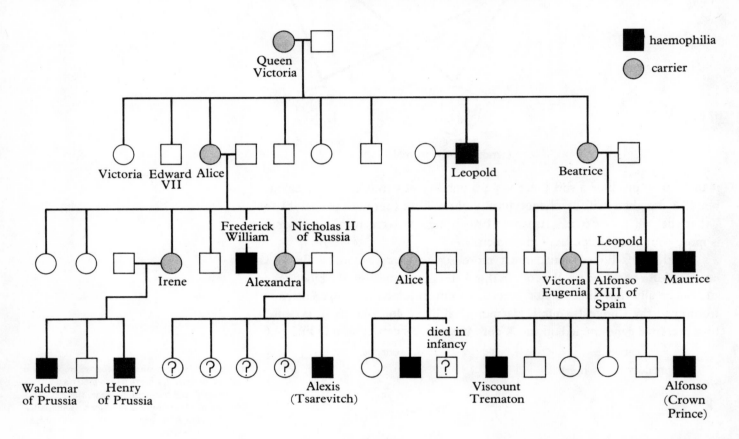

spread to the different European countries as a result of marriages between the various Royal families. Beatrice transmitted her defective genes to the Spanish Royal family and Alice transmitted hers to Russia. One of the most famous haemophiliacs on the Russian side was Alexis Nicholaievitch, the last Tsarevitch. He was badly afflicted with the disease and was constantly at death's door. However, the monk Rasputin had strange powers over the boy and is alleged to have cured him of his bleeding attacks. Through this means Rasputin gained great influence in the Russian court, to such an extent that for a short period just before the revolution he was virtually the ruler of Russia.

SEX LINKAGE AND THE **Y** CHROMOSOME

One conclusion which is drawn from the inheritance of sex-linked traits is that sex-linked genes are absent from the **Y** chromosome. There is ample evidence from breeding experiments that in *Drosophila* the **Y** chromosome is

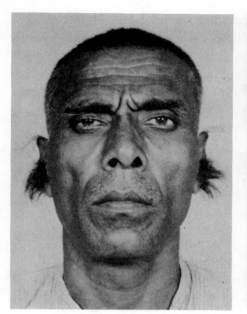 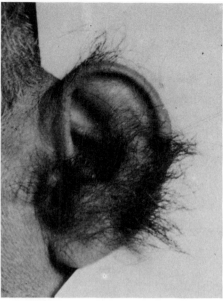

Fig 29.8 Hairy ears, a possible case of **Y**-borne inheritance. Explanation in text. (*Curt Stern, University of California*)

genetically virtually empty, but is this also true of man? If a sex-linked trait is associated exclusively with the **Y** chromosome, it would be expected to show itself in men but not in women. Are there in fact any conditions which show this?

One of the most notable cases of possible **Y**-borne inheritance was the famous 'porcupine man' of eighteenth century England. This unfortunate character had patches of hard, spiny skin rather like a porcupine and is alleged to have transmitted this condition to all his sons. However, much doubt has been cast on the validity of the records and there is little evidence that the condition was associated with the **Y** chromosome, or indeed with the sex chromosomes at all. A better documented case concerns the condition shown in Fig. 29.8: hairy ears. This trait, which shows itself as a tuft of hair sprouting from the pinna of the ear, is particularly common in parts of India and seems to be exclusive to the male sex. However, the surveys which have been made by no means prove that the **Y** chromosome is implicated.

CROSSING-OVER

We have seen that linked genes are carried on the same chromosome. This being so, we would expect that linked genes would always stay together and never separate. This is not however the case. When studying the inheritance of linked genes it is unusual to find complete linkage. More often than not, a certain proportion of the offspring show new combinations, as in independent assortment. We will illustrate this with maize which, like *Drosophila*, is a favourite in genetical research.

In maize a single 'ear' is covered with several hundred kernels. Each kernel is a seed, the product of a single fertilization. The kernels show several clear-cut characteristics including colour and shape. If a maize plant homozygous for kernels which are coloured and smooth is crossed with one having colourless and shrunken kernels, the F1 produces nothing but the coloured smooth variety. Now consider what happens if one of the F1 plants is testcrossed with the double recessive. If the genes for colour and shape were on separate chromosomes, we would expect the F2 to produce all four combinations of characteristics in approximately equal numbers. If they are carried on the same chromosome, the F2 should produce only two types of kernel: coloured smooth and colourless shrunken. But in fact neither of these possibilities turns out to be the case. What we get in the F2 generation are mainly coloured smooth and colourless shrunken kernels, but there are a few coloured shrunken and colourless smooth ones as well:

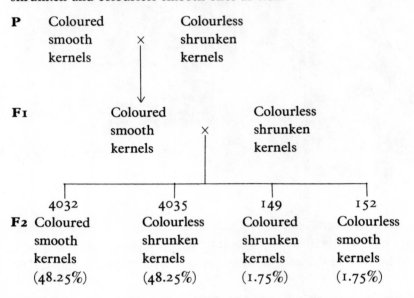

In other words we do get all four combinations of characteristics but not in the proportions we would expect by assuming that the genes are located on separate chromosomes.

How can we explain this? The answer is that the genes are in fact situated on the same chromosome and for the most part they are transmitted together through the gametes to the offspring. However, in the formation of a small proportion of the gametes the genes, instead of staying on their own chromosome, change places. This is called **crossing-over** and it has the effect of separating genes that were previously linked. The result is that, whilst the majority of offspring have the same combination of characteristics as the parents (**parental combinations**), a few show new combinations (**recombinations**). This is summarized in Fig 29.9.

EXPLANATION OF CROSSING-OVER

In seeking an explanation for crossing-over two basic facts must be remembered: (1) In meiosis (the type of cell division which occurs in the formation of the gametes) homologous chromosomes become intimately wrapped round one another, and (2) allelic genes normally occupy the same relative positions on their respective chromosomes and therefore lie alongside each other when the chromosomes come together.

During prophase of the first meiotic division when homologous chromosomes become intertwined, the chromatids of homologous chromosomes are seen to be in contact with each other at certain points along their length (see p. 376). At these points, known as **chiasmata** (singular: **chiasma**), the chromatids break and rejoin as shown in Fig 29.10. The result is that portions of the chromatids belonging to the two homologous chromosomes change places, taking their genes with them. These chromatids, with their new

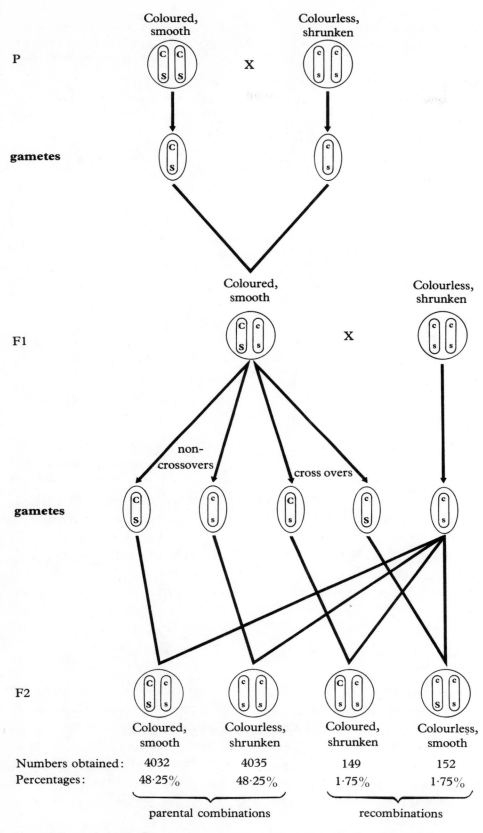

Fig 29.9. Diagram showing the result of crossing maize homozygous for coloured, smooth kernels with maize having colourless, shrunken kernels. The genes controlling colour and texture of the kernels are linked. The small percentage of recombinations in the F2 can be explained by postulating that crossing-over takes place between homologous chromosomes during gamete formation. Gene symbols: coloured, **C**; colourless, **c**; smooth kernel, **S**; shrunken kernel, **s**. **C** is dominant to **c**, and **S** to **s**.

	Coloured, smooth	Colourless, shrunken	Coloured, shrunken	Colourless, smooth
Numbers obtained:	4032	4035	149	152
Percentages:	48·25%	48·25%	1·75%	1·75%
	parental combinations		recombinations	

complements of genes, are termed **crossovers**. Eventually they finish up in separate gametes and, after fertilization, give rise to new combinations of genes in the offspring.

The number of chiasmata formed in a bivalent during meiosis, and therefore the amount of crossing-over, varies from one pair of homologous chromo-

Fig 29.10 Meiosis provides the explanation of crossing over. In prophase when homologous chromosomes become intimately associated with one another, adjacent chromatids break and rejoin as shown below. The result is that previously linked genes become separated and new combinations are established.

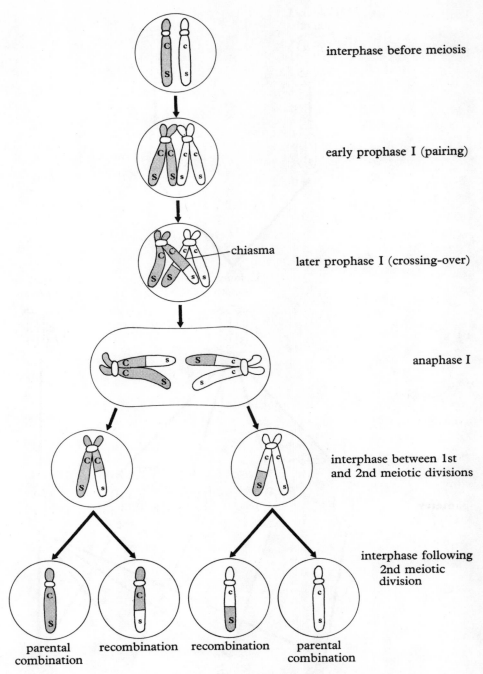

interphase before meiosis

early prophase I (pairing)

chiasma

later prophase I (crossing-over)

anaphase I

interphase between 1st and 2nd meiotic divisions

interphase following 2nd meiotic division

parental combination recombination recombination parental combination

somes to another. One or several chiasmata may be formed, and of course the longer the chromosomes the greater is the number of chiasmata likely to be. The number of possible combinations of genes in the gametes will depend on the number and positioning of the chiasmata relative to the sequence of genes. The importance of crossing-over is that by establishing new gene combinations it is an important instigator of genetic variety. The significance of this is discussed in Chapter 35.

LOCATING GENES ON CHROMOSOMES

To a biologist crossing-over has a special significance for it enables him to do something that would otherwise be impossible, namely to work out the relative positions of genes on a chromosome. To see how this works let us imagine two hypothetical linked genes **A** and **B**. The further apart **A** and **B** are on their chromosome, the more likely it is that the chromosome will break and rejoin at some point between them. Conversely, the closer they are, the less likely it is that breakage and crossing-over will occur between them. So if in a breeding experiment we find that a relatively large percentage of the offspring show separation of **A** from **B**, we conclude that these genes are relatively far apart on the chromosome. On the other hand if a low percentage show separation, we conclude that the genes are relatively close together. The percentage of offspring which show separation of the genes represents the **cross-over value** (**cov**) and for convenience of cov of 1 per cent is taken to represent a distance of 1 unit on the chromosome.

But crossover data can take us further than this. Not only can we deduce the relative distances between genes but we can also work out the order in which they occur on the chromosome.

Imagine that in a breeding experiment it is found that the cov of **A** and **B** is 4 per cent. This means that **A** and **B** are 4 units apart on their chromosome, thus:

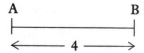

We then go on to investigate the transmission of each of these genes (**A** and **B**) with a third linked gene **C**. If we find that **A** and **C** give a cov of 10 per cent whereas **B** and **C** give only 6 per cent, we conclude that **B** is located somewhere between **A** and **C** as follows:

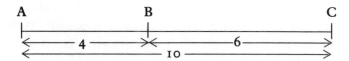

If, on the other hand, **A** and **C** give a cov of 2 per cent, and **B** and **C** 6 per cent, we conclude that **C** lies to one side of A:

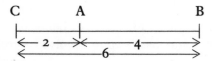

Another possibility might be that **A** and **C**, and **B** and **C** both give covs of 2 per cent. From this we would conclude that **C** lies midway between **A** and **B**:

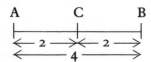

One pitfall should be mentioned. If two genes are widely separated on a chromosome it is possible that breakage may take place at two points between them, resulting in a **double crossover**. Unless a third gene, situated between the other two, is included in the analysis, the double crossover will go undetected, and the number of recombinations will not be a true reflection of the amount of crossing-over. The problem can generally be overcome by choosing genes which, judging by recombination figures, lie close together on the chromosome.

From data collected over many years from numerous breeding experiments, **chromosome maps**, showing the positions of the various genes on the appropriate chromosomes, have been built up for several organisms notably *Drosophila* and maize. The position occupied by a gene is known as the **gene locus**. In *Drosophila* literally hundreds of gene loci have been established in this way. Even in man a start has been made in mapping genes though the data on which this is based are obtained not from breeding experiments (!) but from the study of human pedigrees and the distribution of inherited characteristics in large populations.

CAN GENES BE SEEN?

The chromosome maps described in the previous section are based entirely on indirect evidence. At the time when these chromosome maps were being constructed on the basis of breeding experiments, the chromosomes themselves were visible under the microscope only as uniform threads and could not be seen to contain structures that could be remotely described as genes. In fact geneticists were resigned to the idea that genes were invisible entities hidden within the structure of the chromosomes. Even today, using the best possible techniques of microscopy, chromosomes generally show little internal differentiation and appear unwilling to reveal their secrets.

However in 1934 a discovery was made which has had an enormous impact on the study of genes. It was found that the chromosomes in the salivary glands of *Drosophila* larvae are relatively enormous, about a hundred times larger than those found elsewhere in the body. Moreover, far from being uniform in appearance, these **giant chromosomes** are traversed by numerous darkly-staining bands spaced at irregular intervals and of characteristic shape (Fig 29.11). Some are narrow and rather difficult to detect, others are thicker and clearer. Each type of chromosome has its own characteristic pattern of banding and this is constant in all specimens.

The question is: do these bands represent the genes? In the last thirty years attempts have been made to relate the bands to the chromosome maps of *Drosophila*. The techniques have mainly involved correlating genetic abnormalities detected in breeding experiments with structural abnormalities seen in the chromosomes. The fact that the normal arrangement of the bands is characteristic for each chromosome makes it possible to detect quite small changes in the structure of an individual chromosome. These abnormalities are discussed in Chapter 35 but one example will be mentioned here to illustrate the principle. Sometimes a chromosome breaks and the detached piece rotates through 180° and joins up again the wrong way round. The result is that the normal pattern of banding for that section of the chromosome is reversed, resulting in a change in the relative positions of the genes in the chromosome map. In some instances it is possible to correlate such structural abnormalities detected in the chromosomes with genetic abnormalities observed in breeding experiments. This kind of analysis has made it possible to relate individual bands in the chromosomes to specific gene loci in the chromosome maps (Fig. 29.12). But do the bands actually represent individual genes?

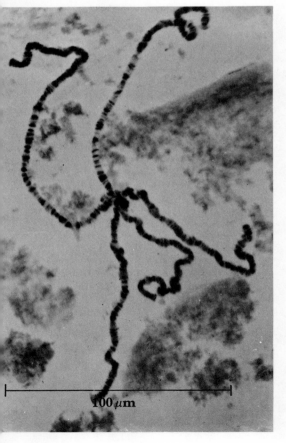

Fig 29.11 Giant chromosomes from the salivary gland of *Drosophila melanogaster*. (*H. G. Callan, University of St. Andrews*)

100 μm

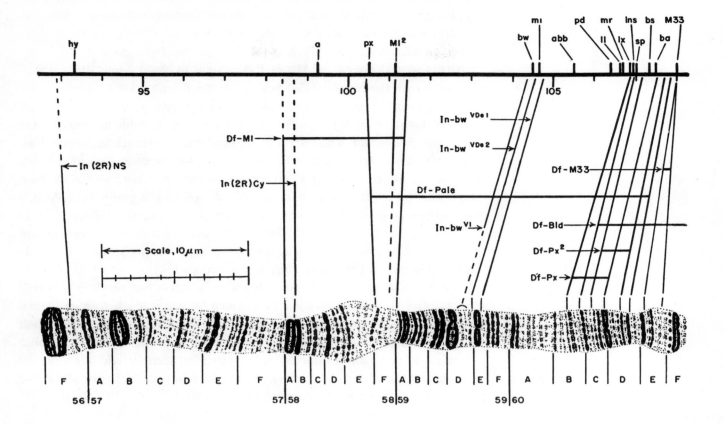

The answer is no. Seeing a chromosome band is not the same as seeing a gene which, as we shall see in Chapter 30, is several orders of magnitude smaller. The salivary chromosomes merely allow us to locate the *positions* of the genes with rather greater precision than is the case with ordinary chromosomes. Nevertheless we have here an example of how information derived from breeding experiments agrees with that obtained from direct observation of cells, an agreement which supports the contention that genes are carried in a linear sequence on the chromosomes.

Fig 29.12 Possible correlations between linkage map of *Drosophila melanogaster* and corresponding part of the second giant salivary gland chromosome. The figures on the chromosome map correspond to map distances. The symbols **hy**, **a**, **px**, etc. correspond to the positions of the various genes as deduced from crossing-over data. Each symbol stands for a particular characteristic: **hy** stands for humpy body, **px** for plexus type of wing venation, **bw** for brown eyes, **sp** for speckled body, **ba** for balloon-shaped wings, etc. (*after Bridges, from Sinnott, Dunn and Dobzhansky*, Principles of Genetics, *McGraw-Hill*)

MULTIPLE ALLELES

Normally, as we have seen, a phenotypic characteristic is controlled by a pair of alleles occurring at a specific locus. Sometimes, however, more than two alleles may be responsible for controlling the same characteristic. All the alleles in the series occur at the same locus but plainly only two can occupy positions on a pair of homologous chromosomes at any one time.

A good example of such **multiple alleles** is provided by the genes controlling the **ABO blood group system** in man. The ABO system is controlled by three alleles which we can designate **A**, **B** and **O**. The physiology of the ABO system is explained in Chapter 17 and here we are only concerned with the genes responsible for determining the different blood groups, and their transmission.

The **A** gene is responsible for the production of type A antigen in the person's red blood cells, the **B** gene for producing type B antigen. The third gene in the series, **O**, produces neither antigen. As only two of the three alleles can be present at any one time, an individual may possess any of the following combinations: **AA, AO, BB, BO, AB** or **OO**. The **A** and **B** alleles show equal dominance with respect to one another (sometimes called **co-dominance**) but both are dominant to **O**. Thus:

A person with the genotype **AA** or **AO** belongs to blood group A

A person with the genotype **BB** or **BO** belongs to blood group B

A person with the genotype **AB** belongs to blood group AB

A person with the genotype **OO** belongs to blood group O.

The fact that there are more than two alleles responsible for determining the blood groups makes no difference to their transmission, which takes place in a normal Mendelian fashion. Thus a mating between two individuals both of group O must result in a child who is also group O (note why). However, the result of a mating between individuals of group A and group O will depend on the genotype of the A parent. If he or she is **AA**, the children will all be group A. On the other hand if the parent is **AO**, children belonging to groups A and O will be produced in equal proportions (i.e. the probability of any one child belonging to either one of these two groups is $\frac{1}{2}$).

An interesting aspect of multiple alleles is that it can result in offspring which differ from both parents. This is seen for example when an individual of group AB mates with group O. Half the children belong to group A, half to group B, none of them belong to the same group as either parent. A similar situation is seen when a mating takes place between individuals with the genotypes **AO** and **BO**. In this case the children can belong to any of the four groups. Draw diagrams to illustrate these crosses.

GENETICS AND THE LAW

Our knowledge of blood groups and their inheritance can be useful in legal questions of paternity. To give a hypothetical example, an unmarried girl gives birth to a child and accuses a well known politician of being the father. The mother belongs to group A and the baby to group O. The accused man, however, belongs to group AB which indicates unequivocally that he cannot be the father. The real father must be A, B or O.

Of course not all paternity suits are as clear-cut as this. If our politician had turned out to belong to group A, B, or O this alone would not be sufficient to convict him; many men belong to these groups and other evidence would have to be obtained, for example from other blood groups such as the MN system, to corroborate that supplied by the ABO system. But it can be very useful in positively eliminating a suspect.

DEGREES OF DOMINANCE

From Mendel's experiments we might conclude that a so-called dominant gene is always completely dominant to its recessive allele. Such a conclusion would be based on the observation that in heterozygous individuals the dominant allele expresses itself fully in the phenotype whereas the recessive allele does not express itself at all. However, the inheritance of the ABO blood group system shows that this is not always true. Sometimes, as in the case of the **A** and **B** genes, two alleles show no dominance with respect to each other, both expressing themselves equally in the phenotype.

In cases where dominance is absent, a mating between two individuals results in offspring which are identical with neither parent but are intermediate between the two. For example a snapdragon with red flowers crossed with white gives plants with pink flowers in the F_1, a result which contrasts sharply with those obtained by Mendel in garden peas. Similarly in shorthorn cattle crossing individuals with red and white coats gives offspring whose coats are composed of a mixture of red and white hairs, the **roan** condition. Another well known example is provided by the **Andalusian** breed of fowls. A black fowl crossed with a splashed-white one (white with small black patches on it) gives F_1 chicks with a blue sheen, the true Andalusian condition. In these cases what will be the result of crossing the F_1 hybrids amongst themselves?

Between the two extremes of complete dominance and no dominance at all are all shades of **partial dominance**. In such cases the offspring fail to resemble either parent exactly, but are closer to one than the other. It is therefore clear that the concept of dominance is by no means as cut and dried as it might at first sight appear. Genes do not interact in an all-or-nothing manner but show varying degrees of intermediate expression.

THE SICKLE-CELL GENE

In man this intermediate expression is seen in the inheritance of a blood condition known as **sickle-cell anaemia** which afflicts American Negroes and the peoples of Mediterranean countries and northern Africa. In sickle-cell anaemia the normal haemoglobin in the red blood cells is entirely replaced by an abnormal haemoglobin known as **haemoglobin S**. Haemoglobin S is much less soluble than normal haemoglobin and it begins to crystallize when the oxygen concentration falls, as it does in the capillaries of the tissues. This causes the red blood cells, normally biconcave disc-shaped, to assume the shape of a crescent or sickle. With their abnormal haemoglobin, the sickled red cells are far less efficient at carrying oxygen. The severe anaemia which results is generally fatal. Fortunately the condition is not very common; about 0.2 per cent of American Negroes suffer from it.

More common, however, is a related condition known as **sickle cell trait**. In this the abnormal haemoglobin S is present only to the extent of about 30 or 40 per cent, the remaining haemoglobin being normal, and the red blood cells are of the normal shape. The anaemia suffered by a person with the sickle-cell trait is slight and generally not sufficient to interfere with normal activity.

How can we explain the relationship between sickle-cell anaemia and sickle-cell trait? We can do so by postulating that the gene responsible for producing normal blood is only partially dominant to its allele producing the sickle-cell condition. A person homozygous for the defective gene suffers from sickle-cell anaemia, the full disease. Heterozygous individuals, however, only develop the milder sickle-cell trait. Because of the partial dominance of the normal allele, the sickle-cell gene is prevented from expressing itself fully in the heterozygote.

GENE ACTION

The sickle-cell story can be taken a step further for it gives us an insight into how genes act in producing their effects in the phenotype. Following the discovery of how the sickle-cell condition is transmitted, intensive research was centred on finding the precise chemical difference between haemoglobin S and normal haemoglobin. The difference turned out to be extremely slight: in haemoglobin S there is one position in each of the two β polypeptide chains in the haemoglobin molecule where the amino acid valine takes the place of glutamic acid. By causing this seemingly trivial change, the defective gene has far-reaching effects on the individual's functioning. This suggests that genes exert their effects, and express themselves in the phenotype, by influencing the organism's biochemical processes, sometimes in a very slight and subtle way. But this is anticipating a topic that is discussed in much more detail in Chapter 30.

LETHAL GENES

It is probably true that in genetical studies more has been learned about genes from abnormal ratios than from normal ones. Consider, for example, the inheritance of fur colour in mice. In mice yellow fur (**Y**) is dominant to grey (**y**). Now if a pair of yellow mice are mated the result is always the same: two thirds of the offspring are yellow, one third grey. In other words yellow to grey occur in a ratio of 2 : 1. The fact that the offspring include both types of fur colour means that both parents must be heterozygous (**Yy**). If this is so why do we not get a 3 : 1 ratio in the F1? There are various theoretical ways of explaining this but the simplest explanation is to postulate that individuals homozygous for the yellow gene (i.e. **YY**) die before birth. In other words the genotype **YY** represents a lethal combination of genes. As can be seen from Fig 29.13, the death of such individuals in the embryonic state would have the effect of removing a quarter of the offspring from the litter. The dead embryos would represent a third of the potential yellow offspring thus reducing the proportions of yellow offspring from what would have been $\frac{3}{4}$ to $\frac{2}{3}$.

Fig 29.13 Inheritance of fur colour in mice. The gene for yellow coat (**Y**) is dominant to grey (**y**). The **YY** combination of genes is lethal so the normal 3:1 ratio is upset. For further details see text.

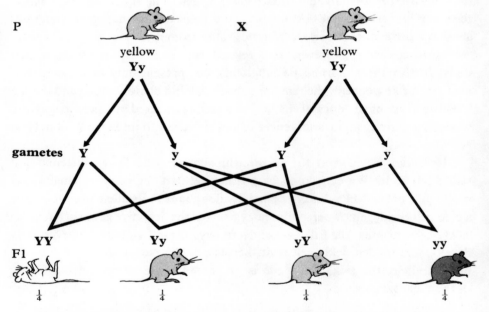

That this hypothesis is correct is confirmed by two further observations.

(1) Crossing yellow with yellow never produces exclusively yellow offspring; in other words it is impossible for a pair of yellow mice to breed true. This can only be explained by assuming that yellow mice are always heterozygous and that a living homozygous yellow mouse is an impossibility.

(2) The expected number of dead embryos, i.e. a quarter of the total

offspring, are found in the uteri of yellow mothers who have been mated with yellow males. No such dead embryos are found in the uteri of grey mothers (or of yellow mothers mated with grey males) because none of the offspring resulting from such matings would be **YY**.

The evidence is thus overwhelming that the **YY** combination is lethal. In other words the **Y** gene not only affects colour but also viability. Should we describe this gene as dominant or recessive? The answer is that it is both depending on how we look at it. As the gene controlling fur colour it is plainly dominant, but as a lethal gene affecting viability it is recessive, exerting its lethal effects only when in the homozygous state.

Lethal genes are known to exist in a wide range of organisms including man. In most cases it is obviously impossible to observe the homozygous lethal as such because it perishes at the embryonic stage. However, its existence can be deduced from unusual ratios of living offspring together with the fact that the total number of offspring will be lower then expected.

In certain cases the existence of a lethal gene can be detected from the fact that the heterozygous condition, though not lethal, has an observable effect on the individual's phenotype. In such cases the lethal gene is not completely recessive but exerts partial dominance so that it expresses itself to a slight extent in the heterozygote.

INTERACTION OF GENES

So far we have dealt only with cases where a characteristic is controlled by one pair of alleles. Sometimes, however, a single characteristic is controlled by two or more pairs of alleles interacting with one another. This applies, for example, to the shape of comb in different breeds of poultry (Fig 29.14).

If a **pea-combed** fowl is crossed with a **rose-combed** fowl, all the F1 offspring display a quite different kind of comb known as **walnut**. If two of these are crossed, the F2 produces birds with all three types of comb (pea, rose and walnut) plus a further fourth type known as the **single comb**.

At first sight a cross in which a totally new character crops up in both the F1 and F2 generations seems inexplicable. However, these results make sense if we assume that the four types of comb are produced by the interaction of two pairs of alleles, which we can designate **Pp** and **Rr**. A pea-comb develops when **P** alone is present without **R**; a rose-comb when **R** is present without **P**; a walnut-comb when **P** and **R** are present together; and a single-comb when **p** and **r** are present alone in the absence of **P** and **R**. The genotypes responsible for the different combs may therefore be summarized as follows:

Pea: **PPrr** or **Pprr**
Rose: **ppRR** or **ppRr**
Walnut: **PPRR, PPRr, PpRR** or **PpRr**
Single: **pprr**

The two pairs of alleles are transmitted in a normal Mendelian manner. They show free assortment and give the same ratios as in ordinary dihybrid crosses. But although they are transmitted independently they are interdependent in the way they express themselves in the phenotype.

In this chapter we have seen how studies on inheritance provide information on the relationship between chromosomes and genes and on the way genes exert their action. In the next chapter we shall analyse genes in more detail.

Fig 29.14 Different types of comb in poultry. Each comb type is produced by the interaction of two genes, each of which is transmitted independently in a Mendelian fashion. See text for details.

pea comb

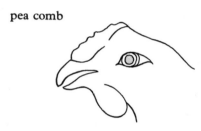

rose comb

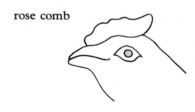

walnut comb

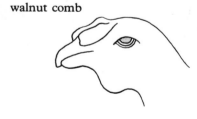

single comb

30 The Nature of the Gene

In this chapter we shall explore the chemical nature of the gene, and we shall see that in its molecular structure lies the explanation of how it works. This is a comparatively new field of biology known as molecular genetics. But before getting down to details let us see what general predictions we can make about the nature of the gene.

Genes, as we have seen, determine an organism's characteristics, and through them these characteristics are handed on to the offspring. This means that genes must contain **information**, a set of instructions if you like, telling the organism how to develop. It also follows that genes must be able to reproduce themselves, or **replicate**, without losing any of this information, otherwise the instructions they carry will be progressively diluted in successive generations.

In searching for the chemical nature of the gene we must therefore look for a molecule that contains information and is capable of replication. For many years protein was considered the only possible candidate because it alone seemed to have the structural diversity necessary for carrying genetic information. However, in 1944 Oswald Avery of the Rockefeller Institute in New York produced evidence that **nucleic acid** rather than protein is the carrier of genetic information. Let us look briefly at what he had done.

EVIDENCE THAT NUCLEIC ACID IS INVOLVED

The bacteria that cause pneumonia, *Pneumococcus*, exist in two different strains, capsulated and non-capsulated. The former are characterized by the cells being surrounded by a thick capsule. This contrasts with the non-capsulated strain in which a capsule is completely missing. Avery and his colleagues prepared a medium containing disrupted capsulated bacteria; it contained no living bacterial cells, just the chemicals extracted from them. Into this medium non-capsulated bacteria were introduced and allowed to grow. The remarkable thing was that the non-capsulated bacteria developed capsules, as did their descendants. The conclusion to be drawn is that a chemical constituent in the extract of capsulated bacteria is capable of bringing about a permanent and inheritable change in the non-capsulated strain. The question is: what particular component of the extract does the trick? By systematically isolating, purifying and testing the different chemical constituents, Avery and his colleagues were able to show that nucleic acid is the active agent. With the subsequent confirmation of Avery's results by similar bacterial transformations the attention of biologists was riveted on nucleic acids.

THE STRUCTURE OF NUCLEIC ACIDS

Although the chemical building blocks of nucleic acids have been known since the turn of the century, it is only recently that we have come to understand how they are fitted together to form a nucleic acid molecule. We now know that nucleic acids occur in all living cells and in all organisms including bacteria and viruses. Unravelling their structure has been one of the most exciting adventures in modern science. It has brought us to a closer understanding of heredity and development, and has opened the doors to understanding the nature of life itself.

Two principal types of nucleic acid are found in cells: **deoxyribonucleic acid (DNA)**, and **ribonucleic acid (RNA)**. By means of chemical tests it has been shown that DNA is confined to the nucleus, whereas RNA is found mainly, but not exclusively, in the cytoplasm. Like proteins, nucleic acids are long chain molecules, but the chains are much longer than those of proteins and the sub-units more complicated than amino acids.

The building blocks of a nucleic acid are called **nucleotides** which are themselves complex molecules. A nucleotide consists of three molecules linked together: a **5-carbon sugar (pentose)**, an **organic base**, and **phosphoric acid**. Let us consider these three constituents in turn.

First, the **sugar**. You will recall that a hexose sugar such as glucose consists of a six-membered ring of five carbon atoms and one oxygen atom to which various side groups are attached. A pentose sugar is basically similar except that there is one less carbon atom in the ring with the result that the molecule is constructed as follows:

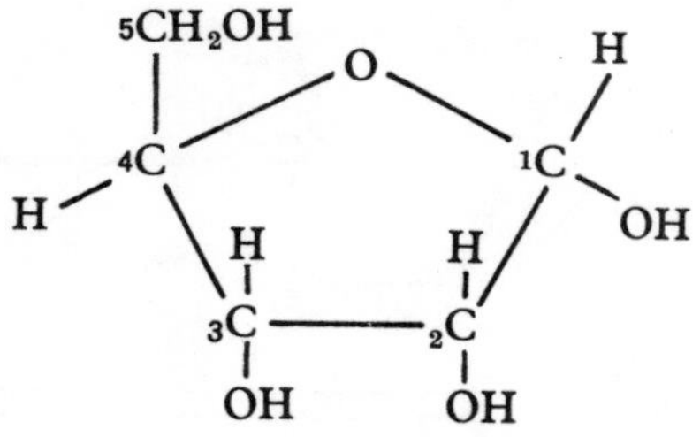

This particular sugar is **ribose**. Deoxyribose differs from it only in that the hydroxyl side group at position 2 is replaced by a hydrogen atom. In other words deoxyribose has one less oxygen atom than ribose:

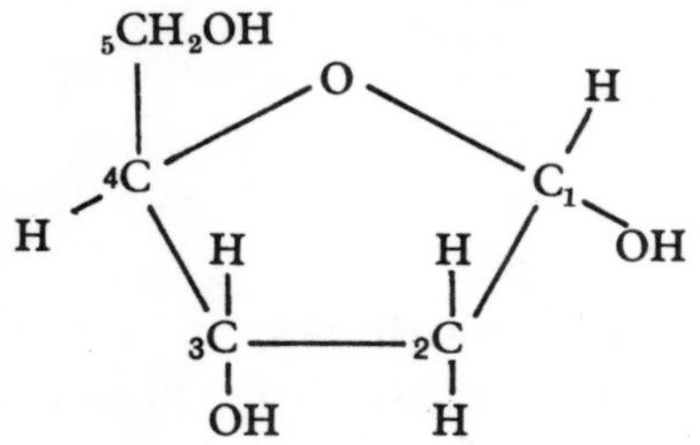

The second constituent of a nucleotide is **phosphoric acid** (H_3PO_4) which has the structural formula:

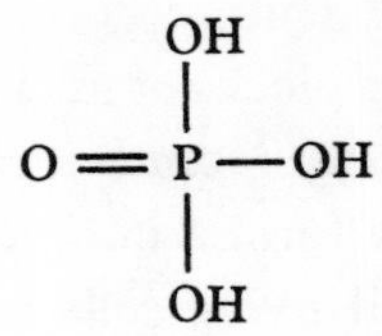

The third component is the **organic base**. This can be any one of five: **adenine, guanine, thymine, cytosine** or **uracil,** which we will abbreviate to **A, G, T, C** and **U** respectively. All these are complex ring compounds, the rings being composed of carbon and nitrogen atoms which for simplicity we will represent thus:

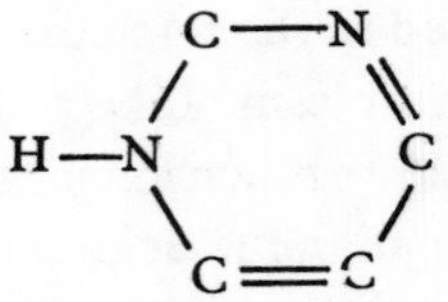

Do not worry about their detailed structure for the moment, but notice the hydrogen atom projecting from the nitrogen on the left: this plays an important part in the building up of a nucleotide.

A

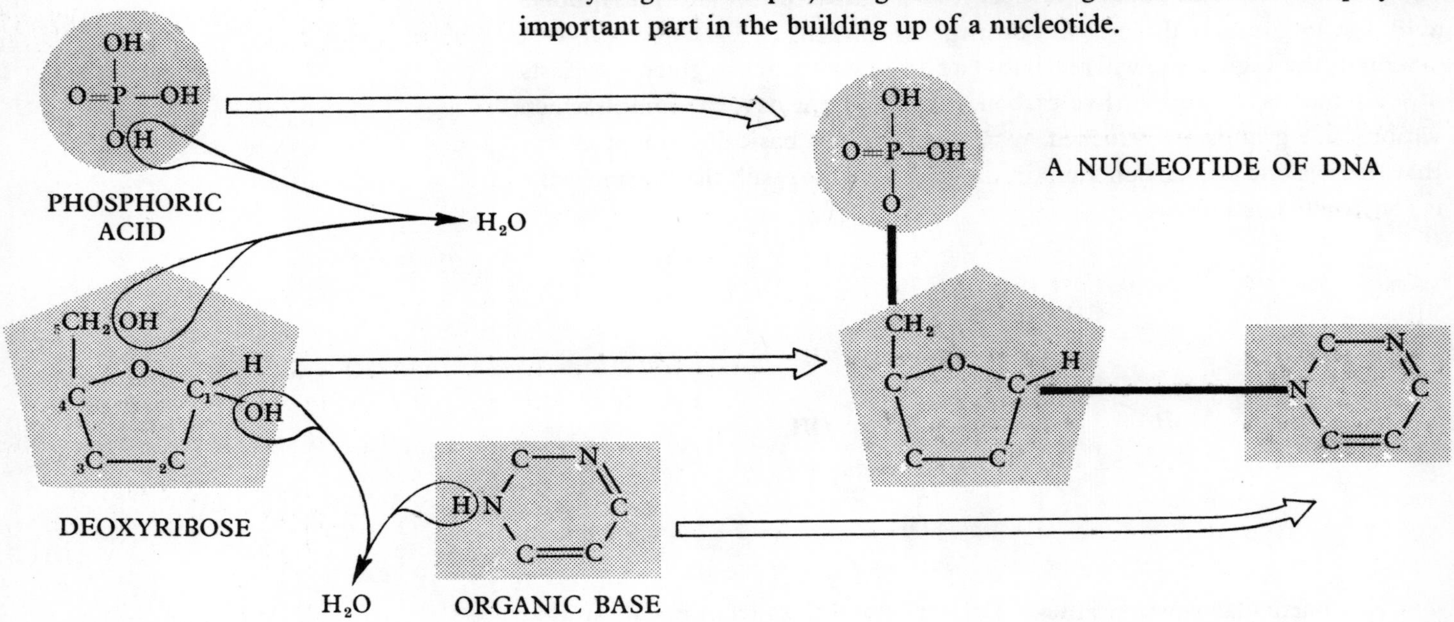

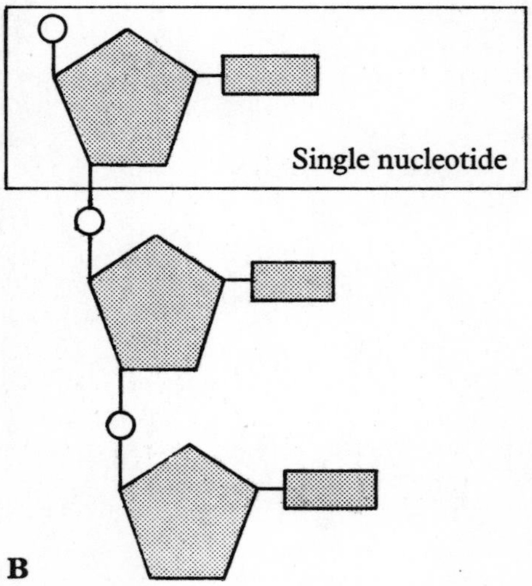

B

Fig 30.1 How a nucleic acid is built up. A) A pentose sugar, in this case deoxyribose, unites with a phosphoric acid molecule and an organic base to form a nucleotide. B) Nucleotides join through their phosphate groups to form a polynucleotide chain which is the basis of a nucleic acid.

How do the three components link up to form a nucleotide? As in the construction of other organic compounds the process is one of condensation involving the removal of water (Fig 30.1A). A sugar ring attaches itself to a phosphoric acid molecule at position 5, and to one of the five bases at position 1. Two molecules of water are removed in the process.

The nucleotides are then strung together to form a nucleic acid molecule.

This also involves condensation. The elements of water are removed from the sugar of one nucleotide and the phosphate radical of another, thereby forming a **dinucleotide**. The addition of further nucleotides produces a long **polynucleotide chain** whose backbone consists of alternating sugar and phosphate groups with the bases projecting out sideways from the sugars (Fig 30.1B).

The sugar and phosphate groups are identical all the way along the chain; in other words the backbone is absolutely stereotyped and shows no variation. In RNA all the sugars are ribose; in DNA they are all deoxyribose. The phosphate groups are identical in both RNA and DNA. The bases, however, show no such uniformity. We have already noted that there are five different ones: **A, G, C, T** and **U**. DNA contains **A, G, C** and **T**; in RNA **T** is replaced by **U**. Now the sequence in which these bases occur along the length of the nucleic acid chain varies from species to species and from individual to individual. It is in this sequence of bases that the nucleic acid carries information controlling the organism's development. It may seem strange that all the instructions required to produce an organism as complicated as, say, a human being, should be conveyed by only four kinds of base, but when we bear in mind that a single nucleic acid molecule may contain a total of five million nucleotides, the number of possible combinations is almost infinite. The bases are, in fact, like letters in a four-letter alphabet. From these four letters an almost infinite number of words can be constructed which, though meaningless in our own language, may convey meaningful information in the cell.

But we are leaping ahead of the story, for although the chemical components of nucleic acids were known at the time that Avery did his experiments on *Pneumococcus*, their significance was not appreciated. In fact even after Avery's work some biologists were reluctant to accept that nucleic acids were of genetic importance.

THE WATSON-CRICK HYPOTHESIS

In 1953 the whole situation was changed when James Watson and Francis Crick, working together at the Cavendish Laboratory in Cambridge, put forward a possible structure for the DNA molecule. This was the turning point in our concept of the gene. Let us briefly consider their proposition.

Some years before it had been found by direct chemical analysis of DNA that, irrespective of the source of the DNA, the ratio of adenine to thymine, and of guanine to cytosine, is always one. In other words it appears that there are always the same number of adenine and thymine molecules, and of guanine and cytosine molecules. However, no such fixed relationship was found between adenine and cytosine or guanine and thymine. To Watson and Crick this suggested that DNA might consist of two parallel strands held together by pairs of bases; adenine being paired with thymine, and guanine with cytosine.

Meanwhile attempts had been made to work out the structure of the DNA molecule by X-ray diffraction analysis (see p. 77). Despite various technical difficulties, mainly caused by DNA's reluctance to crystallize, some reasonably clear X-ray photographs were obtained, notably by Maurice Wilkins and his colleagues at King's College, London. One such photograph is illustrated in Fig 30.3. Interpreting X-ray diffraction patterns is no easy matter at the best of times and for a large and complex molecule like DNA it is particularly difficult, especially when the patterns are rather diffuse. However, Watson

Fig 30.2 Francis Crick (left) and James Watson in front of King's College Chapel, Cambridge. (*From* The Double Helix *by James Watson, courtesy of Weidenfeld and Nicolson Ltd.*)

Fig 30.3 An X-ray diffraction photograph of DNA. (*Maurice Wilkins, King's College, London*)

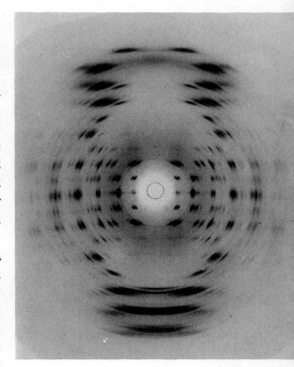

and Crick set to work on the X-ray patterns and came to the conclusion that the DNA molecule consists of two parallel chains twisted to form a **double helix**. From the relative positions of certain spots in the X-ray photographs they concluded that the two chains are cross-linked at regular intervals corresponding to the nucleotides and that there are ten nucleotides for one complete turn of the helix.

Watson and Crick thus envisaged the DNA molecule as a kind of twisted ladder, the two uprights consisting of chains of alternating sugar and phosphate groups, the rungs as pairs of bases sticking inwards towards each other and linked up in a specific relationship: **A** with **T** and **C** with **G** (Fig 30.4).

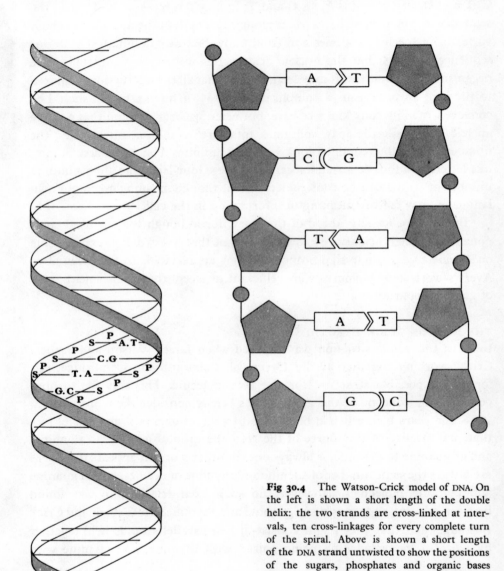

Fig 30.4 The Watson-Crick model of DNA. On the left is shown a short length of the double helix: the two strands are cross-linked at intervals, ten cross-linkages for every complete turn of the spiral. Above is shown a short length of the DNA strand untwisted to show the positions of the sugars, phosphates and organic bases (**A**, **G**, **C** and **T**). Note that **A** pairs with **T** and **C** with **G**.

They then proceeded to build molecular models based on this idea. Accurate models of the four different types of nucleotides with all their atoms and bonds in the right places were fitted together in various ways. By trial and error it was found that the only satisfactory arrangement was to have the two chains running in opposite directions, and the bases linked **A** with **T** and **C** with **G** by hydrogen bonds. When their model of the DNA molecule was complete they worked out what its X-ray diffraction pattern should be and found it to agree very closely with the pattern actually obtained by Wilkins. Fig 30.5 is an atomic model of DNA.

The complementary relationship between the pairs of bases has been confirmed by studying their 3-dimensional shape or stereochemistry. Thymine and cytosine are **pyrimidine** bases consisting of a single hexagonal ring. Adenine and guanine, on the other hand, are **purine** bases consisting of a hexagonal ring joined to a pentagonal ring. The rungs in the DNA ladder can only be formed by linking purine with pyrimidine bases: their relative sizes are such that there would be insufficient room for two purines and too much room for two pyrimidines. The bases are held together by hydrogen bonds and the positions of the hydrogen atoms in relation to the shape of the molecule

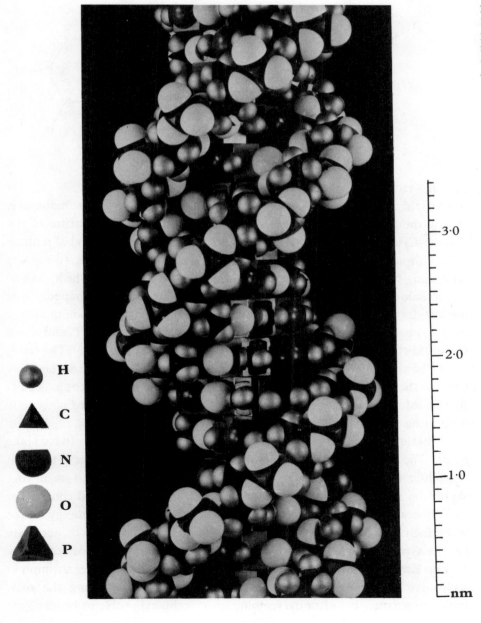

H

C

N

O

P

3·0

2·0

1·0

nm

Fig 30.5 An atomic model of DNA showing the hydrogen, carbon, nitrogen, oxygen and phosphorus atoms in their correct positions. Can you follow the two strands as they coil round each other? (*Maurice Wilkins, King's College, London*)

ensures that **A** can only link with **T**, and **C** with **G** rather like the fitting together of complementary pieces in a jig-saw puzzle. (Fig 30.6)

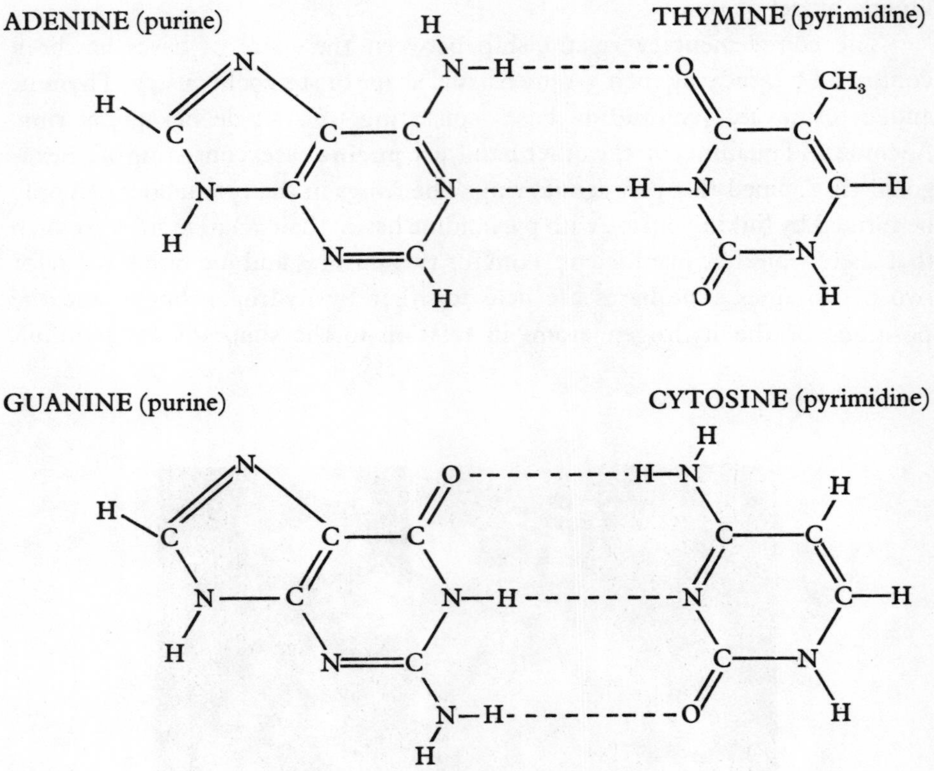

Fig 30.6 Diagram showing the structural formulae of the four organic bases in DNA and how they link up with one another. The rungs in the DNA ladder are formed by linking purine with pyrimidine bases: their relative sizes are such that there would be insufficient room for two purines and too much room for two pyrimidines. The configuration of the molecules is such that two hydrogen bonds are formed between **A** and **T**, three between **C** and **G**. In RNA uracil replaces thymine: **U** is identical with **T** except that the CH$_3$ group is replaced by H. It therefore links with **A** just as thymine does.

ADENINE (purine)   THYMINE (pyrimidine)

GUANINE (purine)   CYTOSINE (pyrimidine)

REPLICATION OF THE DNA MOLECULE

It was said at the outset that an essential property of the genetic material is that it should be able to replicate accurately. An attractive feature of the Watson-Crick model is that it provides us with a possible method of replication. This is illustrated in Fig 30.7. It is not difficult to imagine the two chains separating from each other rather like undoing a zip-fastener. The hydrogen bonds linking the base pairs are weak so breakage could easily happen. Free nucleotides might then come along and take up positions in relation to each of the two chains. The complementary relationship between **A** and **T**, and **C** and **G**, would ensure that the bases aligned themselves in the right way. The result would be two double helices identical with the original one. The sequence of bases in the daughter molecules is exactly the same as in the parent molecule, so the information content of the DNA is unchanged, a feature of genetic replication which we agreed earlier to be absolutely essential.

What evidence is there that accurate replication of DNA does indeed take place? The way biochemists tackle this kind of problem is to see if they can reproduce the process in vitro outside the cell. To synthesize DNA all that should be necessary is to provide the necessary ingredients and conditions, i.e. a liberal sprinkling of free nucleotides of all four types, an energy source (ATP), the required enzyme, and a quantity of intact DNA to act as the template for the synthesis of new DNA. The first successful attempt was carried out by Arthur Kornberg, then at the University of Washington, another triumph in molecular biology. Kornberg's greatest difficulty was isolating the DNA-building enzyme, but after overcoming great technical problems he managed

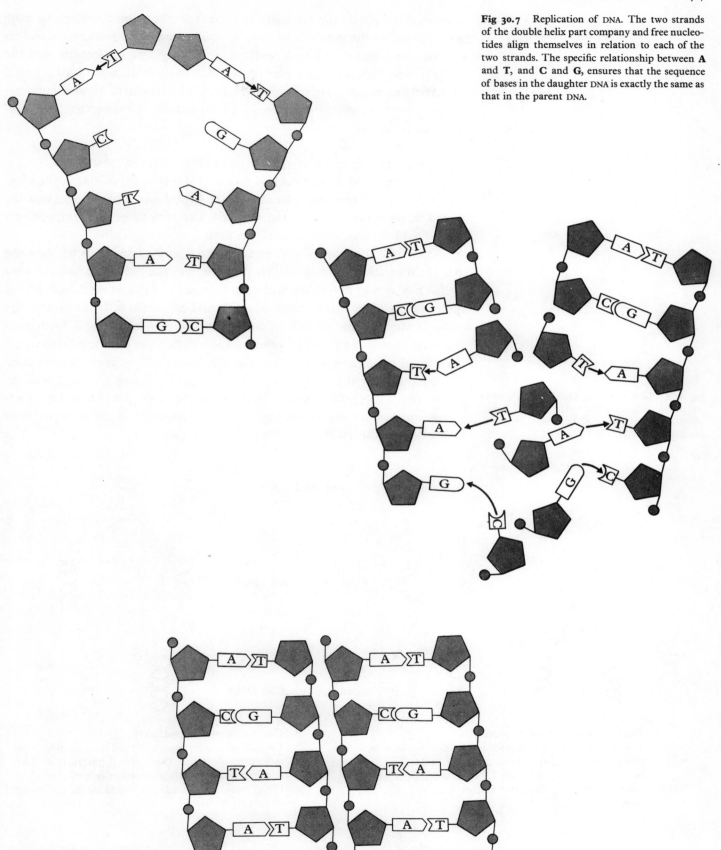

Fig 30.7 Replication of DNA. The two strands of the double helix part company and free nucleotides align themselves in relation to each of the two strands. The specific relationship between **A** and **T**, and **C** and **G**, ensures that the sequence of bases in the daughter DNA is exactly the same as that in the parent DNA.

to extract and purify the necessary enzyme from the colon bacillus. The stage was now set for the synthesis of DNA. Kornberg and his co-workers found that if intact DNA was added to a solution containing free nucleotides and the enzyme, new DNA molecules were formed. On analysis the new DNA was found to have the same relative proportions of adenine-thymine and guanine-cytosine as the original parent DNA, a strong indication that accurate replication had occurred.

CONSERVATIVE VERSUS SEMI-CONSERVATIVE REPLICATION

The zip-fastener idea is a neat and economical way of explaining replication, but it is not the only one. An alternative hypothesis is to suppose that the double helix remains intact and in some way stimulates the synthesis of a second double helix identical with the first.

These two hypotheses were put to the test in the late 1950s by Meselson and Stahl in an experiment which demonstrates the 'scientific method' very well. They argued as follows. If the zip-fastener hypothesis is correct none of the products of successive replications should contain exactly the same molecules as the original double helix; one of the two strands will be derived from the parent DNA, but the other strand will be new. On the other hand, if the second hypothesis is correct one of the double helices in every generation will have exactly the same molecules as the parental one – in both strands. These two hypotheses, summarized in Fig 30.8, are known as the **semi-conservative** and **conservative** hypotheses respectively[1] and they were tested by an elegant experiment involving the colon bacillus.

Fig 30.8 Diagrams comparing the two ways in which DNA might replicate. The original parent DNA is shown as thin lines, the new DNA strands as thick lines.

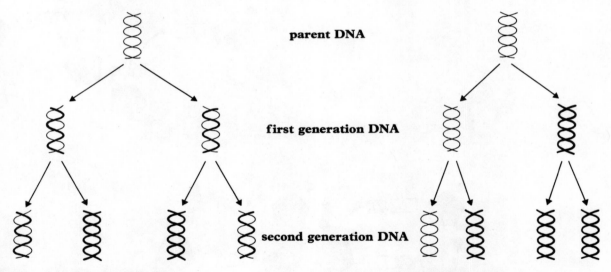

parent DNA

first generation DNA

second generation DNA

A) Semi-conservative hypothesis
The two strands unzip and new nucleotides come into position in relation to each of the strands. The result is that the daughter DNA molecules never contain precisely the same nucleotides as the original parent DNA.

B) Conservative hypothesis
Instead of unzipping, an entire new double helix is formed alongside the original parent one. The result is that in each generation one of the DNA molecules is exactly the same as the original parent molecule.

Cells of the colon bacillus (*E. coli*) were grown for many generations on a medium in which normal nitrogen ^{14}N was replaced with the heavy isotope ^{15}N. This was continued until it was certain that heavy nitrogen had been incorporated into all the DNA, after which the heavy nitrogen was replaced

[1]Why do you think these terms are used for the two hypotheses?

by normal nitrogen. Samples of bacteria were then taken at intervals, and the relative amounts of the two types of nitrogen were estimated by a technique which depends on the fact that compounds containing ^{15}N are very slightly heavier than those containing ^{14}N. The light and heavy DNA were separated by means of an extremely high speed centrifuge, and the DNA was then detected by its absorption of ultraviolet light.

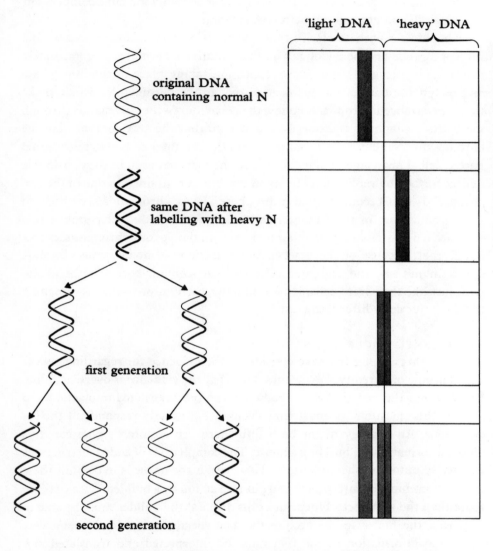

'light' DNA   'heavy' DNA

original DNA containing normal N

same DNA after labelling with heavy N

first generation

second generation

Fig 30.9 The Meselson-Stahl experiment. On the right are shown ultraviolet absorption diagrams of bacterial DNA after treatment with the heavy isotope of nitrogen ^{15}N. A band to the left of the centre line indicates normal 'light' DNA; a band to the right indicates 'heavy' DNA, i.e. DNA containing ^{15}N; a band situated right on the centre line indicates equal amounts of 'light' and 'heavy' DNA. On the left is the interpretation in terms of DNA replication: the 'heavy' DNA is shown in thick lines, the 'light' DNA in thin lines. Note that the interpretation, the only one that satisfactorily explains the absorption diagrams, involves postulating semi-conservative replication. What would the absorption diagrams look like if replication were conservative?

The results, shown in Fig 30.9, give unequivocal support to the semi-conservative (zip-fastener) hypothesis. In the first generation after being switched back to normal nitrogen, the DNA was found to have a density midway between what it would have if it contained only ^{14}N or only ^{15}N: in other words it contained equal amounts of each. In the second generation two sorts of DNA were detected: half was identical with that obtained in the first generation; the other half was light, i.e. it contained no heavy nitrogen at all. These results fit in perfectly with what we would expect from the semi-conservative hypothesis.

DNA REPLICATION AND MITOSIS
We know from special staining techniques and isotope labelling that in all but the simplest organisms DNA is confined to the chromosomes. In fact

a chromosome can be regarded as an enormously long strand of DNA coiled on itself. From the now well established fact that the amount of DNA in the nucleus doubles during interphase, it is thought that replication takes place before mitosis, i.e. before the chromatids make their appearance. Each chromatid therefore contains a double helix derived from replication of the DNA belonging to the original chromosome. By separating the chromatids, mitosis ensures that the products of cell division contain exactly the same complement of DNA as the parent cell did before replication.

Should you wish to clarify this, construct a series of diagrams showing the appearance and behaviour of a representative length of DNA immediately before and during mitosis. Show replication taking place during interphase and assign the correct complement of DNA to the chromatids as they make their appearance in prophase. Follow the behaviour of the chromatids through metaphase, anaphase and telophase and note that the DNA content, like the chromosome content, is exactly the same in the daughter cells as in the original parent cell. This, you will agree, is due to the fact that each of the two double helices formed by replication during interphase is contained in one of the two chromatids that become visible in prophase. These chromatids then become the chromosomes of the daughter cells. Now consider the behaviour of two separate lengths of DNA belonging to a pair of homologous chromosomes and notice that, because of the independent behaviour of homologous chromosomes in mitosis, the daughter cells come to contain both lengths of the DNA, just like the parent cells. How would the distribution of the DNA amongst the daughter cells differ in meiosis?

THE ESSENTIAL ROLE OF DNA

So far in this chapter we have considered the evidence for regarding DNA as the molecule of heredity. Few would deny that the evidence is overwhelming. But how are the instructions embodied in the DNA translated into action? To answer this question we must first decide what exactly responds to the instructions. An analogy might be helpful here. An architect's plan contains all the information for building a house. The complete set of coded instructions fits neatly onto one sheet of paper. However, if the house is to be built these instructions must be interpreted and put into action by a builder. With nothing more than the architect's blueprint at his disposal the builder must be able to construct the house to the architect's exact specifications. In the same way the coded information in the DNA must be interpreted and translated into action in the cell.

DNA AND PROTEIN SYNTHESIS

There is every reason to believe that the essential link is provided by **enzymes**. In Chapter 6 we saw that everything a cell does — what it breaks down, what it synthesizes, what it develops into — is controlled by enzymes. We can therefore postulate that the rôle of DNA is to tell the cell what enzymes to make.

This is not mere speculation. It was shown as long ago as the 1940s by George Beadle and Edward Tatum that genes control the production of enzymes, at least in lower organisms. Beadle and Tatum, then on the staff at Stanford University in California, were interested in the genetics of the bread mold *Neurospora crassa*. They found that this mold would thrive on a minimal medium containing nothing but salts, sucrose and the vitamin biotin; it appeared that it could synthesize all its own amino acids. They then subjected samples of the mold to radiation treatment and found as a consequence that some of their progeny failed to live successfully on the minimal medium (Fig 30.10). By systematically testing the growth of these mutants on different media, they discovered that they had developed an inability to synthesize one specific amino acid. Moreover, crossing these mutants with the normal 'wild type' revealed this inability to be transmitted in a normal

Fig 30.10 Beadle and Tatums' experiment on the bread mold *Neurospora crassa*, on which the one gene-one enzyme hypothesis is based. Irradiation with X-rays produces mutant strains which are incapable of growing on the minimal medium because they cannot synthesize one of the required amino acids. The particular amino acid is determined by systematically testing the mutant's growth on a series of media each of which is supplemented with one of the amino acids. In the case illustrated here the mutant grows successfully on the minimal medium to which the amino acid **T** has been added, indicating that **T** is the particular amino acid which it cannot make for itself. Crossing this mutant with the normal 'wild type' mold produces spores, half of which are capable of growing on the minimal medium whilst the other half only grow successfully on the minimal medium supplemented with **T**. The defect is therefore transmitted in a Mendelian fashion, suggesting that the enzyme responsible for the synthesis of **T** is controlled by a single gene.

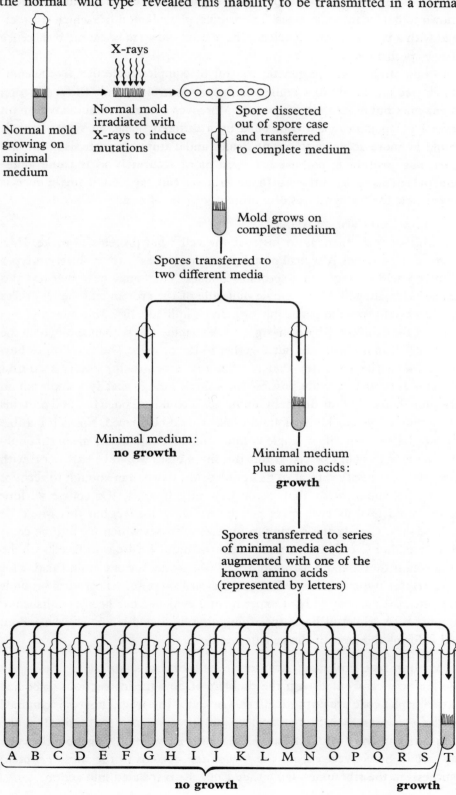

X-rays

Normal mold growing on minimal medium

Normal mold irradiated with X-rays to induce mutations

Spore dissected out of spore case and transferred to complete medium

Mold grows on complete medium

Spores transferred to two different media

Minimal medium: **no growth**

Minimal medium plus amino acids: **growth**

Spores transferred to series of minimal media each augmented with one of the known amino acids (represented by letters)

A B C D E F G H I J K L M N O P Q R S T

no growth   **growth**

Mendelian manner. On the assumption that the inability to synthesize the amino acid was caused by the absence of a specific enzyme, Beadle and Tatum put forward the hypothesis that a single gene controls the production of a single enzyme, the **one gene-one enzyme hypothesis**. Numerous experiments on the biochemistry and genetics of micro-organisms have confirmed the general validity of this idea.

Nor is this concept confined to lower organisms. Various human disorders, known to be due to single genes (for example, phenylketonuria which is associated with a type of mental deficiency) have been shown to be caused by absence of one specific enzyme.

From studies on the genetic control of complex molecules like haemoglobin (see p. 511) we now realize that genes control not only the production of enzymes but other proteins as well. Moreover a single protein may contain several polypeptides each controlled by a different gene. In view of this it would be more accurate to restate Beadle and Tatums' hypothesis as the **one gene—one protein hypothesis**, or even more accurately as the **one gene—one polypeptide hypothesis**. In general we can say that a single gene is responsible for the synthesis of a protein, or part of one.

DNA AS A CODE FOR PROTEINS

The rôle of DNA, then, is to instruct the cell what proteins to make. How might this be done? A typical protein contains at least twenty different types of amino acid arranged in a specific sequence. There may be as many as 500 amino acids altogether in a single molecule of the protein and the way these are arranged gives the particular protein its individuality. Somehow the DNA with its four different bases has got to determine the sequence in which the 20 types of amino acid are put together in the protein. The question is: how can a four-letter code (for that is what DNA amounts to) specify a 20-unit protein? It is obviously not possible for a single base to specify a single amino acid, for then only four different amino acids could be coded for, and proteins containing only four kinds of amino acids would be formed. Nor is it feasible for two bases to specify a single amino acid since this would mean that only 16 amino acids could be coded for. But three bases would be sufficient: with these we can specify a total of 64 amino acids, more than enough to account for the 20 amino acids that occur naturally in cells.[1] Of course all four bases would give us even more possibilities (how many?) but this would be unnecessary. Clearly the minimum number of bases which have to be combined together to code for 20 amino acids is three. We are thus faced with the proposition that a combination of three bases codes for one amino acid. This **base triplet hypothesis** was first put forward on purely theoretical grounds but since then a firm body of experimental evidence has been established to support it. A triplet of bases specifying an amino acid is known as a **codon**, and it forms the basis of the genetic code.

HOW DOES DNA COMMUNICATE WITH THE CYTOPLASM?

We are now confronted with the problem of how the instructions embodied in this triplet code are put into effect. Look at it this way. The DNA is confined to the **chromosomes** in the nucleus but proteins are synthesized in the **ribosomes** in the cytoplasm (see p. 18). This means that somehow the code has got to be conveyed from the DNA in the nucleus to the sites of protein synthesis in the ribosomes. Only then can it be translated into action.

[1] If you have any difficulty in following this you should reason it out for yourself. The number of different combinations that can be made with two out of four bases is $4 \times 4 = 16$; with three : $4 \times 4 \times 4 = 64$.

How might this happen? There are really only two possibilities. One is that the DNA itself, or part of it, moves out of the nucleus into the cytoplasm. The other is that the DNA stays in the nucleus and another molecule, acting as a 'messenger', carries instructions from the DNA to the cytoplasm. The first hypothesis may be discounted on the grounds that DNA cannot be detected in the cytoplasm, which leaves us with the second.

Evidence from a variety of sources suggests that the DNA in the nucleus somehow causes the production of another nucleic acid molecule, RNA. As this is the molecule which conveys the instructions to the cytoplasm it is called **messenger RNA**. Messenger RNA is similar to DNA except that it consists of only one strand instead of two, and it contains the sugar ribose instead of deoxyribose, and the base uracil instead of thymine.

FORMATION OF MESSENGER RNA

How is messenger RNA formed? Before attempting to answer this question let us think what messenger RNA has to do. A typical protein contains about 200 amino acid sub-units. If the triplet hypothesis is correct the instructions required for assembling such a protein will involve at least 600 bases in DNA. In terms of the one gene—one protein concept this would constitute a single gene. At a conservative estimate there are at least a million such genes in a human cell (even a bacterial cell contains about a thousand) so a tiny fraction of the full DNA complement is involved in the production of one particular protein. What all this boils down to is that a given messenger RNA molecule is required to carry instructions from only a very short section of a DNA molecule. The important thing is that the messenger RNA should copy exactly the sequence of bases in this short section of the DNA.

The simplest way of explaining this is to assume that the appropriate part of the DNA molecule serves as a template for the formation of messenger RNA. It is envisaged that the double helix of DNA unzips in the relevant region and that free RNA nucleotides align themselves opposite one of its two strands. Because of the complementary relationship between the bases, cytosine pairs with guanine, and adenine with uracil. Once assembled, the messenger RNA molecule peels off its DNA template and moves out of the nucleus, probably through the pores in the nuclear membrane, into the cytoplasm. Meanwhile the unzipped section of the DNA zips up again. Clearly the sequence of bases in the completed RNA molecule will be identical with one of the two strands of the DNA, the one which did not act as the template.

THE ASSEMBLY OF A PROTEIN

When the messenger RNA gets out into the cytoplasm it attaches itself to the surface of a ribosome where it causes the assembly of a protein. This it does through the intermediacy of yet another kind of nucleic acid called **transfer RNA**. This is a comparatively short molecule, single-stranded like messenger RNA but with the strand double-backed on itself and twisted into a helix. One part of the molecule has three unpaired bases projecting from it and these vary from one molecule to another.

Now it is known that a cell possesses a pool containing as many types of transfer RNA molecules as there are types of amino acids. What appears to happen is that one end of the transfer RNA molecule links up with a specific amino acid and draws it to the messenger RNA on the ribosome. The three unpaired bases at the other end then link up with the appropriate triplet in the messenger RNA molecule. In this way the amino acids are lined up in an order corresponding to the sequence of base triplets in the messenger RNA. As the latter is determined by the sequence of base triplets in the original DNA, it follows that the base sequence in the DNA determines the order in which amino acids line up on the ribosome.

Once aligned, peptide bonds are established between adjacent amino acids

(1) Messenger RNA (——) synthesised alongside DNA
strand in nucleus

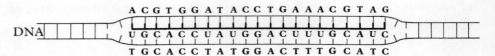

(2) Messenger RNA passes
out of nucleus into cytoplasm

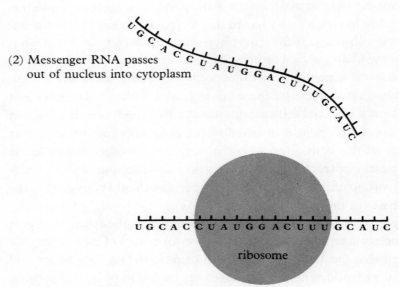

(3) Messenger RNA becomes attached to ribosome

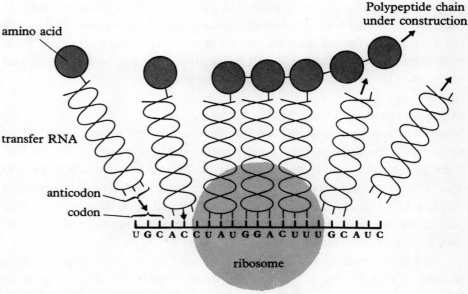

Fig 30.11 Diagram summarizing how DNA in the nucleus controls the assembly of a polypeptide in the cytoplasm. For explanation see text.

(4) Amino acids drawn into position by transfer RNA molecules

and a polypeptide chain is formed. This process of assembly is thought to start at one end of the chain, the end with a free amino group, and to proceed, amino acid by amino acid, to the other end, the end with a free carboxyl group. As the amino acids join up, the completed polypeptide chain peels off from the transfer RNA molecules. Their job complete, the latter detach themselves from the messenger RNA and return to the transfer RNA pool in the cytoplasm, from which presumably they can be drawn again when required. The entire sequence of events is summarized in Fig 30.11.

THE ROLE OF THE RIBOSOMES

Clearly one function of the ribosomes is to provide a suitable surface for the attachment of messenger RNA and the assembly of protein. But there is more to it than this. It is now known that ribosomes occur in groups, or chains, called **polyribosomes** or **polysomes**. In the electron microscope a polysome is seen to consist of from 5 to 50 individual ribosomes: the exact number varies from one type of cell to another. Special staining techniques show these to be connected by a thin thread 1.0 to 1.5 nm in diameter, which is about the thickness of a single strand of RNA. That this strand really is RNA is suggested by the fact that it is readily dissolved by ribonuclease, an enzyme which splits RNA into its constituent nucleotides.

How does a polysome function in protein synthesis? The available evidence suggests that a ribosome attaches itself to one end of a messenger RNA strand and then progresses towards the other end (Fig 30.12). As the ribosome passes a triplet of bases the appropriate transfer RNA molecule takes up position, bringing its amino acid with it. The ribosome then moves on to the next section of the messenger RNA strand and another amino acid is drawn into position, and so on. As the ribosome moves along the messenger RNA, more and more amino acids are added to the growing polypeptide chain. Meanwhile other ribosomes follow suit so that several ribosomes may move along the messenger RNA strand simultaneously, each synthesizing a polypeptide chain as it does so.

Fig 30.12 How a polyribosome works. It is envisaged that a series of ribosomes move along the messenger RNA strand in convoy, each synthesising a polypeptide chain as it does so. Transfer RNA molecules draw the appropriate amino acids into position as the ribosome passes each triplet of bases in the messenger RNA. When the ribosome reaches the end of the RNA strand it releases its polypeptide chain and returns to the beginning. (*based on Rich*)

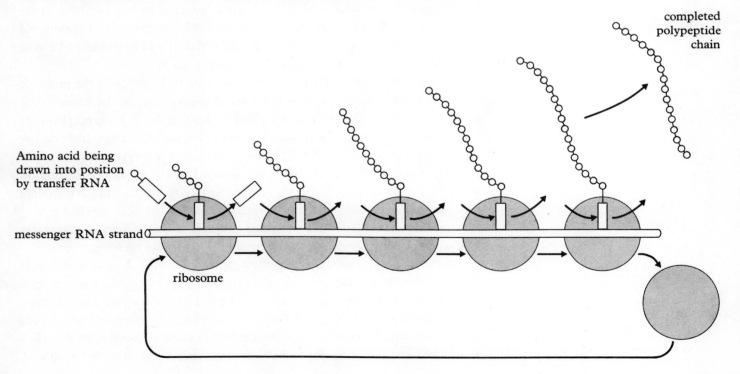

completed polypeptide chain

Amino acid being drawn into position by transfer RNA

messenger RNA strand

ribosome

This enables a large number of polypeptides to be assembled on a single messenger RNA strand in a comparatively short period. It has been estimated that in red blood cells, for example, the time required for a single ribosome to travel the full length of a messenger RNA strand and produce a completed polypeptide chain is one minute. In bacterial cells the time may be much less, perhaps as little as 10 seconds. On reaching the end of the messenger RNA strand the ribosome drops off and liberates its polypeptide chain.

It should be emphasized that no one has ever watched this process actually taking place. The evidence, much of it indirect, is based on very sophisticated experiments involving, for example, separating the individual ribosomes of a polysome and then estimating the amount of polypeptide associated with each one. You should consult recent articles in scientific magazines for further details.

Ribosomes have been likened to computer-controlled machine tools. Each individual ribosome may be regarded as a single machine tool for assembling a polypeptide chain from amino acid building blocks. Just as the product of a programmed machine tool depends on the instructions contained in the computer tape, so the particular polypeptide turned out by a ribosome is determined by the instructions encoded in the messenger RNA strand. Messenger RNA is therefore analogous to computer tape. In a factory each machine tool has its own tape but in the cell a group of identical ribosomes all share the same messenger RNA, thereby increasing efficiency and achieving maximum economy of function within the cell.

EVIDENCE FOR DNA'S CONTROL OF PROTEIN SYNTHESIS

In this brief account of protein synthesis we have glossed over many of the difficulties. How much solid evidence is there to support it? In answer to this let it be said at once that the evidence is certainly not complete. On the other hand it would be quite wrong to dismiss it as fanciful theorizing. The evidence supporting it, some of it based on experiments of great elegance and ingenuity, gets weightier every day, and so far no data has come to hand which contradicts it. In a book of this sort it would be impossible to review all the evidence in detail, but several pointers will be mentioned if only to give some indication of the kind of approach which has been made.

For example, the synthesis of RNA from free nucleotides in the presence of DNA has been achieved, and the relative proportions of the bases in the synthesized RNA was found to correspond to those of the DNA. Direct chemical analysis of cell components, together with the use of isotope tracers, has substantially proved the rôle of the ribosomes in protein synthesis. Transfer RNA has been positively identified and shown to have an affinity for amino acids. More than ten different types of naturally occurring transfer RNA are known to occur in cells and their primary structure has been worked out.

Nothing we know of its structure conflicts with the idea that transfer RNA serves as an intermediary between amino acids and messenger RNA.

If this evidence seems rather negative the following is more positive. It has been shown that the position taken up by an amino acid in a polypeptide chain is determined not by the amino acid but by the transfer RNA molecule which carries it. This has been shown by isolating an amino acid attached to its transfer RNA before the latter had attached itself to messenger RNA. The amino acid was then transformed by catalytic action into another amino acid.

It was found that the latter became incorporated into the polypeptide as if it were the original one.

Thanks to the discovery of the necessary enzyme it is now possible to make synthetic RNA in the laboratory. Using synthetic RNA it has proved possible to induce the formation of unnatural polypeptides. For example, if a synthetic RNA consisting of nothing but uracil nucleotides (called poly-U) is added to a cell-free preparation of ribosomes, it directs the synthesis of a polypeptide chain consisting of nothing but the one amino acid phenylalanine. Similar experiments with other synthetic RNA, for example poly-A and poly-C, has made it possible to say exactly which triplets specify particular amino acids.

TRANSCRIPTION AND TRANSLATION OF THE GENETIC CODE

The control of protein synthesis in a cell can be looked upon as a process in which a coded message is first **transcribed** and then **translated**. Specifically, the coded message contained in DNA is transcribed into messenger RNA, and the coded message embodied in the RNA is then translated by the ribosomes into protein structure. The nucleotide triplets in the DNA and RNA can be regarded as code words. Knowing the amino acids that correspond to each code word, it is possible to construct a **genetic dictionary** showing the relationship between the triplets in DNA, messenger RNA and each amino acid (Table 30.1).

Code words for all 20 amino acids have now been worked out, but the order of nucleotides in each triplet is not yet known. It has been found that two or more triplets, differing from each other in one of their three nucleotides, may code for the same amino acid. It seems that generally only two of the three nucleotides is required to specify a particular amino acid. So the code contains more potential information than is actually used by the cell: to use the cybernetic term, the code is **degenerate.**

Some triplets do not code for any known amino acid: they are called **nonsense triplets**. But although they do not code for amino acids, some of them may be important in ensuring that the construction of a polypeptide chain starts, and ends, at a fixed point. In other words, they tell the ribosomes where to start, and stop, reading the code.

Having described the mode of operation of the genetic code, we turn now to a situation where we can see the controlling influence of DNA in action.

DNA'S CONTROL OVER THE CELL

DNA exerts a potent influence over the biochemical processes which take place in cells. Much of the evidence for this has come from the study of viruses that attack bacteria. These are known as **bacterial viruses, bacteriophages** or just **phages**. (The word bacteriophage is derived from the Greek and means 'eater of bacteria'). One type of bacteria readily infected by phage is the colon bacillus *Escherichia coli* which lives in the mammalian large intestine and can be quite easily cultured in the laboratory.

Partly because of their possibilities as agents for destroying disease-causing bacteria, tremendous interest has been centred on the structure and reproduction of bacterial viruses. The structure of one particular strain is shown in Fig 30.13. Chemical analysis and electron microscopy show it to consist of DNA surrounded by a protein coat. A very long thread of DNA is packed into a **head** from which a short **tail** projects. Though somewhat larger than the majority of viruses they are nevertheless exceedingly small, the head having a total width of approximately 60 nm.

DNA triplet	transcription	m RNA triplet	translation	Amino acid
CGA	→	GCU	→	Alanine
CGG	→	GCC		
GCA	→	CGU	→	Arginine
GCG	→	CGC		
TTA	→	AAU	→	Asparagine
TTG	→	AAC		
CTA	→	GAU	→	Aspartic acid
CTG	→	GAC		
ACA	→	UGU	→	Cysteine
ACG	→	UGC		
CTT	→	GAA	→	Glutamic acid
CTC	→	GAG		
GTT	→	CAA	→	Glutamine
GTC	→	CAG		
CCA	→	GGU	→	Glycine
CCT	→	GGA		
GTA	→	CAU	→	Histidine
GTG	→	CAC		
TAA	→	AUU	→	Isoleucine
TAG	→	AUC		
AAT	→	UUA	→	Leucine
AAC	→	UUG		
TTT	→	AAA	→	Lysine
TTC	→	AAG		
TAT	→	AUA	→	Metathione
TAC	→	AUG		
AAA	→	UUU	→	Phenylalanine
AAG	→	UUC		
GGA	→	CCU	→	Proline
GGG	→	CCC		
AGA	→	UCU	→	Serine
AGG	→	UCC		
TGG	→	ACC	→	Threonine
TGT	→	ACA		
ACT	→	UGA	→	Tryptophan
ACC	→	UGG		
ATA	→	UAU	→	Tyrosine
ATG	→	UAC		
CAA	→	GUU	→	Valine
CAG	→	GUC		

Table 30.1 Table showing the relationship between DNA, messenger RNA (mRNA) and amino acids in protein synthesis. Each DNA triplet acts as a template for the formation of the corresponding mRNA triplet. By this means the information in DNA is transcribed to mRNA. Each mRNA triplet codes for a particular amino acid. In this way the information embodied in the mRNA is translated into protein structure. RNA triplets corresponding to all 20 amino acids are now known, as is shown in the genetic dictionary above. Not all the triplets corresponding to each amino acid are given here: in some cases as many as four different triplets code for the same amino acid.

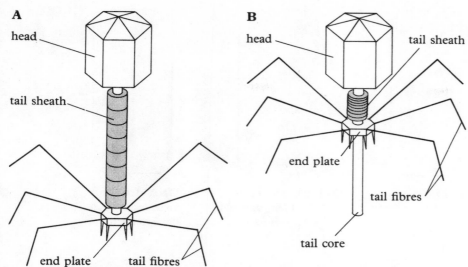

A

head

tail sheath

end plate   tail fibres

B

head

tail sheath

end plate

tail fibres

tail core

Fig 30.13 Structure of bacteriophage (A) before discharge, (B) after discharge. The head, a bipyramidal hexagonal prism, contains a long coiled thread of DNA. The tail consists of a hollow tube (core) surrounded by a helical sheath. Attached to the distal end of the sheath is a hexagonal end plate from which six tail fibres project. On discharge of the DNA thread, the tail sheath contracts, thrusting the inner tube into the body of the bacterium. Seven different strains of phage, known as T1 to T7, all attack the colon bacillus. The one illustrated here is T2. A1 and B1 show the type of electron micrographs on which the diagrams are based. (*Electron micrographs* A1 *and* B1: *R. W. Horne, John Innes Institute*)

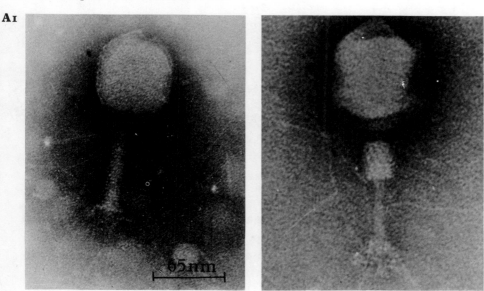

A1

65 nm

B1

When a phage attacks a bacterium it adheres to the bacterial surface by its tail (Fig 30.14). By a process which resembles the action of a hypodermic syringe, the DNA thread is injected into the bacterium where it proceeds to replicate prolifically. Under the influence of the viral DNA, new virus heads and tails are manufactured and assembled within the bacterial cell. After about 30 minutes the bacterium bursts open, releasing some 300 viruses which then repeat the process in other bacterial cells. The process can be followed in electron micrographs, examples of which are shown in Fig 30.15.

That the viral DNA alone is responsible for directing operations in the bacterial cell was demonstrated in 1952 by Alfred Hershey and Martha Chase in America. In order to follow the process in detail they labelled the DNA fraction of the phage with one radioactive isotope and the protein fraction with another. This is made possible by the fact that DNA contains phosphorus but not sulphur, whereas protein contains sulphur but not phosphorus. The phage DNA can therefore be labelled with radioactive ^{32}P, and the protein with ^{35}S. In one set of experiments, viruses were grown in a bacterial medium containing labelled phosphates. The suspension of infected bacteria was then shaken up in a Waring blender, a device similar to that used for stirring milk shakes. This treatment had the effect of stripping most of the virus coats from the

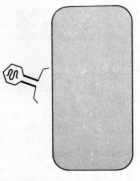

(1) Virus approaches bacterial cell

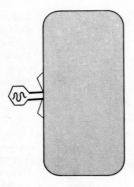

(2) Virus adheres to surface of bacterium

(3) Inner core of tail thrust through wall of bacterium

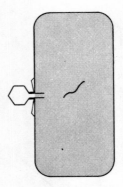

(4) DNA strand of virus injected into bacterial cell

Fig 30.14 Life cycle of a bacterial virus. Only the viral DNA enters the bacterial cell, the protein coat remaining outside. The viral DNA contains all the necessary instructions for directing the synthesis of new viruses within the bacterium. Further explanation in text.

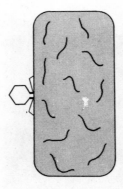

(5) Virus DNA replicates inside bacterial cell

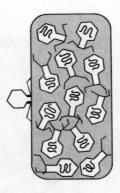

(6) New viruses assembled

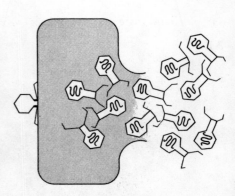

(7) New viruses released from bacterium

surface of the bacteria. The virus coats and infected bacteria were then isolated from one another by means of a high speed centrifuge. On analysis the infected bacteria, thrown down to the bottom of the centrifuge tubes, were found to contain a considerable amount of ^{32}P whereas the supernatant fluid, consisting of viral coats, contained very little.

In a series of complementary experiments, phage was grown in a medium containing labelled sulphates. After separation, most of the ^{35}S was found to be confined to the virus coats and very little was present in the infected bacterial cells.

The conclusion to be drawn from these results is that on infection only viral DNA enters the bacterial cell, the protein coats being left behind. So the DNA alone must be responsible for directing the synthesis of new viruses within the bacterial cell. We now know that a very small quantity of protein is injected into the bacterium along with the DNA, and this consists of the enzymes required to start the process going. The viral DNA takes over the entire metabolic machinery of the bacterium, suppressing its normal control processes and causing it to manufacture new protein coats and nucleic acid threads identical with those of the invading phage. This monstrous process has been likened to the occupation of a country by a foreign power, the latter reorganizing the nation's entire way of life to its own ends.

Although our knowledge of virus reproduction is based mainly on bacteriophage, other viruses are believed to behave similarly, and it is therefore easy to see why they are always associated with disease. Viruses which actively attack and proliferate in cells are described as **virulent**. Sometimes, however,

A

B

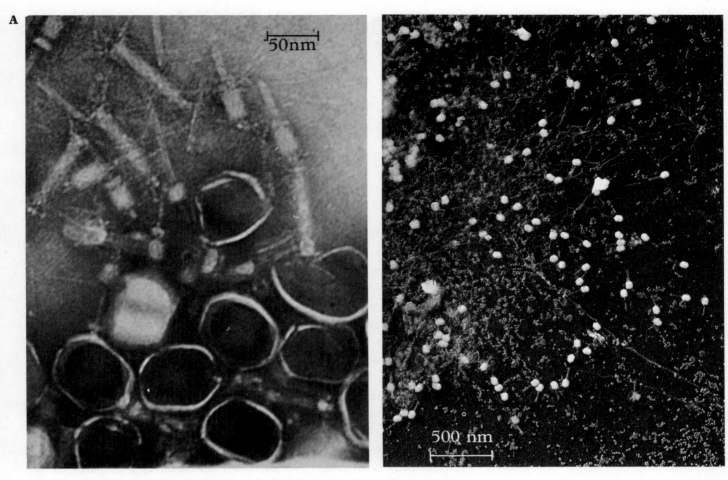

50nm

500 nm

a virus may infect a cell and then remain in a quiescent state for an indefinite period of time. Instead of replicating, the viral DNA joins the cell's own chromosome complement, exerting no influence over the cell and replicating only when the host's own DNA replicates. In this **temperate** state the virus is known as a **provirus**. We know that the host cell does not destroy the viral DNA because from time to time it loses its benign nature and gives rise to a crop of new viruses which burst out of the cell.

One theory of cancer is that cells may harbour foreign genetic material, possibly derived from viruses or proviruses, which exists in cells in temperate form and then suddenly becomes virulent. It is envisaged that in this case the genetic material does not manufacture new viruses but instead causes the host cell to become malignant, allowing it to escape the body's normal suppressive influences over cell division so that it proliferates into a tumour or growth. What this suggests is that although DNA may exert a powerful influence in a cell it can be suppressed or inactivated, at least temporarily. This is very important because if all the DNA present in a cell was active all the time chaos might result. We shall come back to this in the next chapter.

From the point of view of molecular genetics phage provides us with a marvellous tool for studying the action of DNA. Its rapid and prolific reproduction, together with its sensitivity to various mutagenic agents, have made it possible to study the precise effects exerted by different sections of its DNA thread. The results of this research, too detailed to go into here, have gone a long way towards confirming the Watson-Crick hypothesis and establishing the mode of action of DNA.

Fig 30.15 Electron micrographs of bacteriophage. A) shows disrupted T2 viruses: notice the separate heads, tails and fibres. B) shows the release of newly assembled viruses following rupture of the bacterial cell: notice the phage particles lying amongst the remains of the bacterial cytoplasm. (A: *R. W. Horne, John Innes Institute;* B: *E. Kellenberger, University of Geneva*)

31 Genes and Development

In following the development of a complex organism from the fertilized egg to the adult one is struck by two things. Firstly, development involves not only an increase in size, but also a progressive addition of visible complexity as cell differentiation proceeds in the embryo. Secondly, it involves a highly ordered sequence of events in which the various cells, interacting with one another, differentiate at just the right time and in precisely the right place within the embryo. For this to be so there must be an elaborate mechanism by which development is controlled both in space and time. To elucidate the nature of this mechanism is one of the central problems of developmental biology at the present time, and in this chapter we can do no more than touch on some of the basic issues involved.

THE IMPORTANCE OF THE NUCLEUS

If a fertilized egg is deprived of its nucleus it fails to develop properly; in fact the enucleated egg usually dies unless the nucleus is returned within a short period of time. This in itself shows that the nucleus plays an important part in directing the development of the fertilized egg into an embryo.

The nucleus is, of course, required for cell division, but it also determines the type of structure which the cell eventually develops into. This has been demonstrated by the German biologist J. Hämmerling, using the marine alga *Acetabularia*. The great advantage of this organism is that although it is a single cell it is unusually large. The body consists of a 'head', a stalk about 40 mm in length, and a base of root-like rhizoids. The nucleus is situated at the base of the stalk. In *A. mediterranea* the head is cup-shaped, but in another closely related species, *A. crenulata*, it consists of a circlet of finger-like processes.

Now let us examine Hämmerling's experiments and draw such conclusions from them as we can. The experiments are summarized in Fig 31.1. In one of his early experiments he cut through the stalk of *A. mediterranea* so as to separate the head from the base (Fig 31.1, experiment 1). He found that the top half (without a nucleus) died, but the bottom half (with the nucleus) regenerated a new cup-shaped head. It looked as if some kind of chemical substance might be passing from the nucleus into the cytoplasm of the stalk stimulating the formation of a new head.

To test this latter hypothesis Hämmerling cut across the stalk in two places, just below the head and just above the base (Fig 31.1, experiment 2). The isolated stalk was then returned to sea water and after a time it regenerated a new head. At first sight this might appear to contradict the notion that

A) Experiment 1

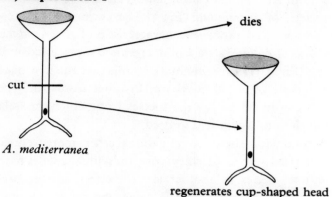

dies

cut

A. mediterranea

regenerates cup-shaped head

Fig 31.1 Hämmerling's experiments with the unicellular green alga *Acetabularia*. Further explanation on pp 502–4.

B) Experiment 2

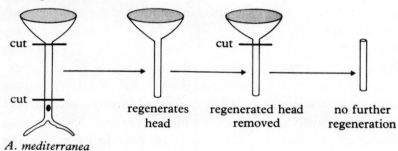

cut

cut

cut

A. mediterranea

regenerates head

regenerated head removed

no further regeneration

C) Experiment 3

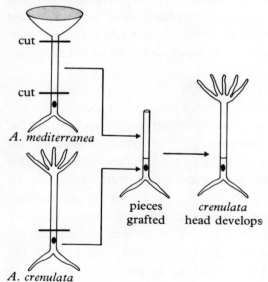

cut

cut

A. mediterranea

A. crenulata

pieces grafted

crenulata head develops

D) Experiment 4

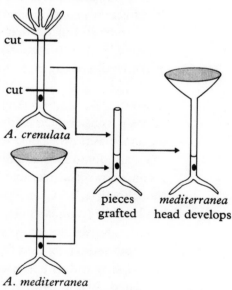

cut

cut

A. crenulata

A. mediterranea

pieces grafted

mediterranea head develops

the nucleus is required for development of a head, but consider the next step in the experiment. After regeneration was complete the new head was removed. This time it was found that the isolated stalk failed to regenerate another head. How can we interpret this result? One interpretation is to assume that the supposed chemical substance produced by the nucleus is present in the stalk. There appears to be sufficient for the formation of one new head, but insufficient for the formation of further heads.

By further ingenious experiments Hämmerling was able to show that the chemical substance produced by the nucleus also determines the type of head developed. A stalk of *A. mediterranea* was cut out and grafted onto an isolated base, containing the nucleus, of *A. crenulata* (Fig 31.1, experiment 3). A new head was formed with finger-like processes characteristic of *A. crenulata*. The reverse experiment was also done: a stalk of *A. crenulata* was grafted onto a base of *A. mediterranea* (Fig 31.1, experiment 4). In this case the new head was cup-shaped, characteristic of *A. mediterranea*. In other words the type of head that develops is determined not by the species from which the stalk is obtained but by the nucleus to which it is grafted.

Since Hämmerling conducted these investigations other more refined experiments have been carried out on *Acetabularia*. Individual nuclei have been transplanted from one species to another and the effect on regeneration noted. All these experiments point to the same general conclusion, namely that the nucleus plays a central rôle in initiating and guiding morphogenesis. In view of modern genetical theory this is not surprising for the nucleus contains the genetic material responsible for laying down the characteristics of the organism.

THE PROBLEM OF MULTICELLULAR ORGANISMS

Acetabularia provides us with a pleasingly simple case of development: a single cell, one nucleus and a comparatively uncomplicated structure. But how do we explain the development of a complex multicellular organism like those considered in Chapter 26? In particular how do we explain the fact that the various cells of the embryo differentiate into distinct types of cell despite the fact that they are all derivatives of one original cell, the fertilized egg? What makes this even more remarkable is the fact that the embryo is effectively sealed off from the outside world. Only respiratory gases pass across the membranes surrounding it and we are forced to conclude that everything required for the formation of a differentiated multicellular embryo is present in the fertilized egg right from the start.

THE ROLE OF THE CYTOPLASM

There is considerable evidence that, however important the nucleus may be, the cytoplasm also plays a significant part in the process of differentiation. One has only to look at a mature egg (see Fig 24.3) to appreciate that the cytoplasm shows considerable differentiation even before fertilization occurs. Mitochondria, RNA, and pigment granules are found towards the metabolically active animal pole, whereas the heavier yolk granules are concentrated towards the more inert vegetal pole. This polarity determines the longitudinal axis of the future embryo, its anterior end lying towards the animal pole and its posterior end towards the vegetal pole.

There is also evidence, derived mainly from work on amphibians, that the dorso-ventral axis of the embryo is established as the result of a redistribution of cytoplasmic granules after fertilization. Such redistribution results in further differentiation within the cytoplasm, some regions possessing potentials for development that other regions lack. This is borne out by the fact that eggs from which certain regions of the cytoplasm have been removed give rise to incomplete embryos. However, removal of other regions of the cytoplasm has no apparent effect on subsequent development.

A particularly interesting discovery is that in certain organisms, notably plants and unicells like *Paramecium*, specific characteristics are determined by genes located not in the nucleus but in the cytoplasm. The genes responsible for this **cytoplasmic inheritance** are replicating units, though they are not transmitted in a normal Mendelian manner. Little is known about the mode of action of these cytoplasmic genes but their existence suggests that the rôle of the cytoplasm in determining an organism's characteristics may in some cases involve more than merely carrying out instructions received from the nucleus.

THE FATE OF CELLS IN THE BLASTULA

In chordate embryos the origin of the three germ layers, ectoderm, endoderm and mesoderm, can be traced back to certain formative cells in the blastula. The way this is done is to mark specific cells of the blastula or gastrula with an **intra-vitam dye**, so-called because the dye is applied to the living cells. The movements of the stained cells can be followed during subsequent development, and their fate determined. This kind of study was first carried out by the German embryologist W. Vogt who constructed **fate maps** to illustrate the eventual destiny of the different regions in the wall of the blastula and early gastrula of amphibians (Fig 31.2). These regions are called **presumptive areas** and the cells in each area may be expected, under normal conditions, to give rise to predictable tissues in later stages of development. The question we have to answer is whether the fate of a particular area is fixed and unalterable from the start, or dependent on influences arising within the embryo itself as development proceeds.

TISSUE CULTURE AND GRAFTING

The answer to these questions has come from two different types of experiment: **tissue culture** and **grafting**. In tissue culture a small piece of presumptive tissue, say ectoderm, is isolated from the blastula and the cells kept alive in a medium containing the necessary materials. In amphibians each cell has a sufficient food supply for about three weeks from its enclosed yolk. In this case the medium merely has to provide water, salts and a buffer.

If conditions are satisfactory the cells of the **explant** (as it is called) divide and development occurs in isolation from other tissues. The results of such experiments indicate that in general the majority of presumptive tissues develop along fixed lines even when they are isolated from other tissues. As a broad generalization it can be said that presumptive ectoderm develops into epidermal tissue, presumptive mesoderm into somites etc., presumptive chorda into notochord, and presumptive endoderm into gut tissues. In this last case the part of the gut formed may depend on the exact piece of presumptive endoderm isolated. So it seems that for most tissues their fate is sealed, or **determined** to use the technical term, as far back as the blastula stage. However, there is one outstanding exception to this and that is the **neural plate**, the region of ectoderm destined to become the neural tube. When isolated and grown in tissue culture, presumptive neural plate develops not into nervous tissue but into epidermis. It seems, therefore, that unless it is subjected to some kind of influence from neighbouring tissue, the tendency is for the neural plate to develop into ordinary epidermis.

Which particular tissue in the intact embryo exerts an influence on the neural plate causing it to develop into nervous tissue? This is where grafting experiments come in. By careful surgery it is possible to cut out small pieces

Fig 31.2 Vogt's fate map showing the presumptive areas in a late blastula of the frog, A) in side view; B) in dorsal view. The map is based on experiments in which small areas of the blastula are coloured with intra-vitam dyes which are then followed into the later stages of development. The arrows indicate the future antero-posterior axis of the embryo. The black object in the endoderm marks the position of the blastopore where invagination occurs (see p. 416). *(based on Vogt)*

A

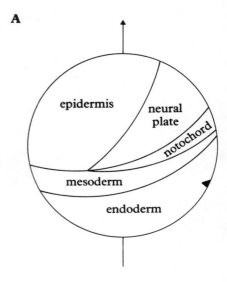

B

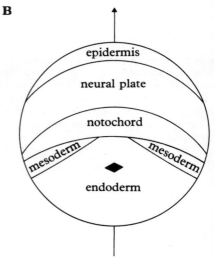

of presumptive tissue from one embryo (the **donor**) and graft them into another (the **host**). This has been done mainly with newts and the results are tolerably clear cut. In the majority of cases the fate of the transplanted tissue is altered so that it develops into tissue appropriate to the host. Thus, for example, if a small piece of presumptive ectoderm is grafted into the mesoderm of a host embryo, the graft develops not into epidermis but into appropriate mesodermal tissue. Somehow the 'normal' development of the graft is suppressed by the surrounding mesoderm, and its fate is changed in accordance with its new position. However, there are two situations in which precisely the reverse takes place, the donor tissue influencing the host's tissue to develop into structures appropriate to the donor. The two tissues capable of doing this are **mesoderm** and prospective **notochord (chorda tissue)**. If either of these are grafted into another tissue, the latter is transformed accordingly. It is in fact chorda tissue which causes the adjacent ectoderm to develop into nervous tissue.

SPEMANN AND MANGOLDS' EXPERIMENT

The powerful influence exerted by chorda and mesoderm on neighbouring tissues was dramatically demonstrated in 1924 by the famous German embryologist Hans Spemann and his co-worker Otto Mangold. Their experiment, one of the landmarks in developmental biology, has profound implications. What they did was to take a small piece of tissue from the dorsal lip of one embryo and transplant it into the ventral side of an early gastrula (Fig 31.3). The result of this operation was that the ectoderm on the ventral side of the

Fig 31.3 Spemann and Mangolds' experiment showing the organizing power of the dorsal lip of an amphibian embryo. A) The operation; B) The result, showing a secondary embryo attached to the ventral side of the primary one; C) Transverse section of the twin embryos: note the two neural tubes.

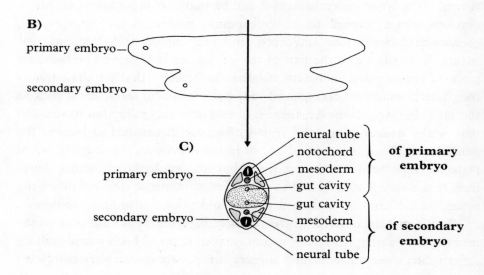

host invaginated to form a second ventrally placed neural tube. On sectioning the embryo, it was found that the other structures such as gut and mesodermal somites, consisting of mixed host and graft tissues, had also been formed on the ventral side of the host. In some cases an almost complete **secondary embryo** was present.

In interpreting the result of Spemann and Mangolds' experiment it is important to appreciate the composition and fate of the dorsal lip. As is explained in Chapter 26, it is made up of cells which are mainly destined to become the notochord and mesodermal somites. As the distinction between these two types of cell is not very clear-cut, it is best to refer to the whole area as **chorda-mesoderm**. In the early gastrula the chorda-mesoderm lies immediately above the dorsal lip of the blastopore where invagination takes place. As gastrulation proceeds this area invaginates to form the roof of the archenteron. Once it has got into this position, and provided that certain other conditions are satisfied, it induces the overlying ectoderm to invaginate and form a neural tube.

ORGANIZERS

The importance of Spemann and Mangold's experiment is that it provides a vivid demonstration of the profound influence which one tissue can have on a neighbouring tissue during development. Such tissues are called **organizers** and they are said to **induce** adjacent cells to develop in a particular way. The process by which one tissue influences another is called **induction**. A tissue which is capable of responding to the stimulus of an organizer is described as **competent**[1].

Powerful though it is, chorda-mesoderm is by no means the only organizer encountered during development. Most tissues are capable of exerting some degree of induction at some stage in their history. Typically the inductive power of an organizer changes as development proceeds, so that a tissue which has a strong influence in, say, the gastrula, may lose its inductive ability in later stages of development.

It is also true that a tissue which is readily induced by an adjacent organizer at one stage of development may become incapable of being induced at a later stage, i.e. it loses its competence. This applies, for example, to the neural plate. The period during which the ectoderm responds to the inductive influence of underlying chorda tissue is quite short. Once the neural tube has been formed, the ectoderm loses its capacity to respond and will only develop into epidermis.

So, as development proceeds, the capacity of different tissues to induce or respond changes. The chorda-mesoderm is the first clearly demonstrable organizer encountered in the course of development, and it is responsible for determining the basic organization of the embryo. For this reason we call it the **primary organizer**. The primary organizer induces the formation of certain structures which in turn act as **secondary organizers** inducing the development of further structures, and so on. There is, as it were, a hierarchy or organizers (primary, secondary, tertiary etc.) each responsible for inducing the formation of certain structures in the embryo.

This hierarchical system would doubtless appear very complex if one were to consider the development of an entire organism with all its interrelated organ systems. However, the basic principles can be seen clearly if

[1] Cells may, of course, influence one another in less drastic ways than is involved in the transformation of one tissue into another. For example, the influence may simply be that the cells attract one another so as to hold the organism together. This is seen in the way isolated sponge cells come together after being separated; or the amoeboid cells of slime molds, after leading a solitary existence, aggregate to form a multicellular reproductive body, as shown in the photograph on the title page of this book (pp. ii–iii).

Fig 31.4 Transverse sections through the head of a vertebrate embryo showing the development of the CNS and eye. The formation of the CNS is shown below, and the development of the eye on the opposite page.

(1)

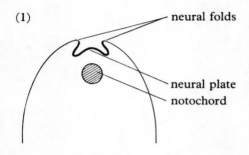

Neural folds grow over neural plate on the dorsal side of the gastrula.

(2)

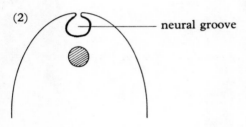

Neural folds meet in the midline thus closing off the neural tube.

(3)

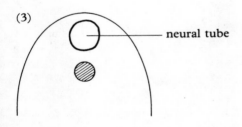

(4)

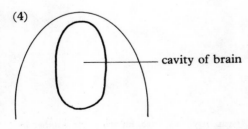

The neural tube expands anteriorly to form the brain.

we confine ourselves to the development of one organ, the eye. The development of the vertebrate eye is well understood, and the system of organizers controlling it has been analysed in some detail. Let us see what happens.

The basic story is outlined in Fig 31.4. As the eye is intimately associated with the brain, the first step in its development is the formation of the **neural tube** which, as we have already seen, is induced by the underlying notochord. The neural tube becomes the central nervous system which expands anteriorly to form the **brain**. The next step is the development of a pouch-like evagination, the so-called **optic vesicle**, on each side of the forebrain. Later the optic vesicles invaginate to become cup-shaped, but in the meantime they induce the adjacent ectoderm to form the **lens rudiment**. The lens then acts as a further organizer, inducing the adjacent epidermis to develop into the **cornea**.

The precise sources of all the inductive influences required to produce a complete eye are still a matter of controversy, but one thing seems certain: it is the optic vesicle which is mainly responsible for the formation of the lens. This has been shown by two experiments. In the first, the optic vesicle on one side of the brain is carefully removed without disturbing the epidermis immediately alongside it. The result is that a lens is not formed on that side of the head though, as would be expected, a normal eye develops on the other intact side.

The second experiment is spectacular if somewhat bizarre. The optic vesicle is removed from one side of the brain and a tiny slit is made in the epidermis further down the body, in the flank region for example. The optic vesicle is then pushed into this slit so that it comes to lie immediately beneath the epidermis. The epidermis itself suffers no undue damage as a result of this procedure; it quickly heals up after the optic vesicle has been implanted. Now normally the embryonic epidermis in this part of the body would give rise to the superficial layer of the skin, but under the influence of the optic vesicle it forms a lens. The end result is the formation of a structurally complete eye in the animal's flank. This eye is described as **atopic**, meaning 'in the wrong place'. Such eyes have no innervation and are therefore non-functional as they cannot 'see'.

THE CHEMICAL NATURE OF ORGANIZERS

In the light of such experiments it would be difficult to deny the validity of the organizer concept. Even quite simple experiments indicate unequivocally that different tissues have a profound effect on each other, and the pattern of development is to a large extent moulded by these influences. This being so, the question arises as to what chemical constituent of the organizer is the effective agent in influencing neighbouring cells. This question has been tackled by extracting the chemical components of an organizer (chordamesoderm for example) and testing their effects on other tissues. The organizer tissue is ground up, centrifuged and the chemical constituents separated and purified. Each constituent is then tested by placing it in contact with a tissue that is known to be competent. The effect of the particular chemical substance on the development of the tissue is then followed. What conclusions can be drawn from the results of these experiments? It would be nice to be able to say that such-and-such a chemical substance causes active cell differentiation, but unfortunately, at the present time, this is simply not possible. The results of the experiments which have been done so far are inconsistent. The anomalies are caused partly by inadequate isolation of the chemical substances, and consistent results are not likely to be forthcoming until the techniques of extraction and purification are improved. The results to date tend to favour **protein** as the most likely candidate though **nucleic acid** is claimed by some to be the active principle in organizers. What may eventually emerge is that induction is brought about by protein, the latter being synthesized in the cytoplasm under the influence of nucleic acid.

THE ROLE OF DNA IN DEVELOPMENT

We come now to the most vexing question of all, namely the part played by the genetic code during development. Briefly the problem is this. The cells of an adult animal, even an embryo, are highly specialized in structure and function: some develop into epithelium, others into muscle, nerve, and so on. And yet all these cells are derived from an original single cell, the fertilized egg. In Chapter 30 we saw that the characteristics of a cell are determined by its genetic code, that is by the sequence of nucleotides in the DNA within its nucleus. Now if we assume, as we did in Chapter 30, that the DNA replicates accurately prior to cell division, then it follows that all the cells of the adult should have identical genetic constitutions. So we are faced with a dilemma. The cells of the adult differ from each other despite the fact that their genetic constitutions are the same.

There are two ways of getting round this difficulty. One is to assume that, although the replication of DNA appears to be accurate, changes do in fact occur in the genetic code as development proceeds. Alternatively we can postulate that the code remains unchanged but that different parts of it are used in different cells at the various stages of development.

In order to decide between these two alternative hypotheses an experiment can be done which is elegant in its simplicity and directness. Let us consider the theory behind the experiment before we get involved with the practical details. If it is true that the genetic code remains unaltered during development, it follows that all the nuclei of the embryo, and indeed the adult, must contain all the necessary genetical information required for producing a complete adult organism. This being so, it follows that if a nucleus is extracted from an embryo or adult and implanted into an egg which has had its own nucleus removed, the egg, if it is capable of developing at all, should give rise to a normal individual. All this assumes, of course, that such a finicky operation can be done without unduly damaging the nucleus, or the egg into which it is implanted.

By now you will have gathered that the ingenuity of experimental embryologists is almost boundless and it will therefore not surprise you to know that this experiment has been done, and indeed is now a commonplace procedure in embryological laboratories. The operation was first performed successfully in 1952 by two American biologists, Robert Briggs of the University of Indiana and Thomas King of the Institute for Cancer Research in Philadelphia. They used frogs for their experiments.

Briefly what they did (Fig 31.5) was to take an embryo at the late blastula stage and immerse it in a fluid which, though harmless to the cells themselves, caused them to dissociate from each other and fall apart. A nucleus is then extracted from one of the cells by means of a micro-pipette. The secret is to

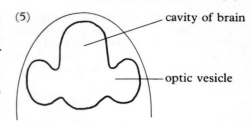

(5)

cavity of brain

optic vesicle

Optic vesicles grow out on each side of the brain.

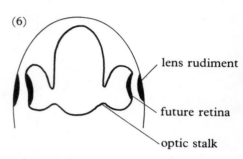

(6)

lens rudiment

future retina

optic stalk

Lens rudiment starts to develop from the epidermis, and the optic vesicle starts to invaginate.

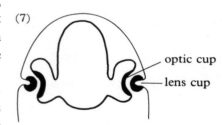

(7)

optic cup

lens cup

Lens rudiment invaginates and the optic vesicle has become cup-shaped.

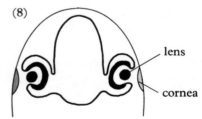

(8)

lens

cornea

Lens complete and partially enveloped by optic cup. Cornea develops from epidermis.

Fig 31.5 Briggs and Kings' method of transplanting nuclei in amphibians. A nucleus is extracted from a blastula cell and injected into an enucleated egg. The latter is found to develop normally in a significant number of cases.

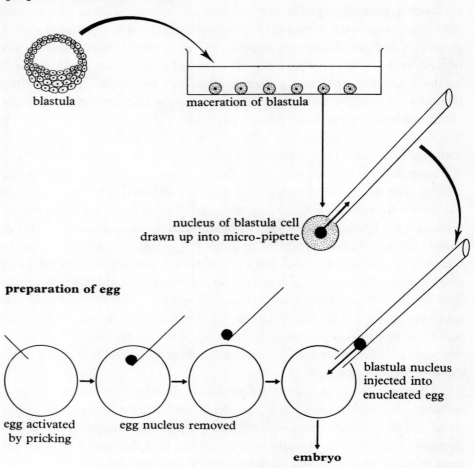

preparation of donor nucleus

blastula

maceration of blastula

nucleus of blastula cell drawn up into micro-pipette

preparation of egg

egg activated by pricking

egg nucleus removed

blastula nucleus injected into enucleated egg

embryo

use a pipette whose opening is just large enough to allow the nucleus to enter but too narrow to admit the rest of the cell. Meanwhile an egg is prepared for receiving the nucleus. It is activated by pricking its surface, or applying a brief electric shock, and its nucleus is flicked out by means of a fine glass needle. This last procedure is made easier by the fact that at this stage the nucleus lies close to the edge of the egg cell. Now it only remains for the donor nucleus to be implanted into the enucleated egg. This is done by carefully injecting it into the cytoplasm of the egg near the animal pole. The egg, with its own nucleus replaced by a nucleus taken from a blastula, is now left to develop.

This experiment has been repeated many times since Briggs and King first did it in the early 1950s. Despite a high mortality rate at first, the results are now reasonably consistent. In general, nuclei obtained from late blastulae are capable of sustaining normal embryos which, on occasions, develop right through to adults. The conclusion is that at least up to the late blastula stage no irreversible changes occur in the genetic information contained within the nucleus.

But might it not be possible that nuclear changes occur after the blastula stage? The most obvious way of testing this possibility is to repeat the experiment with nuclei taken from later stages. Unfortunately this is difficult due to the small size of the cells, but it has been done. The results are variable, depending on the region of the body and the embryological stage from which the nucleus is taken. Also different species of amphibia give different results. In some cases nuclei taken from late gastrulae give normal tadpoles and even adults, and in one of the most spectacular cases nuclei from tumour cells of an adult were found to be capable of producing normal tadpoles. Such results are typical of the African clawed toad, *Xenopus laevis*, which has been much used for this kind of experimentation. On the other hand, nuclei taken from gastrulae of the common frog, *Rana pipiens*, give a high proportion of abnormal embryos. Moreover it has been shown that these abnormalities are stable and heritable, suggesting that they are caused by changes in the genetic material within the nuclei.

What are we to conclude from all this? The fact that a nucleus taken from a differentiated cell can bring about normal development can only be adequately explained by assuming that the genetic material in the nucleus is unchanged at that point in time. The production of abnormal embryos, however, indicates that in some cases the genetic constitution of the nucleus does change, at least during the later stages of development. These two conclusions are not necessarily in conflict: there is good reason to suppose that both are correct, differentiation being brought about partly by changes occurring in the genetic code and partly by different parts of the code being used at different times as development proceeds.

GENE SWITCHING

Let us assume for the moment that different parts of the genetic code come into operation at different times as differentiation takes place. This means that there must be a mechanism which ensures that the right parts of the code operate at the correct time: a mechanism which, in other words, switches the appropriate genes on or off, as and when they are required.

Before considering the evidence for this, let us look at an example to illustrate the basic idea. During human development, two types of **haemoglobin** are formed. In the adult, the red blood corpuscles contain a type of haemoglobin consisting of a haem group surrounded by two α and two β polypeptide chains (see p. 164). In the foetus, however, the central haem group is surrounded by two α polypeptide chains and two γ chains. The differences between the β and γ chains, though slight, are sufficient to confer on these two forms of haemoglobin the markedly different oxygen-carrying powers discussed in Chapter 11. At birth the production of foetal haemoglobin (with the γ chain) ceases and is replaced by adult haemoglobin (with the β chain). Now there is strong evidence from genetical studies that the formation of each of these polypeptide chains is controlled by a single gene. This evidence comes mainly from studying abnormal haemoglobin, such as that which occurs in sickle-cell anaemia. Sickle-cell anaemia is caused by a single gene, and it has been traced to a single incorrect amino acid in the β chain. Other abnormal haemoglobins, attributable to single genes, have been traced to defects in one or other of the polypeptide chains. From these studies a hypothesis can be put forward to explain in genetical terms the replacement of foetal with adult haemoglobin at birth. Briefly the idea is this: the gene which specifies the α chain works all the time, that is during adult as well as embryonic life. The gene which specifies the γ chain works only during embryonic life, and is then switched off. The gene specifying the β chain is inactive during embryonic life, and is switched on at birth. We can also predict the existence of two other genes, one for switching off the γ gene and one for switching on the β gene at birth.

THE JACOB-MONOD HYPOTHESIS OF GENE ACTION

The idea of genes being switched on and off may seem rather fanciful but in recent years spectacular evidence has come to hand supporting it. During the late 1950s François Jacob and Jacques Monod of the Pasteur Institute in Paris carried out a series of brilliant researches on the genetic control of enzyme-synthesis in the bacterium *Escherichia coli*. Briefly what they found was this. *E. coli* will only synthesize enzymes if and when it is appropriate to do so. For example, it produces enzymes for breaking down sugar for the production of energy. One such sugar is lactose. If lactose is present in the nutrient medium in which the bacteria are growing, the bacteria produce an enzyme, β **galactosidase**, to break it down. But if lactose is absent from the medium, this enzyme is not produced. If lactose is added to a medium which previously lacked it, then, and only then, will the bacteria start synthesizing the enzyme.

From these experiments, together with other investigations into the genetics and biochemistry of *E. coli*, Jacob and Monod have put forward a hypothesis to explain how the gene responsible for the production of β galactosidase is regulated. The basic scheme is outlined in Fig 31.6. The section of the DNA strand which codes for the enzyme represents what Jacob and Monod have called the **structural gene**; through the intermediacy of messenger RNA it brings about synthesis of the enzyme. Situated close to the structural gene is another section of the DNA strand called the **operator gene**. At times when the enzyme is needed the operator gene activates the structural gene, causing it to promote synthesis of the enzyme. Now the operator gene, according to Jacob and Monod's theory, comes under the influence of yet another gene, the **regulator gene**, situated further along the DNA chain. At times when the enzyme is not required the regulator gene causes the synthesis of a **repressor substance** which inhibits the operator gene thereby preventing the structural gene from doing its stuff. When the enzyme is required the repressor substance is inhibited, the operator gene becomes functional, the structural gene is activated, and the enzyme is produced.

Fig 31.6 The Jacob–Monod theory of gene action. **R**, regulator gene; **O**, operator gene; **S**, structural gene; + , activation; − , inhibition. Explanation in text.

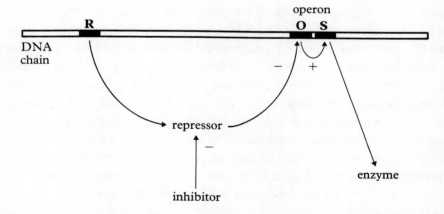

It is an attractive theory and explains many of the known facts of genetics and embryology. But to what extent is it supported by direct experimental evidence? As yet no one has disproved it, nor has anyone put forward a more plausible hypothesis. On the positive side facts keep emerging which strongly support it. For example, a repressor substance has been identified and it has been found that it can bind with the operator gene, thereby preventing synthesis of the enzyme. It has also been demonstrated that lactose itself can inactivate the repressor by combining with it. In this way the sugar ensures its own breakdown by removing the block that would otherwise prevent synthesis of the enzyme. It remains to be seen if the theory will stand up to further rigorous testing. And of course we need to know if it applies to higher organisms as well as to bacteria.

THE RÔLE OF THE ENVIRONMENT

The last section will have made it clear that the environment can influence development by affecting gene action. Environment in that context means the immediate chemical surroundings of the individual cell. We must now briefly consider the extent to which more general factors of the environment, temperature, light and so on, affect development.

A well known case of environmental control is provided by the development of fur colour in the Himalayan rabbit. This rabbit has a white body with black ears, nose, feet and tail (Fig 31.7). At first glance it might be thought that this pattern is under purely genetical control, but a simple experiment shows that this is not the case. If a cold pad is fixed to the rabbit's back during development, black hair develops beneath the pad. We now know that heat prevents the development of the black pigment; only in those parts of the body which are cool enough, i.e. the extremities, will the black pigment develop.

There are many other cases of the environment influencing an organism's development. For example, in plants, chlorophyll will only develop if light is available, flowers if the day length is right and the temperature suitable, and so on. These and other examples are discussed in Chapter 27. Generally environmental influences tend to be cruder than internal ones. They affect the total appearance or otherwise of a particular feature rather than the subtler aspects of differentiation.

CONCLUSION

One of the most interesting facts to emerge from the experiments on gene action in *Escherichia coli* is that lactose itself can turn on the switch leading to synthesis of the enzyme. In other words a chemical molecule which is not itself part of the nuclear material can influence gene action. This fact is of great importance and reflects the belief held by most embryologists and geneticists that development is controlled by a complex interaction between the nucleus, cytoplasm and environment. If the nucleus contains the genetic blueprint responsible for producing a highly differentiated multicellular organism from a fertilized egg, the cytoplasm and environment determine how, where and when this information should be used. Only by postulating such interaction can we explain induction, gene switching and the other facts of development and morphogenesis.

Fig 31.7 Influence of environment on development. In the Himalayan rabbit black fur develops on the ears, feet, nose, and tail due to the lower temperature of these exposed parts of the body. Verified by attaching a cold pad to the side of the body during development: black fur develops underneath pad.

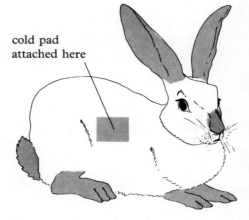

cold pad
attached here

It is generally believed that organisms have evolved in such a way that they are adapted to their environments. Evolution and environment are therefore intimately related, which is why they are discussed together in this part of the book. They are left to the end because an appreciation of their significance depends on an understanding of the principles and concepts discussed in earlier chapters. So in many respects this final part of the book is concerned with integrating the principles outlined earlier, and drawing the threads together.

The basic principles of ecology are presented in Chapter 32, and this is followed in Chapter 33 by an account of the different kinds of association that can be established between organisms of the same or different species, these being an aspect of ecology.

All these ecological situations have arisen as a result of evolution. The way evolution reveals itself in the biology of organisms is outlined in Chapter 34, and the mechanism by which it is believed to have come about is discussed in Chapter 35. The final chapter is devoted to a discussion of some of the more important steps in evolution, from the origin of life to the emergence of man.

Reconstruction of Permian amphibian *Diplo-vertebron*. (*American Museum of Natural History*)

32 The Organism and its Environment

In the course of this book we have constantly referred to the places where organisms live. The word 'place' can be taken to mean many different things, of which geographical location is but one. In this chapter we shall explore some of the meanings of this word.

Studies of this sort come under the general heading of **ecology**, the study of organisms in relation to their environment. In a sense it is artificial to segregate ecology from other aspects of biology. Previous chapters will have made it clear that an organism's structure, physiology, behaviour, and evolution are all inextricably bound up with its environment. In this chapter we shall be relating the themes of other chapters to the general concept of the environment.

FROM BIOSPHERE TO ECOLOGICAL NICHE

Ecology comes from the Greek word *oikos* meaning home, so it literally means studying the 'homes' of organisms. In its broadest sense an organism's 'home' is that part of the earth and its atmosphere inhabitable by living things. This is called the **biosphere**. The biosphere can be divided into a series of **biogeographical regions**, each inhabited by distinctive species of animals and plants. These organisms are able to move freely from place to place within each region, but not from one region to another, this being prevented by various natural barriers. Such biogeographical regions include Eurasia, South America, Africa, and Australia.

Although they may possess a different fauna and flora, two such geographical regions may be virtually identical in their climatic conditions, and this is reflected in similar adaptations shown by their respective inhabitants. These equivalent areas, cutting right across different continents, are lumped together as **biomes**. Each has its own unique set of conditions, and each supports a particular type of flora and fauna. Such areas include the tundra of the Arctic and Antarctic regions; grasslands, desert, and the different kinds of forest (coniferous, temperate, and tropical). The oceans can be regarded as a single biome (the marine biome), though coral reefs are generally regarded as a separate biome. Within each biome only those organisms with the necessary adaptations for surviving the physical conditions are found. Thus the tundra and coniferous forest biomes only have organisms capable of withstanding long periods of extreme cold. Desert organisms, on the other hand, must be able to cope with intense heat and drought. Marine organisms must be able to thrive in salt water, and so on. Needless to say there is no sharp line between adjacent biomes.

Biomes are subdivided into small units called **zones**, each with its own particular set of physical conditions. For example, a forest biome can be divided into ground zone and canopy zone; the marine biome into surface, abyssal and intertidal (littoral) zones; the desert into surface and subterranean zones. Individual organisms may be specially adapted to live in one of these different zones, in some cases providing them with a means of avoiding the extremes of climatic conditions met with in the biome as a whole. For example, many desert animals avoid the heat of the midday sun by burrowing into the subterranean zone.

Within each biome are numerous **habitats**, specific localities each with a particular set of conditions and an appropriately adapted community of organisms. Typical habitats include fresh-water ponds, slow-flowing streams, rock pools, and hedgerows.

All the ecological units mentioned so far are essentially localities of varying sizes. Cutting across all of them is the concept of **environment**. This is a collective term embracing all the conditions in which an organism lives, for example light, temperature, water, and other organisms. We shall return to this later.

Within a particular habitat an organism is generally confined to a restricted situation. For example in a slow-flowing stream, crayfish live in burrows in the bank, flatworms and shrimps under stones, fishes in the open water, and so on. This idea is embodied in the term **ecological niche**. But when we speak of an organism's ecological niche we mean more than just the precise place where it lives. We also include what it does there, i.e. its behaviour, feeding habits, etc. In other words the term ecological niche embraces the organism's entire way of life, its 'profession', so to speak. As such it may include more than one habitat, as is illustrated by amphibians such as frogs and toads which exploit both aquatic and terrestrial habitats. An ecological niche may even span several biomes, as is the case with migratory birds and butterflies. For example, in the course of its seasonal migration the golden plover may cover thousands of miles, taking it through as many as five different biomes (Fig. 32.1).

THE PHYSICAL ENVIRONMENT

An organism's environment, i.e. the conditions in which it lives, may be subdivided into the **physical** and **biotic environments**. Both profoundly affect the distribution of organisms in different habitats.

The physical environment embraces everything that is not associated directly with the presence of other organisms.

(1) Temperature. The narrowness of the temperature range over which biochemical processes can function efficiently means that organisms, wherever they live, must have the necessary physiological or behavioural adaptations to combat or avoid extremes of environmental temperature (Chapter 15).

(2) Water. The extent to which an organism is dependent on an abundant water supply depends on its requirements and its ability to conserve it in adverse conditions. Organisms living in dry habitats generally have good water conservation (see Chapter 14).

(3) Light. This is essential for all green plants and photosynthetic bacteria, and for all animals dependent on the plants. Plants have numerous adaptations for obtaining optimum illumination (see Chapter 10).

Fig 32.1 Migration route of the Atlantic golden plover *Pluvialis dominica*. This bird migrates in flocks from its summer nesting grounds in the Arctic to its winter quarters in Argentina, a distance of over 7,000 miles. Its journey spans more than five major biomes. Notice that its southbound and northbound routes are different.

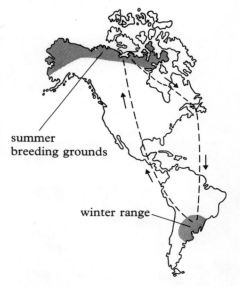

summer
breeding grounds

winter range

Fig 32.2 The effect of the environment on the appearance of an organism is here illustrated by the famous Jeffrey pine on Sentinel Dome in Yosemite National Park, California. The fact that this tree has withstood years of buffetting by screeching winds speaks well of its toughness, but its shape has been much influenced by the environment. (*US National Park Service*)

Fig 32.3 The holdfast of *Laminaria* by which this seaweed or kelp grips onto pieces of rock for firm attachment. (*B. J. W. Heath, Marlborough College*)

(4) **Humidity**. This is important because it can affect the rate at which water evaporates from the surface of an organism, which in turn influences its ability to withstand drought.

(5) **Wind and air currents**. This particularly applies to plants. Only plants with strong root systems and tough stems can live in exposed places where winds are fierce (Fig. 32.2). Wind is also instrumental in the dispersal of spores and seeds.

The above factors are all directly climatic. The following are not necessarily associated with climate as such, though they may be the indirect result of it. Those exclusively connected with soil are known as **edaphic factors**.

(6) **pH**. This influences the distribution of plants in soil and fresh-water ponds. Some plants thrive in acidic conditions, others in neutral or alkaline conditions. Most are highly sensitive to changes in pH.

(7) **Mineral salts and trace elements**. These particularly affect the distribution of plants in the soil. Plants living in soil deficient in a particular element must have special methods of obtaining it. These methods include the harbouring of nitrogen-fixing bacteria and the carnivorous habit, both of which are discussed in previous chapters.

(8) **Water currents,** particularly in rivers and streams. Only organisms capable of stemming or avoiding strong currents can survive. For this reason animals incapable of actively swimming generally live under stones or in burrows and crevices in the bank.

(9) **Salinity.** The importance of this is seen in the sharp distinction between marine and fresh-water species. It also influences the distribution of estuarine animals: these have special physiological or behavioural adaptations for withstanding the daily fluctuations in salinity that accompany the tidal rhythm (see Chapter 14).

(10) **Wave action**. This particularly affects organisms living in the intertidal zone. To survive periodical buffeting by waves, and exposure to air, special adaptations are required. These include the sessile habit of animals like limpets and sea anemones, burrowing by shrimps and sand hoppers, and the firm attachment to rocks and general toughness of sea weeds such as *Fucus* and *Laminaria* (Fig. 32.3).

(11) **Topography**. When looking at the distribution of organisms one finds all the difference between, for example, the centre and edge of a stream, the top-side and under-side of a stone, a north-facing and south-facing wall. This may be explained by differences in illumination, temperature, moisture, etc. Minor topographical differences may be just as important in influencing the distribution of organisms as wide geographical separation.

(12) **Background**. The distribution of organisms whose shape or coloration are such that they are camouflaged when viewed against a particular background is related to the general texture and pattern of the environment. A good example of cryptic coloration is provided by the white and black varieties of the peppered moth which are protected against light and dark coloured tree trunks respectively (see p. 589). There are many other cases of protective camouflage, particularly amongst the insects.

THE BIOTIC ENVIRONMENT

An organism's biotic environment is made up of all the other organisms with which it comes into regular contact. With certain of these organisms it has a special relationship which may profoundly influence its distribution and abundance. Here are some examples.

(1) **Predation.** An organism may feed on, or be fed on by, other organisms. In such cases the distribution of feeder and fed are related. Thus herbivores are only found where there is suitable plant food; carnivorous plants where there are suitable insects etc; predators where there is suitable prey.

An example of the effect that a feeding relationship can have on distribution is provided by the cactus moth *Cactoblastis* and the prickly pear cactus. The caterpillars of this moth feed on the cactus thereby imposing strict limits on its distribution. Such was the case in Australia in the 1920s when *Cactoblastis* was introduced from its native Argentina in order to control the spread of prickly pear which was overrunning vast areas of land in Queensland. Within a comparatively short time the rapid spreading of the cactus was brought under control (Fig 32.4).

A

B   C

(2) **Competition.** Organisms frequently compete with one another for such commodities as food, light, water, shelter, mate or resting site. Competition has resulted in all manner of adaptations, structural and behavioural, which enable individuals to win out in the inevitable 'struggle for existence'. Competition exists both between individuals of the same species (**intraspecific competition**) and between individuals of different species (**interspecific competition**), and the closer the ecological niches of the competing organisms the fiercer is the competition.

(3) An organism may use another as a **habitat**. This specialized aspect of the biotic environment is seen in parasitism and other related associations discussed in Chapter 33.

(4) Certain plants rely on insects and other small animals for **pollination** and/or **dispersal**. Pollination sometimes involves a highly elaborate reciprocal relationship between a certain plant and a specific insect (see p. 400).

(5) In the course of evolution some animals have come to bear a striking and often detailed resemblance to part of a plant, thereby affording them-

Fig 32.4 The extreme influence of the biotic environment is here illustrated by the effect of the cactus moth *Cactoblastis* on the prickly pear cactus. Native to parts of North and South America, the prickly pear was introduced into Australia where it spread rapidly over large areas. Following the discovery of the cactus moth in Argentina, eggs were introduced into Australia. In a comparatively short time the caterpillars, which bore through and feed on the cactus, had destroyed vast numbers of cacti, thus making the land available for farming.   A) *Cactoblastis* caterpillars on the surface of a prickly pear cactus.   B) Dense prickly pear in the Chinchilla area, Queensland, in 1962.   C) The same area three years later after complete destruction of the cacti by *Cactoblastis*. This case illustrates how man can make use of one organism to control another harmful one. (*Australian News and Information Bureau*)

Fig 32.5 Protective resemblance in insects. A) Insect pupa resembling a dead leaf. B) Insect pupa like a thorn. C) Frog with shape and colour of a leaf. (*A and C: M. W. F. Tweedie; B: British Museum – Natural History*)

A

B

insect

selves protection from attack by predators. Certain insects furnish spectacular examples of this, resembling such objects as leaves, twigs, sticks or thorns (Fig 32.5). In such cases the plant species concerned forms a most important part of the insect's biotic environment.

(6) Some animals closely resemble another species that happens to be unpalatable to a predator, a phenomenon known as **mimicry**. The unpalatable species generally possesses distinctive colours or markings (**warning colora-tion**). Predators learn to recognize these signs and avoid attacking this parti-cular species. Armed with similar markings the mimic is also protected from attack.

In some cases the mimic is also unpalatable. At first sight this might appear to be pointless, but it is in fact advantageous to both mimic and mimicked. Can you explain this?

(7) One of the most powerful biotic factors is **man**. This can be illustrated by returning to the prickly pear in Australia (Fig 32.4). Prior to 1840 there were no prickly pear cacti at all in Australia, but in that year a certain Dr Carlyle brought some specimens from America and planted them in his garden. They quickly ran wild and by the turn of the century had spread over more than 10 million acres of land, rendering it unsuitable for agricultural use. By 1910 the cactus was spreading at a rate of a million acres per year. However, the introduction of the cactus moth put a hasty end to this, thus providing an example of **biological control**.

C

The control of the prickly pear cactus in Australia provides a happy example of man's control over nature, but all too often his influence is destructive and harmful. At his most destructive, man completely anihilates habitats, as can be seen in the flooding of valleys to create reservoirs, the pollution of lakes and rivers, and the general spread of industrialization and urban development. But man is also responsible for changing habitats, and creating new ones. Railway embankments and cuttings have long been a botanist's paradise, and nowadays the verges at the sides of newly constructed highways provide a virgin habitat for many species of animals and plants. Even a bombed site provides a habitat for exploitation by any species that can thrive there.

From this brief analysis of the environment it can be seen that organisms do not exist in isolation but are continually influencing each other. This fact is embodied in the word **ecosystem** which we shall now consider.

THE CONCEPT OF THE ECOSYSTEM

An ecosystem is a natural unit composed of living and non-living components whose interactions result in a stable, self-perpetuating system. It is made up of communities of organisms which interact with one another and with non-living constituents of the environment.

It follows that to analyse an ecosystem involves studying natural communities. Such studies are known as **synecology**. The community may exist in a fresh water pond, an oakwood, or a rock pool on the sea shore. Whatever the situation, it is first necessary to identify the organisms living there. This can be done with the aid of systematic keys. It is then necessary to determine the distribution of the different species quantitatively, and to correlate this with the physical and biotic factors of the environment. This is not always easy and generally involves carrying out statistical analysis on the distribution data. Parallel studies on the behaviour of the different organisms show how they interact with one another to form a stable and balanced community.

SUCCESSION

Communities do not remain stable indefinitely. A slight change in the environment may upset the balance and result in a change in the structure of the community. Typically communities evolve from simple beginnings, gradually becoming more complex. Think, for example, of an unexploited rocky terrain. The first organisms to inhabit such an area would be lichens, the only organisms that can subsist on bare rock (see p. 536). The lichens break up the rock surface, enabling dust particles and humus to accumulate in the crevices. This provides a foothold for mosses and, later, ferns and grasses. With the further breakdown of the rock and build-up of humus, small shrubs take root. In the course of time trees move in: conifers at first, particularly those that can thrive on relatively poor soil, and later the more exacting deciduous trees. At each stage in this gradual evolution a **dominant species** can be recognized. This then influences the environment in such a way as to make it suitable for another species. This then replaces the former as the dominant species, and so on. Eventually a more or less stable **climax community** comes into existence: there is no further influx of new species to replace those already established. However, it only requires a change in the environment to alter this state of affairs. A forest fire, a prolonged drought, or the introduction of a new species by man can knock a dominant species off its pedestal and return the community to a more primitive state.

THE CYCLING OF MATTER AND FLOW OF ENERGY

Whatever the situation, an ecosystem consists of three nutritional groups:

(1) **Producers:** autotrophic organisms (mainly green plants) so-called because they are responsible for synthesizing organic food.

(2) **Consumers:** heterotrophic organisms (mainly animals).

(3) **Decomposers:** saprophytic organisms (bacteria and fungi).

Producers, consumers and decomposers are related as shown in Fig 32.6. Organic materials synthesized by the producers are eaten and assimilated by

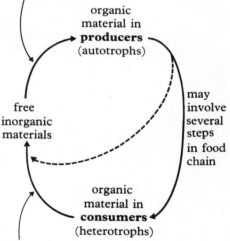

Fig 32.6 Cycle illustrating the interdependence of producers and consumers in an ecosystem. The photosynthetic and chemosynthetic activities of the producers (green plants, etc.) produce organic materials which are fed on by the consumers. The saprophytic activities of the decomposers (bacteria, etc.) frees inorganic materials from the dead bodies of the producers and consumers, thereby ensuring a continual supply of raw materials for the producers. The carbon and nitrogen cycles are detailed applications of this general principle.

the consumers. With the aid of the decomposers, all the organic materials incorporated into the bodies of the consumers are eventually broken down into inorganic materials. These are then rebuilt into organic compounds by the synthetic activities of the producers. So it can be seen that matter circulates in nature: though it constantly changes its form, there is no overall loss or gain. The carbon and nitrogen cycles (pp. 141 and 160 respectively) are detailed applications of this general principle.

Although matter circulates, energy flows in a straight line. The photosynthetic producers convert the radiant energy of sunlight into the energy of chemical bonds in plant carbohydrates. By their respiratory activities, the producers, consumers and decomposers incorporate this energy into ATP, whose subsequent breakdown provides energy for the cells' vital activities. Both in the formation of ATP, and in its subsequent usage, a proportion of the energy is 'lost' as heat, and ultimately it is all converted into heat. The continual trapping of light energy by green plants makes good this loss and maintains the uninterrupted flow of energy in an ecosystem.

FOOD CHAINS AND FOOD WEBS

In the cycle shown in Fig 32.6 the consumers are depicted as feeding directly on the producers. In practice the situation is generally more complicated, the producers being eaten by herbivores (**primary consumers**) which are then eaten by carnivores (**secondary consumers**). Moreover, there may be several carnivores in the series, the first being preyed upon by a second and so on. The resulting nutritional sequence is known as a **food chain** (Fig. 32.7). In general a food chain is a series of organisms existing in any ecosystem through which energy is transferred. Each organism in the series feeds on, and therefore derives energy from, the preceding one. It in turn is consumed

Fig 32.7 A generalized food chain is shown on the right and, below, an indication of the biomass and number of individuals at each level. The drop in biomass and numbers at each level gives the so-called pyramid of biomass and pyramid of numbers respectively.

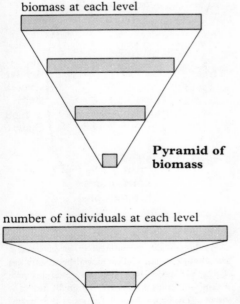

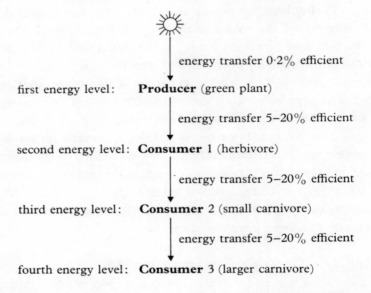

by, and provides energy for, the one following it. Not all food chains are as long as the one illustrated in Fig 32.7. Some are very short, such as grass ⟶ cow ⟶ man, or grass ⟶ deer ⟶ jaguar. There are rarely more than six links in the chain.

The food chain shown in Fig 32.7 is a simple linear series. This means that if the green plant (producer) disappears from the community, the herbivores and carnivores (consumers) will quickly follow suit. Except in communities where there are relatively few interacting organisms (e.g. the desert), such simple food chains are rare. More often the range of plant species is sufficiently great for a herbivore to have several alternative sources of food, and the herbivore in turn may be preyed upon by several different predators. The nutritional relationships between such organisms consitute a **food web**, the basic principle of which is illustrated in Fig 32.8. As a practical example, the food web in a typical fresh water pond is given in Fig 32.9.

Can you work out from your own daily diet, the different food chains in which you yourself participate? As an omnivore, man consumes both plants and animals. When we eat plants we are primary consumers, when we eat the flesh of a herbivore such as a cow or sheep we are secondary consumers, and when we eat the flesh of a carnivore such as a fish we are tertiary consumers. Man is thus the final link in a number of food chains, though on some unfortunate occasions he may suddenly find himself the penultimate link.

Man's rôle in the ecosystem is of great importance when it comes to solving the world food problem. When energy flows through a food chain, only a small proportion of the energy taken up by each link is transferred to

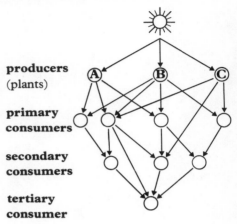

Fig 32.8 A hypothetical food web showing the complex feeding relationships that may exist between different animal and plant species in an ecological community (ecosystem). Note that the dying out of one species does not necessarily affect the well-being of the others. For example if plant A dies the community can still remain stable since the primary consumers that fed on A have an alternative source of food in B and C.

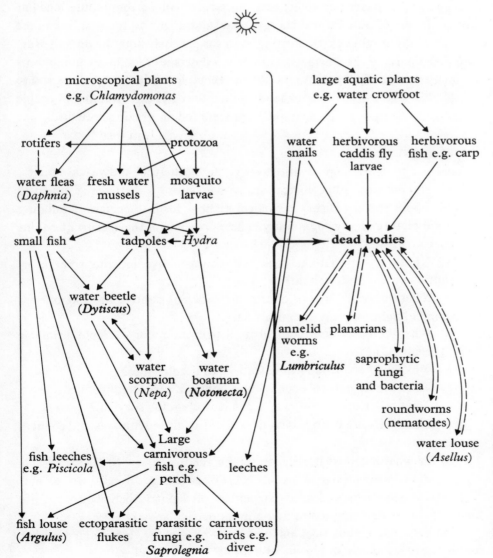

Fig 32.9 Food web in a typical fresh water pond. Microscopical animals and plants living suspended in water constitute plankton which also includes the juvenile stages of larger animals. Most organisms that consume the plankton are filter feeders, e.g. water fleas and mussels. Despite its complexity this ecosystem can be divided into producers, primary, secondary and tertiary consumers and so on just like the simple food chain discussed earlier.

the next step. The reason for this is that at each transfer most of the energy is lost as heat. In fact the efficiency of the energy transformation when an animal eats a plant (or another animal) is generally less than 20 per cent. In other words only 20 per cent of the energy contained in the food material becomes incorporated into chemical bonds in the tissues of the consumer. The rest is lost as heat. This progressive loss of energy at each level of a food chain puts a natural limit on the total weight of living matter that can exist at each level. This is called the **biomass,** and clearly it decreases at each successive level in the chain. So does the total number of individuals: in fact the drop in numbers at each level is even more noticeable than the drop in biomass, because the animals at the end of a food chain tend to be larger than their predecessors. The progressive drop in numbers is shown in Fig 32.7 and is described as a **pyramid of numbers.** There is also, of course, a **pyramid of biomass** corresponding to the drop in biomass at each level. The steady loss of energy explains why there are generally not more than six links in a chain.

It was said earlier that man represents the end of a number of food chains. The size of any human population is limited by the length of such chains. The longer the chain, the less energy can be derived from it, and the smaller will be the resulting population. Conversely, the shorter the chain, the more energy can be derived from it, and the larger will be the population. For minimum loss of transformed energy, man should feed on the first link in the chain, i.e. the producers. Interestingly, this is precisely what he does in over-populated parts of the world where there is a shortage of food. The inhabitants of such countries feed predominantly on vegetable matter: rice, wheat and so on. In this way a given area of land can support a greater number of people than would be the case if the population were fed on grazing animals.

AUTECOLOGY

In contrast to synecology is **autecology,** the ecology of individual species. In undertaking any autecological study one aims to discover as much as possible about the organism's ecological niche. This is important in building up a total picture of any community. These are some of the specific questions one tries to answer.

(1) What type of habitat does the organism live in, and what is its distribution within that habitat?

(2) What physical factors of the environment control its distribution, for example, light, oxygen, chemicals, etc?

(3) What biotic factors control its distribution, for example, available food, predators, hosts, etc?

(4) What food chains and food webs does it enter into?

(5) What are its natural enemies and how is it protected from them?

(6) What other organisms is it dependent upon for survival?

(7) What is its life history, and how does this fit in with seasonal changes in the environment?

(8) By what means is it dispersed and spread?

(9) How does it survive unfavourable periods such as drought, winter, etc?

(10) To what extent does its structure, physiology, behaviour and life history make it well adapted to living in its particular habitat?

(11) To what extent does its structure, physiology, behaviour and life history limit its ability to exploit its habitat?

Obviously one does not necessarily tackle these questions in the order outlined above. After all, some of them overlap with each other. Usually the answers gradually emerge as one investigates the general biology of the organism. Autecological studies have the great merit of drawing together the many different branches of biology, because the survival of any organism in its natural habitat necessarily involves its structure, physiology, biochemistry, behaviour, indeed every aspect of its biology. Specific examples of such autecological studies will not be described here since they are best carried out by personal observation in the field, laboratory studies and specialized reading.

MAN AND HIS ENVIRONMENT

Man influences his natural environment in two main ways: he replaces it with his own artificial environment (towns, cities, reservoirs etc), and he poisons it by releasing toxic chemical substances into it. Urban spread is inevitable, and will continue so long as human populations soar as they have done in the past (see p. 586). It can only be hoped that further urbanization will be carried out with due regard for the conservation of nature.

The second problem is in many ways more difficult for it is bound up with man's economic survival. To take an example, the use of pesticides is essential in modern agriculture. However, there is abundant evidence that some pesticides get into the food chains of animals, either killing them outright or affecting their reproductive success. Such pesticides include the organic chlorine insecticides such as DDT. In Britain in the early 1960s it was found that a marked fall in the numbers of certain wild birds coincided in space and time with the use of these insecticides. Following this observation, a large-scale survey was carried out on the occurrence of organic chlorine residues in the muscles, fat tissue and eggs of a large sample of birds obtained from a wide range of habitats. Variable amounts were found in the different species investigated (Fig 32.10). Interestingly, the highest concentrations were found to occur in predatory birds such as sparrowhawks and herons which, being carnivorous, do not have direct access to pesticides. This suggests that the pesticides get incorporated into food chains and accumulate in the tissues of the consumers: as the poisons pass through the food chains they get more and more concentrated. It also means that even though their use may be localized, they rapidly become dispersed. This, together with their persistence, makes them a danger to wild life, and a potential hazard to man. Similar con-

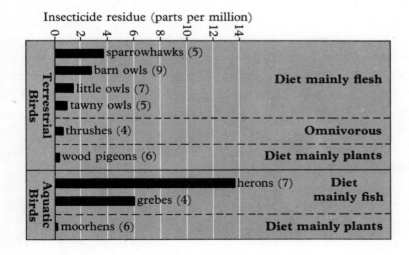

Insecticide residue (parts per million)

Fig 32.10 Average concentration of organic chlorine insecticide residue in the breast muscle of different types of birds. The number of species analysed is given in brackets after the name of each type. (*data by Moore and Walker, 1964*)

clusions have been drawn from independent investigations in North America.

What effects do these organic chlorine pesticides have on the body? Levels above 30 to 40 parts per million are known to be lethal to certain birds, but the amount does not have to be lethal to have an adverse effect. Sub-lethal doses at levels as low as 1.0 parts per million can reduce the fertility of the eggs and viability of the chicks. Furthermore, organic chlorine residues stored in the body fat may upset the metabolism of the birds when their food reserves are mobilized during starvation.

Happily, the use of DDT has now been banned in Britain and America, but every new pesticide presents a potential hazard until its effects become clear. Herbicides also take their toll, for in removing weeds from agricultural land they may upset the balance of nature by depriving animals of their natural habitats or removing essential links in food chains.

Even more serious to man are the various substances that are released into the atmosphere, rivers, lakes, and sea. Hydrocarbons and other organic compounds emitted in motor car exhaust can have an adverse effect on health, and the potential dangers of radioactive contamination have been widely publicized. Apart from its more immediate effects such as causing leukaemia and bone tumours, radiation increases the mutation rate in the germ cells, leading to genetic abnormalities in future generations (see p. 574).

Unfortunately the true facts about the effects of radioactive contamination and chemical pollutants are often obscured by those who wish to justify them. Nowhere can this be better seen than in the hotly debated subject of the genetic damage caused by radioactive fallout from atomic or thermonuclear explosions. It has been stated quite truthfully that the accumulated fallout before the test ban treaty of 1963 amounted to not much more that 1.0 per cent of the natural radiation to which all of us are exposed anyway. This is equivalent to the amount of radiation each person would normally receive in the course of about 14 weeks. However, it has been estimated that even this tiny increase, which is well below anything that would normally be classified as dangerous, may result in as many as 100,000 cases of leukaemia and 12,000 genetically defective babies.

The only general answer to these problems lies in our gaining a thorough understanding of the short and long term effects on the environment of man's activities. Most scientific advances have good and bad effects, and decisions as to their use have to be made by weighing up their relative merits and dangers. It has been estimated that the radiations derived from peaceful uses such as X-ray diagnosis, radiotherapy, nuclear reactors, and luminous watches amount to a general increase of 50 to 100 per cent over that which existed previously, and yet few people would suggest that these things should be stopped forthwith. More important is an awareness of the problem and an avoidance of indiscriminate use of scientific discoveries. The 1963 test ban treaty, the banning of DDT, and the setting up of government bodies to promote conservation, are signs that world leaders are beginning to take these problems seriously.

33 Associations between Organisms

There are many situations in which organisms form close associations with one another. Such associations may occur between individuals of the same species (**intraspecific associations**) or different species (**interspecific associations**). Either way, the environment of each individual is profoundly influenced by the presence of the other. This phenomenon is therefore an aspect of ecology and we shall look at it from this point of view. Intraspecific associations form the basis of social organization which will be discussed towards the end of the chapter. The first part will be concerned with interspecific associations.

In interspecific associations the body of one organism generally provides the habitat for another. Such associations are a specialized aspect of the biotic environment, though the relationship is generally closer and more permanent than in ordinary ecological situations. The extent to which organisms live on one another's bodies is prodigious. For example a single loggerhead sponge may contain as many as 17,000 organisms, belonging to ten different species. It was this kind of thing that prompted a humorous poet to reflect:

> Big fleas have little fleas
> Upon their backs to bite 'em;
> And little fleas have lesser fleas,
> And so ad infinitum.

Not that organisms living in an association always bite one another.

Three types of interspecific association are recognized:

(1) **Parasitism.** This is an association between two organisms in which one, the parasite, lives temporarily or permanently in or on the other, the host, deriving benefit from it and causing harm to it. In other words the parasite gains, but the host loses.

(2) **Commensalism.** In this association one of the two organisms, the commensal, gains whilst the host neither loses nor gains. Commensalism literally means 'eating at the same table'.

(3) **Symbiosis.** Here the association is mutually beneficial, both participants, symbiont and host, gaining from the relationship.

A word of warning: there is some confusion over the use of the word symbiosis. Its literal meaning is 'living together', and some biologists prefer to use it as a blanket term covering all three types of relationship. In such an arrangement, mutually beneficial associations are known as **mutualism.**

There is no sharp dividing line between these three types of association. It is sometimes difficult to decide if an organism living in or on another

Fig 33.1 Parasitic fungi such as *Phytophthora infestans*, the cause of potato blight, are partly intercellular and partly intracellular. The fungus consists of an intercellular network (mycelium) of slender filaments (hyphae) which grow through the plant sending short side branches into the cells. The latter represent the intracellular part of the parasite. They penetrate the cellulose walls of the host's cells by secreting cellulase at their tips. Once inside the cell they develop finger-like haustoria which secrete digestive enzymes. The soluble products are absorbed and sent back to the rest of the mycelium.

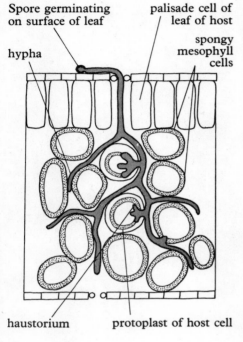

Spore germinating on surface of leaf

palisade cell of leaf of host

spongy mesophyll cells

hypha

haustorium

protoplast of host cell

actually inflicts harm on its host, or not. And in any case what do we mean by 'harm'? The general tendency is to label such organisms parasites, but if it can be demonstrated that no harm is done, they must be regarded as commensals. Similar difficulties arise when trying to decide whether an association is commensalism or symbiosis. The benefit gained by a host may be tenuous to say the most, and sometimes what starts as a mutually beneficial relationship may degenerate into a harmful one. We shall return to this later.

PARASITISM

To what extent do parasites conform to the general definition given above? To arrive at an answer it is necessary to investigate the detailed nature of the relationship. It must be demonstrated convincingly

(1) that the parasite really does live in or on its host,

(2) that the parasite benefits, and

(3) that the host suffers harm.

Let us look briefly at each of these criteria.

(1) **Living in or on the host**. The association between parasite and host can be divided into spatial and temporal relationships. By spatial relationship is meant the part of the host's body that is exploited by the parasite. Some parasites live on the surface of the host (**ectoparasites**) whilst others live inside it (**endoparasites**); and there is much variation as to how far in they live. Thus some parasites live in, or just beneath, the skin; others in the gut, others in the tissue fluid between the cells (**intercellular parasites**), and so on. The most 'endo-' of all parasites are those that penetrate individual cells. The malarial parasite and certain parasitic fungi are examples of such **intracellular parasites** (Fig 33.1).

By temporal relationship we mean the time that the parasite spends in or on its host. Here all gradations are found. Some parasites never leave their hosts; one might call them 'permanent lodgers'. Gut parasites like the tapeworm provide a good example. At the other end of the scale are certain ectoparasites that attach themselves to the host only when they feed. The mosquito and tsetse fly furnish examples of such 'occasional visitors' (Fig 33.2). Can you think of any intermediate conditions between these two extremes?

Fig 33.2 The tsetse fly *Glossina palpalis* is an ectoparasite that draws blood by means of a slender proboscis and suctorial pharynx. The tsetse fly is the vector for a number of endoparasites including the trypanosomes, cause of African sleeping sickness. (*Wellcome Museum of Medical Science*)

(2) Benefits gained. The two main benefits gained by the parasite are shelter and food. For maximum shelter, endoparasites are in the best situation: they are ensured of a stable environment from which nourishment can be derived with the minimum of effort. In some cases this involves no more than soaking up the host's body fluids. Admittedly gut parasites have a special problem in that they must protect themselves from being digested by the host's digestive enzymes, but this can be achieved by the production of appropriate inhibitors. These inactivate the enzymes in the immediate vicinity of the parasite's body.

(3) Harm suffered. In extreme cases, where the parasite feeds on the tissues, much harm, even death, may be suffered as a result of sheer damage to the cells. Such is the case with parasitic fungi such as *Phytophthora infestans*, the cause of potato blight, and *Pythium de baryanum* which causes damping-off in seedlings. The same may be said of the malarial parasite which attacks the host's red blood cells, and the hookworm which rasps at the lining of the small intestine (Fig 33.3). Some parasites do no direct damage to the cells but harm the host by liberating toxic substances into its bloodstream. Such is the case with the trypanosomes, a group of flagellated protozoons that cause African sleeping sickness (Fig 33.4). Transmitted by the tsetse fly, they live and multiply in the bloodstream where they absorb nutrients from the plasma. However they do not invade the cells.

The least harmful parasites are those that feed on the host's digested food without damaging the tissues. Such gut parasites include the tapeworm *Taenia*, a ribbon-like cestode whose total length may exceed four metres. *Taenia* clings to the wall of the small intestine by a head, or **scolex**, armed with hooks and

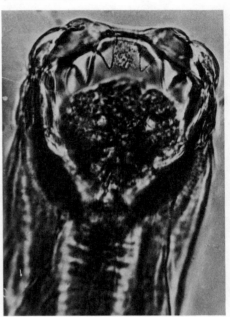

Fig 33.3 Head of the nematode hookworm *Ancylostoma duodenale*, parasite of man, showing the teeth with which this parasite rasps at the lining of the small intestine. The continual loss of blood results in anaemia and general lethargy. It has been estimated that over 600 million people have hookworms, and it is probably the greatest single contributor to the misery and debility of peoples in tropical countries. (*Wellcome Museum of Medical Science*)

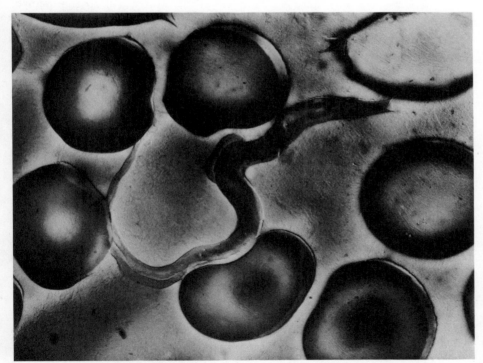

Fig 33.4 Electron micrograph of a trypanosome in a human blood smear. This particular species is *T. rhodesiense* which causes an acute form of sleeping sickness in man and antelope. Trypanosomes are flagellated unicells that live in the bloodstream. Harm is caused mainly by toxic substances liberated by the parasites. Death follows once they get into the cerebrospinal fluid surrounding the brain and spinal cord. (*K. O. Wood and R. G. Bird, London School of Hygiene and Tropical Medicine*)

suckers. The host's digested food is absorbed straight across the integument. The damage done by the insertion of the scolex into the intestinal lining is minimal; the only other harm it does is to steal a proportion of the host's food, an imposition which some people might find beneficial. The only major

Fig 33.5 *Ascaris lumbricoides* is a large parasitic nematode that inhabits the human small intestine, where it feeds on digested food. It does not attack the tissues but may occur in such large numbers that it blocks the gut. (*Wellcome Museum of Medical Science*).

Fig 33.6 A comparatively mild case of elephantiasis caused by the filaria worm *Wuchereria bancrofti*, a parasitic nematode of tropical countries. (*Wellcome Museum of Medical Science*)

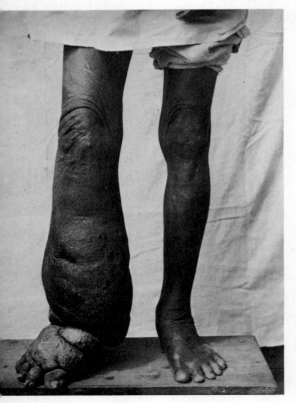

harm caused by gut parasites is that sometimes they block the intestine. This does not happen with *Taenia* since no human host ever harbours more than one at a time, but it can happen with *Ascaris lumbricoides*, a roundworm that inhabits the human small intestine (Fig 33.5). *Ascaris* can reach a length of 300 mm, and a hundred this size are enough to block the intestine and kill the host. The gut of one unfortunate individual is reputed to have contained 5,000 worms.

Sometimes a parasite has very strange effects on its host. This is well illustrated by the filaria worm, *Wuchereria bancrofti*, a nematode that lives in the lymphatic system of man. The tiny larvae, transmitted by the mosquito, invade the lymphatics where they mature into the adult worms. The latter are about 80 mm long and may occur in such numbers as to block the lymph vessels. This, together with excessive growth of the tissues in the infected areas, results in enormous enlargement of the extremities, particularly the legs, breasts and scrotum, a condition known as **elephantiasis** (Fig 33.6). The filaria worm demonstrates the profound influence that a parasite may have on the visible structure of its host. An even more remarkable case is seen in *Sacculina*, a parasitic crustacean that attacks the shore crab *Carcinus*. When the crab moults following infection, it changes its sexual features: if it is a male it develops female characteristics; if it is a female it changes towards a juvenile type lacking gonads. Little is known about the cause of this so-called **parasitic castration**, except that the parasite is clearly interfering with the host's hormonal balance.

ADJUSTMENT BETWEEN PARASITE AND HOST

In parasitic associations there is a complete gradation from almost immediate death to virtually no harm at all. This reflects differences in the degree of adjustment of parasites towards their hosts. It is not in the best interests of a parasite to kill its host: if it does so it destroys itself as well. A well adjusted parasite inflicts minimum harm on its host, so that the latter survives and reproduces normally. It is generally thought that recently evolved parasitic associations tend to be rougher on the host than older ones, the point being that in older associations parasite and host have had longer to adjust to one another. This may explain why trypanosomes cause sleeping sickness in man, but exert no ill effects on wild animals. Possibly they have been parasites of wild animals for many millions of years, but invaded man only comparatively recently: the adjustment between parasite and host, long since established in wild animals, has not yet been achieved in the human species.

A badly adjusted parasite that summarily kills its host must reproduce smartly and spread to another host. The only alternative would be to feed on its dead host. This latter possibility is exploited by *Pythium*: such is the rate of growth of this fungus that it quickly kills its host, thereafter feeding saprophytically on the dead remains.

The amount of harm done is to some extent related to the parasite's **specificity**. Some parasites can only survive in one particular species of host. Presumably the affinity between the parasite and its host is a biochemical one, and it often goes hand in hand with a marked tolerance of the host for the parasite. This is not to imply that such parasites do no harm at all, but at least they do not kill the host outright, which is really all that matters as far as the survival of the parasite is concerned. Many parasites are less specific, attacking a wide range of host species. Such parasites generally give their hosts a rougher time than the more specific ones.

PARASITIC ADAPTATIONS

From the ecological point of view parasitism presents certain problems. For example, the parasite must be able to cling onto, or gain entrance into, its host, and it must also have means of spreading from one host to another. To these ends parasites are specially adapted in a number of ways.

(1) They show **degeneration** or total loss of unwanted organs. For example most endoparasites lack sense organs, particularly eyes, and frequently have a reduced nervous system; gut parasites like the tapeworm lack an alimentary canal (though their free-living relatives have one); and parasites that wallow in their host's intercellular fluids have no osmoregulatory devices.

(2) Depending on how they are transferred from one host to another, they may have **penetrative devices** for gaining entrance into the host and its cells. The miracidium larva of the liver fluke, for example, has a slender tip onto which open a group of glands which secrete lytic enzymes (Fig 33.7). By softening the tissues, these glands enable the larva to bore into the foot of a freshwater snail, the intermediate host. On the plant side, a number of parasitic fungi secrete **cellulase** at the tips of the hyphae, thereby enabling penetration into the host's cells.

(3) Many parasites (especially ectoparasites) have **attachment devices** enabling them to cling onto the host. A large range of trematodes inhabit the gill passages of fishes. The constant flow of water over the gills would sweep them away were it not for an assembly of **suckers, hooks** and **anchors** which enable them to cling to the epithelium. Hooks and suckers are also responsible for attachment of the scolex of the tapeworm *Taenia solium* to the wall of the gut. Some of these devices are illustrated in Fig 33.8.

(4) Gut parasites have **protective devices** which prevent the body being harmed by the host's digestive processes. These devices include the possession of a thick protective **cuticle**, the secretion of large quantities of mucus, and the production of **inhibitor substances** which inactivate the host's digestive enzymes.

(5) One of the greatest problems facing any parasite is getting from one host to another. At best this is a risky business and parasites have evolved various means of ensuring success in this respect. Many employ a **secondary** or **intermediate host** which disperses the parasite over a wide area, and may convey it directly from host to host. An excellent example is the mosquito *Anopheles* which transfers the malarial parasite *Plasmodium* from one host to another. The parasites are drawn up into the mosquito's gut when it sucks the blood of an infected person, and are transferred to another host when the mosquito injects saliva into the blood prior to sucking. Here the intermediate host is functioning as a carrier or **vector**.

(6) Despite the use of vectors infection of new hosts is a hazardous business, and to raise the chance of success vast numbers of offspring are produced. The reproductive powers of many parasites are phenomenal. Countless millions of tiny wind-dispersed spores may be produced by a single parasitic fungus, and a glance at Fig 33.9 will give you some idea of the reproductive powers of the malarial parasite. You will see that reproduction occurs at no less than four stages in the life cycle, two of them in the human host and two in the

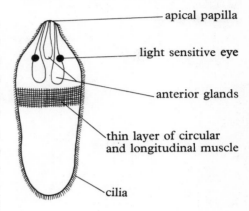

Fig 33.7 Miracidium larva of the liver fluke *Fasciola hepatica*. The cilia enable the larva to swim through water to the intermediate host, a snail. The protrusible apical papilla adheres to the snail's foot, and a secretion from the anterior glands dissolves the flesh, thereby permitting the larva to penetrate. The larva wriggles through the tissues by contraction of its circular and longitudinal muscles. Only a small portion of the muscle is shown in the diagram below.

apical papilla

light sensitive eye

anterior glands

thin layer of circular and longitudinal muscle

cilia

Fig 33.8 Attachment devices of 3 representative parasites. *Gyrodactylus*, an ectoparasitic fluke, lives attached to the gills of freshwater fishes by a circlet of hooks and claw-like anchors at the posterior end. *Sphyranura*, another fluke, is a parasite of newts to which it is attached by means of muscular suckers aided by hooks and anchors. *Taenia solium*, the pork tapeworm, inhabits the human small intestine: the head (scolex) is buried in the gut wall which it grips by means of its hooks and suckers (A: *after Mueller & Van Cleave;* B: *after Alvey*)

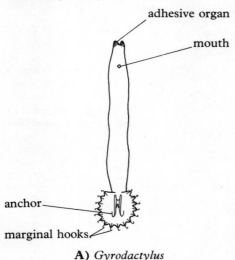

A) *Gyrodactylus*

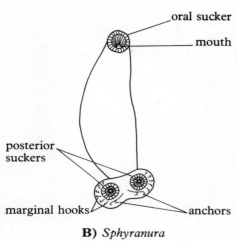

B) *Sphyranura*

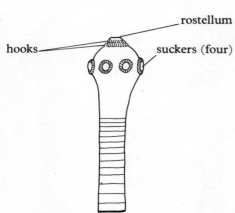

C) *Taenia solium*

mosquito. In the type of malaria caused by *Plasmodium vivax*, the number of parasites in the blood may exceed 30,000/mm³.

The life cycle of a parasite often consists of a series of structurally distinct larval stages, each responsible for rapid asexual multiplication. This is illustrated by the liver fluke *Fasciola hepatica* whose life cycle is summarized in Fig 33.10. This parasite serves to illustrate another important principle, namely the importance of having a **dormant resistant stage** in the life cycle. Provided they are immersed in water, the **encysted cercariae** of the liver fluke remain viable for up to a year, though the survival time is reduced to a few weeks if they are exposed to the air. Many other parasites, plant as well as animal, have stages that remain in a dormant, yet viable, state until a suitable host is forthcoming.

PARASITES AND MAN

The eradication of parasites harmful to man and his domestic animals and crops is a major preoccupation of scientists. Many diseases, like malaria and African sleeping sickness, that only a few decades ago were major killers, are now being brought under control. The same applies to plant parasites such as the eelworm *Heterodera* which attacks the roots of many useful crop plants, and the potato blight fungus *Phytophthora infestans*.

This would not be the place to go into details of how these different parasites are eliminated, but the principles underlying the control measures should be understood. First and foremost, it is essential that the autecology of both the parasite and its host should be understood. Once the detailed life cycle of the parasite and its host are known, it is possible to attack the parasite at its 'weak points'. Take the malarial parasite for example. This has the mosquito *Anopheles* as its vector, so if you destroy *Anopheles* you destroy the parasite. But how does one set about destroying the mosquito? To some extent this can be done by spraying infested areas with insecticides. But more foolproof methods immediately spring to mind when one realizes that the mosquito has an **aquatic larva** which hangs onto the surface film through which it absorbs oxygen into its tracheal system. This is the 'weak point' in its life history. Spreading oil over the surface of the water lowers its surface tension, with the result that the larvae can no longer cling to the surface, so they sink and drown (Fig 33.11). This technique, coupled with the thorough draining of swamps, or stocking them up with fish that feed on mosquito larvae, has gone a long way towards reducing the incidence of malaria in countries like Africa.

Knowledge of a parasite's life cycle often opens up the possibility of quite simple methods of eradication. For example, knowing that the potato blight fungus is spread by wind-dispersed conidia leads to the simple but effective expedient of not planting potatoes down the pathways followed by prevailing winds. Again, the knowledge that the ciliated miracidium larva of the liver fluke requires water in which to swim to its intermediate host, opens up the possibility of eliminating it by draining wet pastureland or restricting grazing to sloping ground.

Understanding the life cycle of a parasite also means understanding the nature of the disease caused by it. Only in this way can effective treatment be devised. In the case of malaria, various **drugs** have been found that destroy the parasite at different stages in its asexual and sexual cycles: some attack it in the liver cells, others while it is in the bloodstream; others go for the gameto-

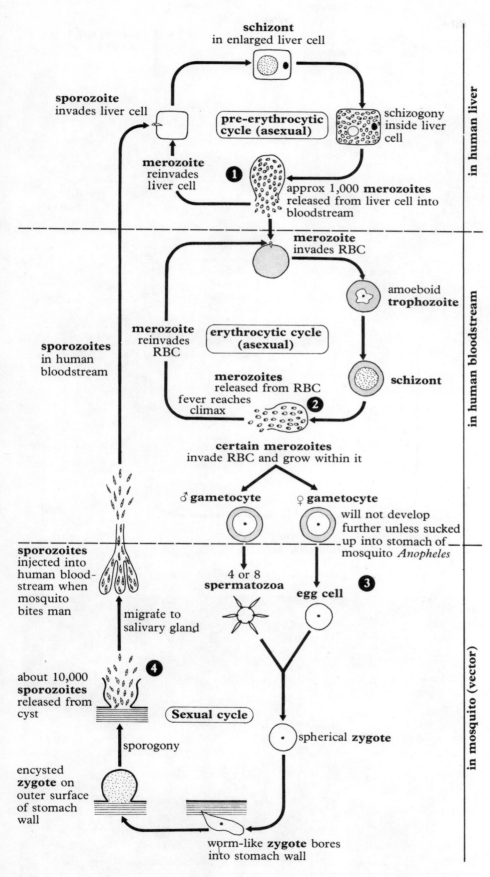

schizont
in enlarged liver cell

sporozoite
invades liver cell

**pre-erythrocytic
cycle (asexual)**

schizogony
inside liver
cell

merozoite
reinvades
liver cell

❶

approx 1,000 **merozoites**
released from liver cell into
bloodstream

in human liver

merozoite
invades RBC

amoeboid
trophozoite

merozoite
reinvades
RBC

**erythrocytic cycle
(asexual)**

schizont

sporozoites
in human
bloodstream

merozoites
released from RBC
fever reaches
climax

❷

in human bloodstream

certain merozoites
invade RBC and grow within it

♂ **gametocyte**　　♀ **gametocyte**

will not develop
further unless sucked
up into stomach of
mosquito *Anopheles*

sporozoites
injected into
human blood-
stream when
mosquito
bites man

4 or 8
spermatozoa

egg cell

❸

migrate to
salivary gland

about 10,000
sporozoites
released from
cyst

❹

Sexual cycle

spherical **zygote**

sporogony

in mosquito (vector)

encysted
zygote on
outer surface
of stomach
wall

worm-like **zygote** bores
into stomach wall

Fig 33.9 Life cycle of the malarial parasite showing the four stages where reproduction occurs (numbered 1 to 4). Following the bite by an infected mosquito, the parasite, in the form of a slender sporozoite, attacks and reproduces within the liver cells, forming numerous so-called merozoites. Some of these merozoites invade the red blood cells (RBC's), producing further off-spring. The release of the merozoites from the red blood cells generally occurs at regular 48 or 72 hour intervals and is accompanied by attacks of high fever. After several generations some of the merozoites develop into sexual forms (gameto-cytes) which develop no further until taken up by a mosquito of the genus *Anopheles*. In the mosquito's stomach, eggs and motile sperm are liberated and fertilization takes place. The zygotes, spherical at first, acquire a vermiform shape which enables them to penetrate into the stomach wall, on whose outer surface they form wart-like cysts. Within each cyst rapid asexual multiplication takes place forming thousands of slender sporozoites which, on rupture of the cyst, migrate to the mosquito's salivary glands. On injection into the human bloodstream the sporo-zoites invade the liver, thus completing the cycle. The terms schizont and schizogony are explained on p. 402.

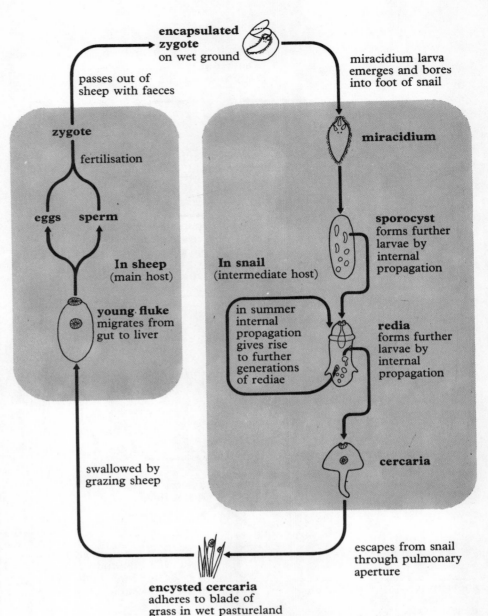

Fig 33.10 Life cycle of the liver fluke *Fasciola hepatica*. The adult fluke lives in the bile passages of sheep where it reproduces sexually producing numerous encapsulated zygotes. The ciliated miracidium larva emerges from the capsule and, swimming through water on the ground, penetrates the fleshy foot of the snail *Limnaea truncatula* where it turns into a sporocyst. Inside the sporocyst special propagatory cells divide to form rediae larvae which, armed with a muscular collar and a pair of posterior processes, burrow into the liver of the snail, feeding on the tissues. Inside each redia, propagatory cells give rise to either more rediae or numerous cercariae larvae. The latter usually work their way to the mantle cavity and leave the snail via its pulmonary aperture, encysting on blades of grass. If and when it is consumed by a sheep, the cyst bursts open releasing a small immature fluke which migrates from the host's gut to its liver. Here it feeds and grows to maturity, thus completing the cycle.

The labelled elements in the figure read:

encapsulated zygote on wet ground · passes out of sheep with faeces · miracidium larva emerges and bores into foot of snail · zygote · fertilisation · miracidium · eggs · sperm · sporocyst forms further larvae by internal propagation · **In sheep** (main host) · **In snail** (intermediate host) · young fluke migrates from gut to liver · in summer internal propagation gives rise to further generations of rediae · redia forms further larvae by internal propagation · swallowed by grazing sheep · cercaria · escapes from snail through pulmonary aperture · encysted cercaria adheres to blade of grass in wet pastureland

Fig 33.11 Applying oil to a swamp from a hand-operated pneumatic sprayer during anti-malarial control operations in Mombasa. The oil lowers the surface tension of the water thereby preventing the mosquito larva from hanging onto the surface film. (*Shell photograph*)

cytes. Again the underlying principle is that the biology of the parasite must be known in all its aspects. This applies to all parasites. The singular success that has marked man's endeavours to control diseases like diphtheria, pneumonia and poliomyelitis, has been largely owing to his ever-increasing knowledge of the causative agents, namely bacteria and viruses. Eradication of diseases like influenza and the common cold, will be achieved only by understanding the basic life processes of viruses.

To a considerable extent the elimination of organisms that plague man depends on personal hygiene and public health. Many parasites—tapeworms, hookworms and blood flukes, to mention but a few—pass out of the host with the faeces. Proper disposal of human faeces breaks the life cycle, thereby preventing reinfection of other hosts. In backward parts of the world, particularly where human faeces are used as fertilizer, the long-term solution to the parasite problem is more likely to be achieved through education than through medicine.

COMMENSALISM

For a relationship to be described as truly commensalitic it must be shown that the host neither loses nor gains. As has already been stressed, this is not always easy to decide. A well established case of commensalism is the association between the **colonial hydroid** *Hydractinia echinata* and the **hermit crab** *Pagurus bernhardus*. (Do not worry about the specific names: they serve to emphasize that this particular association is highly specific.) The hermit crab is a curious creature whose asymmetrical abdomen fits snugly into the coils of an empty whelk shell. *Hydractinia* is one of several organisms commonly found attached to shells occupied by hermit crabs. In this particular association the hydroid obtains food particles from the crab and, more importantly, it is taken into regions which otherwise would be unsuitable for it because of the softness of the substratum. The hydroid can exist on empty shells or rocks, but not on soft mud. The crab gains nothing from the association: the hydroid has stinging cells (like those of *Hydra*), but as far as is known they provide no defence for the crab. Nor, when selecting an empty shell, does the crab show any preference for a shell covered with the hydroid. This is therefore a rather loose relationship from which only one of the two partners benefits.

SYMBIOSIS

For an association to qualify as symbiotic it must be possible to demonstrate that both partners benefit from the relationship. There are all possible grades of symbiosis, ranging from rather loose associations in which the two organisms gain relatively little from each other, to associations so intimate that the two partners may be regarded as a single organism.

Returning to the hermit crab, certain species of *Pagurus* always have the **sea anemone** *Adamsia palliata* attached to their shells. The anemone is attached in such a way that its mouth and tentacles are close to the mouth of the hermit crab. It is not difficult to see that this association is mutually beneficial: the anemone obtains scraps of food from the hermit crab, not to mention free transport; the hermit crab gains protection from the anemone's stinging cells. This is an obligatory association: neither partner can do without the other. If the crab is taken out of its shell, the anemone drops off and dies; if the anemone is removed, the crab will find another one and place it on its shell

Fig 33.12 The large hermit crab *Eupagurus bernhardus* in a whelk shell to which is attached the sea anemone *Calliactis parasitica*. (*Douglas P. Wilson*)

Fig 33.13 Lichens, formed by a symbiotic union between an alga and a fungus, are the hardiest of plants and are generally the first to colonize barren areas. The lichen shown in A is growing on rock. B) Section of a lichen showing the algal cells amongst the fungal hyphae. (*A: B. J. W. Heath, Marlborough College*)

A)

B)

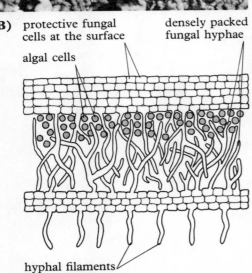

protective fungal cells at the surface

densely packed fungal hyphae

algal cells

hyphal filaments for anchorage

with its pincers. A slightly looser association is found between hermit crabs and the sea anemone *Calliactis* (Fig 33.12). Both partners benefit but the association is not obligatory, each organism being perfectly capable of leading an independent existence.

In many symbiotic associations, one partner lives inside the other. We have come across examples of this in previous chapters; for instance, in the association between cellulase-secreting micro-organisms and herbivorous mammals (see p. 135). The ultimate in intimacy is achieved by one of the two organisms living inside the cells of the other. A well known example of this, and one that can be easily seen in the laboratory, is provided by the **green hydra** *Chlorohydra viridissima* which harbours large numbers of the unicellular green alga *Chlorella* (in this case referred to as zoochlorella) in its endodermal cells. Although it is impossible experimentally to deprive a hydra of its zoochlorellae, various lines of evidence indicate that this association is mutually beneficial, if not obligatory. For example, it has been found that the zoochlorellae survive when *Chlorohydra* is kept in the dark, suggesting that, although unable to photosynthesize, they derive essential nourishment from the host. Conversely, specimens of *Chlorohydra* kept in the light but deprived of organic food, survive for longer than similarly starved species of *Hydra* lacking zoochlorellae. What, then, are these two organisms gaining from each other? In a nutshell, the alga is afforded shelter and protection, whilst at the same time obtaining carbon dioxide from its host's respiration, and nitrogen compounds from its excretory waste. In return, the hydra obtains oxygen and carbohydrates from the alga's photosynthesis.

A similar metabolic relationship is found in those extraordinary plants, **lichens** (Fig 33.13). A lichen is the result of a union between a green alga and a fungus. The algal cells are surrounded by fungal hyphae. The fungus gains oxygen and carbohydrates from the alga, whilst the alga obtains water and mineral salts from the fungus, as well as protection from drying out. The alga is also provided with a means of attachment since the fungal hyphae penetrate into the substratum. The result is an astonishingly hardy organism that can thrive in the most unlikely places, for example on exposed rocks at high altitudes and in Arctic and Antarctic regions. Always the first plants to move into barren areas, they flourish in situations where no other vegetation exists and where algae and fungi alone could not possibly survive.

Lichens provide an example of an extremely well balanced relationship. The two partners do nothing but good for one another. However, not all symbiotic partnerships are as harmonious as this. In some cases the benefits enjoyed by one or other of the participants may be marginal, and sometimes a beneficial relationship may deteriorate into a harmful one. Such is the case with the association between herbivorous mammals like the cow and their rumen bacteria. Initially the bacteria receive, in return for digesting cellulose, shelter and protection; but later the bacteria pass out of the rumen chamber into the more posterior parts of the gut where they are digested by the host.

SOME QUESTIONS CONCERNING THESE ASSOCIATIONS

In the more intimate symbiotic associations, such as that between *Chlorohydra* and its zoochlorella, the two organisms are transmitted together from one generation to the next, subsequently developing side by side. However, in other associations it must be supposed that the two participants are brought together by one (or both) being attracted to the other. What, for example, makes the miracidium larva of the liver fluke *Fasciola hepatica* bore into the foot of the snail *Limnaea truncatula*? What causes a hermit crab to pick up the appropriate species of sea anemone and place it on its shell? In some cases the sea anemone takes the initiative, climbing onto the shell: what causes it to do this?

Many experiments have been conducted to see if **chemical attractants** might be operating in these situations. The basic principle underlying these experiments is that chemical substances are extracted from different host species, and the responses to them of commensals and parasites investigated. These experiments show definitely that chemical attraction, often of a highly specific nature, is involved. However, little is known of the exact substances involved or how they operate in releasing the appropriate behaviour. This is not to imply that other types of stimulation are not important. It is probable that in many instances chemical attraction is aided by touch.

Another fundamental question concerns the evolution of these associations. Do we assume that commensalism is the most primitive association and that parasitism and symbiosis followed later, or is there some other interpretation of the relationship between them? It is easy to imagine that an organism which has been living as a commensal (or even a symbiont), might suddenly start exploiting its host in some way, thus becoming a parasite. Equally, a parasitic association might gradually become so well balanced that no harm at all is inflicted on the host, thus turning it into a commensalitic relationship. It is possible that many different sequences have occurred in the evolution of different associations.

SOCIAL ANIMALS

So far we have been dealing with associations between different species (interspecific associations). We turn now to associations between members of the same species: **intraspecific associations**.

Except when they breed, and possibly for a short time afterwards, most individuals of a species have little or nothing to do with each other during their lives. However, certain animals live in groups or colonies showing **social organization**. The two groups to have developed this to a particularly high degree are insects and mammals.

From a purely descriptive point of view we know much about these social groups, particularly in insects such as ants, termites and bees. Within each colony are two or more types of individual, or **castes**, each with distinct functions to perform. In the **honey bee colony** the **queen** is solely responsible for laying eggs, the **drones** for fertilizing her, and the **workers** for gathering food and performing sundry duties in the hive. Each caste is adapted for its particular job: thus the queen is a fertile female, the drones fertile males, and the workers sterile females with well developed mouthparts and other structural adaptations for collecting nectar and pollen. What determines this rigid caste system? The particular type of individual an egg develops into is determined partly by its genetic constitution and partly by the diet supplied to the larvae (Fig 33.14). The sexes are determined genetically. When the queen lays eggs she can control whether or not they are fertilized by sperm from her sperm receptacle. Unfertilized (haploid) eggs develop into males (drones), fertilized (diploid) eggs into females. Whether a female becomes a queen or a worker depends on its diet. If a larva is fed on nothing but **royal jelly**, a rich food solution secreted by special glands into the pharynx of the workers, a queen eventually emerges from the pupa. If, however, royal jelly is replaced after the first few days by a diet of honey and predigested pollen, a worker is formed.

Fig 33.14 Summary of the caste system, and how it arises, in a honey bee colony. The structure and behaviour of each type of individual, determined by its chromosome constitution and upbringing, are adapted to the needs of the community as a whole.

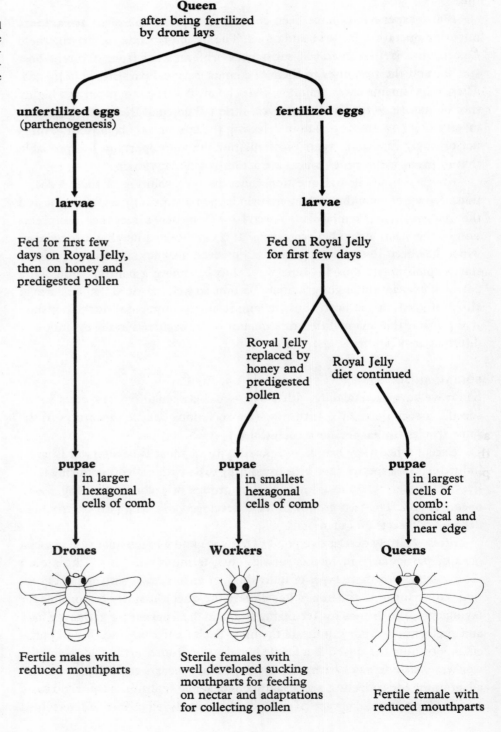

Moreover, the workers change their duties as their age increases. They start by feeding the larvae on regurgitated food mixed with special secretions from the salivary glands; then they ventilate and cool the hive by vigorously fanning it with their wings; then they clean the hive by behaving as scavengers; and finally they leave on food-collecting expeditions. There is thus a strict division of labour, each caste performing the functions which its genetic constitution, upbringing and age command it to perform. It is all very rigid but devastatingly efficient.

In certain species of **ants** social organization involves the exploitation of other species. For example, one group of ants belonging to the Myrmicinae cultivate fungi which they grow in special parts of the nest. The fungus is used as food for both larvae and adults. Even more interesting are ants which form an association with aphids. Aphids suck plant juices and since they generally consume far more than they need, they constantly exude the surplus as a sweet sticky substance called **honeydew**. The so-called **dairying ants** exploit this situation by keeping aphids in their nests. On being touched by an ant, the aphid exudes honeydew which the ant then consumes. Dairying ants often guard their aphid 'cows', and may even provide them with succulent roots on which to feed. The association is therefore akin to symbiosis.

Other ants exploit other species of ants. The most fantastic case of this is provided by the red ant *Formica sanguinea*. This aggressive insect periodically raids the nests of its brown relative *Formica fusca* and steals its pupae. Workers emerging from these pupae are then used as **slaves** performing all the routine duties in the nest. The red ant is not completely dependent on its slaves: its own workers can perform the tasks equally well if required to do so. However, another ant, a South American species of *Polyergus*, is absolutely dependent on having slaves. Its enormous fang-like jaws, though admirable weapons for offence and defence, are totally unsuitable for feeding the larvae, a job which must be performed by its slaves. These it obtains from *Formica fusca*, a persecuted species if ever there was one.

INTERACTION BETWEEN INDIVIDUALS

In any social structure it is necessary that the individuals in the colony should be held together, so that they do not disperse and become solitary. How is this achieved? The full answer is far from understood, though there is evidence that mutual **chemical stimulation** between individuals plays an important part. For example, in both bees and ants the larvae are fed by the workers continuously and ceaselessly until they pupate. Time not spent feeding the larvae is spent licking them, presumably to obtain an attractive substance produced by them.

Also important in social organization is **communication**. The efficiency of the colony can be greatly increased if its members can transmit information to one another in some way. The most thoroughly investigated species in this respect is the honey bee, the subject of a series of classical studies by the German zoologist Karl von Frisch. By marking bees and following their movements in special observation hives, von Frisch concluded that a foraging worker informs other bees as to the whereabouts of food by means of dances (Fig 33.15). On returning to the hive from a food source, two kinds of dance may be performed. If the food is less than approximately 100 metres distant the bee performs a **round dance** in which it moves round and round in a tight

Fig 33.15 Dances performed by the workers in a honey bee colony. A) The round dance and figure-of-eight waggle dance. B) Von Frisch's theory as to how the waggle dance provides information on the direction of food in relation to the position of the hive. Further details in text.

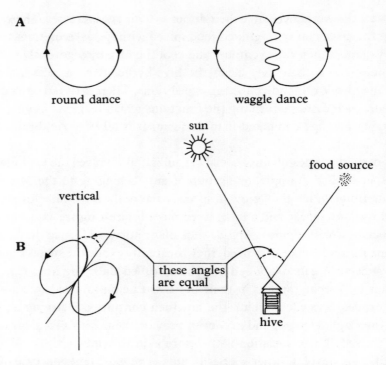

circle. The other bees cluster round and, presumably when they get the message, fly off in search of the food. According to von Frisch, the round dance only tells the bees that there is food somewhere within 100 metres of the hive. No information is given as to its exact location. If however, the food is more than 100 metres away, another dance is performed which tells other bees exactly where the food is. In this so-called **waggle dance** the bee moves through a tight figure of 8, its abdomen waggling vigorously from side to side during the cross-piece of the 8. Before reading on, how do you think this simple dance might contain information telling other bees where they should fly to get the food?

Von Frisch's theory is that two essential pieces of information are contained in the waggle dance:

(1) the distance of the food source from the hive, and

(2) the direction of the food relative to the hive.

Let us look at each in turn.

It is claimed that the distance of the food from the hive is inversely proportional to the speed of the dance, i.e. the distance is proportional to the time taken by the bee to complete one circuit of the dance. Moreover it is said that the bee can take into account headwinds encountered during its flight to the food. If it experienced a strong headwind it does the dance correspondingly slower, as if the distance were greater. So the speed of the waggle dance would seem to be related to the energy expenditure in the outgoing flight. It is thought that this also applies to the round dance, its speed being inversely proportional to the distance of the food from the hive within the 100 metres limit.

Von Frisch claims that the direction of the food source is indicated by the orientation of the waggle dance. Generally a returning forager performs its dance on a vertical wall in the hive. It is said that the angle between the vertical and the cross-piece of the dance is equal to the angle between a line drawn from the hive to the sun and a line from the hive to the food source (Fig 33.15B). When a bee is doing its waggle dance other bees cluster round in an excited manner, touching it with their antennae and following its movements in a crude kind of way. Presumably capable of appreciating the subtle meaning of the dance, they then fly off to the food source.

The dances of bees are fascinating, but during the last decade serious doubt has been thrown on their significance. It is claimed by A. M. Wenner, an American scientist, that bees communicate by sound. Using a sensitive recording device, Wenner has shown that bees emit a large repertoire of sounds, some of which appear to be related to the position of food sources encountered on foraging trips. This is not to imply that bees do not dance. They certainly do, but their dances may not necessarily be the means by which they communicate.

In this brief discussion on social animals we have concentrated on insects where organization of the community, with its caste system and strict division of labour, is rigid and stereotyped. Looking at the occupants of a bee hive, an ant hill, or a termite nest, one marvels at their industry but shudders at the monotony.

In **mammalian societies**, baboons for example, we find a hierarchical system in which (generally) a large powerful male dominates the group. Like insects, there is division of labour (some baboons act as sentinels while others forage for food, etc.), but here one is struck by the fact that the behaviour of individuals is far less rigid than that of social insects. Watching a troop of baboons, it is far more difficult to predict how an individual will behave than it is with, say, a bee colony. From what you know of animal behaviour, can you suggest explanations for this difference?

Flexibility in behaviour is also the hallmark of human societies, and one cannot help wondering what life would have been like for the human species if evolution had taken a different turn and had resulted in man's brain and behaviour being more like those of insects!

34 Evolution in Evidence

There are approximately two million species of animals and plants living today. The question to which all biologists seek an answer is: how did this tremendous diversity of life come to exist on this planet?

DARWIN AND THE THEORY OF EVOLUTION

The man whose name is most closely associated with this question is Charles Darwin (1809–1882). Ever since his Cambridge days Darwin had been a keen naturalist, devoting much of his time to collecting animals and plants. But his real opportunity came in 1832 when he was offered a berth on HMS *Beagle*, a man-of-war which was to sail round the world on a map-making survey. The journey took nearly five years and the many stops, some of them several months long, gave Darwin an unparalleled opportunity to explore the flora and fauna in many different parts of the world. He was particularly struck by the remarkable nature of the animals and plants on the Galapagos Islands in the Pacific Ocean. During his long voyage Darwin was forced further and further towards the conclusion that animals and plants have arisen by a process of slow and gradual change over successive generations, this being brought about by **natural selection**. For twenty years after returning from his voyage Darwin consolidated his data and filled in the details of his theory, seeking support for his ideas from geology, embryology and other branches of biology. In 1859 his book *On the Origin of Species by means of Natural Selection* was published, the first theory of evolution to be supported by fully documented evidence. This was not in fact the first announcement of the theory, as the previous year Darwin had published a short paper with Alfred Russell Wallace entitled *A Theory of Evolution by Natural Selection*. Independently of Darwin, Wallace had come to the same conclusions though with less factual backing. For various reasons it was agreed that they should announce their conclusions together rather than publish them separately. However, their joint paper attracted relatively little attention and today, perhaps rather unfairly, Darwin is generally credited with the theory.

Nowadays most people take evolution for granted, but it should be remembered that the climate of opinion in the middle of the nineteenth century far from favoured the notion. Although the idea had been in the air for some time, most people preferred to believe that animals and plants had arisen spontaneously by an act of special creation – each species being formed separately. It had even been calculated by the ingenious Dr Lightfoot, vicar of the University Church at Cambridge, that the world was created at exactly 9 a.m. on 23 October 4004 BC. Darwin therefore found himself immediately

Fig 34.1 Charles Darwin at the age of 51. This portrait was made a year after the publication of *The Origin of Species*. (*Royal College of Surgeons of England*)

at variance with the Church. The debates that followed are now part of the annals of biology. The most famous confrontation was between Thomas Henry Huxley and the Bishop of Oxford, Dr Samuel Wilberforce (whose suavity earned him the nickname of 'Soapy Sam'). In a public debate the bishop asked Huxley if he traced his descent from an ape through his grandfather or grandmother; whereupon Huxley replied that he would rather be descended from an ape than from an intelligent being 'who uses his gifts to discredit and crush humble seekers after truth'. But despite these stormy beginnings Darwin's theory has stood the test of time and, with minor modifications, is fully accepted today.

WHAT DID DARWIN DO?

In essence Darwin did two things. First, he marshalled powerful evidence supporting the **fact of evolution**, i.e. that species have not remained unaltered through time, but have changed. This was by no means an entirely new idea, it can be detected in the writings of earlier naturalists and philosophers, but Darwin was the first person to put it on a sound scientific basis. Much of his evidence was based on the geographical distribution of the animals and plants he had observed during his five year voyage. He then went on to argue that if species have arisen by gradual change it should be possible to learn something about their ancestry from the similarities and differences between them, and from their embryology and fossil record. This he set out to do with painstaking care and thoroughness. The result was that he gave a new purpose and direction to biological enquiry. The new disciplines of comparative anatomy and palaeontology grew up. No longer purely descriptive, these subjects became concerned with tracing the ancestral history of organisms and establishing the validity of Darwin's theory.

The second thing he did was to put forward a plausible hypothesis explaining the **mechanism of evolution**. As a naturalist Darwin was enormously impressed with the exquisite way animals and plants are adapted to their surroundings (see Chapter 32). Darwin explained this by postulating that individuals of a species differ from each other in the degree to which they are suited to their environment. The poorly adapted ones, argued Darwin, perish whereas the well adapted ones survive and hand on their beneficial characteristics to their offspring. This is what is meant by natural selection, nature as it were selecting the 'fit' and rejecting the 'unfit', and to Darwin it provided an explanation of the extinct forms seen so clearly in the fossil record. We shall return to this in Chapter 35. In the present chapter we shall be concerned with the way evolution reveals itself in different areas of biology. The areas we shall discuss are: **distribution**, **comparative anatomy**, **taxonomy**, **embryology**, **cell biology**, and **palaeontology**.

Distribution Studies

Travellers to foreign countries are often struck by the fact that places with similar climates and conditions may have widely different animals and plants. To illustrate this let us look at the general distribution of animals in the continents of the world.

CONTINENTAL DISTRIBUTION

First consider **Africa** and **South America**. Both occupy approximately the same range of latitude, and both have much the same variety of habitats, humid jungles, dry plains, high mountain ranges and so forth, and yet both support quite different faunas. True, the same major groups are found on both continents, but for the most part the individual species are different. Thus Africa has short-tailed (Old World) monkeys, anthropoid apes, elephants, camels, antelopes, giraffes and lions, to mention but a few. South America, however, has none of these. Instead we find animals like long-tailed (New World) monkeys, llamas, tapirs, panthers and jaguars.

If we take **Australia** into consideration as well, we find even greater differences despite the fact that this, too, lies on the same latitude as South America and Africa. In Australia we find the pouch mammals or **marsupials**, like the kangaroo, which are totally absent from Africa and are represented in South America only by the opossum. It is here that we find the spiny anteater and duckbilled platypus, the only living representatives of the **monotremes**, a group of primitive egg-laying mammals found nowhere else in the world. On the other hand Australia has very few placental (eutherian) mammals except those that have been introduced by man, their place being taken by the marsupials.

Africa, South America and Australia are all in the Southern Hemisphere. If, however, we examine the continents of the Northern Hemisphere we find that the differences are far less pronounced. In fact the faunas of North America and Eurasia are strikingly similar: elk, reindeer, bison (buffalo), bears, beavers, lynxes, hares, conies, mountain sheep, and goats, are all found on both continents, along with many other animals.

EXPLANATION OF CONTINENTAL DISTRIBUTION

We are left, then, with the general picture of the two great continents of the Northern Hemisphere having a more or less uniform fauna, whereas the three southern continents have sharply contrasting faunas. How can we explain this? A glance at a map of the world will remind you that whereas South America, Africa and Australia are separated from one another by great bodies of water, North America and Asia are separated by a shallow strait (the Bering Strait), less than 100 km wide. Moreover there is reliable evidence that in the geological past a continuous land bridge linked the two continents with one another across what is now the strait.

One theory, summarized in Fig 34.2, is that the main groups of modern mammals arose in the northern hemisphere and subsequently migrated in three major directions: to South America via the Isthmus of Panama; to Africa via the strait of Gibraltar which, like the Bering Strait, has been bridged in past geological time; and to Australia via South East Asia to which it was at one time connected by land. Presumably the shallowness of the Bering Strait would have made the passage of animals between the two northern continents a relatively easy matter, and it explains the present-day similarity between the two faunas. But once they had got right down into the southern continents, they presumably became isolated from each other by various types of barrier. The submerging of the Isthmus of Panama is thought to have isolated the South American fauna; the Mediterranean Sea, and more recently the North African desert, would at least

A)

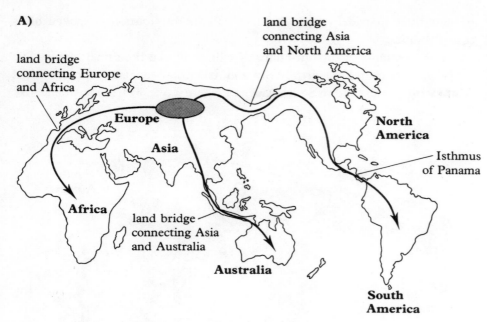

Fig 34.2 It is possible to explain the fact that Africa, Australia and South America have widely different faunas by postulating that ancestral stocks of animals originated in Asia or North America and then migrated southwards into the southern continents via natural land bridges (A). They then became cut off and underwent adaptive radiation independently in their respective continents (B).

B)

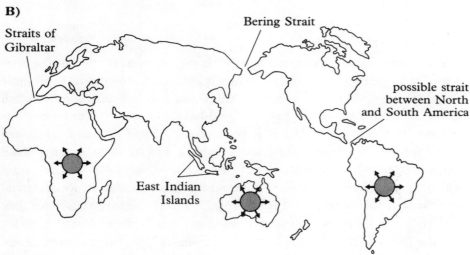

partially isolate the African fauna; and the submerging of the original connection between Australia and South East Asia would isolate the Australian fauna.

Once isolated, it is thought that the animals in each continent proceeded to evolve along their own lines. From the original stocks which invaded these continents arose many different forms filling every available habitat. This is called **adaptive radiation**, and it has occurred independently in each of the three southern continents, resulting in their now possessing markedly different faunas.

EVIDENCE FOR MIGRATION AND ISOLATION

What evidence is there for such mass migrations followed by independent evolution? An important indirect piece of evidence is that although the animals of widely separated regions may be different in detail they have many basic similarities. What this means in practice is that continents like South America and Africa have different genera and species, but these belong to the same orders and families. When a group of animals or plants has re-

Fig 34.3 Support for the hypothesis outlined in Fig 34.2 comes from studying the distribution of modern and fossil forms. The map shows the distribution of the camel family. The black disc in N. America represents the centre of origin of the Camelidae, where the oldest fossils are found (Tertiary). Light grey shows the distribution during pleistocene times, and dark grey the present-day range. Today the Camelidae are represented by the camel in Afro-Asia, and the llama in South America. Both are fundamentally similar, but different in detail. This is an example of discontinuous distribution. The arrows show probable migration routes. (*based on Matthew*)

presentatives in widely separated localities its distribution is spoken of as **discontinuous**.

As an example of discontinuous distribution take the camel family. The Camelidae are found in North Africa and Asia where the family is represented by the camel itself, and in South America where it is represented by the llama.

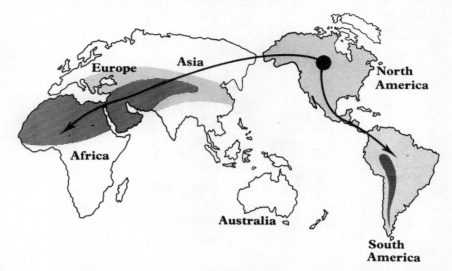

There are, of course, a number of possible explanations of this but the most plausible hypothesis is that the camel family arose in the Northern Hemisphere and then migrated southwards into Africa and South America as described above. If the matter ended here it could hardly be expected to receive unanimous support, but in fact it is borne out by fossil evidence. The fossil record for the Camelidae is admirably complete, and the distribution of fossil forms indicates that the evolution of camels started in North America, from which they migrated across the Bering Straits into Asia and thence to Africa, and through the Isthmus of Panama into South America. Once isolated they evolved along their own lines, giving the modern camel in Asia and Africa and the llama in South America. The evidence for this theory is summarised in Fig 34.3. The fossil record of other animals such as tapirs and horses tells a similar kind of story.

THE AUSTRALIAN FAUNA

In the foregoing section it was said that isolation is followed by independent adaptive radiation. This is illustrated particularly well by the Australian

Fig 34.4 The three mammals shown here are all non-placental species found exclusively in Australia and Tasmania where they occupy ecological niches occupied in other continents by the eutherian (placental) mammals. A) The spiny anteater *Echidna*, an egg-laying monotreme. B) The koala bear, an arboreal marsupial. C) The marsupial wolf *Thylacinus cynocephalus*. (*Australian News and Information bureau*)

A

B

C

fauna. As was mentioned earlier, Australia is the home of the marsupials whose ancestors presumably got there by migration over a continuous land bridge represented today by the Malayan peninsula and Indonesian islands. The dearth of eutherian mammals, living or fossil, in Australia suggests that Australia became cut off before the placental mammals had a chance to get there. Free from competition, the marsupials have evolved to fill all the ecological niches which in other continents are occupied by placental mammals. Thus, the **great red kangaroo** is equivalent in the life it leads to fast-running herbivores like deer and antelope; the **tree kangaroo** and **koala bear** are equivalent to arboreal monkeys and sloths; and the **Tasmanian wolf** *Thylacinus* is a savage carnivore equivalent to the wolf of northern regions. There are also marsupial 'moles', 'hares' and 'squirrels', and even a marsupial anteater with a long snout and extensible tongue just like the placental anteaters of other continents. As was mentioned earlier Australia is also the home of the egg-laying monotremes and this group also includes an ant-eating representative, *Echidna*, the spiny anteater (Fig 34.4A).

OCEANIC ISLANDS

If the effects of geographical isolation are well shown by the three great southern continents, they are even more clearly seen in the faunas of oceanic islands. An oceanic island is one which has never had any connection with the mainland. Most of them are formed by a submerged volcanic mountain erupting and pushing up above the surface of the sea. This has happened at intervals throughout geological time, and is still happening in certain parts of the Pacific Ocean. Once it has broken the surface, the exposed mountain peak cools down and in the fullness of time becomes an island capable of exploitation by any animals and plants which are able to get there. How organisms reach such oceanic islands is a matter of conjecture: they may be blown there by wind, as spores for example; carried by ocean currents, perhaps clinging to drift wood; or of course they may fly there if capable of doing so.

THE GALAPAGOS ISLANDS

Such is the case with the Galapagos islands, situated on the equator some 900 km west of Ecuador in South America. Formed by volcanic action, these islands were at first devoid of life but later became colonized by animals and plants from the mainland. These then evolved along their own lines into species which, though fundamentally similar to the mainland forms, differ from them in detail.

Today the Galapagos islands are a biologist's paradise, for they demonstrate the effect that isolation can have on the subsequent evolution of animals and plants. Darwin visited the islands in 1835, and what he saw there probably shaped his ideas about evolution more forcibly than any other single experience during his five year voyage on the *Beagle*. The Galapagos fauna is unique and, seen against the background of volcanic rock and twisted lava, it makes a visit to the islands seem like stepping back into a past geological age.

The most prominent inhabitants of the islands at the time Darwin visited them were the **giant iguana lizards** (Fig 34.5). About $1\frac{1}{4}$ metres long, there are two species, one terrestrial and the other marine. The marine form is the only known aquatic lizard, and is adapted for living in water by having webbed feet and a laterally-flattened tail which it uses for propelling itself like a newt. The terrestrial species was extremely abundant at the time Darwin visited the

Fig 34.5 Giant marine iguana lizard of the Galapagos, A) basking in the sun, and B) swimming. (A: *Alfred M. Bailey.* B: *Robert I. Bowman, San Francisco State College, California*) **A**

B

islands, and it is thought that the marine form evolved from it as a result of overcrowding and competition for food on land. Both species are herbivorous, the land form feeding on leaves and other forms of vegetation, the marine form on seaweed.

Further inland, where the vegetation becomes more lush and plentiful, **giant land tortoises** are found. Unique to the Galapagos, these extraordinary reptiles reach a metre in height and weigh up to 225 kg. Whereas on the mainland such cumbersome and slow-moving creatures would have long since been ousted by the more agile mammals, on the Galapagos islands, where there are virtually no mammals, they have flourished. Unfortunately man, with his notable propensity for interfering with nature, has put an end to this: it has been estimated that over the last three centuries more than ten million have been exterminated, resulting in the virtual extinction of this magnificent animal. The trend towards large size is also shown by the plants of the Galapagos. For example, in the total absence of palms and conifers so prevalent on the mainland, the normally low-growing prickly pear cactus has assumed the proportions of a tree.

DARWIN'S FINCHES

But from the scientific point of view by far the most interesting inhabitants of the Galapagos islands are the birds. Darwin was particularly interested in the **Galapagos finches** for he could see in them the key to understanding the evolutionary process. The mainland finches are all of one type, possessing short straight beaks for crushing seeds. On the Galapagos islands, however, thirteen species fall into six main types, each having a beak specially adapted for dealing with a particular kind of food (Fig 34.6). The Galapagos finches exploit a wide range of ecological niches which on the mainland are occupied by other groups of birds. Thus the **large ground finch**, the closest to the mainland finches in form and habit, has a typical finch-like beak for crushing seeds. In contrast, the **cactus ground finches** have a long straight beak and split tongue for getting nectar out of the flowers of the prickly pear cactus. The **vegetarian tree finch**, on the other hand, has a curved parrot-like beak with which it feeds on buds and fruits. The **insectivorous tree finches** have a similar beak which they use for feeding on beetles and other small insects. Then there is the **warbler finch** which is so like a true warbler that at first it was thought to be one. It uses its slender beak for feeding on small insects which it catches on the wing like a true warbler. But the most remarkable of all the Galapagos birds is the **woodpecker finch** (Fig 34.7). This resembles true woodpeckers in its ability to climb up vertical tree trunks and bore holes in wood in search of insects. But whereas the true woodpecker uses its long tongue to seek out the insect, the woodpecker finch, not having a long tongue, picks up a small stick or cactus spine in its beak and pokes this into the hole. When the insect emerges the bird drops the stick and devours the insect. Quite apart from its evolutionary implications, this is remarkable for being one of the few cases of tool-using by an animal other than man.

The Galapagos finches, known with justification as 'Darwin's Finches', afford an excellent example of adaptive radiation. It is assumed that a stock of ancestral finches reached the islands from the mainland and then, in the absence of competition, evolved to fill all the empty ecological niches occupied on the mainland by other species. It is comparable to the adaptive radiation

Fig 34.6 An evolutionary tree showing how the different types of finches found on the Galapagos Islands might be related to one another. The beaks have evolved to suit different kinds of diet, thus adapting their possessors to a wide range of ecological niches.

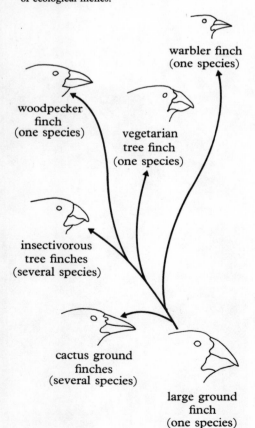

warbler finch
(one species)

woodpecker finch
(one species)

vegetarian tree finch
(one species)

insectivorous tree finches
(several species)

cactus ground finches
(several species)

large ground finch
(one species)

of the marsupials in Australia or the placental mammals in South America.

Within the Galapagos archipelago there are interesting differences between islands which lend further support to this idea. Take the insectivorous tree finches for example. These are not all identical but show small differences from island to island. On the central islands there is a large pale form with a curved parrot-like beak (the type illustrated in Fig 34.6). On Hood Island, the southernmost in the group, the large pale form occurs together with a much smaller, darker form with a straighter beak. In the northwest is a third type somewhat intermediate between the other two. These three forms do not interbreed and are therefore regarded as separate species. It seems that the ancestral stock became split up and isolated on separate islands, each then evolving along its own lines.

ISOLATING BARRIERS

It is clear from this brief review of continental and island faunas that the distribution of animals and plants, and the form they take in different localities, depends to a considerable extent on migration and isolation. These in turn depend on the establishment of **natural barriers** which restrict movement in one direction or another. Barriers limiting the movement of terrestrial organisms include water, mountain ranges, deserts, temperature and other climatic factors. Aquatic organisms are limited by such factors as salinity of the water, currents and tides. These barriers do not remain static throughout geological time but change from one period to another. Routes that have served as a pathway for the migration of animals during one period may become closed in a subsequent period, thus isolating groups of organisms from each other and allowing them to evolve independently.

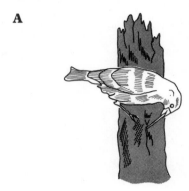

Fig 34.7 A) The woodpecker finch, most remarkable of all the Galapagos finches, lacks a long tongue but instead uses a twig or cactus spine to probe for insects in the bark of trees. (*After Lack, 1947*). B) shows a woodpecker finch using a tool in the laboratory. (*Robert I. Bowman, San Francisco State College, California*)

A

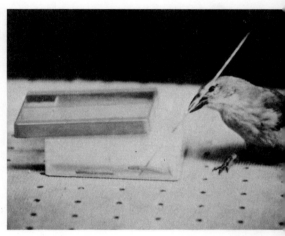

B

Comparative Anatomy

If it is true that widely separated groups of animals and plants share a common ancestry, as their geographical distribution suggests is the case, we would expect them to have certain basic structural features in common. In fact the degree of resemblance between them should indicate how closely related they are in evolution: groups with little in common are assumed to have diverged from a common ancestor much earlier in geological history than groups which have a lot in common. Establishing evolutionary relationships on the basis of structural similarities and differences is the business of **comparative anatomy**.

In deciding how closely related two animals are, a comparative anatomist looks for structures which, though they may serve quite different functions in the adult, are fundamentally similar, suggesting a common origin. Such structures are described as **homologous**. In deciding whether or not two structures are homologous many considerations must be taken into account: their relative positions, gross morphology, histological appearance and so on. In cases where the structures serve different functions in the adult, it may be necessary to trace their origin and development in the embryo if any fundamental similarity is to be discerned.

Fig 34.8 A generalized pentadactyl limb. This kind of limb is possessed by all terrestrial groups of vertebrates. The preaxial border (i. e. the edge of the limb which generally points towards the main axis of the body) is to the left, the postaxial border to the right. The fore- and hindlimbs both conform to the pattern illustrated: the nomenclature used for each is shown to the left and right of the diagram respectively. (*after Grove and Newell*, Animal Biology, *University Tutorial Press*)

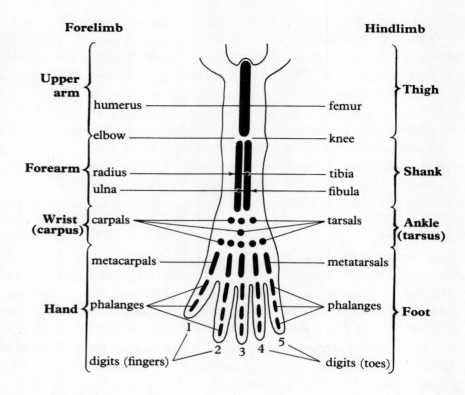

THE PENTADACTYL LIMB

One of the clearest examples of homology is shown by the **pentadactyl limb**, so called because typically it has five digits. The pentadactyl limb is found in all four classes of terrestrial vertebrates (amphibians, reptiles, birds and mammals) and can even be traced back to the fins of certain fossil fishes from which the first amphibians are thought to have evolved. Throughout the land forms the structure of the pentadactyl limb is fundamentally the same, conforming to a greater or lesser extent to the generalized pattern shown in Fig 34.8. But in the course of evolution the limbs of different vertebrates have become adapted for different purposes, in some cases involving severe structural modifications. Something of this can be seen in Fig 34.9 which shows how the mammalian forelimb is modified for grasping, walking, running, digging, swimming and flying in a selection of mammals.

In **monkeys** the fingers are elongated to form a grasping hand admirably suited for swinging from branch to branch. In the **pig** the first digit is lost and the second and fifth reduced; the remaining two are longer and stouter than the rest and bear a hoof for supporting the body. In **horses** this process is carried further; the limb is adapted for support and running by great elongation of the third metacarpal which becomes the cannon bone; the remaining digits are lost though the second and fourth metacarpals, much reduced, persist as the splint bones. In complete contrast, the **mole** has a short, spade-like forelimb for burrowing; an unusual feature is that all nine carpals are present, one being displaced to the side. A rather similar device is seen in the **anteater** of tropical South America which uses its enlarged third digit for tearing down ant hills and termite nests. In the **whale** the forelimb becomes a flipper for steering and maintaining equilibrium during swimming. But the most striking modification is seen in the **bat** whose forelimb has been turned into a wing by great elongation of four of the digits: the hook-like first digit remains free and is used for hanging from trees; the remainder support the skin of the

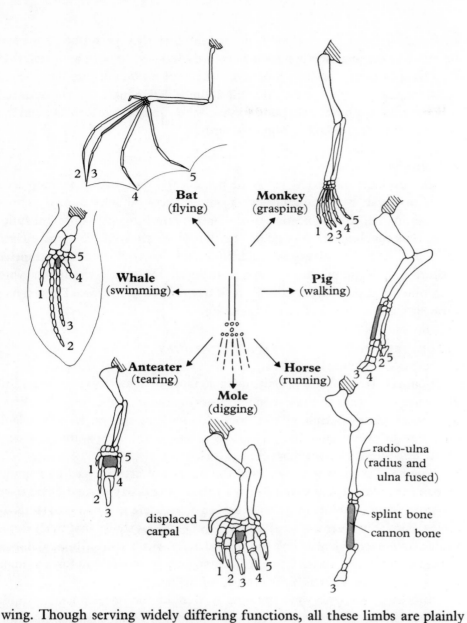

Fig 34.10 Modification of the pentadactyl limb for flight in three different vertebrates. The wing is supported in the bat by four elongated digits; in the bird by the entire limb minus digits 4 and 5; and in the pterodactyl by an elongated 5th digit. (*after Colbert*, The Dinosaur Book, *1951, McGraw-Hill*)

A) Bat

B) Bird

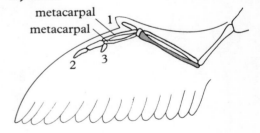

C) Pterodactyl

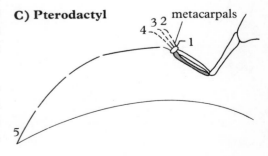

wing. Though serving widely differing functions, all these limbs are plainly recognizable as modifications of the idealized pentadactyl limb, and are clearly homologous.

Comparative studies on the pentadactyl limb can usefully be extended beyond the mammals to other vertebrate groups such as reptiles and birds. Compare, for example, the forelimb of the **bat** with that of a **bird** and a **pterodactyl**. (Fig 34.10). All three are adapted for flight but the way this has been accomplished is different in each case. In the bat, as we have seen, the wing is supported by four greatly elongated digits; in the bird it is supported by the entire limb minus certain of the digits; and in the pterodactyl, an extinct flying reptile contemporary with the dinosaurs, the wing was supported by a single elongated digit. We have here an example of homologous structures becoming adapted in different ways to perform the same basic function.

In some cases evolution has resulted in extreme reduction, even total loss, of a structure. In the bird's wing, for instance, the third digit is very much reduced and the fourth and fifth are missing altogether. Structures which are thus reduced are known as **vestigial**, and their existence has been used as

Fig 34.11 Much can be learned about evolution by studying the comparative anatomy of the vertebrate heart and arterial arches. The single circulation of fishes is gradually replaced by a double circulation with complete separation of oxygenated and deoxygenated bloodstreams. The diagrams are all ventral views and, for clarity, the heart has been deflected back through 180° so that the auricles appear to lie posterior to the ventricles. Arrows in the blood vessels indicate direction of blood flow. **A**, atrium; **V**, ventricle; **R**, right; **L**, left. (*based on Goodrich*)

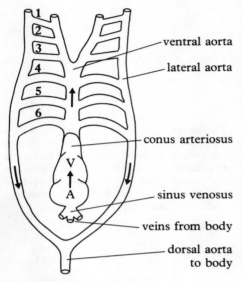

A) Embryonic pattern
Six arterial arches link the ventral aorta with the lateral aortae. Fishes have this arrangement except that arterial arch 1 (and sometimes 2) are missing in the adult. In fishes the arterial arches serve the gills.

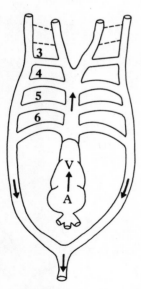

B) Tadpole
Arterial arches 1 and 2 disappear, leaving four functional arches, which serve the gills.

strong evidence for evolution: it is thought that they performed a normal function in the ancestor but have since been reduced to such an extent that they have lost their original function. Well known examples are the reduced pelvic girdle of the whale (the hind limbs being absent), the reduced pelvic girdle and hind limb 'buds' in the python and boa constrictor, and the wings of the kiwi and other flightless birds.

DIVERGENT EVOLUTION

We can explain the variations in the pentadactyl limb within a group such as the mammals by postulating that from an ancestral stock numerous lines of evolution led to modification of the basic pattern to serve different functions, thus enabling the descendants to fill a wide variety of ecological niches. This is described as **divergent evolution**, and clearly it results in **adaptive radiation**. The end products have certain structural features in common with each other and with the ancestral stock from which they arose. These structural similarities are the basis of **homology**.

RECONSTRUCTING AN EVOLUTIONARY PATHWAY: THE VERTEBRATE HEART AND ARTERIAL ARCHES

Comparative anatomy permits us to do more than merely relate a handful of diverse organisms to a common ancestor. In certain situations it also allows us to work out **evolutionary pathways**. To see how this works we will look briefly at the evolution of the vertebrates as seen in the comparative anatomy of the heart and arterial arches (Fig 34.11).

The basic plan of the heart, and the pattern of arteries leading from it, is seen in vertebrate embryos. The heart, receiving deoxygenated blood from the great veins, consists of a single **atrium, ventricle** and **conus arteriosus** which leads into a **ventral aorta**. A study of embryos shows that in all vertebrates six **arterial arches** link the ventral aorta with a pair of **lateral dorsal aortae** on each side of the body. The latter unite posteriorly to form a single median **dorsal aorta** which takes blood to the body.

Although this embryonic pattern is seen in the development of most vertebrates, the only animals to show this primitive arrangement in the adult stage are **fishes**, and even they have modified it to some extent. In adult fishes the first one or two arterial arches disappear and the remainder serve the gills. Fishes have a **single circulation**, blood entering the heart from the great veins whence it is pumped through the gills for oxygenation and thence to the dorsal aorta.

In **amphibians** certain fundamental changes have taken place which can be best understood if we first consider the **tadpole**: this serves as a bridge between the fish and adult amphibian and provides an excellent example of how embryological development can help us to understand an evolutionary pathway. The tadpole has a single circulation: the first two arterial arches disappear early in development but the remainder serve the gills as in fishes. At metamorphosis, when the tadpole forsakes its aquatic existence for dry land, drastic changes occur in its circulation associated with the sudden change from gill to lung respiration. The part of each lateral dorsal aorta between the third and fourth arterial arches constricts to become the **ductus caroticus**; the fifth arterial arch disappears altogether, and the outer portion of the sixth constricts to become the **ductus arteriosus**. This leaves arterial arches

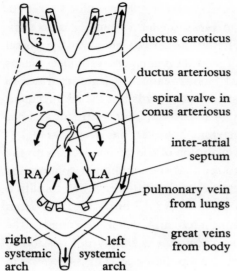

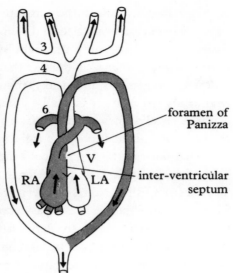

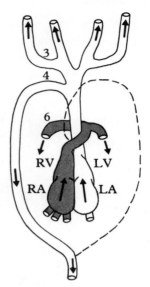

C) Adult amphibian
Lateral aorta between arches 3 and 4 constricts to become the ductus caroticus. Arch 5 disappears. Distal portions of arch 6 constrict to become the ductus arteriosus. This leaves:
Arch 3 → carotid arch (serving head)
Arch 4 → systemic arch (serving body)
Arch 6 → pulmonary arch (serving lungs)
Atrium becomes divided into left and right atria by inter-atrial septum.

D) Crocodile
Inter-ventricular septum divides ventricle into right and left halves. Conus arteriosus of amphibian replaced by separate vessels arising direct from ventricle. But note that left systemic arch carries deoxygenated blood.

E) Bird
Left systemic arch disappears.
Right systemic takes blood to body.

3, 4 and 6: the third becomes the **carotid arch** taking blood to the head; the fourth the **systemic arch** conveying blood to the dorsal aorta; and the sixth becomes the **pulmonary arch** conveying blood to the newly developed lungs. Meanwhile the atrium becomes subdivided into right and left halves which receive blood from the body and lungs respectively. The ventricle, however, remains undivided, the three arterial arches arising from the single conus arteriosus. This being so, we must assume that deoxygenated and oxygenated blood get mixed before they enter the arterial arches. It is difficult to see how this could be avoided, though experimental evidence suggests that the amount of mixing may not be as great as it seems. The **spiral valve** in the conus may play some part in keeping the two blood-streams separate.

In **reptiles** there is variation in the structure of the heart, but we will confine our attention to the crocodile since it is thought to be closely related to the ancestors of higher vertebrates. In the crocodile the single conus arteriosus is replaced by three separate vessels arising directly from the ventricle. These are the pulmonary arch, the left systemic arch, and right systemic arch to which the carotids are attached. The ventricle is completely divided though there is a small hole, the **foramen of Panizza**, connecting the base of the two systemics. We have here a double circulation with virtually complete separation of deoxygenated and oxygenated blood.

The trouble with the crocodile is that the interventricular septum is placed in such a position that the left systemic arch receives predominantly deoxygenated blood. Since this arch leads to the dorsal aorta it follows that a proportion of the blood pumped to the trunk and hind quarters will be

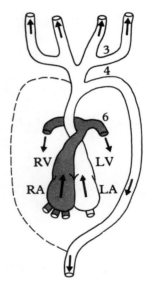

F) Mammal
Right systemic arch disappears.
Left systemic takes blood to body.

deoxygenated. In view of the apparent inefficiency in the crocodile's arrangement, it is not surprising to find that in **birds** the left systemic arch is completely lost, leaving only the right systemic to convey blood to the dorsal aorta. This, together with the fact that the heart is completely divided, means that the oxygenated and deoxygenated bloodstreams are kept separate. In **mammals** the heart is basically similar to that of birds but the arrangement of the arterial arches is quite different: The left systemic is present and has the carotids attached to it; the right systemic, however, has disappeared.

What does all this tell us about the evolution of the vertebrates? With the help of the tadpole it is not difficult to see that the amphibian condition is derived from fishes. Nor is it a very great step to derive the reptilian condition from the amphibian. It is also easy enough to derive the bird circulation from the crocodile by loss of the left systemic arch. But when we come to the mammal we find ourselves in difficulties because it is simply not possible to derive the mammalian circulation from the reptilian one. This is because in reptiles the carotids arise from the right systemic arch whereas in mammals they arise from the left systemic, the right one having disappeared. Under these circumstances it is easiest to derive the mammalian condition from the amphibian. Thus, on the basis of the comparative anatomy of the heart and arterial arches, the evolutionary sequence would appear to be as follows:

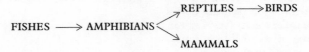

However, we know from fossil evidence that mammals did not arise from amphibians but from a group of reptiles. In fact the evolutionary sequence based on the fossil record is as follows:

$$FISHES \longrightarrow AMPHIBIANS \longrightarrow REPTILES \begin{array}{c} \nearrow BIRDS \\ \searrow MAMMALS \end{array}$$

How can we reconcile these apparently conflicting pathways? The fossil evidence is irrefutable; after all it is based on direct examination of the actual ancestors. The first sequence, however, is derived from a comparison of modern forms and may not necessarily represent what actually occurred in evolutionary history. What we can say is that whereas the group of reptiles from which the birds arose could well have had the kind of circulation seen in present-day crocodiles, the group which gave rise to the mammals are unlikely to have had such an arrangement. The latter are more likely to have had a circulation similar to present-day amphibians, from which the mammalian system could have evolved.

What all this tells us is that deductions based on comparative anatomy alone can be misleading, but comparative anatomy taken in conjunction with fossil data enables us to build up a more accurate picture of the evolutionary history of animals.

CONVERGENT EVOLUTION

This kind of evolution produces seemingly similar structures in organisms having quite different ancestral origins. Such structures are described as **analogous**. Generally they perform the same function, though possibly in very different ways. Any anatomical similarity between them is caused not by a common ancestry but by the fact that structures performing the same function are bound to resemble one another to a degree. Examples of analogous structures are the legs of insects and mammals, and the wings of butterflies and birds. It can easily be seen that the legs of insects and mammals, though performing the same tasks, have a quite different structural organization. Similarly, the wings of butterflies and birds, though both used for flight, are constructed quite differently and operate on different principles.

In these two cases of analogy a cursory glance is enough to tell us that the structures bear no fundamental resemblance to each other. But this is not always the case. The eyes of octopuses and vertebrates, for example, are remarkably similar even down to small points of detail, and an unbiased observer might well conclude that they were homologous. There is, however, one telling difference between them: the retina of the vertebrate eye is inverted, i.e. the light-sensitive cells lie beneath the nerve fibres that unite to form the optic nerve, whereas the retina of the octopus is non-inverted. For this reason the vertebrate eye has a 'blind spot' where the optic nerve emerges from it, but no such structure is found in the octopus eye. The explanation of this difference lies in the way the two eyes develop, and it tells us at once that they are the products of two distinct lines of evolution, resembling each other merely as a result of convergent evolution.

Taxonomy – the Classification of Organisms

The most obvious reason for classifying organisms is convenience. With over two million species on record it produces order out of chaos, and facilitates identification. But there is more to taxonomy than mere pigeon-holing. A natural classification is an expression of evolutionary relationships between different groups. It is assumed that organisms in the same group are closely related whilst those in separate groups are more distantly related. Moreover the size of a group reflects the closeness of the organisms in it. Thus a large unit of classification such as a phylum may contain a wide range of organisms with perhaps only one fundamental feature in common, whereas a smaller unit such as an order or family will contain a much more closely-knit group of organisms with numerous similarities. It would seem that the members of a phylum diverged from each other at a much earlier point in time than the members of a family. Clearly taxonomy is not so much evidence for evolution as the inevitable result of it.

THE BASIS OF A NATURAL CLASSIFICATION
Since the purpose of a natural classification is to reflect evolutionary, or **phylogenetic**, relationships it must be based on homologous and not on analogous structures. Thus it would be highly misleading to base a classification on the possession or otherwise of wings, for butterflies and birds would appear in the same group. This may seem a fatuous example, but there was a time when all worm-shaped animals were placed in one group called the Vermes, which finished up containing animals as diverse as earthworms and

snakes! The point is that shape, wings, legs and so on are highly adaptive features which have evolved independently in many different groups. Classifications based on such criteria are of no scientific value.

A natural classification must therefore be based on homologous structures whose fundamental similarities demonstrate a common ancestry, rather than on analogous structures which happen to bear a superficial resemblance to each other as a result of convergent evolution. To illustrate this let us look at the classification of the chordates.

CLASSIFICATION OF THE CHORDATES

The phylum **Chordata** contains all five classes of vertebrates plus three invertebrate groups. This wide range of animals are held together by one fundamental feature which they all have in common, the possession at some stage in their life history of a **notochord**. This is a flexible rod composed of tightly packed vacuolated cells situated immediately beneath the neural tube on the dorsal side of the body. *Amphioxus* is the only chordate that keeps its notochord in an unmodified form throughout life. In the **vertebrates** it is replaced during development by the vertebral column; in the burrowing **acorn worms** it is reduced to a short rod in the proboscis, and in the sessile **sea squirts** it is present only in the larva. Nevertheless, on the generally agreed assumption that the notochord is of phylogenetic significance, all these animals are placed in the same phylum, the implication being that they share a common ancestry (Fig 34.12). This is borne out by the fact that they share other fundamental features in common. For example they have a **hollow CNS (dorsal nerve tube)** immediately above the notochord; the pharynx is perforated by **pharyngeal slits** which connect its lumen with the outside world; a **tail** extends beyond the anus, and there are **segmental muscle blocks (myotomes)** on either side of the body. These fundamental features are known as the **chordate characters** and are illustrated in Fig 34.13.

Embryology

In the previous section it was mentioned that sea squirts only show a notochord in their larvae. This illustrates the general principle that in order to establish phylogenetic relationships it may be necessary to study an organism's embryological development.

THE TADPOLE LARVA OF SEA SQUIRTS

Nothing could look less like a chordate than an adult sea squirt: a sessile filter feeder, it has hardly any of the features found in other chordates, and there is no trace of a notochord. But a detailed examination of its larva makes its affinity with the chordates abundantly clear (Fig 34.14). Not only is there a notochord but in addition there are pharyngeal clefts, a dorsal tubular nerve cord, segmental myotomes and a post-anal tail, all characteristic of chordates. In fact this little creature is so chordate-like that it is called the **tadpole larva,** and indeed it does bear a striking resemblance to a little tadpole. Eventually it attaches itself to a rock or piece of

A

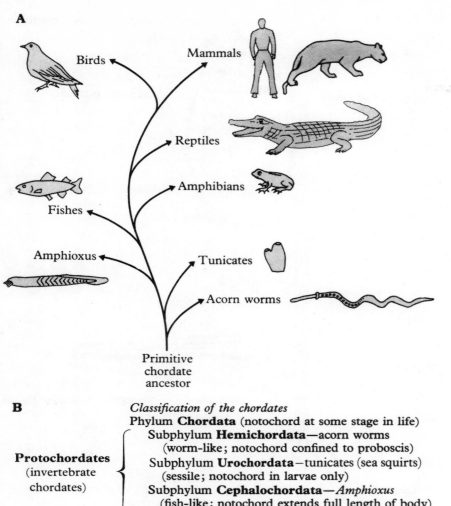

Fig 34.12 The classification of the chordates endeavours to reflect the phylogenetic relationship between the different groups. A) Evolutionary tree for the chordates based on comparative anatomy, embryology and fossil evidence. B) Classification of the chordates based on their evolutionary relationships. (*Evolutionary tree based on Romer*)

B

Classification of the chordates
Phylum **Chordata** (notochord at some stage in life)

Protochordates (invertebrate chordates)
    Subphylum **Hemichordata**—acorn worms
      (worm-like; notochord confined to proboscis)
    Subphylum **Urochordata**—tunicates (sea squirts)
      (sessile; notochord in larvae only)
    Subphylum **Cephalochordata**—*Amphioxus*
      (fish-like; notochord extends full length of body)
    Subphylum **Vertebrata**
    (notochord replaced by backbone during development)
      Class **Pisces** (fishes)
      Class **Amphibia**
      Class **Reptilia**
      Class **Aves** (birds)
      Class **Mammalia**

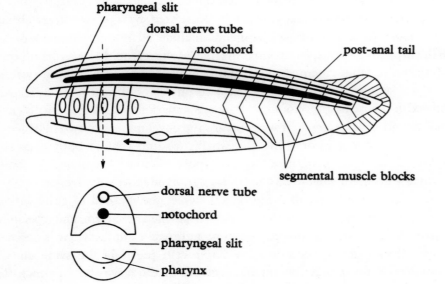

Fig 34.13 A generalized chordate showing all the 'chordate characters'. Only the 'chordate characters' are labelled. Major blood vessels are shown and the direction of blood flow is indicated by arrows. It is characteristic of chordates that blood flows forward in the ventral vessel and backwards in the dorsal vessel.

Fig 34.14 The adult sea squirt (A) bears little resemblance to a chordate. There is no trace of a notochord and, except for the perforated pharynx, none of the other 'chordate characters' are seen. But its phylogenetic affinity with the chordates is given away by its free-swimming 'tadpole' larva (B) which shows all the 'chordate characters'. The sessile adult is a filter feeder: water is drawn by cilia through the inhalent opening whence it passes through the pores in the wall of the pharynx into the atrial cavity, and so to the exterior via the exhalent opening; small particles are trapped in mucus which is drawn by cilia lining the pharynx into the stomach. (*based on Buchsbaum*)

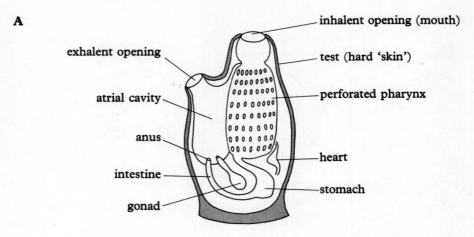

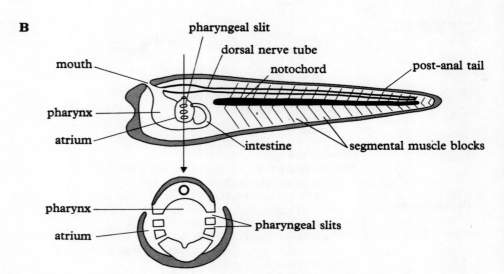

weed by its head and undergoes metamorphosis into the sessile adult, thereby obscuring its phylogenetic relationships.

THE CONNECTION BETWEEN ANNELIDS AND MOLLUSCS

In order to establish phylogenetic relationships it is sometimes necessary to compare the development of the two groups concerned, a field of study known as **comparative embryology**. This is illustrated by the two invertebrate phyla **annelids** and **molluscs**. Annelids include all the segmented worms: earthworms, ragworms, leeches and the like. Molluscs, on the other hand, contain the totally unsegmented snails and their allies, the whelks, clams, oysters, etc. On grounds of adult structure no two phyla could look more different. And yet in their early development they are remarkably similar, even down to small details. For example, if we compare the development of two marine representatives, the ragworm and the whelk, we find that in both cases the zygote cleaves to give rise to strikingly similar blastulae, in both of which a group of small micromeres are arranged in a characteristic pattern above a smaller number of larger macromeres; the method of gastrulation is the same in both, as is the formation of the mesoderm and coelomic cavities. The mesoderm affords the most impressive similarity: it is derived entirely from the proliferation of a single cell; precisely the same one in both cases. If this is not convincing, a glance at the larva should clinch the matter. Both animals share a so-called **trochophore larva** (Fig 34.15), a little creature with a curved gut, a characteristic girdle of cilia and a number of other diagnostic features.

EVOLUTION AND EMBRYOLOGY: A WARNING

Although embryology can be indispensible in establishing phylogenetic relationships, there has been a tendency in the past to take it too far. Thus Ernst Haeckel (1843–1919) suggested that during its embryological development an organism repeats its ancestral history or, to use Haeckel's own way of putting it, 'ontogeny recapitulates phylogeny'. There is a grain of truth in this: the presence of branchial grooves and segmental myotomes in the human embryo, for example, bears witness to a fish ancestry. But it is quite wrong to assume that an animal literally 'climbs up its family tree' during its development. After all, many of the steps in its ancestral history will have no usefulness in embryological development and will have long since been dropped out. Besides, there is every reason to suppose that embryological stages, particularly later ones, have themselves evolved and may be markedly different now from what they were in the past. This is particularly well illustrated by insects, whose larvae show almost as much diversity of form as the adults.

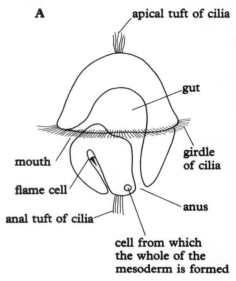

Fig 34.15 A) Generalized trochophore larva showing its characteristic shape and distribution of cilia. This type of larva is possessed by annelids and molluscs and indicates a close phylogenetic relationship between these otherwise sharply contrasting phyla. B) and C) scanning electron micrographs of trochophore larvae of the polychaete *Harmothöe*. Notice the girdle of cilia characteristic of the trochophore larva. (A: *modified after Barrington, 1967*. B and C: *A. Boyde, University College, London*)

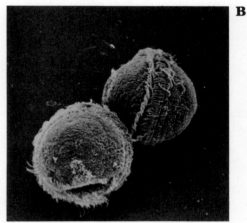

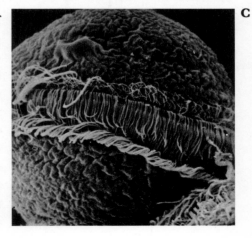

Cell Biology

From the research carried out on the structure and function of cells in recent years, one particularly interesting fact emerges, namely that in many details of their biochemistry and fine structure cells of different organisms are remarkably alike. Indeed chemicals such as **nucleic acids**, ATP and **cytochrome**, and organelles like **mitochondria, endoplasmic reticulum** and **ribosomes** appear to be of almost universal occurrence. This strongly supports the view that all living things have had a common ancestry. Even viruses, which at first sight appear very different from other organisms, possess nucleic acid.

However, many structures and chemical substances are not ubiquitous but are confined to specific groups of organisms. For example, most plants contain **chlorophyll, cellulose** and **starch**, all of which are absent from the tissues of animals; vertebrates possess **adrenaline** and **thyroxine**, neither of which are found in other groups; and the brown algae are the exclusive possessors of the orange pigment **fucoxanthin** which is not found in any other group of plants. Organisms sharing the same chemical characteristics are considered to be more closely related than those lacking such affinities. This principle, known as **biochemical homology**, has been used in recent years to confirm phylogenetic relationships. Let us look at an example in detail.

DISTRIBUTION OF PHOSPHAGENS

It has long been suspected, on grounds of comparative anatomy and embryology, that the animal kingdom can be divided, phylogenetically speaking, into two great stocks: the **chordates** and **echinoderms** on the one hand;

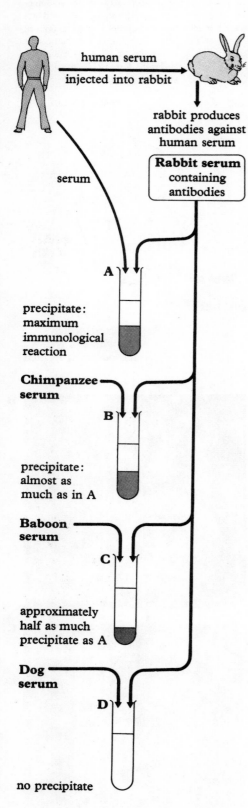

Fig 34.16 The principle underlying serological tests as a means of establishing evolutionary relationships. (*after Moody*, Introduction to Evolution, *Harper, 1962*)

human serum
injected into rabbit

rabbit produces
antibodies against
human serum

Rabbit serum
containing
antibodies

serum

A

precipitate:
maximum
immunological
reaction

**Chimpanzee
serum**

B

precipitate:
almost as
much as in A

**Baboon
serum**

C

approximately
half as much
precipitate as A

**Dog
serum**

D

no precipitate

and the **annelids, molluscs** and **arthropods** on the other. It is thought that these two stocks diverged from a common ancestor many hundreds of millions of years ago and that each has radiated along its own particular lines. The morphological evidence for this, though weighty, is by no means conclusive. However, it is given considerable support by the distribution of **phosphagens**. Present in the muscles of all animals, the function of phosphagens is to provide phosphate groups for the synthesis of ATP. Now phosphagens occur in two forms: **phosphocreatine** and **phosphoarginine**. Chemical analysis of muscle tissue obtained from a wide cross-section of animals has shown that phosphocreatine is found almost exclusively in chordates and echinoderms, whereas phosphoarginine is found in most of the other phyla.

In drawing conclusions from this kind of data one must of course beware of convergent evolution resulting in chemical similarities between groups that are in fact quite unrelated. It so happens, for example, that the muscles of certain annelids contain phosphocreatine instead of phosphoarginine, but this does not mean that we immediately put these animals with the chordates. The freak occurrence of phosphocreatine in the 'wrong group' might be merely an evolutionary quirk of no phylogenetic significance. Again, the occurrence of uric acid as the nitrogenous waste in birds and insects is not indicative of any phylogenetic affinity between these far-removed groups. It is simply the result of convergent evolution arising from the necessity for both groups to conserve water.

DISTRIBUTION OF BLOOD PIGMENTS

Another example of biochemical homology is seen in the distribution of blood pigments in the animal kingdom. There are four such pigments: **haemoglobin, haemoerythrin** and **chlorocruorin**, all of which contain iron in their prosthetic groups, and **haemocyanin** which contains copper. Analysis of the bloods of different animals has shown that haemoglobin is very widely distributed, being found in all vertebrates and many invertebrates including the earthworm. Chlorocruorin, however, is confined to polychaetes, and haemoerythrin to several closely related minor groups such as sipunculids and brachiopods. Haemocyanin, the second most abundant pigment after haemoglobin, occurs in many molluscs and crustaceans. The point is that these pigments do not follow a haphazard distribution but occur in phylogenetically related groups.

SEROLOGICAL TESTS

A useful method of demonstrating possible evolutionary affinities is to compare the blood proteins of different animals. It is argued that the closer two species are phylogenetically, the more alike their blood proteins are liable to be. Demonstrating the closeness of animals on this basis involves the use of **serological tests,** the theory behind which is as follows (Fig 34.16). Suppose we want to find out the affinity between man and certain other mammals. If we inject some human serum into, say, a rabbit, the latter responds by producing antibodies against the antigens in the serum. If some of these antibodies are then mixed with human serum, an immunological reaction takes place with the formation of a precipitate. Now if such antibodies are mixed with serum obtained from a whole series of animals, the amount of precipitation gives an indication of the degree of similarity between the different sera. The greater the amount of precipitation the closer is the serum to that of the human. The complete absence of a precipitate indicates a wide difference between the two sera. What do the results of such tests indicate? They tell us, for example, that human blood is much closer to that of the great apes like the chimpanzee and gorilla than it is to, say, the baboon; from which it may be tentatively inferred that man has a closer affinity with the chimpanzee than with other primate groups. To some extent this conclusion is supported by other information, for example the distribution of blood group systems. However, the data is by no means conclusive and is open to alternative explanations. Such data should only be used to confirm inferences drawn from other evidence, and should not by itself form the basis of hard and fast conclusions.

Palaeontology

The evidences for evolution presented so far are all indirect, based on studying animals and plants living to-day. Direct evidence comes from studying the animals and plants of the past as seen in the fossil record, a branch of biology known as **palaeontology.**

HOW FOSSILS ARE FORMED

Fossils are generally preserved in **sedimentary rock**, which is formed by the deposition of silt, sand, or calcium carbonate over thousands or millions of years. Silt deposits give **shales**, sand gives **sandstone**, and calcium carbonate (which may be precipitated from solution or derived from the shells of animals) gives **limestone**.

The most common method of fossilization involves the ultimate conversion of the hard parts of the body (bones, teeth, shells) into rock. What happens is that when an animal such as a bony vertebrate dies, the organic material in its bones gradually decays away, resulting in the bone becoming porous. If the animal subsequently becomes buried in mud, mineral particles infiltrate into the bones and gradually fill up the pores. If and when the mud turns into rock, the bones harden and are preserved for ever. Sometimes dead animals become covered by wind-blown sand, and if this is subsequently turned into mud by heavy rain or flooding, the same process of mineral infiltration may occur. Fortunately sedimentary rock is comparatively soft – this is why it is so quickly and easily eroded – and it is comparatively easy for a palaeontologist, armed with hammer and chisel, to separate the fossil from the softer rock or **matrix** surrounding it.

Although we generally associate fossils with bones and shells, the process is by no means confined to animals. Given the right circumstances plants can also become fossilized. For example, there are instances of entire tree trunks being preserved as in the 'Petrified Forest' of Arizona (Fig 34.17). In this case the original organic matter of the wood became replaced, particle by particle, by silica, which was carried into the logs by water from the sediment in which the trees were buried. The result was that the wood was literally turned into

Fig 34.17 View of the 'Petrified Forest' in north eastern Arizona. These great logs are the remains of araucaria-like trees that flourished in this area in the triassic period about 160 million years ago. Killed by fires from lightning and possibly fungus attacks, the trees were washed downstream in flood waters and rapidly buried, their wood subsequently becoming replaced by silica. (*M. B. V. Roberts*)

rock, a process called **petrifaction**. So detailed is this transformation that the individual xylem elements and annual rings can be clearly seen.

Sometimes the organism, particularly when buried in rapidly hardening mud, decays completely and the space it occupied becomes filled with another kind of material. This process results in the formation of a **mould**. Sometimes, as in the case of many shells, the mould is composed of silica while the surrounding rock is of limestone. Separation can then be carried out by chemical means, the limestone being removed by hydrochloric acid.

Occasionally footprints made by an animal walking on mud become covered over by successive layers of deposit which subsequently harden. There is a line of weakness between successive layers which, if split, reveals these fossil **impressions**. Although the information they provide may be fragmentary, such impressions can tell us quite a lot about the nature of the fauna at different stages in geological history. Impressions have also provided useful information about leaves, scales, feathers and other surface structures. An example is given in Fig 34.18.

In certain circumstances an organism may be preserved by being immersed in some kind of natural preservative. A spectacular example is provid-

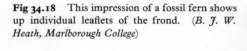

Fig 34.18 This impression of a fossil fern shows up individual leaflets of the frond. (*B. J. W. Heath, Marlborough College*)

ed by insects preserved in **amber**. These creatures were caught in the resin exuding from coniferous trees. On hardening, the resin turned to amber, preserving the animals intact. Subsequent serial sectioning of the embedded specimens has enabled their internal anatomy to be described in detail. Another natural preservative is the asphaltic oil found in **tar pits** and **asphalt lakes** in various parts of the world. For example, complete skeletons of the sabre-toothed tiger and other extinct mammals have been obtained from the tar pits at Rancho La Brea in Southern California. Yet another means of preservation is **ice**. Complete specimens of a mammoth and woolly rhinoceros have been found frozen in the ice in northern Siberia. The mammoth's flesh is said to have been in such a good state of preservation that it was eaten as steaks by the local people.

From the foregoing account the impression may have been given that fossilization is a very common process. On the contrary it is extremely capricious, requiring just the right conditions and circumstances. In fact a comparatively small percentage of animals become preserved after they have died and this, coupled with the fact that many fossils get destroyed by erosion or are inaccessible, means that the geological record is for the most part scanty and incomplete. Nevertheless there are certain places, such as the famous **Ludlow Bone Bed** in Shropshire and the great **dinosaur beds** in Utah and Colorado, where large numbers of fossils have been found concentrated together in a comparatively small area (Fig 34.19). It is thought that in these cases large populations of animals met a sudden end as the result of a cataclysmic change in the environment. The Ludlow Bone Bed contains innumerable fragments of armoured fishes which are thought to have perished as a result of a sudden change in the salinity of the water. In the case of the American dinosaur beds it is assumed that the animals were drowned and then washed downstream after a sudden flood. Excavation of these fossils and careful reconstruction has enabled palaeontologists to build up a reasonably complete picture of the flora and fauna that has existed on this planet during past ages and of the environmental changes which accompanied their evolution.

RECONSTRUCTING EVOLUTIONARY PATHWAYS FROM THE FOSSIL RECORD
It is obviously important to arrange extinct animals and plants into some kind of chronological sequence, and to demonstrate how, and when, one group evolved into another. The fact that fossils are formed in sedimentary rock helps us to do this. In the formation of sedimentary rock a layer of mud hardens and then another layer is formed on top of it, and so on. The resulting rock consists of a series of horizontal strata, each containing fossils characteristic of the time when it was laid down. The oldest rock, and therefore the earliest fossils, are contained in the lowest strata; the youngest rock, and most recent fossils, are in the highest strata. It follows that we should be able to trace the evolutionary history of a particular group by examining their fossil remains in successive strata. But how do we get at the lowest layers? Fortunately nature has come to our aid by eroding sedimentary rock in such a way that the different strata are exposed. Nowhere can this be seen better than in the Grand Canyon in Arizona. Almost two kilometres deep, the canyon has been cut by the eroding action of the Colorado River which has exposed a series of strata spanning some 500 million years of geo-

Fig 34.19 Dinosaur bones exposed in the famous quarry in north-eastern Utah, USA. Parts of over 300 individual dinosaurs, belonging to ten different species, have been excavated from the quarry. Over 20 of them are virtually complete skeletons which are now mounted in various American museums. The dinosaurs discovered in the quarry are all Jurassic forms approximately 140 million years old. (*US National Park Service*)

logical history. Studying the fossil inhabitants of these different strata, not only in the Grand Canyon but in many other parts of the world, has made it possible to reconstruct the evolution of successive groups of animals and plants during geological time. This is summarized in the geological chart given in Table 34.1.

Era (and Duration)	Period	Estimated Time since Beginning of each Period (in Millions of Years)	Fauna and Flora
Cenozoic (age of mammals: lasted about 70 million years)	Quaternary	1	Modern species of mammals, extinction of large forms, such as mammoth; dominance of man.
	Tertiary	70	Rise of birds and placental mammals.
Mesozoic (age of reptiles: lasted about 120 million years)	Cretaceous	120	Dominance of flowering plants commences; extinction of large reptiles and ammonites by end of period.
	Jurassic	155	Reptiles dominant on land, sea, and in air; first birds; archaic mammals.
	Triassic	190	First dinosaurs, turtles, ichthyosaurs, plesiosaurs; cycads and conifers dominant.
Palaeozoic (lasted about 360 million years)	Permian	215	Radiation of reptiles, which displace amphibians as dominant groups; widespread glaciation.
	Carboniferous	300	Ferns as dominant plant group; sharks and crinoids abundant; radiation of amphibians; first reptiles.
	Devonian	350	Age of fishes (mostly fresh-water); first trees and first amphibians.
	Silurian	390	Invasion of the land by plants and arthropods; brachiopods; primitive jawless vertebrates.
	Ordovician	480	Appearance of vertebrates (armoured fishes); brachiopods and cephalopods dominant.
	Cambrian	550	Appearance of all invertebrate phyla and many classes; dominance of trilobites and brachiopods; diversified algae.

Table 34.1 Table of geological periods together with a brief summary of the animals and plants characteristic of each period as inferred from the fossil record. The Proterozoic and Archaeozoic eras, predating the Palaeozoic, are omitted since fossils were not abundant at that time. The Periods are subdivided into Epochs which for the Tertiary are the Palaeocene, Eocene, Oligocene, Miocene, and Pliocene; and for the Quaternary: Pleistocene and Recent. The first monkeys appeared during the Oligocene, and man during the Pleistocene. (*from Romer*, The Vertebrate Story, *University of Chicago Press, 1959*)

DATING FOSSIL REMAINS

In the geological chart shown in Table 34.1 you will notice that approximate times are given for the different eras. These figures are based on estimating the ages of the various rocks. How is this done? Very rough estimates can be made on the basis of how long it takes for sedimentary rocks to be laid down, but such estimates may be very wide of the mark. More accurate methods are based on the fact that the older the rock, the less radioactive it is. This decline in radioactivity is due to radioactive elements decaying to their non-radioactive isotopes. By estimating the rate at which uranium decays to lead, or potassium to argon, it is possible to make reasonably accurate datings of rocks and fossils. The potassium-argon method is particularly useful because potassium is a common element found in all sorts of rock, and it decays into its argon isotope extremely slowly. It is therefore possible to date very old rocks, over three billion years old in fact, in this way. For younger fossils which still contain some organic material, radioactive **carbon dating** can be used. This method is based on the fact that after an organism dies the radioactive carbon (^{14}C) it contains gradually disintegrates into normal carbon. As disintegration is rapid, this method can only be used for dating recent fossils not more than about 70,000 years old, but it is an accurate method and has been used to make precise datings not only of fossils but also of archaeological remains.

What kind of evolutionary sequences have been established from the fossil record? There are obviously far too many for us to be able to look at all of them, so we shall choose two for detailed discussion: the evolution of horses and the vertebrate ear ossicles.

THE EVOLUTION OF HORSES

The history of horses is summarized in Fig 34.20. It starts with a little animal called *Hyracotherium* (old name: *Eohippus*) which lived in North America in the Eocene epoch about 50 million years ago and then spread across to Europe and Asia. Fossil remains of *Hyracotherium* obtained from Eocene rocks in North America show it to have differed from modern horses in three important respects: it was very much smaller, had well developed digits (four on the forefoot, three on the hindfoot), and possessed low-crowned molar teeth lacking the serrated surface typical of modern horses. We know that *Hyracotherium* lived in rather marshy, well-wooded country in which its spreading toes would have afforded it much better support than a hoof. It probably fed on soft vegetation and fruit, for which its non-grinding molars would have been perfectly adequate.

The subsequent evolution of horses can be followed through a complex series of fossils obtained from successively younger rock strata. The series finishes up with the modern horse *Equus*, and there are at least a dozen intermediate forms, not to mention many off-shoots from the main line. The record is pretty complete and shows several changes to have taken place. But before considering these it must be pointed out that they did not take place in the beautifully graded fashion suggested by Fig 34.20. It was much more complicated than this, changes taking place, not gradually, but irregularly and sporadically.

First, then, we see an increase in size. *Hyracotherium* was barely 0.4 m tall. The later fossils show a progressive increase in size, finishing up with the modern horse which stands a good 1.5 m off the ground. Second, the third digit gets stouter and longer; meanwhile the remaining ones are withdrawn so that by the Pliocene period they have disappeared altogether. The lengthening of the third digit is brought about by great elongation of the metacarpal and metatarsal, this being accompanied by conversion of the distal phalange (or rather its nail) into a hoof. Thus the modern horse stands on the tip of its third digit. The third change concerns the teeth. From being low-crowned in *Hyracotherium*, the molar teeth acquire higher crowns with a complete covering of cement. Wearing of the surface results in the serrated structure typical of the modern horse. These changes are accompanied by a transformation of the premolars into molar-type teeth.

The three changes just described can be directly related to changes in the environment. The fossil plants found in the different strata tell us that the marshy, wooded country in which *Hyracotherium* lived was gradually replaced by a drier type, and the Miocene descendants of *Hyracotherium* found them-

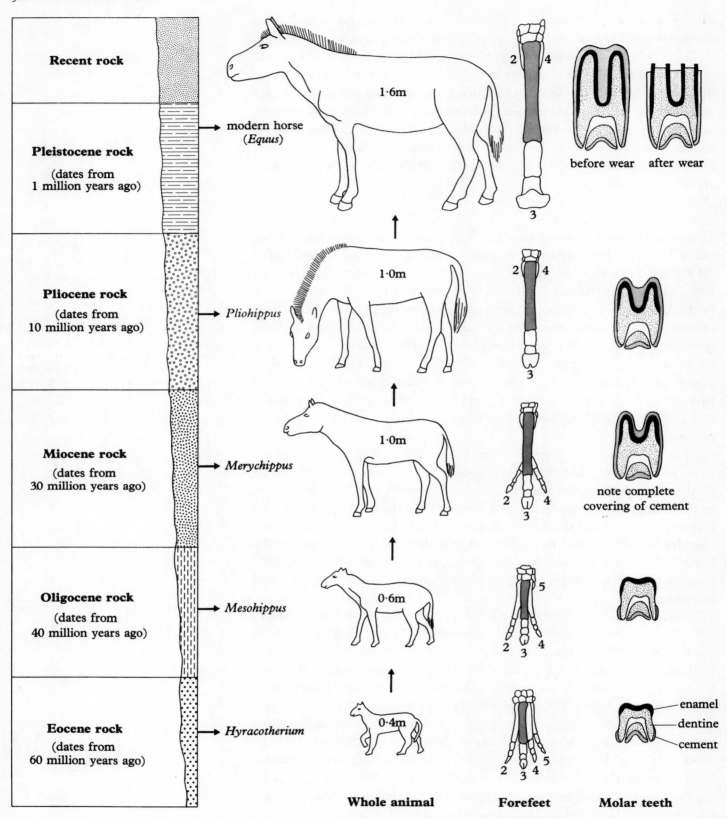

Fig 34.20 Evolution of the horse showing re-constructions of the fossil species obtained from successive rock stata. The foot diagrams are all front views of the left forefoot. The 3rd meta-carpal is shaded throughout. The teeth are shown in longitudinal section. *(based on de Beer, 1964)*

selves living in open prairie offering little concealment. Survival now depended on the head being in an elevated position for gaining a good view of the sur-rounding countryside, and on a high turn of speed for escape from predators. Hence the increase in size and the replacement of the splayed-out foot by the hoofed foot. The drier, harder ground would make the original splayed-out foot unnecessary for support. The changes in the teeth can be explained by assuming that the diet changed from soft vegetation to grass.

EVOLUTION OF THE VERTEBRATE EAR OSSICLES

This story, also well documented by fossils, is extraordinary because it illustrates how structures can radically change their function in the course of evolution. The sequence is summarized in Fig 34.21. Fishes, both living and fossil, have no middle ear. The back of the jaws are held in position by a brace, the **hyomandibular**, which runs from the point of articulation between the upper and lower jaws to the back of the cranium just by the auditory capsule. In amphibians and reptiles the first of a series of remarkable changes takes place. The upper jaw, previously slung from the cranium by ligaments, becomes fused to the floor of the cranium. The hyomandibular, now released from its

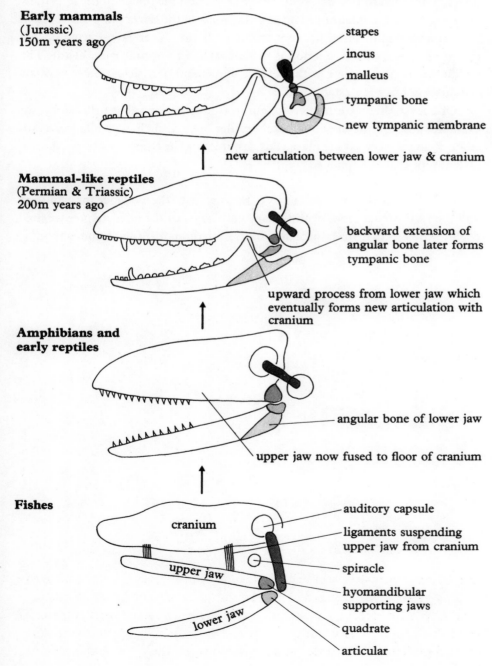

Early mammals
(Jurassic)
150m years ago

stapes
incus
malleus
tympanic bone
new tympanic membrane

new articulation between lower jaw & cranium

Mammal-like reptiles
(Permian & Triassic)
200m years ago

backward extension of angular bone later forms tympanic bone

upward process from lower jaw which eventually forms new articulation with cranium

Amphibians and early reptiles

angular bone of lower jaw

upper jaw now fused to floor of cranium

Fishes

cranium

auditory capsule
ligaments suspending upper jaw from cranium
spiracle
hyomandibular supporting jaws
quadrate
articular

upper jaw

lower jaw

Fig 34.21 Evolution of the vertebrate ossicles. The diagrams are based on fossils which show the incorporation of the hyomandibular of fishes into the amphibian middle ear cavity as the stapes. In the evolution of the mammal-like reptiles the stapes is joined by the articular and quadrate bones from the back of the jaws. These become the malleus and incus respectively. Hyomandibular and stapes in dark tone; quadrate and incus in medium tone; articular and malleus in lighter tone. (*modified after Simpson, 1957*)

function of supporting the jaws, goes into the newly-developed middle ear chamber as the **stapes**. Its function is now to transmit sound waves from the tympanic membrane to the inner ear.

In amphibians, reptiles and birds the stapes is the only ossicle in the middle ear. But in mammals two others are present: the **malleus** and **incus** (see p. 311). Where do they come from? Fossil evidence, supported by comparative anatomy and embryology, indicates that they are derived from, of all places, the upper and lower jaws. In amphibians and reptiles the jaws articulate between the **quadrate** bone at the back of the upper jaw and the **articular bone** at the back of the lower jaw. A very complete series of fossil mammal-like reptiles shows that the articular and quadrate gradually became 'pinched off' the back of the lower and upper jaws and incorporated into the middle ear as the malleus and incus respectively. Meanwhile a new jaw articulation is set up between an upward process from the lower jaw and the region of the upper jaw just in front of the quadrate. The **angular bone** of the lower jaw grows back in an arc and finally loses its connection with the lower jaw, becoming the **tympanic bone**, whose function is to support the **tympanic membrane**. The middle ear is now replete with its full complement of ossicles, and a new jaw articulation has been established.

Of course there are numerous unresolved questions about this story; for example, how did the 'mammal-like' reptiles hear, and chew, while these fantastic changes were taking place? But despite such functional problems there is little doubt that it happened.

This chapter has been primarily concerned with the evidence for evolution and the reconstruction of evolutionary pathways. In the next chapter we shall explore the mechanism by which evolution is thought to have come about.

35 The Mechanism of Evolution

Of all the many things that Darwin saw during his voyage on the *Beagle* none impressed him more than the distribution and habits of the birds on the Galapagos islands, particularly the finches discussed in the last chapter. It was this more than anything else that led him to put forward his theory as to how evolution has come about. In this chapter we will start by looking at the theory in outline and then explore its various propositions in detail.

SUMMARY OF DARWIN'S THEORY

You have only to look about you to appreciate that the different individuals of a species, including offspring of the same parents, show considerable **variation** from one another. In addition to visible phenotypic variation there is much hidden variation caused by concealed genetic differences between individuals. Some of the visible variation results from environmental influences, particularly diet, and this plays no direct part in the evolutionary process. The kind of variation which is important in evolution is that which the organism is able to hand on to its offspring, transmitted, as we now know, through the genes. Such variation may show a smooth gradation through the population (**continuous variation**) or a sudden transition between two or more extremes (**discontinuous variation**). Either way, it results in some individuals being better adapted to the environment than others. As organisms generally produce far more offspring than the environment can support, there always tends to be a **struggle for existence** due to over-crowding and competition between individuals. It follows that those individuals with favourable variations will stand a better chance of survival in this struggle. This can be expressed as the **survival of the fittest**.[1] The result is that well adapted individuals reach reproductive age and hand on their favourable characteristics to their offspring, whereas less well adapted individuals fail to do so. Nature, as it were, selects those individuals which are sufficiently well adapted and allows them to survive, and rejects those that are poorly adapted. The latter usually perish before they reach sexual maturity. Darwin coined the term **natural selection** to describe this weeding-out process. As environmental conditions are constantly changing, natural selection is for ever favouring the emergence of new forms, or, as Darwin put it, the **origin of species**.

LAMARCK'S THEORY

A central tenet of Darwin's argument is that the variations which form the 'raw material' for natural selection, arise spontaneously. They are in no way

[1] This graphic phrase, commonly attributed to Darwin, was in fact introduced by the philosopher Herbert Spencer in 1867, eight years after the publication of Darwin's *Origin of Species*. However, the idea embraced by it was Darwin's and it is customary to use it in summarizing Darwin's theory.

dictated by the environment or purposefully geared towards making the possessor better adapted to it. This idea is in direct contrast to an alternative theory put forward in 1809 (the year of Darwin's birth) by the French naturalist Jean-Baptiste de Lamarck. Lamarck proposed, with little supporting evidence, that when an organism develops a need for a particular structure, this induces its appearance. This idea was based on the observation that structures which are subjected to constant use become well developed, whereas those that are not used tend to degenerate. This in itself is not an unreasonable proposition; after all everyone knows the effect that exercise and training can have on the development of muscles in a weight-lifter. But Lamarck went on to suggest that these beneficial characteristics, acquired during an individual's lifetime, could be handed on to the progeny. In other words, evolutionary change could be achieved by the **transmission of acquired characters**. This implies that a weight-lifter who has developed enormous arm muscles produces children who are born with specially large arm muscles.

THE DARWINIAN AND LAMARCKIAN THEORIES COMPARED

To illustrate the difference between the Darwinian and Lamarckian theories consider the case of the giraffe. We know from the fossil record that the ancestors of the modern giraffe had short necks, and that in the course of geological history the neck got gradually longer and longer. How do we explain this? The Lamarckian explanation would go something like this. The ancestors of the giraffe fed on the leaves of bushes and trees and, competing with each other for food, stretched their necks in order to reach the higher branches. This condition was transmitted to the offspring who therefore started with a somewhat longer neck and repeated the process. Hence the descendants of the original stock acquired progressively longer necks.

The Darwinian explanation is based on a quite different premise, namely that the development of a long neck was a spontaneous occurrence and not the result of environmental need. In the ancestral population some individuals *happened* to have longer necks than others. Since they could reach the leaves on higher branches than their shorter-necked contemporaries, such individuals would be expected to win out in the 'struggle for existence'. Assuming the long-necked condition to have a genetic basis, these individuals would hand on this beneficial characteristic to their progeny.

In just the same way Darwin postulated that the original population of finches on the Galapagos islands produced variants which were favoured in the process of natural selection, so that they eventually evolved to fill the different ecological niches that on the mainland are occupied by other birds.

How do the Darwinian and Lamarckian theories stand today? In the course of the last hundred years a mass of evidence, much of it admittedly indirect, has been marshalled in support of Darwin's theory. On the other hand no decisive evidence whatsoever has been found in favour of Lamarck's theory. This is not to say that the environment plays no part at all in directing the course of evolution. It certainly does, though not by dictating what structures an organism should or should not develop, but by selecting those individuals which *by chance* happen to be better adapted to it.

Having briefly reviewed Darwin's theory we must now look at its propositions in detail. Although the theory is still adhered to today it has been greatly expanded largely as a result of our increasing knowledge of genetics.

You must remember that Darwin formulated his ideas with no knowledge of heredity; the work of Mendel was quite unknown to him. The Darwinian theory reappraised in terms of modern genetics is sometimes called **neo-darwinism**.

Variation

The pattern of continuous variation in a population can be studied by measuring a chosen variable in a large sample of individuals. The results can be conveniently summarized in the form of a **frequency distribution histogram**. As an example, Fig 35.1 shows how height varies in a human population. Different heights are represented along the horizontal axis, the numbers of individuals sharing each height on the vertical axis. The bell-shaped curve is typical of the results obtained when any clear-cut feature in a population of animals or plants is measured. It is known as a **curve of normal distribution** or **Gaussian curve** (after the German mathematician Karl Friedrich Gauss who first established its theoretical basis). The highest point on the graph, or **mode**, represents the most frequent or 'popular' height. Typically a normal distribution curve is symmetrical on either side of the mode, but in practice this is not always so. Sometimes frequency curves are distinctly asymmetrical. Can you think of reasons for this?

Both environmental and hereditary variations usually give normal distribution curves if measured separately. In nature, however, both operate together, and variations caused by genetic differences can only be distinguished clearly from those attributable to the environment by making measurements in an absolutely constant environment, a well-nigh impossible task.

It has already been mentioned that from the evolutionary point of view genetic variation, i.e. the kind that can be transmitted from parents to offspring, is much more important than environmental variation. With this in mind let us look into the causes of genetic variation and assess their rôle in evolution.

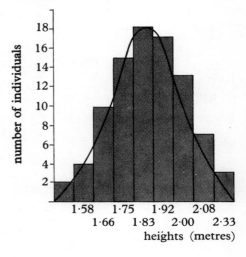

Fig 35.1 Histogram of heights in an adult human population. The size of each block indicates the number of individuals sharing that particular height. The curve is a curve of normal distribution.

THE CAUSES OF GENETIC VARIATION

Before explaining the causes of genetic variation it is necessary to clear up a problem that may be bothering you. In Chapter 28 we saw that in accordance with Mendel's laws of particulate inheritance there is generally no smooth gradation between contrasting features shown by an organism. Thus pea plants are either tall or short, coloured or white; *Drosophila* has red or white eyes, grey or black body. There are no intermediate conditions. How is it, then, that many characteristics, human skin pigmentation for example, show a complete gradation from one extreme to another? The answer is that an organism's characteristics are generally controlled not by single genes but by the interaction of numerous genes. A characteristic produced by the cumulative effect of a number of genes is called a **polygenic character**. Clearly if genes making up a polygenic complex get separated their effects are weakened. Sometimes the effect produced by one pair of genes may be hidden, or accentuated, by another pair of genes. The latter, sometimes called **modifier genes**,

Fig 35.2 Diagram showing how different kinds of gametes can be formed, depending on how homologous chromosomes segregate during meiosis. The parent cells show the different ways chromosomes can arrange themselves on the spindle prior to separating. The gametes show the different combinations of chromosomes and genes resulting from division of the parent cells. Note that eight different gametes are produced, 2 to the power of the haploid number.

parent cells

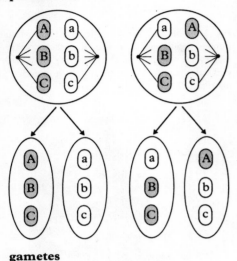

gametes

parent cells

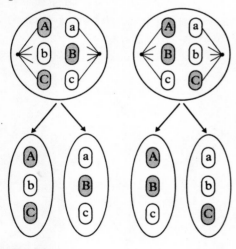

gametes

clearly affect the expression of other genes. Obviously if they become separated the individual's characteristics will be modified accordingly. Such is the behaviour of chromosomes in sexual reproduction that genes are continually being separated and recombined in sundry different ways, thus generating the variation seen in typical frequency distribution histograms.

We can take this idea a step further. With the exception of identical twins no two human beings are ever genetically identical. They may share the same height, or the same colour of eyes, but if all their features are taken into consideration the two individuals are certain to differ in some respect. Each individual possesses a unique combination of characteristics which is not repeated in any other individual.

This has the same general explanation as the distribution frequencies described earlier. An individual possesses a very large number of different characteristics controlled by a vast number of genes. The reshuffling of genes that occurs in sexual reproduction ensures a wide range of variation in the population as a whole. The truth of this is seen in the fact that human offspring are often markedly different from their parents, and even brothers may be surprisingly unlike each other.

RESHUFFLING OF GENES

How does this genetic variation come about? Opportunity for reshuffling of genes is provided by **meiosis**, which is described in Chapter 23. During metaphase of the first meiotic division homologous chromosomes come together in pairs, and subsequently segregate into the daughter cells, independently of each other. The result of this **independent segregation** is the production of a wide variety of different gametes depending on which particular chromosomes are combined in the daughter cells. This in turn depends on the way the different chromosomes line up on the spindle prior to separating. The total number of possible combinations depends on how many pairs of chromosomes there are in the parent cell. To illustrate this, consider three pairs of chromosomes carrying the alleles **Aa**, **Bb** and **Cc** respectively. From Fig 35.2 it can be seen that the various ways they segregate gives a total of eight different combinations in the gametes. In general terms, the number of combinations is 2^n where n is the haploid number of chromosomes, in this case three. In man, with a haploid number of 23, the number of combinations is $2^{23} = 8,388,608$. No wonder brothers never look exactly alike.

The mechanism described above, though important in promoting genetic variety, can only mix up genes carried on different chromosomes. It can play no part in separating and recombining genes located on the same chromosome. The opportunity for this kind of reshuffling is provided by **crossing-over**.

In prophase of the first meiotic division, when homologous chromosomes are in intimate contact with one another, the chromatids of homologous chromosomes break and rejoin at certain points called **chiasmata** (see p. 374). The number and position of the chiasmata relative to the sequence of genes determines the different combinations possible in the gametes. This is made clear in Fig 35.3 and no further elaboration is required here. The important point to note is that it is another important source of genetic variety.

Still further variety is achieved on **fertilization**. Union of gametes permits parental genes to be brought together in different combinations. In this way good, and bad, qualities of the parents can be combined in the offspring.

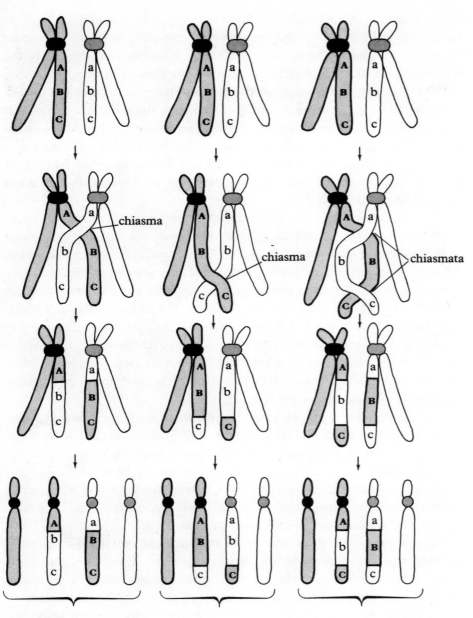

Fig 35.3 Diagram showing how crossing-over between the chromatids of homologous chromosomes can lead to genetic variety. The new combinations possible in the gametes depends on the number of genes involved and their positions on the chromosomes relative to the chiasmata. Crossing-over takes place during prophase of the first meiotic division.

THE ROLE OF RESHUFFLING OF GENES IN EVOLUTION

The cellular mechanisms just described provide the basis of continuous variation in a population. However, although they promote genetic variety they play only a limited rôle in evolution. The reason is that independent segregation of chromosomes and crossing-over may establish a new combination of genes in one generation but the same processes can just as readily 'undo' it in later generations. Thus genetic novelty produced this way is generally nonpersistent and therefore plays only a limited part in effecting long-term evolutionary change. A notable exception is seen when two genes are brought into very close proximity as a result of crossing-over. This lessens the chance of a chiasma being formed between them in later generations with the result that they may remain together indefinitely. But for the most part mere reshuffling of genes produces only short-term changes. Much more profound initiators of genetic variety from the evolutionary point of view are **mutations,** to which we now turn.

WHAT ARE MUTATIONS?

Every now and again a natural population of animals or plants throws up an individual with some characteristic strikingly different from the rest of the population, for example, haemophilia and cystic fibrosis in man, white eyes and vestigial wings in *Drosophila*, resistance of bacteria to penicillin or flies to DDT – the list is unending. It can be shown from studies of inheritance that these conditions are generally recessive and are transmitted in a Mendelian fashion. The genetical mechanism producing the change is known as a **mutation**. Any descendants of an individual with a mutation, showing the new characteristic, are called **mutants**. Mutations are the basis of discontinuous variation in populations.

It is characteristic of mutations that they are comparatively rare, at least in terms of the rate at which gametes are produced. In most organisms the mutation rate varies between one and 30 per million gametes. The rarity of mutations can also be seen in their occurrence in a population, the **mutation frequency**. At any one gene locus at each generation a mutation occurs in only one individual out of approximately 500,000. The exact number is variable since the genes at different loci have different mutation rates. Expressed in this way the mutation rate certainly seems to be extremely low. However, it must be remembered that each individual has a very large number of loci at which it can happen. The overall mutation rate is therefore much greater than is realized by considering a single locus.

The low rate of mutation made it difficult to study them until the discovery in 1927 by the American geneticist H. J. Muller that the mutation rate of *Drosophila* can be greatly accelerated by irradiation with X rays. Since then it has been found that other factors also speed up the mutation rate. These **mutagenic agents** include gamma rays, ultraviolet light and a variety of chemicals including mustard gas. This has made it possible to study the development and transmission of mutations in some detail.

From these studies three salient facts emerge. First, mutations arise spontaneously and are undirected by the environment. Environmental influences can affect the mutation rate but they cannot induce a particular mutation to take place. From the evolutionary standpoint, the only direct part played by the environment is in selecting mutants that happen to possess advantageous characteristics. Second, mutations, though not necessarily permanent, are relatively persistent. A mutant gene may be transmitted through many generations without further change, after which it may mutate again producing yet another novel characteristic, or it may change back to its original condition. But the fact that it can persist unaltered for at least several generations makes it possible for natural selection to get to work on it. Third, the vast majority of mutations confer disadvantages on the organisms that inherit them. The occurrence of a useful mutation is an extremely rare event indeed. In the course of evolutionary history infinitely more harmful mutations have been selected against than beneficial mutations selected for. This being so, evolution would seem to be a wasteful process, and it certainly is in terms of the lives of individual organisms. In the long history of animals and plants there have been innumerable 'bosh shots' and relatively few successes. But by transmitting favourable mutations to their offspring it is the successes that have been responsible for the wide variety of highly adapted forms which populate our planet today.

THE SPREAD OF MUTATIONS THROUGH A POPULATION

If they are to promote evolutionary change, beneficial mutant genes must be able to spread through a population. In the rare event of such a gene being dominant it will be selected for at once and will spread quickly through the population, bringing about rapid evolutionary change. But what will happen if, as is usually the case, the mutant gene is recessive? It will only show its effects in organisms homozygous for it. The chances of two such genes being brought together into the homozygous state are slight, and many generations may pass without it happening. In the meantime, or course, the gene may be lost from the population altogether. At best, fully recessive mutant genes spread very slowly.

The whole situation can, however, be altered if the mutant gene, recessive at first, becomes dominant in the course of time. This might occur if the gene confers a beneficial effect, however slight, on the heterozygote.[1] Selection will then favour the heterozygote, and the gene will spread rapidly through the population. Moreover there will be selection for a genetic background favourable to the fuller expression of the mutant gene in the heterozygote, so that eventually it assumes full dominance. Needless to say if the gene confers a disadvantage on the heterozygote it will become fully recessive and selection will favour its complete elimination from the population.

THE TYPES OF MUTATION

In general two kinds of mutation are recognized: **chromosome mutations** and **gene mutations**. The former involve changes in the gross structure of chromosomes and are generally detectable under the microscope. The latter are chemical changes in individual genes and cannot be detected by microscopical means. Let us look briefly at each.

CHROMOSOME MUTATIONS

During meiosis when chromosomes become intertwined there is plenty of opportunity for various kinds of structural aberration to take place (Fig 35.4). For example, a chromosome may break in two places and the section in between may drop out, taking all its genes with it. The two ends then join up, giving a shorter chromosome with a chunk missing in the middle. This is called a **deletion** and insofar as it leads to the total loss of genes it can have a profound effect on the development of an organism: in fact all but the very shortest deletions are lethal. Another kind of chromosome abnormality occurs if a chromosome breaks in two places and the middle piece then turns round and joins up again so that the normal sequence of genes is reversed. This is called **inversion**. Sometimes a section of one chromosome breaks off and becomes attached to another chromosome. This is rather like crossing-over except that it occurs between non-homologous chromosomes; it is known as **translocation**. Yet another abnormality occurs when a section of a chromosome replicates so that a set of genes is repeated: this is **duplication**.

In all these cases the sequence of genes is altered. This is important because it can bring into close proximity genes whose combined effect produces a beneficial characteristic. The closer together such genes are on the same chromosome the less likely they are to be separated by crossing over, and the more persistent will be the resultant characteristic. All the abnormalities described so far are visible when homologous chromosomes pair during meiosis.

[1] Under these circumstances the mutant gene is not fully recessive. Because it exerts some effect in the heterozygous condition it should be described as semi-dominant (see p. 476).

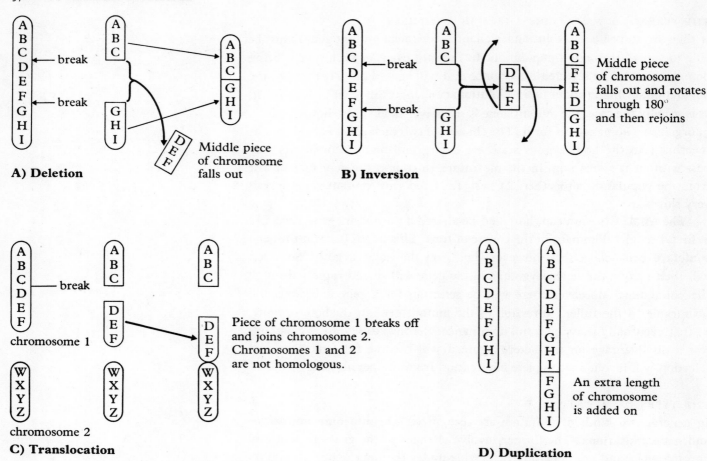

A) Deletion

Middle piece of chromosome falls out

B) Inversion

Middle piece of chromosome falls out and rotates through 180° and then rejoins

C) Translocation

chromosome 1

chromosome 2

Piece of chromosome 1 breaks off and joins chromosome 2. Chromosomes 1 and 2 are not homologous.

D) Duplication

An extra length of chromosome is added on

Fig 35.4 Diagrams showing different types of chromosome mutation. Note that each type of mutation results in an alteration in the number and/or sequence of genes (represented by letters) on the chromosome.

Since equivalent loci lie alongside each other, any change in the sequence of genes will prevent the chromosomes lying parallel with one another. Thus deletion will cause a loop, inversion a twist, and so on.

Another kind of chromosome abnormality is caused by the addition or loss of one or more whole chromosomes. To understand how this comes about we must return to meiosis for a moment. Normally in meiosis homologous chromosomes come together and then segregate into separate cells, so that the gametes finish up with only one of each type of chromosome instead of the normal two. However, on some occasions the two homologous chromosomes, instead of separating, go off into the same gamete. This phenomenon is known as **non-disjunction** and it results in half the gametes having two of the chromosomes whilst the other half have none (Fig 35.5). The fusion of the first kind of gamete with a normal gamete of the other sex will give an individual with three such chromosomes, i.e. the normal pair plus an extra one. Fusion of the second kind of gamete with a normal gamete will give an individual with only one of this particular type of chromosome.

Quite how important non-disjunction has been in generating useful genetic novelty is uncertain, but there is no doubt that it can have a profound effect on an organism's development. For example, **mongolism** (better called **Down's syndrome** after the clinician who first described it) is now known to be caused by the presence of an extra chromosome in the cells, chromosome number 21 to be precise (Fig 35.6). This is one of the smallest of all the human chromosomes, and yet its presence plays havoc with the individual's normal development. Sufferers from Down's syndrome, if they survive at all, have a

A) Normal meiosis

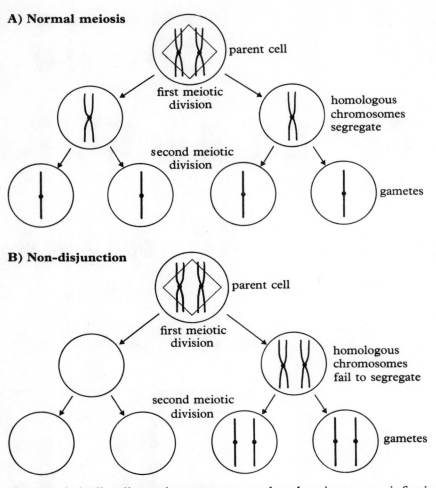

B) Non-disjunction

Fig 35.5 Diagrams comparing normal meiosis (A) with non-disjunction (B), for one pair of chromosomes. In normal meiosis the homologous chromosomes segregate, so the gametes each contain one chromosome. In non-disjunction the chromosomes fail to segregate, so half the gametes contain two chromosomes each and the other half contain no chromosomes at all. Generally non-disjunction takes place with respect to just one pair of homologous chromosomes, the rest behaving normally.

characteristically slit-eyed appearance, reduced resistance to infection and are always mentally deficient, all because of one extra chromosome.

Also caused by non-disjunction are various sex-chromosome abnormalities. For example, some unfortunate individuals have the genetic constitution **XXY (Klinefelter's syndrome)** caused by failure of the **X** chromosomes to separate during oögenesis in the mother. Normally such individuals are outwardly males but have some female characteristics and fail to manufacture sperm. The reverse situation, i.e. where an **X** chromosome is missing, gives a condition called **Turner's syndrome**. Such individuals have the genetic constitution **XO** and are sterile females. Many other sex-chromosome anomalies are now known, and all arise as a result of abnormal behaviour of the chromosomes during meiosis. Some of these conditions are associated with mental deficiency. Recently it has been suggested that the **XYY** combination may be associated with psychopathic traits and a criminal tendency. This is claimed on the grounds that maximum security prisons contain a significantly higher percentage of **XYY** individuals than the general population.

POLYPLOIDY

Sometimes all the chromosome pairs undergo non-disjunction simultaneously, resulting in half the gametes having two of each type of chromosome (i.e. diploid), the rest having none. If a diploid gamete fuses with a normal haploid gamete the resulting individual is **triploid**, i.e. it has three of each type of

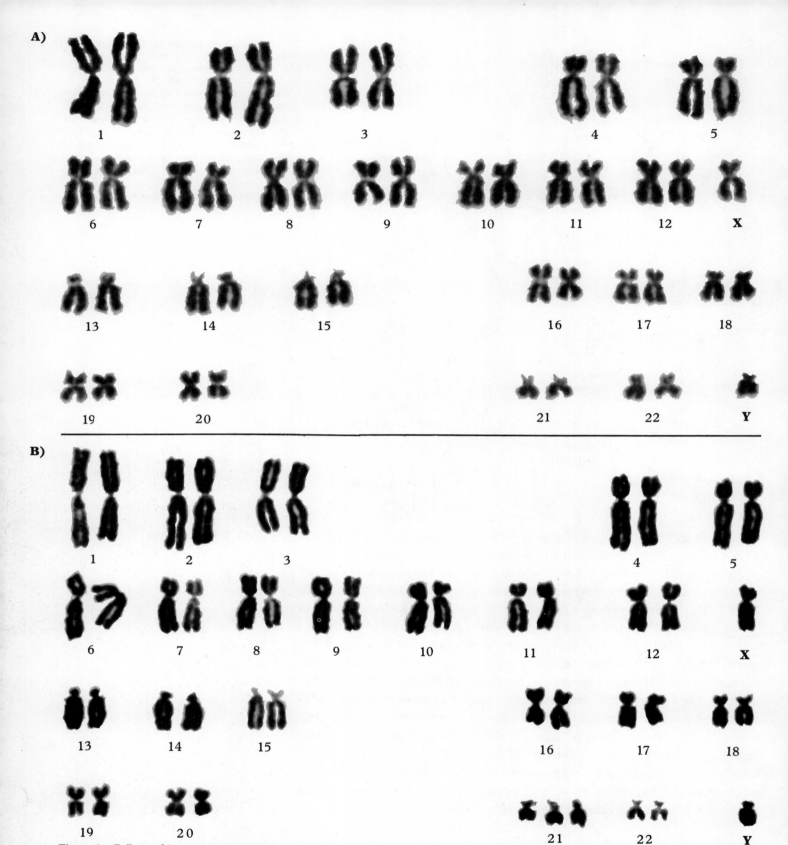

Fig 35.6 Full set of human chromosomes arranged in sequence. Each chromosome can be seen to consist of a pair of chromatids joined at the centromere. The chromosomes differ from one another in size and position of centromere. A) Normal male. B) Mongol male: notice the extra chromosome 21. (A: *E. H. R. Ford, University of Cambridge*. B: *G. E. Roberts, Ida Darwin Hospital, Fulbourn*).

chromosome. If two diploid gametes fuse, a **tetraploid** individual results. It is thus possible for an organism to acquire one or more complete extra sets of chromosomes, a phenomenon called **polyploidy**.

Polyploidy is rare in animals but common in plants where it is often associated with advantageous characteristics: greater hardiness, resistance to disease, and so on. To some extent these advantages are offset by its often being associated with lowered fertility, a fact which may have considerably reduced its importance as an instigator of useful evolutionary change. Nevertheless there is no doubt that in plants polyploidy has played a significant part in the formation of new species.

As an example of how polyploidy can contribute to evolutionary change consider the 'rice grass' *Spartina townsendii* which was mentioned in Chapter 14 in connection with its ability to thrive in salt water. This species is the hybrid formed by crossing *S. stricta* (diploid chromosome number 56) with *S. alterniflora* (diploid number 70). If the hybrid were formed by the fusion of normal haploid gametes one would expect it to have a diploid number of 63. However, it seems that in the formation of this particular hybrid no reduction division occurred, resulting in the formation of a polyploid with a chromosome number of 126. What is the significance of this in evolution? In the nineteenth century the prevalent species on the English south coast was *S. stricta*. Sometime during that century accidental hybridization took place between the English species and *S. alterniflora* introduced from North America. *S. townsendii* first appeared in Southampton Water around 1870. A tough and highly fertile halophyte, the new species spread rapidly, eventually replacing the old species. Today *Spartina townsendii* is the dominant species on the south coast and indeed certain areas are choked up with it (Fig 35.7). This illustrates the general principle that the offspring resulting from a cross between two genetically dissimilar lines often possess beneficial characteristics (including higher fertility) not shown by either of the parents. This is termed **hybrid vigour**. Generally it is caused by increased heterozygosity as a result of gene mixing, but in the case of *Spartina* it is clearly associated with the polyploid condition. Polyploidy is particularly common amongst grasses and is responsible for the ability of certain species to survive in arctic and desert conditions.

Doubling of the chromosome number can be induced experimentally by heat or cold shock or by various chemical agents, notably **colchicine**, an alkaloid substance extracted from the crocus *Colchicum*. Applied in the correct concentration, colchicine prevents spindle-formation during mitosis. The chromosomes replicate in the usual way and the chromatids separate. However, the cell fails to divide and when the nuclear membrane reforms it envelops twice the normal number of chromosomes, i.e. the cell becomes tetraploid.

Colchicine and other agents have been used to induce polyploidy in various crops, including wheat, thereby increasing their hardiness and vigour. This is an example of man altering the evolution of a species to his own ends, a topic we shall return to at the end of this chapter.

GENE MUTATIONS

Gene mutations are thought to have played by far the most important part in generating evolutionary change. A gene mutation arises as a result of a chemical change in an individual gene. An alteration in the sequence of nucleotides in that part of a DNA molecule corresponding to a single gene will change the order of amino acids making up a protein, and this can have far-reaching consequences on the development of an organism. In some cases the proper functioning of some vital protein may be completely prevented. If this results in the early death of the organism, the genes are described as **lethal genes** (see p. 478).

Fig 35.7 Rice grass, *Spartina townsendii*, a polyploid formed by hybridization between *S. stricta* and *S. alterniflora*. A vigorous halophyte, it is here seen flourishing in the salt marshes around Poole Harbour, Dorset. (*B. J. W. Heath, Marlborough College*)

Drastic effects can sometimes be produced by a seemingly trivial change in the nucleotide sequence of a gene. This can be seen, for example, in the formation of **haemoglobin**. In the inherited disease known as **sickle-cell anaemia** (see p. 477) the red blood cells, normally biconcave discs, are sickle-shaped and the victim suffers all the symptoms of extreme oxygen shortage: weakness, emaciation, kidney and heart failure etc. The anaemia is not caused by the distorted shape of the red blood cells as such, but by the fact that they contain an abnormal haemoglobin, **haemoglobin S**, which is inefficient at carrying oxygen. It has been shown that haemoglobin S differs from normal haemoglobin only in that each of its two β polypeptide chains contains the amino acid valine instead of glutamic acid at *one* position along the length of the chain. Such a 'mistake' is caused by a change in as few as two nucleotides in the DNA responsible for haemoglobin-formation. This is a seemingly trivial change from the molecular point of view, yet one has only to look at someone suffering from sickle-cell anaemia to realize what a profound effect it can have on the normal functioning of the body.

DIFFERENT KINDS OF GENE MUTATION

The kind of gene mutation that causes sickle cell anaemia is called a **substitution** because an incorrect nucleotide is substituted for the correct one. If we look upon DNA as a conveyor of coded information, we can see at once that the slightest change is enough to completely alter its information content. As an analogy consider the wrong information which might be conveyed by a misprint in a telegram in which a single incorrect letter is substituted for the correct one:

Intended message: FLOSSIE NOW ARRIVING BY AIR FROM NEW YORK.
Actual message: FLOSSIE NOT ARRIVING BY AIR FROM NEW YORK.

Such a trivial error could alter a person's destiny, just as gene mutations have undoubtedly altered the course of evolution.

Let us consider some other misprints in telegrams. Here is an example of an extra letter creeping into a message. A geneticist would call it an **insertion**:

Intended message: PLEASE SAY WHERE YOU ARE.
Actual message: PLEASE STAY WHERE YOU ARE.

On the other hand a letter might be left out (**deletion**) as in the following example:

Intended message: WILL SEND FRIEND TO COLLECT JEWELLERY.
Actual message: WILL SEND FIEND TO COLLECT JEWELLERY.

Another possibility is that two letters might be printed the wrong way round, resulting in an **inversion**:

Intended message: GUERILLAS SENDING ARMS TO AID RIOTERS.
Actual message: GUERILLAS SENDING RAMS TO AID RIOTERS.

Sometimes an inversion may involve more than two letters as in this example:

Intended message: BRING THERMOS ON OUTING.
Actual message: BRING MOTHERS ON OUTING.

Substitutions, insertions, deletions and inversions all occur in nature and can be induced in laboratory organisms by various mutagenic agents. The mistake in the genetic code may be very small and yet it can have far-reaching consequences out of all proportion to the size of the mistake itself.

THE IMPORTANCE OF MUTATIONS IN EVOLUTION

What kind of mutations are likely to have been important in evolution? Although chromosome mutations such as deletions and translocations are important in the genetics of individuals, it is likely that the most significant processes from the evolutionary point of view are those that result in an increase in the amount of genetic material. It is true that the marked inferiority of 'mongols' makes one doubt the evolutionary importance of non-disjunction, but in the long run how can we account for the different chromosome numbers and the very varied amounts of genetic material present in different species except by postulating that there has been an increase in genetic material? Although the addition of whole chromosomes may have been disastrous more often than it has been helpful, it is highly likely that the addition of new genes within individual chromosomes (insertions) has been extremely important in promoting evolutionary novelty.

The Struggle for Existence

This graphic phrase is much used, but there is of course no conscious struggle and to that extent the expression is misleading. However, it does convey the idea that organisms are constantly in a state of conflict with each other and with what can at times be a hostile environment. The conflict stems from predation and competition, which themselves are related to the way populations grow.

POPULATION GROWTH

Consider what happens if a few individuals of a species enter a previously unoccupied area. Assuming that there is no shortage of food and no predators, reproduction will occur and the number of individuals will increase as shown in Fig 35.8. Having got underway, the population grows exponentially and finally reaches an equilibrium. The steepest part of the growth curve is the exponential phase, and it represents the maximum possible growth rate under ideal environmental conditions. The absence of predators and competition for food and space is summed up by saying that there is no **environmental resistance**. Under these circumstances the struggle for existence is low, survival is high and the species realizes its full **reproductive potential**. If this were to continue indefinitely the population curve would continue to rise without limit and gross overcrowding would ensue. What stops this happening and causes the growth curve to flatten out is environmental resistance. The environment becomes, as it were, saturated with this particular species and it can support no more. When this point is reached the birth and death rates exactly balance each other. This pattern of population growth is the same irrespective of the species: it applies equally to bacteria on an agar plate, yeast cells in a flask of cider, or rabbits in Australia.

Fig 35.8 Generalized graph of population growth. When the curve reaches a plateau the environment is 'saturated': it can support no more individuals.

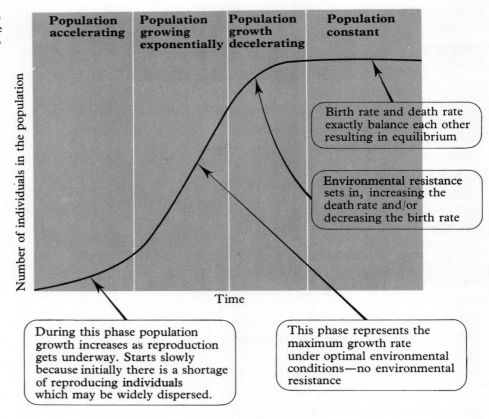

ENVIRONMENTAL RESISTANCE

The form that environmental resistance takes depends on the species in question. In general the following factors all limit population growth.

(1) Shortage of food, water or oxygen. These are obvious checks to population growth and affect all organisms to a greater or lesser extent. Organisms compete for these commodities and the greater the population the more severe is the competition.

(2) Lack of light. This is particularly important in the growth of plant populations because they rely on light for photosynthesis. Plants compete with each other for light as can be seen in any woodland or forest.

(3) Predators. The presence of animals that prey on a particular species plays a critical part in keeping down the population of that species. We shall return to this point shortly.

(4) Lack of shelter. This may be shelter from predators or shelter from physical factors of the environment such as excessive heat. Again organisms may compete for this.

(5) Disease. This can be one of the most potent forces in checking the uncontrolled growth of populations. Overcrowding facilitates the rapid spread of disease within a population.

(6) Psychological factors. Sheer physical overcrowding may lead to a reluctance on the part of individual members of the population to breed. In some animals, particularly birds, overcrowding is offset by territorial behaviour in which each individual acquires and defends its own piece of territory.

All these factors, and more besides, make up environmental resistance and keep populations from growing indefinitely. The first person to realize the importance of this was Thomas Malthus, an English clergyman, who in 1798 wrote *An Essay on the Principles of Population*. He argued that while the human population increases geometrically, the food supply only grows arithmetically; in other words the population always tends to outstrip the available food supply leading to 'famine, pestilence and war'. It was a grim doctrine and cut little ice with Malthus' contemporaries, largely because England was enjoying a period of prosperity and rapid population growth at the time. However, Darwin was much impressed with Malthus' argument and could see that it applied even more forcibly to other organisms besides man. In the Malthusian doctrine Darwin could see his 'struggle for existence' and the seeds of an evolutionary process by natural selection.

MAINTENANCE OF POPULATIONS

When a population reaches equilibrium it does not remain absolutely constant but fluctuates because of variations in the environmental resistance. For example, if we have a population of yeast cells multiplying in a nutrient medium and we add some *Paramecium*, the exponential growth of the yeast is arrested and the two populations are found to fluctuate as shown in Fig 35.9. You will notice that the fluctuations in the *Paramecium* population exactly mirror, but lag slightly behind, the fluctuations in the yeast. The explanation is that the yeast population drops as soon as *Paramecium* starts feeding on it. The *Paramecium* population then grows until the yeast population falls so low that the predator itself begins to suffer. With a decline in the number of predators the yeast population begins to grow again, and the cycle is repeated.

These observations, which have been found to apply to many other relationships besides that between *Paramecium* and yeast, can be put into a general statement. For a given species in a particular environmental situation there is a certain optimum population which the environment can support. This we can call the **norm** or **set point**. If the population rises above the set point, competition or predation takes place to such an extent that the population falls. If it falls below the set point environmental resistance is temporarily relieved so that the population rises again (Fig 35.10). In the normal course of events populations fluctuate on either side of the set point, but the regulatory processes described prevent the fluctuations being excessive. Clearly we have here an example of **homeostasis** which we have already met in a physiological context (see Part III). An increase or decrease in population sets into motion processes which keep the population on an even keel, a clear case of **negative feedback**.

SUDDEN CHANGES IN POPULATIONS

The homeostatic control of a population breaks down if some factor of the environment is suddenly changed. If, for example, the *Paramecia* were suddenly removed from the situation depicted in Fig. 35.9, the yeast population would no longer be held in check and would rise exponentially as it did originally. This we would describe as **positive feedback**, an increase in population providing the impetus for further increase.

This kind of thing certainly happens in the wild, though mainly as a result of man's interference with nature. A particularly dramatic example is furnished by the deer that live on the Kaibab plateau just north of the Grand Canyon in Arizona. Prior to 1907 the plateau had a stable population of about

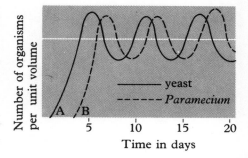

Fig 35.9 The effect on a population of yeast cells of adding *Paramecium*. A few yeast cells are introduced into a nutrient medium (A), and a few *Paramecia* are added three days later (B). Note the fluctuations in the *Paramecium* population follow the same patterns as fluctuations in the yeast population. (*data after Simpson et al*)

Fig 35.10 Scheme summarizing the homeostatic control of populations. The optimum population (norm) is the set point in negative feedback process by which the population is kept more or less constant. Only by altering the environment can the population be changed. Further details on pp. 583–4.

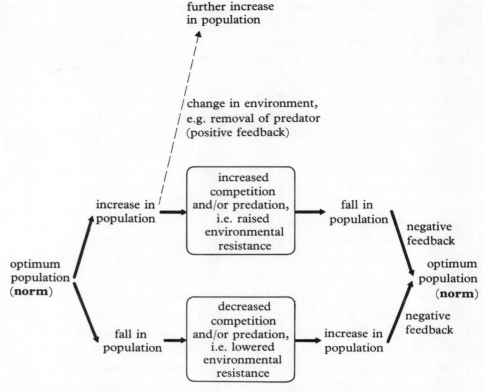

4,000 deer which was maintained at a constant level as a result of predation by wolves, coyotes and pumas. In 1907 a vast number of predators were exterminated by man, with the result that in the course of the next 20 years the deer population rose from about 4,000 to over 100,000. The predators were removed in order to free the deer from their natural enemies, but in fact the result was disastrous. By 1925 the deer population was far greater than the environment could support, resulting in mass starvation and death, so the population quickly dropped to a new equilibrium point. In general terms what happened was that a sudden change in the environment caused a shifting of the set point. The original set point of about 4,000 was maintained by predation. After the removal of the predator the new set point was maintained by an equally powerful force, competition for food. The effect of this can be seen in Fig 35.11.

Shifting of the set point can also be seen in the growth of human populations. Figures for early human populations are hard to come by, but there is good reason to suppose that in the course of human history there have been three major population explosions, each corresponding to a major change in the environment (cultural revolution) enabling a larger number of individuals to survive. The first explosion, occurring about 20,000 years ago, was probably brought about by the use of tools allowing improvements in hunting and food-gathering techniques (tool-making revolution). The second, about 6,000 years ago, was brought about by improvements in farming (agricultural revolution). And the third, which really got underway about 300 years ago, has been caused by improvements in food production, industry and medicine (scientific-industrial revolution). Each of these three bursts is associated with a substantial shift in the population set point: the birth rate rises, the death rate falls, and the population surges up exponentially to a new level.

In the case of human populations the environmental changes responsible

A

B

for raising the set point have been created by man himself, and inevitably it brings with it new problems such as food shortage and sheer physical overcrowding. In other words man changes the environment and moves the set point which then, paradoxically, influences his own adjustment to the environment. We are in the midst of the third great population explosion at the moment (Fig. 35.12). The question is, how long will the human population continue to rise exponentially, and when will a new equilibrium, imposed either by man or by nature, be reached?

Fig 35.11 Before man exterminated the predators, the deer on the Kaibab Plateau were sufficiently few in number for ground vegetation to provide sufficient food for the whole population. After extermination of the predators the numbers rose so high, and competition became so fierce, that the deer had to reach for fodder from the trees (A). B) Junipers on the Kaibab Plateau in normal unbrowsed condition C) The same trees after severe browsing by the deer.

C

POPULATIONS AND EVOLUTION

How does this help our understanding of evolution? In the first place it is clear that there is usually a tendency for a species to produce more individuals than the environment can support. This results in a high death rate. However, the mortality that keeps populations stable is not random, i.e. it does not affect all individuals equally. On the contrary, it strikes more fiercely at those individuals that are least well adapted for survival. In other words the population is kept in check by a process of **differential mortality**. To revert to the deer on the Kaibab Plateau, the individuals which used to lose out in the struggle for existence prior to 1907 were those that were insufficiently fleet of foot to avoid the predators. The ones that lost out after removal of the predators were those that could not hold their own in the fierce competition for fodder. In this way differential mortality continued to weed out the 'weak' from the 'strong', but the environmental definition of weakness and strength, i.e. of what is 'fit' and 'unfit' – had changed. Thus, as well as keeping the popul-

Fig 35.12 Like any other population which is not restrained by the normal checks, the human population grows exponentially. Plotted on a logarithmic scale, the graph shows how the world population has risen since 1800. The dotted line **A** represents the projected population growth over the next 120 years assuming that the 1960 – 69 growth rate is maintained. Notice that whereas the population increased by less than 1,500 million during the century from 1850 to 1950, it may well rise by more than 15,000 million in the next hundred years. In the years to come, will the growth rate remain unchanged, giving curve **A**, or will it continue to increase as it has done in the past, giving a curve that approximates to **B**? Or will checks imposed either by men or nature cause it to level off as in the curve **C**? It has been estimated that if the population were to increase as in **B**, there would be 'standing room only' within 600 years. (*based on United Nations data*)

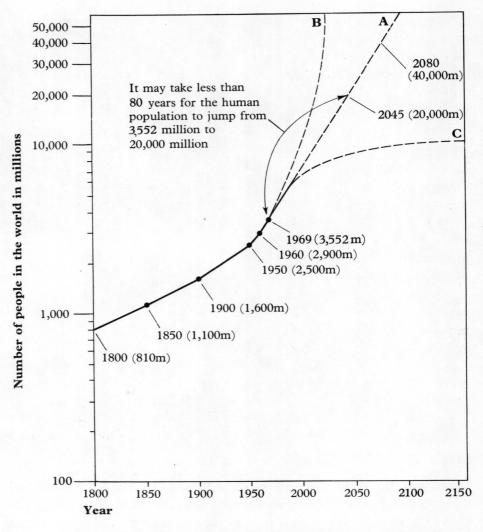

ation on an even keel, differential mortality favours the perpetuation of beneficial characteristics. This is the essence of natural selection, to which we now turn.

Natural Selection

Natural selection is not so much concerned with the survival or death of individuals as with the perpetuation or elimination of the genes carried by them.

THE ACTION OF NATURAL SELECTION ON GENES

From the evolutionary point of view differential mortality is important because, if it occurs before the individual has had a chance to reproduce, it eliminates unfavourable genes from the population. However, death is not the only way of achieving this. Any process that encourages the transmission of favourable genes and blocks the transmission of unfavourable genes can be said to contribute towards evolutionary progress. If, for example, an animal with unfavourable characteristics is somehow prevented from breeding, either by natural infertility or losing in the struggle for a mate, then this individual's genes will be prevented from going through to the next generation.

Although this does not involve the death of the organism it is just as effective at eliminating its genes from the population. Sometimes called **genetic death**, it is caused not by differential mortality but by differential fertility.

It must not be thought that bad genes are always quickly weeded out of a population by natural selection. Sometimes a gene which carries a major disadvantage in the homozygous state confers an advantage in the heterozygous condition. This will favour its survival, and such genes may show a surprisingly high frequency in a population. Such is the case, for example, with the gene that causes **sickle-cell anaemia** (see p. 477). You will remember that sickle-cell anaemia develops when the gene is in the homozygous state, and it is mercifully rare. Much more common is the milder **sickle-cell trait** which develops when the gene is in the heterozygous state. It has been found that the high incidence of this latter condition is caused by the heterozygote being more resistant to **malaria**. This advantage is enough to ensure the survival of the sickle-cell gene.

Other examples of this kind of thing could be given. The important principle is that owing to heterozygote advantage, individual organisms and populations always harbour a certain number of bad genes. This is sometimes referred to as the **genetic load**.

NATURAL SELECTION AS AN AGENT OF CONSTANCY AS WELL AS CHANGE
A central feature of Darwin's theory is that species change. In proposing his theory a major difficulty Darwin had to contend with was that species appear to remain remarkably constant. This was seized upon by some of his critics as a weakness in the argument.

But in fact natural selection is responsible both for maintaining the constancy of species and for changing them. To see how this works we must return to the frequency distribution curves discussed earlier (p. 571). If we construct a frequency diagram based on all the individuals born into a population and compare this with a diagram based only on those which reproduce, we find that so long as the environment remains constant the curve for the breeding individuals is narrower but has exactly the same mean as that for the original population. The mean represents the type which is best adapted to the environment. Differential mortality favours this 'ideal type' and is biased against the extremes. We can look upon the mean as the set point in a homeostatic control mechanism. Natural selection is most destructive against those individuals that deviate most widely from the mean, and favours those that conform most closely to it. This kind of selection is called **stabilizing selection** because it maintains the constancy of species over the generations. It can be seen in the first three generations in Fig 35.13.

Stabilizing selection normally occurs only when the environment remains constant. What happens, then, if the environment suddenly changes? The effect of this is shown in the fourth and fifth generations in Fig 35.13. Let us suppose that the environment changes in such a way as to make it an advantage to be slightly taller than before. This new height becomes the 'ideal', and selection will be biased in its favour. The result is that the mean in the frequency curve of the breeding individuals becomes shifted to the right. This can be seen happening in the fourth and fifth generations in Fig 35.13 and its effect is to achieve an appreciable overall increase in the mean height of the population. What has happened is that the set point in the homeostatic process

Fig 35.13 Diagram showing stabilizing and progressive selection. Stabilizing selection perpetuates the normal distribution curve from generation to generation, and is therefore responsibile for maintaining the constancy of a species. Progressive selection causes a shift in the normal distribution curve by favouring a size which hitherto has been discriminated against in the selection process. In each generation the frequency curve for all individuals born is shown as a bold line. The curve for breeding individuals is shown as a broken line.

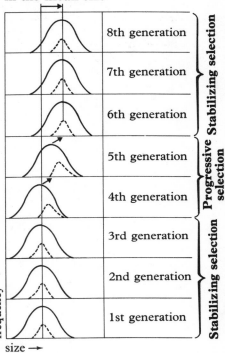

Overall shift in the mean size

8th generation
7th generation
6th generation
5th generation
4th generation
3rd generation
2nd generation
1st generation

Stabilizing selection
Progressive selection
Stabilizing selection

frequency ↑
size →

Fig 35.14 Diagram summarizing the relative frequencies of the dark and light forms of the peppered moth *Biston betularia* in different parts of Britain. In each disc the white sector represents the light form and the black sector the dark form. The dark form predominates in areas which are blackened by industrialization. (*data from H. B. D. Kettlewell*)

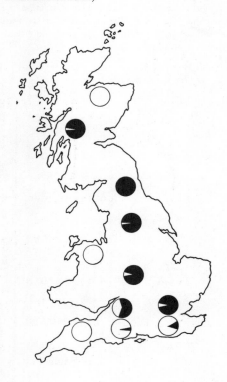

has been shifted by a change in the environment. This kind of selection clearly favours the emergence of new forms and is called **progressive** or **directional selection**. Once the new mean has been established it is maintained over successive generations by stabilizing selection, as can be seen in the sixth, seventh and eighth generations in Fig 35.13. To summarize: natural selection perpetuates constancy so long as the environment remains stable, but promotes the emergence of new forms if and when the environment changes.

NATURAL SELECTION IN ACTION

Another weakness in Darwin's original theory was that he was unable to demonstrate natural selection actually taking place. This is hardly surprising as it normally takes a long time for its effects to become apparent. To-day, however, we have ample evidence of natural selection in action. Biochemical mutants in micro-organisms such as *Neurospora* and *Escherichia coli*, resistance to DDT in flies and to penicillin in bacteria can be shown to spread through their respective populations by natural selection. But one of the most striking examples of natural selection taking place 'in front of our very eyes' is furnished by the peppered moth *Biston betularia* which has been investigated by Bernard Kettlewell and his colleagues at Oxford.

The peppered moth is very common in England and normally rests on the trunks and branches of trees where it depends on its cryptic coloration to blend in with the background. The normal form of the moth is speckled white, but occasionally mutants appear which are very much darker. This is called the **melanic** form. The first melanic moths were reported in 1848 near Manchester. Since then the number has increased prodigiously in various parts of Britain. Kettlewell has conducted an extensive survey on the relative abundance and distribution of the normal and melanic forms in different parts of the country (Fig 35.14), and the interesting fact emerges that the melanic form abounds in industrial regions where smoke and soot from factory chimneys has blackened the bark of trees. Around Manchester, for example, the frequency of the melanic form may exceed 95 per cent. In non-polluted areas, however, the light form predominates. In the north of Scotland and the extreme south-west of England it may even reach 100 per cent.

How can we explain this distribution? The peppered moth is predated upon by birds, such as thrushes, which peck them off the trees. In polluted areas the dark form is almost invisible against the darkened tree trunks whereas the light form stands out like a beacon (Fig 35.15). In clean areas the reverse is true: the light form is admirably camouflaged against the background of unsooted lichens, but the dark form is very clearly seen. Kettlewell has shown that in polluted areas such as Birmingham far more light moths are picked off the trees by birds than the better camouflaged dark forms, with the result that the frequency of dark moths is significantly higher. In non-polluted areas, however, it is mainly the dark moths that fall prey to the birds, so the frequency of light moths is higher.

In each case differential mortality is achieved by **selective predation**. In homeostatic terms, the darkening of trees with the coming of the industrial revolution shifted the set point in industrial areas so that dark body colour was favoured over light, a clear case of progressive selection. Within each area stabilizing selection has kept the frequency of the two forms more or less constant.

The peppered moth provides an example of **polymorphism**, the existence within a species of two or more distinct forms which occur in approximately constant proportions in a given population. Many other examples of polymorphism are known, for instance the banding pattern and colour of the shells of the snail *Cepaea*, and the different castes in social insects such as bees and ants (see p. 537). The different forms, or **morphs** as they are called, need not necessarily be visibly different. For example, the human species is polymorphic with respect to blood groups, the difference in this case being a biochemical one.

NATURAL SELECTION AND POPULATION GENETICS

To be effective in producing long-term evolutionary change the forces of natural selection do not merely act on individual genes, but on populations.

A more or less genetically isolated unit of population is known as a **deme**. A good example is provided by the deer on the Kaibab Plateau in Arizona mentioned earlier. Other examples might be the carp in a pond, beech trees in a copse, or *Paramecium* in a ditch. Although there may be considerable movement of individuals to and from an individual deme, a deme generally perpetuates itself by interbreeding of the individuals within it. It thus represents a genetic unit, and what happens to it in evolutionary terms depends on the genes it contains. The genetic constitution of a deme, i.e. the sum total of all the different genes in the population, is known as the **gene pool**. Just as the future of an individual organism depends on its genetic constitution, so the evolutionary future of a deme depends on its gene pool.

The frequency of any given gene in a population, relative to all its alleles at the same locus, is known as the **gene frequency.** It was proved mathematically by G. H. Hardy in Britain and W. Weinberg in Germany in 1908 that, provided there are no disruptive influences such as mutation or selection, the frequency of genes in a population remains constant generation after generation. This is known as the **Hardy-Weinberg principle**. Of course there will be a continual movement of genes, or **gene flow**, within the population as a result of breeding, but the overall gene frequencies remain constant. This stability is referred to as **genetic equilibrium**. Under these circumstances a deme remains unchanged in its overall characteristics, and no evolution occurs. Evolution only takes place if and when the genetic equilibrium is upset as a result of mutation, environmental change and natural selection. A change in the environment alters the selection pressure, favouring a particular gene (or genes) whose frequency increases accordingly. This is the genetic basis of progressive selection.

Two other factors can upset the genetic equilibrium, thereby bringing about evolutionary change. First, a gene may be lost altogether from a population as a result of pure chance. For example, in a small population it may happen that none of the individuals carrying a particular gene reproduce successfully. The result will be the total loss of that particular gene. This is an extreme example, but although total loss of a gene may be comparatively rare it is quite likely that in a small population a sufficient proportion may be lost to bring about a significant drop in the gene frequency. Such variations in gene frequency are known as **genetic drift**. Of course the chances of this happening become smaller the larger the population; thus genetic drift can only be expected to produce significant changes in small populations.

Fig 35.15 The light and dark forms of the peppered moth, *Biston betularia*, at rest on tree trunks. A) On soot-covered oak trunk near Birmingham. B) On lichen-covered trunk in unpolluted countryside in Dorset. (*H. B. D. Kettlewell, University of Oxford*)

A

B

The second factor which may upset the genetic equilibrium depends on the fact that neighbouring populations may not necessarily be totally isolated from each other. Generally a population is sufficiently close to an adjacent one for individuals to move from one to the other, or for interbreeding to take place at the boundary. In this way a population may gain or lose genes, thereby altering its genetic composition. Sometimes adjacent populations actually join and if interbreeding takes place they merge into a single population with a new gene pool.

The Origin of Species

We are now in a position to discuss the formation of new species. For purposes of the present discussion we will define a species as a group of organisms all sharing the same morphological features, which are potentially able to breed amongst themselves but not with any other species.

THE IMPORTANCE OF ISOLATION

As a first step in the development of a new species it is necessary for the population to become split up into two or more separate demes each with its own gene pool. These demes must be isolated from one another because if genes are exchanged between them they will effectively behave as one population, and any genetic differences which might arise between them will not be perpetuated. If they are isolated, mutation and selection can take place independently in the two populations and each can develop into a distinct species.

ISOLATING MECHANISMS

The kinds of isolating mechanisms that can lead to the emergence of new species are as follows:

(1) **Ecological isolation**. In this are included all the environmental barriers that keep populations apart. Thus the populations may be widely separated geographically or divided by impenetrable barriers such as mountain ranges and rivers. Even if they occupy the same locality they may be separated by having a preference for slightly different habitats. Again, migratory populations may be effectively isolated from each other by the fact that they occupy the same habitat during different seasons of the year.

(2) **Reproductive isolation**. Even though two populations may not be ecologically separated, they may be effectively isolated by the fact that they cannot interbreed. This might be caused by lack of attraction between males and females or by physical non-correspondence of genitalia. In the case of flowering plants it may be due to the fact that pollination is impossible between the two populations. In animals with elaborate behaviour patterns it may be because the courtship behaviour of one fails to stimulate the other.

(3) **Genetic isolation**. Even though mating may be possible, fundamental differences in genetic constitution may prevent reproduction being successful. Thus the gametes may be prevented from fusing, for example by the pollen grain of one population of plants failing to germinate on the stigmas of the

other. Even if fertilization does occur the zygotes may be inferior in some way and fail to develop properly. Sometimes offspring are produced but the hybrids may be adaptively inferior, living for only a short time. Or they may be sterile, as in the famous case of the mule, formed by interbreeding between horses and donkeys.

THE EMERGENCE OF NEW SPECIES

It is reasonable to suppose that in the formation of new species some kind of isolation is a necessary first step. Geographically isolated populations are known as **allopatric** which literally means 'different countries'. Such isolation provides the opportunity for each population to evolve along its own lines. If the two populations subsequently come together, each may have changed to such an extent that for physiological or genetic reasons interbreeding is impossible. The two populations may merge geographically (forming what is called a **sympatric** population) but, as they are now reproductively and/or genetically isolated, they retain their genetic individuality. They have in fact become separate species.

EVOLUTION IN RETROSPECT

Quite understandably students often find it hard to believe that random mutations, totally undirected by the environment, could be responsible for producing an organism as complex as man, or an organ as elaborate as, for example, the ear. Such doubts are by no means confined to beginners; they are shared by many professional biologists as well. The idea seems all the more implausible when one discovers that mutation rates are low, and relatively few are beneficial.

There is no easy answer to this problem. The only thing one can say is that evolution has had a very long time in which to occur. The fossil record goes back for over a thousand million years, and in all probability life has existed for very much longer than that, possibly as much as three thousand million years. The history of eutherian mammals alone spans 60 million years, and man has been in existence for over a million. It should also be remembered that although man's mutation rate seems slow, it occurs much faster in organisms with higher rates of reproduction. Nor should it be assumed that mutation rates have always been as slow as they are to-day. It is possible that in past geological ages greater intensities of ultraviolet light and natural radiation may have resulted in faster mutation rates, at least in the early stages of evolution.

This brings us to the question of what limits the rate of evolution. On this there is some doubt. In some circumstances it may well be the mutation rate, but it appears that in many species the rate of mutation is much greater than can be 'made use of' as it were. In other words a factor other that the mutation rate holds up the rate of evolution. The bottleneck in the process would seem to be the rate of change of the environment, i.e. the forces of natural selection. For the most part an organism's environment remains pretty constant, but now and again an event occurs that radically alters this situation. Think, for example, of the time when the first animals crawled out of water onto land: immediately a vast wealth of new ecological niches became available, radically altering the pattern of natural selection. As radiation into these niches took place, more niches were created by the evolving organisms themselves. On the greatly condensed geological time scale it is as if thousands of new species came into being simultaneously.

ARTIFICIAL SELECTION

For hundreds of years man has altered the evolution of species by imposing on them his own process of selection. This is known as **artificial selection** and it forms the basis of animal and plant breeding. Man selects the males and females with the characteristics he wants, and allows them to interbreed. Individuals lacking the desired qualities are prevented from mating by extermination, segregation or sterilization. By rigorous selection over many generations, the quality of the stock can be improved and special breeds developed for particular purposes. To take cattle as an example: Jersey, Guernsey and Ayrshire breeds are all dairy cattle developed for milk production, whereas Herefords and Aberdeen-Angus are beef cattle developed for meat production. Other animals that have been subjected to artificial selection include sheep for meat and wool, horses for racing and hauling, pigs for bacon and lard production, pigeons for flight capacity and plumage, poultry for egg and meat production, and dogs for hunting, retrieving, racing and appearance. Amongst plants, many crops like wheat, barley and potatoes have been bred for higher yield, greater resistance to disease, ability to withstand adverse weather conditions and so on. In all these cases the different varieties, each with its distinctive characteristics, have arisen by a process of artificial selection in which man plays the rôle normally performed by nature.

Could the same kind of selection be imposed on the human race? To most people this idea seems repugnant, conjuring up visions of restricted marriages and compulsory birth control. But in fact it already happens to a slight degree as when, for example, a couple with a history of abnormality decide, on the advice of a genetic counsellor, not to have any children. There is nothing inately evil about wanting to improve the quality of the human race, especially when it involves eliminating from the population deleterious genes such as those that cause haemophilia and mental defects. The theory and practice of improving the human race by genetical means is known as **eugenics**, and despite its sinister undertones it should not be automatically dismissed as totally unacceptable.

36 Some Major Steps in Evolution

Julian Huxley has described man as 'the spearhead of the evolutionary process'. In this final chapter we shall take a look at some of the events which have led to the emergence of man as the dominant species. We start by going back some 3,000 million years to the time when it is thought that life first arose on this planet.

REFLECTIONS ON THE ORIGIN OF LIFE

In 1953 Stanley Miller, then at the University of Chicago, succeeded in synthesizing amino acids, the building blocks of proteins, by putting a spark across a mixture of simple gases in a closed system. This was a momentous achievement, not because he synthesized amino acids as such, but because he managed to produce them under the primitive conditions that we believe might have existed on this planet some 3×10^9 years ago. We will return to this later, but in the meantime let us see why his achievement hit the scientific headlines.

Ever since the beginning of recorded history man has speculated on the origin of life. In early times it was generally believed that organisms were generated spontaneously from non-living matter. Thus the ancient Egyptians believed that snakes arose from mud, and the ancient Greeks believed that rats came from garbage. Renaissance scientists knew better than this, but in the seventeenth and eighteenth centuries it was widely believed that bacteria (discovered in 1676) were spontaneously generated. It was not until 1862 that Louis Pasteur disproved, once and for all, the theory of **spontaneous generation**. What had attracted people to this idea was that when a nutrient broth was allowed to stand, micro-organisms quickly developed in it. Pasteur found that if the neck of a flask containing broth was drawn out into a long S-shaped tube, and the broth then boiled, micro-organisms failed to develop, at any rate so quickly. If, however, the tube was broken off close to the flask, micro-organisms quickly appeared. Pasteur reasoned that the micro-organisms had entered the flask from the atmosphere, but that in the first case they had become trapped on the walls of tube and so failed to reach the broth.

From this grew the idea that all life comes from pre-existing life, and all notions of spontaneous generation died a quiet death. In fact the latter theory became so unfashionable that some people could not even accept that the first organisms to evolve on this planet arose by spontaneous generation, preferring to believe that life was brought here by meteorites from other planets. For obvious reasons this is most unlikely, and we are therefore forced to conclude that life originally arose by spontaneous generation, even though the process appears not to be repeatable today.

Fig 36.1 Louis Pasteur in his laboratory in Paris. One of his famous long-necked flasks can be seen in the centre of his laboratory bench. (*Wellcome Historical Medical Museum*)

Before discussing how such an event may have taken place, we must decide what form the first organisms might have taken. It goes without saying that, however simple they may have been, they must have reproduced themselves. But what would they have used as a source of energy for this? In other words how did they feed? There are really only two possibilities: they must have fed either **heterotrophically** or **autotrophically**. At first sight autrotrophism would seem to be the most likely method: autotrophs are structurally less complex than heterotrophs, and generally seem to be on a lower level of organization. But a moment's reflection tells us that, though this might appear to be the case when one looks at the whole organism, at the biochemical level it is far from true. The metabolic equipment required to synthesize organic substances from inorganic raw materials, which is what an autrotroph does, is far more complex than that needed by a heterotroph feeding on ready-made organic matter. On these grounds it would seem more likely that the first organisms were heterotrophs.

This immediately raises a whole host of questions. If the first organisms were heterotrophs what were they like and how did they originate? How did the organic substances, on which they were dependent for food, come into being? How did they metabolize these organic substances to produce energy? And so on! The hypothesis accounting for these events, since it is based on the premise that the first organisms were heterotrophic, is called the **heterotroph hypothesis**. Let us examine its propositions, step by step.

SYNTHESIS OF ORGANIC MOLECULES

The estimated age of the earth is around 5×10^9 years, and the first indications of primitive life occurred about 2×10^9 years ago. In the early stages of its existence the earth would have been too hot for life to exist, and we can therefore narrow down the origin of life to about 3×10^9 years ago, give or take a few hundred million years! Geochemical evidence suggests that at this time the earth's atmosphere consisted of four simple gases: **methane** (CH_4), **ammonia** (NH_3), **hydrogen** (H_2) and **water vapour**. Oxygen gas could not have existed as such since the atmosphere was too hot: it would have combined with other materials (iron, silicon etc.) whose oxides form much of the earth's crust. Water vapour is thought to have been formed mainly from volcanic activity: about 10 per cent of the material released in a volcanic eruption is water.

The first step towards the origin of life must have been the synthesis of simple organic molecules. Many suggestions have been put forward to explain how this might have happened, but the most generally accepted theory is that they were formed by the action of lightning, ultraviolet radiation, or possibly gamma radiation, on the four simple gases mentioned above. Evidence for this rests on the fact that scientists have been able to perform such syntheses in the laboratory. Stanley Miller's experiment, with which we started this discussion, is illustrated in Fig 36.2. Since then several other steps in the synthesis of organic molecules have been carried out under primitive earth conditions. Miller himself synthesized amino acids from the four gases using an electric spark. From the same raw materials Melvin Calvin, using gamma radiation, managed to produce a mixture of amino acids, simple 6-carbon sugars, and also purines and pyrimidines which, you will recall, enter into the composition of nucleic acids. Since then more complex organic molecules have been made:

polypeptides have been synthesized from amino acids simply by heating to melting point and then cooling; and nucleic acids have been formed from nucleotides merely by heating under pressure. These laboratory syntheses do not, or course, prove that similar events took place millions of years ago, but they do show that such events could have taken place.

Assuming that the synthesis of organic molecules took place in the atmosphere, it is argued that they were subsequently brought down to the earth's surface in heavy rains and that, in the course of time, they accumulated in primitive oceans and lakes. One can imagine that these great bodies of water must have been teaming with organic molecules, a kind of 'organic soup'.

THE FIRST ORGANISMS

The next step is crucial, but one on which there is precious little evidence. Somehow or other, certain of these organic compounds came together to form the first living organisms, i.e. they must have combined in such a way as to produce a stable and integrated chemical system capable of releasing energy and replicating itself. This is an enormous step and is full of unanswered questions. Were random collisions responsible for bringing the right molecules together? How were the molecules held together once they had collided? Was some sort of membrane formed round them, and if so what form did it take? Do you think modern studies on cells and cell membranes can help us to answer these questions?

However it was put together, there are certain things that this simple organism must have been able to do if it was to qualify as a living being. First and foremost it must have been able to obtain energy, presumably by the breakdown of organic molecules. Since there was no oxygen gas present at this time, its respiration must have been **anaerobic**. With respiration proceeding all the time, it would need some way of replenishing its supplies of organic compounds. We have already dismissed the idea that it could synthesize these for itself (autotrophic nutrition), which means that it must have fed **heterotrophically**, presumably on the 'organic soup' surrounding it. Here again questions immediately spring to mind. Were these organic compounds absorbed by diffusion, or was there some kind of active uptake? Presumably the ingestion of solid particles by pinocytosis and phagocytosis would come later, but who knows?

Secondly, it must have been able to reproduce itself. At this early stage of evolution reproduction must have been a simple **asexual** process involving nothing more than the replication of macromolecules. To what extent do you think we can gain some insight into this by our knowledge of viruses and bacteria? Do you think our understanding of the way DNA replicates can help us?

Once the first reproducing organisms had originated, rapid evolution would be expected to have produced more complex and better adapted forms. Assuming the Darwinian theory to be sound, this would mean that the offspring resulting from reproduction of our primitive organism would show transmissible variations. In other words replication of its macromolecules would not be absolutely accurate, but would show deviations. Our knowledge of gene mutations helps us to understand how such deviations may have arisen. Further variation would result from **sexual reproduction**, and we must assume that sex, involving the transfer of material from one individual to an-

Fig 36.2 The apparatus used by Stanley Miller to synthesize organic compounds under primitive earth conditions. A) Photograph of Miller in his laboratory in 1953. B) Diagram of the apparatus. The apparatus was first evacuated, then hydrogen, methane and ammonia admitted to the desired pressures. The water in the flask was then boiled to promote circulation and to supply the necessary water vapour for oxygenation of compounds in the sparking chamber. The products of the discharge were condensed in the tube running through the cooling jacket, and collected in the U tube, which also serves to prevent circulation in the wrong direction. By this means Miller synthesized some 15 amino acids which were subsequently separated by paper chromatography. The identifiable ones included glycine, alanine, glutamic, and aspartic acids. (A: *United Press International;* B: *after Miller, Journal of the American Chemical Society, Vol. 77, 1955).*

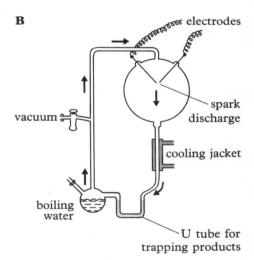

other, arose at a very early stage of evolution. Bacteria give us a possible clue as to how this might have taken place.

DEVELOPMENT OF AUTOTROPHS

Such was the primitive scene as envisaged by the heterotroph hypothesis. What happened next? It is quite likely that the supply of organic molecules, originally present in vast quantities in the primitive oceans, was gradually exhausted by the ever-growing population of heterotrophs. The competition must have been increasingly fierce, and this would have placed a premium on any organisms capable of an alternative method of feeding. It is thought that there now evolved certain organisms capable of synthesizing their own organic food from inorganic materials (**autotrophic nutrition**). Independent of the rapidly declining organic food supply, these autrotrophs would have been at a distinct advantage over their heterotrophic contemporaries. Of course we have no idea what kind of autrotrophic feeding they indulged in. It may have been a process of **chemosynthesis** akin to that used by certain present day bacteria. If it involved **photosynthesis** it may have been similar to that used by present-day green and purple sulphur bacteria which, living in anaerobic conditions, use hydrogen sulphide as the source of hydrogen for reducing carbon dioxide. They have a type of chlorophyll chemically simpler than that of more highly evolved autrotrophs. Sooner or later, however, we must assume that autotrophs arose that could photosynthesize like modern green plants.

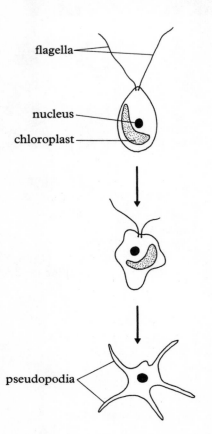

Fig 36.3 The green flagellate *Ochromonas* showing conversion of the autotrophic flagellated form into the heterotrophic amoeboid form. (*after Pascher*)

flagella

nucleus

chloroplast

pseudopodia

AEROBIC RESPIRATION AND THE DEVELOPMENT OF SECONDARY HETEROTROPHS

The arrival of photosynthesizing autotrophs would have resulted in oxygen accumulating in the atmosphere. This was important for two reasons. Firstly it resulted in the formation of a layer of ozone (O_3) high in the earth's atmosphere. This forms a barrier to the sun's radiation and would have prevented any further synthesis of organic compounds in the atmosphere, and in later stages of evolution it would have protected organisms from exposure to harmful rays. Secondly, it is envisaged that some organisms evolved the ability to utilize this free oxygen for **aerobic respiration**, thereby achieving a more thorough breakdown of organic substances with the release of additional energy. You will remember that the aerobic breakdown of sugar, involving Krebs' citric acid cycle as well as glycolysis, provides about ten times as much energy as anaerobic respiration.

The development of aerobic respiration was a great step forward, for, with the greater metabolic rate that it permitted, all manner of new possibilities opened up: greater synthetic powers for manufacturing new materials and structures, a more active life, and so on. If mutations resulted in the development of more complex structures, the energy was there for making full use of them.

Meanwhile the original heterotrophs were rapidly declining in numbers. At this point, organisms are thought to have evolved which were capable of feeding on the autotrophs. How they arose is just another unanswered, perhaps unanswerable, question. What we can safely say is that these secondary heterotrophs must have acquired the ability to ingest and digest other organisms, presumably by **phagocytosis**. Examination of present-day protozoons like *Amoeba* and *Paramecium* shows us how they might have achieved this.

So we finish up with **autotrophs** and **heterotrophs** coexisting side by side. These provide the basic components of a balanced ecosystem, the autotrophs providing oxygen and organic compounds for the heterotrophs, the heterotrophs providing carbon dioxide for the autotrophs. Both were presumably unicellular at this early stage of evolution. It is thought that the autotrophs gave rise to the **plant kingdom**, the heterotrophs to the **animal kingdom**.

THE DIVERGENCE OF ANIMALS AND PLANTS

The theory just propounded assumes that animals and plants had a common ancestry, a belief that is substantiated by their many biochemical similarities. But if this is so, might we not expect to see, even to-day, unicellular organisms capable of feeding autotrophically and heterotrophically? In fact such organisms do exist and belong to a group of protists known as the **phytoflagellates** ('plant flagellates'). The phytoflagellates are generally green (because they contain chlorophyll) and have one or sometimes two flagella. Many of them feed exclusively by photosynthesis and will die if kept in the dark. However, some of them, more than was originally thought, thrive in darkness. This is because they can feed heterotrophically by absorbing soluble organic matter through the cell wall. In some cases phagocytosis of organic particles takes place across a specialized part of the cell membrane.

Some green flagellates can change into colourless forms that lack chlorophyll and feed on organic matter. Such is the case with *Ochromonas* illustrated in Fig 36.3. If conditions become unfavourable for photosynthesis it loses its chloroplast, withdraws its flagella and becomes amoeboid, feeding heterotrophically by phagocytosis. An example of a permanently colourless form is *Peranema* which has developed the heterotrophic feeding habit to the point that it can attack and swallow other flagellates. This tendency of green flagellates to become heterotrophic may give us a clue as to how heterotrophs might have arisen from autotrophs in the evolutionary past.

ORIGIN OF MULTICELLULAR ORGANISMS

Like so many other aspects of early evolution, this is a controversial matter and we can do no more than touch on the possibilities. One obvious possibility is that the first multicellular organisms may have arisen as a result of a unicell dividing and the daughter cells failing to separate. We see this in the colonial **collar flagellates** which represent a step in the development of the sponges (see p. 45). On the plant side we see it in the *Volvox* colony in which the constituent cells are interconnected by thin cytoplasmic threads and show a reasonable degree of integration. Of course, neither *Volvox* nor sponges, both of which are colonial organisms, could represent the direct ancestors of multicellular plants and animals, but some biologists feel that they can give us an insight into what primitive multicellular organisms may have been like and how the multicellular condition may have arisen.

An alternative possibility is that the nucleus of an ancestral unicellular organism may have divided repeatedly to give a **multinucleate cell**. If cell membranes were to form between the nuclei, a simple multicellular organism would result (Fig 36.4). The attractiveness of this theory is that it permits us to derive a properly integrated multicellular organism direct from a protist, rather than via a questionable colonial intermediate. Moreover it is supported by evidence. Multinucleate protists do exist, and indeed this condition is quite

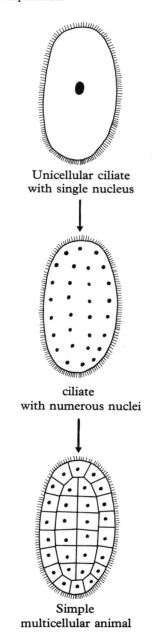

Fig 36.4 Diagrams illustrating how a simple multicellular animal may have arisen in evolution from a multinucleate protozoon. See text for further explanation.

Unicellular ciliate
with single nucleus

ciliate
with numerous nuclei

Simple
multicellular animal

Fig 36.5 Fossil evidence suggests that migration from water to land took place in the Devonian period about 350 million years ago when a group of fresh water fishes, the crossopterygians, gave rise to the first amphibians. The crossopterygians had lobed fins, constricted at the base, which in some cases show a marked tendency towards the pentadactyl limb. It is thought that these were used for locomotion on land. The Devonian was a period of prolonged seasonal droughts and any fishes that were capable of struggling overland from one body of water to another would have been at a distinct advantage. A) and B) pectoral fins of two crossopterygians showing their likeness to the pentadactyl limb. C) forelimb of an early amphibian for comparison. D) artist's reconstruction of the lobe-finned crossopterygian *Eusthenopteron* clambering out of water. This is the fish whose pectoral fin is shown in A. (A–C based mainly on Romer; D: *American Museum of Natural History*)

A) *Eusthenopteron* (crossopterygian)

B) *Sauripterus* (crossopterygian)

C) Early amphibian

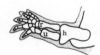

h humerus
r radius
u ulna

common amongst ciliates. Furthermore, there are certain multicellular animals, the **acoel flatworms**, in which the cell membranes separating adjacent nuclei are incompletely formed. Like other free-living flatworms, the acoels are ciliated, and it is just possible that they may have arisen from a multinucleate ciliate.

LARVAL FORMS AND EVOLUTION

In trying to reconstruct evolutionary pathways one generally endeavours to derive an organism from the adult form of a more primitive group. For example, in searching for the ancestor of the vertebrates it is tempting to accept *Amphioxus* as the most likely candidate: its basic organization is very like that of a primitive vertebrate, and moreover the ammocoete larva of jawless fishes like the lamprey is very similar to *Amphioxus* in many points of detail. However, there is good reason to believe that on more than one occasion during evolutionary history, a group may have evolved not from an adult ancestor, but its larva. On first consideration this may seem absurd; after all, larvae do not normally reproduce. However, there is evidence that in the course of evolution certain larval forms may have acquired sexual maturity, and then given rise to higher forms. The adult would thus become redundant and disappear from the life cycle. On this basis it is possible that the vertebrates originated not from an *Amphioxus*-like creature but from the 'tadpole' larva of the sea squirts (see p. 557). The evidence for this is, of course, circumstantial and indirect, but it is certainly true that larvae can, on occasions, become sexually mature. Such is the case with the axolotl, a native of certain Mexican lakes. A favourite feature in many aquaria, this giant tadpole is the larval form of the salamander *Ambly-*

D

stoma. More than 15 cm long, it reaches sexual maturity and is able to reproduce, without undergoing metamorphosis into the adult. This process is called **neoteny**, and it makes one suspect that evolution from a larval form may not be as unlikely as it appears at first sight.

THE COLONIZATION OF DRY LAND

We have seen that in all probability life began in the ocean. Many millions of years later, armed with the appropriate adaptations, organisms began to forsake their aquatic home and move onto dry land. This has happened several times in evolutionary history, in both plants and animals. The best known example, well documented by fossils, is the migration onto land by a group of fishes in Devonian times some 350 million years ago. Known as the crossopterygians, they and the early amphibians to which they gave rise, show a gradual transformation of paired fins into pentadactyl limbs typical of modern tetrapods (Fig 36.5). But rather than get involved with the details of this particular migration, let us briefly consider the anatomical and physiological changes that would be required for any animal to move from water onto dry land. In reviewing such changes we are greatly helped by modern amphibians like the frog, which have to make this migration in the course of their life history. Some of the changes necessary for an animal, would also have to be made by a plant undergoing the same kind of transition.

The changes can be summarized as follows:

(1) Air being a much rarer medium than water, there would have to be new means of breathing and support. The latter particularly applies to land-dwelling vertebrates, where, in contrast with fishes, the skeleton functions as a cantilever holding the head and abdomen clear of the ground (see Chapter 21). New problems of stability also arise, these being solved by mechanical means in conjunction with complex vestibular reflexes.

(2) Terrestrial organisms are liable to lose water by evaporation. As explained in Chapter 14, both active and passive mechanisms have been evolved for coping with this. The passive technique of developing an impermeable cuticle is common to animals and plants.

(3) Water, particularly the ocean, is a comparatively stable medium, not liable to the wide fluctuations in temperature which characterize the terrestrial environment. Land animals must have structural, physiological, or behavioural means of maintaining a comparatively constant body temperature, irrespective of external fluctuations (see Chapter 15).

(4) The greater rarity of the medium means there is less resistance to movement, so higher speeds can be achieved. The salmon, a comparatively fast moving fish, can just manage 37 km/h, but only for short distances, and the dolphin, despite its streamlining and strong propulsive tail, cannot exceed 40 km/h. Terrestrial animals can do much better: 65 km/h in a good racehorse, 97 km/h in the antelope and dragonfly, and over 140 km/h in a homing pigeon! Even man, despite various physical shortcomings, can top 36 km/h. What this all boils down to is that the general pace of living is greater on land than in water: both prey and predator move more swiftly. If terrestrial animals are to survive they must be adapted to meet this situation. It does not of course follow that all terrestrial animals should be fast movers: some animals, such as worms and snails, have 'opted out' by retiring to highly protected habitats.

(5) Coping with the faster pace of living means developing an improved

musculo-skeletal system and better neuro-sensory mechanisms. The evolution of elaborate sense organs (the compound eye of insects for example) and rapid reflexes, has made it possible for animals to respond quickly to the slightest changes in the environment. Of course such developments are not confined to land animals, nor do all land animals show them, but they are essential pre-requisites for thorough exploitation of the terrestrial environment.

(6) In many aquatic animals fertilization is external, eggs and sperm being shed into the surrounding water. As this would be impossible on land, terrestrial animals must have some means of introducing sperm into the female, thereby bringing about internal fertilisation. The only alternative is to return to water for breeding purposes, which is what amphibians do. With internal fertilization comes the possibility of internal development and viviparity. This is an enormous advantage to a terrestrial animal, for the environmental hazards that plague the adult can have an even more devastating effect on its helpless eggs and young. It is interesting that those forms which have not developed viviparity lay encapsulated eggs in which the embryo is surrounded by some kind of protective covering. Think, for example, of the chick embryo in its fluid-filled amniotic cavity surrounded by extra-embryonic membranes and hard shell. But undoubtedly viviparity is the best solution, with the embryo nourished in the uterus by means of a placenta.

EVOLUTIONARY TRENDS IN ANIMALS AND PLANTS
By and large, evolution leads to an increase in complexity. When we look at the phylogenetic tree of the animal kingdom we see a general trend towards more complex nervous, circulatory and feeding systems, not to mention musculo-skeletal arrangements, sense organs and so on. The animal kingdom can be split into two great stocks which are thought to have diverged from each other at an early stage in evolution. One stock contains the **annelids, molluscs, arthropods** and various other related groups; the other contains the **echinoderms, protochordates** and **vertebrates**, including man. Some of the reasons for relating animals in this way are discussed in Chapter 34. Within each of these two great stocks the evolutionary trends mentioned above can be detected.

To say that evolution tends to produce more complex forms is not to imply that it is never regressive. There are many instances of what are thought to have been quite active animals taking up a sedentary or perhaps parasitic existence. Resorting to such passive modes of life generally involves the degeneration or even total loss of many organs, which puts them in sharp contrast with their free-living relatives. Can you think of some specific examples of this?

Because of the scantiness of the fossil record, it is not easy to reconstruct the evolutionary history of the plant kingdom. However, studying the structure and development of modern forms enables us to detect certain trends. In the more primitive green algae all the cells are alike and all have the capacity to reproduce sexually. Such is the case with the simpler members of the colonial Volvocales and the filamentous *Spirogyra*. But in more advanced forms the cells have become differentiated into those that reproduce (**gonadic cells**) and those that carry out purely vegetative functions like photosynthesis (**somatic cells**). The rudiments of such differentiation are seen even in *Volvox*, but the process is carried much further in higher algae. This separation of 'sex and soma' is one of the earliest and most fundamental evolutionary trends in the plant kingdom.

In Chapter 25 we saw that in the majority of plants there is a distinction between sporophyte and gametophyte generations. The diploid sporophyte undergoes meiosis to produce haploid spores; and the haploid gametophyte produces gametes. Amongst higher plants there is a trend towards suppression of the gametophyte until eventually, in flowering plants, it becomes incorporated into the body of the sporophye itself (see p. 410). This is associated with the development of the seed habit characteristic of these plants.

In both animals and plants there have been trends towards larger size. Some of the problems attendant on this are mentioned in Chapter 4 (p. 48). Increase in size generally goes hand in hand with the development of ever more complex tissues and organs. A tendency towards increased size has occurred independently in many different groups. California, which has extremes in everything, boasts of having the world's largest and smallest trees. The largest are the giant redwoods (*Sequoia sp.*): the coastal species can be 120 m tall. The smallest is a tiny alpine willow, *Salix petrophila*, only about 200 mm tall.

In the course of evolution many bizarre experiments have been tried, most of them leading to extinction. One of the classic examples is provided by the **dinosaurs**, a group of reptiles that carried size to its ultimate conclusion. These great animals were the dominant fauna during the Mesozoic era about 120 million years ago (Fig 36.6). Some, like *Tyrannosaurus rex*, were savage carnivores; others, like *Brontosaurus*, were vegetarian. Their fossil remains tell us that they achieved fantastic sizes: *Diplodocus*, probably the largest animal ever to roam this earth, was over 30 metres long, and *Brachiosaurus* weighed 50,000 kg!

The main problem facing such a large walking animal is how to support the body. The weight of an animal varies with the cube of its linear dimensions. In other words, if the animal's size doubles in all directions (length, height, etc.), its weight increases by about eight times. But the strength of its legs varies with their cross-sectional area, which increases in squares. So if an animal's dimensions double, its weight increases by eight times, but the strength of its legs only increases by four times. This means that for such an animal to support itself, the size of its legs must increase at a greater rate than the rest of the body. Judging by the fossils, the legs of the dinosaurs were indeed extremely thick, (cf. elephants), but even so some zoologists are doubtful if they could have supported such heavy bodies. For this reason the suggestion has been put forward that they spent much of their time wallowing in lakes and lagoons where the water would buoy up their massive bodies.

By the end of the Mesozoic era none of these mighty animals remained: they had all died. What led to their extinction? We do not know, but their large size may well have had something to do with it. There is good evidence that the end of the Mesozoic saw marked changes in climate: land levels changed, mountain ranges were thrown up, and there was a pronounced drop in temperature, at any rate in North America which was the main centre of dinosaur evolution. Their large size would have made temperature regulation difficult (why?), and their eggs would have been exposed to a hostile environment. So the changing climate may have killed them off. And with the passing of the vegetarians the carnivores would have perished too. Another factor contributing to their extinction may have been changes in vegetation. It has even been suggested that the change from ferns, with their distinctly laxative action, to flowering plants may have caused them to die of constipation!

During the Mesozoic era, other land animals that were in any way conspicuous readily fell foul of the carnivorous dinosaurs. They were in a literal sense the 'ruling reptiles'. However, in the course of their reign, a group of rather inconspicuous, sharp toothed animals appeared on the scene. These were the first mammals. Scanty though their fossil record is, we know that for upwards of 80 million years these little mammals remained small and insignificant, probably living in caves and trees and feeding on small insects and buds. But as soon as the great age of dinosaurs came to an end, these animals blossomed forth and in a comparatively short time gave rise to the many mammalian groups that we know today.

THE EMERGENCE OF MAN

Thanks to the work of L. S. B. Leakey in East Africa, there is good reason to believe that the origin of man as a tool-using primate goes back well over a million years. However, his origins from lower primates, and his exact relationship with fossil 'men' such as *Proconsul*, *Australopithecus* and Neanderthal man, are controversial matters on which there is much argument. Yet, the general trend in the skeleton of these forms is towards the humanoid condition, and many now accept that some branch of the primates gave rise to man.

Although the fossil story is full of perplexing problems, our knowledge of modern man and of lower primates permits us to specify the advances that must have marked man's emergence from a lower primate stock. In the first place there must have been a profound increase in **mental ability**. What exactly this meant in terms of evolution of the brain is uncertain, though it must have involved more than mere increased cranial capacity. Man's increased brain power enabled him to think rationally, foresee the outcome of his actions, and solve problems. But this increased brain power would have been valueless without the wherewithal to use it. In this respect the evolution of **hands** must have been of paramount importance. Probably originating in man's ancestors as prehensile grasping devices for climbing trees, they became efficient manipulative devices for making tools and hurling weapons. Today, of course, they provide the very means by which man shows his creativeness

Fig 36.6 Life in the Mesozoic era, from a mural at Marlborough College painted by Alan MacKichan. The animals and plants depicted in the mural are based on reconstructions by Joseph Augusta and Zdenĕk Burian. The large dinosaur on the left is *Brachiosaurus*, and behind it, slightly to the right, is *Diplodocus*. Both were herbivores. The carnivorous *Tyrannosaurus rex* is just to the right of centre. Several heavily armoured herbivorous species are shown towards the right, and *Pteranodon*, a flying reptile with a wing span of over 10 metres, can be seen overhead. Notice the plants: the Mesozoic saw the rise in dominance of the sporophyte. Tree ferns and conifers abounded at the beginning of the period, with flowering plants appearing later.

and transmits his achievements down the generations.

Full use of the hands would not have been possible had it not been for man's ancestors becoming **bipedal**. Necessitating major changes in the musculo-skeletal system, particularly the vertebral column and pelvis, this freed the hands from their former function of locomotion, allowing them to be used for other purposes. Man owes much to his bipedal stance.

In the emergence of man as a dominant species the development of **speech** has been extremely important. It is likely that primitive man, and his immediate ancestors, had long been social animals, living and hunting in packs. Doubtless starting as little more than inarticulate grunts, speech gradually became a means of communicating complex information between individuals. Speech depends on the production of sound by the **larynx**. In the sound-production process two elastic strands of tissue, the **vocal cords**, are vibrated by blasts of air emitted from the lungs. Pitch is determined by the tension on the vocal cords, together with the size of the glottis aperture; loudness by the strength of the blast of air. Production of sound this way is not of course confined to man, but what is unique to man is his ability to produce a wide repertoire of sounds by subtle movements of his lips and tongue. This is achieved by complex neural mechanisms intiated by the brain.

Today, in the more advanced human societies natural selection is not the potent force that it is amongst animals, or indeed used to be in human communities. Modern science and medicine have put an end to that. Although this makes man's lot a happier one, it presents problems, of which the population explosion and over-exploitation of the environment are the most pressing. Does this mean that man is no longer evolving? The answer is no. Natural selection still operates, though on a limited scale and in more subtle ways than in past ages. But, more important, man has entered a new phase in his evolution, a phase in which advances depend not so much on structural changes in his body as on the transmission of accumulated experience from one generation to the next. This could be described as cultural evolution or, to use Julian Huxley's term, psycho-social evolution. Surely this holds the key to man's future.

Phylum **Protozoa** (single-celled)	*Euglena, Amoeba, Paramecium,* trypanosomes, malarial parasite.
Phylum **Porifera** (porous)	Sponges.
Phylum **Coelenterata** (sac-like, two layers of cells separated by non-cellular mesogloea; stinging cells (nematoblasts))	*Hydra,* colonial hydroids, jellyfish, sea anemones, corals.
Phylum **Platyhelminthes** (flatworms)	planarians, parasitic flukes, tapeworms.
Phylum **Nematoda** (round worms)	*Ascaris,* hookworms, *Wuchereria* (filaria).
Phylum **Annelida** (ringed worms, segmented)	
POLYCHAETA	*Nereis* (ragworm), fanworms, lugworm.
OLIGOCHAETA	earthworms.
HIRUDINEA	leeches.
Phylum **Mollusca** (soft-bodied typically with shell)	
GASTROPODA	whelks, limpets, snails, slugs.
LAMELLIBRANCHIATA (bivalves)	mussels, clams, oysters.
CEPHALOPODA	squids and octopuses.
Phylum **Arthropoda** (jointed limbs, hard exoskeleton)	
CRUSTACEA	water fleas, *Sacculina,* barnacles, shrimps, woodlice, prawns, crayfish, lobsters, crabs.
MYRIAPODA	centipedes and millipedes.
ARACHNIDA	eurypterids (fossils), horseshoe crab (*Limulus*), scorpions, mites, ticks, spiders.
INSECTA	cockroach, locust, aphids, lice, mosquitoes, flies, fleas, bees, bugs (e.g. *Rhodnius*), butterflies and moths.
Phylum **Echinodermata** (spiny-skinned, pentaradiate)	starfish, brittle stars, sea urchins, sea cucumbers, sea lilies.
Phylum **Chordata** (possess notochord at some stage in life history)	
PROTOCHORDATA (invertebrate chordates)	sea squirts, acorn worms, *Amphioxus.*
VERTEBRATA (notochord replaced by vertebral column)	
CYCLOSTOMATA (jawless fishes)	lampreys and hagfishes.
PISCES (true fishes)	elasmobranchs (modern cartilaginous fishes) e.g. dogfish, rays, sharks; teleosts (modern bony fishes) e.g. cod.
AMPHIBIA	newts, salamanders, frogs, toads.
REPTILIA	dinosaurs, lizards, crocodiles, turtles, snakes.
AVES (birds)	pigeons, gulls, kiwi, ostrich, etc.
MAMMALIA	
MONOTREMES (egg-laying)	duck-billed platypus, spiny anteater.
MARSUPIALS (pouch mammals with rudimentary non-allantoic placenta)	opossums, Tasmanian wolf (*Thylacinus*), koala bear, kangaroos.
EUTHERIANS (have true placenta)	the many groups include the rodents, great cats, and primates (e.g. baboons, chimpanzee, man).

Note

All the above groups up to and including the protochordata lack a backbone (vertebral column) and are therefore classified as **Invertebrates.**

Appendix II

Outline Classification of the Plant Kingdom

The Thallophyta and Bryophyta are non-vascular plants, having no specialised tissues for transport of water and food materials; the Pteridophyta and Spermatophyta are vascular plants and may be combined into one group, the **Tracheophyta.**

The Angiosperms are referred to as the **Flowering plants** because of their conspicuous flowers.

The existence of unicellular phytoflagellates that can be either autotrophic or heterotrophic makes the distinction between protozoons and unicellular algae artificial. Some authorities consider that they should be put together in a separate kingdom called the **Protista.**

Bacteria are of doubtful affinity with the fungi and perhaps they too should be placed in a separate kingdom. **Viruses** certainly cannot be classified in either the animal or plant kingdoms.

Phylum **Thallophyta** (includes single-celled and simple multicellular plants)		
	ALGAE (photosynthetic)	unicellular and colonial green plants like *Chlamydomonas* and *Volvox* respectively; filamentous pond weeds like *Spirogyra* and *Oedogonium*; brown seaweeds like *Fucus* and *Laminaria*.
	FUNGI (non-photosynthetic— saprophytic and parasitic)	molds (e.g. *Neurospora*, *Penicillium*); *Phytophthora* (potato blight); *Pythium*, yeast, mushrooms, toadstools, puff balls, (bacteria ?).
	LICHENS (dual alga-fungus)	*Xanthoria*.
Phylum **Bryophyta** (more complex green plants usually with simple leaves)		
	HEPATICAE (not differentiated into roots, stem and leaves)	liverworts.
	MUSCI (stems and leaves but no true roots)	mosses.
Phylum **Pteridophyta** (green plants with stem, leaves, true roots and vascular system, but no flowers)		Ferns, bracken, horsetails, club mosses.
Phylum **Spermatophyta** (produce seeds).		
	GYMNOSPERMS ('naked' seeds— i.e. not enclosed within a fruit)	Chinese maidenhair tree, *Cycads*; Conifers (have cones instead of flowers)
		pine, larch, fir, redwood (*Sequoia*), etc.
	ANGIOSPERMS (flowering plants, seeds enclosed in some sort of fruit)	
	DICOTYLEDONS (2 seed-leaves)	buttercup, pea, rose, oak, elm, etc. etc.
	MONOCOTYLEDONS (1 seed-leaf)	grasses, orchids, lily, etc.

Appendix III
Size scale

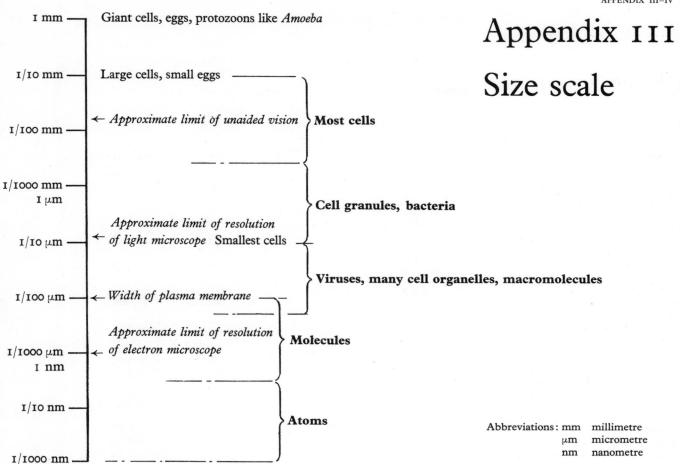

1 mm	Giant cells, eggs, protozoons like *Amoeba*
1/10 mm	Large cells, small eggs
1/100 mm	← *Approximate limit of unaided vision* } **Most cells**
1/1000 mm 1 μm	**Cell granules, bacteria**
1/10 μm	*Approximate limit of resolution of light microscope* Smallest cells
1/100 μm	**Viruses, many cell organelles, macromolecules** ← *Width of plasma membrane*
1/1000 μm 1 nm	*Approximate limit of resolution of electron microscope* **Molecules**
1/10 nm	**Atoms**
1/1000 nm	

Abbreviations: mm millimetre
μm micrometre
nm nanometre

Appendix IV
Conversions to S.I. units

Length

1 in	25.4 mm
1 ft	305 mm
1 yd	0.91 m
1 mile	1.61 km

Microscopical and submicroscopical lengths

1 μ	1 μm
1 mμ	1 nm
1 Å	0.1 nm

Area

1 in²	645 mm²
1 ft²	0.09 m²
1 yd²	0.836 m²
1 mile²	2.6 km²

Volume

1 in³	16 387 mm³
1 ft³	0.03 m³
1 UK gal	4.5 dm³

(Note 1 dm³ = 1 litre—symbol l)

Velocity

| 1 ft/s | 0.3 m/s |
| 1 mile/h | 0.45 m/s |

Mass

| 1 lb | 454g |
| | 0.45kg |

Density

| 1 lb/in³ | $2.8 \times 10^4 \text{kg/m}^3$ |
| 1 lb/ft³ | 16 kg/m³ |

Force

| 1 pdl | 0.14 N |
| 1 lbf | 4.45 N |

Pressure

| 1 mm Hg | 133 N/m² |
| 1 lbf/in² | 6.9 kN/m² |

Energy (work, heat)

| 1 cal | 4.19 J |
| | 0.00419 kJ |

Power

| 1 hp | 745.7 W |

Temperature

$t°F = \frac{5}{9}(t - 32)°C$

Frequency

| 1 c/sec | 1 Hz |

Abbreviations that may be unfamiliar

m	metre
μm	micrometre
nm	nanometre
dm	decimetre
s	second
g	gramme
N	newton
J	joule
W	watt
Hz	hertz

The former units micron (μ), millimicron (mμ) and Angstrom Unit (Å) have now been discontinued.

Appendix v

Formulae of the naturally occurring amino acids

The general formula of amino acids is:

$$\overset{\displaystyle R}{\underset{\displaystyle NH_2—CH—COOH}{|}}$$

The R group is variable; the remainder of the molecule, i.e. the part responsible for forming peptide linkages, is common to all the amino acids. In the following list the R groups are given, the rest of the molecule being represented thus:

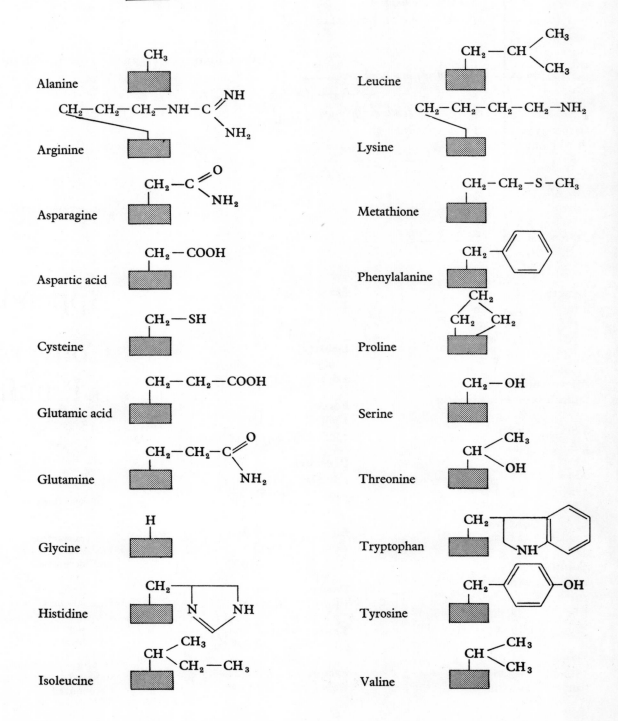

Alanine

Arginine

Asparagine

Aspartic acid

Cysteine

Glutamic acid

Glutamine

Glycine

Histidine

Isoleucine

Leucine

Lysine

Metathione

Phenylalanine

Proline

Serine

Threonine

Tryptophan

Tyrosine

Valine

Suggestions for Further Reading

What follows is not an exhaustive bibliography but a selection of readings which I personally recommend for use with this book. I have included books of general interest as well as works of reference, and have added a brief explanatory note on each. For convenience the readings are divided into traditional categories, viz: cell biology, biochemistry, physiology, etc. Within each category elementary books are mentioned first, more advanced or specialized ones later. Articles from periodicals are included only in cases where they offer something unique that is not available in a book. All such articles are obtainable as offprints from the publishers (see p. 617).

GENERAL BIOLOGY TEXTS

D. G. Mackean, *Introduction to Biology* (John Murray, 4th edition 1969)
Sir Francis Knowles, *Man and Other Living Things* (Harrap, 2nd edition 1959)
These well illustrated texts will provide you with a convenient source of background information if you have done no biology before.

M. Abercrombie, C. J. Hickman and M. L. Johnson, *A Dictionary of Biology* (Penguin, 5th edition 1966)
A source of potted information on technical terms used in biology. A useful book for your bookshelf.

Garrett Hardin, *Biology: its Principles and Implications* (Freeman, 2nd edition 1966)
One of the most balanced American texts for first year college students. Written with the needs of the general citizen in mind, it contains ideas rather than detailed facts and stresses human aspects of biology.

A. E. Vines and N. Rees, *Plant and Animal Biology* (Pitman, 2 volumes, 3rd edition 1969)
Large (volume one alone is over 1000 pages), but packed full of the kind of factual information that a sixth former might want to refer to. Volume one deals mainly with organisms, volume two with physiology, genetics, evolution and ecology.

SCIENTIFIC METHOD

W. I. B. Beveridge, *The Art of Scientific Investigation* (Heinemann Educational Books, 3rd edition 1968)
An excellent account by an animal pathologist of how hypotheses are formulated and tested, amply illustrated with case histories.

P. B. Medawar, *The Art of the Soluble* (Methuen, 1967; Pelican, 1969)
In the last chapter the author, who is director of the National Institute for Medical Research in London, explodes the myth that the scientific method is endowed with a special mystique.

WHO'S WHO AMONGST ANIMALS AND PLANTS

A. S. Romer, *The Procession of Life* (Weidenfeld and Nicolson, 1968)
Like all Romer's books, compulsive reading. Traces the evolution of the animal kingdom, summarizing the salient characteristics of each group.

Ralph Buchsbaum, *Animals without Backbones* (Pelican, 2 volumes 1951)
Deals exclusively with the invertebrates, which are reviewed in an entertaining way. This has been a bestseller for years.

A. S. Romer, *The Vertebrate Story* (University of Chicago Press, 1959)
This revised edition of Romer's earlier book *Man and the Vertebrates* (Pelican, 2 volumes 1954) is a thoroughly readable account of the evolution of the vertebrates, including man.

E. J. H. Corner, *The Life of Plants* (Weidenfeld and Nicolson, 1964)
Belonging to the same series as Romer's *The Procession of Life*, this book discusses evolutionary trends in the plant kingdom.

V. A. Greulach and J. E. Adams, *Plants: an Introduction to Modern Botany* (Wiley, 2nd edition 1967)
An excellent general botanical text, easy to read and full of good illustrations. Contains a useful survey of the main plant groups.

THE WORLD OF MICRO-ORGANISMS

René Dubos, *The Unseen World* (USA: Rockefeller Institute; UK: Oxford University Press, 1962)
Based on the first series of Rockefeller Institute Christmas lectures for High School students, this very readable book traces the historical development of microbiology from Pasteur to the present day.

K. M. Smith, *Viruses* (Cambridge University Press, 1962)
A very good book, quite short and easy to read, on the structure, reproduction, and diseases caused by viruses, written by one of the men who discovered them.

W. M. Stanley and E. G. Valens, *Viruses and the Nature of Life* (USA: Dutton, 1961; UK: Methuen, 1962)
A popular but authoritative account of viruses by experts in this field. Stanley was the first person to successfully crystallize a virus in pure form.

W. R. Sistrom, *Microbial Life* (Holt, Rinehart and Winston, 2nd edition 1969)
A brief review of micro-organisms from viruses to protists. Deals mainly with their physiological processes.

GENERAL CELL BIOLOGY

W. D. McElroy and C. P. Swanson, *Modern Cell Biology* (Prentice Hall, 1968)
Deals concisely with cell structure and function, including metabolism and protein synthesis.

E. J. Ambrose and D. M. Easty, *Cell Biology* (Nelson, 1970)
An up-to-date textbook of cell biology, clearly written and well illustrated. Accounts of protein synthesis, cell respiration and photosynthesis are particularly useful.

Donald Kennedy (editor), *The Living Cell* (W. H. Freeman, 1965)
A collection of articles from *Scientific American* on cell biology. The articles span the period from 1958–64 and include several from the special issue of *Scientific American* (Sept 1961) on the cell. The articles cover such topics as fine structure, respiration, photosynthesis, the genetic code, cell division and specialization, transmission of nerve impulses, cell movement and contraction of muscle.

J. A. Ramsay, *The Experimental Basis of Modern Biology* (Cambridge University Press, 1965)
A discussion of mainly cellular topics, particularly useful in that it emphasises the experimental evidence on which modern theories of cell function are based. Now available as a paperback.

A. C. Giese, *Cell Physiology* (Saunders, 1968)
This standard American textbook is very full but by no means unreadable, and it has good reference lists.

FINE STRUCTURE OF CELLS

S. W. Hurry, *The Microstructure of Cells* (John Murray, 1965)
Compiled specially for British sixth forms, this slim volume is amply illustrated with electron micrographs. The principal organelles are dealt with in turn. Text brief and to the point.

T. P. O'Brien and M. E. McCully, *Plant Structure and Development* (Macmillan, 1969)
Like Hurry, but on plants, and includes photomicrographs as well as electron micrographs.

D. W. Fawcett, *An Atlas of Fine Structure* (Saunders, 1966)
Lavishly illustrated survey of animal cells with superb electron micrographs, many of them made by the author.

F. A. L. Clowes and B. E. Juniper, *Plant Cells* (Blackwell, 1968)
An authoritative and well illustrated account of the fine structure of plant cells with a strong functional slant. The introduction contains useful information on the preparation of electron micrographs.

STRUCTURE OF TISSUES AND ORGANS

W. H. Freeman and Brian Bracegirdle, *An Atlas of Histology* (Heinemann Educational Books, 1966)
Compiled specially for British sixth forms, this modestly sized book contains clear black-and-white photomicrographs of mammalian tissues and organs with corresponding drawings. A great help in interpreting microscopical sections.

A. C. Shaw, S. K. Lazell and G. N. Foster, *Photomicrographs of the Flowering Plant* (Longmans Green, 1965).
Similar kind of thing to Freeman and Bracegirdle, but on the flowering plant.

W. Bloom and D. W. Fawcett, *A Textbook of Histology* (Saunders, 8th edition 1962)
Updated version of Maximow and Bloom's famous histology text. The later editions include fine structure of specialized mammalian cells which is integrated with the traditional histology.

A. W. Ham, *Histology* (USA: Lippincott; UK: Pitman; 5th edition, 1965)
Standard work of reference in which histological structure is related to function. Full of useful information.

CHEMISTRY OF LIFE

Steven Rose, *The Chemistry of Life* (Pelican, 1966)
If you have an aversion to chemistry try this little book. The sections on enzyme action, energy release, and biochemical control in the cell are particularly good.

K. Harrison, *A Guide-Book to Biochemistry* (Cambridge University Press, 2nd edition 1965)
E. Baldwin, *The Nature of Biochemistry* (Cambridge University Press, 2nd edition 1967)
Two short introductions to biochemistry. Baldwin is more general and makes few demands on one's knowledge of chemistry. Harrison is slightly more detailed on the chemistry. Both stress basic principles and are ideal for general biologists.

J. S. Fruton and S. Simmonds, *General Biochemistry* (Wiley, 2nd edition 1958)
H. R. Mahler and E. H. Cordes, *Biological Chemistry* (Harper & Row/Weatherhill, 1966)
Both standard biochemistry textbooks, heavy going for the non-chemist but good for reference.

SPECIFIC BIOCHEMICAL TOPICS

D. W. Moss, *Enzymes* (Oliver and Boyd, 1968)
Only 100 pages: a short discussion of enzymes, their structure and how they work. Not much detailed chemistry.

John Marks, *The Vitamins in Health and Disease* (Churchill, 1968)
A mine of useful information, attractively presented, on the chemistry, sources, metabolism, and deficiency diseases of the vitamins required by man and domestic animals.

A. L. Lehninger, 'How Cells Transform Energy' (*Scientific American*, Sept 1961, Offprint no. 91).
This article from *The Living Cell* is a useful summary of the chemistry of respiration and photosynthesis.

HUMAN PHYSIOLOGY

A. C. Guyton, *Function of the Human Body* (Saunders, 1969)
This attractive American textbook is up to date and just right for British A level students who want to broaden their knowledge of human physiology.

O. C. J. Lippold and F. R. Winton, *Human Physiology* (Churchill, 6th edition 1968)
This new edition of Winton and Bayliss' textbook is rather too detailed for the needs of the average sixth former, but is readable and full of useful information.

H. Davson and M. G. Eggleton (editors), *Principles of Human Physiology* (Churchill, 14th edition, 1968)
New edition of Starling and Lovatt Evans' textbook, the chapters written by different contributors. An excellent reference book.

GENERAL AND COMPARATIVE ANIMAL PHYSIOLOGY

P. T. Marshall and G. M. Hughes, *The Physiology of Mammals and Other Vertebrates* (Cambridge University Press, 1965).
Written with sixth forms in mind, this book relates structure to function and is particularly useful on respiration and circulation.

J. A. Ramsay, *A Physiological Approach to the Lower Animals* (Cambridge University Press, 2nd edition 1968).
Ramsay's crisp and lucid writing makes this one of the most readable biology books. Fundamental principles and generalizations are emphasized throughout. Highly recommended.

W. B. Yapp, *An Introduction to Animal Physiology* (Oxford University Press, 3rd edition, 1970).
Human, invertebrate, and vertebrate physiology brought together into one compact volume. Contains much factual information useful to sixth formers, particularly scholarship candidates

E. Florey, *An Introduction to General and Comparative Physiology* (Saunders, 1966).
Advanced, but good for reference particularly on reception of stimuli and related topics.

C. L. Prosser and F. A. Brown, *Comparative Animal Physiology* (Saunders, 2nd edition 1961)
Hardly bedtime reading, but a useful survey of the research literature on the physiology of a very wide range of animals. Extensive reference lists.

SPECIFIC TOPICS IN ANIMAL PHYSIOLOGY

G. M. Hughes, *Comparative Physiology of Vertebrate Respiration* (Heinemann Educational Books, 1963)
In this little book the author discusses respiration in fishes, amphibians, reptiles, birds, and mammals in a way that can be readily understood by sixth formers.

E. Baldwin, *An Introduction to Comparative Biochemistry* (Cambridge University Press, 4th edition 1964).
This short and very readable book is particularly useful on the comparative aspects of osmoregulation, excretion, and respiratory pigments.

K. Schmidt-Nielsen, *Animal Physiology* (Prentice-Hall, 2nd edition 1964)
A simple general physiology text, recommended for its chapters on water and temperature control.

A. P. M. Lockwood, *Animal Body Fluids and their Regulation* (Heinemann Educational Books, 1964)
Assumes a rather wide knowledge of zoology, but provides a useful source of information on osmotic and ionic regulation in animals.

E. J. Holborow, *An ABC of Modern Immunology* (Lancet, 1968).
Reprinted from 1967 issues of *The Lancet*, this short book provides a useful modern account of the immune response.

Sir Macfarlane Burnet, 'Current Theories of Immunology' (*Science Journal*, Sept 1965. Reprint Ref: 65/50).
In this article the author discusses his clonal selection theory in the light of other theories of immunity.

R. D. Keynes, 'The Nerve Impulse and the Squid' (*Scientific American*, Dec 1958, Offprint no. 58)
Describes the pioneering experiments of Hodgkin and Huxley which led to elucidation of the nature of the nerve impulse.

'The Human Brain' (*Science Journal*, May 1967)
This special issue of the *Science Journal* includes articles on memory, synaptic transmission, and the biochemistry of behaviour.

H. E. Huxley, 'The Contraction of Muscle' (*Scientific American*, Nov 1958, Offprint no. 19)
H. E. Huxley, 'The Mechanism of Muscular Contraction' (*Scientific American*, Dec 1965, Offprint no. 1026).
In the first of these two articles Huxley describes the fine structure of skeletal muscle and relates this to the mechanism of contraction. The second article deals with the more biochemical aspects of muscle action.

J. Gray, *How Animals Move* (Cambridge University Press, 1953; Pelican, 1959)
Based on Sir James Gray's Royal Institution Christmas lectures for schools, this little book outlines in non-mathematical terms the mechanism of locomotion in aquatic and terrestrial animals.

J. Gray, *Animal Locomotion* (Weidenfeld and Nicolson, 1968)
This could be described as an advanced version of *How Animals Move*. If you are keen on a mathematical approach, it is all here. A magnificent volume.

C. R. Austin, *Fertilization* (Prentice-Hall, 1965).
A useful account in less than 150 pages of the significance, mechanics and consequences of fertilization.

'Human Reproduction' (*Science Journal*, June 1970)
This special issue of *Science Journal* contains articles on the sperm cell, fertilization, sex hormones, and early development of the embryo.

F. H. Marshall and A. S. Parkes (editors), *Physiology of Reproduction* (Longmans Green; vol. 1, Parts 1 and 2, 3rd editions 1960; vol. 2, 3rd edition 1952; vol. 3, 3rd edition 1966).
A large tome, full of facts and figures on the functional aspects of reproduction. Excellent for reference.

GENERAL PLANT PHYSIOLOGY

W. M. M. Baron, *Organization in Plants* (Edward Arnold, 2nd edition 1967)
Quite a short book in which the classical experiments of plant physiology are related to recent advances and to ecology. Highly recommended.

F. C. Steward, *Plants at Work* (Addison-Wesley, 1964)
Less readable but rather fuller than Baron, with more biochemical detail.

B. S. Meyer, D. B. Anderson and R. H. Böhning, *Introduction to Plant Physiology* (Van Nostrand, 1960)
Detailed but well within the scope of the sixth former who wants to extend his knowledge of plant physiology.

Walter Stiles, *An Introduction to the Principles of Plant Physiology* (Methuen, 3rd edition 1969)
An up to date treatise containing over 600 pages. Good for reference.

SPECIFIC TOPICS IN PLANT PHYSIOLOGY

D. I. Arnon, 'The Rôle of Light in Photosynthesis' (*Scientific American*, Nov 1960, Offprint no. 75)
J. A. Basham, 'The Path of Carbon in Photosynthesis' (*Scientific American*, June 1962, Offprint no. 122)
These two articles give a clear account, with experimental backing, of the light and dark stages of photosynthesis. (Both are included in *The Living Cell*)

W. M. M. Baron, *Physiological Aspects of Water and Plant Life* (Heinemann Educational Books, 1967)
A concise summary of the rôle of water in the physiology of plants, including turgor and transpiration.

J. F. Sutcliffe, *Mineral Salts Absorption in Plants* (Pergamon, 1962)
This book contains a useful discussion on possible mechanisms of active transport.

M. Richardson, *Translocation in Plants* (Edward Arnold, 1969)
An up-to-date summary and evaluation of the experimental data on the ascent of sap and movement of organic substances in plants.

ANIMAL BEHAVIOUR

Konrad Lorenz, *King Solomon's Ring* (Methuen, 1952)
Light relief, but written by a great zoologist who knows what he is talking about.

J. D. Carthy, *The Study of Behaviour* (Edward Arnold, 1966)
A useful introduction, particularly good on orientation and the rôle of stimuli. Only 57 pages.

Aubrey Manning, *An Introduction to Animal Behaviour* (Edward Arnold, 1967)
Quite readable for anyone who wants to take this subject further. Has some good accounts of behaviour experiments and a generally modern outlook. Good reference lists.

R. A. Hinde, *Animal Behaviour* (McGraw Hill, 2nd edition 1970)
This is more than just a survey of the research literature: it links behaviour studies with physiology and comparative psychology. But it is strong meat!

Desmond Morris, *The Naked Ape* (Cape, 1967; Corgi Books, 1969)
Human behaviour interpreted in terms of man's animal origins. Worth reading if only because of its controversial aspects. Its sequel *The Human Zoo* (Cape, 1969; Corgi Books, 1971) is concerned with human society. If you read these books you should also read:

John Lewis and Bernard Towers, *Naked Ape or Homo sapiens?* (Garnstone Press, 1969)
This is a reply by an anthropologist and a human anatomist to *The Naked Ape*.

CELL DIVISION

J. McLeish and B. Snoad, *Looking at Chromosomes* (Macmillan, 1962)
A short account of mitosis and meiosis illustrated with very clear photomicrographs.

ANIMAL GROWTH AND DEVELOPMENT

J. Cohen, *Living Embryos* (Pergamon, 1963)
A factual summary of the structural changes that occur in the development of the main animal groups.

W. H. Freeman and Brian Bracegirdle, *An Atlas of Embryology* (Heinemann Educational Books, 1963)
Photomicrographs of frog and chick embryos with corresponding drawings. Useful for interpreting sections and whole mounts.

C. W. Bodemer, *Modern Embryology* (Holt, Rinehart and Winston, 1968).
A large up to date textbook with emphasis on experimental studies. Good for reference.

PLANT GROWTH AND DEVELOPMENT

J. G. Torrey, *Development in Flowering Plants* (Macmillan, 1967)
A readable summary of the structural and physiological aspects of plant development. Contains a good account of plant hormones.

A. C. Leopold, *Plant Growth and Development* (McGraw-Hill, 1965)
A standard work of reference reviewing many of the experiments that have been carried out in recent years on the control of plant growth and development.

HEREDITY

A. M. Winchester, *Heredity, an Introduction to Genetics* (USA: Barnes and Noble, 1961; UK: Harrap, 1964). A fairly short, readable book on heredity with emphasis on human aspects. Interesting problems are posed at the ends of the chapters and there is a valuable section on the hazards of atomic radiation.

A. M. Srb and R. D. Owen, *General Genetics* (W. H. Freeman, 2nd edition 1965)
This general textbook includes chapters in which the principles of genetics are related to animal and plant breeding.

Curt Stern, *Principles of Human Genetics* (W. H. Freeman, 2nd edition 1960)
Quite advanced but compulsive reading. The last part includes chapters on the hazards of radiation and the genetic aspects of race.

K. R. Lewis and B. John, *The Matter of Mendelian Heredity* (Churchill, 1964)
A sophisticated book in which the discoveries of Mendel are looked at in the light of modern discoveries about chromosomes and genes. (Second edition due for publication Autumn 1971).

A. M. Srb, R. D. Owen and R. S. Edgar (editors), *Facets of Genetics* (W. H. Freeman, 1970)
A collection of *Scientific American* articles covering the years 1948–70. Topics include Mendel, the genetic basis of evolution and differentiation, and the genetic code.

GENES AND GENE ACTION

George and Muriel Beadle, *The Language of Life* (Gollancz, 1966)
Popular book meant for non-scientists, but well worth reading. The authors put molecular genetics into historical perspective.

John Kendrew, *The Thread of Life* (Bell, 1966)
Based on a series of BBC television lectures given by the author in 1964. The last five chapters deal with the genetic code in a way that can be readily understood by the general reader.

Navin Sullivan, *The Message of the Genes* (Routledge and Kegan Paul, 1968)
A simple and lucid account of the structure of DNA and its significance. A useful introduction for someone just starting molecular biology.

J. D. Watson, *The Double Helix* (Weidenfeld and Nicolson, 1965; Penguin, 1970)
The co-discoverer of the structure of DNA describes how the genetic code was cracked. If you think all scientists are stuffy old men in white coats you should read this book.

R. H. Haynes and P. C. Hanawalt (editors), *The Molecular Basis of Life* (W. H. Freeman, 1968)
A collection of articles from *Scientific American* on molecular biology spanning the years from 1948–68. Information on molecular genetics is mainly in Section III, which contains articles by Marshall Nirenberg and Francis Crick on the genetic code.

M. Ptashne and W. Gilbert, 'Genetic Repressors' (*Scientific American*, June 1970)
Isolation of chemical repressor substances goes some way towards confirming the Jacob-Monod theory of gene action.

OTHER BOOKS ON DEVELOPMENTAL BIOLOGY

M. Sussman, *Growth and Development* (Prentice-Hall, 2nd edition 1964)
Discusses in a fairly simple way the functional aspects of cell growth and differentiation.

J. A. Moore, *Heredity and Development* (Oxford University Press, 1963)
Cell division, fertilization, Mendelian genetics, chromosomes, genes, and embryology all in one volume of less than 250 pages. Good value but it will help if you already know something about these topics.

N. J. Berrill, *Growth, Development and Pattern* (W. H. Freeman, 1961)
A lucid discussion on the control of growth, morphogenesis, and regeneration in animals and plants. Advanced but worth reading.

P. C. Clegg, *Introduction to Mechanisms of Hormone Action* (Heinemann Educational Books, 1969)
A small book explaining clearly how hormones may influence cells by affecting their enzymes or their genes.

ECOLOGY

E. P. Odum, *Ecology* (Holt, Rinehart and Winston, 1963).
A broad introduction to the principles of ecology. Brief and to the point.

D. P. Bennett and D. A. Humphries, *Introduction to Field Biology* (Edward Arnold, 1965)
A good book for schools, practically orientated, with emphasis on methods and analysis of results. A good bibliography lists several identification keys suitable for school use.

W. H. Dowdeswell, *Practical Animal Ecology* (Methuen, 1959)
Another valuable book, ideal for sixth forms. Though it deals specifically with animal communities basic principles are stressed. Bibliography includes references to books and pamphlets on specific animal groups suitable for someone starting an autecological project.

Maurice Ashby, *Introduction to Plant Ecology* (Macmillan, 2nd edition 1969)
A readable account of plant environments and communities with a useful section on soil. Bibliography includes books on experimental techniques.

T. Lewis and L. R. Taylor, *Introduction to Experimental Ecology* (Academic Press, 1967)
45 ecological exercises suitable for school and college use. Includes information on apparatus, quantitative methods and identification.

Helen Mellanby, *Animal Life in Fresh Water* (Methuen, 6th edition 1963)
Fits into the average pocket. Enables quick identification of the more common fresh water animals and summarizes the general features of each group.

A. C. Allee, A. E. Emmerson, T. Park and K. P. Schmidt, *A Fundamental Treatise on Ecology* (Saunders, 1949)
A standard work of reference packed full of information.

'The Biosphere' (*Scientific American*, September 1970) Special issue of *Scientific American* devoted to the biosphere. Includes articles on the cycling of matter and energy, human food production etc.; eminently suitable for sixth forms.

ASSOCIATIONS

R. A. Wilson, *An Introduction to Parasitology* (Edward Arnold, 1967)
This useful introduction stresses basic principles, including physiology, and contains some notes on practical techniques.

E. R. Noble and G. A. Noble, *Parasitology: the Biology of Animal Parasites* (USA: Lea and Febiger, 2nd edition 1964; UK: Henry Kimpton, 1964)
A large textbook packed full of information on the structure, life histories, and medical effects of animal parasites. Nicely illustrated and, despite size, readable.

E. J. W. Barrington, *Invertebrate Structure and Function* (Nelson, 1967).
This general textbook of invertebrate zoology includes a useful discussion on invertebrate associations, particularly symbiosis and commensalism.

V. B. Wigglesworth, *The Life of Insects* (Weidenfeld and Nicolson, 1964)
This fascinating account of insect life by a leading insect physiologist contains a valuable discussion on social insects.

EVOLUTION

Charles Darwin, *The Origin of Species* (John Murray, 1959)
A best seller for over 100 years but hardly ever read! Its worth dipping into if you can stomach the heavy Victorian prose.

Alan Moorehead, *Darwin and the Beagle* (Hamish Hamilton, 1969)
An exciting account of Darwin's voyage. More of a travelogue than a biological treatise, but great fun.

P. A. Moody, *Introduction to Evolution* (Harper, 1962)
A general textbook on evolution, summarizing the evidences and mechanism in a way that can be readily understood by sixth formers.

Garrett Hardin, *Nature and Man's Fate* (Mentor, 1959)
A popular summary of Darwin's theory emphasising the historical events leading up to it, and its consequences.

W. H. Dowdeswell, *The Mechanism of Evolution* (Heinemann Educational Books, 3rd edition 1963)
A useful introduction to the theory of natural selection in less than 150 pages. Suitable for sixth forms.

John Maynard Smith, *The Theory of Evolution* (Pelican, 1958).
This book assumes a considerable background knowledge but is well worth reading for its excellent discussions on variation, population genetics, and speciation.

Sir Gavin de Beer, *Atlas of Evolution* (Nelson, 1964).
An expensive volume, superbly illustrated and succinctly written. Touches on all aspects of evolution including genes, development, behaviour, blood groups etc. Excellent on adaptation.

ENQUIRIES IN BIOLOGY

Biological investigations involve asking questions and formulating hypotheses. The art of enquiry is particularly fostered by some of the problems presented in the Nuffield A level biology *Study Guide* (Penguin Education, 1970), by the questions at the ends of the chapters in Hardin's *Biology: its Principles and Implications* (see p. 609), and in the following books:

Joseph J. Schwab, *BSCS Biology Teacher's Handbook* (Wiley, 1963)
This book contains 44 'Invitations to Enquiry' involving the formulation of hypotheses and interpretation of data in a wide range of investigations.

S. W. Hurry and D. G. Mackean, *Enquiries in Biology* (John Murray, 1968)
A slim volume (only 40 pages) with teacher's guide, containing seven investigations all involving interpretation of experimental evidence. The investigations cover muscle action, breathing, photosynthesis, and moulting in insects.

QUANTITATIVE BIOLOGY

O. N. Bishop, *Statistics for Biology* (Longmans Green, 1966)
The results of biological investigations, particularly in genetics and ecology, often need to be analyzed statistically. This book explains in terms that do not make great demands on one's mathematical knowledge how such analyses are carried out.

J. M. Eggleston, *Problems in Quantitative Biology* (English Universities Press, 1968)
If you are keen on a mathematical approach you will find this book challenging and exciting. The problems cover a wide range of biological topics, including physiology and genetics.

Laboratory Work

ANATOMY

On vertebrate and invertebrate anatomy there is nothing to beat H. G. Q. Rowett's dissection guides, published in the UK by John Murray (1950–53) and in the USA by Holt, Rinehart and Winston. These cover the dogfish, frog, rat, and selected invertebrates, and are available separately or combined together in a single volume. The diagrams and instruction notes are admirably concise and clear.

If you want instructions on a wider range of animals I would recommend:

W. S. Bullough, *Practical Invertebrate Anatomy* (Macmillan, 2nd edition 1958)
Contains instructions on the examination of over 100 invertebrate species including microscopical ones.

J. T. Saunders and S. M. Manton, *Practical Vertebrate Morphology* (Oxford University Press, 4th edition 1969).
Includes dissection of the lamprey, skate, bony fish, salamander, lizard, grass snake, and pigeon.

EXPERIMENTAL

The BSCS texts,[1] and the Nuffield A level biology *Laboratory Guides* (Penguin Education, 1970), provide a wide range of experiments suitable for sixth forms. Other recommended books:

J. Creedy and G. D. Brown, *Experimental Biology Manual* (Heinemann Educational Books, 1970)
Specially devised to accompany a modern A level biology course, this new manual includes experiments in microbiology, genetics, biochemistry, physiology, and behaviour.

P. Abramoff and R. G. Thompson, *An Experimental Approach to Biology* (W. H. Freeman, 1967)
Contains many experiments that do not require expensive equipment and are suitable for schools.

W. D. Zoethout, *Laboratory Experiments in Physiology* (USA: Mosby; UK: Henry Kimpton; 6th edition 1963)
An account of experiments in mammalian physiology, many of which are well within the scope of school sixth forms.

W. M. M. Baron, *Organization in Plants* (Edward Arnold, 2nd edition 1967)
Written for sixth forms. In the appendix there is a very useful section on experimental procedures.

G. Wald et al, *Twenty-six Afternoons of Biology* (Addison-Wesley, 2nd edition 1967)
An account of experiments featuring in the introductory biology course at Harvard University. Full of useful ideas but many of the experiments are demanding of apparatus. The experiments can be done equally well in the morning.

Useful Journals and Periodicals

Scientific American (monthly)
Contains articles on almost all aspects of biology, many of which are available as separate offprints. Full list of offprints can be obtained from the publisher, W. H. Freeman and Company: 660 Market Street, San Francisco, California 94104; or 58 Kings Road, Reading RGI 3AA.

Endeavour (published quarterly by ICI)
Contains many articles in biology, often on rather off-beat topics. Usually well illustrated and not too technical.

New Scientist (weekly, merged with *Science Journal* in 1971.)
Looking through the *New Scientist* each week is a good way of keeping up with current developments. In addition to special articles (often on topical subjects), the early pages contain brief reports of recent discoveries.

In addition the following more specialized journals frequently contain useful articles, though in most cases of more use to teachers than to students: *Nature, Nature New Biology, Science, Science Progress, The Advancement of Science, The School Science Review, The Journal of Biological Education,* and *Biological Reviews.*

[1] Biological Sciences Curriculum Study of the American Institute of Biological Sciences.

Index